Nonequilibrium Quantum Transport Theory of Spinful and Topological Systems

A New Perspective and Foundation for Topotronics

Nonequilibrium Quantum Transport Theory of Spinful and Topological Systems

A New Perspective and Foundation for Topotronics

Felix A Buot
University of San Carlos, Cebu, Philippines

World Scientific

NEW JERSEY • LONDON • SINGAPORE • BEIJING • SHANGHAI • HONG KONG • TAIPEI • CHENNAI • TOKYO

Published by

World Scientific Publishing Co. Pte. Ltd.
5 Toh Tuck Link, Singapore 596224
USA office: 27 Warren Street, Suite 401-402, Hackensack, NJ 07601
UK office: 57 Shelton Street, Covent Garden, London WC2H 9HE

Library of Congress Cataloging-in-Publication Data
Names: Buot, Felix A. (Felixberto Alcudia), 1938–   author.
Title: Nonequilibrium quantum transport theory of spinful and topological systems : a new perspective and foundation for topotronics / Felix A. Buot, University of San Carlos, Cebu, Philippines.
Description: New Jersey : World Scientific, [2024] | Includes bibliographical references and index.
Identifiers: LCCN 2023054135 | ISBN 9789811264719 (hardcover) |
ISBN 9789811264726 (ebook for institutions) | ISBN 9789811264733 (ebook for individuals)
Subjects: LCSH: Quantum field theory. | Nanostructured materials--Transport properties. |
Condensed matter. | Hilbert space. | Chirality.
Classification: LCC QC174.45 .B86 2024 | DDC 620/.50153143--dc23/eng/20231222
LC record available at https://lccn.loc.gov/2023054135

British Library Cataloguing-in-Publication Data
A catalogue record for this book is available from the British Library.

For any available supplementary material, please visit
https://www.worldscientific.com/worldscibooks/10.1142/13096#t=suppl

Desk Editors: Nimal Koliyat/Joseph Ang

Typeset by Stallion Press
Email: enquiries@stallionpress.com

To my parents; to my wife Evelina, to Max and Eugene;
to Anna Charlotte, James Maxwell, and Daniel Francis

Preface

This book gives a new perspective of topological materials. This enables kinetic transport formulation and simulations of open systems for nanodevice designs, high-speed performance and reliability analyses. And opens novel THz simulations of spinful and topological materials. This book is a sequel to my book on *Nonequilibrium Quantum Transport Physics in Nanosystems.*

Research in condensed matter physics has led to functional new materials, nanoelectronics, spintronics, topotronics, spin and pseudospin caloritronics, electrocaloric effect in nanoscale antiferroelectric and ferroelectric phases, microfluidics, photonics, nano-optoelectronics, plasmonics, phononics, quantum computing, quantum information, topological phases of matter, and generally, in quantum science and technology. Topological analyses also has a crucial role in fluid dynamics, such as, the earth's oceans, plasmas, the biological cells in fluid-like *active* matter, etc.

The use of topology in quantum physics started in the guise of the Bohr–Sommerfeld quantization condition. Classically, it started long before, with the *vector potential* in the Maxwell's theory of electromagnetism. Maxwell's theory leads to Einstein's geometric spacetime in general relativity theory, and to Dirac relativistic theory of positrons and electrons, with particle velocities limited within the light cone. The next geometrical step in physics is the Yang–Mills gauge theory of bosons in standard model of particle physics. A purely *quantum* mechanical step comes with the importance of the relative phases of wavefunctions in the Aharonov–Bohm effect. This is followed by Berry who underscores the physical role of the geometric phases of the wavefunctions. The Berry connections is exemplified by the *effective* (*vector potential*) "magnetic" Hamiltonian of *nuclear motion* in the Born–Oppenheimer approximation. Strong interest in topological techniques develops after the computational discovery in 1976 of the Hofstadter butterfly spectrum. This triggers developments of Mandelbrot fractals in mathematics and physics, as well as in engineering, e.g., in iterated maps on a voltage interval of a ring-oscillator microelectronic circuit. Several Nobel prize winning works followed: the integer quantum Hall effect (IQHE), the fractional quantum Hall effect (FQHE), newly discovered magnetic phases of materials — the theory of which goes well beyond Ginzburg–Landau symmetry breaking phase-transition paradigm, etc.

Further advances in the use of topology in CMP and QFT comes with the use of Chern–Simons theory developed by pure mathematicians, S.-S. Chern and J.H. Simons. The use of a complex topological space, namely, the 6D Calabi–Yau manifold, has been applied in supersymmetric string theory of fermions and bosons (includes gravitons) in QFT. It will surely find analogous and fruitful applications in CMP. A purely geometric theory of entanglement is now being sought, e.g., in black holes entanglement linked by Einstein–Rosen bridge.

For physicists and engineers engaged in the analyses of experimental results, physical modeling, nanodevice designs, and numerical simulations of active nanodevices of topological materials and spin systems, there is a need for a broader and deeper background knowledge of the theoretical foundation and relevant quantum physics. Essentially, this book provides a new perspective of chiral systems in topological phases of matter from a nonequilibrium quantum transport and lattice Weyl transform techniques. Thus, it uniquely serves as a foundation for implementing numerical kinetic transport simulations of nonlinearly active spintronic, pseudospintronic, and topotronic nanodevices.

This book consists of *eight* parts. Parts 1, 2, and 3 provide the necessary background knowledge, including recent results by the author, to make the book self-contained. A comprehensive view of mixed Hilbert-space representations, which generalizes the coherent-state representation, is given in Part 1. Part 2 deals with quantization of chiral degrees of freedom, i.e., U(1) gauge theory of the Bohr–Sommerfeld quantization. Part 3 summarizes the author's non-equilibrium quantum superfield transport physics. Readers interested mainly in quantum transport theory of spin, pseudospin, and topological systems can go directly to Parts 4–8. In Part 4, the kinetic quantum transport theory of spin and spin-charge coupling, and multi-band spinful Bloch electrons are given. For intuitive understanding, a new perspective on entanglement-of-formation entropy, concurrence, entropic-distance entanglement measures, and joint distributions as entanglements is given in Part 5.

Parts 6–8 deal with a nonequilibrium transport theory of topological systems.

An alternative *quantum transport theory of topological phases of matter* is treated in Part 6. This brings the theory to numerical simulation and engineering community. This part gives an *entirely new perspective of* several chiral degrees of freedom and their entanglements. In Part 7, the nonequilibrium quantum transport theory of IQHE and FQHE is also given. In Part 8, we introduced a Chern–Simons sub-band-gap mapping of the Hofstadter butterfly spectrum, and deduced the IQHE and FQHE regions, where the FQHE is due to *a widely-separated subband* in the fractal-like spectrum at very high-magnetic fields window. Thus, the Hofstadter butterfly spectrum lends support to the derivation of FQHE in Part 7, which is based on restructuring of a widely-separated sub-band in high-fields window.

I thank the publishers for inviting me to write a book based on my chapter in *Topics in Nanoscience*, vol. 1 (2022), edited by W. Schommers. I am grateful to my wife, Belen, for her constant support.

Felix A. Buot
C&LB Research Institute,
Carmen, Cebu, March 2024

About the Author

Dr. Felix A. Buot obtained his M.S.E.E. from Stanford University and his Ph.D. in Condensed Matter Physics from the University of Oregon. He was awarded Postdoctoral Research Associateships/ Fellowships in the following institutions: (1) Department of Mathematics (Westfield College/Bedford College), University of London, and as a Visiting Researcher at Cambridge University (2) International Center for Theoretical Physics, Trieste Italy; (3) McGill University, Montreal, Canada; (4) St. Francis Xavier University, Nova Scotia, Canada; (5) Stanford University, Stanford, CA; (6) Laboratory of Atomic and Solid State Physics and then jointly with National Center for Submicron Structures, Cornell University, Ithaca, NY. He worked as a Research Physicist at the U.S. Naval Research Laboratory for more than 23 years and received several NRL-wide research publication awards. He served in the Editorial Board and as Associate Editor of several Scientific Research Journals, as well as organized special journal issues of emerging research areas. He was a Research Professor at George Mason University, Fairfax, VA. He was a UNDP-TOKTEN Professor at the University of the Philippines. Dr. Buot was also a U.S. State Department Fulbright Professor at the University of San Carlos. He served as President of the Philippine-American Academy of Science and Engineering (PAASE). He is the author of a best-seller book: *Non-equilibrium Quantum Transport Physics in Nanosystems* published in 2009 by World Scientific.

Contents

Part 1

Condensed Matter Physics and Quantum Field Theory

Chapter 1

Prologue

This part of the book is intended to delve deeper and more comprehensive into the discrete phase-space formalism, generalized Pauli-matrix operators, phase-space distribution functions (functionals), and functional path integrals of quantum physics. Here, we focus on a unified treatment and parallelism between a particle theory of condensed matter physics (CMP) and a pure field formalism of quantum field theory (QFT).[1] We should also point out that quantum phase-space formalisms are crucial in the formulation of quantum *kinetic* transport equations of technological importance in active nanodevices performance simulations.

The parallelism between the *particle* formalism of CMP and a *pure field* formalism of QFT focuses on mixed Hilbert space representations, better known as the phase-space representations, of either the lattice site and crystal momentum or the canonically-conjugate quantum fields.[2] In short, a *particle* and a *pure field* formalism of quantum physics are analogous

[1]By a *pure* field formalism, we mean a dynamical theory based on distribution functionals and their dynamics. Contemporary quantum field theories calculations make use of the particle concepts, their scatterings and Feynman diagrams, with concomitant renormalization techniques.

[2]Parallelism in quantum physics is ubiquitous. For example, statistical physics at equilibrium or steady-state makes parallel formal developments with many-body quantum field theory at zero temperature, giving anti-periodic Green's function with respect to periodic time and yielding to Matsubara frequencies in Fourier transform space. We shall see later that the formal developments of real-time non-equilibrium transport theory of Part 3 also makes parallel developments with many-body zero temperature formalism. Other kind of parallelism is in the form of dualities, such as the duality of low and high-temperature statistical physics pioneered by Kramers and Wannier [1]. Even for nanotechnology, duality exists between type I and type II resonant tunneling heterostructures [2].

in all respects. Moreover, a pure field theory of bosons and fermions are also analogous on all accounts. This will be made clear in this part of the book.

The canonical conjugate variables formulation of either classical or elementary quantum physics has been thoroughly established and seems to characterize both classical and quantum dynamics. The extension to the discrete quantum mechanics and mixed representations, namely, the lattice Weyl–Wigner (W–W) quantum physics has been developed in CMP [3–7] with its various successful applications [8–10], as well as in the formulation of non-gradient quantum transport kinetic equations [11–14]. Here, we further extend the formalism to non-Hermitian operators of QFT, where translation operators or quantum state generators are well-defined, as well as transition function between the two Hilbert spaces. A forerunner[3] of this has been given by the author in his 1986 paper [16]. The present general formalism unifies all mixed-space representations on quantum mechanics, Hermitian and non-Hermitian operators and includes coherent state (CS) representation [17] as a special case. Mixed-space representations has been employed by the author in calculating the magnetic susceptibility of interacting many-body Bloch electrons [18], and in direct construction of bosonic and fermionic path integrals [16].

The case of non-equilibrium quantum transport of many-body physics in CMP, which not only involves non-Hermitian operators but also time-reversal symmetry breaking, deserves a separate treatment. This is discussed in Part 3 of this book. By virtue of the *doubling* of degrees of freedom, there it is expected that the emergent dual sector consist of chronological and anti-chronological solutions of the quantum Liouville equation, acquiring prominence over dual spaces made through ordinary field operator conjugation in many-body physics of equilibrium or steady-state quantum field theory.

The present unification naturally leads us to propose an *alternative* formulation for a *pure field* formulation of QFT based on the Q-distribution functionals of canonically-conjugate fields, which is structurally[4] expected to be free of infinities due to singularities, and hence no need of renormalizations or artificially imposed cut-off ubiquitous in QFT. Moreover, a *third quantization scheme* is introduced suitable for formulating a pure field non-equilibrium quantum transport theory in QFT, i.e., to obtain

[3]To the author's knowledge, the eigenvalues of quantum fields was first explored by Schwinger [15].
[4]Similar to lattice gauge theory, assuming non-degenerate countable states.

the quantum transport equation of the quantum distribution functionals of canonically-conjugate fields. Consequently, time-dependent problems can then be dealt with as a non-equilibrium QFT transport physics of a third quantized S-matrix theory. It is worthwhile to point out that the use of functionals in quantum physics has proven to be more physically revealing theoretically and more powerful computationally. It is heralded by Feynman functional path integrals in QFT and by the Kohn–Sham density functional theory in CMP.

First, we give an introduction on the physical basis of the discrete W–W non-equilibrium quantum transport theory. One of the important aspects of the discrete formalism is the bijective pairing of the elements of the dual spaces, known in mathematics as Pontryagin duality, which allows generalization to other dual spaces.

1.1 Physical models for discrete phase-space quantum mechanics

Historically, there are already *existing quantum models in physics* that have physically guided the discrete phase-space physics of the lattice W–W formulation, namely, (a) *Localized Wannier function* and *extended Bloch function* for discrete lattices in solid state physics, obeying the Born–von Karman boundary condition (strictly speaking, obeying *modular arithmetic* based on *finite fields*, akin to a group theory of integers). (b) Dirac delta function and plane waves in the continuum limit, i.e., only for *uncompacted continuous* coordinate space (non-cyclical domains) can one have *unquantized continuous* momentum space (this bijective quantum-mechanical canonical variables is disregarded in some of the W–W formalism of lattice models, where cyclical *discrete lattice* coordinates is unphysically paired with *compact continuous* momentum space [19–21]).

In fact, it is mathematically inconsistent for momentum to be continuous for a translationally symmetric discrete lattice, or for compactified continuous coordinate with periodic boundary condition. For all these *cyclical* domains, momentum or frequencies are always quantized or discrete.

In both (a) and (b), we have the eigenvector for positions (e.g., discrete lattice position), $|q\rangle$, and eigenvector for momentum (or discrete crystal momentum), $|p\rangle$. Of course, all respective eigenspaces go to continuum spaces in the limit of non-cyclical systems with lattice constant goes to zero as in (b).

These respective eigenvectors, $|q\rangle$ and $|p\rangle$, in (a) and (b) are bijectively related by Fourier transformation, $|q\rangle$ to $|p\rangle$ via the transition function, $\langle q|p\rangle$, and often produces results *akin* to quantum uncertainty principle in their probabilistic continuous coordinate components. Precisely, to construct a physically-based discrete phase space or discrete W–W quantum physics one must be guided by the following principles.

1.1.1 *Discrete phase space on finite fields*

The original formulation of lattice-space discrete Weyl transform and discrete Wigner distribution function (WDF) [3] is based on crystalline solid with inversion symmetry, and hence based on an odd number of discrete lattice points, (q, p), obeying the Born–von Karman boundary condition. In terms of mathematical rigor [22], this formulation is generally based on a finite fields represented by a finite prime number, N, of lattice points obeying modular arithmetic, closed under addition and multiplication.[5] The presence of multiplicative inverses in the formulation assumes that the finite number of lattice points is a prime number, since primality is required for the non-zero elements to have multiplicative inverses. Indeed, as the number of lattice points becomes large, the lattice point coordinates obeying the Born–von Karman boundary condition, assumes the domain of prime integers.[6]

1.1.2 *Transition functions between dual Hilbert spaces*

The existence of a transition function between dual spaces can be easily established for Hermitian operators leading to bijective discrete Fourier transformation. The physically-based construction of lattice W–W formalism in CMP is guided by well-known aspects of solid-state physics [3], namely, (a) the invariance of this physical and canonical scheme of *complete and orthogonal* set of $\{|q\rangle\}$ and *complete and orthogonal set* of $\{|p\rangle\}$, in going from discrete to continuum physics, and (b) in the compactification of the solid-state lattice, the number of discrete lattice points (and hence the number of

[5]There are *infinitely* many prime numbers. Examples of prime numbers are: 2, 3, 5, 7, 11, 13, 17, 19, 23, 29, 73, 79, 83, 89, 97, 101, 103, 107, 179, 181, 191, 193, 197, 199, 211, 223, 227, and 229, etc. When written in base 10, all prime numbers except 2 and 5 end in 1, 3, 7 or 9.

[6]A review article by Kasperkovitz and Peev, *Ann. Phys.* **230**, 21 (1994) and more recently by Fialkovsky *et al.* [19] makes misleading statements by failing to recognize the finite-field aspects of the theory.

discrete momentum points) must be a prime number for obvious inversion symmetry reason and mathematical definiteness. Moreover, all arithmetic operations on this group of numbers must be closed, i.e., all arithmetic operation must be a modular arithmetic with *prime number modulus* (*akin to a group operation on prime number of integers*). In short, all arithmetic operation is a modular arithmetic based on finite fields, since only for finite fields with prime number modulus does every non-zero element have well-defined multiplicative inverse, and hence modular division operation also provides closure. Thus, these two set of Hilbert spaces, $\{|q\rangle\}$ and $\{|p\rangle\}$, is clearly connected by a transition function which defines the lattice bijective Fourier transform.

1.1.3 *The direct and reciprocal lattice in CMP and SQFT*

In CMP and statistical quantum-field theories (SQFTs), compactification of space and time, respectively, into periodic dimensions and segments can both be mathematically expressed in analogous terminology.

We first summarize the Bravais lattice vectors and their corresponding reciprocal lattice vectors, since this points to symmetry properties of the discrete W–W formulation in CMP. In 2D lattice, we have the reciprocal lattice vectors, $\vec{b}_1$ and $\vec{b}_2$ given by the matrix,

$$(\vec{b}_1\ \vec{b}_2) = \frac{1}{\hat{n}\cdot(\vec{a}_1\times\vec{a}_2)}(\vec{a}_2\times\hat{n}\ \hat{n}\times\vec{a}_1), \tag{1.1}$$

which geometrically means that $\vec{b}_1$ is perpendicular to $\vec{a}_2$ and $\vec{b}_2$ perpendicular to $\vec{a}_1$. The $\hat{n}$ is the unit vector normal to the 2D lattice plane.

In 3D lattice, the reciprocal lattice vectors, $\vec{b}_1$, $\vec{b}_2$, and $\vec{b}_3$ are given by

$$(\vec{b}_1\ \vec{b}_2\ \vec{b}_3) = \left(\frac{\vec{a}_2\times\vec{a}_3}{\vec{a}_1\cdot(\vec{a}_2\times\vec{a}_3)}\ \frac{\vec{a}_3\times\vec{a}_1}{(\vec{a}_3\times\vec{a}_1)\cdot\vec{a}_2}\ \frac{\vec{a}_1\times\vec{a}_2}{(\vec{a}_1\times\vec{a}_2)\cdot\vec{a}_3}\right). \tag{1.2}$$

The above formulas are independent of any choosen coordinate system. We have

$$(\vec{a}_1\ \vec{a}_2\ \vec{a}_3)(\vec{b}_1\ \vec{b}_2\ \vec{b}_3)^T = I. \tag{1.3}$$

Therefore in the compactification of the direct lattice space into N lattice sites with the Born–von Karman periodic-boundary condition, the reciprocal lattice is correspondingly broken up into Brillouin zones with highest quantized wavelength given by $2a$ (of wavevector $\pm\frac{2\pi}{\tilde{N}a}\frac{\tilde{N}}{2} = \pm\frac{\pi}{a}$), along $\tilde{N}$

lattice vectors, where a is the direct lattice constant. The Brillouin zone is the corresponding compactification of the reciprocal lattice sites into quantized wave-vector Bloch states on one-to-one correspondence with discrete N states within the N direct lattice sites, i.e., the N localized Wannier functions in a *separated* energy band.

Similarly, in an equilibrium or steady-state statistical system characterized by temperature T, the compactification of continuous time τ by a period $\beta\hbar$, i.e., $0 \leq \tau \leq \beta\hbar$, has the corresponding reciprocal "Matsubara–Brillouin zone" with quantized *angular* frequency given by

$$\omega_n = \frac{2\pi n}{\hbar\beta} \qquad \text{for bosons,}$$

$$\omega_n = \frac{(2n+1)\pi}{\hbar\beta} \qquad \text{for fermions,} \tag{1.4}$$

where $\beta = \frac{1}{k_B T}$, k_B is the Boltzmann constant. Here, one may consider the following relation for the continuous-time compactification by a period $\beta\hbar$ as

$$\text{Lim}_{N \Longrightarrow \infty,\ a \Longrightarrow 0}\ Na = \hbar\beta. \tag{1.5}$$

The quantized-quantum number, $-\infty \leq n \leq \infty$, is *unlimited or* yields infinite highest angular frequency of the "Matsubara–Brillouin zone". The reason for the expression Eq. (1.4) is the *anti-periodicity* of fermions as function of time, e.g., the Matsubara Green's function. Very simply put, the physical reason for Eq. (1.4) is that fermions rotate by odd numbers of π compatible with *anti-commutation* relations and Pauli exclusion principle, e.g., see Section 3.2.2 on particles.

1.1.3.1 *Fourier transform space: Quantized momentum and frequency*

In summary, for all translationally symmetric or *cyclical* domains, either continuous or discrete, the Fourier transform space or reciprocal space is, strictly speaking, *quantized.* For discrete lattice the quantized momentum is limited to the region of the first Brillouin zone of a separated energy band. Likewise, a frequency cut-off implies discreteness of the temporal domain. For all cyclical domains, either continuous or discrete, the use of continuous momentum or frequency is at best only an approximation, sometimes employed for calculational and mathematical convenience.

1.1.4 *Translation symmetry*

The one eclectic concept in solid state physics is the concept of translation symmetry along any symmetry directions of the lattice. Inversion symmetry of the lattice structure is also among the common symmetry properties. Thus, the discrete phase space W–W formalism in CMP [3] is compatible with any of the lattice structures defined by Eqs. (1.1) and (1.2), i.e., not limited to cubic lattice structures only.

1.1.5 *Generalization to other discrete quantum and classical systems*

The guidance of (a) and (b) allow us to generalize discrete phase space based on finite fields to be useful when the quantum numbers specifying the quantum states, are discrete configurations other than the particle position and momentum [12], even to a system describable by a smallest prime number 2. A simplest example is that of quantum bit or qubit, like a two-level atomic systems. Another example lies in the two-state entangled diagrams of multi-qubit systems [23], where the entangled basis states are formed by the Hadamard lattice transform. Moreover, entangled qubits have emergent qubit properties.

1.1.6 *Generalized operator basis*

Another generalization has to do in the construction of the WDF. This is done through the generalized projection operators, sometimes referred to as the generalized Pauli spin operators for both orthogonal and biorthogonal spaces. This will be made clear later. Indeed, the construction can be generalized to the spaces of non-Hermitian operators or biorthogonal systems. The power of using finite fields is that one can also generalized the discrete Wigner distribution construction to the algebraic concept of finite fields, which are extension of prime fields, where q and p are field elements (*mod irreducible polynomial*), useful in quantum computing, visualization, and communication/information sciences. Here we can have p^n elements for some prime p and some integer $n > 1$ useful for constructing the WDF for spin-$\frac{1}{2}$ systems [12, 22]. Moreover, frequency–time phase-space Wigner function has become an important analytical tool in digital-signal processing.

1.1.7 *CMP mapping of discrete momentum and lattice position spaces*

Each energy band corresponds to the splitting of an energy level of one atomic site into N levels where N is the number of lattice sites in a *compactified* Bravais lattice obeying the Born–von Karman boundary condition in a given symmetry direction. This is the basis of a separated energy band quantum dynamics [24]. Therefore, the number of crystal momentum states in each band (in a Brillouin zone) is exactly equal to the number of lattice sites. Hence, there has to be a bijective mapping between number of discrete lattice sites, in compactified direct lattice, and the number of discrete crystal momentum states in the corresponding Brillouin zone. This is the essence of the powerful theoretical concept of localized function around each lattice site, the *Wannier function* $|q\rangle$, and the extended function over all lattice points, the *Bloch function* $|p\rangle$, related through the bijective discrete Fourier transformation. The bijectivity and finite fields aspects are crucial to the mathematical manipulations, for mathematical definiteness which avoid ambiguities.

It is worth cautioning that the use of discrete lattice models coupled with the *compact continuous* momentum space in some W–W formulations [19] renders an ill-defined transition function, by incurring non-bijective canonical conjugate dynamical variables, e.g., absence of bijective transformation between momentum $|p\rangle$ and coordinate $|q\rangle$. This, and *together* with the lack of modular arithmetic on finite fields not invoked at all, is adding to several ambiguities. As mentioned before, for translationally symmetric domains, continuous momentum or frequency is at best only an approximation, for calculational and mathematical convenience.

1.1.7.1 *Bijective Fourier transformation (Pontryagin duality)*

In short, the crystal momentum space is essentially discrete and yield a bijective mapping to the discrete lattice sites through a discrete Fourier transformation. Moreover, dual Hilbert spaces related by generalized Fourier transform also endows quantum uncertainty relations, well known in classical probability theory [25].

The Bloch function space is orthogonal and complete and so is the Wannier function space. Observe that these eigenfunctions of phase-space operators are well established for gapped energy band structures, or energy band far removed from the other energy bands. Generalized Wannier function can also be defined for coupled energy bands, using a decoupling scheme like the Foldy–Wouthuysen transformation applied to relativistic Dirac electrons. Indeed, it has been shown that counterparts of Wannier function and Bloch function exist for the decoupled positive energy states of relativistic electrons, with the *Dirac–Wannier function* localization about the size of Compton wavelength [5, 8]. Electric Wannier function and magnetic Wannier function also exist, as well as their respective Bloch functions, for uniform external electromagnetic fields [24]. This physical idea has been extended to formally construct the discrete phase space quantum mechanics *based on finite fields* for cases where q and p are not position and momentum variables, useful for quantum computing [12].

Although every finite field, with p^n elements for some prime p and some integer $n \geq 1$, often deals with irreducible polynomials over ring $\mathbb{Z}$ of integers, or over field $\mathbb{Q}$ of rational numbers, or over field $\mathbb{R}$ of real numbers, or over field $\mathbb{C}$ of complex numbers, the role of irreducible polynomials can be played by prime numbers themselves for $n = 1$: prime numbers (together with the corresponding negative numbers of equal modulus) are the irreducible integers. They exhibit many of the general properties of the concept "irreducibility" that equally apply to irreducible polynomials, such as the essentially unique factorization into prime or irreducible factors: Every polynomial $p(x)$ in ring of polynomials with coefficients in F, denoted by $F[x]$, can be factorized into polynomials that are irreducible over F. This factorization is unique up to permutation of the factors and the multiplication of constants from F to the factors.[7]

In general, the discrete phase space in CMP is based on the mathematics of finite fields [22] and holds for any prime number of lattice points obeying the Born–von Karman boundary condition, i.e., with prime modulus. It even holds for the most elementary prime number 2 of lattice points, to yield the 2×2 Pauli spin matrices and the well-known *Hadamard transformation* between "Wannier function" and "Bloch function", i.e., discrete Fourier transformation of two points in "phase space" leading to *transformation of qubits* [12]. Here the "Wannier function" and "Bloch function" has acquired the status of a simple theoretical device for discrete quantum

[7]The simplest case of interest in Buot's CMP discrete W–W formulation is when $n = 1$. In this case the finite field $GF(p)$ is the ring $\frac{Z}{pZ}$. This is a finite field with p elements, usually labeled $0, 1, 2, \ldots, p-1$, where arithmetic is performed modulo p, where non-zero elements have multiplicative inverses.

physics. Indeed, the CMP discrete phase-space formalism [3] also gives the generalized Pauli spin operators for any given prime number of lattice points, which can be generalized to spaces of non-Hermitian operators in CS representation. It has also yielded all the entangled basis states, e.g., for two, three, four and multi-qubits, to be discussed in Chapter 23, crucial to the physics of quantum teleportation [12]. In other words, bijective pairing of different Hilbert spaces is crucial to various generalization discussed here.

Chapter 2

Mixed Space Representations in CMP

The mixed q-p representations in the particle formalism is very useful in making analogy with classical dynamics in terms of position and momentum eigenvalues. This has also ushered a better understanding of the path integral formulations of quantum field theory [16] to make analogy with classical partition function of statistical physics, especially with the compactification of time by the aid of Matsubara technique for steady-state systems or temporally translation-invariant systems. This quantum field mixed representation theory aspects will be discussed in later chapters.

2.1 The q-p representation

The mixed q-p representation basically start by expanding any quantum operator, $\hat{A}$, in terms of mutually unbiased basis states, namely the eigenvector of position operator, $\hat{Q}$, and the eigenvector of momentum operator, $\hat{P}$. We have

$$\begin{aligned} \hat{A} &= \sum_{p,q} |q\rangle \langle q| \hat{A} |p\rangle \langle p|, \\ &= \sum_{p,q} \langle q| \hat{A} |p\rangle \; |q\rangle \langle p|. \end{aligned} \tag{2.1}$$

2.1.1 *The completeness of mixed-space projector*

Thus, the set $\{|q\rangle \langle p|\}$ is the basis operator for the mixed q-p representation. From the completeness relations of the unbiased basis states, $\{|q\rangle\}$ and $\{|p\rangle\}$, the set $\{|q\rangle \langle p|\}$ obeys the completeness relation,

$$\sum_{q,p} \langle q| |p\rangle |q\rangle \langle p| = 1. \tag{2.2}$$

Substituting the expression for $\langle q| |p\rangle$,

$$\langle q| |p\rangle = C_o e^{\frac{i}{\hbar}\vec{p}\cdot\vec{q}}, \tag{2.3}$$

$$\langle p| |q\rangle = C_o^* e^{\frac{-i}{\hbar}\vec{p}\cdot\vec{q}}, \tag{2.4}$$

we obtained, for the completeness relation,

$$(N\hbar^3)^{-\frac{1}{2}} \sum_{q,p} e^{\frac{i}{\hbar}\vec{p}\cdot\vec{q}} |q\rangle \langle p| = 1, \tag{2.5}$$

where C_o is choosen as

$$C_o = \left(N\hbar^3\right)^{-\frac{1}{2}}. \tag{2.6}$$

Equations (2.5) or (2.2) can be rewritten as

$$\sum_{q,p} \frac{|q\rangle \langle p|}{\langle p| |q\rangle} = 1. \tag{2.7}$$

Here we use the transformation identities in the mixed q-p representations,

$$|p\rangle = \sum_{q} \langle q| |p\rangle \ |q\rangle, \tag{2.8}$$

$$\langle p| = \sum_{q} \langle q| \ \langle p| |q\rangle, \tag{2.9}$$

$$|q\rangle = \sum_{p} \langle p| |q\rangle \ |p\rangle, \tag{2.10}$$

$$\langle q| = \sum_{q} \langle p| \ \langle q| |p\rangle, \tag{2.11}$$

$$\langle p| |q\rangle = \exp\left(-\frac{i}{\hbar} p \cdot q\right). \tag{2.12}$$

2.1.2 *Covariant state vector and contravariant components*

The position operator, $\hat{Q}$, and momentum operator, $\hat{P}$, obey the following commutation relations. When operating on components, we have,

$$\left[\hat{Q},\hat{P}\right]\langle x|\,|q\rangle = i\hbar\ \langle x|\,|q\rangle = i\hbar\ \psi(x,q). \tag{2.13}$$

On the other hand, when operating on the eigenvector, $|q\rangle$,

$$\left[\hat{Q},\hat{P}\right]|q\rangle = -i\hbar\ |q\rangle. \tag{2.14}$$

The translation operator, $T\left(q'\right)$, is defined as

$$\exp\left(-\frac{i}{\hbar}q'\cdot\hat{P}\right)|q\rangle = \left|q+q'\right\rangle, \tag{2.15}$$

where

$$\hat{P}\,|p\rangle = p\,|p\rangle, \tag{2.16}$$

$$\hat{P}\,|q\rangle = i\hbar\frac{\partial}{\partial q}\,|q\rangle, \tag{2.17}$$

whereas,

$$\exp(ip'\cdot\hat{Q})\,|p\rangle = |p+p'\rangle, \tag{2.18}$$

where

$$\hat{Q}\,|q\rangle = q\,|q\rangle, \tag{2.19}$$

$$\hat{Q}\,|p\rangle = -i\hbar\frac{\partial}{\partial p}\,|p\rangle. \tag{2.20}$$

Equations (2.15) and (2.18) can be deduced from Eqs. (2.8)–(2.12). We will talk more concerning Eqs. (2.14)–(2.20) below for emphasis of its relevance in our theoretical developments [e.g., see Eqs. (3.1)–(3.9)] below.

Chapter 3

Physics of Commutators and Anti-Commutators

Here, we proposed a couple of intuitive physical interpretation of commutators and anti-commutators in quantum physics. To the author's knowledge, only Schwinger [15] seems to have first hinted the physical implications for bosons and fermions.

3.1 Conjugate field operators and quantum state dynamics

Indeed, conjugate operators in CMP and QFT serve to generate new quantum states or states dynamics. This is part of the reasons why time variables do not have any role as a quantum operator but just as a simple parameter.

3.1.1 *Space translation, time evolution, and state generation*

It is not often emphasized in the literature, as well as in physics textbooks, that commutation and anti-commutation relation of canonically conjugate operators do carry very important dynamical meanings [15], i.e., quantum state generation or state dynamics, and imply either space translation (Wannier state generation) or time evolution, and in QFT it implies energy or excitation-number state generation.

3.1.2 *States generators*

We have

$$P\,|p\rangle = p\,|p\rangle, \tag{3.1}$$

$$\begin{aligned} Q\,|p\rangle &= Q\sum_q e^{\frac{ip\cdot q}{\hbar}}\,|q\rangle \\ &= -i\hbar\nabla_p\,|p\rangle, \end{aligned} \tag{3.2}$$

and

$$P\,|p\rangle = p\,|p\rangle, \tag{3.3}$$

$$\begin{aligned} P\,|q\rangle &= P\sum_p e^{\frac{-ip\cdot q}{\hbar}}\,|p\rangle \\ &= i\hbar\nabla_q\,|q\rangle. \end{aligned} \tag{3.4}$$

We have, independent of the "coordinates" or eigenvector basis, the commutation relation,

$$[Q, P] = -i\hbar. \tag{3.5}$$

We denote by

$$X_{q'} = \exp\left\{-\frac{i}{\hbar}q'\cdot P\right\} \tag{3.6}$$

as the generator of $|q\rangle$ states, and

$$Z_{p'} = \exp\left\{\frac{i}{\hbar}p'\cdot Q\right\} \tag{3.7}$$

as the "rotation" or phase operator for $|q\rangle$ states.

Correspondingly, we denote

$$Z_{p'} = \exp\left\{\frac{i}{\hbar}p'\cdot Q\right\} \tag{3.8}$$

as the generator of $|p\rangle$ states, and

$$X_{q'} = \exp\left\{-\frac{i}{\hbar}q'\cdot P\right\} \tag{3.9}$$

as the "rotation" or phase operator for $|p\rangle$ states, a sort of reverse roles from that of $|q\rangle$ "coordinates".

We have for localized Wannier states at lattice position q, Eq. (2.15),

$$|q\rangle = \exp\left\{-\frac{i}{\hbar}q\cdot P\right\}|0\rangle\,, \tag{3.10}$$

where the operator $P = i\hbar\nabla_q$ is operating on the basis eigenvector $|q\rangle$, since $P = -i\hbar\nabla_q$ acting on the components, $\psi(q)$, or wavefunctions behaves contravariantly. Here, we basically made an assumption in the above expression that there exist continuous function of q having an infinite radius of convergence, which are equal to $|q\rangle$ at the lattice points, i.e., we expand the exponential in Eq. (3.10) as a well-defined Taylor series.

3.1.3 *Canonical conjugate operators implies state generators*

All these results are reminiscent of well-known relations with other canonical conjugate variables operating on wavefunctions, for example, we have

$$\Psi(t+t_0) = \exp\left(-i\frac{t\,\mathcal{H}}{\hbar}\right)\Psi(t_0), \tag{3.11}$$

where

$$\mathcal{H} = i\hbar\frac{\partial}{\partial t},$$

and

$$\Psi(q+q_0) = \exp\left(i\frac{q\cdot\hat{P}}{\hbar}\right)\Psi(q_0). \tag{3.12}$$

In other words, canonical conjugate variables acts as displacement operators acting on its dual eigenvector space, respectively. This operation is also known as generation of eigenvectors. This principle is useful when we deal with QFT later. In fact we shall see that in QFT, the commutation relation of annihilation and creation operator given by

$$\left[\hat{a},\hat{a}^{\dagger}\right] = 1, \tag{3.13}$$

implies the generation of eigenvector of $\hat{a}$ as follows:

$$\exp(\alpha\hat{a}^{\dagger})|0\rangle = |\alpha\rangle, \tag{3.14}$$

where α is an eigenvalue of annihilation operator $\hat{a}$. This implication toward quantum state generator holds for commutator and anti-commutator relations, either fermions or bosons [15].

As a side remark, time in Eq. (3.11) cannot be elevated to an energy displacement operator in quantum mechanics although it appears with energy variable in Poisson bracket operator on the same level as the position and momentum variables. Time can only serve as a parameter, it is simply inconceivable to have an unbiased quantum state $|t=0\rangle$, independent of dynamics since this violates Einstein special relativity theory i.e., time is simply not an independent dynamical variable but simply act as an "attached" parameter in both classical and quantum dynamical processes. Indeed, energy or generally quantum state displacement operations are ubiquitous in QFT and many-body problems in CMP, but these are implemented by the use of canonically-conjugate annihilation and creation operators.

3.2 Anti-commutator and commutator in quantum transport

3.2.1 *Entanglement-induced localization and delocalization*

In quantum transport theory of interacting *chiral* (axial or pseudovectors) degrees of freedom discussed in Part 6 of this book, e.g., Landau orbits, spin and pseudospins, the anti-commutator of scattering terms are identified as a localization or non-diffusive, characterized by a fixed point in phase space. Whereas commutator terms are associated with delocalization or quantum diffusion. There, the concept of locality and non-locality have been extended to entangled chiral degrees of freedom in non-equilibrium quantum transport physics, leading to an entirely new perspective of topological phases of matter.

3.2.2 *Particles and forces: Fermions and bosons*

Indeed, anti-commutation is associated with a wavefunction property of a rotating spinors where excitations have spatial sizes, or inherent non-zero masses, and behaving like billiard balls upon exchange, resulting in the wavefunction rotation about a point and phase change of π. This is the root of the Matsubara frequencies for fermions at equilibrium or steady-state, at some specified temperature. This is illustrated in Fig. 3.1.

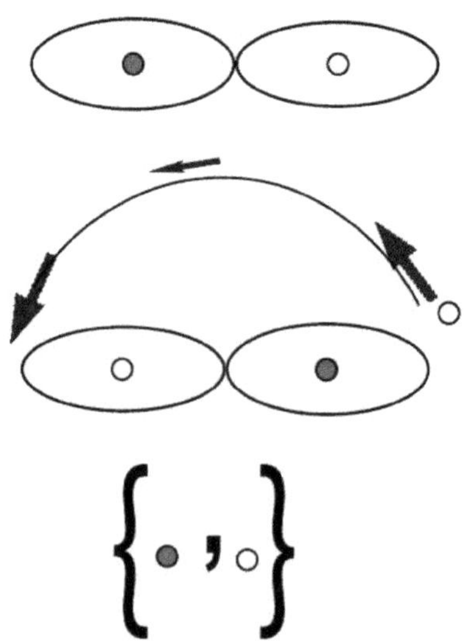

Fig. 3.1. A very simple intuitive explanation of spinors. The anti-commutator represents local or non-diffusive term in quantum transport since repetition of the anti-commutation operation just result in pure rotation of Ψ with a phase change of π. Fermions are spinors and behave like billiard balls under exchange, as indicated in the figure.

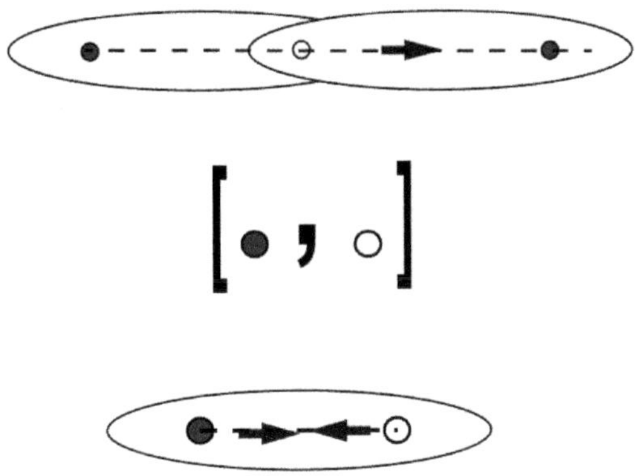

Fig. 3.2. A qualitative explanation and schematic representation of bosons as obeying a non-local commutation relation. Bosons can pass through each other so the wavefunction does not have to rotate to affect an exchange of particles. Thus, there is no wavefunction phase change. The commutation terms in quantum transport usually implies unitary evolution as well as a non-local diffusive terms, since repeated commutation will result in diffusion as schematically indicated in the figure. Note that the boson particle can pass through the other boson particle in straight line, unlike the fermions which must go around since the cannot occupy the same state as indicated in Fig. 3.1. This is the reason that bosons are often spatially non-local excitations with generally zero masses. There is also an implied overlap of their wavefunctions which allows for quantum diffusion. The wavefunctions maintain their phase after exchange with effective diffusion. The lower graph is a naïve interpretation of zero-point dynamics of non-local bosons. Bosons have quantized energy and momentum but spatially non-local or cannot be localized even for Bose condensate which cannot be a solid.

On the other hand, a commutator of conjugate operators implies spatial non-locality of excitation wavefunctions, where energy is the one quantized with generally indefinite non-local spatial dimensions, inherent zero masses, as depicted in Fig. 3.2. The exception are the bosons of short-range forces which can have masses, such as the W and Z bosons, through the Higgs mechanism. Even the zero-point energy is still non-local in space, which is the main reason why the lowest energy Bose condensate cannot form a solid. Clearly, commutator cannot be characterized by a point, i.e., it is non-local and excitations can conceptually pass through each other in the process of exchange [26–28]. Indeed, Fig. 3.2 seems to suggest that only boson quasiparticles can move independently in one dimensions.

3.2.3 *Particles versus quantized force fields: Nambu–Goldstone bosons and the Higgs bosons*

The spatial size or finite mass implications of the anti-commutation relations is the reason why elementary particles are generally fermions whereas

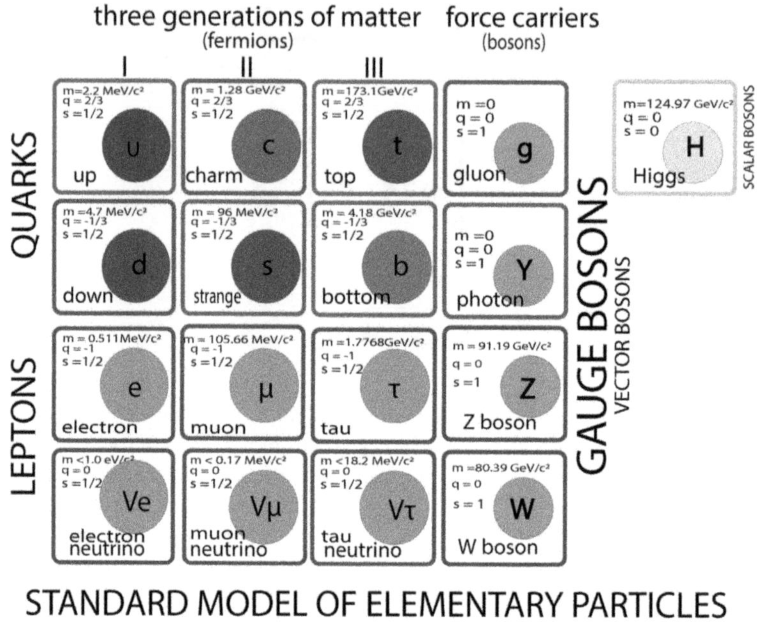

Fig. 3.3. In the standard model of particle physics, there are four major types of gauge bosons namely, photons, W bosons, Z bosons, and gluons. Photons are particles that carry electromagnetic interactions, while W and Z bosons tend to carry weak interactions, and gluons can carry strong interactions.

Source: Redrawn from: https://en.wikipedia.org/wiki/Standard_Model.

quantized energy of excitations of force fields are bosons, Figs. 3.3 and 3.4. At the early days of the standard model, the generation of mass of weak forces (bosons) has become problematic. But newly discovered Higgs fields are thought to provide the short-range force field and very heavy short-range bosons. Therefore, Higgs fields are also responsible for giving masses to particles [29]. In the Standard Model of elementary particle physics, the phrase "Higgs mechanism" refers to the generation of masses for the $W_{\pm}$, and Z weak gauge bosons through electroweak symmetry breaking. The Nambu–Goldstone bosons (NGBs) is generally for CMP as the massive Z bosons are for elementary particle physics. The Nambu–Goldstone mode is a long-wavelength fluctuation of the order parameter, e.g., in superconductors. In particle and CMP, Goldstone bosons or NGBs are bosons that appear in models exhibiting spontaneous breakdown of continuous symmetries. On the other hand, although symmetry breaking excitation solitons (bosons) in condensed matter do not generally interact and pass through each other without interacting, the bosons of strong forces in elementary particle physics

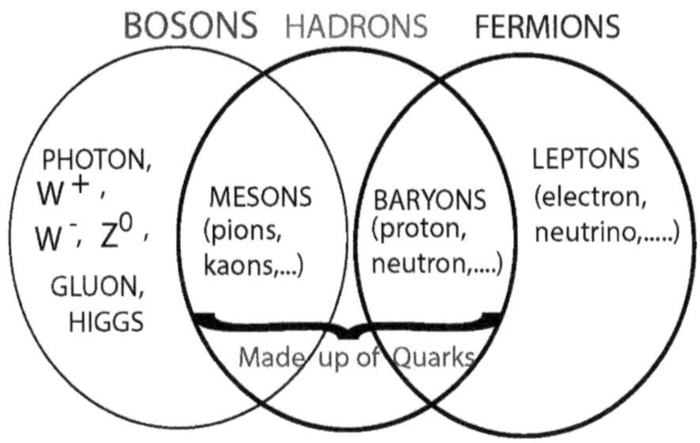

Fig. 3.4. All meson particles are unstable. These particles tend to decay, forming electrons and neutrinos if the meson has a charge. But uncharged mesons undergo decay forming photons. The mesons have an integer spin (baryons have half-integer spin).

Source: Redrawn from: https://en.wikipedia.org/wiki/Boson.

are highly-formed bunches which do not decrease with distance. This strong long-range forces are referred to as gluons.

The charge of U(1) gauge bosons maybe determined in similar way as in the determination of the charge of Dirac magnetic monopole discussed in Section 13.6. Therefore interacting fermions carry boson quantized charges, e.g., protons have positive charge while electrons have negative charge. In the standard model, fermions are *classified* according to what charges they carry or how they interact.

3.2.4 *Fermionization and bosonization: Symmetry breaking excitations, spinons, magnons*

Fermionization and bosonization are discussed in several works [30–34]. The Jordan–Wigner transformation is discussed in Refs. [35, 36]. Indeed, there is duality between free bosons and free fermions, or free Bloch fermions in CMP.

Symmetry breaking excitations of ground states can generally pass through each other, even in lower dimensions, and are characterized as bosons. In fact spinons are bosons which carry spin-$\frac{1}{2}$ in 1D and 2D antiferromagnets [37–41]. Magnons are also bosons [42]. Within superconductivity in CMP and electroweak symmetry breaking in particle physics, some of the intriguing phenomena can be understood within the regime of spontaneous breaking of symmetries, i.e., the ground state of a quantum system being less symmetric than the corresponding Hamiltonian. A characteristic sign of broken symmetries are massless bosonic excitations, which generally

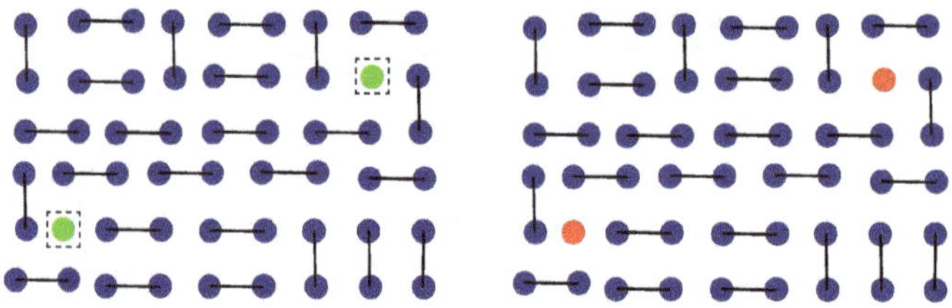

(a) (b)

Fig. 3.5. An illustration of spin-charge separation in spin singlet dimers. (a) A system with two spinons, made by unpairing two of the electrons and no longer in singlet state. The collective excitations must not have electric charge because the system still has the same charge. (b) The system has two electrons removed resulting in collective excitations called holons. The system has spin zero but now has charge.

Source: Redrawn from Ref. [45].

Table 3.1. Differences between baryons and mesons.

	Baryons	Mesons
Definitions	Subatomic particles	Hadronic particles
Category	Fermions	Bosons
Spin	Half-integer spin	Integer spin
Quarks	Has three quarks	Has a pair of quarks antiquarks
Interactions	Participate in strong interactions	Participate in strong & weak interactions
Examples	Protons and neutrons	Heavier mesons decay to lighter mesons

can pass through each other, known as NGBs. The number of such distinct bosonic excitations are taken to be a measure of how many of the original symmetries are dynamically broken. However, a general formula that predicts the number of different NGBs in a given dynamical system with broken symmetries is still not resolved [43, 44].

Fermions in one dimension cannot move independently, but only collide with each other resulting in density fluctuations. An interesting instability of quasiparticles in CMP is illustrated by the transformation of an electron into spinon, holon, as well as orbiton in quasi 1D sample of strontium cuprate [40, 41]. Spinons and holons are illustrated in Fig. 3.5. Spin-charge separations occurs in 1D Tomonaga–Luttinger (TL) liquids.

Table 3.1 illustrate the difference between baryons and mesons as subatomic quasiparticles in matter. The key difference between baryons and mesons is that baryons consist of a combination of three quark particles, whereas mesons consist of a pair of quark–antiquark particles.

Table 3.1 shows the marked difference between baryons and mesons. In the third column, the lighter mesons will ultimately decay to stable electrons, neutrinos, photons, etc.

3.2.5 *The Tomonaga–Luttinger liquids and Hubbard models*

An iconic bosonization of a 1D interacting fermi system occurs in the so-called TL liquids and Hubbard models. Although the bare system of electrons in a quantum wire consists of fermions, the low-energy elementary excitations in a TL liquid are bosons, these are sound waves of the 1D electron. In one dimension, a fermion (electron) that tries to propagate has to push its neighbors because of electron–electron interactions and Pauli exclusion principle. No individual motion in possible. Any individual excitation has to be a collective one, e.g., density fluctuation quasiparticles. The peculiarities of a 1D system of interacting fermions makes bosonization of low-energy excitation spectrum so effective. The low-energy excitation is well described by the TL liquid theory.

Moreover, the 1D Hubbard model in a magnetic field is equivalent under renormalization-group transformation to a multicomponent TL model. The numerical evaluations of the correlation functions and the analytic results [46–48] indicated clearly that the 1D Hubbard Model is a TL liquid.

3.2.5.1 *"Decay" of fermion into spinon, holon, and orbiton in CMP*

Remarkably, the TL model exhibits spin-charge separation, in which the spin and charge of the fermions possess independent dynamics, so the velocities of the spin and charge excitations of a TL liquid are distinct. These quasi-particles are the so-called spinons, holons, and in some cases orbitons, e.g., in *quasi* 1D sample of strontium cuprate [49, 50]. Since spinons are bosons, we expect spinons move faster than holons and separates.

Chapter 4

The Expansion of Any Operators in CMP: Phase-Space Point Projector

Any operator, $\hat{A}$, can be can be expanded in terms of mixed-space projector by Eq. (2.1), which we rewrite as

$$A = \sum_{p'',q'} |p''\rangle \langle p''| A |q'\rangle \langle q'| , \tag{4.1}$$

$$A = \sum_{p'',q'} \langle p''| A |q'\rangle \; |p''\rangle \langle q'| . \tag{4.2}$$

We wish to express $\langle p''| A |q'\rangle$ and $|p''\rangle \langle q'|$ in terms of the position eigenstate matrix elements and momentum space projectors.

4.1 Change of units for a unification of CMP and QFT

In preparations of our discussions using creation and annihilation operators on the unification with quantum field theory, we let $\hbar = 1$ in what follows unless specified. Using, Eqs. (2.8)–(2.12), we write

$$\begin{aligned} \langle p''| A |q'\rangle &= \frac{1}{\sqrt{N}} \sum_{q''} e^{-ip''\cdot q''} \langle q''| A |q'\rangle , \\ &= \sum_{q''} \frac{\langle q''| A |q'\rangle}{\langle q''| |p''\rangle}, \end{aligned} \tag{4.3}$$

$$
\begin{aligned}
|p''\rangle \langle q'| &= \frac{1}{\sqrt{N}} \sum_{p'} e^{ip' \cdot q'} |p''\rangle \langle p'|, \\
&= \sum_{p'} \frac{|p''\rangle \langle p'|}{\langle p'| |q'\rangle}, \qquad (4.4)
\end{aligned}
$$

with completeness relation, characteristic of dual Hilbert spaces notably given by Eq. (4.6)

$$
\frac{1}{\sqrt{N}} \sum_{p,q} e^{-ip \cdot q} |p\rangle \langle q| = 1, \qquad (4.5)
$$

$$
\sum_{p,q} \frac{|p\rangle \langle q|}{\langle q| |p\rangle} = 1. \qquad (4.6)
$$

Introducing the notation in Eq. (4.4),

$$
\vec{p}' = \vec{p} + \vec{u}, \quad \vec{q}' = \vec{q} + \vec{v}, \qquad (4.7)
$$

$$
\vec{p}'' = \vec{p} - \vec{u}, \quad \vec{q}'' = \vec{q} - \vec{v}, \qquad (4.8)
$$

we have, upon substituting in Eq. (4.2),

$$
\begin{aligned}
A &= \sum_{p'', q'} \langle p''| A |q'\rangle \, |p''\rangle \langle q'|, \\
&= \frac{1}{N} \sum_{p,q,u,v} \left(e^{i2\vec{p} \cdot \vec{v}} \langle \vec{q} - \vec{v}| A |\vec{q} + \vec{v}\rangle \right) \left(e^{i2\vec{u} \cdot \vec{q}} |\vec{p} - \vec{u}\rangle \langle \vec{p} + \vec{u}| \right). \qquad (4.9)
\end{aligned}
$$

4.1.1 *Mixed space operator basis*

We write the last result as an expansion in terms of *phase-space point projector*, $\hat{\Delta}(p, q)$, defined as the *lattice Weyl transform* of a projector, by

$$
\hat{\Delta}(p, q) = \frac{1}{\sqrt{N}} \sum_{u} e^{i2\vec{u} \cdot \vec{q}} |\vec{p} - \vec{u}\rangle \langle \vec{p} + \vec{u}|, \qquad (4.10)
$$

and the coefficient of expansion, the so-called *lattice Weyl transform* of matrix element of operator, $A(p, q)$, defined by

$$
A(p, q) = \frac{1}{\sqrt{N}} \sum_{v} e^{i2\vec{p} \cdot \vec{v}} \langle \vec{q} - \vec{v}| A |\vec{q} + \vec{v}\rangle. \qquad (4.11)
$$

Clearly, for a density matrix operator $\hat{\rho}$, the lattice Weyl transform obeys,

$$\sum_{p,q} \rho(p,q) = 1. \tag{4.12}$$

If one accounts for other extra discrete quantum labels like spin and energy-band indices, we can incorporate this simply by adding discrete indices and performing summation in a form of a trace.

Thus, we eventually have any operator expanded in terms of $\hat{\Delta}_{\lambda'\lambda}(p,q)$,

$$\hat{A} = \sum_{p,q,\lambda,\lambda'} A_{\lambda\lambda'}(p,q)\ \hat{\Delta}_{\lambda'\lambda}(p,q), \tag{4.13}$$

and hence it follows:

$$A_{\lambda\lambda'}(p,q) = \mathrm{Tr}(\hat{A}\hat{\Delta}_{\lambda'\lambda}(p,q)). \tag{4.14}$$

The validity of Eq. (4.14) readily follows from the important rule established in the next chapter that the trace of binary product of operators equals to product of its respective lattice Weyl transforms. The lattice Weyl transform of $\hat{\Delta}_{\lambda'\lambda}(p,q)$ is

$$\begin{aligned}
\hat{\Delta}_{\lambda'\lambda}(p,q)[p'q']_{\alpha\beta} &= \sum_{u'} e^{2iq'\cdot u'} \langle p'+u',\alpha \,|\, \hat{\Delta}_{\lambda'\lambda}(p,q) \,|\, p'-u',\beta\rangle \\
&= \frac{1}{N}\sum_{u',u} e^{2i(q'\cdot u'+u\cdot q)} \langle p'+u',\beta||\vec{p}-\vec{u},\lambda'\rangle \\
&\quad \times \langle \vec{p}+\vec{u},\lambda|\,|p'-u',\alpha\rangle \\
&= \frac{1}{N}\sum_{u',u} e^{2i(q'\cdot u'+u\cdot q)} \delta_{p,p'}\delta_{u.-u'}\delta_{\beta,\lambda'}\delta_{\lambda,\alpha} \\
&= \frac{1}{N}\sum_{u'} e^{2i(q'-q)\cdot u'} \delta_{p,p'}\delta_{\beta,\lambda'}\delta_{\lambda,\alpha} \\
&= \delta_{q,q'}\delta_{p,p'}\delta_{\beta,\lambda'}\delta_{\lambda,\alpha}.
\end{aligned} \tag{4.15}$$

Therefore,

$$\begin{aligned}
A_{\lambda\lambda'}(p,q) &= \mathrm{Tr}\left[\hat{A}\hat{\Delta}_{\lambda'\lambda}(p,q)\right] \\
&= \sum_{p'q',\alpha,\beta} A(p'q')_{\alpha\beta}\hat{\Delta}_{\lambda'\lambda}(p,q)[p'q']_{\beta\alpha}
\end{aligned}$$

$$= \sum_{p'q',\alpha,\beta} A(p'q')_{\alpha\beta}\delta_{q,q'}\delta_{p,p'}\delta_{\beta,\lambda'}\delta_{\lambda,\alpha}$$

$$= A(p,q)_{\lambda\lambda'}.$$

Upon similar procedure based on Eq. (4.2), an equivalent expression can be obtained for $A_{\lambda\lambda'}(p,q)$ and $\hat{\Delta}_{\lambda'\lambda}(p,q)$, namely,

$$A_{\lambda\lambda'}(p,q) = \frac{1}{\sqrt{N}}\sum_{\bar{u}} e^{2i\bar{q}\cdot\bar{u}}\langle\bar{p}+\bar{u},\lambda|\,\hat{A}|\bar{p}-\bar{u},\lambda'\rangle, \tag{4.16}$$

$$\hat{\Delta}_{\lambda'\lambda}(p,q) = \frac{1}{\sqrt{N}}\sum_{\bar{v}} e^{2i\bar{p}\cdot\bar{v}}|\bar{q}+\bar{v},\lambda\rangle\langle\bar{q}-\bar{v},\lambda'\,|. \tag{4.17}$$

Chapter 5

Phase-Space Trace Formula: Weyl Transform of Matrix Elements

5.1 Trace as a classical-like averaging in phase space

Here, we will discuss in more detail the trace of binary product of operators in phase-space, since this is crucial to the development of the nonequilibrium quantum transport theory of IQHE and FQHE in Parts 7 and 8. Let an operator $\hat{A}$ be given in the mixed representation as ($\hbar \Longrightarrow 1$),

$$\hat{A} = \sum_{p_1,q_1,\alpha,\beta} A_{\alpha\beta}\left(p_1, q_1\right) \Delta_{\alpha\beta}\left(p_1, q_1\right). \tag{5.1}$$

Similarly, we have for the operator $\hat{B}$ as

$$\hat{B} = \sum_{p_2,q_2,\gamma\mu} B_{\gamma\delta}\left(p_2, q_2\right) \Delta_{\gamma\mu}\left(p_2, q_2\right). \tag{5.2}$$

Upon taking the trace of the binary product of $\hat{A}$ and $\hat{B}$, we have

$$\begin{aligned} \mathrm{Tr}(\hat{A}\hat{B}) = & \sum_{p_1,q_1,\alpha,\beta,p_2,q_2,\gamma\mu} A_{\alpha\beta}(p_1, q_1) B_{\gamma\mu}(p_2, q_2) \\ & \times \mathrm{Tr}(\Delta_{\alpha\beta}(p_1, q_1)\Delta_{\gamma\mu}(p_2, q_2)). \end{aligned} \tag{5.3}$$

Thus, the trace operation amounts to taking the trace of the binary product of two Δ's. Then we have, using the expressions

$$\Delta_{\alpha\beta}\left(p_1, q_1\right) = \frac{1}{\sqrt{N}} \sum_{u_1} e^{(2i/\hbar)q_1.u} \left|p_1 - u_1, \alpha\right\rangle \left\langle p_1 + u_1, \beta\right|, \tag{5.4}$$

$$\Delta_{\gamma\mu}\left(p_2, q_2\right) = \frac{1}{\sqrt{N}} \sum_{u_2} e^{(2i/\hbar)q_2.u_2} \left|p_2 - u_2, \gamma\right\rangle \left\langle p_2 + u_2, \mu\right|. \tag{5.5}$$

We have

$$\begin{aligned}
&\Delta_{\alpha\beta}(p_1, q_1)\Delta_{\gamma\mu}(p_2, q_2)\\
&\quad = \frac{1}{N}\sum_{u_1} e^{(2i/\hbar)q_1.u_1} \sum_{u_2} e^{(2i/\hbar)q_2.u_2}\, |p_1 - u_1, \alpha\rangle\\
&\qquad \times\, \langle p_1 + u_1, \beta|\, |p_2 - u_2, \gamma\rangle\, \langle p_2 + u_2, \mu|,\\
&\quad = \frac{1}{N}\sum_{u_1 u_2} e^{(2i/\hbar)(q_1.u + q_2.u_2)}\, |p_1 - u_1, \alpha\rangle\, \langle p_2 + u_2, \mu|\;\; \delta_{p_1+u_1, p_2-u_2}\delta_{\beta\gamma}.
\end{aligned} \tag{5.6}$$

Taking the trace yields

$$\begin{aligned}
&= \frac{1}{N}\sum_{u_1 u_2, p, \alpha'} e^{(2i/\hbar)(q_1.u_1+q_2.u_2)}\langle p, \alpha'||p_1 - u_1, \alpha\rangle\langle p_2 + u_2, \mu|\, |p, \alpha'\rangle\\
&\quad \times \delta_{p_1+u_1, p_2-u_2}\delta_{\beta\gamma},\\
&= \frac{1}{N}\sum_{u_1 u_2, p, \alpha'} e^{(2i/\hbar)(q_1.u_1+q_2.u_2)}\delta_{p1-u_1, p}\delta_{p2+u_2, p}\;\delta_{p_1+u_1, p_2-u_2}\delta_{\beta\gamma}\delta_{\alpha'\alpha}\delta_{\mu\alpha'},\\
&= \frac{1}{N}\sum_{p, \alpha'} e^{(2i/\hbar)(q_1.(p_1-p)+q_2.(p-p_2))}\;\delta_{p_1+(p_1-p), p_2-(p-p_2)}\delta_{\beta\gamma}\delta_{\alpha'\alpha}\delta_{\mu\alpha'},\\
&= \frac{1}{N}\sum_{p, \alpha'} e^{(2i/\hbar)(q_1.(p_1-p)+q_2.(p-p_2))}\;\delta_{2p_1-p, 2p_2-p}\delta_{\beta\gamma}\delta_{\alpha'\alpha}\delta_{\mu\alpha'},\\
&= \frac{1}{N}\sum_{p, \alpha'} e^{(2i/\hbar)(q_1.(p_1-p)+q_2.(p-p_2))}\;\delta_{p_1, p_2}\delta_{\beta\gamma}\delta_{\alpha'\alpha}\delta_{\mu\alpha'}.
\end{aligned} \tag{5.7}$$

Therefore,

$$\begin{aligned}
&\mathrm{Tr}(\Delta_{\alpha\beta}(p_1, q_1)\Delta_{\gamma\delta}(p_2, q_2))\\
&\quad = \frac{1}{N}\sum_{\eta, p}\sum_{u_1 u_2} \exp\left[\frac{2i}{\hbar}(q_1 \cdot u_1 + q_2 \cdot u_2)\right]\delta_{p_1-u_1, p}\\
&\qquad \times \delta_{p_2+u_2, p}\;\delta_{p_1+u_1, p_2-u_2}\;\delta_{\eta\alpha}\;\delta_{\beta\gamma}\;\delta_{\mu\eta},\\
&\quad = \frac{1}{N}\sum_{\eta, p}\exp\left[\frac{2i}{\hbar}p\cdot(q_2 - q_1) + q_1 \cdot p_1 - q_2 \cdot p_2\right] \times \delta_{p_1 p_2}\;\delta_{\eta\alpha}\;\delta_{\beta\gamma}\;\delta_{\mu\eta},
\end{aligned}$$

$$= \frac{1}{N}\sum_{\eta,p} \exp\left[\frac{2i}{\hbar}(p-p_1)\cdot(q_2-q_1)\right] \delta_{\eta\alpha}\ \delta_{\beta\gamma}\ \delta_{\mu\eta},$$

$$= \sum_{\eta,u} \exp\left[\frac{2i}{\hbar}u\cdot(q_2-q_1)\right] \delta_{\eta\alpha}\ \delta_{\beta\gamma}\ \delta_{\mu\eta},$$

$$= \frac{1}{N}\sum_{\eta}(N\hbar^3)\delta(q_2-q_1)\delta_{\eta\alpha}\ \delta_{\beta\gamma}\ \delta_{\mu\eta}. \tag{5.8}$$

Hence, we have

$$\begin{aligned}\mathrm{Tr}(\hat{A}\hat{B}) &= (N)^{-1} \sum_{p_1,q_1,\alpha,\beta,p_2,q_2,\gamma\mu} A_{\alpha\beta}(p_1,q_1)B_{\gamma\mu}(p_2,q_2) \\ &\quad \times \mathrm{Tr}(\Delta_{\alpha\beta}(p_1,q_1)\Delta_{\gamma\mu}(p_2,q_2)), \\ &= (N\hbar^3)^{-1} \sum_{p_1,q_1,\alpha,\beta,\gamma\mu} A_{\alpha\beta}(p_1,q_1)B_{\gamma\mu}(p_2,q_2)\sum_{\eta}(N\hbar^3)\delta_{q_2q_1}\delta_{\eta\alpha}\ \delta_{\beta\gamma}\ \delta_{\mu\eta}, \\ &= \sum_{p_1,q_1,\gamma\eta} A_{\eta\gamma}(p_1,q_1)B_{\gamma\eta}(p_1,q_1),\ \text{(recalling } \hbar\to 1) \\ &= \sum_{p,q,\alpha\beta} A_{\alpha\beta}(p,q)B_{\beta\alpha}(p,q).\end{aligned} \tag{5.9}$$

Similar results can be obtained if we use the alternate expression for the Δ operators, namely,

$$\Delta_{\alpha\beta}(p_1,q_1) = \frac{1}{\sqrt{N}}\sum_{v_1} e^{(2i/\hbar)p_1\cdot v_1}\left|q_1+v_1,\alpha\right\rangle\left\langle q_1-v_1,\beta\right|, \tag{5.10}$$

$$\Delta_{\gamma\mu}(p_2,q_2) = \frac{1}{\sqrt{N}}\sum_{v_2} e^{(2i/\hbar)p_2\cdot v_2}\left|q_2+v_2,\gamma\right\rangle\left\langle q_2-v_2,\mu\right|, \tag{5.11}$$

where, just as before

$$\hat{A} = \sum_{p_1,q_1,\alpha,\beta} A_{\alpha\beta}(p_1,q_1)\Delta_{\alpha\beta}(p_1,q_1), \tag{5.12}$$

and

$$\mathrm{Tr}(\hat{A}\hat{B}) = \sum_{p_1,q_1,\alpha,\beta,p_2,q_2,\gamma\mu} A_{\alpha\beta}(p_1,q_1)B_{\gamma\mu}(p_2,q_2)\,\mathrm{Tr}(\Delta_{\alpha\beta}(p_1,q_1)\Delta_{\gamma\mu}(p_2,q_2)). \tag{5.13}$$

Now with the alternate expression for Δ,

$$\begin{aligned}
&\Delta_{\alpha\beta}(p_1,q_1)\Delta_{\gamma\mu}(p_2,q_2)\\
&\quad=\frac{1}{N}\sum_{v_1,v_2} e^{(2i/\hbar)(p_1\cdot v_1+p_2\cdot v_2)}\,|q_1+v_1,\alpha\rangle\langle q_1-v_1,\beta|\,|q_2+v_2,\gamma\rangle\langle q_2-v_2,\mu|\,,\\
&\quad=\frac{1}{N}\sum_{v_1,v_2} e^{(2i/\hbar)(p_1\cdot v_1+p_2\cdot v_2)}\,|q_1+v_1,\alpha\rangle\langle q_2-v_2,\mu|\,\delta_{q_1-v_1,q_2+v_2}\delta_{\beta\gamma}.
\end{aligned}\tag{5.14}$$

We have,

$$\begin{aligned}
&\mathrm{Tr}(\Delta_{\alpha\beta}(p_1,q_1)\Delta_{\gamma\mu}(p_2,q_2))\\
&\quad=\frac{1}{N}\sum_{v_1,v_2,q,\eta} e^{(2i/\hbar)(p_1\cdot v_1+p_2\cdot v_2)}\,\langle q,\eta|\,|q_1+v_1,\alpha\rangle\langle q_2-v_2,\mu|\\
&\qquad\times|q,\eta\rangle\,\delta_{q_1-v_1,q_2+v_2}\delta_{\beta\gamma},\\
&\quad=\frac{1}{N}\sum_{v_1,v_2,q,\eta} e^{(2i/\hbar)(p_1\cdot v_1+p_2\cdot v_2)}\delta_{q_1+v_1,q}\delta_{q_2-v_2,q}\delta_{q_1-v_1,q_2+v_2}\delta_{\beta\gamma}\delta_{\eta\alpha}\delta_{\mu\eta},\\
&\quad=\frac{1}{N}\sum_{v_1,v_2,q,\eta} e^{(2i/\hbar)(p_1\cdot v_1+p_2\cdot v_2)}\delta_{q-q_1,v_1}\delta_{q_2-q,v_2}\delta_{q_1-v_1,q_2+v_2}\delta_{\beta\gamma}\delta_{\eta\alpha}\delta_{\mu\eta},\\
&\quad=\frac{1}{N}\sum_{q,\eta} e^{(2i/\hbar)((p_1-p_2)\cdot(q-q_1))}\delta_{q_1,q_2}\delta_{\beta\gamma}\delta_{\eta\alpha}\delta_{\mu\eta}\\
&\quad=\frac{1}{N}\sum_{v,\eta} e^{(2i/\hbar)((p_1-p_2)\cdot v)}\delta_{q_1,q_2}\delta_{\beta\gamma}\delta_{\eta\alpha}\delta_{\mu\eta},\\
&\quad=\frac{1}{N}\sum_{\eta}(N)\delta_{p_1,p_2}\delta_{q_1,q_2}\delta_{\beta\gamma}\delta_{\eta\alpha}\delta_{\mu\eta}.
\end{aligned}\tag{5.15}$$

Therefore,

$$\begin{aligned}
\mathrm{Tr}(\hat{A}\hat{B})&=(N)^{-1}\sum_{p_1,q_1,\alpha,\beta,p_2,q_2,\gamma\delta} A_{\alpha\beta}(p_1,q_1)B_{\gamma\mu}(p_2,q_2)\\
&\quad\times\mathrm{Tr}(\Delta_{\alpha\beta}(p_1,q_1)\Delta_{\gamma\mu}(p_2,q_2)),\\
&=(N\hbar^3)^{-1}\sum_{p_1,q_1,\alpha,\beta,p_2,q_2,\gamma\delta} A_{\alpha\beta}(p_1,q_1)B_{\gamma\mu}(p_2,q_2)
\end{aligned}$$

$$\times \sum_{\eta}(N\hbar^3)\delta_{p_1,p_2}\delta_{q_1,q_2}\delta_{\beta\gamma}\delta_{\eta\alpha}\delta_{\mu\eta},$$

$$= \sum_{p,q,\eta,\alpha} A_{\eta\alpha}(p,q)B_{\alpha\eta}(p,q), \tag{5.16}$$

yielding the same expression as before, Eq. (5.9). Note we restored $\hbar$ in the second line of Eq. (5.16).

Thus, in the mixed representation, the product of the lattice Weyl transforms of two operators, sum over the (p,q) phase space, and taking the trace over discrete indices η, α yields the trace of binary product of any quantum operators.

Example 5.1. As well-known application of the formula, consider a more familiar quantum averaging of any operator $\hat{A}$ using the density matrix operator, $\hat{\rho}$. We have the familiar expression,

$$\langle \hat{A} \rangle = \mathrm{Tr}(\hat{A}\hat{\rho}). \tag{5.17}$$

By applying our formula for the expression of trace in mixed representation, we have

$$\langle \hat{A} \rangle = \sum_{p,q,\eta,\alpha} A_{\eta\alpha}(p,q)W_{\alpha\eta}(p,q), \tag{5.18}$$

where $A_{\eta\alpha}(p,q)$ is the lattice Weyl transform of $\hat{A}$. Here, $W_{\alpha\eta}(p,q)$, the Wigner distribution function, is the lattice Weyl transform of the density matrix operator, $\hat{\rho}$. Except for the trace over discrete indices or quantum labels, Eq. (5.18) is entirely of the form of the familiar classical averaging in phase space in classical statistical mechanics, although the Wigner distribution function, $W_{\alpha\eta}(p,q)$, is not positive definite, unlike classical Boltzmann distribution functions.

This trace relation in the mixed representation is crucial in our non-equilibrium quantum transport approach to IQHE and FHQE, discussed in more detail in later chapters, in Part 7, where the discrete indices correspond to energy-band indices. Section 5.2[1] outlines these derivations.

---

[1]The next two examples maybe omitted by the reader and proceed to the next chapter. These examples are treated in more detail in Parts 7 and 8 of the book.

5.2 Example: Quantized Hall electrical conductivity

Consider an interesting case of quantum Hall electrical conductivity[2] with the electric field in the x-direction,

$$\sigma_{yx} = \left(\frac{e^2}{h}\right)\frac{1}{(2\pi\hbar)}\sum_{\alpha\beta}\int\int\int d\vec{\mathcal{K}}_x d\vec{\mathcal{K}}_y dt$$

$$\times\left[\frac{\partial^{(a)}}{\partial\vec{\mathcal{K}}_x}\frac{\partial^{(b)}}{\partial\vec{\mathcal{K}}_y} - \frac{\partial^{(a)}}{\partial\vec{\mathcal{K}}_y}\frac{\partial^{(b)}}{\partial\vec{\mathcal{K}}_x}\right] H^{(a)}_{\alpha\beta}(\vec{\mathcal{K}},\mathcal{E})(-iG^{<(b)}_{\beta\alpha}(\vec{\mathcal{K}},\mathcal{E})), \quad (5.19)$$

where the lattice Weyl transforms are effectively given by,

$$H^{(a)}_{\alpha\beta}(\vec{p},\vec{q};E,t) \Longrightarrow H_{\alpha\beta}(\vec{\mathcal{K}},\mathcal{E}), \quad (5.20)$$

$$G^{<(b)}_{\beta\alpha}(\vec{p},\vec{q};E,t) \Longrightarrow G^{<(b)}_{\beta\alpha}(\vec{\mathcal{K}},\mathcal{E}). \quad (5.21)$$

Here, the expected dynamical variables in the phase space including the time variable occurs as particular combinations in $\vec{\mathcal{K}}$ and $\mathcal{E}$. This situation occurs when a crystalline solid is subject to a uniform electric field. Therefore, besides the crystal momentum varying in time as

$$\vec{\mathcal{K}} = \vec{p} + e\vec{F}t, \quad (5.22)$$

the energy variable vary with $\vec{q}$ as

$$\mathcal{E} = E + e\vec{F}\cdot\vec{q}, \quad (5.23)$$

where $\vec{F}$ is the electric field strength. A rigorous physical justification of Eqs. (5.22) and (5.23) is given in Chapter 31.

5.2.1 *"Pull back" of the lattice Weyl transformation of matrix elements*

The pull-back process means we have to undo the effect of lattice Weyl transformation operator, $\mathcal{W}$, which maybe viewed as a *Fourier transformation of matrix elements*. The results of the pull-back process after integration with respect to the time variable, given in more details in Part 7, Eq. (5.19) can

[2]The derivation of this equation is given in later chapters dealing with IQHE and FQHE.

be cast in the form,

$$
\begin{aligned}
&\langle \sigma_{yx}(\omega) \rangle \\
&= i\frac{e^2}{h}\frac{\delta(\omega)}{(2\pi)} \int\int dk_x dk_y \\
&\quad \times \sum_{\alpha} \left\{ \begin{array}{l} -\left\langle \alpha, \frac{\partial}{\partial k_x}\vec{\mathcal{K}}, \mathcal{E} \right| \sum_{\beta} \left| \beta, \vec{\mathcal{K}}, \mathcal{E} \right\rangle f(E_\beta) \left\langle \beta, \vec{\mathcal{K}}, \mathcal{E} \right| \left| \alpha, \frac{\partial}{\partial k_y}\vec{\mathcal{K}}, \mathcal{E} \right\rangle \\ + \left\langle \alpha, \frac{\partial}{\partial k_y}\vec{\mathcal{K}}, \mathcal{E} \right| \sum_{\beta} \left| \beta, \vec{\mathcal{K}}, \mathcal{E} \right\rangle f(E_\beta) \left\langle \beta, \vec{\mathcal{K}}, \mathcal{E} \right| \left| \alpha, \frac{\partial}{\partial k_x}\vec{\mathcal{K}}, \mathcal{E} \right\rangle \end{array} \right\} \\
&\quad + \left\{ \begin{array}{l} \left\langle \alpha, \frac{\partial}{\partial k_x}\vec{\mathcal{K}}, \mathcal{E} \right| \sum_{\beta} \left| \beta, \vec{\mathcal{K}}, \mathcal{E} \right\rangle \left\langle \beta, \vec{\mathcal{K}}, \mathcal{E} \right| \left| \alpha, \frac{\partial}{\partial k_y}\vec{\mathcal{K}}, \mathcal{E} \right\rangle \\ - \left\langle \alpha, \frac{\partial}{\partial k_y}\vec{\mathcal{K}}, \mathcal{E} \right| \sum_{\beta} \left| \beta, \vec{\mathcal{K}}, \mathcal{E} \right\rangle \left\langle \beta, \vec{\mathcal{K}}, \mathcal{E} \right| \left| \alpha, \frac{\partial}{\partial k_x}\vec{\mathcal{K}}, \mathcal{E} \right\rangle \end{array} \right\} f(E_\alpha).
\end{aligned}
\tag{5.24}
$$

The first curly bracketed term in Eq. (5.24) contributes to zero.

Proof. Assuming Hermitian density matrix for $\hat{\rho}$, consider the first curly bracketed term in Eq. (5.24),

$$
\begin{aligned}
&-\left\{ \left\langle \alpha, \frac{\partial}{\partial k_x}\vec{\mathcal{K}}, \mathcal{E} \right| \left| \beta, \vec{\mathcal{K}}, \mathcal{E} \right\rangle f(E_\beta) \left\langle \beta, \vec{\mathcal{K}}, \mathcal{E} \right| \left| \alpha, \frac{\partial}{\partial k_y}\vec{\mathcal{K}}, \mathcal{E} \right\rangle \right\}^{\dagger} \\
&\qquad = -\left\langle \alpha, \frac{\partial}{\partial k_y}\vec{\mathcal{K}}, \mathcal{E} \right| \left| \beta, \vec{\mathcal{K}}, \mathcal{E} \right\rangle f(E_\beta) \left\langle \beta, \vec{\mathcal{K}}, \mathcal{E} \right| \left| \alpha, \frac{\partial}{\partial k_x}\vec{\mathcal{K}}, \mathcal{E} \right\rangle.
\end{aligned}
\tag{5.25}
$$

Therefore,

$$
\begin{aligned}
&\sum_{\alpha,\beta} \left\{ \begin{array}{l} -\left\langle \alpha, \frac{\partial}{\partial k_x}\vec{\mathcal{K}}, \mathcal{E} \right| \left| \beta, \vec{\mathcal{K}}, \mathcal{E} \right\rangle f(E_\beta) \left\langle \beta, \vec{\mathcal{K}}, \mathcal{E} \right| \left| \alpha, \frac{\partial}{\partial k_y}\vec{\mathcal{K}}, \mathcal{E} \right\rangle \\ + \left\langle \alpha, \frac{\partial}{\partial k_y}\vec{\mathcal{K}}, \mathcal{E} \right| \left| \beta, \vec{\mathcal{K}}, \mathcal{E} \right\rangle f(E_\beta) \left\langle \beta, \vec{\mathcal{K}}, \mathcal{E} \right| \left| \alpha, \frac{\partial}{\partial k_x}\vec{\mathcal{K}}, \mathcal{E} \right\rangle \end{array} \right\} \\
&\quad = 2i\,\mathrm{Im} \sum_{\alpha,\beta} \left\langle \alpha, \frac{\partial}{\partial k_y}\vec{\mathcal{K}}, \mathcal{E} \right| \left| \beta, \vec{\mathcal{K}}, \mathcal{E} \right\rangle f(E_\beta) \left\langle \beta, \vec{\mathcal{K}}, \mathcal{E} \right| \left| \alpha, \frac{\partial}{\partial k_x}\vec{\mathcal{K}}, \mathcal{E} \right\rangle.
\end{aligned}
\tag{5.26}
$$

Assuming $\langle \alpha, \frac{\partial}{\partial k_y}\vec{\mathcal{K}}, \mathcal{E}|$ and $|\alpha, \frac{\partial}{\partial k_x}\vec{\mathcal{K}}, \mathcal{E}\rangle$ are expanded in terms of the orthogonal basis states, then since $\hat{\rho}$ is a projector, we must have,

$$
\mathrm{Im} \sum_{\alpha,\beta} \left\langle \alpha, \frac{\partial}{\partial k_y}\vec{\mathcal{K}}, \mathcal{E} \right| |\beta, \vec{\mathcal{K}}, \mathcal{E}\rangle f(E_\beta) \langle \beta, \vec{\mathcal{K}}, \mathcal{E}| \left| \alpha, \frac{\partial}{\partial k_x}\vec{\mathcal{K}}, \mathcal{E} \right\rangle = 0, \tag{5.27}
$$

being a diagonal Hermitian density matrix. □

Taking the limit $\omega \Longrightarrow 0$ in Eq. (5.24) and summing over the states β, we readily obtain the total conductivity, σ_{yx}, sum over the α band index,

$$\sigma_{yx} = \frac{e^2}{h}\sum_{\alpha} f(E_\alpha)\frac{i}{(2\pi)}\int\int dk_x dk_y \begin{bmatrix} \left\langle \alpha, \frac{\partial}{\partial k_x}\vec{\mathcal{K}}, \mathcal{E}\right|\left|\alpha, \frac{\partial}{\partial k_y}\vec{\mathcal{K}}, \mathcal{E}\right\rangle \\ -\left\langle \alpha, \frac{\partial}{\partial k_y}\vec{\mathcal{K}}, \mathcal{E}\right|\left|\alpha, \frac{\partial}{\partial k_x}\vec{\mathcal{K}}, \mathcal{E}\right\rangle \end{bmatrix}, \tag{5.28}$$

where the integration variable is transformed with Jacobian unity as

$$\vec{\mathcal{K}} = \vec{p} + e\vec{F}t \Longrightarrow \vec{p} = \hbar\vec{k}. \tag{5.29}$$

At low temperature, we can just write Eq. (5.28) as

$$\begin{aligned}(\sigma_{yx})_{\alpha\alpha} &= \frac{e^2}{2\pi\hbar}\frac{1}{(2\pi)}\sum_{\alpha}\int\int_{\text{occupiedBZ}} dk_x dk_y \left[\nabla_{\vec{k}} \times \left\langle \alpha, \vec{k}\right| i\frac{\partial}{\partial \vec{k}}\left|\alpha, \vec{k}\right\rangle\right]_{\text{plane}}, \\ &= \frac{e^2}{2\pi\hbar}\frac{1}{(2\pi)}\sum_{\alpha}\oint dk_c \left[\left\langle \alpha, \vec{k}\right| i\frac{\partial}{\partial k_c}\left|\alpha, \vec{k}\right\rangle\right]_{\text{contour or Wilson loop over BZ}}, \\ &= \left(\frac{e^2}{h}\right)\sum_{\alpha}\frac{\Delta\phi_{\text{total}}}{2\pi} = \sum_{\alpha}\left(\frac{e^2}{h}\right)n_\alpha, \end{aligned} \tag{5.30}$$

over all occupied bands α, where $n_\alpha \in \mathbb{Z}$ is the topological first Chern (or *winding*) number for each band. Thus, the the Hall conductivity is quantized in units of $(\frac{e^2}{h})$ as derive with the aid from the trace formula applied to the quantum transport method used here.

5.3 Example: Kubo current–current correlation formula

In the time-dependent linear response perturbation of the Kubo current–current correlation, a time varying electric field is indirectly used. To get to IQHE the limiting case of $\omega \Longrightarrow 0$ is taken after Fourier transformation of a convolution integral of the linear response scheme. We start with the RHS of σ_{yx}, properly expressed as a trace, before any time integration,

$$\sigma_{yx}$$
$$= \frac{e^2}{h}\frac{1}{(2\pi\hbar)}\int\int\int d\vec{\mathcal{K}}_x d\vec{\mathcal{K}}_y dt$$

$$\times \sum_{\alpha,\beta} \left[\begin{array}{c} (E_\beta(\vec{\mathcal{K}},\mathcal{E}) - E_\alpha(\vec{\mathcal{K}},\mathcal{E})) \\ \times \left\{ \begin{array}{c} \left\langle \alpha, \frac{\partial}{\hbar \partial k_x} \vec{\mathcal{K}}, \mathcal{E} \right| \left| \beta, \vec{\mathcal{K}}, \mathcal{E} \right\rangle \left\langle \beta, \vec{\mathcal{K}}, \mathcal{E} \right| \left| \alpha, \frac{\partial}{\hbar \partial k_y} \vec{\mathcal{K}}, \mathcal{E} \right\rangle \\ - \left\langle \alpha, \frac{\partial}{\hbar \partial k_y} \vec{\mathcal{K}}, \mathcal{E} \right| \left| \beta, \vec{\mathcal{K}}, \mathcal{E} \right\rangle \left\langle \beta, \vec{\mathcal{K}}, \mathcal{E} \right| \left| \alpha, \frac{\partial}{\hbar \partial k_x} \vec{\mathcal{K}}, \mathcal{E} \right\rangle \end{array} \right\} \\ \times\ f(E_\alpha) e^{i(\omega_{\alpha\beta})t} \end{array} \right]. \tag{5.31}$$

Denoting $E_\beta(\vec{\mathcal{K}},\mathcal{E}) - E_\alpha(\vec{\mathcal{K}},\mathcal{E}) = \hbar\omega_{\beta\alpha}$, we have,

$$\sigma_{yx} = \frac{e^2}{h} \frac{1}{(2\pi\hbar)} \int\int\int d\vec{\mathcal{K}}_x d\vec{\mathcal{K}}_y dt \times \sum_{\alpha,\beta} \left[\begin{array}{c} \hbar\omega_{\beta\alpha} \\ \times \left\{ \begin{array}{c} \left\langle \alpha, \frac{\partial}{\hbar \partial k_x} \vec{\mathcal{K}}, \mathcal{E} \right| \left| \beta, \vec{\mathcal{K}}, \mathcal{E} \right\rangle \left\langle \beta, \vec{\mathcal{K}}, \mathcal{E} \right| \left| \alpha, \frac{\partial}{\hbar \partial k_y} \vec{\mathcal{K}}, \mathcal{E} \right\rangle \\ - \left\langle \alpha, \frac{\partial}{\hbar \partial k_y} \vec{\mathcal{K}}, \mathcal{E} \right| \left| \beta, \vec{\mathcal{K}}, \mathcal{E} \right\rangle \left\langle \beta, \vec{\mathcal{K}}, \mathcal{E} \right| \left| \alpha, \frac{\partial}{\hbar \partial k_x} \vec{\mathcal{K}}, \mathcal{E} \right\rangle \end{array} \right\}, \\ \times f(E_\alpha) e^{i(\omega_{\alpha\beta})t} \end{array} \right] \tag{5.32}$$

$$\sigma_{yx} = \frac{e^2}{h} \frac{\hbar}{(2\pi\hbar)\hbar^2} \int\int\int d\vec{\mathcal{K}}_x d\vec{\mathcal{K}}_y dt \times \sum_{\alpha,\beta} \left[\begin{array}{c} \omega_{\beta\alpha} \\ \times \left\{ \begin{array}{c} \left\langle \alpha, \frac{\partial}{\partial k_x} \vec{\mathcal{K}}, \mathcal{E} \right| \left| \beta, \vec{\mathcal{K}}, \mathcal{E} \right\rangle \left\langle \beta, \vec{\mathcal{K}}, \mathcal{E} \right| \left| \alpha, \frac{\partial}{\partial k_y} \vec{\mathcal{K}}, \mathcal{E} \right\rangle \\ - \left\langle \alpha, \frac{\partial}{\partial k_y} \vec{\mathcal{K}}, \mathcal{E} \right| \left| \beta, \vec{\mathcal{K}}, \mathcal{E} \right\rangle \left\langle \beta, \vec{\mathcal{K}}, \mathcal{E} \right| \left| \alpha, \frac{\partial}{\partial k_x} \vec{\mathcal{K}}, \mathcal{E} \right\rangle \end{array} \right\} \\ \times f(E_\alpha) e^{i(\omega_{\alpha\beta})t} \end{array} \right]. \tag{5.33}$$

We make use of the general relations,

$$\langle \alpha, \vec{p}|\, \vec{v}\, |\beta, \vec{p}\rangle \equiv \omega_{\beta\alpha} \left\langle \alpha, \nabla_{\vec{k}} \vec{p} \right| |\beta, \vec{p}\rangle, \tag{5.34}$$

$$\left\langle \alpha, \nabla_{\vec{k}} \vec{p} \right| |\beta, \vec{p}\rangle = \frac{\langle \alpha, \vec{p}|\, \vec{v}_g\, |\beta, \vec{p}\rangle}{\omega_{\beta\alpha}}. \tag{5.35}$$

Similarly, we have

$$\langle \beta, \vec{p}|\, \vec{v}\, |\alpha, \vec{p}\rangle \equiv \omega_{\beta\alpha} \langle \beta, \vec{p}|\, \nabla_{\vec{k}}\, |\alpha, \vec{p}\rangle, \tag{5.36}$$

$$\langle \beta, \vec{p}|\, \nabla_{\vec{k}}\, |\alpha, \vec{p}\rangle = \frac{\langle \beta, \vec{p}|\, \vec{v}_g\, |\alpha, \vec{p}\rangle}{\omega_{\beta\alpha}}. \tag{5.37}$$

Substituting in Eq. (5.31), we then have the convolution integral of the *Hall current response function* incorporated in $v_{g,y}(t-t')$, with respect to time,

$$\text{RHS} = \frac{e^2}{h}\frac{1}{(2\pi\hbar)\hbar}\int\int\int d\vec{\mathcal{K}}_x d\vec{\mathcal{K}}_y dt' \times \sum_{\alpha,\beta}\left[\left\{\begin{array}{c} \frac{1}{\omega_{\beta\alpha}}\left\langle \alpha,\vec{\mathcal{K}},\mathcal{E}\right| v_{g,x}\left|\beta,\vec{\mathcal{K}},\mathcal{E}\right\rangle \\ \times\left\langle \beta,\vec{\mathcal{K}},\mathcal{E}\right| v_{g,y}(t-t')\left|\alpha,\vec{\mathcal{K}},\mathcal{E}\right\rangle \\ -\left\langle \alpha,\frac{\partial}{\partial k_y}\vec{\mathcal{K}},\mathcal{E}\right| v_{g,y}(t-t')\left|\beta,\vec{\mathcal{K}},\mathcal{E}\right\rangle \\ \times\left\langle \beta,\vec{\mathcal{K}},\mathcal{E}\right| v_{g,x}\left|\alpha,\vec{\mathcal{K}},\mathcal{E}\right\rangle \frac{1}{\omega_{\beta\alpha}} \end{array}\right\} \times e^{i(\omega_{\alpha\beta})t'} f(E_\alpha)\right]. \quad (5.38)$$

Consider the following Fourier transformation:

$$\frac{1}{\sqrt{2\pi}}\int_{-\infty}^{\infty} e^{i\omega t}F(t)dt = \frac{1}{\sqrt{2\pi}}\int_{-\infty}^{\infty} e^{i\omega t}dt\int_{-\infty}^{0} f(t-t')g(t')dt'. \quad (5.39)$$

Making the substitution,

$$t - t' = \xi \Longrightarrow dt = d\xi, \quad (5.40)$$

$$t = (t' + \xi), \quad (5.41)$$

we have,

$$\frac{1}{\sqrt{2\pi}}\int_{-\infty}^{\infty} e^{i\omega t}F(t)\,dt = \frac{1}{\sqrt{2\pi}}\int_{-\infty}^{\infty} e^{i\omega t'}g(t')dt'\int_{-\infty}^{0} e^{i\omega\xi}f(\xi)d\xi. \quad (5.42)$$

The first factor changes the $\frac{1}{\omega_{\beta\alpha}}$ to $\frac{1}{\omega}$. For the second factor, we can transform the range of integration as follows,

$$\begin{aligned}\int_{-\infty}^{0} e^{i\omega\xi}f(\xi)d\xi &= \int_{\infty}^{0} e^{-i\omega\xi}f(-\xi)(-d\xi) = \int_{0}^{\infty} e^{-i\omega\xi}f(-\xi)d\xi, \\ &= \int_{0}^{\infty} e^{-i\omega\alpha}f^{\dagger}(\xi)d\xi = \int_{0}^{\infty} e^{-i\omega\xi}f(\xi)d\xi, \end{aligned} \quad (5.43)$$

since $f(\xi) = j(\xi)$ is the observable current density and hence self-adjoint. We can apply this result in what follows. Defining the current density as

$$j_x = \frac{ev_{g,x}}{a^2}, \quad (5.44)$$

we obtain, after Fourier transforming the convolution integral as

$$
\begin{aligned}
&\text{RHS}\\
&= \frac{a^2}{\hbar\omega}\int\int\left(\frac{a}{(2\pi\hbar)}\right)^2 d\vec{\mathcal{K}}_x d\vec{\mathcal{K}}_y \int_0^\infty dt\\
&\quad\times\sum_{\alpha,\beta}\left[\begin{Bmatrix}\left\langle \alpha,\vec{\mathcal{K}},\mathcal{E}\right| \frac{ev_{g,x}}{a^2}\left|\beta,\vec{\mathcal{K}},\mathcal{E}\right\rangle\left\langle \beta,\vec{\mathcal{K}},\mathcal{E}\right| \frac{ev_{g,y}(t)}{a^2}\left|\alpha,\vec{\mathcal{K}},\mathcal{E}\right\rangle\\ -\left\langle \alpha,\vec{\mathcal{K}},\mathcal{E}\right| \frac{ev_{g,y}(t)}{a^2}\left|\beta,\vec{\mathcal{K}},\mathcal{E}\right\rangle\left\langle \beta,\vec{\mathcal{K}},\mathcal{E}\right| \frac{ev_{g,x}}{a^2}\left|\alpha,\vec{\mathcal{K}},\mathcal{E}\right\rangle\end{Bmatrix}\atop \times\, e^{-i(\omega-i\eta)t}f(E_\alpha)\right],\\
&= \frac{a^2}{\hbar\omega}\int_0^\infty dt\int\int\left(\frac{a}{(2\pi\hbar)}\right)^2 d\vec{\mathcal{K}}_x d\vec{\mathcal{K}}_y\\
&\quad\times\sum_{\alpha,\beta}\left[\begin{Bmatrix}\left\langle \alpha,\vec{\mathcal{K}},\mathcal{E}\right| j_x(0)\left|\beta,\vec{\mathcal{K}},\mathcal{E}\right\rangle\left\langle \beta,\vec{\mathcal{K}},\mathcal{E}\right| j_y(t)\left|\alpha,\vec{\mathcal{K}},\mathcal{E}\right\rangle\\ -\left\langle \alpha,\vec{\mathcal{K}},\mathcal{E}\right| j_y(t)\left|\beta,\vec{\mathcal{K}},\mathcal{E}\right\rangle\left\langle \beta,\vec{\mathcal{K}},\mathcal{E}\right| j_x(0)\left|\alpha,\vec{\mathcal{K}},\mathcal{E}\right\rangle\end{Bmatrix}\atop \times\, e^{-i(\omega-i\eta)t}f(E_\alpha)\right],\\
&= \frac{a^2}{\hbar\omega}\int_0^\infty dt\ \text{Tr}\rho_0\left\{[j_x(0),j_y(t)]\right\}e^{-i(\omega-i\eta)t},
\end{aligned}
\tag{5.45}
$$

where η is just a regularization exponent at ∞. Therefore, the Kubo formula for the conductivity is given by

$$
\sigma_{yx}(\omega) = \frac{a^2}{\hbar\omega}\int_0^\infty dt\ \text{Tr}\rho_0\left\{[j_x(0),j_y(t)]\right\}e^{-i(\omega-i\eta)t}. \tag{5.46}
$$

This is the Kubo current–current correlation formula for the Hall conductivity.

Chapter 6

Fully Symmetric Translation Operators

The state,

$$\exp\{-iq\cdot P\}\ |0\rangle = |q\rangle\,, \tag{6.1}$$

is an eigenstate of the position operator with displaced eigenvalue by q. However, if undisplaced $|0\rangle \Rightarrow |q'\rangle$ then the state

$$\exp\{-iq\cdot P\}\ \left|q'\right\rangle = \left|q'+q\right\rangle \tag{6.2}$$

is an eigenstate of the position operator, Q, with eigenvalue $(q'+q)$.

We can symmetrize the translation operator of $|0\rangle$ by inserting $\exp\{ip\cdot Q\}$ in front of $|0\rangle$ which effectively insert unity. We thus have

$$|q\rangle = \exp\{-iq\cdot P\}\exp\{ip\cdot Q\}\,|0\rangle\,. \tag{6.3}$$

By the use of the Campbell–Baker–Hausdorff operator identity, we obtained

$$\begin{aligned}
&\exp\{-iq\cdot P\}\exp\{ip\cdot Q\}\\
&= \exp\{-i(q\cdot P-p\cdot Q)\}\exp\left\{\frac{[-iq\cdot P, ip\cdot Q]}{2}\right\},\\
&= \exp\{-i(q\cdot P-p\cdot Q)\}\exp\left\{i\frac{p\cdot q}{2}\right\}.
\end{aligned} \tag{6.4}$$

where in the last line we made use of Eq.(2.14). Therefore we have the symmetric form for the displacement operator generating the state $|q\rangle$ from $|0\rangle$ given by

$$T(q)^{\text{sym}} = \exp\left\{i\frac{p\cdot q}{2}\right\}\exp\{-i(q\cdot P-p\cdot Q)\}\,. \tag{6.5}$$

The displacement operator may also be interpreted as an operator for the preparation of the quantum eigenstate out of the "vacuum", $|0\rangle$,[1] basis eigenstate. Actually Eq. (6.5) incorporates generation of $|q\rangle$ and $|p\rangle$ states as the case maybe.

6.1 The generalized Pauli-matrix operator basis

The symmetric operator factor,[2]

$$\begin{aligned} Y_{p,q} &= (\exp\{-i(q\cdot P - p\cdot Q)\}), \\ &= \exp\left\{-i\frac{p\cdot q}{2}\right\}\exp\{-iq\cdot P\}\exp\{ip\cdot Q\}, \end{aligned} \tag{6.6}$$

is referred to here as the generalized mixed-space projector or sometimes referred to as the generalized Pauli-matrix operator [12]. It can be considered the universal form of projector in mixed representations of CMP. We shall see that this also holds with dual spaces of non-Hermitian operators in QFT.

6.1.1 *Algebra of generalized Pauli-matrix operator*

When operating on Wannier wave-vectors, $|q\rangle$, we call the operator $X_{q'} = \exp\left\{-\frac{i}{\hbar}q'\cdot P\right\}$ as the translation operator $X_{q'}$ and the operator $Z_{p'} = \exp\left\{\frac{i}{\hbar}p'\cdot Q\right\}$ as the phase operator $Z_{p'}$. They exchange roles when operating on Bloch wave-vectors, $|p\rangle$. We have

$$X_{q'}|q\rangle = |q+q'\rangle, \tag{6.7}$$

$$X_{q'}|p\rangle = \exp\left\{-\frac{i}{\hbar}q'\cdot p\right\}|p\rangle \tag{6.8}$$

$$Z_{p'}|q\rangle = \exp\left\{\frac{i}{\hbar}p'\cdot q\right\}|q\rangle, \tag{6.9}$$

$$Z_{p'}|p\rangle = |p+p'\rangle. \tag{6.10}$$

[1]Here, the concept of a vacuum state does not have a special meaning since $|0\rangle$ represent arbitrary reference lattice-site position. It is introduced simply to bring analogy with zero-eigenvalue of non-Hermitian operators in later chapters, there the state $|0\rangle$ has a distinguished position.

[2]The materials of this section are taken from the author's book [12]. This is given here for completeness.

6.1.1.1 *Symmetric translation operator*

The *shift* operator, $X_{q'}$, and *phase* operator, $Z_{p'}$, obey the following equations:

$$X_{q'}\left|q\right\rangle = \exp\left\{-\frac{i}{\hbar}q'\cdot P\right\}\left|q\right\rangle = \left|q+q'\right\rangle, \tag{6.11}$$

$$Z_{p'}\left|q\right\rangle = \exp\left\{\frac{i}{\hbar}p'\cdot Q\right\}\left|q\right\rangle = \exp\left\{\frac{i}{\hbar}p'\cdot q\right\}\left|q\right\rangle. \tag{6.12}$$

The product of $X_{q'}$ and $Z_{p'}$ is thus given by[3]

$$\begin{aligned} X_{q'}Z_{p'} &= \exp\left\{-\frac{i}{\hbar}q'\cdot P\right\}\exp\left\{\frac{i}{\hbar}p'\cdot Q\right\} = T\left(q',p'\right)^{\text{sym}} \\ &= \exp\left\{-\frac{i}{\hbar}(q'\cdot P - p'\cdot Q)\right\}\exp\left\{\frac{\left[-\frac{i}{\hbar}q'\cdot P, \frac{i}{\hbar}p'\cdot Q\right]}{2}\right\} \\ &= \exp\left\{\frac{i}{\hbar}\frac{p'\cdot q'}{2}\right\}Y_{q'p'}. \end{aligned} \tag{6.13}$$

where $Y_{q'p'}$ is the generalized spin-matrix operator.[4] Clearly we have, from Eq. (6.6),

$$\begin{aligned} \hat{Y}^{\dagger}_{q'p'} &= \exp\left\{\frac{i}{\hbar}\left(q'\cdot\hat{P}^{\dagger} - p'\cdot\hat{Q}^{\dagger}\right)\right\} \\ &= \exp\left\{\frac{i}{\hbar}\left(q'\cdot\hat{P} - p'\cdot\hat{Q}\right)\right\} = \hat{Y}_{-q'-p'}. \end{aligned} \tag{6.14}$$

6.1.1.2 *Commutation relations and products of* $Y_{q'p'}$

The use of Campbell–Baker–Hausdorff operator identity allows us to evaluate the following products of $Y_{q'p'}$, operating on wavevectors, $|q\rangle$.

$$Y_{q'p'}Y_{q''p''} = \exp\left\{-\frac{i}{\hbar}\left(q'\cdot P - p'\cdot Q\right)\right\}\exp\left\{-\frac{i}{\hbar}\left(q''\cdot P - p''\cdot Q\right)\right\}.$$

---

[3]The result may be considered a version of rotation and translation operation.

[4]Since $X_{q'}$, $Z_{p'}$, and $Y_{q'p'}$ are exponential function of the canonical operators, their eigenvalues for a qubit system is the set $\{1,-1\}$ instead of the set $\{0,1\}$ for the canonical operators themselves. Not the prefactor, $\exp\left\{\frac{i}{\hbar}\frac{p'\cdot q'}{2}\right\}$, is because we are operating on wavevector $|q\rangle$, by Eq. (2.14).

Therefore, for $\vec{Q}$ and $\vec{P}$ in the same direction, as in 1D system, then

$$Y_{q'p'}Y_{q''p''} = Y_{q'+q''p'+p''} \exp\left\{\frac{i}{2\hbar}\left[q'p'' - p'q''\right]\right\}. \tag{6.15}$$

Hence, we have the commutation relation,

$$\left[Y_{q'p'}, Y_{q''p''}\right] = Y_{q'+q''p'+p''} \begin{bmatrix} \exp\left\{\frac{i}{2\hbar}\left[q'p'' - p'q''\right]\right\} \\ - \exp\left\{\frac{i}{2\hbar}\left[q''p' - p''q'\right]\right\} \end{bmatrix}. \tag{6.16}$$

Note that $Y_{q'p'}$ and $Y_{q''p''}$ commute if and only if the symplectic product $[q'p'' - p'q''] = 0$, i.e., the two vectors $\binom{q'}{p'}$ and $\binom{q''}{p''}$ are collinear. Similarly, we have

$$\begin{aligned} Y_{q'p'}Y_{q''p''}Y_{q'''p'''} &= Y_{q'+q''p'+p''}Y_{q'''p'''} \exp\left\{\frac{i}{2\hbar}\left[q'p'' - p'q''\right]\right\} \\ &= Y_{q'+q''+q'''p'+p''+p'''} \exp\left\{\frac{i}{2\hbar}\left[q'p'' - p'q''\right]\right\} \\ &\quad \times \exp\left\{\frac{i}{2\hbar}\left[\left[q' + q''\right]p''' - \left[p' + p''\right]q'''\right]\right\} \\ &= Y_{q'+q''+q'''p'+p''+p'''} \exp\left\{\frac{i}{2\hbar}\left[q'\left[p'' + p'''\right] - p'\left[q'' + q'''\right]\right]\right\} \\ &\quad \times \exp\left\{\frac{i}{2\hbar}\left[q''p''' - p''q'''\right]\right\}. \end{aligned} \tag{6.17}$$

This can be generalized as

$$\begin{aligned} &Y_{q_1p_1}Y_{q_2p_2}Y_{q_3p_3}\cdots Y_{q_np_n} \\ &= Y_{\sum\limits_{j=1}^{n} q_j \sum\limits_{l=1}^{n} p_l} \exp\left\{\frac{i}{2\hbar}\left[q_1\sum_{j=2}^{n} p_j - p_1\sum_{j=2}^{n} q_j\right]\right\} \\ &\quad \times \exp\left\{\frac{i}{2\hbar}\left[q_2\sum_{j=2+1}^{n} p_j - p_2\sum_{j=2+1}^{n} q_j\right]\right\} \end{aligned}$$

$$\times \cdots\cdots$$

$$\exp\left\{\frac{i}{2\hbar}\left[q_{n-1}p_n - p_{n-1}q_n\right]\right\}. \tag{6.18}$$

If all the q_i's are equal to q and all p_l's are equal to p, then we have

$$(Y_{qp})^n = Y_{nq,np}\exp\left\{\frac{i\;n(n-1)}{4\hbar}[qp - pq]\right\}. \tag{6.19}$$

We can use Eq. (6.17) to evaluate the following expression:

$$\begin{aligned}
&Y_{q'p'}(Y_{qp})^n Y^\dagger_{q'p'} \\
&= Y_{q'p'}(Y_{qp})^n Y_{-q'-p'} \\
&= Y_{q'p'}Y_{nq,np}Y_{-q'-p'}\exp\left\{\frac{i\;n(n-1)}{4\hbar}\left[qp - pq\right]\right\} \\
&= Y_{q'+nq-q',p'+np-p'}\exp\left\{\frac{i}{2\hbar}\left[q'(np-p') - p'(nq-q')\right]\right\} \\
&\quad\times\exp\left\{-\frac{i}{2\hbar}\left[nq(p') - np(q')\right]\right\} \\
&= Y_{nq,np}\exp\left\{\frac{in}{\hbar}\left(q'p - p'q\right)\right\}.
\end{aligned} \tag{6.20}$$

6.1.2 *Pauli matrices from generalized Pauli-matrix operator*

We shall see below that these operators, X, Z, and Y, lead to the generalized Pauli matrices. Note that these results hold for discrete eigenvalues q and p belonging to finite fields, i.e., whose number of elements are power of a prime number.

In the orthonormal *Wannier function* basis states the matrix elements of X, Z, and Y, are calculated as follows:

$$\langle q''|\,X_{q'}\,|q\rangle = \langle q''|\exp\left\{-\frac{i}{\hbar}q'\cdot P\right\}|q\rangle = \delta_{q'',q+q'}, \tag{6.21}$$

$$\langle q''|\,Z_{p'}\,|q\rangle = \langle q''|\exp\left\{\frac{i}{\hbar}p'\cdot Q\right\}|q\rangle = \exp\left\{\frac{i}{\hbar}p'\cdot q\right\}\delta_{q'',q}, \tag{6.22}$$

$$\langle q''|\,Y_{q'p'}\,|q\rangle = \exp\left\{-\frac{i}{\hbar}\frac{p'\cdot q'}{2}\right\}\langle q''|\,X_{q'}Z_{p'}\,|q\rangle$$

$$= \exp\left\{-\frac{i}{\hbar}\frac{p'\cdot q'}{2}\right\}\exp\left\{\frac{i}{\hbar}p'\cdot q\right\}\langle q''|\,X_{q'}\,|q\rangle$$

$$= \exp\left\{-\frac{i}{\hbar}\frac{p'\cdot q'}{2}\right\}\exp\left\{\frac{i}{\hbar}p'\cdot q\right\}\delta_{q'',q+q'}. \tag{6.23}$$

Thus, we may express the generalized spin matrix operator expanded in terms of the *Wannier function* basis states as

$$Y_{q'p'} = \sum_{q'',q}\exp\left\{-\frac{i}{\hbar}\frac{p'\cdot q'}{2}\right\}\exp\left\{\frac{i}{\hbar}p'\cdot q\right\}\delta_{q'',q+q'}\,|q''\rangle\langle q|$$

$$= \exp\left\{-\frac{i}{\hbar}\frac{p'\cdot q'}{2}\right\}\sum_{q}\exp\left\{\frac{i}{\hbar}p'\cdot q\right\}|q+q'\rangle\langle q|\,. \tag{6.24}$$

In the orthonormal *Bloch wave-vector* basis states the matrix elements of X, Z, and Y, are calculated as follows:

$$\langle p''|\,X_{q'}\,|p\rangle = \langle p''|\exp\left\{-\frac{i}{\hbar}q'\cdot P\right\}|p\rangle = \exp\left\{-\frac{i}{\hbar}q'\cdot p\right\}\delta_{p'',p}, \tag{6.25}$$

$$\langle p''|\,Z_{p'}\,|p\rangle = \langle p''|\exp\left\{\frac{i}{\hbar}p'\cdot Q\right\}|p\rangle = \delta_{p'',p+p'}, \tag{6.26}$$

$$\langle p''|\,Y_{q'p'}\,|p\rangle = \exp\left\{-\frac{i}{\hbar}\frac{p'\cdot q'}{2}\right\}\langle p''|\,X_{q'}Z_{p'}\,|p\rangle$$

$$= \exp\left\{-\frac{i}{\hbar}\frac{p'\cdot q'}{2}\right\}\exp\left\{-\frac{i}{\hbar}q'\cdot p\right\}\langle p''|\,Z_{p'}\,|p\rangle$$

$$= \exp\left\{-\frac{i}{\hbar}\frac{p'\cdot q'}{2}\right\}\exp\left\{-\frac{i}{\hbar}q'\cdot p\right\}\delta_{p'',p+p'}. \tag{6.27}$$

6.1.2.1 *The prime number 2*

Now let us take the smallest field of prime number consisting of two elements, obeying modular arithmetic. Then the eigenvalues q, may be designated by the number 0 and 1, and the eigenvalues p designated by $\frac{2\pi\hbar n}{N}$, $\{n = 0, 1 \text{ and } N = 2\}$. Observe that the product of the form $pq = \frac{2\pi\hbar nm}{N}$, where n and m are integers, the number N must be a prime number such as the number 2. In this discrete quantum mechanics, we may still designate the *Wannier function* basis states by the Dirac notation $|q\rangle$, in this case by

$|0\rangle$ and $|1\rangle$ for $N = 2$. Substituting the possible values of q and p, namely, $\{0, 1\}$ and $\{\frac{2\pi\hbar 0}{2}, \frac{2\pi\hbar 1}{2}\}$, respectively, and using modular arithmetic, then from Eqs. (6.21)–(6.23), we have,

$$X_0 = Z_0 = Y_{0,0} = \begin{pmatrix} 1 & 0 \\ 0 & 1 \end{pmatrix} = I, \tag{6.28}$$

$$X_1 = \begin{pmatrix} 0 & 1 \\ 1 & 0 \end{pmatrix} = \sigma_x, \tag{6.29}$$

$$Z_1 = \begin{pmatrix} 1 & 0 \\ 0 & -1 \end{pmatrix} = \sigma_z, \tag{6.30}$$

$$\begin{aligned} Y_{01} &= \exp\left\{\frac{i}{\hbar}p' \cdot q\right\} \delta_{q'',q+q'} \\ &= \begin{pmatrix} 1 & 0 \\ 0 & -1 \end{pmatrix} = Z_1 = \sigma_z, \end{aligned} \tag{6.31}$$

$$Y_{10} = \delta_{q'',q+q'} = \begin{pmatrix} 0 & 1 \\ 1 & 0 \end{pmatrix} = X_1 = \sigma_x, \tag{6.32}$$

$$\begin{aligned} Y_{1,1} &= -i \exp\left\{\frac{i}{\hbar}p' \cdot q\right\} \delta_{q'',q+q'} \\ &= \begin{pmatrix} 0 & i \\ -i & 0 \end{pmatrix} = -\sigma_y, \end{aligned} \tag{6.33}$$

where σ_x, σ_y, and σ_z are the well-known Pauli matrices, the transformation matrices for particles with spin $\frac{1}{2}$.

Likewise, in the orthonormal *Bloch wave-vector* basis states the matrix elements of X, Z, and Y, can be calculated from the matrix expression,

$$\begin{aligned} \langle p''|\, \tilde{Y}_{p'q'}\, |p\rangle &= \exp\left\{\frac{i}{\hbar}\frac{p' \cdot q'}{2}\right\} \langle p''|\, \tilde{X}_{p'} \tilde{Z}_{q'}\, |p\rangle \\ &= \exp\left\{\frac{i}{\hbar}\frac{p' \cdot q'}{2}\right\} \exp\left\{-\frac{i}{\hbar}q' \cdot p\right\} \langle p''|\, \tilde{X}_{p'}\, |p\rangle \\ &= \exp\left\{\frac{i}{\hbar}\frac{p' \cdot q'}{2}\right\} \exp\left\{-\frac{i}{\hbar}q' \cdot p\right\} \delta_{p'',p+p'}, \end{aligned} \tag{6.34}$$

where σ_x and σ_y exchange role, with only $\tilde{Y}_{1,1}$ differing by a phase,

$$\tilde{Y}_{1,1} = \begin{pmatrix} 0 & -i \\ i & 0 \end{pmatrix} = \sigma_y. \tag{6.35}$$

We have the following relations for the Pauli matrices, the anticommutation is

$$\{\sigma_i, \sigma_j\} = 0 \quad \text{for all } i, j \text{ in } 0, 1$$

while the commutation obeys the "coordinate" cross-product rule,

$$[\sigma_i, \sigma_j] = 2i\ e_{ijk}\ \sigma_k.$$

On the other hand, a coordinate-free relations is given in the following section.

6.1.3 *Coordinate-free products and commutation properties*

For the products, we from Eq. (6.15)

$$Y_{q'p'}Y_{q''p''} = Y_{q'+q''p'+p''} \exp\left\{\frac{i}{2\hbar}\left[q'p'' - p'q''\right]\right\}. \tag{6.36}$$

The square of Y_{ij} gives,

$$\begin{aligned} Y_{ij}Y_{ij} &= Y_{2i,2j} \exp\left\{\frac{i}{2}\left[\pi j - j\pi\right]\right\} \\ &= Y_{0,0} = I. \end{aligned} \tag{6.37}$$

We have (noting that two-state systems obey modular arithmetic [*mod* (2)]),

$$\begin{aligned} Y_{10}Y_{11} &= Y_{1+1,0+1} \exp\left\{\frac{i}{2}\left[1\pi - 01\right]\right\} \\ &= iY_{0,1} \Longrightarrow [Y_{10}, Y_{11}] = 2iY_{0,1}, \end{aligned} \tag{6.38}$$

$$\begin{aligned} Y_{11}Y_{01} &= Y_{1+0,1+1} \exp\left\{\frac{i}{2}\left[1\pi - \pi 0\right]\right\} \\ &= iY_{1,0} \Longrightarrow [Y_{11}, Y_{01}] = 2iY_{1,0}, \end{aligned} \tag{6.39}$$

$$\begin{aligned} Y_{01}Y_{10} &= Y_{0+1,1+0} \exp\left\{\frac{i}{2}\left[00 - \pi 1\right]\right\} \\ &= -iY_{1,1} \Longrightarrow [Y_{01}, Y_{10}] = -2iY_{11}, \end{aligned} \tag{6.40}$$

where the expression for $Y_{1,1}$ is given by Eq. (6.33).

Expanding on the commutation properties, we have from Eq. (6.16), we have the general operator expressions,

$$\left[Y_{q'p'}, Y_{q''p''}\right] = Y_{q'+q''p'+p''} \begin{bmatrix} \exp\left\{\frac{i}{2\hbar}\left[q'p'' - p'q''\right]\right\} \\ -\exp\left\{\frac{i}{2\hbar}\left[q''p' - p''q'\right]\right\} \end{bmatrix}. \tag{6.41}$$

We have, therefore,

$$\begin{aligned} [Y_{10}, Y_{01}] &= Y_{1+0,0+1} \begin{bmatrix} \exp\left\{\frac{i}{2}\left[1\pi - 00\right]\right\} \\ -\exp\left\{\frac{i}{2}\left[00 - \pi 1\right]\right\} \end{bmatrix} \\ &= 2iY_{1,1}, \end{aligned} \tag{6.42}$$

$$\begin{aligned} [Y_{01}, Y_{11}] &= Y_{0+1,1+1} \begin{bmatrix} \exp\left\{\frac{i}{2}\left[0\pi - \pi 1\right]\right\} \\ -\exp\left\{\frac{i}{2}\left[1\pi - \pi 0\right]\right\} \end{bmatrix} \\ &= -2iY_{1,0}, \end{aligned} \tag{6.43}$$

$$\begin{aligned} [Y_{11}, Y_{10}] &= Y_{1+1,1+0} \begin{bmatrix} \exp\left\{\frac{i}{2}[10 - \pi 1]\right\} \\ -\exp\left\{\frac{i}{2}[1\pi - 01]\right\} \end{bmatrix} \\ &= -2iY_{0,1}, \end{aligned} \tag{6.44}$$

and

$$\begin{aligned} [Y_{01}, Y_{10}] &= Y_{0+1,1+0} \begin{bmatrix} \exp\left\{\frac{i}{2}\left[00 - \pi 1\right]\right\} \\ -\exp\left\{\frac{i}{2}\left[\pi 1 - 00\right]\right\} \end{bmatrix} \\ &= -2iY_{11}. \end{aligned} \tag{6.45}$$

6.1.4 *Symmetric form of* $\hat{\boldsymbol{\Delta}}_{\lambda'\lambda}\,(p,q)$

Consider Eq. (4.17). The following identity can be verified

$$|\bar{q} + \bar{v}, \lambda\rangle = \exp[-2i\hat{P}\cdot\bar{v}]\,|\bar{q} - \bar{v}, \lambda\rangle\,. \tag{6.46}$$

We also have,

$$\begin{aligned} &|\bar{q} - \bar{v}, \lambda\rangle\,\langle\bar{q} - \bar{v}, \lambda'| \\ &= (N)^{-1}\sum_{\bar{u}}\exp[-2i(\bar{q} - \bar{v} - \hat{Q})\cdot\bar{u}]\sum_{q}|q, \lambda\rangle\,\langle q, \lambda'|\,, \\ &= \sum_{q}\delta(\bar{q} - \bar{v} - \hat{Q})\,|q, \lambda\rangle\,\langle q, \lambda'| \\ &= \sum_{q}\delta\left(\bar{q} - \bar{v} - q\right)\;|q, \lambda\rangle\,\langle q, \lambda'|\,, \end{aligned} \tag{6.47}$$

$$\Omega_{\lambda\lambda'} = \sum_{q} |q, \lambda\rangle \langle q, \lambda'| \,. \tag{6.48}$$

Substituting the expressions, Eqs (6.46) and (6.47) in Eq. (4.17), we obtain a completely symmetric expression of $\hat{\Delta}_{\lambda'\lambda}(p, q)$,

$$\begin{aligned}
&\hat{\Delta}_{\lambda'\lambda}(p, q) \\
&\quad = \frac{1}{\sqrt{N}} \sum_{\bar{v}} e^{2i\bar{p}\cdot\bar{v}} |\bar{q} + \bar{v}, \lambda\rangle \langle \bar{q} - \bar{v}, \lambda'| \\
&\quad = \frac{1}{\sqrt{N}} (N)^{-1} \sum_{\bar{v},\bar{u}} e^{\frac{2i}{\hbar}\bar{p}\cdot\bar{v}} \exp[-2i\hat{P}\cdot\vec{v}] \exp[-2i(\bar{q} - \bar{v} - \hat{Q})\cdot\bar{u}] \\
&\qquad \times \sum_{q} |q, \lambda\rangle \langle q, \lambda'| \,, \\
&\quad = \frac{1}{\sqrt{N}} (N)^{-1} \sum_{\bar{v},\bar{u}} \exp\{-2i(P\cdot v - Q\cdot u)\} \exp 2i \{p\cdot v - q\cdot u\} \Omega_{\lambda\lambda'}.
\end{aligned} \tag{6.49}$$

Thus, we have changed the seemingly asymmetric expression of the first line into a symmetric form of the last line.[5] We can combine the exponential operators to obtain

$$\hat{\Delta}_{\lambda'\lambda}(p, q) = \frac{1}{\sqrt{N}} (N)^{-1} \sum_{v,u} \exp -2i[(\hat{P} - p)\cdot v - (Q - q)\cdot u]\Omega_{\lambda\lambda'}, \tag{6.50}$$

where,

$$\Omega_{\lambda\lambda'} = \sum_{q'} |q', \lambda\rangle \langle q', \lambda'| \,, \tag{6.51}$$

We therefore have,

$$\begin{aligned}
A(p, q) &= \mathrm{Tr}(\hat{A}\hat{\Delta}) \\
&= \frac{1}{\sqrt{N}} (N)^{-1} \left(\begin{array}{c} \sum_{v,u} \exp 2i\,[p\cdot v - q\cdot u] \\ \times \mathrm{Tr}\left\{ \hat{A} \exp\left\{ -2i \left[\hat{P}\cdot v - Q\cdot u \right] \right\} \Omega_{\lambda\lambda'} \right\} \end{array} \right).
\end{aligned} \tag{6.52}$$

[5]The exponential terms containing $\bar{v}\cdot\bar{u}$ reduces to unity under the summation or integral operation.

Therefore, the characteristic distribution of $A(p,q)$ is identified as

$$A_{\lambda\lambda'}(u,v) = (N)^{-1}\mathrm{Tr}\{\hat{A}\exp(-2i[\hat{P}\cdot v - Q\cdot u])\Omega_{\lambda\lambda'}\}.$$

By using this characteristic distribution function, we can write $A_{\lambda\lambda'}(p,q)$,

$$A_{\lambda\lambda'}(p,q) = \left(\frac{1}{N}\right)^{\frac{1}{2}} \sum_{u,v} A_{\lambda\lambda'}(u,v)e^{-2i(p\cdot v - q\cdot u)}, \tag{6.53}$$

with the inverse,

$$A_{\lambda\lambda'}(u,v) = \left(\frac{1}{N}\right)^{\frac{1}{2}} \sum_{p,q} A_{\lambda\lambda'}(p,q)\, e^{2i(p\cdot v - q\cdot u)}. \tag{6.54}$$

We can consider Eq. (6.55) simply like a Fourier transform (caveat: Fourier transform to *operator space*) of the characteristic function of the lattice Weyl transform of the operator $\hat{A}$,

$$\hat{A} = (N)^{-1} \sum_{u,v,\lambda,\lambda'} A_{\lambda\lambda'}(u,v)\exp[(-2i(\hat{P}\cdot v - \hat{Q}\cdot u))]\Omega_{\lambda\lambda'}. \tag{6.55}$$

In continuum approximation, we have,

$$\begin{aligned}\hat{\Delta}_{\lambda\lambda'}(p,q) &= (2\pi)^{-3}\int du\, dv\; e^{-i[(\hat{P}-\bar{p})\cdot\vec{v}-(Q-\bar{q})\cdot\bar{u}].u}\Omega_{\lambda\lambda'},\\ &= (2\pi)^{-3}\int du\, dv\; e^{(i)(p\cdot v - q\cdot u)}e^{(-i)(P.v-Q.u)}\Omega_{\lambda\lambda'}.\end{aligned} \tag{6.56}$$

From Eq. (6.53), we also have,

$$\begin{aligned}A_{\lambda\lambda'}(p,q) = \mathrm{Tr}\left(\hat{A}\hat{\Delta}\right) &= (2\pi)^{-3}\left(\int du\, dv\; e^{i(p\cdot v-q\cdot u)}\mathrm{Tr}\left(\hat{A}e^{(-i)(P.v-Q.u)}\right)\right),\\ &= (2\pi)^{-1}\left(\int du\, dv\; e^{(i)(p\cdot v-q\cdot u)}A_{\lambda\lambda'}(u,v)\right).\end{aligned} \tag{6.57}$$

6.2 Smoothing of the characteristic distributions

In general, we can have different expression for the characteristic function depending on what is referred to in corresponding CS formalism[6] as the normal and anti-normal expressions,

$$e^{(-i)(P.v-Q.u)} = \exp\left\{-i\frac{u\cdot v}{2}\right\}\exp\{-iv\cdot P\}\exp\{iu\cdot Q\},$$
$$= \exp\left\{i\frac{u\cdot v}{2}\right\}\exp\{iu\cdot Q\}\exp\{-iv\cdot P\}, \tag{6.58}$$

so that with $\exp\left\{i\frac{u\cdot v}{2}\right\}e^{(-i)(P.v-Q.u)}$ as our reference, we have the so-called normal and anti-normal order expressions given by

$$\exp\{-iv\cdot P\}\exp\{iu\cdot Q\} = \exp\left\{i\frac{u\cdot v}{2}\right\}e^{(-i)(P.v-Q.u)}, \tag{6.59}$$

$$\exp\{iu\cdot Q\}\exp\{-iv\cdot P\} = \exp\left\{-i\frac{u\cdot v}{2}\right\}e^{(-i)(P.v-Q.u)}, \tag{6.60}$$

respectively, yielding the following different expressions for $A_{\lambda\lambda'}(u,v)$, namely,

$$A^{w}_{\lambda\lambda'}(u,v) = (N)^{-1}\,\mathrm{Tr}\left(\hat{A}\exp\left(-i\right)\left(P.v-Q.u\right)\right), \tag{6.61}$$

$$A^{n}_{\lambda\lambda'}(u,v) = \mathrm{Tr}\ \left\{\hat{A}\ \exp\{-iv\cdot P\}\exp\{iu\cdot Q\}\right\},$$
$$= (N)^{-1}\exp\left\{i\frac{u\cdot v}{2}\right\}\mathrm{Tr}\left(\hat{A}\exp\left(-i\right)\left(\ P.v-Q.u\right)\right), \tag{6.62}$$

$$A^{a}_{\lambda\lambda'}\left(u,v\right) = \mathrm{Tr}\ \left\{\hat{A}\ \exp\{ip\cdot Q\}\exp\{-iq\cdot P\}\right\},$$
$$= (N)^{-1}\exp\left\{-i\frac{u\cdot v}{2}\right\}\mathrm{Tr}\left(\hat{A}\exp(-i)\left(P.v-Q.u\right)\right). \tag{6.63}$$

Although, Eqs. (6.62) and (6.63) only amounts to difference in the phase factors, the corresponding quantities in non-Hermitian dual spaces gives a very different distributions often referred to as smooth-out distributions. Examining Eqs. (6.59) and (6.60), and the fact that in non-Hermitian mixed representation, $\left\{i\frac{u\cdot v}{2}\right\}$ is a real quantity that resembles a Gaussian function,

[6]For creation and annihilation operators in many-body quantum physics, the proof relies on the use of normal or anti-normal ordering of canonical operators, which can then be treated like $\mathbb{C}$-numbers in expansion of exponentials. The exponential in Eq. (6.55) is sometimes referred to as the generalized Pauli-spin operator.

Eqs. (6.59) clearly represent some smoothing of the Wigner-distribution characteristic function and hence smoothing of the Wigner distribution function itself. In Eqs. (6.62) and (6.63) no informations are lost.

Indeed, more *general* phase-space distribution functions, $f^{(g)}(p,q,t)$, can be obtained from the expression

$$\begin{aligned} f^{(g)}(p,q,t) &= \left(\frac{1}{2\pi\hbar}\right)^3 \int dudve^{-i(qu+pv)} \\ &\quad \times [\{\mathrm{Tr}\ \hat{\rho}\ \exp(-i)(P.v - Q.u)\}\, g(u,v)], \\ &= \frac{1}{2\pi\hbar}\int dudve^{-\frac{i}{\hbar}(qu+pv)}[C^{(w)}(u,v,t)\ \ g(u,v)], \end{aligned} \tag{6.64}$$

where $g(u,v)$ is some choosen phase function.

6.2.1 *Alternative symmetrization of for* $\hat{\Delta}_{\lambda'\lambda}(p,q)$

We can also take up Eq. (4.10) to symmetrize $\hat{\Delta}_{\lambda'\lambda}(p,q)$. We have

$$\begin{aligned} \hat{\Delta}(p,q) &= \frac{1}{\sqrt{N}}\sum_u e^{i2\vec{u}\cdot\vec{q}}\,|\vec{p}-\vec{u}\rangle\,\langle\vec{p}+\vec{u}|\,, \\ &= \frac{1}{\sqrt{N}}\sum_u e^{-i2\vec{u}\cdot\vec{q}}\,|\vec{p}+\vec{u}\rangle\,\langle\vec{p}-\vec{u}|\,, \\ &= \frac{1}{\sqrt{N}}\sum_u e^{i2(Q-q)\cdot u}\,|\vec{p}-\vec{u}\rangle\,\langle\vec{p}-\vec{u}|\,, \end{aligned} \tag{6.65}$$

where

$$|\vec{p}-\vec{u}\rangle\,\langle\vec{p}-\vec{u}| = (N)^{-1}\sum_{\vec{v},\vec{p}'}\exp[2i(\vec{p}-\vec{u}-\hat{P})\cdot\vec{v}]\ \left|\vec{p}'\right\rangle\left\langle\vec{p}'\right|. \tag{6.66}$$

Therefore,

$$\begin{aligned} \hat{\Delta}(p,q) &= \frac{1}{\sqrt{N}}(N)^{-1}\sum_{u,v,\vec{p}'} e^{i2(Q-q)\cdot u}\exp\left[-i2\left(P-p\right)\cdot v\right] \\ &\quad \times \exp\left[-i2\vec{u}\cdot v\right]\left|\vec{p}'\right\rangle\left\langle\vec{p}'\right|, \end{aligned} \tag{6.67}$$

where

$$\begin{aligned} e^{i2Q\cdot u}e^{-i2P\cdot v} &= e^{-2i(P\cdot v - Q\cdot u)}e^{\frac{4}{2}[Q,P]\vec{u}\cdot v}, \\ &= e^{-2i(P\cdot v-Q\cdot u)}e^{-2i\vec{u}\cdot v}. \end{aligned} \tag{6.68}$$

Therefore,

$$\hat{\Delta}(p,q) = \frac{1}{\sqrt{N}}(N)^{-1}\sum_{u,v}\exp 2i(p\cdot v - q\cdot u) \times \exp[-2i(P\cdot v - Q\cdot u)]\sum_{\vec{p}'}\left|\vec{p}'\right\rangle\left\langle\vec{p}'\right|. \quad (6.69)$$

Thus,

$$\hat{A} = \sum_{p,q,\lambda,\lambda'} A_{\lambda\lambda'}(p,q)\;\hat{\Delta}_{\lambda'\lambda}(p,q)$$
$$= \frac{1}{\sqrt{N}}(N)^{-1}\sum_{p,q,\lambda,\lambda',u,v}\left[\begin{array}{c} A_{\lambda\lambda'}(p,q)\;\exp 2i(p\cdot v - q\cdot u) \\ \times\exp[-2i(P\cdot v - Q\cdot u)]\sum_{\vec{p}'}\left|\vec{p}'\lambda\right\rangle\left\langle\vec{p}'\lambda'\right|, \end{array}\right], \quad (6.70)$$

with characteristic function given by

$$A(u,v) = \frac{1}{\sqrt{N}}\sum_{p,q} A_{\lambda\lambda'}(p,q)\;\exp 2i(p\cdot v - q\cdot u), \quad (6.71)$$

and

$$\hat{A} = (N)^{-1}\sum_{\lambda,\lambda',u,v,\vec{p}_0} A(u,v)\exp[-2i(P\cdot v - Q\cdot u)]\;\left|\vec{p}_0\lambda\right\rangle\left\langle\vec{p}_0\lambda'\right|. \quad (6.72)$$

with

$$A(u,v) = (N)^{-1}\,\mathrm{Tr}\left\{\hat{A}\exp[-2i(P\cdot v - Q\cdot u)]\;\left|\vec{p}_0\lambda\right\rangle\left\langle\vec{p}_0\lambda'\right|\right\} \quad (6.73)$$

The algebra of $\exp\{-i(q'\cdot P - p'\cdot Q)\}$ is discussed in Section 6.1.1 for completeness, while its relevance to the physics of two-state systems, spin systems, quantum computing, entanglements [23] and teleportation, are also discussed in the author's book [12].

Chapter 7

Direct Construction of Path Integral in CMP

We are here interested in expressing the transition amplitude as a lattice path integral in solid-state physics. A well-known procedure to implement this is to decomposed the evolution operator, $\hat{U}(t, t_0) \equiv \exp \frac{-i}{\hbar}(t - t_0)\mathcal{H}$,[1] in terms of infinitesimal time increments, $\epsilon = t_j - t_{j-1}$. First, we decomposed in time increments as

$$\hat{U}(t, t_0) = \operatorname*{Lim}_{n \Longrightarrow \infty} \prod_{j=1}^{n+1} \hat{U}(t_j, t_{j-1}), \tag{7.1}$$

where $t_{n+1} = t$. As a wandering incremental paths between t_{n+1} and t_0 in the crystal lattice, we make use of the Wannier-function wavevector as the pointer to the lattice sites and its completeness relations, as

$$\begin{aligned}
&\langle q\,(t_{n+1})|\,\hat{U}\,(t, t_0)\,|q\,(t_0)\rangle \\
&\quad = \langle q\,(t_{n+1})|\,\hat{U}\,(t_{n+1}, t_n) \sum_{q_n} |q\,(t_n)\rangle\,\langle q\,(t_n)|\,\hat{U}\,(t_n, t_{n-1}) \\
&\qquad \times \sum_{q_{n-1}} |q\,(t_{n-1})\rangle\,\langle q\,(t_{n-1})| \\
&\qquad \times \cdots \sum_{q_1} |q\,(t_1)\rangle\,\langle q\,(t_1)|\,\hat{U}\,(t_1, t_0)\,|q\,(t_0)\rangle, \\
&\quad = \sum_{q_1} \cdots \sum_{q_n} \prod_{j=1}^{n+1} \langle q\,(t_j)|\,\hat{U}\,(t_j, t_{j-1})\,|q\,(t_{j-1})\rangle,
\end{aligned} \tag{7.2}$$

[1]Here we restore $\hbar \neq 1$ to invite familiarity of the succeeding equations. Later we again let $\hbar = 1$ to bring out parallelism with QFT.

where the end coordinates, namely, $q(t_{n+1})$ and $q\left(t_0\right)$ are fixed. By making the time intervals infinitely small or by letting $\mathrm{Lim}_{n \Longrightarrow \infty}$, we can take advantage of the linearity of the small time evolution operators to calculate the matrix elements after which it can be recomposed in exponential form but now cast as $\mathbb{C}$-numbers. We have

$$\hat{U}\left(t_j, t_{j-1}\right) \simeq 1 - \frac{i}{\hbar}\left(t_j - t_{j-1}\right)\mathcal{H} + O\left(t_j - t_{j-1}\right)^2 + \cdots . \tag{7.3}$$

Now comes our big advantage in calculating the matrix elements in mixed-space representation. We now expand the operator $\hat{U}\left(t_j, t_{j-1}\right)$ in terms of the phase-space point projector $\hat{\Delta}\left(p, q\right)$ in $\mathcal{H}$. We have,

$$\mathcal{H} = \sum_{p,q} H\left(p, q\right)\hat{\Delta}\left(p, q\right). \tag{7.4}$$

Then

$$\begin{aligned}
&\langle q\left(t_j\right)|\,\hat{U}\left(t_j, t_{j-1}\right)|q\left(t_{j-1}\right)\rangle \\
&\quad = \langle q\left(t_j\right)|\left(1 - \frac{i}{\hbar}\left(t_j - t_{j-1}\right)\mathcal{H}\right)|q\left(t_{j-1}\right)\rangle, \\
&\quad = 1 - \frac{i}{\hbar}\left(t_j - t_{j-1}\right)\sum_{p,q} H\left(p, q\right)\ \{\langle q\left(t_j\right)|\,\hat{\Delta}\left(p, q\right)|q\left(t_{j-1}\right)\rangle\},
\end{aligned} \tag{7.5}$$

where for intermediate points,

$$\{\langle q\left(t_j\right)|\,\hat{\Delta}\left(p, q\right)|q\left(t_{j-1}\right)\rangle\} = \exp\left(\frac{i}{\hbar}p_j \cdot \left(q_j - q_{j-1}\right)\right)\delta\left(2q - [q_j + q_{j-1}]\right), \tag{7.6}$$

and for the end points, we have

$$\{\langle q\left(t_{n+1}\right)|\,\hat{\Delta}\left(p, q\right)|q\left(t_n\right)\rangle\} = \exp\left(\frac{i}{\hbar}p_n \cdot \left(q_{n+1} - q_n\right)\right)\delta\left(2q - [q_{n+1} + q_n]\right), \tag{7.7}$$

$$\{\langle q(t_1)|\,\hat{\Delta}\left(p, q\right)|q\left(t_0\right)\rangle\} = \exp\left(\frac{i}{\hbar}p_1 \cdot \left(q_1 - q_0\right)\right)\delta(2q - [q_1 + q_0]), \tag{7.8}$$

so that we can write to first order in time increment, $(t_j - t_{j-1})$, as

$$\begin{aligned}
&\langle q\left(t_j\right)|\,\hat{U}\left(t_j, t_{j-1}\right)|q\left(t_{j-1}\right)\rangle \\
&\quad = \left(N\hbar^3\right)^{-1}\sum_{p_j, q, q_j}\exp\left(\frac{i}{\hbar}p_j \cdot \left(q_j - q_{j-1}\right)\right)
\end{aligned}$$

$$\times \exp\left(\frac{-i}{\hbar}\left(t_j - t_{j-1}\right) H\left(p, q\right)\right) \delta\left(2q - \left[q_j + q_{j-1}\right]\right),$$

$$= \left(N\hbar^3\right)^{-1} \sum_{j=1}^{n} \exp\left[\frac{i}{\hbar}\left(t_j - t_{j-1}\right)\left(\begin{array}{c} p_j \cdot \frac{(q_j - q_{j-1})}{(t_j - t_{j-1})} \\ -H\left(p, \frac{q_j + q_{j-1}}{2}\right)\end{array}\right)\right]. \tag{7.9}$$

Substituting in Eq. (7.2), we obtained

$$\left\langle q\left(t_{n+1}\right)\right| \hat{U}\left(t, t_0\right) \left|q\left(t_0\right)\right\rangle$$

$$= \sum_{q_1} \cdots \sum_{q_n} \sum_{p_1} \cdots \sum_{p_n} \prod_{j=1}^{n} \left(N\hbar^3\right)^{-1}$$

$$\times \exp\left[\frac{i}{\hbar}\left(t_j - t_{j-1}\right)\left(p \cdot \frac{(q_j - q_{j-1})}{(t_j - t_{j-1})} - H\left(p, \frac{q_j + q_{j-1}}{2}\right)\right)\right],$$

$$= \prod_{j=1}^{n} \sum_{q_j} \prod_{j=1}^{n} \sum_{p} \left(N\hbar^3\right)^{-1} \exp\left[\frac{i}{\hbar}\left(t_j - t_{j-1}\right)\left(\begin{array}{c} p \cdot \frac{(q_j - q_{j-1})}{(t_j - t_{j-1})} \\ -H\left(p, \frac{q_j + q_{j-1}}{2}\right)\end{array}\right)\right]. \tag{7.10}$$

Now since $\sum_p$ is performed for each term labeled by $j = 1 \cdots n$, we can also label summation over p by summation over p_j, i.e., $\sum_{p_j}$ giving us a more symmetrical form,

$$\left\langle q\left(t_{n+1}\right)\right| \hat{U}\left(t, t_0\right) \left|q\left(t_0\right)\right\rangle$$

$$= \prod_{j=1}^{n} \sum_{q_j} \prod_{j=1}^{n} \sum_{p_j}$$

$$\times \left(N\hbar^3\right)^{-1} \exp \sum_{j=1}^{n} \left[\frac{i}{\hbar}\left(t_j - t_{j-1}\right)\left(\begin{array}{c} p_j \cdot \frac{(q_j - q_{j-1})}{(t_j - t_{j-1})} \\ -H\left(p_j, \frac{q_j + q_{j-1}}{2}\right)\end{array}\right)\right]. \tag{7.11}$$

7.1 Continuum limit

If we take the continuum limit, i.e., $N \Longrightarrow \infty$,and $\Delta t = t_j - t_{j-1} \Longrightarrow 0$, and defining,

$$\int \mathcal{D}p \equiv \prod_{j=1}^{n} \sum_{p_j}, \tag{7.12}$$

$$\int_{q_i = q(t_0)}^{q_f = q(t_n)} \mathcal{D}q \equiv \prod_{j=1}^{n} \sum_{q_j}. \tag{7.13}$$

We have in the exponential,

$$S = \sum_{j=1}^{n} \left[\frac{i}{\hbar}\epsilon \left(p_j \cdot \frac{(q_j - q_{j-1})}{\epsilon} - H\left(p_j, \frac{q_j + q_{j-1}}{2} \right) \right) \right]$$

$$\Longrightarrow \frac{i}{\hbar}\int_{t_i}^{t_f} dt\ [p(t) \cdot \dot{q}(t) - H(p(t), q(t))], \tag{7.14}$$

where the *time-integral* in S is identified as the Hamiltonian action. Then we end up with

$$\langle q(t_f)| \hat{U}(t, t_0) |q(t_i)\rangle = \left(2\pi\hbar^3\right)^{-N} \iint \mathcal{D}q\mathcal{D}p\ \exp(S). \tag{7.15}$$

This last result is often called functional integral since now the set, $\{p(t), q(t)\}$, defines phase-space path functions of time.

Chapter 8

General Mixed Space Representation in QFT

Let the operators $\hat{\psi}$ and $\hat{\psi}^\dagger$ be non-Hermitian operators. These operators obey either commutation or anti-commutation relations, i.e.,

$$[\hat{\psi}, \hat{\psi}^\dagger]_\eta = 1, \tag{8.1}$$

where the subscript $\eta = +$ is for the anti-commutation and $\eta = -$ stands for commutation relation. These non-Hermitian operators have *distinguished* left and right eigenvectors. We have

$$\hat{\psi} |\alpha\rangle = \alpha |\alpha\rangle, \tag{8.2}$$

$$\langle\alpha| \hat{\psi}^\dagger = \langle\alpha| \alpha^*, \tag{8.3}$$

$$\hat{\psi}^\dagger |\beta\rangle = \beta |\beta\rangle, \tag{8.4}$$

$$\langle\beta| \hat{\psi} = \langle\beta| \beta^*. \tag{8.5}$$

This means the left eigenvector of $\hat{\psi}$ is $\langle\beta|$, with eigenvalue β^*, whereas the left eigenvector of $\hat{\psi}^\dagger$ is $\langle\alpha|$, with eigenvalue α^*. We have,

$$\begin{aligned}\langle\beta| \hat{\psi} |\alpha\rangle &= \alpha \langle\beta| |\alpha\rangle, \\ &= \beta^* \langle\beta| |\alpha\rangle, \\ (\alpha - \beta^*) \langle\beta| |\alpha\rangle &= 0, \end{aligned} \tag{8.6}$$

and similarly

$$\begin{aligned}\langle\alpha|\,\hat{\psi}^{\dagger}\,|\beta\rangle &= \alpha^{*}\,\langle\alpha|\,|\beta\rangle,\\ &= \beta\,\langle\alpha|\,|\beta\rangle,\\ (\alpha^{*}-\beta)\,\langle\alpha|\,|\beta\rangle &= 0.\end{aligned} \tag{8.7}$$

From Eqs. (8.6) and (8.7), if $\alpha_n \neq \beta_m^*$, $\langle\beta_m|\,|\alpha_n\rangle = 0$, imitating orthogonal Hermitian Hilbert space. However, if $\alpha_n = \beta_m^*$, then $\langle\beta_m|\,|\alpha_n\rangle \neq 0$, amenable to probabilistic interpretation. Moreover, with $|\alpha\rangle$ and $\langle\beta|$ we have the general projection given by

$$\sum_n \frac{|\alpha_n\rangle\,\langle\beta_n|}{\langle\beta_n|\,|\alpha_n\rangle} = 1, \tag{8.8}$$

which project only within the paired eigenspace of $\{|\alpha\rangle\,,\langle\beta|\}$. Similarly,

$$\sum_n \frac{|\beta_n\rangle\,\langle\alpha_n|}{\langle\alpha_n|\,|\beta_n\rangle} = 1, \tag{8.9}$$

only projects within the paired space, $\{\langle\alpha|\,,|\beta\rangle\}$. Equation (8.8) reminds us of Eq. (2.7) of CMP.

The general proof of Eq. (8.8) lies in the following expansion of $|\Psi\rangle$, where $|\Psi\rangle$ is a complete orthonormal eigenvector. Let,

$$|\Psi\rangle = \sum_{n'} c^{n'}\,|\alpha_{n'}\rangle, \tag{8.10}$$

then we have,

$$\begin{aligned}\sum_n \frac{|\alpha_n\rangle\,\langle\beta_n|\,|\Psi\rangle}{\langle\beta_n|\,|\alpha_n\rangle} &= \sum_{n,n'} \frac{|\alpha_n\rangle\,\langle\beta_n|\,|\alpha_{n'}\rangle\,c^{n'}}{\langle\beta_n|\,|\alpha_n\rangle},\\ &= \sum_n c^n\,|\alpha_n\rangle = |\Psi\rangle.\end{aligned} \tag{8.11}$$

confirming Eq. (8.8) as a simple projector.

8.1 Mixed Hilbert-space construction

Thus, from Eqs. (8.6) and (8.7), we have a well-defined Hermitian-like operation in terms of the paired eigenvector set $\{\langle\alpha|\,,|\beta\rangle\}$ and $\{\langle\beta|\,,|\alpha\rangle\}$ as dual eigenspaces. Indeed, assuming non-degenerates countable states, we can pair the states so that $\alpha_n = \beta_m^*$ and rewrite the pairs with same quantum subscript, i.e., $\alpha_n = \beta_n^*$, which would be consistent with the relation, $[\hat{\psi}_n, \hat{\psi}_n^\dagger]_\eta = 1$ leading to quantum-state generation in what follows.

Just as the dot product $\langle p| |q\rangle \neq 0$ in condensed matter physics discussions, so $\langle \alpha_n| |\alpha_n\rangle \neq 0$, means that $\langle \alpha_n| |\alpha_n\rangle$ cannot be made equal to zero, since these belongs to separate paired set, $\{\langle\alpha|, |\beta\rangle\}$ and$\{\langle\beta|, |\alpha\rangle\}$, respectively. In some sense, the $|\alpha\rangle$, and $|\beta\rangle$ respective spaces are reminiscent of the *q*-*p* phase space, where $\langle p| |q\rangle \neq 0$ is the transition function between the two spaces. This conclusion can be made quite general in what follows.

8.2 Pairing algorithm and Hermitianization

Thus, in order to work with Hermitian-like operators for convenience, we want the eigenvalues $\alpha = \beta^*$ and $\alpha^* = \beta$. For non-degenerate countable finite system, this type of pairing of the different $|\alpha\rangle$ and $|\beta\rangle$ spaces is well defined. For convenience in what follows, we relabel the $|\alpha\rangle$ and $|\beta\rangle$ notations and their adjoints to reflect the Hermitian-like new spaces, sort of *renormalize* new dynamical vector spaces.

From Eqs. (8.6) and (8.7), it seems trivial just like for the Hermitian operators to prove orthogonality or more appropriately, biorthogonality, in terms of the $\{\langle\alpha|, |\beta\rangle\}$ and $\{\langle\beta|, |\alpha\rangle\}$ "dual" eigenspaces. This seems to suggest the following Hermitian-like relations:

$$\langle\alpha| \hat{\psi}^\dagger |\beta\rangle \equiv \alpha^* \langle\alpha| |\beta\rangle, \tag{8.12}$$

$$\langle\beta| \hat{\psi} |\alpha\rangle \equiv \alpha \langle\beta| |\alpha\rangle, \tag{8.13}$$

where $\langle\alpha|$ and $|\beta\rangle$ are the left and right eigenvectors, respectively of $\hat{\psi}^\dagger$, whereas, $\langle\beta|$ and $|\alpha\rangle$ are the left and right eigenvectors, respectively, of $\hat{\psi}$. The above pairing allows us to form dual eigenvectors to simulate the Hilbert-spaces of Hermitian operators.

We now denote the $\langle\alpha|$ and $|\beta\rangle$ left and right eigenvectors, respectively of $\hat{\psi}^\dagger$ as making up the α^*-Hilbert space, and we will adopt a new consistent labels, $\langle\alpha| \implies \langle\mathfrak{p}|$ and $|\beta\rangle \implies |\mathfrak{p}\rangle$. Similarly, we relabel the $\langle\beta|$ and $|\alpha\rangle$ left and right eigenvectors, respectively, of $\hat{\psi}$ as making up the α-Hilbert space, with $\langle\beta| \implies \langle\mathfrak{q}|$ and $|\alpha\rangle \implies |\mathfrak{q}\rangle$. The Fraktur font label of $\mathfrak{q}$ and $\mathfrak{p}$-eigenspaces constitute our newly-formed quantum label for Hilbert spaces for $\hat{\psi}$ and $\hat{\psi}^\dagger$, respectively. So Eqs. (8.12) and (8.13) becomes,

$$\langle\mathfrak{p}| \hat{\psi}^\dagger |\mathfrak{p}\rangle \equiv \alpha^* \langle\mathfrak{p}| |\mathfrak{p}\rangle, \tag{8.14}$$

$$\langle\mathfrak{q}| \hat{\psi} |\mathfrak{q}\rangle \equiv \alpha \langle\mathfrak{q}| |\mathfrak{q}\rangle, \tag{8.15}$$

8.2.1 CMP-like operators $\hat{\mathcal{P}}$ and $\hat{\mathcal{Q}}$

To make parallel developments with those of CMP, it is more convenient to switch operators from $\hat{\psi}^\dagger$ and $\hat{\psi}$ to $\hat{\mathcal{P}}$ and $\hat{\mathcal{Q}}$, where $\hat{\mathcal{P}} = i\hat{\psi}^\dagger$and $\hat{\mathcal{Q}} = -i\hat{\psi}$. Then we have,[1]

$$\langle \mathfrak{p}| \hat{\mathcal{P}} |\mathfrak{p}\rangle \equiv i\alpha^* \langle \mathfrak{p}| |\mathfrak{p}\rangle, \tag{8.16}$$

$$\langle \mathfrak{q}| \hat{\mathcal{Q}} |\mathfrak{q}\rangle \equiv -i\alpha \langle \mathfrak{q}| |\mathfrak{q}\rangle, \tag{8.17}$$

From the last two equations, we may set the eigenvalues of $\hat{\mathcal{P}}$ as $\mathfrak{p} = i\alpha^*$, and similarly, the eigenvalues of $\hat{\mathcal{Q}}$ as $\mathfrak{q} = -i\alpha$. Then, we can recast Eqs. (8.16) and (8.17) as

$$\hat{\mathcal{P}} |\mathfrak{p}\rangle = \mathfrak{p} |\mathfrak{p}\rangle, \quad \hat{\mathcal{Q}} |\mathfrak{q}\rangle = \mathfrak{q} |\mathfrak{q}\rangle$$
$$\langle \mathfrak{p}| \hat{\mathcal{P}} = \langle \mathfrak{p}| \mathfrak{p}, \quad \langle \mathfrak{q}| \hat{\mathcal{Q}} = \langle \mathfrak{q}| \mathfrak{q}.$$

8.2.2 Completeness relations

In the new dual space representation, Eq. (8.8) becomes simply a completeness relation,

$$\sum_n \frac{|\alpha_n\rangle \langle \beta_n|}{\langle \beta_n| |\alpha_n\rangle} = 1 \Longrightarrow \sum_{\mathfrak{q}} \frac{|\mathfrak{q}\rangle \langle \mathfrak{q}|}{\langle \mathfrak{q}| |\mathfrak{q}\rangle} = 1,$$
$$\Longrightarrow \sum_{\mathfrak{q}} |\mathfrak{q}\rangle \langle \mathfrak{q}| = 1, \tag{8.18}$$

which yields the completeness relation, with normalized $\langle \mathfrak{q}| |\mathfrak{q}\rangle \equiv 1$. Thus, the $|\mathfrak{q}\rangle$ and $|\mathfrak{p}\rangle$ eigenstates obey the completeness relations,[2]

$$\sum_{\mathfrak{q}} |\mathfrak{q}\rangle \langle \mathfrak{q}| = 1, \tag{8.19}$$

$$\sum_{\mathfrak{p}} |\mathfrak{p}\rangle \langle \mathfrak{p}| = 1. \tag{8.20}$$

[1]Note that $\hat{\psi}^\dagger$ and $\hat{\mathcal{P}}$ commutes and therefore share the same eigenvectors, $|\mathfrak{p}\rangle$. And similarly holds for $\hat{\psi}$ and $\hat{\mathcal{Q}}$, with shared eigenvector, $|\mathfrak{q}\rangle$.

[2]The power of the use of functionals in quantum physics has been demonstrated before. It is pioneered by the use of Feynman functional path integrals in QFT, by the Kohn–Sham density functional theory in CMP, and by the coherent state formulation in nonlinear quantum optics.

Then it becomes trivial to see the transformation between the dual spaces, $|\mathfrak{q}\rangle$ and $|\mathfrak{p}\rangle$ eigenstates,

$$|\mathfrak{q}\rangle = \sum_{\phi} |\mathfrak{p}\rangle \langle\mathfrak{p}| \, |\mathfrak{q}\rangle, \tag{8.21}$$

$$|\mathfrak{p}\rangle = \sum_{\theta} |\mathfrak{q}\rangle \langle\mathfrak{q}| \, |\mathfrak{p}\rangle, \tag{8.22}$$

with transformation function between elements of new dual spaces given by $\langle\mathfrak{p}| \, |\mathfrak{q}\rangle$ and $\langle\mathfrak{q}| \, |\mathfrak{p}\rangle$, respectively. From Eqs. (8.6) and (8.7), we have

$$\langle\mathfrak{q}_m| \, |\mathfrak{q}_n\rangle = \delta_{m,n}, \tag{8.23}$$

$$\langle\mathfrak{p}_m| \, |\mathfrak{p}_n\rangle = \delta_{m,n}, \tag{8.24}$$

firmly defining the complete and orthogonal dual Hilbert spaces, $\{|\mathfrak{q}\rangle\}$ and$\{|\mathfrak{p}\rangle\}$.

8.2.3 *Generation of states*

In what follows, partial derivatives of field variables means partial functional derivatives. Equation (8.1), defines the generation of state $|\mathfrak{q}\rangle$,

$$\begin{aligned} |\alpha\rangle &= C_o \exp \alpha\hat{\psi}^\dagger \; |0\rangle, \\ |\alpha^*\rangle &= C_o \exp \alpha^*\hat{\psi} \; |0\rangle, \end{aligned} \tag{8.25}$$

which translates to our operator, $\hat{\mathcal{P}}$ as

$$|-i\alpha\rangle = C_o \exp(-i\alpha) \cdot \left(-i\hat{\mathcal{P}}\right) \; |0\rangle,$$

$$|\mathfrak{q}\rangle = C_o \exp -i\mathfrak{q}\cdot\hat{\mathcal{P}} \; |0\rangle.$$

$$\hat{\psi}^\dagger |\mathfrak{q}\rangle = \hat{\psi}^\dagger \exp q \; \hat{\psi}^\dagger \; |0\rangle = \frac{\partial}{\partial\mathfrak{q}} \; |\mathfrak{q}\rangle, \tag{8.26}$$

$$\hat{\mathcal{P}} |\mathfrak{q}\rangle = \hat{\mathcal{P}} C_o \exp -i\mathfrak{q}\cdot\hat{\mathcal{P}} \; |0\rangle = i\frac{\partial}{\partial\mathfrak{q}} \; |\mathfrak{q}\rangle. \tag{8.27}$$

Similarly,

$$\hat{\mathcal{Q}} |\mathfrak{p}\rangle = \hat{\mathcal{Q}} C_o \exp i\mathfrak{p} \cdot \hat{\mathcal{Q}} \; |0\rangle = -i\frac{\partial}{\partial\mathfrak{p}} \; |\mathfrak{p}\rangle. \tag{8.28}$$

Inserting the term $\exp\{\alpha^*\hat{\psi}\}$ right in front of $|0\rangle$ of Eq. (8.25), which has the effect of multiplying by unity, we obtain a fully symmetric form as

$$|\alpha\rangle = \exp \alpha\hat{\psi}^\dagger \exp\{\alpha^*\hat{\psi}\}\ |0\rangle. \tag{8.29}$$

which also translates to

$$|\mathfrak{q}\rangle = \exp\left(-i\mathfrak{q}\cdot\hat{\mathcal{P}}\right)\exp\{i\mathfrak{p}\cdot\hat{\mathcal{Q}}\}\ |0\rangle$$

We have,

$$\hat{\psi}^\dagger\,|\mathfrak{q}\rangle = \frac{\partial}{\partial\mathfrak{q}}\,|\mathfrak{q}\rangle = -i\hat{\mathcal{P}}\ |\mathfrak{q}\rangle, \tag{8.30}$$

$$\hat{\mathcal{P}}\,|\mathfrak{p}\rangle = \ \mathfrak{p}\ |\mathfrak{p}\rangle. \tag{8.31}$$

To calculate $\langle\mathfrak{p}|\,|\mathfrak{q}\rangle$, we proceed as follows:

$$\langle\mathfrak{p}|\,\hat{\psi}^\dagger\,|\mathfrak{q}\rangle = \langle\mathfrak{p}|\left(\frac{\partial}{\partial\mathfrak{q}}\right)|\mathfrak{q}\rangle, \tag{8.32}$$

$$-i\mathfrak{p}\,\langle\mathfrak{p}|\,|\mathfrak{q}\rangle = \frac{\partial}{\partial\mathfrak{q}}\,\langle\mathfrak{p}|\,|\mathfrak{q}\rangle, \tag{8.33}$$

$$\frac{\frac{\partial}{\partial\mathfrak{q}}\,\langle\mathfrak{p}|\,|\mathfrak{q}\rangle}{\langle\mathfrak{p}|\,|\mathfrak{q}\rangle} = -i\mathfrak{p}, \tag{8.34}$$

$$\frac{\partial}{\partial\mathfrak{q}}\ln\langle\mathfrak{p}|\,|\mathfrak{q}\rangle = -i\mathfrak{p}, \tag{8.35}$$

$$\langle\mathfrak{p}|\,|\mathfrak{q}\rangle = \exp\left(-i\ \mathfrak{p}\cdot\mathfrak{q}\right). \tag{8.36}$$

Similarly, we have[3]

$$\hat{\psi}\,|\mathfrak{p}\rangle = \frac{\partial}{\partial\mathfrak{p}}\ \ |\mathfrak{p}\rangle = i\hat{\mathcal{Q}}\ |\mathfrak{p}\rangle, \tag{8.37}$$

$$\hat{\mathcal{Q}}\ |\mathfrak{q}\rangle = \mathfrak{q}\ |\mathfrak{q}\rangle, \tag{8.38}$$

and

$$\langle\mathfrak{q}|\,\hat{\psi}\,|\mathfrak{p}\rangle = \langle\mathfrak{q}|\,\frac{\partial}{\partial\mathfrak{p}}\ \ |\mathfrak{p}\rangle, \tag{8.39}$$

$$i\mathfrak{q}\,\langle\mathfrak{q}|\,|\mathfrak{p}\rangle = \frac{\partial}{\partial\mathfrak{p}}\,\langle\mathfrak{q}|\,|\mathfrak{p}\rangle, \tag{8.40}$$

$$\langle\mathfrak{q}|\,|\mathfrak{p}\rangle = \exp\left(i\ \mathfrak{p}\cdot\mathfrak{q}\right). \tag{8.41}$$

[3]Likewise $\hat{\psi}$ and $\hat{\mathcal{Q}}$ commutes and therefore share the same eigenvectors.

So far, all the above developments holds for fermions and bosons. However, note that for fermions the eigenvalues corresponding to $\mathfrak{q}$ and $\mathfrak{p}$ are elements of the Grassmann algebra. Hence, Eqs. (8.36) and (8.41) are not obvious for fermions. To demonstrate its validity for fermions, the transition functions for fermions will be derived in a more specific details in Section 10.1 of Chapter 10. First we will treat the case of mixed space representation of boson quantum field theory.

Chapter 9

Mixed Space Representation in QFT: Bosons

9.1 Commutation relations of $\hat{\mathcal{Q}}$ and $\hat{\mathcal{P}}$

We have in eigenvector basis,

$$\hat{\mathcal{Q}} = \sum_{\mathfrak{q}} \mathfrak{q} \ \ |\mathfrak{q}\rangle \langle \mathfrak{q}|, \tag{9.1}$$

whereas we have,

$$\begin{aligned}\hat{\mathcal{P}} &= \sum_{\mathfrak{q},\mathfrak{q}'} \langle \mathfrak{q}'| \, P \, |\mathfrak{q}\rangle \ \ |\mathfrak{q}'\rangle \langle \mathfrak{q}| \\ &= \sum_{\mathfrak{q},\mathfrak{q}'} \langle \mathfrak{q}'| \left(i \frac{\partial}{\partial \mathfrak{q}} \right) |\mathfrak{q}\rangle \ \ |\mathfrak{q}'\rangle \langle \mathfrak{q}|, \end{aligned} \tag{9.2}$$

where $\frac{\partial}{\partial \mathfrak{q}} |\mathfrak{q}\rangle$ denotes a a functional derivative operation. Then, we have

$$\begin{aligned}\hat{\mathcal{Q}}\hat{\mathcal{P}} &= \sum_{\mathfrak{q}} \mathfrak{q} \ \ |\mathfrak{q}\rangle \langle \mathfrak{q}| \sum_{\mathfrak{q}',\mathfrak{q}''} \langle \mathfrak{q}'| \left(i \frac{\partial}{\partial \mathfrak{q}} \right) |\mathfrak{q}''\rangle \ \ |\mathfrak{q}'\rangle \langle \mathfrak{q}''| \\ &= \sum_{\mathfrak{q}} \mathfrak{q} \langle \mathfrak{q}| \left(i \frac{\partial}{\partial \mathfrak{q}} \right) |\mathfrak{q}\rangle, \end{aligned} \tag{9.3}$$

since $\hat{\psi}^{\dagger}$ acts like $\frac{\partial}{\partial \mathfrak{q}}$ acting on $|\mathfrak{q}\rangle$ basis. On the other hand, we have

$$\begin{aligned}\hat{\mathcal{P}}\hat{\mathcal{Q}} &= \sum_{\mathfrak{q}',\mathfrak{q}''} \langle \mathfrak{q}'| \left(i \frac{\partial}{\partial \mathfrak{q}''} \right) |\mathfrak{q}''\rangle \ \ |\mathfrak{q}'\rangle \langle \mathfrak{q}''| \sum_{\mathfrak{q}} \mathfrak{q} \ \ |\mathfrak{q}\rangle \langle \mathfrak{q}| \\ &= \sum_{\mathfrak{q}} \mathfrak{q} \langle \mathfrak{q}| \left(i \frac{\partial}{\partial \mathfrak{q}} \right) |\mathfrak{q}\rangle + i. \end{aligned} \tag{9.4}$$

These entail the commutation relation,

$$\left[\hat{\mathcal{Q}}, \hat{\mathcal{P}}\right] = -i. \tag{9.5}$$

The seemingly strange relation of Eq. (9.5) arise since with respect to the eigenvector $|\mathfrak{q}\rangle$ basis, $\hat{\mathcal{Q}}$ acts like $\mathfrak{q}$ whereas $\hat{\mathcal{P}}$ acts like $i\frac{\partial}{\partial \mathfrak{q}}$. Hence, essentially Eq. (9.5) looks like

$$\left[\mathfrak{q}, i\frac{\partial}{\partial \mathfrak{q}}\right] = -i.$$

Compare this with

$$\left[\hat{\psi}, \hat{\psi}^{\dagger}\right] = 1,$$

which acts on wavefunction basis like,

$$\left[\mathfrak{q}, -\frac{\partial}{\partial \mathfrak{q}}\right] = 1.$$

Similarly, with respect to $|\mathfrak{p}\rangle$ basis $\hat{\mathcal{P}}$ acts like $\mathfrak{p}$ whereas $\hat{\mathcal{Q}}$ acts like $-i\frac{\partial}{\partial \mathfrak{p}}$. All these results are in line with Eqs. (2.13)–(2.20) dealing with covariant and contravariant basis and their operators.

Since the commutator for bosons, is a $\mathbb{C}$-number (i.e., not an operator), we can readily make use of the Campbell–Baker–Hausdorff operator identity, namely,

$$\begin{aligned}
&\exp\left\{-i\mathfrak{q}\cdot\hat{\mathcal{P}}\right\}\exp\left\{i\mathfrak{p}\cdot\hat{\mathcal{Q}}\right\} \\
&= \exp\left(-i\left\{\mathfrak{q}\cdot\hat{\mathcal{P}} - \mathfrak{p}\cdot\hat{\mathcal{Q}}\right\}\right)\exp\left\{\frac{\left[\mathfrak{q}\cdot\hat{\mathcal{P}}, \mathfrak{p}\cdot\hat{\mathcal{Q}}\right]}{2}\right\}, \\
&= \exp\left(-i\left\{\mathfrak{q}\cdot\hat{\mathcal{P}} - \mathfrak{p}\cdot\hat{\mathcal{Q}}\right\}\right)\exp\left\{i\frac{\mathfrak{q}\cdot\mathfrak{p}}{2}\right\}.
\end{aligned} \tag{9.6}$$

Therefore, the generator of $|\mathfrak{q}\rangle$ states as

$$\begin{aligned}
|\mathfrak{q}\rangle &= \exp\left\{i\frac{\mathfrak{q}\cdot\mathfrak{p}}{2}\right\}\exp\left(-i\left\{\mathfrak{q}\cdot\hat{\mathcal{P}} - \mathfrak{p}\cdot\hat{\mathcal{Q}}\right\}\right)|0\rangle, \\
&= \mathcal{D}(\mathfrak{q})\,|0\rangle.
\end{aligned} \tag{9.7}$$

Thus, $\mathcal{D}(\mathfrak{q})$ is the displacement operator that generates the state $|\mathfrak{q}\rangle$ from the vacuum, $|0\rangle$. Note that $\exp(i\mathfrak{p}'\cdot\hat{\mathcal{Q}})\,|\mathfrak{p}\rangle = \exp\left(\mathfrak{p}'\cdot\frac{\partial}{\partial \mathfrak{p}}\right)|\mathfrak{p}\rangle$ when operating on $|\mathfrak{p}\rangle$ states.

9.2 Reformulation as mixed space representation

We can now formulate the non-Hermitian operators to Hermitian-like dual space representation, similar to the q-p dual space representation of CMP, Section 2.1 in Chapter 2.

9.2.1 *Isomorphism between* $\mathfrak{q}$-$\mathfrak{p}$ *and* q-p *dual spaces*

We now demonstrate the remarkable isomorphism of annihilation and creation operators, $\mathcal{Q}$ and $\mathcal{P}$, and their dual Hilbert spaces, $\mathfrak{q}$-$\mathfrak{p}$, with the position and momentum operators, Q and P, and their dual Hilbert spaces, q-p. This is clearly seen by the following bijective and homomorphic or isomorphic transformation, i.e., the harmonic oscillator transformation from position and momentum operators to annihilation and creation operators,

$$\begin{pmatrix} \hat{\psi} \\ \hat{\psi}^\dagger \end{pmatrix} = \begin{pmatrix} i\mathcal{Q} \\ -i\mathcal{P} \end{pmatrix} = \frac{1}{\sqrt{2}} \begin{pmatrix} 1 & 1 \\ 1 & -1 \end{pmatrix} \begin{pmatrix} \hat{Q} \\ i\hat{P} \end{pmatrix}, \tag{9.8}$$

$$= \frac{1}{\sqrt{2}} \begin{pmatrix} \hat{Q} + i\hat{P} \\ \hat{Q} - i\hat{P} \end{pmatrix}, \tag{9.9}$$

where $\hat{Q}$ and $\hat{P}$ are dimensionless canonical operators, with eigenvalues, q and p, respectively. Then the transformation of Eq. (9.8) induces a transformation of Eq. (9.10) as

$$\begin{aligned} \mathcal{D}_{\text{sym}}(\mathfrak{q}, \mathfrak{p}) &= \exp\{-i(\mathcal{P}\cdot\mathfrak{q} - \mathcal{Q}\cdot\mathfrak{p})\}, \\ &= \exp\{-i\mathcal{P}\cdot\mathfrak{q} + \mathrm{i}\mathcal{Q}\cdot\mathfrak{p}\}, \\ &= \exp\left\{\hat{\psi}^\dagger\cdot\mathfrak{q} + \hat{\psi}\cdot\mathfrak{p}\right\}, \end{aligned} \tag{9.10}$$

where the last line makes use of Eq. (8.37). We have for CS,

$$\begin{aligned} \hat{\psi}^\dagger\cdot q - \hat{\psi}\cdot p &= \frac{1}{2}[(\hat{Q} - i\hat{P})\cdot(v + iu) - (\hat{Q} + i\hat{P})\cdot(v - iu)], \\ &= \frac{1}{2}[(\hat{Q} - i\hat{P})\cdot(v + iu) + (\hat{Q} + i\hat{P})\cdot(-v + iu)], \\ &= -i(\hat{P}\cdot v - \hat{Q}\cdot u), \end{aligned} \tag{9.11}$$

However, in QFT Eq. (9.10) gives

$$Y(\mathfrak{q}, \mathfrak{p}) = \exp\{-i(\mathcal{P}\cdot\mathfrak{q} - \mathcal{Q}\cdot\mathfrak{p})\}, \tag{9.12}$$

whereas in CS, we have,

$$Y(q, p) = \exp[-i(\hat{P}\cdot v - \hat{Q}\cdot u)], \tag{9.13}$$

as the symmetric translation operator in position-momentum, q-p phase space. Therefore, the dual space $\{\mathfrak{q}, \mathfrak{p}\}$ or $\{\alpha, \alpha^*\}$ in CS is isomorphic to $\{q, p\}$ dual space.

9.3 The $\mathfrak{q}$-$\mathfrak{p}$ representations and lattice Weyl transform

The mixed $\mathfrak{q}$-$\mathfrak{p}$ representation basically start by expanding any quantum operator, $\hat{A}$, in terms of mutually unbiased basis states, namely the eigenvector of annihilation operator, $\hat{\psi}$ or $\mathcal{Q}$, and the eigenvector of creation operator, $\hat{\psi}^\dagger$ or $\mathcal{P}$. We have

$$\begin{aligned}\hat{A} &= \sum_{\mathfrak{p},\mathfrak{q}} |\mathfrak{q}\rangle\langle\mathfrak{q}|\hat{A}|\mathfrak{p}\rangle\langle\mathfrak{p}|, \\ &= \sum_{\mathfrak{p},\mathfrak{q}} \langle\mathfrak{q}|\hat{A}|\mathfrak{p}\rangle\ |\mathfrak{q}\rangle\langle\mathfrak{p}|. \end{aligned}\tag{9.14}$$

9.3.1 *The completeness of mixed spaces*

The set $\{|\mathfrak{q}\rangle\langle\mathfrak{p}|\}$ is the basis operators for the mixed $\mathfrak{q}$-$\mathfrak{p}$ representation. From the completeness relations of the unbiased basis states, $\{|\mathfrak{q}\rangle\}$ and $\{|\mathfrak{p}\rangle\}$, the set $\{|\mathfrak{q}\rangle\langle\mathfrak{p}|\}$ obeys the completeness relation,

$$\sum_{\mathfrak{q},\mathfrak{p}} |\mathfrak{q}\rangle\langle\mathfrak{q}||\mathfrak{p}\rangle\langle\mathfrak{p}| = 1, \tag{9.15}$$

$$\sum_{\mathfrak{q},\mathfrak{p}} \langle\mathfrak{q}||\mathfrak{p}\rangle\ |\mathfrak{q}\rangle\langle\mathfrak{p}| = 1. \tag{9.16}$$

Substituting the expression for $\langle\mathfrak{q}||\mathfrak{p}\rangle$,

$$\langle\mathfrak{q}||\mathfrak{p}\rangle = C_0 \exp(i\mathfrak{p}\cdot\mathfrak{q}), \tag{9.17}$$

$$\langle\mathfrak{p}||\mathfrak{q}\rangle = C_0 \exp(-i\mathfrak{p}\cdot\mathfrak{q}), \tag{9.18}$$

we obtained, for the completeness relation,

$$C_0 \sum_{\mathfrak{q},\mathfrak{p}} \exp\left(i\mathfrak{p}\cdot\mathfrak{q}\right) |\mathfrak{q}\rangle \langle\mathfrak{p}| = 1, \tag{9.19}$$

where, assuming *non-degenerates countable states*, C_o can be choosen as

$$C_o = (N)^{-\frac{1}{2}}. \tag{9.20}$$

Equation (9.19) can be rewritten as

$$\sum_{\mathfrak{q},\mathfrak{p}} \frac{|\mathfrak{q}\rangle \langle\mathfrak{p}|}{\langle\mathfrak{p}| |\mathfrak{q}\rangle} = 1, \tag{9.21}$$

and similarly,

$$\sum_{\mathfrak{q},\mathfrak{p}} \frac{|\mathfrak{p}\rangle \langle\mathfrak{q}|}{\langle\mathfrak{q}| |\mathfrak{p}\rangle} = 1. \tag{9.22}$$

Here we use the transformation identities in the mixed $\mathfrak{q}$-$\mathfrak{p}$ representation,

$$|\mathfrak{p}\rangle = \sum_{\mathfrak{q}} \langle\mathfrak{q}| |\mathfrak{p}\rangle\ |\mathfrak{q}\rangle, \tag{9.23}$$

$$\langle\mathfrak{p}| = \sum_{\theta} \langle\mathfrak{p}| |\mathfrak{q}\rangle\ \langle\mathfrak{q}|, \tag{9.24}$$

$$|\mathfrak{q}\rangle = \sum_{\phi} \langle\mathfrak{p}| |\mathfrak{q}\rangle\ |\mathfrak{p}\rangle, \tag{9.25}$$

$$\langle\mathfrak{q}| = \sum_{\phi} \langle\mathfrak{q}| |\mathfrak{p}\rangle\ \langle\mathfrak{p}|, \tag{9.26}$$

with transformation functions given by Eqs. (9.17)–(9.18). Note the similarities of Eqs. (9.23)–(9.26), combined with Eqs. (9.17)–(9.18), with those of Eqs. (2.8)–(2.12) of CMP

9.3.2 *The expansion of any operators in a mixed space*

Any operator, $\hat{A}$, can be expressed in terms of the eigenvector spaces, $|\mathfrak{q}'\rangle$ and $|\mathfrak{p}''\rangle$, respectively, in a mixed representation by Eq. (9.14), which we rewrite for convenience as

$$\begin{aligned}\hat{A} &= \sum_{\mathfrak{p}'',\mathfrak{q}'} |\mathfrak{p}''\rangle \langle\mathfrak{p}''| A |\mathfrak{q}'\rangle \langle\mathfrak{q}'|, \\ &= \sum_{\mathfrak{p}'',\mathfrak{q}'} \langle\mathfrak{p}''| A |\mathfrak{q}'\rangle\ |\mathfrak{p}''\rangle \langle\mathfrak{q}'|. \end{aligned} \tag{9.27}$$

We wish to express $\langle \mathfrak{p}''| A |\mathfrak{q}'\rangle$ and $|\mathfrak{p}''\rangle \langle \mathfrak{q}'|$ in terms of the $|\mathfrak{q}\rangle$-eigenstate matrix elements and $|\mathfrak{p}\rangle$-space projectors, respectively. Using, Eqs. (9.23)–(9.26), we write

$$\langle \mathfrak{p}''| A |\mathfrak{q}'\rangle = \frac{1}{\sqrt{N}} \sum_{\mathfrak{q}''} e^{-i\mathfrak{p}''\cdot\mathfrak{q}''} \langle \mathfrak{q}''| A |\mathfrak{q}'\rangle, \tag{9.28}$$

$$|\mathfrak{p}''\rangle \langle \mathfrak{q}'| = \frac{1}{\sqrt{N}} \sum_{\mathfrak{p}'} e^{i\mathfrak{p}'\cdot\mathfrak{q}'} |\mathfrak{p}''\rangle \langle \mathfrak{p}'|, \tag{9.29}$$

with completeness relation, using the transformation function characteristic of dual spaces,

$$\frac{1}{\sqrt{N}} \sum_{\mathfrak{q},\mathfrak{p}} \exp\left(i\mathfrak{p}\cdot\mathfrak{q}\right) |\mathfrak{q}\rangle \langle \mathfrak{p}| = 1 = \sum_{\mathfrak{p}} |\mathfrak{p}\rangle \langle \mathfrak{p}|. \tag{9.30}$$

Introducing the notation in Eq. (9.29),

$$\mathfrak{p}' = \mathfrak{p} + \mathfrak{u}, \quad \mathfrak{q}' = \mathfrak{q} + \mathfrak{v}, \tag{9.31}$$

$$\mathfrak{p}'' = \mathfrak{p} - \mathfrak{u}, \quad \mathfrak{q}'' = \mathfrak{q} - \mathfrak{v}. \tag{9.32}$$

Then, upon substituting in Eq. (9.27), we end up with

$$\begin{aligned} A &= \sum_{\mathfrak{p}'',\mathfrak{q}'} |\mathfrak{p}''\rangle \langle \mathfrak{p}''| A |\mathfrak{q}'\rangle \langle \mathfrak{q}'|, \\ &= \frac{1}{N} \sum_{\mathfrak{p},\mathfrak{q},\mathfrak{u},\mathfrak{v}} e^{i2(\mathfrak{p}\cdot\mathfrak{v}+\mathfrak{u}\cdot\mathfrak{q})} \langle \mathfrak{q}-\mathfrak{v}| A |\mathfrak{q}+\mathfrak{v}\rangle \; |\mathfrak{p}-\mathfrak{u}\rangle \langle \mathfrak{p}+\mathfrak{u}|. \end{aligned} \tag{9.33}$$

9.3.3 *Mixed space operator basis,* $\hat{\Delta}(\mathfrak{p}, \mathfrak{q})$

We write the last result as an expansion in terms of *mixed-phase point projector*, $\hat{\Delta}(\mathfrak{p}, \mathfrak{q})$, defined as the *Weyl transform* of a projector, by

$$\begin{aligned} \hat{\Delta}(\mathfrak{p}, \mathfrak{q}) &= \frac{1}{\sqrt{N}} \sum_{\mathfrak{u}} e^{-i2\mathfrak{u}\cdot\mathfrak{q}} |\mathfrak{p}+\mathfrak{u}\rangle \langle \mathfrak{p}-\mathfrak{u}|, \\ &= \frac{1}{\sqrt{N}} \sum_{\mathfrak{u}} e^{-i2\mathfrak{u}\cdot\mathfrak{q}} e^{2iQ\cdot u} |\mathfrak{p}-\mathfrak{u}\rangle \langle \mathfrak{p}-\mathfrak{u}|, \end{aligned}$$

$$= \frac{1}{\sqrt{N}} (N)^{-1} \sum_{\mathfrak{u}} e^{-i2\mathfrak{u}\cdot\mathfrak{q}} e^{2i\mathcal{Q}\cdot\mathfrak{u}} \sum_{\mathfrak{v}} e^{2i(\mathfrak{p}-\mathfrak{u}-\mathcal{P})\cdot\mathfrak{v}} |\mathfrak{p}_0\rangle \langle\mathfrak{p}_0|,$$

$$= \frac{1}{\sqrt{N}} (N)^{-1} \sum_{\mathfrak{u},\mathfrak{v}} e^{2i(\mathcal{Q}-\mathfrak{q})\cdot\mathfrak{u}} e^{-2i(\mathcal{P}-\mathfrak{p})\cdot\mathfrak{v}} e^{-2i\mathfrak{u}\cdot\mathfrak{v}} |\mathfrak{p}_0\rangle \langle\mathfrak{p}_0|,$$

$$= \frac{1}{\sqrt{N}} (N)^{-1} \sum_{\mathfrak{u},\mathfrak{v}} e^{2i(\mathfrak{p}\cdot\mathfrak{v}-\mathfrak{q}\cdot\mathfrak{u})} e^{-2i(\mathcal{P}\cdot\mathfrak{v}-\mathcal{Q}\cdot\mathfrak{u})} \sum_{\mathfrak{p}_0} |\mathfrak{p}_0\rangle \langle\mathfrak{p}_0|, \quad (9.34)$$

and the coefficient of expansion, the so-called *Weyl transform* of matrix element of operator, $A(\mathfrak{p},\mathfrak{q})$, defined by

$$A(\mathfrak{p},\mathfrak{q}) = \frac{1}{\sqrt{N}} \sum_{\mathfrak{v}} e^{i2\mathfrak{p}\cdot\mathfrak{v}} \langle\mathfrak{q}-\mathfrak{v}| A |\mathfrak{q}+\mathfrak{v}\rangle. \quad (9.35)$$

Clearly, for a density matrix operator $\hat{\rho}$, the Weyl transform functional or Wigner distribution functional obeys,

$$\sum_{\mathfrak{p},\mathfrak{q}} \rho(\mathfrak{p},\mathfrak{q}) = 1. \quad (9.36)$$

If one accounts for other extra discrete quantum labels like spin, isospin, and energy-band indices, we can incorporate this in the summation in a form of a trace.

Thus, we eventually have any operator expanded in terms of mixed-space operator basis, $\hat{\Delta}(\mathfrak{p},\mathfrak{q})$,

$$\hat{A} = \sum_{\mathfrak{p},\mathfrak{q}} A(\mathfrak{p},\mathfrak{q})\ \hat{\Delta}(\mathfrak{p},\mathfrak{q}),$$

$$= \frac{1}{\sqrt{N}} (N)^{-1} \sum_{\mathfrak{u},\mathfrak{v}} \left(\sum_{\mathfrak{p},\mathfrak{q}} A(\mathfrak{p},\mathfrak{q})\, e^{2i(\mathfrak{p}\cdot\mathfrak{v}-\mathfrak{q}\cdot\mathfrak{u})} \right) e^{-2i(\mathcal{P}\cdot\mathfrak{v}-\mathcal{Q}\cdot\mathfrak{u})} |\mathfrak{p}_0\rangle \langle\mathfrak{p}_0|. \quad (9.37)$$

We have

$$A(\mathfrak{p},\mathfrak{q}) = \mathrm{Tr}(\hat{A}\ \hat{\Delta}(\mathfrak{p},\mathfrak{q}))$$

$$= \frac{1}{\sqrt{N}} \sum_{\mathfrak{u},\mathfrak{v}} e^{2i(\mathfrak{p}\cdot\mathfrak{v}-\mathfrak{q}\cdot\mathfrak{u})} \mathrm{Tr}((N)^{-1} \hat{A} e^{-2i(\mathcal{P}\cdot\mathfrak{v}-\mathcal{Q}\cdot\mathfrak{u})} |\mathfrak{p}_0\rangle \langle\mathfrak{p}_0|),$$

$$= \frac{1}{\sqrt{N}} \sum_{\mathfrak{u},\mathfrak{v}} e^{2i(\mathfrak{p}\cdot\mathfrak{v}-\mathfrak{q}\cdot\mathfrak{u})} A(\mathfrak{u},\mathfrak{v}), \quad (9.38)$$

where $A(\mathfrak{u}, \mathfrak{v})$ is the characteristic function of $A(\mathfrak{p}, \mathfrak{q})$ distribution. Upon similar procedure based on Eq. (9.27), an equivalent expression can be obtain for $A(\mathfrak{p}, \mathfrak{q})$ and $\hat{\Delta}(\mathfrak{p}, \mathfrak{q})$, namely,

$$A(\mathfrak{p}, \mathfrak{q}) = \frac{1}{\sqrt{N}} \sum_{\mathfrak{u}} e^{i2\mathfrak{u}\cdot\mathfrak{q}} \langle \mathfrak{p} + \mathfrak{u}| \hat{A} |\mathfrak{p} - \mathfrak{u}\rangle, \tag{9.39}$$

$$\hat{\Delta}(\mathfrak{p}, \mathfrak{q}) = \frac{1}{\sqrt{N}} \sum_{\mathfrak{v}} e^{i2\mathfrak{p}\cdot\mathfrak{v}} |\mathfrak{q} + \mathfrak{v}\rangle \langle \mathfrak{q} - \mathfrak{v}|. \tag{9.40}$$

9.3.4 *Symmetrization of the generator of eigenstates*

Generator of eigenstates occupy a central role in non-Hermitian quantum dynamics. We can also define this in the Hermitian mixed $\mathfrak{q}$-$\mathfrak{p}$ representation. We have for $\mathfrak{q}$,

$$|\mathfrak{q}\rangle = \exp\{-i\mathfrak{q}\cdot\mathcal{P}\}|0\rangle, \tag{9.41}$$

where again the operator $\mathcal{P} = i\nabla_{\mathfrak{q}}$ is operating on the basis eigenvector $|\mathfrak{q}\rangle$, since $\mathcal{P} = -i\nabla_x$ acting on the x-components behaves contravariantly. Here, we again made an assumption in the above expression that there exist continuous functional of $\mathfrak{q}$ having an infinite radius of convergence, which are equal to $|\mathfrak{q}\rangle$ at the "lattice" functional points, i.e., we expand the exponential in Eq. (9.41) as a well-defined Taylor series.

9.3.5 *Fully symmetric translation operators*

The state,

$$|\mathfrak{q}\rangle = \exp\{-i\mathfrak{q}\cdot\mathcal{P}\}\ |0\rangle, \tag{9.42}$$

is an eigenstate with displaced eigenvalue by $\mathfrak{q}$. However, if $|0\rangle \Rightarrow |\mathfrak{q}'\rangle$ then the state $[\exp\{-i\mathfrak{q}\cdot\mathcal{P}\}|\mathfrak{q}'\rangle] = |\mathfrak{q}' + \mathfrak{q}\rangle$ is an eigenstate of the "position operator" with eigenvalue $(\mathfrak{q}' + \mathfrak{q})$.

We can symmetrize the translation operator by inserting $\exp\{i\mathfrak{p}\cdot\mathcal{Q}\}$ in front of $|q = 0\rangle$, in Eq. (9.42), which effectively insert unity. We thus have

$$|\mathfrak{q}\rangle = \exp\{-i\mathfrak{q}\cdot\mathcal{P}\}\exp\{i\mathfrak{p}\cdot\mathcal{Q}\}|0\rangle. \tag{9.43}$$

By the use of the Campbell–Baker–Hausdorff operator identity, we obtained

$$\begin{aligned}
&\exp\{-i\mathfrak{q}\cdot\mathcal{P}\}\exp\{i\mathfrak{p}\cdot\mathcal{Q}\}\\
&\quad=\exp\{-i\left(\mathfrak{q}\cdot\mathcal{P}-\mathfrak{p}\cdot\mathcal{Q}\right)\}\exp\left\{\frac{\left[-i\mathfrak{q}\cdot\mathcal{P},i\mathfrak{p}\cdot\mathcal{Q}\right]}{2}\right\},\\
&\quad=\exp\{-i\left(q\cdot P-p\cdot Q\right)\}\exp\left\{i\frac{\mathfrak{p}\cdot\mathfrak{q}}{2}\right\}.
\end{aligned}\tag{9.44}$$

Therefore, we have the symmetric form for the displacement operator generating the state $|\mathfrak{q}\rangle$ from $|0\rangle$ given by

$$T\left(\mathfrak{q},\mathfrak{p}\right)_{\mathrm{sym}}=\exp\left\{i\frac{\mathfrak{p}\cdot\mathfrak{q}}{2}\right\}\exp\left\{-\frac{i}{\hbar}\left(\mathfrak{q}\cdot\mathcal{P}-\mathfrak{p}\cdot\mathcal{Q}\right)\right\},\tag{9.45}$$

where we used the symbol $T\left(\mathfrak{q},\mathfrak{p}\right)_{\mathrm{sym}}$ because of isomorphism demonstrated in Eq. (9.45). The displacement operator may also be interpreted as an operator for the preparation of the quantum eigenstate out of the "vacuum", $|0\rangle$, basis eigenstate.[1]

9.3.6 *The symmetric operator basis for mixed representations*

The ubiquitous appearance of the symmetric operator factor,

$$\begin{aligned}
Y_{\mathfrak{p},\mathfrak{q}}&=\exp\left(-i\left(\mathfrak{q}\cdot\mathcal{P}-\mathfrak{p}\cdot\mathcal{Q}\right)\right),\\
&=\exp\left\{i\frac{\mathfrak{p}\cdot\mathfrak{q}}{2}\right\}\exp\{-i\mathfrak{q}\cdot\mathcal{P}\}\exp\{i\mathfrak{p}\cdot\mathcal{Q}\},
\end{aligned}\tag{9.46}$$

suggest that this operator is the basis operator for mixed space representation theory. This is referred to here as the generalized projector or as "generalized Pauli-matrix operator" [12]. It can be considered the universal form of projector in mixed representations of quantum physics, either dealing with Hermitian or with non-Hermitian operators.

[1]Here, the concept of a vacuum state does have a special meaning in quantum field theory, i.e., no excitations or quasiparticles.

Symmetric form of $\hat{\Delta}_{\lambda'\lambda}(\mathfrak{p},\mathfrak{q})$ If we consider Eq. (9.40) as the expression for $\hat{\Delta}_{\lambda'\lambda}(\mathfrak{p},\mathfrak{q})$, the following identities can be verified,

$$|\mathfrak{q}+\mathfrak{v},\lambda\rangle = \exp\left[-2i\mathcal{P}\cdot\mathfrak{v}\right]|\mathfrak{q}-\mathfrak{v},\lambda\rangle. \tag{9.47}$$

Then

$$\begin{aligned}
&|\mathfrak{q}-\mathfrak{v},\lambda\rangle\left\langle\mathfrak{q}-\mathfrak{v},\lambda'\right| \\
&= (N)^{-1}\sum_{\mathfrak{u},\mathfrak{q}_0}\exp\left[-2i\left(\mathfrak{q}-\mathfrak{v}-\mathcal{Q}\right)\cdot\mathfrak{u}\right]|\mathfrak{q}_0,\lambda\rangle\left\langle\mathfrak{q}_0,\lambda'\right|, \\
&= \sum_{\mathfrak{q}_0}\delta\left(\mathfrak{q}-\mathfrak{v}-\hat{\mathcal{Q}}\right)|\mathfrak{q}_0,\lambda\rangle\left\langle\mathfrak{q}_0,\lambda'\right| \\
&= \sum_{\mathfrak{q}_0}\delta\left(\mathfrak{q}-\mathfrak{v}-\mathfrak{q}_0\right)\ |\mathfrak{q}_0,\lambda\rangle\left\langle\mathfrak{q}_0,\lambda'\right|,
\end{aligned} \tag{9.48}$$

$$\Omega_{\lambda\lambda'} = |\mathfrak{q}_0,\lambda\rangle\left\langle\mathfrak{q}_0,\lambda'\right|. \tag{9.49}$$

Substituting the expressions, Eqs (9.47) and (9.48) in Eq. (9.40), we obtain a completely symmetric expression of $\hat{\Delta}(\mathfrak{p},\mathfrak{q})$,

$$\begin{aligned}
&\hat{\Delta}(\mathfrak{p},\mathfrak{q}) \\
&= \frac{1}{\sqrt{N}}\sum_{\mathfrak{v}}e^{i2\mathfrak{p}\cdot\mathfrak{v}}\ |\mathfrak{q}+\mathfrak{v}\rangle\langle\mathfrak{q}-\mathfrak{v}| \\
&= \frac{1}{\sqrt{N}}(N)^{-1}\sum_{\bar{\mathfrak{v}},\bar{\mathfrak{u}}}e^{i2\mathfrak{p}\cdot\mathfrak{v}}\exp[-2i\mathcal{P}\cdot\mathfrak{u}]\exp[-2i(\mathfrak{q}-\mathfrak{v}-\hat{\mathcal{Q}})\cdot\mathfrak{u}] \\
&\quad\times\sum_{\mathfrak{q}_0}|\mathfrak{q}_0,\lambda\rangle\left\langle\mathfrak{q}_0,\lambda'\right|.
\end{aligned} \tag{9.50}$$

Then we have

$$\begin{aligned}
&\exp\left\{-i\left(\mathcal{P}-\mathfrak{p}\right)\cdot 2\mathfrak{v}\right\}\exp\left\{2i\left(\hat{\mathcal{Q}}-\mathfrak{q}\right)\cdot\bar{\mathfrak{u}}\right\}\exp 2i\mathfrak{v}\cdot\mathfrak{u} \\
&\quad= \exp\left\{-i\left(\mathcal{P}\cdot\mathfrak{v}-\mathcal{Q}\cdot\mathfrak{u}\right)\right\}\exp\left\{2i\left(\mathfrak{p}\cdot\mathfrak{v}-\mathfrak{q}\cdot\mathfrak{u}\right)\right\},
\end{aligned} \tag{9.51}$$

which is the same as Eq. (9.34). Equation (9.51) is a symmetric form.

We can also combine the exponential operators to obtain,

$$\frac{1}{\sqrt{N}}(N)^{-1}\sum_{\bar{\mathfrak{v}},\bar{\mathfrak{u}}} e^{2i\mathfrak{p}\cdot\mathfrak{v}}\exp\left[-2i\mathcal{P}\cdot\mathfrak{v}\right]\exp\left[2i\left(\mathfrak{q}-\mathfrak{v}-\mathcal{Q}\right)\cdot\bar{\mathfrak{u}}\right]\sum_{\mathfrak{q}_0}\left|\mathfrak{q}_0,\lambda\right\rangle\left\langle\mathfrak{q}_0,\lambda'\right|$$

$$=\frac{1}{\sqrt{N}}(N)^{-1}\sum_{\bar{\mathfrak{v}},\bar{\mathfrak{u}}}\exp -2i\left[\left(\mathcal{P}-\mathfrak{p}\right)\cdot\mathfrak{v}+\left(\mathcal{Q}-\mathfrak{q}\right)\cdot\mathfrak{u}\right]\Omega_{\lambda\lambda'}, \tag{9.52}$$

where

$$\Omega_{\lambda\lambda'}=\sum_{\mathfrak{q}_0}\left|\mathfrak{q}_0,\lambda\right\rangle\left\langle\mathfrak{q}_0,\lambda'\right|=\sum_{\mathfrak{p}_0}\left|\mathfrak{p}_0,\lambda\right\rangle\left\langle\mathfrak{p}_0,\lambda'\right|, \tag{9.53}$$

and

$$A\left(\mathfrak{p},\mathfrak{q}\right)=\mathrm{Tr}\left(\hat{A}\hat{\Delta}\right)$$
$$=\frac{1}{\sqrt{N}}\left(\begin{array}{c}\sum\limits_{\bar{\mathfrak{v}},\bar{\mathfrak{u}}}\exp 2i\left[\mathfrak{p}\cdot\mathfrak{v}-\mathfrak{q}\cdot\mathfrak{u}\right]\\ \times\,\mathrm{Tr}\left\{(N)^{-1}\hat{A}\exp\left\{-2i\left[\mathcal{P}\cdot\mathfrak{v}-\mathcal{Q}\cdot\mathfrak{u}\right]\right\}\Omega_{\lambda\lambda'}\right\}\end{array}\right). \tag{9.54}$$

Therefore, the characteristic distribution for $A\left(\mathfrak{p},\mathfrak{q}\right)$ is identically,

$$A_{\lambda\lambda'}\left(\mathfrak{u},\mathfrak{v}\right)=(N)^{-1}\,\mathrm{Tr}\{\hat{A}\exp\left\{-2i\left[\mathcal{P}\cdot\mathfrak{v}-\mathcal{Q}\cdot\mathfrak{u}\right]\right\}\Omega_{\lambda\lambda'}\}, \tag{9.55}$$

like before. By using the characteristic distribution function for $A_{\lambda\lambda'}\left(\mathfrak{p},\mathfrak{q}\right)$

$$A_{\lambda\lambda'}\left(\mathfrak{u},\mathfrak{v}\right)=\left(\frac{1}{N}\right)^{\frac{1}{2}}\sum_{p,q}A_{\lambda\lambda'}\left(\mathfrak{p},\mathfrak{q}\right)\exp -2i\left[\mathfrak{p}\cdot\mathfrak{v}-\mathfrak{q}\cdot\mathfrak{u}\right], \tag{9.56}$$

with inverse

$$A_{\lambda\lambda'}\left(\mathfrak{p},\mathfrak{q}\right)=\left(\frac{1}{N}\right)^{\frac{1}{2}}\sum_{u,v}A_{\lambda\lambda'}\left(\mathfrak{u},\mathfrak{v}\right)\exp\left\{2i\left[\mathfrak{p}\cdot\mathfrak{v}-\mathfrak{q}\cdot\mathfrak{u}\right]\right\}, \tag{9.57}$$

then we can write Eq. (9.37) simply like a Fourier transform (caveat: Fourier transform to *operator space*) of the characteristic function of the lattice Weyl transform of the operator $\hat{A}$,

$$\hat{A}=(N)^{-1}\sum_{\bar{u},\bar{v},\lambda,\lambda'}A_{\lambda\lambda'}\left(\mathfrak{u},\mathfrak{v}\right)\exp\left\{-2i\left[\mathcal{P}\cdot\mathfrak{v}-\mathcal{Q}\cdot\mathfrak{u}\right]\right\}\Omega_{\lambda'\lambda}, \tag{9.58}$$

where the inverse can be written as

$$A_{\lambda\lambda'}\left(\mathfrak{u},\mathfrak{v}\right)=(N)^{-1}\,\mathrm{Tr}\{\hat{A}\exp\{-2i[\mathcal{P}\cdot\mathfrak{v}-\mathcal{Q}\cdot\mathfrak{u}]\}\}\Omega_{\lambda\lambda'}. \tag{9.59}$$

In continuum approximation, we have from Eq. (9.52),

$$\hat{\Delta}(\mathfrak{p},\mathfrak{q}) = (2\pi)^{-1}\int d\mathfrak{u}\wedge d\mathfrak{v}\ e^{2i[((\mathfrak{p}-\mathcal{P}).v-(\mathfrak{q}-\mathcal{Q})\cdot u)]},$$

$$= (2\pi)^{-1}\int d\mathfrak{u}\wedge d\mathfrak{v}\ e^{2i[\mathfrak{p}\cdot\mathfrak{v}-\mathfrak{q}\cdot\mathfrak{u}]}e^{(-2i)(\mathcal{P}\cdot\mathfrak{v}-\mathcal{Q}\cdot\mathfrak{u})}. \tag{9.60}$$

9.3.7 *On the counting of states: Coherent state formulation*

The measure of counting of states in Eq. (9.60) can be shown to be related to original annihilation and creation operator in the case of harmonic oscillator. We have

$$(2\pi)^{-1}\int d\mathfrak{u}\wedge d\mathfrak{v} \Longrightarrow (2\pi i)^{-1}\int d\mathfrak{u}\wedge d\mathfrak{v},$$

$$= (2\pi i)^{-1}\int \frac{1}{2}d\left(\bar{\mathfrak{q}}-i\bar{\mathfrak{p}}\right)\wedge d\left(\bar{\mathfrak{q}}+i\bar{\mathfrak{p}}\right),$$

$$= (2\pi)^{-1}\int (d\bar{\mathfrak{q}}\wedge d\bar{\mathfrak{p}}),$$

$$= \frac{1}{\pi}\int (d\,\mathrm{Re}\,\alpha\wedge d\,\mathrm{Im}\,\alpha), \tag{9.61}$$

since $d\,\mathrm{Re}\,\alpha = \frac{d\bar{\mathfrak{q}}}{\sqrt{2}}$ and $d\,\mathrm{Im}\,\alpha = \frac{d\bar{\mathfrak{p}}}{\sqrt{2}}$.

9.4 Characteristic distribution of lattice Weyl transform

We have the identity for the characteristic function, $A_{\lambda\lambda'}(\mathfrak{u},\mathfrak{v})$,

$$A_{\lambda\lambda'}(\mathfrak{u},\mathfrak{v}) = (N)^{-1}\mathrm{Tr}\left\{\hat{A}\exp\{-2i\left[\mathcal{P}\cdot\mathfrak{v}-\mathcal{Q}\cdot\mathfrak{u}\right]\}\right\}\Omega_{\lambda\lambda'}. \tag{9.62}$$

Thus, the characteristic function exist for all function of canonical quantum operators, either Hermitian or non-Hermitian, spinor (fermions) or boson operators.[2]

9.4.1 *Implications on CS formulation*

In what follows, for a moment we will drop the discrete indices λ and λ' to make contact with CS formulation of quantum physics. In general, we can

[2]For creation and annihilation operators in many-body quantum physics, the proof relies on the use of normal or anti-normal ordering of canonical operators, which can then be treated like $\mathbb{C}$-numbers in expansion of exponentials. The exponential in Eq. (9.62) is sometimes referred to as the generalized Pauli-spin operator.

have different expression for the characteristic function depending on the use of, what is often referred to, in corresponding CS formalism, as the normal and anti-normal expressions. This is made clear by restoring the annihilation and creation operators, $\hat{\psi}$ and $\hat{\psi}^\dagger$, thus,

$$\begin{aligned}
&\exp\{-2i[\mathcal{P}\cdot\mathfrak{v}-\mathcal{Q}\cdot\mathfrak{u}]\}\\
&\quad\to\exp\{2[\hat{\psi}^\dagger\cdot\mathfrak{v}+\hat{\psi}\cdot\mathfrak{u}]\}\\
&\quad=\exp\left(2\mathfrak{v}\cdot\mathfrak{u}\right)\exp\{2\mathfrak{v}\cdot\hat{\psi}^\dagger\}\exp\left\{2\mathfrak{u}\cdot\hat{\psi}\right\},\\
&\quad=\exp\{2\alpha\cdot\alpha^*\}\exp\left\{2\mathfrak{v}\cdot\hat{\psi}^\dagger\right\}\exp\left\{2\mathfrak{u}\cdot\hat{\psi}\right\}.
\end{aligned}\tag{9.63}$$

so that

$$\exp\left\{-2i\mathfrak{v}\cdot\mathcal{P}\right\}\exp\left\{2i\mathfrak{u}\cdot\mathcal{Q}\right\}\Longrightarrow\exp\left\{-2\alpha\cdot\alpha^*\right\}\exp\left\{2\left[\hat{\psi}^\dagger\cdot\mathfrak{v}+\hat{\psi}\cdot\mathfrak{u}\right]\right\},\tag{9.64}$$

$$\exp\left\{2i\mathfrak{u}\cdot\mathcal{Q}\right\}\exp\left\{-2i\mathfrak{v}\cdot\mathcal{P}\right\}\Longrightarrow\exp\left\{2\alpha\cdot\alpha^*\right\}\exp\left\{2\left[\hat{\psi}^\dagger\cdot\mathfrak{v}+\hat{\psi}\cdot\mathfrak{u}\right]\right\},\tag{9.65}$$

yielding, upon restoring indices λ and λ', the following different expressions for $A_{\lambda\lambda'}(u,v)$, namely,

$$A^{w}_{\lambda\lambda'}(u,v)=(N)^{-1}\,\mathrm{Tr}\left(\hat{A}\exp\left(-2i\right)\left(\mathcal{P}.\mathfrak{v}-\mathcal{Q}.\mathfrak{u}\right)\right),\tag{9.66}$$

which is the characteristic function for the Wigner distribution function. We also have the so-called normal characteristic distribution function,

$$\begin{aligned}
A^{n}_{\lambda\lambda'}(u,v)&=(N)^{-1}\,\mathrm{Tr}[\hat{A}\exp\{-2i\mathfrak{v}\cdot\mathcal{P}\}\exp\{2i\mathfrak{u}\cdot\mathcal{Q}\}],\\
&=(N)^{-1}\exp\{2i\mathfrak{u}\cdot\mathfrak{v}\}\mathrm{Tr}(\hat{A}\exp(-2i)(\mathcal{P}.\mathfrak{v}-\mathcal{Q}.\mathfrak{u})),
\end{aligned}\tag{9.67}$$

and the anti-normal characteristic distribution function given by

$$\begin{aligned}
A^{a}_{\lambda\lambda'}(u,v)&=(N)^{-1}\,\mathrm{Tr}\,\{\hat{A}\;\exp\{2i\mathfrak{u}\cdot\mathcal{Q}\}\exp\{-2i\mathfrak{v}\cdot\mathcal{P}\}\},\\
&=(N)^{-1}\exp\{-2i\mathfrak{u}\cdot\mathfrak{v}\}\mathrm{Tr}(\hat{A}\exp(-2i)(\mathcal{P}.\mathfrak{v}-\mathcal{Q}.\mathfrak{u})).
\end{aligned}\tag{9.68}$$

Although, Eqs. (9.67) and (9.68) only amounts to difference in the phase factors in the case of canonical position-momentum q-p ordinary mixed space representation, here the similar quantities in non-Hermitian dual spaces gives a real exponents giving very different distributions often referred to as smooth-out distributions. Examining Eqs. (9.64) and (9.65), and the fact that in non-Hermitian mixed representation, for example in CS,

$$\begin{aligned}\exp\{2\mathfrak{u}\cdot\mathfrak{v}\} &\equiv \exp(2\bar{\alpha}^*\bar{\alpha}) = \exp[(\bar{q}-i\bar{p})(\bar{q}+i\bar{p})],\\ &= \exp\left[\left(\mathrm{Re}\,\bar{\alpha}^2+\mathrm{Im}\,\bar{\alpha}^2\right)\right], \qquad (9.69)\\ \exp\{-2\mathfrak{u}\cdot\mathfrak{v}\} &\equiv \exp(-2\bar{\alpha}^*\bar{\alpha}) = \exp[-(\bar{q}-i\bar{p})(\bar{q}+i\bar{p})],\\ &= \exp\left[-\left(\mathrm{Re}\,\bar{\alpha}^2+\mathrm{Im}\,\bar{\alpha}^2\right)\right], \qquad (9.70)\end{aligned}$$

in original notation is a real quantity that resembles a Gaussian function, Eqs. (9.65) clearly represent some smoothing of the Wigner distribution characteristic functional and hence smoothing of the Wigner distribution functional itself. Moreover, in Eqs. (9.67) and (9.68) no informations are lost.

Indeed, more *general* phase-space distribution functionals, $f^{(g)}(\mathfrak{p},\mathfrak{q},t)$, can be obtained from the expression

$$\begin{aligned}f^{(g)}(\mathfrak{p},\mathfrak{q},t) &= \left(\frac{1}{N}\right)^{\frac{1}{2}}\sum_{u,v} A_{\lambda\lambda'}(\mathfrak{u},\mathfrak{v})\exp\{-2i[\mathfrak{p}\cdot\mathfrak{v}-\mathfrak{q}\cdot\mathfrak{u}]\}\,g(\mathfrak{u},\mathfrak{v}),\\ &= \frac{1}{2\pi}\int dudve^{-i([\mathfrak{p}\cdot\mathfrak{v}-\mathfrak{q}\cdot\mathfrak{u}])}[C^{(w)}(\mathfrak{u},\mathfrak{v},t)\;\; g(\mathfrak{u},\mathfrak{v})], \qquad (9.71)\end{aligned}$$

where $g(\mathfrak{u},\mathfrak{v})$ is some choosen smoothing function.

9.4.2 *P- and Q-function: Husimi distribution and smoothing*

Generally all distribution functionals will become meaningful under the functional integral sign, thus these have the properties of generalized functional distributions. The distribution function $f^a(\mathfrak{p},\mathfrak{q},t)$ that one obtain from Eq. (9.69) is known as the P-functional and that obtain from Eq. (9.70) is also known as the Q-functional or the Husimi distribution in quantum optics. A detailed discussion of these two distribution is given by the author's book [12] on the topic of coherent state formulation, and will not be repeated here for lack of space.

The identical algebra of the CMP's $\exp\{-i(q'\cdot P-p'\cdot Q)\}$ as well as its relevance to the physics of two-state systems, spin systems, quantum computing, entanglements (discussed in Part 5 of this book) and teleportation, are discussed in the author's book [12] as well.

9.5 Path integral for bosons

The path integral for bosons follows from the above $\mathfrak{q}$-$\mathfrak{p}$ mixed space representation. Except for the strictly functional integral [16], the procedure and results is analogous to that of CMP path integral before. We will just give the result as

$$\begin{aligned}
&\langle \mathfrak{q} | U(t, t_o) | \mathfrak{q}_o \rangle \\
&= \frac{1}{(2\pi\hbar)^{N\lambda}} \lim_{n \Longrightarrow \infty} \int .. \int \prod_{i=1}^{n} d^{N\lambda}\mathfrak{q}_i \prod_{i=1}^{n+1} d^{N\lambda}\mathfrak{p}_i \\
&\times \exp\left\{ \frac{i}{\hbar} \sum_{j=1}^{n} \epsilon \begin{bmatrix} \mathfrak{p}_j \cdot \frac{\mathfrak{q}_j - \mathfrak{q}_{j-1}}{\epsilon} \\ -H\left(\mathfrak{p}_j, \frac{\mathfrak{q}_j + \mathfrak{q}_{j-1}}{2}\right) \end{bmatrix} \right\},
\end{aligned} \tag{9.72}$$

where $N\lambda$ is the number of countable states of $\mathfrak{q}_i$ and $\mathfrak{p}_i$, $\epsilon = (t_j - t_{j-1})$ and $H\left(\mathfrak{p}_j, \frac{\mathfrak{q}_j + \mathfrak{q}_{j-1}}{2}\right)$ comes from the Weyl transform of the Hamiltonian.

Chapter 10

Mixed Space Representation in QFT: Fermions

The canonical field operators satisfy the following anticommutation relations,

$$\left[\psi_\mu^\dagger, \psi_\nu^\dagger\right]_+ \equiv \left\{\psi_\mu^\dagger, \psi_\nu^\dagger\right\} = 0, \tag{10.1}$$

$$[\psi_\mu, \psi_\nu]_+ \equiv \{\psi_\mu, \psi_\nu\} = 0, \tag{10.2}$$

$$\left[\hat{\psi}, \hat{\psi}^\dagger\right]_+ \equiv \left\{\hat{\psi}, \hat{\psi}^\dagger\right\} = 1. \tag{10.3}$$

We have, as before write, $\hat{\psi}^\dagger = -i\hat{\mathcal{P}}$, $\hat{\psi} = i\hat{\mathcal{Q}}$ as before. To avoid confusion, we set the respective eigenvalues of $\hat{\mathcal{Q}}$ and $\hat{\mathcal{P}}$ as $\mathfrak{q}$ and $\mathfrak{p}$, respectively. As before, we have the left and right eigenvector defined by

$$\hat{\mathcal{P}}\,|\mathfrak{p}\rangle = \mathfrak{p}\,|\mathfrak{p}\rangle = |\mathfrak{p}\rangle\,\mathfrak{p}, \tag{10.4}$$

$$\langle\mathfrak{p}|\,\hat{\mathcal{P}} = \langle\mathfrak{p}|\,\mathfrak{p} = \mathfrak{p}\,\langle\mathfrak{p}|\,. \tag{10.5}$$

Likewise, we have,

$$\hat{\mathcal{Q}}\,|\mathfrak{q}\rangle = \mathfrak{q}\,|\mathfrak{q}\rangle = |\mathfrak{q}\rangle\,\mathfrak{q}, \tag{10.6}$$

$$\langle\mathfrak{q}|\,\hat{\mathcal{Q}} = \langle\mathfrak{q}|\,\mathfrak{q} = \mathfrak{q}\,\langle\mathfrak{q}|\,. \tag{10.7}$$

Equations (10.4)–(10.7) follows from,

$$\langle\mathfrak{p}|\,\hat{\mathcal{P}}\,|\mathfrak{p}\rangle = \mathfrak{p}$$

$$\langle\mathfrak{p}|\,\mathfrak{p}\,|\mathfrak{p}\rangle = \mathfrak{p}\,\langle\mathfrak{p}|\,|\mathfrak{p}\rangle \tag{10.8}$$

$$= \langle\mathfrak{p}|\,|\mathfrak{p}\rangle\,\mathfrak{p}. \tag{10.9}$$

However, we have the following differential operator relations:

$$\begin{aligned}\langle \mathfrak{q}| \hat{\mathcal{P}} |\mathfrak{q}\rangle \mathfrak{q} &= \langle \mathfrak{q}| i\overrightarrow{\frac{\partial}{\partial \mathfrak{q}}} |\mathfrak{q}\rangle \mathfrak{q} - \langle \mathfrak{q}| |\mathfrak{q}\rangle i\overrightarrow{\frac{\partial}{\partial \mathfrak{q}}}\mathfrak{q} \\ &= -\mathfrak{q} \langle \mathfrak{q}| i\overrightarrow{\frac{\partial}{\partial \mathfrak{q}}} |\mathfrak{q}\rangle - i. \end{aligned} \tag{10.10}$$

Noting that Eq. (9.4) is negative for fermions, we have the anti-commutation relation,

$$\{\hat{\mathcal{Q}}, \hat{\mathcal{P}}\} = -i. \tag{10.11}$$

For fermions, we can also use Eqs. (8.25)–(8.41) which are common to both bosons and fermions. However, note that for fermions, the eigenvalues are elements of the Grassmann algebra. The following eigenvector generation holds for bosons and fermions, namely,

$$\hat{\psi}^{\dagger} |\mathfrak{q}\rangle = -i\hat{\mathcal{P}} |\mathfrak{q}\rangle = \overrightarrow{\frac{\partial}{\partial \mathfrak{q}}} |\mathfrak{q}\rangle \Longrightarrow \hat{\mathcal{P}} |\mathfrak{q}\rangle = i\overrightarrow{\frac{\partial}{\partial \mathfrak{q}}} |\mathfrak{q}\rangle, \tag{10.12}$$

$$\langle \mathfrak{q}| \overleftarrow{\hat{\psi}^{\dagger}} = \langle \mathfrak{q}| \left(-i\hat{\mathcal{P}}\right) = \langle \mathfrak{q}| \overleftarrow{\left(\frac{\partial}{\partial \mathfrak{q}}\right)} \Longrightarrow \langle \mathfrak{q}| \hat{\mathcal{P}} = \langle \mathfrak{q}| \overleftarrow{\left(i\frac{\partial}{\partial \mathfrak{q}}\right)}, \tag{10.13}$$

$$\hat{\psi} |\mathfrak{p}\rangle = i\hat{\mathcal{Q}} |\mathfrak{p}\rangle = \frac{\partial}{\partial \mathfrak{p}} |\mathfrak{p}\rangle \Longrightarrow \hat{\mathcal{Q}} |\mathfrak{p}\rangle = \overrightarrow{-i\frac{\partial}{\partial \mathfrak{p}}} |\mathfrak{p}\rangle, \tag{10.14}$$

$$\langle \mathfrak{p}| \overleftarrow{\left(i\hat{\mathcal{Q}}\right)} = \langle \mathfrak{p}| \overleftarrow{\left(\frac{\partial}{\partial \mathfrak{p}}\right)} \Longrightarrow \langle \mathfrak{p}| \hat{\mathcal{Q}} = \langle \mathfrak{p}| \overleftarrow{\left(-i\frac{\partial}{\partial \mathfrak{p}}\right)}, \tag{10.15}$$

which is compatible with Eqs. (10.5)–(10.7). We also have

$$|\mathfrak{q}\rangle = \exp\left(-i\mathfrak{q}.\hat{\mathcal{P}}\right) |0\rangle \Longrightarrow \hat{\mathcal{P}} |\mathfrak{q}\rangle = i\overrightarrow{\frac{\partial}{\partial \mathfrak{q}}} |\mathfrak{q}\rangle, \tag{10.16}$$

$$\langle \mathfrak{q}| = \langle 0| \exp\left(i\hat{\mathcal{P}}\cdot\mathfrak{q}\right) \Longrightarrow \langle \mathfrak{q}| \hat{\mathcal{P}} = \langle \mathfrak{q}| \overleftarrow{\left(-i\frac{\partial}{\partial \mathfrak{q}}\right)}, \tag{10.17}$$

$$|\mathfrak{p}\rangle = \exp\left(i\mathfrak{p}.\hat{\mathcal{Q}}\right) |0\rangle \Longrightarrow \hat{\mathcal{Q}} |\mathfrak{p}\rangle = \overrightarrow{-i\frac{\partial}{\partial \mathfrak{p}}} |\mathfrak{p}\rangle, \tag{10.18}$$

$$\langle \mathfrak{p}| = \langle 0| \exp\left(-i\hat{\mathcal{Q}}\cdot\mathfrak{p}\right) \Longrightarrow \langle \mathfrak{p}| \hat{\mathcal{Q}} = \langle \mathfrak{p}| \overleftarrow{\left(i\frac{\partial}{\partial \mathfrak{p}}\right)}, \tag{10.19}$$

which is again compatible with Eqs. (10.5)–(10.7), where the arrows denote the left and right derivatives and the dot product is defined by

$$\psi^\dagger \cdot \mathfrak{q} = \sum_\mu \mathfrak{q}_\mu \psi_\mu^\dagger \equiv \sum_\mu \mathfrak{q}_\mu(-i\hat{\mathcal{P}}_\mu). \tag{10.20}$$

Note that we retained the notation, $\mathfrak{p}$ and $\mathfrak{q}$ for eigenvalues, with the understanding that for fermions, $\mathfrak{p}$ and $\mathfrak{q}$ are elements of Grassmann algebra, not ordinary $\mathbb{C}$-functionals.

10.1 Transformation functions

Here, we derive the transformation functions specific to fermions and thereby justify the use of Eqs. (8.25)–(8.41) universal relations for QFT. Using the identity: $e^A e^B = e^B e^A e^{[A,B]}$, for the case where the commutator $[A, B]$ is a $\mathbb{C}$-number, we have the following expression[1] for the transformation functions

$$\begin{aligned} \langle \mathfrak{p}|\,|\mathfrak{q}\rangle &= \langle \mathfrak{p}=0|\exp\left(i\hat{\mathcal{Q}}\cdot\mathfrak{p}\right)\exp\left(-i\mathfrak{q}.\hat{\mathcal{P}}\right)|\mathfrak{q}=0\rangle\,, \\ &= C_0\,\langle 0|\,|0\rangle\exp\left(-i\mathfrak{p}.\mathfrak{q}\right), \end{aligned} \tag{10.21}$$

since, we have

$$\begin{aligned} \left[i\hat{\mathcal{Q}}\cdot\mathfrak{p}, -i\mathfrak{q}.\hat{\mathcal{P}}\right]_- &= \left[\left(\hat{\mathcal{Q}}\cdot\mathfrak{p}\right)\left(\mathfrak{q}.\hat{\mathcal{P}}\right) - \left(\mathfrak{q}.\hat{\mathcal{P}}\right)\left(\hat{\mathcal{Q}}\cdot\mathfrak{p}\right)\right], \\ &= \sum_{\mu,v} \mathfrak{p}_\mu \mathfrak{q}_\nu \left\{\ddot{\mathcal{Q}}_\mu \ddot{\mathcal{P}}_\nu + \hat{\mathcal{P}}_\nu \hat{\mathcal{Q}}_\mu\right\} \\ &= -i\sum_{\mu,v} \mathfrak{p}_\mu \mathfrak{q}_\nu \delta_{\mu\nu}, \\ &= -i\sum_\mu \mathfrak{p}_\mu \mathfrak{q}_\mu = -i\mathfrak{p}\cdot\mathfrak{q}, \end{aligned} \tag{10.22}$$

where we make use of the variant of BHC formula given by

$$\begin{aligned} e^X e^Y &= e^{\left(Y + [X,Y] + \frac{1}{2!}[X,[X,Y]] + \cdots\right)} e^X \\ &= e^Y e^X e^{[X,Y]}, \end{aligned} \tag{10.23}$$

---

[1]To avoid ambiguities due to Grassmann functional variables, we employ the displacement operators or generation of states for deriving the transition functionals between the Hilbert spaces of fermions.

since $[X, Y]$ commutes with both X and Y. In our case above only the term relating to $e^{[X,Y]}$ remains since the $e^Y e^X$ term yields unity (operating on vacuum states). Note that the last term of Eq. (10.21) agrees with Eq. (8.36) for the general case.

Similarly, we have,

$$\begin{aligned}\langle\mathfrak{q}|\,|\mathfrak{p}\rangle &= \langle\mathfrak{q}=0|\exp(-i\hat{\mathcal{P}}\cdot\mathfrak{q})\exp(i\mathfrak{p}.\hat{\mathcal{Q}})\,|\mathfrak{p}=0\rangle\,,\\ &= C_0\,\langle 0|\,|0\rangle\exp\left(i\mathfrak{p}\cdot\mathfrak{q}\right),\end{aligned} \tag{10.24}$$

since, we have

$$\begin{aligned}[-i\hat{\mathcal{P}}\cdot\mathfrak{q}, i\mathfrak{p}.\hat{\mathcal{Q}}] &= [\hat{\mathcal{P}}\cdot\mathfrak{q}, \mathfrak{p}.\hat{\mathcal{Q}}]\\ &= \hat{\mathcal{P}}\cdot\mathfrak{q}\mathfrak{p}.\hat{\mathcal{Q}} - \mathfrak{p}.\hat{\mathcal{Q}}\hat{\mathcal{P}}\cdot\mathfrak{q}\\ &= -\mathfrak{p}_\mu\mathfrak{q}_\nu\hat{\mathcal{P}}_\nu\hat{\mathcal{Q}}_\mu - \mathfrak{p}_\mu\mathfrak{q}_\nu\hat{\mathcal{Q}}_\mu\hat{\mathcal{P}}_\nu\\ &= -\mathfrak{p}_\mu\mathfrak{q}_\nu\{\hat{\mathcal{P}}_\nu, \hat{\mathcal{Q}}_\mu\}\\ &= i\mathfrak{p}_\mu\mathfrak{q}_\nu\delta_{\nu\mu}\\ &= i\mathfrak{p}\cdot\mathfrak{q}.\end{aligned} \tag{10.25}$$

Note that Eqs. (10.21) and (10.24) agrees with Eqs. (8.36) and (8.41), respectively.

Equation (10.11), defines the generation of state $|\mathfrak{q}\rangle$,

$$|\mathfrak{q}\rangle = \sqrt{C_0}\exp(-i\mathfrak{q}\cdot\hat{\mathcal{P}})\,|0\rangle\,. \tag{10.26}$$

Inserting the term $\exp\{ip\cdot\hat{\mathcal{Q}}\}$ right in front of $|0\rangle$, which has the effect of multiplying by unity, we obtain a fully symmetric form as

$$|\mathfrak{q}\rangle = \sqrt{C_0}\exp(-i\mathfrak{q}\cdot\hat{\mathcal{P}})\exp\{ip\cdot\hat{\mathcal{Q}}\}\,|0\rangle\,. \tag{10.27}$$

We have,

$$\hat{\psi}^\dagger\,|\mathfrak{q}\rangle = \frac{\partial}{\partial\mathfrak{q}}\,|\mathfrak{q}\rangle = -i\hat{\mathcal{P}}\,|\mathfrak{q}\rangle\,, \tag{10.28}$$

$$\hat{\mathcal{P}}\,|\mathfrak{p}\rangle = \mathfrak{p}\,|\mathfrak{p}\rangle\,. \tag{10.29}$$

Note that for fermions the eigenvalues corresponding to $\mathfrak{q}$ and $\mathfrak{p}$ are elements of the Grassmann algebra.

We have from Eqs. (9.15)–(9.16)

$$\sum_{\mathfrak{q},\mathfrak{p}} |\mathfrak{q}\rangle \langle\mathfrak{q}|\,|\mathfrak{p}\rangle \langle\mathfrak{p}| = 1, \tag{10.30}$$

$$C_0 \sum_{\mathfrak{q},\mathfrak{p}} |\mathfrak{q}\rangle \exp\left(i\mathfrak{q}\cdot\mathfrak{p}\right) \langle\mathfrak{p}| = 1, \tag{10.31}$$

$$C_0 \sum_{\mathfrak{q},\mathfrak{p}} \exp\left(i\mathfrak{q}\cdot\mathfrak{p}\right) |\mathfrak{q}\rangle \langle\mathfrak{p}| = 1, \tag{10.32}$$

and from Eqs. (9.23)–(9.26)[2]

$$|\mathfrak{p}\rangle = \frac{1}{\sqrt{N}} \sum_{\mathfrak{q}} \exp\left(i\mathfrak{q}\cdot\mathfrak{p}\right) |\mathfrak{q}\rangle\,, \tag{10.33}$$

$$\langle\mathfrak{p}| = \frac{1}{\sqrt{N}} \sum_{\mathfrak{q}} \langle\mathfrak{q}| \exp\left(-i\mathfrak{q}\cdot\mathfrak{p}\right), \tag{10.34}$$

$$|\mathfrak{q}\rangle = \frac{1}{\sqrt{N}} \sum_{\mathfrak{p}} \exp\left(-i\mathfrak{p}\cdot\mathfrak{q}\right)\ |\mathfrak{p}\rangle\,, \tag{10.35}$$

$$\langle\mathfrak{q}| = \frac{1}{\sqrt{N}} \sum_{\mathfrak{p}} \langle\mathfrak{p}| \exp\left(i\mathfrak{p}\cdot\mathfrak{q}\right). \tag{10.36}$$

Compare Eqs. (10.33)–(10.36) with Eqs. (10.16)–(10.19). Therefore, we have also the completeness relation expressed as

$$\sum_{\mathfrak{q}} |\mathfrak{q}\rangle \langle\mathfrak{q}| = 1, \tag{10.37}$$

$$\sum_{\mathfrak{p}} |\mathfrak{p}\rangle \langle\mathfrak{p}| = 1, \tag{10.38}$$

$$\frac{1}{\sqrt{N}} \sum_{\mathfrak{q},\mathfrak{p}} \exp\left(-i\mathfrak{q}\cdot\mathfrak{p}\right) |\mathfrak{p}\rangle \langle q| = 1, \tag{10.39}$$

$$\frac{1}{\sqrt{N}} \sum_{\mathfrak{q},\mathfrak{p}} \exp\left(i\mathfrak{p}\cdot\mathfrak{q}\right) |\mathfrak{q}\rangle \langle p| = 1. \tag{10.40}$$

Equations (10.39)–(10.40) are the ones used by Halperin *et al.* [51] in following the coherent state formulation.

---

[2]We have again here assuming non-degenerates countable states.

10.2 "Lattice" Weyl transform for fermions

First we will formulate the Weyl transformation based on finite set of eigenvalues discussed in Section 8.1. On the other hand, for continuous eigenvalues or fields, one need to use the Berezin [52] Grassmann calculus involving Berezin integration of anticommuting variables.

We have for finite set of eigenvalues, expand any operator as

$$\begin{aligned}\hat{A} &= \sum_{\mathfrak{p}'',\mathfrak{q}'} \left|\mathfrak{p}''\right\rangle \left\langle\mathfrak{p}''\right| A \left|\mathfrak{q}'\right\rangle \left\langle\mathfrak{q}'\right|, \\ &= \sum_{\mathfrak{p}'',\mathfrak{q}'} \left\langle\mathfrak{p}''\right| A \left|\mathfrak{q}'\right\rangle \left|\mathfrak{p}''\right\rangle \left\langle\mathfrak{q}'\right|, \end{aligned} \tag{10.41}$$

or

$$\begin{aligned}\hat{A} &= \sum_{\mathfrak{p}'',\mathfrak{q}'} \left|\mathfrak{q}''\right\rangle \left\langle\mathfrak{q}''\right| A \left|\mathfrak{p}'\right\rangle \left\langle\mathfrak{p}'\right|, \\ &= \sum_{\mathfrak{p}'',\mathfrak{q}'} \left\langle\mathfrak{q}''\right| A \left|\mathfrak{p}'\right\rangle \ \left|\mathfrak{q}''\right\rangle \left\langle\mathfrak{p}'\right|. \end{aligned} \tag{10.42}$$

Using, Eqs. (2.8)–(2.12), we write

$$\left\langle\mathfrak{p}''\right| A \left|\mathfrak{q}'\right\rangle = \frac{1}{\sqrt{N}} \sum_{\mathfrak{q}''} e^{-i\mathfrak{p}''\cdot\mathfrak{q}''} \left\langle\mathfrak{q}''\right| A \left|\mathfrak{q}'\right\rangle, \tag{10.43}$$

$$\left|\mathfrak{p}''\right\rangle \left\langle\mathfrak{q}'\right| = \frac{1}{\sqrt{N}} \sum_{\mathfrak{p}'} e^{i\mathfrak{q}'\cdot\mathfrak{p}'} \left|\mathfrak{p}''\right\rangle \left\langle\mathfrak{p}'\right|. \tag{10.44}$$

Introducing the notation in Eq. (9.29),

$$\mathfrak{p}' = \mathfrak{p} + \mathfrak{u}, \quad \mathfrak{q}' = \mathfrak{q} + \mathfrak{v}, \tag{10.45}$$

$$\mathfrak{p}'' = \mathfrak{p} - \mathfrak{u}, \quad \mathfrak{q}'' = \mathfrak{q} - \mathfrak{v}. \tag{10.46}$$

then,

$$q' \cdot p' = (\vec{q} + \vec{v}) \cdot (\vec{p} + \vec{u}) = \vec{p} \cdot \vec{q} + \vec{u} \cdot \vec{v} + \vec{p} \cdot \vec{v} + \vec{u} \cdot \vec{q}, \tag{10.47}$$

$$-\vec{p}'' \cdot \vec{q}'' = -(\vec{p} - \vec{u}) \cdot (\vec{q} - \vec{v}) = -\vec{p} \cdot \vec{q} - \vec{u} \cdot \vec{v} + \vec{p} \cdot \vec{v} + \vec{u} \cdot \vec{q}. \tag{10.48}$$

Upon substituting in Eq. (9.27), we end up with

$$A = \sum_{\mathfrak{p}'',\mathfrak{q}'} |\mathfrak{p}''\rangle \langle\mathfrak{p}''| A |\mathfrak{q}'\rangle \langle\mathfrak{q}'| ,$$

$$= \frac{1}{N} \sum_{\mathfrak{p},\mathfrak{q},\mathfrak{u},\mathfrak{v}} e^{i2(\mathfrak{p}\cdot\mathfrak{v}+\mathfrak{u}\cdot\mathfrak{q})} \langle\mathfrak{q}-\mathfrak{v}| A |\mathfrak{q}+\mathfrak{v}\rangle |\mathfrak{p}-\mathfrak{u}\rangle \langle\mathfrak{p}+\mathfrak{u}| . \quad (10.49)$$

Likewise as in previous formulations, we write the last result as an expansion in terms of *phase-space point projector*, $\hat{\Delta}(p,q)$, defined as the fermion *lattice Weyl transform* of a projector, by

$$\hat{\Delta}(p,q) = \frac{1}{\sqrt{N}} \sum_{u} e^{i2\vec{u}\cdot\vec{q}} |\vec{p}-\vec{u}\rangle \langle\vec{p}+\vec{u}| , \quad (10.50)$$

and the coefficient of expansion, the fermion *lattice Weyl transform* of matrix element of operator, $A(p,q)$, defined by

$$A(p,q) = \frac{1}{\sqrt{N}} \sum_{v} e^{i2\vec{p}\cdot\vec{v}} \langle\vec{q}-\vec{v}| A |\vec{q}+\vec{v}\rangle . \quad (10.51)$$

An equivalent expression can also be obtain for $A(\mathfrak{p},\mathfrak{q})$and $\hat{\Delta}(\mathfrak{p},\mathfrak{q})$, namely,

$$A(\mathfrak{p},\mathfrak{q}) = \frac{1}{\sqrt{N}} \sum_{\mathfrak{u}} e^{i2\mathfrak{u}\cdot\mathfrak{q}} \langle\mathfrak{p}+\mathfrak{u}| \hat{A} |\mathfrak{p}-\mathfrak{u}\rangle , \quad (10.52)$$

$$\hat{\Delta}(\mathfrak{p},\mathfrak{q}) = \frac{1}{\sqrt{N}} \sum_{\mathfrak{v}} e^{i2\mathfrak{p}\cdot\mathfrak{v}} |\mathfrak{q}+\mathfrak{v}\rangle \langle\mathfrak{q}-\mathfrak{v}| . \quad (10.53)$$

10.2.1 *Symmetric form of* $\hat{\Delta}_{\lambda'\lambda}(\mathfrak{p},\mathfrak{q})$

In the expression, Eq. (10.53) for $\hat{\Delta}_{\lambda'\lambda}(\mathfrak{p},\mathfrak{q})$, we can use the following identities:

$$|\mathfrak{q}+\mathfrak{v},\lambda'\rangle = \exp\left[-2i\mathcal{P}\cdot\mathfrak{v}\right] |\mathfrak{q}-\mathfrak{v},\lambda'\rangle . \quad (10.54)$$

Then

$$|\mathfrak{q}-\mathfrak{v},\lambda'\rangle \langle\mathfrak{q}-\mathfrak{v},\lambda|$$

$$= (N)^{-1} \sum_{\mathfrak{u},\mathfrak{q}_0} \exp\left[-2i(\mathfrak{q}-\mathfrak{v}-\mathcal{Q})\cdot\mathfrak{u}\right] |\mathfrak{q}_0,\lambda'\rangle \langle\mathfrak{q}_0,\lambda| ,$$

$$= \sum_{\mathfrak{q}_0} \delta\left(\mathfrak{q} - \mathfrak{v} - \hat{\mathcal{Q}}\right) \left|\mathfrak{q}_0, \lambda'\right\rangle \left\langle \mathfrak{q}_0, \lambda\right|$$

$$= \sum_{\mathfrak{q}_0} \delta\left(\mathfrak{q} - \mathfrak{v} - \mathfrak{q}_0\right) \left|\mathfrak{q}_0, \lambda'\right\rangle \left\langle \mathfrak{q}_0, \lambda\right|, \tag{10.55}$$

$$\Omega_{\lambda'\lambda} = \left|\mathfrak{q}_0, \lambda'\right\rangle \left\langle \mathfrak{q}_0, \lambda\right|. \tag{10.56}$$

Substituting the expressions, Eqs. (6.46) and (6.47) in Eq. (9.40), we obtain a completely symmetric expression of $\hat{\Delta}(\mathfrak{p}, \mathfrak{q})$,

$$\begin{aligned}
&\hat{\Delta}_{\lambda'\lambda}(\mathfrak{p}, \mathfrak{q}) \\
&\quad = \frac{1}{\sqrt{N}} \sum_{\mathfrak{v}} e^{i2\mathfrak{p}\cdot\mathfrak{v}} \left|\mathfrak{q} + \mathfrak{v}, \lambda'\right\rangle \left\langle \mathfrak{q} - \mathfrak{v}, \lambda\right|, \\
&\quad = \frac{1}{\sqrt{N}} (N)^{-1} \sum_{\bar{\mathfrak{v}}, \bar{\mathfrak{u}}} e^{i2\mathfrak{p}\cdot\mathfrak{v}} \exp\left[-2i\mathcal{P}\cdot\mathfrak{v}\right] \exp[-2i(\mathfrak{q} - \mathfrak{v} - \hat{\mathcal{Q}})\cdot\mathfrak{u}] \\
&\quad\quad \times \sum_{\mathfrak{q}_0} \left|\mathfrak{q}_0, \lambda'\right\rangle \left\langle \mathfrak{q}_0, \lambda\right|.
\end{aligned} \tag{10.57}$$

Then we have, even accounting for the anticommuting Grassmann variables,

$$\begin{aligned}
&\exp\left\{-i\left(\mathcal{P} - \mathfrak{p}\right)\cdot 2\mathfrak{v}\right\} \exp\{2i(\hat{\mathcal{Q}} - \mathfrak{q})\cdot\bar{\mathfrak{u}}\} \exp 2i\mathfrak{v}\cdot\mathfrak{u} \\
&\quad = \exp\left\{-i\left(\mathcal{P}\cdot\mathfrak{v} - \mathcal{Q}\cdot\mathfrak{u}\right)\right\} \exp\left\{2i\left(\mathfrak{p}\cdot\mathfrak{v} - \mathfrak{q}\cdot\mathfrak{u}\right)\right\},
\end{aligned} \tag{10.58}$$

which is the same as Eq. (9.34). Thus, we have changed the seemingly asymmetric expression of the first line into a symmetric form of the last line. And again, we have,

$$\begin{aligned}
\hat{\Delta}_{\lambda'\lambda}(\mathfrak{p}, \mathfrak{q}) &= \frac{1}{\sqrt{N}} (N)^{-1} \sum_{\mathfrak{v}, \mathfrak{u}} \exp\left\{-i\left(\mathcal{P}\cdot\mathfrak{v} - \mathcal{Q}\cdot\mathfrak{u}\right)\right\} \\
&\quad \times \exp\left\{2i\left(\mathfrak{p}\cdot\mathfrak{v} - \mathfrak{q}\cdot\mathfrak{u}\right)\right\} \sum_{\mathfrak{q}_0} \left|\mathfrak{q}_0, \lambda'\right\rangle \left\langle \mathfrak{q}_0, \lambda\right|, \\
&= \frac{1}{\sqrt{N}} (N)^{-1} \sum_{\mathfrak{v}, \mathfrak{u}} \exp\left\{-i\left(\mathcal{P}\cdot\mathfrak{v} - \mathcal{Q}\cdot\mathfrak{u}\right)\right\} \\
&\quad \times \exp\left\{2i\left(\mathfrak{p}\cdot\mathfrak{v} - \mathfrak{q}\cdot\mathfrak{u}\right)\right\} \Omega_{\lambda'\lambda},
\end{aligned} \tag{10.59}$$

where,

$$\Omega_{\lambda'\lambda} = \sum_{\mathfrak{q}_0} \left|\mathfrak{q}_0, \lambda\right\rangle \left\langle \mathfrak{q}_0, \lambda'\right|. \tag{10.60}$$

Moreover, we also have

$$A(\mathfrak{p},\mathfrak{q}) = \mathrm{Tr}\left(\hat{A}\hat{\Delta}\right)$$
$$= \frac{1}{\sqrt{N}}\begin{pmatrix} \sum_{\bar{\mathfrak{v}},\bar{\mathfrak{u}}} \exp 2i\left[\mathfrak{p}\cdot\mathfrak{v}-\mathfrak{q}\cdot\mathfrak{u}\right] \\ \times\mathrm{Tr}\{(N)^{-1}\hat{A}\exp\{-2i\left[\mathcal{P}\cdot\mathfrak{v}-\mathcal{Q}\cdot\mathfrak{u}\right]\}\Omega\}\end{pmatrix}. \tag{10.61}$$

10.3 Characteristic distribution

Therefore, the characteristic distribution for $A(p,q)$ is identically,

$$A_{\lambda\lambda'}(\mathfrak{u},\mathfrak{v}) = \mathrm{Tr}\left\{(N)^{-1}\hat{A}\exp\{-2i\left[\mathcal{P}\cdot\mathfrak{v}-\mathcal{Q}\cdot\mathfrak{u}\right]\}\Omega_{\lambda\lambda'}\right\}, \tag{10.62}$$

which is formally as before. By using the characteristic distribution function for $A_{\lambda\lambda'}(\mathfrak{p},\mathfrak{q})$,

$$A_{\lambda\lambda'}(\mathfrak{u},\mathfrak{v}) = \left(\frac{1}{N}\right)^{\frac{1}{2}}\sum_{p,q} A_{\lambda\lambda'}(\mathfrak{p},\mathfrak{q})\exp 2i\left[\mathfrak{p}\cdot\mathfrak{v}-\mathfrak{q}\cdot\mathfrak{u}\right], \tag{10.63}$$

with inverse

$$A_{\lambda\lambda'}(\mathfrak{p},\mathfrak{q}) = \left(\frac{1}{N}\right)^{\frac{1}{2}}\sum_{u,v} A_{\lambda\lambda'}(\mathfrak{u},\mathfrak{v})\exp\{-2i\left[\mathfrak{p}\cdot\mathfrak{v}-\mathfrak{q}\cdot\mathfrak{u}\right]\}, \tag{10.64}$$

then we can write $\hat{A}$ as

$$\hat{A} = \sum_{\mathfrak{p},\mathfrak{q}} A_{\lambda\lambda'}(\mathfrak{p},\mathfrak{q})\,\hat{\Delta}_{\lambda'\lambda}(\mathfrak{p},\mathfrak{q}),$$
$$= \sum_{\mathfrak{u},\mathfrak{v},\lambda'\lambda} A_{\lambda\lambda'}(\mathfrak{u},\mathfrak{v})\,e^{-2i(\mathcal{P}\cdot\mathfrak{v}-\mathcal{Q}\cdot\mathfrak{u})}\sum_{\mathfrak{q}_0}\left|\mathfrak{q}_0,\lambda'\right\rangle\left\langle\mathfrak{q}_0,\lambda\right|. \tag{10.65}$$

The first summation of Eq. (10.65) is simply like a Fourier transform (caveat: Fourier transform to *operator space*) of the characteristic function of the lattice Weyl transform of the operator $\hat{A}$,

$$\hat{A} = \sum_{\bar{u},\bar{v},\lambda,\lambda'} A_{\lambda\lambda'}(\mathfrak{u},\mathfrak{v})\exp\{-2i\left[\mathcal{P}\cdot\mathfrak{v}-\mathcal{Q}\cdot\mathfrak{u}\right]\}\Omega_{\lambda'\lambda}, \tag{10.66}$$

where the inverse can be written as

$$A_{\lambda\lambda'}(\mathfrak{u},\mathfrak{v}) = \mathrm{Tr}\{(N)^{-1}\hat{A}\exp\{-2i\left[\mathcal{P}\cdot\mathfrak{v}-\mathcal{Q}\cdot\mathfrak{u}\right]\}\Omega_{\lambda\lambda'}\}. \tag{10.67}$$

10.3.1 *Distribution functions*

We have for the normal characteristic distribution

$$A^{n}_{\lambda\lambda'}(u,v) = \exp\{2iu\cdot\mathfrak{v}\}\,\mathrm{Tr}((N)^{-1}\hat{A}\exp(-2i)(\mathcal{P}.\mathfrak{v}-\mathcal{Q}.\mathfrak{u}))\Omega_{\lambda\lambda'}, \tag{10.68}$$

and the anti-normal characteristic distribution function given by

$$A^{a}_{\lambda\lambda'}(u,v) = \exp\{-2iu\cdot\mathfrak{v}\}\,\mathrm{Tr}((N)^{-1}\hat{A}\exp(-2i)(\mathcal{P}.\mathfrak{v}-\mathcal{Q}.\mathfrak{u})). \tag{10.69}$$

Formally again, a more *general* phase-space distribution functions, $f^{(g)}(\mathfrak{p},\mathfrak{q},t)$, can be obtained from the expression

$$f^{(g)}(\mathfrak{p},\mathfrak{q},t) = \left(\frac{1}{N}\right)^{\frac{1}{2}}\sum_{u,v} A_{\lambda\lambda'}(\mathfrak{u},\mathfrak{v})\exp\{-2i[\mathfrak{p}\cdot\mathfrak{v}-\mathfrak{q}\cdot\mathfrak{u}]\}\,g(\mathfrak{u},\mathfrak{v}), \tag{10.70}$$

where $g(\mathfrak{u},\mathfrak{v})$ is some choosen smoothing function. Thus, for finite system, the formalism exactly maps to that of CMP, as well as that of the QFT of bosons.

10.4 Continuous fields: Calculations of integrals for fermions

With the continuous annihilation and creation field operators for fermions, our pairing algorithm in Section 8.2 is ambiguous, but the results are assumed to be extendable to continuous fields. So far, our discussion on QFT implies a lattice of the functionals of the set $\{\mathfrak{p},\mathfrak{q}\}$. This is akin to the the structurally incorporating the ubiquitous use of cut-of though incorporated structurally from the beginning. This may be related to the use of lattice structure in lattice gauge theory.

The major difference from bosonic case occurs in the functional integration of continuous Grassmann variables. The reason for this is that Berezin integration [52] is actually similar to (standard) functional differentiation. Formally, we can still associate a sort of unification in a symbolic sense for the Grassmann integration.

A sort of a justification is by considering translation invariance of a definite integral and linearity:

$$\int_{-\infty}^{\infty} f(x)\,dx = \int_{-\infty}^{\infty} f(x+y)\,dx, \tag{10.71}$$

where $f(x)$ is at most a linear function of x, a Grassmann variable. We write

$$f(x) = a + bx = \begin{pmatrix} a & b \end{pmatrix} \begin{pmatrix} 1 \\ x \end{pmatrix}, \tag{10.72}$$

or more precisely,

$$\begin{aligned} f &= f_{00} + f_{01}\hat{\psi}^\dagger + f_{10}\hat{\psi} + f_{11}\hat{\psi}^\dagger\hat{\psi}, \\ &= \begin{pmatrix} f_{00} & f_{01} & f_{10} & f_{11} \end{pmatrix} \begin{pmatrix} 1 \\ \hat{\psi}^\dagger \\ \hat{\psi} \\ \hat{\psi}^\dagger\hat{\psi} \end{pmatrix}. \end{aligned} \tag{10.73}$$

Linearity says

$$\int_{-\infty}^{\infty} (a + bf(x))\, dx = \int_{-\infty}^{\infty} adx + \int_{-\infty}^{\infty} bf(x)\, dx \tag{10.74}$$

Thus, we have

$$\begin{aligned} \int_{-\infty}^{\infty} f(x+y)\, dx &= \int_{-\infty}^{\infty} adx + \int_{-\infty}^{\infty} b(x+y)\, dx, \\ &= \int_{-\infty}^{\infty} (a+bx)\, dx + \int_{-\infty}^{\infty} bydx, \\ &= \int_{-\infty}^{\infty} f(x)\, dx + by \int_{-\infty}^{\infty} dx. \end{aligned} \tag{10.75}$$

But by translational invariance,

$$\int_{-\infty}^{\infty} f(x+y)\, dx = \int_{-\infty}^{\infty} f(x)\, dx. \tag{10.76}$$

Therefore, since by is arbitrary, one is forced to equate,

$$\int_{-\infty}^{\infty} dx = 0, \tag{10.77}$$

which does looks like a differentiation of a constant equals to 1. Thus, integration of Grassmann numbers obeys

$$\int dx = 0, \tag{10.78}$$

$$\int xdx = 1. \tag{10.79}$$

The previous relations involving integrals must now be interpreted as standard differentiation with respect to the integration variable. We observe that like before we have the eigenvalue equations for the left and right eigenvector defined by

$$\hat{\mathcal{P}}\,|\mathfrak{p}\rangle = \mathfrak{p}\,|\mathfrak{p}\rangle = |\mathfrak{p}\rangle\,\mathfrak{p}, \tag{10.80}$$

$$\langle\mathfrak{p}|\,\hat{\mathcal{P}} = \langle\mathfrak{p}|\,\mathfrak{p} = \mathfrak{p}\,\langle\mathfrak{p}|\,. \tag{10.81}$$

Likewise we have

$$\hat{\mathcal{Q}}\,|\mathfrak{q}\rangle = \mathfrak{q}\,|\mathfrak{q}\rangle = |\mathfrak{q}\rangle\,\mathfrak{q}, \tag{10.82}$$

$$\langle\mathfrak{q}|\,\hat{\mathcal{Q}} = \langle\mathfrak{q}|\,\mathfrak{q} = \mathfrak{q}\,\langle\mathfrak{q}|\,, \tag{10.83}$$

where the eigenvalues are now continuous Grassmann functional variables.

The completeness relation is given formally as before by

$$\begin{aligned}
1 &= \sum_{\mathfrak{q}} |\mathfrak{q}\rangle\,\langle\mathfrak{q}| \Longrightarrow \int d\,[\mathfrak{q}]\,|\mathfrak{q}\rangle\,\langle\mathfrak{q}|\,,\\
&= \prod\left(\frac{\partial}{\partial\mathfrak{q}'} + \frac{\partial}{\partial\mathfrak{q}''}\right)|\mathfrak{q}'\rangle\,\langle\mathfrak{q}''|\,,\\
&= \int \mathfrak{D}\mathfrak{q}\,|\mathfrak{q}\rangle\,\langle\mathfrak{q}|\,,
\end{aligned} \tag{10.84}$$

where the multicomponent integral $\int \mathfrak{D}\mathfrak{q}$ is only a symbolic meaning for continuous fields, so as to formally resembles that of discrete pairing case before. Likewise,

$$1 = \sum_{\mathfrak{p}} |\mathfrak{p}\rangle\,\langle\mathfrak{p}| \Longrightarrow \int \mathfrak{D}\mathfrak{p}\,|\mathfrak{p}\rangle\,\langle\mathfrak{p}|\,. \tag{10.85}$$

10.4.1 *The transformation function and resolution of identity*

The transformation function is the same as Eq. (10.30). The fermion Fourier transforms are similar as Eqs. (10.33)–(10.36) but for continuous case, the summation is now replaced by functional integral obeying the Berezin fermion field integrals. Similarly the completeness relations, often referred to as resolutions of unity, are as given by Eqs. (10.30)–(10.32) with summation replaced by Berezin fermion integrals, Eqs. (10.84) and (10.85).

10.4.2 Continuous Weyl transform for fermions

The discrete expression for the mixed representation of any operator given before, Eq. (10.41), now reads in the continuous case as

$$\hat{A} = \sum_{\mathfrak{p}'',\mathfrak{q}'} |\mathfrak{p}''\rangle \langle \mathfrak{p}''| A |\mathfrak{q}'\rangle \langle \mathfrak{q}'| \Longrightarrow \iint \mathfrak{D}\mathfrak{p}''\mathfrak{D}\mathfrak{q}' \langle \mathfrak{p}''| A |\mathfrak{q}'\rangle |\mathfrak{p}''\rangle \langle \mathfrak{q}'|, \tag{10.86}$$

$$\hat{A} = \iint \mathfrak{D}\mathfrak{p}\mathfrak{D}\mathfrak{q} A(\mathfrak{p},\mathfrak{q})\, \hat{\Delta}(\mathfrak{p},\mathfrak{q}), \tag{10.87}$$

where

$$A(\mathfrak{p},\mathfrak{q}) = \int \mathfrak{D}\mathfrak{v} e^{-2\mathfrak{p}.\mathfrak{v}} \langle \mathfrak{q}+\mathfrak{v}| \hat{A} |\mathfrak{q}-\mathfrak{v}\rangle, \tag{10.88}$$

$$\hat{\Delta}(\mathfrak{p},\mathfrak{q}) = \int \mathfrak{D}\mathfrak{u} e^{2\mathfrak{q}.\mathfrak{u}} |\mathfrak{p}+\mathfrak{u}\rangle \langle \mathfrak{p}-\mathfrak{u}|. \tag{10.89}$$

We also have the equivalent expressions,

$$A(\mathfrak{p},\mathfrak{q}) = \int \mathfrak{D}\mathfrak{u} e^{-i2\mathfrak{q}\cdot\mathfrak{u}} \langle \mathfrak{p}+\mathfrak{u}| \hat{A} |\mathfrak{p}-\mathfrak{u}\rangle, \tag{10.90}$$

$$\hat{\Delta}(\mathfrak{p},\mathfrak{q}) = \int \mathfrak{D}\mathfrak{v} e^{i2\mathfrak{p}\cdot\mathfrak{v}} |\mathfrak{q}+\mathfrak{v}\rangle \langle \mathfrak{q}-\mathfrak{v}|. \tag{10.91}$$

in the sense of Berezin fermionic integrals.

Equation (10.59) can now be written in terms of Berezin fermion integral as

$$\hat{\Delta}_{\lambda'\lambda}(\mathfrak{p},\mathfrak{q}) = \iint \mathfrak{D}\mathfrak{v}\mathfrak{D}\mathfrak{u} \exp\{-i(\mathcal{P}\cdot\mathfrak{v} - \mathcal{Q}\cdot\mathfrak{u})\} \exp\{2i(\mathfrak{p}\cdot\mathfrak{v} - \mathfrak{q}\cdot\mathfrak{u})\}\, \Omega_{\lambda'\lambda}, \tag{10.92}$$

and Eq. (10.61) is now,

$$\begin{aligned} A(\mathfrak{p},\mathfrak{q}) &= \mathrm{Tr}\left(\hat{A}\hat{\Delta}\right) \\ &= \begin{pmatrix} \iint \mathfrak{D}\mathfrak{v}\mathfrak{D}\mathfrak{u} \exp 2i\,[\mathfrak{p}\cdot\mathfrak{v} - \mathfrak{q}\cdot\mathfrak{u}] \\ \times \mathrm{Tr}\left\{\hat{A} \exp\{-2i\,[\mathcal{P}\cdot\mathfrak{v} - \mathcal{Q}\cdot\mathfrak{u}]\}\, \Omega\right\} \end{pmatrix}. \end{aligned} \tag{10.93}$$

We also have for Eq. (10.64) the expression,

$$A_{\lambda\lambda'}(\mathfrak{u},\mathfrak{v}) = \iint \mathfrak{D}\mathfrak{p}\mathfrak{D}\mathfrak{q} A_{\lambda\lambda'}(\mathfrak{p},\mathfrak{q}) \exp 2i\,[\mathfrak{p}\cdot\mathfrak{v} - \mathfrak{q}\cdot\mathfrak{u}], \tag{10.94}$$

with inverse

$$A_{\lambda\lambda'}(\mathfrak{p},\mathfrak{q}) = \iint \mathfrak{D}\mathfrak{v}\mathfrak{D}\mathfrak{u}A_{\lambda\lambda'}(\mathfrak{u},\mathfrak{v})\exp\{-2i[\mathfrak{p}\cdot\mathfrak{v}-\mathfrak{q}\cdot\mathfrak{u}]\}. \tag{10.95}$$

10.4.3 *Distribution functions*

$$A^{n}_{\lambda\lambda'}(u,v) = \exp\{2iu\cdot\mathfrak{v}\}\,\mathrm{Tr}(\hat{A}\exp(-2i)(\mathcal{P}.\mathfrak{v}-\mathcal{Q}.\mathfrak{u})), \tag{10.96}$$

and the anti-normal characteristic distribution function given by

$$A^{a}_{\lambda\lambda'}(u,v) = \exp\{-2iu\cdot\mathfrak{v}\}\,\mathrm{Tr}(\hat{A}\exp(-2i)(\mathcal{P}.\mathfrak{v}-\mathcal{Q}.\mathfrak{u})). \tag{10.97}$$

Formally, again a more *general* phase-space distribution functions, $f^{(g)}(\mathfrak{p},\mathfrak{q},t)$, can be obtained from the expression

$$f^{(g)}(\mathfrak{p},\mathfrak{q},t) = \iint \mathfrak{D}\mathfrak{u}\mathfrak{D}\mathfrak{v}A_{\lambda\lambda'}(\mathfrak{u},\mathfrak{v})\exp\{-2i[\mathfrak{p}\cdot\mathfrak{v}-\mathfrak{q}\cdot\mathfrak{u}]\}\,g(\mathfrak{u},\mathfrak{v}), \tag{10.98}$$

where $g(\mathfrak{u},\mathfrak{v})$ is some choosen smoothing function. Finally, we can write Eq. (10.66) as

$$\hat{A} = \sum_{\lambda,\lambda'}\iint \mathfrak{D}\mathfrak{u}\mathfrak{D}\mathfrak{v}A_{\lambda\lambda'}(\mathfrak{u},\mathfrak{v})\exp\{-2i[\mathcal{P}\cdot\mathfrak{v}-\mathcal{Q}\cdot\mathfrak{u}]\}\,\Omega_{\lambda'\lambda}. \tag{10.99}$$

10.4.4 *Fermion path integrals*

We are now equipped to calculate the path integral expression for the evolution operator, $U(t.t_o)$,

$$\begin{aligned}\hat{U}(t.t_o) &= \exp\left(\frac{-i}{\hbar}[t-t_o]\,\mathcal{H}\right),\\ &= \prod_{j=1}^{n+1}\hat{U}(t_j.t_{j-1}).\end{aligned} \tag{10.100}$$

We now make of the following relations for fermions in terms of Berezin Grassmann integrals. Thus for any fermion operator, we have,

$$\hat{A} = \iint \mathfrak{D}\mathfrak{q}\mathfrak{D}\mathfrak{p}A(\mathfrak{p},\mathfrak{q})\,\Delta(\mathfrak{p},\mathfrak{q}), \tag{10.101}$$

where

$$A(\mathfrak{p},\mathfrak{q}) = \int \mathfrak{D}\mathfrak{u}\ e^{-i2\mathfrak{q}\cdot\mathfrak{u}} \langle \mathfrak{p}+\mathfrak{u}|\,\hat{A}\,|\mathfrak{p}-\mathfrak{u}\rangle\,, \tag{10.102}$$

$$\hat{\Delta}(\mathfrak{p},\mathfrak{q}) = \int \mathfrak{D}\mathfrak{v}\ e^{i2\mathfrak{p}\cdot\mathfrak{v}}\,|\mathfrak{q}+\mathfrak{v}\rangle\langle\mathfrak{q}-\mathfrak{v}|\,. \tag{10.103}$$

From the above relations we can transform Eq. (10.100) into a Grassmann path integral. We have from Eq. (10.100),

$$\hat{U}(t_j.t_{j-1}) = \iint \mathfrak{D}\mathfrak{q}\mathfrak{D}\mathfrak{p}\left(U_{t_j,t_{j-1}}(\mathfrak{p},\mathfrak{q})\right)\Delta(\mathfrak{p},\mathfrak{q}), \tag{10.104}$$

where

$$\begin{aligned} U_{t_j,t_{j-1}}(\mathfrak{p},\mathfrak{q}) &= \int \mathfrak{D}\mathfrak{u}\,\mathfrak{e}^{-2i\mathfrak{u}\cdot\mathfrak{q}}\langle\mathfrak{p}+\mathfrak{u}|\,\hat{U}(t_j.t_{j-1})\,|\mathfrak{p}-\mathfrak{u}\rangle, \\ &= \exp\left[-\frac{i}{\hbar}(t_j-t_{j-1})\,H(\mathfrak{p},q)\right], \end{aligned} \tag{10.105}$$

where $H(\mathfrak{p},q)$is the "lattice" Weyl transform of $\mathcal{H}$ for the time interval $(t_j - t_{j-1})$. We also have,

$$\begin{aligned} \langle\mathfrak{q}_j|\,\Delta(\mathfrak{p},\mathfrak{q})\,|\mathfrak{q}_{j-1}\rangle &= \int \mathfrak{D}\mathfrak{v}\,\mathfrak{e}^{2i\mathfrak{p}\cdot\mathfrak{v}}\langle\mathfrak{q}_j|\,|\mathfrak{q}+\mathfrak{v}\rangle\langle\mathfrak{q}-\mathfrak{v}|\,|\mathfrak{q}_{j-1}\rangle, \\ &= \exp\left[2i\mathfrak{p}\cdot(\mathfrak{q}_j-\mathfrak{q}_{j-1})\right]\delta\left(\frac{\mathfrak{q}_j+\mathfrak{q}_{j-1}}{2}-\mathfrak{q}\right). \end{aligned} \tag{10.106}$$

Therefore,

$$\begin{aligned} &\langle\mathfrak{q}_j|\,\hat{U}(t_j.t_{j-1})\,|\mathfrak{q}_{j-1}\rangle \\ &\quad= \iint \mathfrak{D}\mathfrak{q}\mathfrak{D}\mathfrak{p}U_{t_j,t_{j-1}}(\mathfrak{p},\mathfrak{q})\,\Delta(\mathfrak{p},\mathfrak{q}), \\ &\quad= \int \mathfrak{D}\mathfrak{p}\exp\left[2\frac{i}{\hbar}\mathfrak{p}\cdot\left(\frac{\mathfrak{q}_j-\mathfrak{q}_{j-1}}{2}\right)\right] \\ &\qquad\times\exp\left[-\frac{i}{\hbar}(t_j-t_{j-1})\ H\left(\mathfrak{p},\frac{\mathfrak{q}_j+\mathfrak{q}_{j-1}}{2}\right)\right], \\ &\quad= \int \mathfrak{D}\mathfrak{p}\exp\epsilon\frac{i}{\hbar}\left[\mathfrak{p}\cdot\left(\frac{\mathfrak{q}_j-\mathfrak{q}_{j-1}}{\epsilon}\right)-(t_j-t_{j-1})\,H\left(\mathfrak{p},\frac{\mathfrak{q}_j+\mathfrak{q}_{j-1}}{2}\right)\right], \end{aligned} \tag{10.107}$$

where ϵ is the incremental time, $t_j - t_{j-1}$. Therefore the transition amplitude between a state $\Psi(t_0)$ and a state $\Phi(t)$ is given by

$$\begin{aligned}
&\langle \Phi(t) | \, | \Psi(t_0) \rangle \\
&= \int \int \prod_{j=1}^{n+1} \mathfrak{D}\mathfrak{p}_j \mathfrak{D}\mathfrak{q}_j \phi(q_{n+1}, t_{n+1}) \\
&\quad \times \exp \left\{ -\epsilon \sum_{j=1}^{n+1} \begin{bmatrix} \mathfrak{p}_j \cdot \left(\dfrac{\mathfrak{q}_j - \mathfrak{q}_{j-1}}{\epsilon} \right) \\ +H \left(\mathfrak{p}_j, \dfrac{\mathfrak{q}_j + \mathfrak{q}_{j-1}}{2} \right) \end{bmatrix} \right\} \psi(q_1, t_0), \quad (10.108)
\end{aligned}$$

where $\phi(q_{n+1}, t_{n+1}) = \langle \Phi(t) | \, | q_{n+1} \rangle$ and $\psi(q_1, t_0) = \langle q_1 | \, | \Psi(t_0) \rangle$. Another useful result for the path integral is the expression for the partition function, using Matsubara imaginary time [53],

$$\begin{aligned}
\exp(-\beta\Omega) &= \operatorname{Tr} \exp(-\beta(\mathcal{H} - \mu\hat{N})), \\
&= \operatorname{Tr} \exp(-\beta\hat{K}), \qquad (10.109)
\end{aligned}$$

which yields

$$\begin{aligned}
\exp(-\beta\Omega) = \int \int \prod_{j=1}^{n+1} \mathfrak{D}\mathfrak{p}_j \mathfrak{D}\mathfrak{q}_j \exp \left\{ -\epsilon \sum_{j=1}^{n+1} \left[\mathfrak{p}_j \cdot \left(\frac{\mathfrak{q}_j - \mathfrak{q}_{j-1}}{\epsilon} \right) \right. \right. \\
\left. \left. + K \left(\mathfrak{p}_j, \frac{\mathfrak{q}_j + \mathfrak{q}_{j-1}}{2} \right) \right] \right\}. \qquad (10.110)
\end{aligned}$$

The anti-periodic boundary condition for $\mathfrak{p}$ and $\mathfrak{q}$ is precisely what is needed to preserve the anti-periodicity in each time variables (Matsubara time zone for fermions) of the Green's function in the path integral formulation of statistical quantum field theory [54], e.g., $\mathfrak{q}_{n+1} = -\mathfrak{q}_o$ and $\mathfrak{p}_{n+1} = -\mathfrak{p}_o$ with a period β in Matsubara time zone.

Chapter 11

Third Quantization Scheme in QFT

We have seen that both in CMP as well as in QFT, their quantum distribution functionals of field variables are well-defined. This suggests that we can have an alternative formulation of QFT in terms of Q-distribution functional of field variables similar to those of CMP, which is in terms of Wigner distribution functions. Some hints in this direction has been given very early on by Guttinger [55]. More pertinent to our work in this direction has recently been given by Drummond [56, 57], and also by Friederich [58] and by Man'ko *et al.* [59]. However, lacking in all these previous discussions is a topic on how to formulate the distribution functional dynamics in QFT.

11.1 Q-Distribution functional dynamics of QFT

Here, we show that in order to formulate the quantum transport equation of the Q-distribution function, a third quantization scheme is a prerequisite. Clearly, following the route towards the formulation of non-equilibrium superfield theory of quantum transport in CMP [12], we can form a new theory of dissipative nonequilibrium quantum functional field theory based on the quantum transport equation of the Q-distribution functional or the "Wigner distribution function" of conjugate functional field variables. This is a new perspective on the formulation of quantum field theory which we investigate in more details in the next section. A good beginning in this direction has been given by Drummond [56, 57] and more specifically by Friederich [58].

11.2 A third quantization scheme

To formulate non-equilibrium quantum transport theory in terms of quantum distribution functionals of field variables in QFT, we need to devise third-quantize annihilation and creation operators on the Fock space of second quantize field variables. We can then repeat the procedure for obtaining all correlation functions, similar to the correlation functions of the quantum-superfield transport formulation in CMP. We can construct quantum kinetic transport equation in the phase-space domain of functional variables. Thus, in principle, a hierarchy of quantization in quantum physics can be constructed. The real physical meaning and utility of this third quantization constructions has yet to be explored. However, this will surely lead to the *distribution functional transport approach* to QFT. This seems like an uncharted field of investigation, although functional approaches to quantum physics have been heralded by the Feynman functional path integrals in QFT and by the Kohn–Sham density functional theory in CMP.

A third quantization scheme proposed here is the following. We take the eigenfunction of the second quantized annihilation operator to correspond to Wannier function index by $\mathfrak{q}$, sort of *Wannier functional.* Similarly the eigenfunction of the "momentum", $\mathfrak{p}$, sort of *Bloch functional*, non-localized functional in the sense of being a superposition of the "localized" "Wannier functionals", through the transition function between the two separate spaces. In this sense $\mathfrak{p} \cdot \mathfrak{q}$ dot product means summation over the discrete indices, and summation or integration over functional arguments.

Now, the third quantization procedure then consist of coming up with the new $\breve{\psi}\,(\mathfrak{q},t)$ and $\breve{\psi}^{\dagger}\,(\mathfrak{q},t)$, annihilation and creation of second quantized states indicated by $|\mathfrak{q}\rangle$ and its adjoint $\langle\mathfrak{q}|$. With this setup, the algebraic formulation of the non-equilibrium quantum superfield transport equations can be straightforwardly constructed in terms of non-equilibrium Green functional techniques and Wigner distribution functional equation in $(\mathfrak{p},\mathfrak{q})$ functional phase space. This type of formulation is probably suitable in treating chaotic quantum dynamics where several energy or functional states are involved, e.g., fluctuation theory, quantum turbulence and turbulent and reactive flows, cosmology, and explosive dynamics.

The idea of the third quantization, as promulgated by Raptis [60], involves theoretical independence with spacetime manifolds (which carries all sort of ambiguous singularities) in favor of an abstract superspace of canonical functional fields. This has been heuristically discussed by Raptis [60] in his proposed third quantization of gravity and Yang–Mills theories. This means, in our case, avoiding the use of conventional position and momentum dynamical variables, (p, q)-phase space, in favor of canonical functional variables, $(\mathfrak{p}, \mathfrak{q})$-functional phase space, which is crucial in exploring the quantum dynamics of the Q-distribution in $(\mathfrak{p}, \mathfrak{q})$-functional phase space of QFT.

The spirit of the idea of third quantization proposed by Giddings and Strominger [61] somewhat affirm the thinking of Ref. [60]. Our algorithmic procedure presented here follows the ideas brought forth by Raptis [60], Friederich [58], and to some degree with [61].

A different approach proposed by Prosen *et al.* [62, 63] is based on the idea of using a Fock space of density operators of physical states, satisfying the quantum Liouville equation, and adjoint structure of canonical creation and annihilation maps over this space, Majorana operators for fermions. They also called this third quantization, which is geared for solving Langevin transport equation. However their theory is not based on canonical functional variables, i.e., abstract superspace of canonical functional fields, such as $(\mathfrak{p}, \mathfrak{q})$- functional phase space, consistent with spacetime manifolds independence (i.e., in our case, (p, q) independent). In other words, not based on creation and annihilation of second quantized functional states, but specifically based on creation and annihilation operator maps over Fock space of density operators of the Liouville equation, supposedly taking the place of the Schrödinger equation of the second quantization scheme. We believe the theory of Prosen *et al.* [62, 63] should be aptly called *super-quantization*, since it just mimic the conventional second quantization scheme over the wavefunctions space of the Schrödinger equations. Their approach is on Fock space of density operators, which have a lattice Weyl transform in conventional (p, q)-phase space. The lattice Weyl transform of a density operator itself is the conventional Wigner distribution function, $f_w(p, q)$. Thus, this is not the sort of third quantization scheme discussed here. The third quantization in Ref. [64] also mimics the second quantization of QFT, as the wave function of the spacetime and matter fields of the Wheeler–DeWitt (WDW) equation is elevated to an operator, which can then be Fourier expanded through the momentum of the scalar field. The third quantization scheme presented here is more in line with the theory proposed in Refs. [60] and [58] and partly with the theory in Ref. [61].

11.2.1 *Third quantization annihilation and creation field operators*

To construct the third quantization annihilation and creation field operator $\check{\psi}(\mathfrak{q}, t)$ and $\check{\psi}^{\dagger}(\mathfrak{q}, t)$, we decompose a general field operators $\tilde{\Psi}(t)$ in terms of the eigenfunctions, $|\mathfrak{q}\rangle$, of the annihilation operator. We have, i.e.,

$$\tilde{\Psi}(t) = \sum_{\mathfrak{q}} \breve{\psi}(\mathfrak{q}, t) |\mathfrak{q}\rangle, \tag{11.1}$$

$$\tilde{\Psi}^{\dagger}(t) = \sum_{\mathfrak{p}} \breve{\psi}^{\dagger}(\mathfrak{p}, t) \langle\mathfrak{p}|, \tag{11.2}$$

where the presence of x in the argument of the general field operators, $\tilde{\Psi}(t)$, and $|\mathfrak{q}\rangle$ is just suppressed. In Eq. (11.2), we used $\langle\mathfrak{p}|$ since this is associated with the left eigenfunction of the *second* quantization creation, $\hat{\psi}^{\dagger}$. However, we can transform $\langle\mathfrak{p}|$ to $\langle\mathfrak{q}|$. Using Eq. (9.24), we have,

$$\tilde{\Psi}^{\dagger}(t) = \sum_{\mathfrak{p}} \breve{\psi}^{\dagger}(\mathfrak{p}, t) \sum_{\mathfrak{q}} \langle\mathfrak{p}| |\mathfrak{q}\rangle \langle\mathfrak{q}|. \tag{11.3}$$

Upon interchanging the order of summation, we have

$$\begin{aligned}
&= \sum_{\mathfrak{q}} \left\{ \sum_{\mathfrak{p}} \langle\mathfrak{p}| |\mathfrak{q}\rangle \breve{\psi}^{\dagger}(\mathfrak{p}, t) \right\} \langle\mathfrak{q}|, \\
&= \sum_{\mathfrak{q}} \left\{ \sum_{\mathfrak{p}} \exp(-i\mathfrak{p} \cdot \mathfrak{q}) \breve{\psi}^{\dagger}(\mathfrak{p}, t) \right\} \langle\mathfrak{q}|, \\
&= \sum_{\mathfrak{q}} \breve{\psi}^{\dagger}(\mathfrak{q}, t) \langle\mathfrak{q}|.
\end{aligned} \tag{11.4}$$

Thus, $\breve{\psi}^{\dagger}(\mathfrak{q}, t)$ and $\breve{\psi}(\mathfrak{q}, t)$, annihilate and create second quantized fields indicated by $|\mathfrak{q}\rangle$ and its adjoint $\langle\mathfrak{q}|$. This was interpreted as equivalent to second-quantized coupling constants in Ref. [61] in their application of third quantization to cosmology. These third quantized field operators, $\breve{\psi}(\mathfrak{q}, t)$ and $\breve{\psi}^{\dagger}(\mathfrak{q}, t)$, satisfy the equal-time commutation relation for bosons and anticommutation for fermions, namely,

$$\left[\breve{\psi}(\mathfrak{q}), \breve{\psi}^{\dagger}(\mathfrak{q}')\right]_{\pm} = \delta_{\mathfrak{q},\mathfrak{q}'}, \tag{11.5}$$

$$\left[\breve{\psi}(\mathfrak{q}), \breve{\psi}(\mathfrak{q}')\right]_{\pm} = 0, \tag{11.6}$$

$$\left[\breve{\psi}^{\dagger}(\mathfrak{q}), \breve{\psi}^{\dagger}(\mathfrak{q}')\right]_{\pm} = 0, \tag{11.7}$$

where the $+$ subscript is for fermions and the $-$ is bosons.

11.3 Wigner distribution functional of QFT

Following the usual prescription of second quantization procedure, i.e., by writing a *single* particle operator in terms of second-quantize field operators, we have

$$\hat{A}^{(1)} = \int \hat{\Psi}^{\dagger}(x)\, A^{(1)} \hat{\Psi}(x)\, dx. \tag{11.8}$$

Likewise, in the third quantization procedure, we can write in the same way, where $A^{(1)}$ in $\hat{\Psi}^{\dagger}(x)\, A^{(1)}\hat{\Psi}(x)$ is now express in $(\mathfrak{p}, \mathfrak{q})$-space. Moreover, we express $\hat{\Psi}(x)$ in terms of the third quantization annihilation field operator, namely, Eq. (11.1). Now using the expression of $\hat{A}^{(1)}$ in terms of the $\mathfrak{p}$ and $\mathfrak{q}$ field variables of QFT, we write

$$\hat{A} = \sum_{\mathfrak{p},\mathfrak{q}} A(\mathfrak{p}, \mathfrak{q})\, \hat{\Delta}(\mathfrak{p}, \mathfrak{q}), \tag{11.9}$$

where

$$A(\mathfrak{p}, \mathfrak{q}) = \frac{1}{\sqrt{N}} \sum_{\mathfrak{u}} e^{i2\mathfrak{u}\cdot\mathfrak{q}} \langle \mathfrak{p}+\mathfrak{u}|\, \hat{A}\, |\mathfrak{p}-\mathfrak{u}\rangle, \tag{11.10}$$

$$\hat{\Delta}(\mathfrak{p}, \mathfrak{q}) = \frac{1}{\sqrt{N}} \sum_{\mathfrak{v}} e^{i2\mathfrak{p}\cdot\mathfrak{v}} |\mathfrak{q}+\mathfrak{v}\rangle \langle \mathfrak{q}-\mathfrak{v}|. \tag{11.11}$$

We have

$$\begin{aligned}
&\int \hat{\Psi}^{\dagger}(x)\, A^{(1)} \hat{\Psi}(x)\, dx \\
&= \sum_{\mathfrak{p},\mathfrak{q}} A(\mathfrak{p}, \mathfrak{q}) \int \hat{\Psi}^{\dagger}(x)\, \hat{\Delta}(\mathfrak{p}, \mathfrak{q})\, \hat{\Psi}(x)\, dx, \\
&= \frac{1}{\sqrt{N}} \sum_{\mathfrak{p},\mathfrak{q}} A(\mathfrak{p}, \mathfrak{q}) \sum_{\mathfrak{v}} e^{i2\mathfrak{p}\cdot\mathfrak{v}} \left\langle \hat{\Psi}^{\dagger} \middle| \, |\mathfrak{q}+\mathfrak{v}\rangle \langle \mathfrak{q}-\mathfrak{v}| \, \middle| \hat{\Psi}(x) \right\rangle, \\
&= \frac{1}{\sqrt{N}} \sum_{\mathfrak{p},\mathfrak{q},\mathfrak{v},\mathfrak{q}',\mathfrak{q}''} A(\mathfrak{p}, \mathfrak{q})\, e^{i2\mathfrak{p}\cdot\mathfrak{v}} \breve{\psi}^{\dagger}(\mathfrak{q}', t) \langle \mathfrak{q}'|\, |\mathfrak{q}+\mathfrak{v}\rangle \langle \mathfrak{q}-\mathfrak{v}|\, \breve{\psi}(\mathfrak{q}'', t)\, |\mathfrak{q}''\rangle,
\end{aligned} \tag{11.12}$$

and we end up with

$$A^{(1)} = \frac{1}{\sqrt{N}} \sum_{\mathfrak{p},\mathfrak{q},\mathfrak{v}} A(\mathfrak{p}, \mathfrak{q})\, [e^{2i\mathfrak{p},\mathfrak{v}} \breve{\psi}^{\dagger}(\mathfrak{q}+\mathfrak{v})\, \breve{\psi}(\mathfrak{q}-\mathfrak{v})]. \tag{11.13}$$

Then we can write $A^{(1)}$ in terms of third quantization annihilation and creation operators, $\breve{\psi}(\mathfrak{q})$ and $\breve{\psi}^{\dagger}(\mathfrak{q})$, as

$$A^{(1)} = \frac{1}{\sqrt{N}} \sum_{\mathfrak{p},\mathfrak{q},\mathfrak{v}} A(\mathfrak{p}, \mathfrak{q})\, [e^{2i\mathfrak{p},\mathfrak{v}}\ \breve{\psi}^{\dagger}(\mathfrak{q}+\mathfrak{v})\, \breve{\psi}(\mathfrak{q}-\mathfrak{v})], \tag{11.14}$$

in complete analogy to the second quantize version [16]. Here trace over discrete indices is implied. We can more meaningfully write,

$$A^{(1)} = \sum_{\mathfrak{p},\mathfrak{q}} A(\mathfrak{p},\mathfrak{q})\, \check{f}(\mathfrak{p},\mathfrak{q}), \tag{11.15}$$

with

$$\check{f}(\mathfrak{p},\mathfrak{q}) = \frac{1}{\sqrt{N}} \sum_{\mathfrak{v}} e^{2i\mathfrak{p},\mathfrak{v}} \check{\psi}^{\dagger}(\mathfrak{q}+\mathfrak{v})\, \check{\psi}(\mathfrak{q}-\mathfrak{v}), \tag{11.16}$$

where $\check{f}(\mathfrak{p},\mathfrak{q})$ is analogous to the Klimontovich second quantization operator [65] for one-particle phase space distribution function. Here we are talking of QFT $(\mathfrak{p},\mathfrak{q})$-phase space of course, not the ordinary $(p.q)$-phase space. Thus, taking the average, we have,

$$\langle \check{f}(\mathfrak{p},\mathfrak{q}) \rangle = \frac{1}{\sqrt{N}} \sum_{\mathfrak{v}} e^{2i\mathfrak{p},\mathfrak{v}} \langle \check{\psi}^{\dagger}(\mathfrak{q}+\mathfrak{v})\, \check{\psi}(\mathfrak{q}-\mathfrak{v}) \rangle. \tag{11.17}$$

We identify the correlation function,

$$\langle \check{\psi}^{\dagger}(\mathfrak{q}+\mathfrak{v})\, \check{\psi}(\mathfrak{q}-\mathfrak{v}) \rangle = G^{<}(\mathfrak{q}+\mathfrak{v},\mathfrak{q}-\mathfrak{v}), \tag{11.18}$$

where $G^{<}(\alpha,\beta)$ corresponds to the Green function for particle distribution of second quantize quantum transport theory. Since here it pertains to the third quantization scheme, we will denote this by Fraktur style, $\mathfrak{G}^{<}(\mathfrak{q}+\mathfrak{v},\mathfrak{q}-\mathfrak{v})$, so that,

$$\begin{aligned} \langle \check{f}(\mathfrak{p},\mathfrak{q}) \rangle &= \frac{1}{\sqrt{N}} \sum_{\mathfrak{v}} e^{2i\mathfrak{p},\mathfrak{v}} \mathfrak{G}^{<}(\mathfrak{q}+\mathfrak{v},\mathfrak{q}-\mathfrak{v}), \\ &= f_w(\mathfrak{p},\mathfrak{q}). \end{aligned} \tag{11.19}$$

Therefore, $f_w(\mathfrak{p},\mathfrak{q})$ is the lattice Weyl transform of $\mathfrak{G}^{<}$, where $f_w(\mathfrak{p},\mathfrak{q})$ is the Wigner distribution functional in the third quantization, which is akin to the lattice Weyl transform of the Green function, $G^{<}$, for particle distribution in *second* quantized quantum transport theory of Part 3.

11.4 Ballistic transport equation in the third quantization

The ballistic transport equation for $f_w(\mathfrak{p},\mathfrak{q},\mathfrak{t})$ is given by the

$$i\left(\frac{\partial}{\partial t}\right) f_w(\mathfrak{p},\mathfrak{q},\mathfrak{t}) = \mathrm{LWT}_{\mathrm{third}}\left[\hat{H},\hat{\rho}\right], \tag{11.20}$$

which derives from the quantum Liouville equation,

$$i\left(\frac{\partial}{\partial t}\right)\hat{\rho} = \left[\hat{H},\hat{\rho}\right] \equiv: \mathcal{L}\hat{\rho}, \tag{11.21}$$

where $\mathcal{L}$ is the Liouvillian operator and $\mathrm{LWT}_{\mathrm{third}}$ indicates the lattice Weyl transform of the commutator of the Hamiltonian and density matrix operator. Upon carrying out the $\mathrm{LWT}_{\mathrm{third}}$ operation, we obtain,

$$\begin{aligned}\mathrm{LWT}_{\mathrm{third}}\left[\hat{H},\hat{\rho}\right] &= \left[\hat{H},\hat{\rho}\right](\mathfrak{p},\mathfrak{q},\mathfrak{t}),\\ &= \cos\Lambda\left[H(\mathfrak{p},\mathfrak{q},\mathfrak{t})\,f_w(\mathfrak{p},\mathfrak{q},\mathfrak{t}) - f_w(\mathfrak{p},\mathfrak{q},\mathfrak{t})\,H(\mathfrak{p},\mathfrak{q},\mathfrak{t})\right]\\ &\quad - i\sin\Lambda\left\{H(\mathfrak{p},\mathfrak{q},\mathfrak{t})\,f_w(\mathfrak{p},\mathfrak{q},\mathfrak{t}) + f_w(\mathfrak{p},\mathfrak{q},\mathfrak{t})\,H(\mathfrak{p},\mathfrak{q},\mathfrak{t})\right\},\end{aligned} \tag{11.22}$$

where Λ is the Poisson bracket operator [here we put $\hbar \longrightarrow 1$],

$$\Lambda = \frac{1}{2}\left(\frac{\partial^H}{\partial\mathfrak{p}}\cdot\frac{\partial^{f_w}}{\partial\mathfrak{q}} - \frac{\partial^H}{\partial\mathfrak{q}}\cdot\frac{\partial^{f_w}}{\partial\mathfrak{p}}\right). \tag{11.23}$$

The RHS of Eq. (11.22) can also be expressed as an integral which is more suitable for numerical computation. These expressions simplify considerably when the respective lattice Weyl transforms are scalar functions. Furthermore, calculation of currents involves taking the trace, and the first term of Eq. (11.22) does not contribute. Hence, we will consider only the second term of Eq. (11.22). Then we have,

$$[\hat{H},\hat{\rho}](\mathfrak{p},\mathfrak{q},\mathfrak{t}) = -2i\sin\Lambda\left\{H(\mathfrak{p},\mathfrak{q},\mathfrak{t})\;\, f_w(\mathfrak{p},\mathfrak{q},\mathfrak{t})\right\}, \tag{11.24}$$

and in integral form as

$$[\hat{H},\hat{\rho}](\mathfrak{p},\mathfrak{q},\mathfrak{t}) = \frac{1}{(2\pi)^8}\int\mathcal{D}\mathfrak{p}'\mathcal{D}\mathfrak{q}'K_H^s(\mathfrak{p},\mathfrak{q};\mathfrak{p}',\mathfrak{q}')\,f_w(\mathfrak{p}',\mathfrak{q}',t), \tag{11.25}$$

where

$$\begin{aligned}K_H^s(\mathfrak{p},\mathfrak{q};\mathfrak{p}',\mathfrak{q}') &= \int\mathcal{D}\mathfrak{u}\mathcal{D}\mathfrak{v}\exp\left\{\frac{i}{\hbar}\left[(\mathfrak{p}-\mathfrak{p}')\cdot\mathfrak{v} + (\mathfrak{q}-\mathfrak{q}')\cdot\mathfrak{u}\right]\right\}\\ &\quad\times\left[H\left(\mathfrak{p}+\frac{\mathfrak{u}}{2},\mathfrak{q}-\frac{\mathfrak{v}}{2},t\right) - H\left(\mathfrak{p}-\frac{\mathfrak{u}}{2},\mathfrak{q}+\frac{\mathfrak{v}}{2},t\right)\right].\end{aligned} \tag{11.26}$$

Thus, we have the ballistic transport equation in the third quantization scheme given by

$$i\left(\frac{\partial}{\partial t}\right)f_w(\mathfrak{p},\mathfrak{q},\mathfrak{t}) = -2i\sin\Lambda\left\{H(\mathfrak{p},\mathfrak{q},\mathfrak{t})\,f_w(\mathfrak{p},\mathfrak{q},\mathfrak{t})\right\}, \tag{11.27}$$

or

$$i\left(\frac{\partial}{\partial t}\right)f_w(\mathfrak{p},\mathfrak{q},\mathfrak{t}) = \frac{1}{(2\pi)^8}\int\mathcal{D}\mathfrak{p}'\mathcal{D}\mathfrak{q}'K_H^s(\mathfrak{p},\mathfrak{q};\mathfrak{p}',\mathfrak{q}')\,f_w(\mathfrak{p}',\mathfrak{q}',t), \tag{11.28}$$

where $K_H^s(\mathfrak{p},\mathfrak{q};\mathfrak{p}',\mathfrak{q}')$ is given by Eq. (11.26).

In the case of the numerical simulations of the *second* quantization scheme, the form of Eq. (11.28) has proved to be more fruitful for simulating real nanodevices [14]. Equation (11.28) can be generalized to matrix quantities, not just scalar, as well as to energy-dependent quantities. In the above transport equations it is assumed that the energy variable has been integrated out.

In what follows we give a general formulation of the dynamics of the distribution functional in QFT in terms of the *correlation functions* of the third quantization annihilation, $\check{\psi}(\mathfrak{q},t)$, and creation, $\check{\psi}^{\dagger}(\mathfrak{q},t)$, operators. First, we have to cast all quantum operators, especially one-body and two-body operators, in terms of $\check{\psi}^{\dagger}(\mathfrak{q},t)$ and $\check{\psi}(\mathfrak{q},t)$ and show that these have exactly the same form as that of the second quantized many-body formulation [12] of CMP. This then leads us to a parallel formulation of *third* quantum superfield theory of transport physics based on the quantum Liouville equation or von Neumann equation.

11.5 Hamiltonian operators in the third quantization

To construct the dynamics of the QFT distribution in functional phase space, we need to cast the Hamiltonian and all quantum operators in terms of the third quantize annihilation, $\check{\psi}(\mathfrak{q})$, and creation operators, $\check{\psi}^{\dagger}(\mathfrak{q})$, where $\mathfrak{q}$ is the eigenvalue of the second quantize operator, $\hat{\psi}(x)$. Observe that we use the *hat* for second quantize operators, whereas we use *inverted hat* for the *third* quantize operators.

Let $\hat{A}_1$ and $\hat{A}_2$ denote one-body and two-body quantum-mechanical operators, respectively. Following the usual prescription of the many-body quantization procedure, we write $\hat{A}_1$ and $\hat{A}_2$ in terms of the field operators, Ψ and its adjoint $\Psi^{\dagger}$ as

$$\hat{A}_1 = \int \tilde{\Psi}^{\dagger}(x)\, A_1 \tilde{\Psi}(x)\,, \tag{11.29}$$

$$\hat{A}_2 = \iint \tilde{\Psi}^{\dagger}(x)\, \tilde{\Psi}^{\dagger}(x')\, A_2 \tilde{\Psi}(x')\, \tilde{\Psi}(x)\; dx\; dx'. \tag{11.30}$$

As before, we decompose the field operators in terms eigenfunction $|\mathfrak{q}\rangle$ and its adjoint, $\langle\mathfrak{q}|$. We have,

$$\tilde{\Psi}(x) = \sum_{\mathfrak{q}} \check{\psi}(\mathfrak{q})\, |\mathfrak{q}\rangle\,, \tag{11.31}$$

$$\tilde{\Psi}^{\dagger}(x) = \sum_{\mathfrak{q}} \check{\psi}^{\dagger}(\mathfrak{q})\, \langle\mathfrak{q}|\,, \tag{11.32}$$

where we have suppressed summation over other inherent discrete indices if present in the theory. We have also not displayed the argument x in the eigenfunction basis. Equations (11.31) and (11.32) maybe considered the conceptual basis of the Jordan–Schwinger transformation from matrices to field operators.

Thus, $\check{\psi}(\mathfrak{q},t)$ and $\check{\psi}^{\dagger}(\mathfrak{q},t)$, labeled by the second quantized $\mathfrak{q}$ field, annihilate and create (generate) second quantized field eigenvectors, indicated by $|\mathfrak{q}\rangle$ and $\langle\mathfrak{q}|$. These third quantized field operators, $\check{\psi}(\mathfrak{q},t)$ and $\check{\psi}^{\dagger}(\mathfrak{q},t)$, satisfy the equal-time commutation relation for bosons and anticommutation for fermions. We desire to express Eqs. (11.29) and (11.30) in terms of third quantized field operators, $\check{\psi}(\mathfrak{q},t)$ and $\check{\psi}^{\dagger}(\mathfrak{q},t)$, labeled by $\mathfrak{q}$.

We also like to express A_1 and A_2 in terms of $(\mathfrak{p},\mathfrak{q})$-functional phase space. Thus, we need to express A_1 and A_2 in terms of its lattice Weyl transforms and phase-space point projectors. We have using the expression of $\hat{A}_1$ in terms of the $\mathfrak{p}$ and $\mathfrak{q}$ field variables of QFT, we write

$$\hat{A}_1 = \sum_{\mathfrak{p},\mathfrak{q}} A(\mathfrak{p},\mathfrak{q})\,\hat{\Delta}(\mathfrak{p},\mathfrak{q}), \tag{11.33}$$

where

$$A(\mathfrak{p},\mathfrak{q}) = \frac{1}{\sqrt{N}}\sum_{\mathfrak{u}} e^{i2\mathfrak{u}\cdot\mathfrak{q}}\,\langle\mathfrak{p}+\mathfrak{u}|\,\hat{A}\,|\mathfrak{p}-\mathfrak{u}\rangle\,, \tag{11.34}$$

$$\hat{\Delta}(\mathfrak{p},\mathfrak{q}) = \frac{1}{\sqrt{N}}\sum_{\mathfrak{v}} e^{i2\mathfrak{p}\cdot\mathfrak{v}}\,|\mathfrak{q}+\mathfrak{v}\rangle\langle\mathfrak{q}-\mathfrak{v}|\,. \tag{11.35}$$

We have

$$\begin{aligned}
&\int \hat{\Psi}^{\dagger}(x)\,A^{(1)}\hat{\Psi}(x)\,dx\\
&= \sum_{\mathfrak{p},\mathfrak{q}} A(\mathfrak{p},\mathfrak{q})\int \hat{\Psi}^{\dagger}(x)\,\hat{\Delta}(\mathfrak{p},\mathfrak{q})\,\hat{\Psi}(x)\,dx,\\
&= \frac{1}{\sqrt{N}}\sum_{\mathfrak{p},\mathfrak{q}} A(\mathfrak{p},\mathfrak{q})\sum_{\mathfrak{v}} e^{i2\mathfrak{p}\cdot\mathfrak{v}}\langle\hat{\Psi}^{\dagger}||\mathfrak{q}+\mathfrak{v}\rangle\langle\mathfrak{q}-\mathfrak{v}||\hat{\Psi}(x)\rangle,
\end{aligned} \tag{11.36}$$

and substituting Eqs. (11.1) and (11.4), we end up with

$$A^{(1)} = \frac{1}{\sqrt{N}}\sum_{\mathfrak{p},\mathfrak{q},\mathfrak{v}} A(\mathfrak{p},\mathfrak{q})\,[e^{2i\mathfrak{p},\mathfrak{v}}\check{\psi}^{\dagger}(\mathfrak{q}+\mathfrak{v})\,\check{\psi}(\mathfrak{q}-\mathfrak{v})]. \tag{11.37}$$

We can reduce to a more familiar form amenable to field theoretical perturbation technique by integrating with respect to $\mathfrak{p}$ (here summation means functional integration), and indicate the result by $W_1(2\mathfrak{v},2\mathfrak{q})$, i.e.,

$$W_1(2\mathfrak{v},2\mathfrak{q}) = \frac{1}{\sqrt{N}}\sum_{\mathfrak{p}} e^{2i\mathfrak{p},\mathfrak{v}} A(\mathfrak{p},\mathfrak{q}), \tag{11.38}$$

which yields

$$A^{(1)} = \sum_{\mathfrak{q},\mathfrak{v}} W_1\left(2\mathfrak{v}, 2\mathfrak{q}\right) \check{\psi}^{\dagger}\left(\mathfrak{q}+\mathfrak{v}\right) \check{\psi}\left(\mathfrak{q}-\mathfrak{v}\right). \tag{11.39}$$

By transformation of variables, we end up with,

$$A^{(1)} = \sum_{\mathfrak{r},\mathfrak{r}'} W_1\left(\mathfrak{r}-\mathfrak{r}', \mathfrak{r}+\mathfrak{r}'\right) \check{\psi}^{\dagger}\left(\mathfrak{r}\right) \check{\psi}\left(\mathfrak{r}'\right). \tag{11.40}$$

Substituting the expression for $A\left(\mathfrak{p},\mathfrak{q}\right)$, given by Eq. (4.11), in Eq. (11.38) we have

$$\begin{aligned} W_1\left(2\mathfrak{v}, 2\mathfrak{q}\right) &= (N)^{-1} \sum_{\mathfrak{p}} e^{2i\mathfrak{p},\mathfrak{v}} \sum_{\mathfrak{v}'} e^{i2\mathfrak{p}\cdot\mathfrak{v}'} \left\langle \mathfrak{q}-\mathfrak{v}'\right| A \left|\mathfrak{q}+\mathfrak{v}'\right\rangle, \\ &= (N)^{-1} \sum_{\mathfrak{p}} e^{2i\mathfrak{p},\mathfrak{v}} \sum_{\mathfrak{v}'} e^{-i2\mathfrak{p}\cdot\mathfrak{v}'} \left\langle \mathfrak{q}+\mathfrak{v}'\right| A \left|\mathfrak{q}-\mathfrak{v}'\right\rangle, \\ &= (N)^{-1} \sum_{\mathfrak{p},\mathfrak{v}'} e^{2i\mathfrak{p},(\mathfrak{v}-\mathfrak{v}')} \left\langle \mathfrak{q}+\mathfrak{v}'\right| A \left|\mathfrak{q}-\mathfrak{v}'\right\rangle, \\ &= \sum_{\mathfrak{v}'} \delta\left(\mathfrak{v}-\mathfrak{v}'\right) \left\langle \mathfrak{q}+\mathfrak{v}'\right| A \left|\mathfrak{q}-\mathfrak{v}'\right\rangle, \\ &= \left\langle \mathfrak{q}+\mathfrak{v}\right| A \left|\mathfrak{q}-\mathfrak{v}\right\rangle = \left\langle \mathfrak{r}\right| A \left|\mathfrak{r}'\right\rangle. \end{aligned} \tag{11.41}$$

Therefore, we can express Eq. (11.39) as

$$\begin{aligned} A^{(1)} &= \sum_{\mathfrak{r},\mathfrak{r}'} W_1\left(\mathfrak{r},\mathfrak{r}'\right) \check{\psi}^{\dagger}\left(\mathfrak{r}\right) \check{\psi}\left(\mathfrak{r}'\right), \\ &= \sum_{1,2} V\left(1,2\right) \check{\psi}^{\dagger}\left(1\right) \check{\psi}\left(2\right). \end{aligned} \tag{11.42}$$

Thus, we have succeeded in expressing the one-body Hamiltonian in terms of the *third* quantization creation, $\check{\psi}^{\dagger}$, and annihilation, $\check{\psi}$, operators. This is formally the same expression as in the conventional second quantization scheme.

Similarly, for the two-body operator we proceed as follows. We have,

$$A^{(2)} = \iint \Psi^{\dagger}\left(x\right) \Psi^{\dagger}\left(x'\right) A_2 \Psi\left(x'\right) \Psi\left(x\right) dx' dx, \tag{11.43}$$

which yields

$$\begin{aligned} A^{(2)} = (N)^{-1} &\sum_{\mathfrak{p},\mathfrak{p}';\mathfrak{q},\mathfrak{q}',\mathfrak{v}.\mathfrak{v}'} A_2\left(\mathfrak{p},\mathfrak{p}';\mathfrak{q},\mathfrak{q}'\right) \exp\left[2i\left(\mathfrak{p}.\mathfrak{v}+\mathfrak{p}'.\mathfrak{v}'\right)\right] \\ &\times \check{\psi}^{\dagger}\left(\mathfrak{q}+\mathfrak{v}\right) \check{\psi}^{\dagger}\left(\mathfrak{q}'+\mathfrak{v}'\right) \check{\psi}\left(\mathfrak{q}'-\mathfrak{v}'\right) \check{\psi}\left(\mathfrak{q}-\mathfrak{v}\right), \end{aligned} \tag{11.44}$$

where again summation over all other inherent discrete indices, like spins, are suppressed to save space. Again, we can reduce to a more familiar form amenable to field theoretical perturbation technique by integrating with respect to $\mathfrak{p}$ and $\mathfrak{p}'$ (here summation means functional integration), and indicate the result by $W_2(2\mathfrak{v}, 2\mathfrak{v}'; 2\mathfrak{q}, 2\mathfrak{q}')$, i.e.,

$$W_2(2\mathfrak{v},2\mathfrak{v}'; 2\mathfrak{q},2\mathfrak{q}') = (N)^{-1} \sum_{\mathfrak{p},\mathfrak{p}'} A_2(\mathfrak{p}, \mathfrak{p}'; \mathfrak{q}, \mathfrak{q}') \exp\left[2i\left(\mathfrak{p}.\mathfrak{v} + \mathfrak{p}'.\mathfrak{v}'\right)\right], \tag{11.45}$$

where

$$A_2(\mathfrak{p}, \mathfrak{p}'; \mathfrak{q}, \mathfrak{q}') = (N)^{-1} \sum_{v'',v'''} e^{-2ip\cdot v''} e^{-2ip'\cdot v'''} \times \left\langle \mathfrak{q} + \mathfrak{v}''\right| \left\langle \mathfrak{q}' + \mathfrak{v}'''\right| A_2 \left|\mathfrak{q}' - \mathfrak{v}'''\right\rangle \left|\mathfrak{q} - \mathfrak{v}''\right\rangle. \tag{11.46}$$

We have

$$\begin{aligned} W_2(2\mathfrak{v},2\mathfrak{v}'; \mathfrak{q}, \mathfrak{q}') &= (N)^{-2} \sum_{\mathfrak{p},\mathfrak{v}''} e^{2i\mathfrak{p},(\mathfrak{v}-\mathfrak{v}'')} \sum_{\mathfrak{p}',\mathfrak{v}'''} e^{2i\mathfrak{p}',(\mathfrak{v}'-\mathfrak{v}''')} \\ &\quad \times \left\langle \mathfrak{q} + \mathfrak{v}''\right| \left\langle \mathfrak{q}' + \mathfrak{v}'''\right| A_2 \left|\mathfrak{q}' - \mathfrak{v}'''\right\rangle \left|\mathfrak{q} - \mathfrak{v}''\right\rangle, \\ &= \left\langle \mathfrak{q} + \mathfrak{v}\right| \left\langle \mathfrak{q}' + \mathfrak{v}'\right| A \left|\mathfrak{q}' - \mathfrak{v}'\right\rangle \left|\mathfrak{q} - \mathfrak{v}\right\rangle = \left\langle \mathfrak{r}\right| \left\langle \mathfrak{r}'\right| A \left|\mathfrak{r}''\right\rangle \left|\mathfrak{r}'''\right\rangle. \end{aligned} \tag{11.47}$$

From Eq. (11.44), we end up with the expression,

$$A^{(2)} = \sum_{\mathfrak{r},\mathfrak{r}',r'',r'''} \left\langle \mathfrak{r}\right| \left\langle \mathfrak{r}'\right| A \left|\mathfrak{r}''\right\rangle \left|\mathfrak{r}'''\right\rangle \check{\psi}^{\dagger}(\mathfrak{r}) \check{\psi}^{\dagger}(\mathfrak{r}') \check{\psi}(\mathfrak{r}'') \check{\psi}(\mathfrak{r}'''), \tag{11.48}$$

and can be written symbolically as

$$A^{(2)} = \sum_{1,2,3,4} V(1,2,3,4) \check{\psi}^{\dagger}(1) \check{\psi}^{\dagger}(2) \check{\psi}(3) \check{\psi}(4), \tag{11.49}$$

which is formally of the expression as those of second quantize operators for four point potential [12]. Equations (11.42) and (11.49) allow us to write the Hamiltonian in terms of third quantization, $\check{\psi}^{\dagger}$ and $\check{\psi}$, as

$$\check{H} = \sum_{1,2} V(1,2) \check{\psi}^{\dagger}(1) \check{\psi}(2) + \sum_{1,2,3,4} V(1,2,3,4) \check{\psi}^{\dagger}(1) \check{\psi}^{\dagger}(2) \check{\psi}(3) \check{\psi}(4). \tag{11.50}$$

Table 11.1. CMP and QFT distribution-function dynamics.

Dynamical unification of CMP and QFT	
CMP (in second quantization)	QFT (in third quantization)
Bloch function, $\|p\rangle$	Eigenfunction of 2nd quantization creation operator, $\|\mathfrak{p}\rangle$
Wannier function, $\|q\rangle$	Eigenfunction of 2nd quantization annihilation operator, $\|\mathfrak{q}\rangle$
2nd quantization field operators, $\hat{\psi}^{\dagger}(q)$ and $\hat{\psi}(q)$	3rd quantization quantum field operators, $\check{\psi}^{\dagger}(\mathfrak{q})$ and $\check{\psi}(\mathfrak{q})$
2nd quantization field operators, $\hat{\psi}^{\dagger}(p)$ and $\hat{\psi}(p)$	3rd quantization quantum field operators, $\check{\psi}^{\dagger}(\mathfrak{p})$ and $\check{\psi}(\mathfrak{p})$
Correlation functions, typified by $G^{<}(q,q',t,t')$	Correlation functions, typified by $\mathfrak{G}^{<}(\mathfrak{q},\mathfrak{q}',t,t')$
Momentum-position phase space, (p,q)	"Momentum-position" functional phase space, $(\mathfrak{p},\mathfrak{q})$
$\langle\langle 1\| \overline{S}(\infty,-\infty) \|\rho_{eq}\rangle\rangle$ $= \exp\left[\frac{i}{\hbar}W\right]$ generating function	$\langle\langle 1\| \mathcal{S}(\infty,-\infty) \|\rho_{eq}\rangle\rangle$ $= \exp\left[\frac{i}{\hbar}\mathcal{W}\right]$ generating functional
Transport equations of correlation functions	Transport equations of distribution functions

The above developments allow us to map the quantum transport dynamics of CMP to that of QFT, by essentially simple substitution as follows: (a) second quantized operators to third quantize field operators, (b) second quantize correlation functions to third quantize correlation functions, and (c) crystal momentum, p to $\mathfrak{p}$, as well as lattice position q to $\mathfrak{q}$. For example, we can still have a well-defined multi-component quantum superfield operators in the third quantization, namely,

$$\check{\Psi}(1) = \begin{pmatrix} \check{\psi}(1) \\ \widetilde{\psi}^{\dagger}(1) \\ \check{\psi}^{\dagger}(1) \\ \widetilde{\psi}(1) \end{pmatrix} \equiv \left\{\check{\Psi}_{\alpha}(1),\ \alpha = 1,2,3,4\right\}. \tag{11.51}$$

Thus, we obtained a perfect match between the transport variables in CMP and QFT. We summarize the identifications in Table 11.1,

11.6 Third quantization of quantum superfield transport theory

We state here some results without going through similar formal procedure. This similar procedure [12] that can be followed is outlined in Part 3 of this book.

Here, we will simply complete the equation for $\mathfrak{G}^<$ of Eq. (11.27) which is given only for ballistic transport in functional phase space $(\mathfrak{p}, \mathfrak{q})$. We have, upon following similar procedure for second quantize superfield quantum transport formalism, the full equation for the functional distribution correlation function, $\mathfrak{G}^<$, neglecting Cooper pairings terms as

$$\begin{aligned} i\left(\frac{\partial}{\partial t_1}+\frac{\partial}{\partial t_2}\right)\mathfrak{G}^<(1,2) &= \left[H,\mathfrak{G}^<\right](1,2) \\ &\quad + \left[\Sigma^<,\mathrm{Re}\mathfrak{G}^r\right](1,2) \\ &\quad + \frac{i}{2}\left\{\Sigma^<,A\right\}(1,2) \\ &\quad - \frac{i}{2}\left[\Gamma,\mathfrak{G}^<\right](1,2), \end{aligned} \tag{11.52}$$

where $(1,2)$ stands for $(\mathfrak{q}_1, t_1, \mathfrak{q}_2, t_2)$. The quantities in Eq. (11.52) are defined in the third quantization as follows:

$$\begin{aligned} H(1,2) &= \left[H_0(1,2)+V(1,2)\right]\delta(1-2) \\ &\quad + \delta(t_1-t_2)\,\Sigma^{HF}(\mathfrak{q}_1,\mathfrak{q}_2,t_1) + \mathrm{Re}\Sigma^r(1,2), \end{aligned} \tag{11.53}$$

$$\mathrm{Re}\mathfrak{G}^r(1,2) = \frac{-i}{2}\epsilon(t_1-t_2)\,A(1,2), \tag{11.54}$$

$$\mathrm{Re}\Sigma^r(1,2) = \frac{-i}{2}\epsilon(t_1-t_2)\,\Gamma(1,2). \tag{11.55}$$

Upon taking the lattice Weyl transform, neglecting the Cooper pairing terms, a simpler equation,

$$\begin{aligned} &\frac{\partial}{\partial t}\mathfrak{G}^<(\mathfrak{p},E,\mathfrak{q},t) \\ &\quad = 2\sin\left(\check{\Lambda}\right)\left[H(\mathfrak{p},E,\mathfrak{q},t)\,\mathfrak{G}^<(\mathfrak{p},E,\mathfrak{q},t) + \Sigma^<(\mathfrak{p},E,\mathfrak{q},t)\,\mathrm{Re}\mathfrak{G}^r(\mathfrak{p},E,\mathfrak{q},t)\right] \\ &\qquad + \cos\left(\check{\Lambda}\right)\left[\Sigma^<(\mathfrak{p},E,\mathfrak{q},t)\,A(\mathfrak{p},E,\mathfrak{q},t) - \Gamma(\mathfrak{p},E,\mathfrak{q},t)\,\mathfrak{G}^<(\mathfrak{p},E,\mathfrak{q},t)\right]. \end{aligned} \tag{11.56}$$

The interest in integrating Eq. (11.56) with respect to the energy, E, is to obtain an equation for the distribution of quasiparticles,

$$f_w(\mathfrak{p},\mathfrak{q},t) = \int dE\,(-i)\,\mathfrak{G}^<(\mathfrak{p},E,\mathfrak{q},t). \tag{11.57}$$

Equation (11.56) can also be given in integral form [11, 12], this will not be given here [11, 12], the readers are referred to Part 3 of this book.

Chapter 12

Part 1: Concluding Remarks

We see that the parallelism discussed here is rooted in the formally identical expression for the phase-space point projector denoted as $\hat{\Delta}(\mathfrak{p}, \mathfrak{q})$. Here, any quantum operator can be expanded in terms of the $\hat{\Delta}(\mathfrak{p}, \mathfrak{q})$ operator basis. Although $\hat{\Delta}(\mathfrak{p}, \mathfrak{q})$ carries two equivalent expressions as basis operators, this is made a unique and symmetrical expression, which is a again formally identical in the unification discussed here, namely,

$$\hat{\Delta}(\mathfrak{p}, \mathfrak{q}) = (N)^{-1} \sum_{\mathfrak{v},\mathfrak{u}} \exp\{-i(\mathcal{P} \cdot \mathfrak{v} - \mathcal{Q} \cdot \mathfrak{u})\} \exp\{2i(\mathfrak{p} \cdot \mathfrak{v} - \mathfrak{q} \cdot \mathfrak{u})\} \Omega_{\lambda\lambda'}, \tag{12.1}$$

or in terms of continuous Grassmann variables,

$$\hat{\Delta}(\mathfrak{p}, \mathfrak{q}) = \iint \mathfrak{D}\mathfrak{v}\mathfrak{D}\mathfrak{u} \sum_{\mathfrak{v},\mathfrak{u}} \exp\{-i(\mathcal{P} \cdot \mathfrak{v} - \mathcal{Q} \cdot \mathfrak{u})\} \exp\{2i(\mathfrak{p} \cdot \mathfrak{v} - \mathfrak{q} \cdot \mathfrak{u})\} \Omega_{\lambda\lambda'}. \tag{12.2}$$

With the above unification,[1] all physics that follows have corresponding formally identical expressions, such as the distribution in dual space or phase-space and their path integral expressions. These results clearly demonstrate the power of the mathematical language of "lattice" Weyl–Wigner formulation of quantum physics [3], as shown by the authors in previous publications. Indeed, for finite fermion systems, assuming nondegenerate countable states, the creation and annihilation *field* formalism exactly maps to that of

[1] As a side remark, it is worth pointing out that in supersymmetric string theory (superstring theory), another instance of unification of bosons and fermions corresponds to different states of a single string within the Calabi–Yau space.

CMP's *particle* formalism of free Bloch fermions. Indeed, CMP is a renormalized particle position and momentum formalism of Bloch fermions which incorporates the Pauli exclusion principle and Fermi–Dirac distribution.

The fact that,

$$\hat{Y}(\mathfrak{u}, \mathfrak{v}) = \exp[-i(\hat{P} \cdot \mathfrak{v} - \hat{Q} \cdot \mathfrak{u})]\Omega, \tag{12.3}$$

formally span all operators describing fermions, bosons, and spin systems suggests that, in principle, bosonization, fermionization and Jordan–Wigner fermionization for spin systems, as well as the Holstein–Primakoff (and Dyson–Maleev) transformation from boson operators to the spin operator can in principle be performed depending on the physical situations and ease in the calculations. In quantum physics, unitary transformation on the creation and annihilation operators themselves is also employed, as exemplified by the Bogoliubov transformation for quadratic Hamiltonian. Moreover, whenever $\hat{Y}(\mathfrak{u}, \mathfrak{v}) \Longrightarrow M_{ij}$ is already expressed in matrix form, the Jordan–Schwinger transformation is a map from matrices M_{ij} to bilinear expressions of creation and annihilation operators, e.g., of the form, $\hat{a}_i^\dagger M_{ij} \hat{a}_j$, which expedites computation of representations. This follows in spirit with Eqs. (11.31) and (11.32).

The present unification suggests an alternative formulation of QFT in terms of "Wigner distribution" or Q-distribution of conjugate functional-field variables quantum transport equations. Using the phase-space formalism of quantum mechanics, there seems to be a self-similarity of the mathematical structure, through an iterative algorithm, in going from solid-state crystalline phase space (p, q), to second quantization scheme $(\mathfrak{p}, \mathfrak{q})$ phase-space, and further into the third quantization scheme, brought by the third quantization creation, $\check{\psi}^\dagger(\mathfrak{q}, t)$, and annihilation, $\check{\psi}(\mathfrak{q}, t)$, field operators whose eigenvalues will lead to higher quantization schemes. This then will produce similar generating functional and nonequilibrium superfield quantum transport equations. This suggests that quantization *per se* appears to possess a hierarchical and "self-similar" mathematical framework.

This third quantization may have been interpreted as equivalent to second-quantized coupling constants in Ref. [61] in their application of their own third quantization scheme to cosmology. The advantages and utility of the third quantization scheme in quantum field theory needs further research and investigations, e.g., this may have applications in quantum fluctuation theory of systems with many degrees of freedom, e.g., all spatial and energy scales in turbulent flows, in the functional theory of coherent state representation in harmonic oscillators and nonlinear optics), and in cosmology.

It is hope that functional theories and third quantization dynamics of Wigner type distribution of functionals will lead to a more comprehensive perspective of nature in CMP and QFT. The power of the use of functionals in quantum physics has been demonstrated before. It has been heralded by the use of Feynman functional path integrals in QFT and by the Kohn–Sham density functional theory in CMP.

Since these are mentioned above, for completeness we refer the readers to the world wide web/cyberspace/Wikipedia for detailed discussions of the following bosonization and fermionization transformations, namely, the Jordan–Wigner transformation, Holstein-Primakoff transformation, Jordan–Schwinger transformation, and Bogoliubov transformation. Moreover, we mention some other very important transformations in CMP and QFT which deals with the decoupling of the different degrees of freedom. This is exemplified, e.g., by the well-known Foldy–Wouthuysen transformation for relativistic Dirac electrons [66] given in Appendix A, as well as by the perturbative decoupling of energy bands to all orders in the calculation of magnetic susceptibility of Bloch fermions [18].

Part 2

Chiral Quantization: U(1) Gauge Theory of Bohr–Sommerfeld Quantization

Chapter 13

Chiral Degrees of Freedom and Quantization

There are two general themes that will be addressed in this chapter, namely, (a) quantum transport in open systems and (b) vortex/precession-motion quantization of low-dimensional systems, and stationary quantization of *confined motion* in phase space due to oscillatory dynamics or *compactification* of space and time for translationally symmetric and steady-state systems (e.g., particle in a box, Brillouin zone, and Matsubara time-zone or Matsubara quantized frequencies). The former is a recent phenomenon which has become the springboard of modern condensed-matter transport physics research, whereas the latter has an old beginning in quantum mechanics of atomic systems. Although, the latter beginning is old it continues to be interesting and a powerful way of getting a handle of complex problems. Indeed this old beginning has now merited revisits since its innocent looking formalism actually have a rigorous and deep geometric origin, which has only been recently exposed by the author. This is based on the geometrical concept of connection [67]. Our task is to show that the old quantization scheme is the forebear of the new and permeates or pervades the whole of modern topological quantum physics. The geometrical concept of connection also allows us to deduce the quantized Dirac magnetic-monopole charge (i.e., gauge charge) without the use of Dirac semi-infinite string [68].

The quantization of Hall conductance of electrical conductivity in a 2D periodic potential was first explained by Thouless, Kohmoto, Nightingale, and den Nijs (TKNN) [69] using the Kubo current–current correlation. A similar approach was employed by Streda [70]. Earlier, Laughlin[71], and later Halperin [72], study the effects produced by changes in the vector potential on the states at the edges of a finite system, where quantization of the conductance is made explicit, but whether their result is independent of the edge conditions was not clear. In contrast, the use of Kubo formula by TKNN pertains to bulk 2D conductors.

These topological works were motivated by the 1980 Nobel Prize winning experimental discovery of von Klitzing, Dorda, and Pepper [73] on the quantization of the Hall conductance of a 2D electron gas in a strong magnetic field. The strong magnetic field basically provides the gapped energy structure, in form of equally gapped LL, for the experiments of 2D electron gas. On the other hand, there is no magnetic field in the TKNN approach, only a periodic potential in crystalline solid under weak electric field is being treated. A strong magnetic field is not needed to provide the gapped energy structure in their theory, only peculiar, gapped energy-band structures. In principle, in the presence of strong electric field the discrete Landau levels (LLs) is replaced by the *unstable* discrete Stark ladder-energy levels in the absence of scatterings [24, 74] due to a finite energy-band width and compactification of space into a torus, known as the Born-von Karman boundary condition.

Our new approach in exposing the old quantization scheme to integer quantum Hall effect (IQHE) makes use of the real-time superfield and lattice Weyl transform non-equilibrium Green's function (SFLWT-NEGF) quantum transport equation, this is outlined in Part 3 of this book. This is given to first order in the gradient expansion of the quantum transport equation, given in more details in Part 7. This purely nonequilibrium quantum transport approach is in contrast to the use of conventional equilibrium-fluctuation Kubo current–current formula originally employed by TKNN [75]. In the present approach, the topological invariant of quantum transport in $(\vec{p}, \vec{q}; E, t)$-space is thus identified. How the Bohr–Sommerfeld (BS) quantization enters into the analysis is here demonstrated. For a self-contained treatment, we give outline of the derivation in this part of the book, although it is also given in more details in Part 7. The BS quantization also enters in the calculation leading to quantized orbital magnetic moment and edge states.

Another Nobel prize winning theoretical discovery [75], and confirmed experimentally, is the so-called Kosterlitz–Thouless phase transition (KTPT) first discovered in X–Y model of spin systems. This goes well-beyond the well-established theorem, the so-called Mermin–Wagner theorem [76], sometimes referred to as the Mermin–Wagner–Berezinskii theorem, which states that continuous symmetries cannot be spontaneously broken at finite temperature in systems with sufficiently short-range interactions in dimensions $d \leq 2$. The vortex and antivortex solutions of KTPT goes beyond Ginzburg–Landau symmetry breaking phase transition which been quite successful in the past. On the other hand, the vortex solutions of KTPT has a topological content (basically topological defects) and the quantization bears resemblance to the BS quantization condition. The BS quantization also crept into the theory of another novel topological phase of matter, so-called Haldane phase of odd-integer spin chain. The emergence of new topological phases of matter which violates symmetry breaking and are topological in nature has ushered the interaction of theoretical physics and pure mathematics.

More recent experiment reveals the appearance of incompressible fractional quantum Hall effect (FQHE) states in monolayer graphene at $\nu = \pm\frac{1}{2}$, $\pm\frac{1}{4}$ and $\pm\frac{3}{4}$ substituting the compressible Hall metal states at these fillings in the lowest LL in a narrow magnetic field window depending on the sample parameters. Jacak propose [77] an explanation of these observations in terms of homotopy of monolayer graphene consistent with a general theory of correlated states in planar Hall systems, which is beyond the composite fermion model of Jain [78]. An increase of the magnetic field flux quantum is proven for multiloop trajectories (the formal proof goes via the BS rule [69]), resulting in the magnetic flux quantum, $\Phi_k = (2k+1)\frac{h}{e}$, for planar braids with additional k loops [77]. More recently the BS quantization has been employed by Jacak [79] to show that a larger flux quantum for multi-loop cyclotron braid orbits defines larger spatial dimension of such orbits. Generalization of the BS quantization rule shows that the size of a magnetic field flux quantum grows for multiloop orbits like $(2k+1)\ \frac{h}{c}$ with the number of loops k. Utilizing this property for electrons on the 2-D substrate jellium, the author has derived upon the path integration a complete FQHE hierarchy deemed in excellent agreement with experiments [77].

Indeed, the recent developments in the cross-fertilization of pure mathematics and physics is invigorated due to recent discoveries in physics which have strong relevance to pure mathematics. The IQHE and fractional FQHE have recently been geometrically formalized and unified based on Hecke operators and Hecke eigensheaves (mathematical replacement of the term eigenfunctions), i.e., of the geometric Langlands program by Ikeda [260]. In particular the plateaus of the Hall conductance are associated by the Hecke eigensheaves. Moreover, the fractal energy spectrum of a tight-binding Hamiltonian in a magnetic field, known as the Hofstadter's butterfly, as been associated with Langlands duality in the quantum groups. Indeed, the Langlands program in pure mathematics may carry several realizations in theoretical physics.

As it turns out, the old BS quantization rules has recently been realized to have much deeper ramifications in quantum physics. In a much simpler terms in some cases, it has to do with the counting of discrete "$(2\pi\hbar)^3$-voxels" of actions in phase-space volume of generalized canonical variables measured in units of Planck's constant, h, or number of quantum flux in the case of magnetic fields. Contour integral of Berry connection are measured in terms of $2\pi\hbar$-pixels. By counting operation means that the results belong to the domain of integers, $\mathbb{Z}$.

Here we show that the old quantization scheme have bearings on LL degeneracy, massive Dirac magnetic monopole, Aharonov–Bohm effect, discrete vortex charge, quantization of orbital magnetic moment and angular momentum, and BS quantization in superfluidity, superconductivity, and topological band theory (TBT) and topological quantum field theory (TQFT).

13.1 Geometric origin of the Bohr–Sommerfeld quantization

We give here a geometrical point of view of the BS quantization rule given by

$$\frac{1}{2\pi\hbar}\oint_{H(p,q)=E} p_i dq_i = \frac{1}{2\pi\hbar}\oint_{H(p,q)=E} \vec{p}\cdot d\vec{q} = n. \tag{13.1}$$

The energy spectrum of hydrogen atom agreed exactly with observed spectrum as obtained by Sommerfeld. The BS theory was applied with varying success to other systems. In fact, it is remarkable that the energy spectrum of Dirac equation for an electron in an atom obtained by Sommerfeld also agrees exactly [80]. With a minor modifications BS quantization gives the energy spectrum of π meson, which can be obtained by solving the Klein–Gordon equation [81].

Closer examination of Eq. (13.1) suggests that it does not really make any sense if one considers the set $\{p, q\}$ as independent conjugate dynamical variables. Obviously, it is totally inconsistent with the surface integral

$$\frac{1}{2\pi\hbar}\oint_{H(p,q)=E[\partial S]} \vec{p}\cdot d\vec{q} = \frac{1}{2\pi\hbar}\iint_S \nabla_{\vec{q}}\times\vec{p}\cdot d\vec{S} = 0. \tag{13.2}$$

The only plausible quantity that could be applied to give a new meaning to the variable $\vec{p}$ in Eq. (13.1) is the vector potential $A(q)$ in Maxwell's equations. Then the integer n would give the LL degeneracies or the number of quantized vortices in type II superconductors. But in the absence of magnetic fields, in which the BS is often applied, we have to appeal to more general gauge theories beyond Maxwell's vector potential, e.g., the gauge of TQFT via the Chern–Simons gauge theory [82] or the gauge of TBT in terms of Berry connection and Berry curvature.

Here we give a much simpler and intuitive geometric derivation of the BS quantization condition, as a contour integral of the gradient of geometric phase or Berry connection. Let $|n, q\rangle$ be the basis eigenvector of the eigenstates of the Schrödinger equation, parametrized by q,

$$H(q)|n, q\rangle = E_n |n, q\rangle . \tag{13.3}$$

Note from the original quantum mechanical viewpoint, p_i and dq_i in BS quantization are c-numbers not operators. These c-numbers must actually be derived from the Schrödinger wavefunctions, adiabatically, i.e., as diagonal matrix element of the momentum operator, $i\hbar\nabla_q$ when operating on the eigenvector $|n, q\rangle$,

$$\begin{aligned} \frac{1}{2\pi\hbar}\oint p_i dq_i &= \frac{1}{2\pi\hbar}\oint \langle n, q| (i\hbar\nabla_{q_i}) |n, q\rangle \, dq_i, \\ &= \frac{1}{2\pi\hbar}(i\hbar)\oint \langle n, q| \nabla_{q_i} |n, q\rangle \, dq_i. \end{aligned} \tag{13.4}$$

Thus,

$$\begin{aligned} \frac{1}{2\pi\hbar}\oint p_i dq_i &= \frac{i}{2\pi}\oint \langle n, q| \nabla_{q_i} |n, q\rangle \, dq_i, \\ &= \frac{1}{2\pi}\oint d\phi = n, \quad n = 1, 2, 3 \ldots, \end{aligned} \tag{13.5}$$

where we made use of the result of the change of phase in parallel transport of Schrödinger wavefunction in q-space,

$$d\phi = i\langle \alpha, q| \frac{\partial}{\partial q} |\alpha, q\rangle \cdot dq, \tag{13.6}$$

where $n \Longrightarrow \alpha$ denotes the discrete energy level.[1] Furthermore,

$$\nabla\phi = \frac{1}{\hbar}\langle n, q| i\hbar\nabla_q |n, q\rangle = \frac{\vec{p}}{\hbar} \equiv: \frac{e}{\hbar c}\vec{A}, \tag{13.7}$$

[1]Indeed, if $|n, q\rangle = Ce^{-\frac{ip\cdot q}{\hbar}}$ are eigenfunction of momentum operator, such as for free electron Hamiltonian, then we get exactly the c-number p, the independent conjugate dynamical variable as proportional to the connection. BS quantization makes sense by space compactification such as a particle in a box. It is this strictly momentum viewpoint in the literature that when naively applied to general cases makes the BS condition valid only as a limiting semiclassical approximation in phase-space, when in fact it has more general validity as demonstrated here. See also [83].

where we defined $\vec{A}$ as acting like a vector potential. In terms of the introduced "vector potential", $\vec{A}$, the BS quantization condition now reads

$$\frac{1}{2\pi\hbar}\oint \vec{p}\cdot d\vec{q} = \frac{1}{2\pi}\oint \frac{e}{\hbar c}\vec{A}\cdot d\vec{q},$$
$$= \frac{1}{2\pi}\oint \nabla_{\vec{q}}\phi\left(\vec{q}\right)\cdot d\vec{q} = n. \tag{13.8}$$

Therefore, we arrive at the B–S quantization rules for closed orbits purely from a geometric considerations, where $\langle n,q|\,\nabla_{q_i}\,|n,q\rangle \equiv: \frac{e}{c}\vec{A}$ is the Berry connection. It does appear that the BS quantization rules have an exact geometric origin, as well as confirming the Planck's discretization of phase space in units of Planck's constant h. Therefore, it is expected that BS condition will play an important role in TBT, as will be shown here. In what follows we give various examples where the BS geometric quantization condition enters in the analyses.

If one accounts for vacuum fluctuations due to uncertainty principle, then the integer $n \Longrightarrow n+\frac{1}{2}$ since $\frac{1}{2}\hbar$ is the minimal vacuum fluctuation of the action.

13.2 The BS condition and quantization of angular momentum

An immediate case of a "closed orbits" deal with either orbital or spin angular momentum. The Schrödinger equation for angular momentum about the z-axis is given by

$$i\hbar\frac{\partial}{\partial\phi}\left|\psi\left(\phi\right)\right\rangle = \hat{J}_z\left|\psi\left(\phi\right)\right\rangle, \tag{13.9}$$

where $\hat{J}_z$ is the angular momentum operator for the z-component. Therefore, we obtain the BS condition as

$$\hbar\oint\left\langle\psi\left(\phi\right)\right|i\frac{\partial}{\partial\phi}\left|\psi\left(\phi\right)\right\rangle d\phi = \hbar\oint\nabla\theta\left(\phi\right)d\phi = \hbar 2\pi m,$$
$$\oint\left\langle\psi\left(\phi\right)\right|\hat{J}_z\left|\psi\left(\phi\right)\right\rangle d\phi = J_z\oint d\phi = 2\pi J_z$$
$$= 2\pi\hbar m. \tag{13.10}$$

Therefore the eigenvalues of the observable component, $J_z = m\hbar \le \pm|\vec{J}|$, $(m = \pm1, \pm2, 3, \ldots \le \pm|\vec{J}|)$, i.e., quantized in units of Planck action. This BS quantization condition has been used by Haldane [84] in his theory of a new state of matter now known as the Haldane phase.

13.3 The BS phase factor and Schrödinger equation

The Berry phase and Berry curvature was originally derived by Berry [67] using the Schrödinger equation whose Hamiltonian depends on a time-dependent parameter, $R(t)$. This specifically made clear in the case of the Born–Oppenheimer approximation in molecular physics, where the Berry connection exactly acts like a vector potential in the effective Schrödinger equation for the slow nuclear variable [12]. Therefore, the wavefunction evolves both in time and parameter space. However, because of the presence of the dynamical phase factor in the wavefunction, the BS phase factor is not the only phase factor and hence the BS quantization condition, on grounds of wavefunction uniqueness, cannot be implemented. This is left as integral around a contour in parameter R-space, under time *compactification* (i.e., 'time-Brillouin zone' or 'temporal box': $t = 0$ to T) of steady-state condition. We have

$$|\psi(T)\rangle = \exp\left\{\frac{-i}{\hbar}\int_0^T dtE_n(R(t))\right\}\exp\{i\gamma_n(C)\}|\psi(R(0))\rangle, \qquad (13.11)$$

where the BS phase or geometrical phase factor or Berry connection is given by

$$\gamma_n(C) = \oint_C \langle n, R| i\frac{\partial}{\partial \vec{R}} |n, R\rangle \cdot d\vec{R}. \qquad (13.12)$$

13.4 The BS condition and Landau-level degeneracy

Indeed, the quantization condition in terms of the vector potential, Eq. (13.8) materializes in the method of counting of Planck states in phase space for the Landau circular orbits. This is the calculation of the LL degeneracy, LL_D. Classically this may be approximated by

$$LL_D = \frac{A}{\pi r^2} = \frac{\pi R^2}{\pi r^2} = \frac{R^2}{r^2}, \qquad (13.13)$$

where r is the classical radius of Landau orbits in a uniform magnetic fields and the system area is given by $A = \pi R^2$. Equation (13.13) has the dimensional units of total flux divided by the dimensional units of "quantum" flux, $\frac{(\pi R^2)B}{\pi\left(\frac{EL}{e}\right)} \Longrightarrow\sim \frac{\Phi}{\phi_o}$, where EL has the dimension of energy multiplied by the length measure.

Quantum mechanically, the more accurate expression for the degeneracy is $\frac{\Phi}{\frac{2\pi\hbar c}{e}} = \frac{\Phi}{\phi_o}$. This can be inferred simply by counting of Planck states ("pixel" of action) in phase space using Berry's curvature and connection, i.e., magnetic field and vector potential, respectively. In Gaussian units, we have,

$$\begin{aligned}
LL_D &= \frac{1}{2\pi\hbar}\iint \vec{\nabla}\times\vec{K}\cdot d\vec{a}, \\
&= \frac{1}{2\pi\hbar}\iint \vec{\nabla}\times\frac{e}{c}\vec{A}\cdot d\vec{a}, \\
&= \frac{1}{2\pi\hbar}\frac{|e|}{c}\iint \vec{B}\cdot d\vec{a}, \\
&= \frac{\Phi}{\phi_o}\,\epsilon\,\mathbb{Z}, \\
&= \frac{1}{2\pi\hbar}\oint \frac{e}{c}\vec{A}\cdot d\vec{q} = n\,\epsilon\,\mathbb{Z},
\end{aligned} \tag{13.14}$$

where Φ is the total magnetic flux, and ϕ_o is the quantum flux $= \frac{hc}{|e|}$. This is a realization of the BS quantization condition given in Eq. (13.8). The resulting quantization of orbital motion leads to edge states and integer quantum Hall effect under uniform magnetic fields. The general analysis of edge states marks the works of Laughlin [71], and Halperin [72].

13.5 Peierls phase factor and Wilson loops

The above argument on confined or bounded motion can be made precise by noting that localized wavefunctions in energy-band theory with vector potential, either real electromagnetic or Berry connection, always carry the so-called Peierls phase factor. This is well-known since the early days of solid-state physics. Thus, "bringing" (i.e., using *magnetic translation operator* defined below) a localized wavefunction around a closed loop (in modern terminology the so-called Wilson loop or plaquette) would acquire a total phase factor determined by Eq. (13.8),

$$\exp\left(\frac{i}{\hbar}\oint_C \frac{e}{c}A\cdot dq\right) = 1, \tag{13.15}$$

by virtue of uniqueness. With the contour enclosing a magnetic flux and dividing the closed trajectory into halves with each half endowed with opposite chirality, one end up with Aharonov–Bohm effect.

For a Wilson loop, this means that for an electromagnetic vector potential, we have,

$$\frac{i}{\hbar}\oint_C \frac{e}{c} A \cdot dq = 2\pi n, \; n \in \mathbb{Z}, \tag{13.16}$$

$$\frac{i}{2\pi\hbar}\frac{e}{c}\iint \nabla \times A \cdot d\vec{a} = n, \; n \in \mathbb{Z},$$

$$= \frac{\Phi}{\phi_o} \in \mathbb{Z}, \tag{13.17}$$

which again follows from the BS quantization condition.

We note that the Landau orbits are confined in phase space and the amount of "h^3-voxels" or more appropriately the number of h-pixels of phase space enclosed by the orbits in units of Planck's constant is quantized, i.e., $\in \mathbb{Z}$. We can even say that any oscillatory and harmonic motion in phase space entails some interdependence of canonical coordinates so as to enclose an integral number of Planck's states, i.e., h-pixels. In the case of magnetic fields, this translates to the number of quantum fluxes. The minimal coupling of canonical momentum under a uniform magnetic field provides the natural interdependence of the canonical conjugate variables, where the dependence on the coordinates is only provided by the vector potential.

13.6 BS quantization of magnetic charge of a Dirac monopole

The classical Maxwell's equations are

$$\nabla \cdot E = \rho_E, \quad \nabla \times E = -\frac{\partial B}{\partial t}, \tag{13.18}$$

$$\nabla \cdot B = 0, \quad \nabla \times B = \frac{\partial E}{\partial t} + j_E. \tag{13.19}$$

If Dirac magnetic monopole exists, giving a magnetic charge density, ρ_m, then we have a complete duality of electric and magnetic fields,

$$\nabla \cdot E = \rho_E, \quad \nabla \times E = -\frac{\partial B}{\partial t} - j_m, \tag{13.20}$$

$$\nabla \cdot B = \rho_m, \quad \nabla \times B = \frac{\partial E}{\partial t} + j_E. \tag{13.21}$$

The electric-magnetic duality is characterized by the following replacements:

$$E \Longrightarrow B, \quad \rho_E \Longrightarrow \rho_m, \quad j_E \Longrightarrow j_m, \tag{13.22}$$

$$B \Longrightarrow -E, \quad \rho_m \Longrightarrow -\rho_E, \quad j_m \Longrightarrow -j_E. \tag{13.23}$$

The four equations in Eq. (13.21) is condensed into two using complex fields,

$$\nabla \cdot (E + iB) = (\rho_E + i\rho_m), \tag{13.24}$$

$$\nabla \times (E + iB) = i\frac{\partial}{\partial t}(E + iB) + i(j_E + ij_m), \tag{13.25}$$

which manifest additional symmetry of the electric-magnetic duality. Although, classical arguments persist against the existence of magnetic monopole, quantum theory is more favorable of its existence when viewed in small region in space, as will be shown below.

Here, we can continue to make use of the Peierls phase-factor arguments similar to that used by Jackiw [85] to deduce the existence of Dirac magnetic monopole, instead of the use of semi-infinitely-long/thin Dirac string originally employed by Dirac [68]. This done by creating a tetrahedron or bounded 3D domain using the magnetic translation operator to define the surfaces of the bounded region. The corners is served by localized functions, represented by the magnetic Wannier functions, translated between lattice sites to define a tetrahedron-enclosed region in space.

For our purpose, it is convenient to formalize the magnetic translation operator in the presence of vector potential defined by

$$T(\vec{q}) = \exp\left[\frac{i}{\hbar}\left(\frac{\hbar}{i}\nabla_{\vec{r}} + \frac{e}{c}\vec{A}(\vec{r})\right).\vec{q}\right], \tag{13.26}$$

where $\vec{A}(\vec{r})$ is the vector potential, chosen in symmetric gauge, $\vec{A}(\vec{r}) = \frac{1}{2}\vec{B} \times \vec{r}$, and $\vec{q}$ is the crystal lattice vector.[2] $T(\vec{q})$ generates all the magnetic Wannier functions belonging to a band index λ from a given Wannier function centered at the origin. Therefore, operating on the lattice-position eigenvector centered at the origin, we have

$$T(\vec{q})\,|0, \lambda\rangle = |\vec{q}, \lambda\rangle\,. \tag{13.27}$$

Consider

$$T(-\vec{q}_1) = \exp\left[-\frac{i}{\hbar}\left(\frac{\hbar}{i}\nabla_{\vec{r}} + \frac{e}{c}\vec{A}(\vec{r})\right).\vec{q}_1\right]. \tag{13.28}$$

Using the Baker–Campbell–Hausdorff (BCH) formula, we have,

$$\exp\left[-\frac{i}{\hbar}\left(\frac{\hbar}{i}\nabla_{\vec{r}} + \frac{e}{c}\vec{A}(\vec{r})\right).\vec{q}_1\right] = \exp\left[-\frac{i}{\hbar}\left(\frac{e}{c}\vec{A}(\vec{r})\right).\vec{q}_1\right]\exp\left[-\frac{i}{\hbar}\left(\frac{\hbar}{i}\nabla_{\vec{r}}\right).\vec{q}_1\right]. \tag{13.29}$$

[2]The continuum limit of all the results can be taken in the end.

Using the localized wavefunction, $w_\lambda(r,0)$, this means that,

$$T(-\vec{q}_1)\, w_\lambda(r,0) = \exp\left(-\frac{i}{\hbar}\frac{e}{c}\vec{A}(\vec{r})\,.\vec{q}_1\right) w_\lambda(\vec{r}-\vec{q}_1), \qquad (13.30)$$

where now the displaced localized wavefunction is now *centered* at the lattice point $\vec{q}$. We define $\tilde{w}_\lambda(\vec{r}-\vec{q})$ as the magnetic Wannier function centered at $\vec{q}$. Because we are translating a localized Wannier function, we need to perform negative lattice vector shift to obtain a Wannier function centered in a new resultant vector lattice point. The exponential factor of Eq. (13.30) is the so-called *Peierls phase factor*.

The $T(\vec{q})$-algebra is determined from BCH formula,

$$\begin{aligned}T(-\vec{q}_1)\,T(-\vec{q}_2) &= \exp\left[-\frac{i}{\hbar}\left(\frac{\hbar}{i}\nabla_{\vec{r}} + \frac{e}{c}\vec{A}(\vec{r})\right).\vec{q}_1\right]\\ &\quad\times\exp\left[-\frac{i}{\hbar}\left(\frac{\hbar}{i}\nabla_{\vec{r}} + \frac{e}{c}\vec{A}(\vec{r})\right).\vec{q}_2\right],\\ &= \exp\left(-\frac{ie}{\hbar c}\vec{A}(\vec{q}_1)\,.\vec{q}_2\right) T(-(\vec{q}_1+\vec{q}_2)),\\ &= \exp\left(-\frac{ie}{\hbar c}\vec{B}.\frac{1}{2}(\vec{q}_1\times\vec{q}_2)\right) T(-(\vec{q}_1+\vec{q}_2)). \qquad (13.31)\end{aligned}$$

This algebra thus enable us to move the localized Wannier function, carrying *Peierls phase factor*, in different lattice sites. Therefore, consider additional magnetic translation by $\vec{q}_3$ We obtain,

$$\begin{aligned}&[(T(-\vec{q}_1)\,T(-\vec{q}_2))]\,T(-\vec{q}_3)\\ &\quad= \exp\left(-\frac{ie}{\hbar c}\vec{B}.\frac{1}{2}(\vec{q}_1\times\vec{q}_2)\right)\exp\left(-\frac{ie}{\hbar c}\vec{B}.\frac{1}{2}(\vec{q}_1+\vec{q}_2)\times\vec{q}_3\right)\\ &\qquad\times T(-(\vec{q}_1+\vec{q}_2+\vec{q}_3)),\\ &\quad= \exp\left(-\frac{ie}{\hbar c}\vec{B}.\left[\frac{1}{2}(\vec{q}_1\times\vec{q}_2)+\frac{1}{2}(\vec{q}_1+\vec{q}_2)\times\vec{q}_3\right]\right)\\ &\qquad\times T(-(\vec{q}_1+\vec{q}_2+\vec{q}_3)). \qquad (13.32)\end{aligned}$$

On the other hand, using a different bilinear grouping, we can have

$$\begin{aligned}&T(-\vec{q}_1)\,[T(-\vec{q}_2)\,T(-\vec{q}_3)]\\ &\quad= \exp\left(-\frac{ie}{\hbar c}\phi_{23}\right)\exp\left(-\frac{ie}{\hbar c}\phi_{1(2+3)}\right) T(-(\vec{q}_1+\vec{q}_2+\vec{q}_3)). \qquad (13.33)\end{aligned}$$

By virtue of the associative properties of the quantum operators $T(\vec{q})$, The RHS of Eq. (13.32) must be equal to the RHS of Eq. (13.33). We obtain four surfaces enclosing a tetrahedron defined by the phase,

$$T\left(-\left(\vec{q}_1+\vec{q}_2+\vec{q}_3\right)\right)$$
$$=\exp\left(-\frac{ie}{\hbar c}\left[\phi_{12}+\phi_{(1+2)3}-\phi_{23}+\phi_{1(2+3)}\right]\right)T\left(-\left(\vec{q}_1+\vec{q}_2+\vec{q}_3\right)\right), \tag{13.34}$$

where $\left[\phi_{12}+\phi_{(1+2)3}-\phi_{23}+\phi_{1(2+3)}\right]=\Phi$ is the total magnetic flux emerging out of bounded domain enclosed by the surfaces formed by tetrahedron. Writing explicitly the total flux, Φ, emerging out of tetrahedron,

$$\Phi=\vec{B}\cdot\frac{1}{2}\left[\vec{q}_2\times\vec{q}_1+\left(\vec{q}_3\times\left(\vec{q}_1+\vec{q}_2\right)\right)+\left(\vec{q}_2\times\vec{q}_3\right)+\left(\vec{q}_1\times\left(\vec{q}_2+\vec{q}_3\right)\right)\right],$$
$$=\vec{B}\cdot\left[\frac{1}{2}\begin{array}{l}\left(\vec{q}_1+\vec{q}_2\right)\times\vec{q}_1+\left(\left(\vec{q}_1+\vec{q}_2+\vec{q}_3\right)\times\left(\vec{q}_1+\vec{q}_2\right)\right)\\ +\left(\vec{q}_2\times\left(\vec{q}_2+\vec{q}_3\right)\right)+\left(\vec{q}_1\times\left(\vec{q}_1+\vec{q}_2+\vec{q}_3\right)\right)\end{array}\right]. \tag{13.35}$$

This means that the phase of Eq. (13.34) is zero, as expected since application of the resulting total magnetic translation, $T\left(-\vec{q}_1\right)T\left(-\vec{q}_2\right)T\left(-\vec{q}_3\right)$ applied to a localized Wannier function, $w\left(\vec{r},o\right)$, should be equivalent to

$$T\left(-\left(\vec{q}_1+\vec{q}_2+\vec{q}_3\right)\right)w\left(\vec{r},o\right)$$
$$=\exp\left(-\frac{i}{\hbar}\frac{e}{c}A\left(\vec{r}\right).\left(\vec{q}_1+\vec{q}_2+\vec{q}_3\right)\right)w\left(\vec{r}-\left(\vec{q}_1+\vec{q}_2+\vec{q}_3\right)\right),$$
$$=\exp\left(-\frac{i}{\hbar}\frac{e}{c}A\left(\vec{r}\right).\vec{q}\right)w\left(\vec{r}-\vec{q}\right), \tag{13.36}$$

where $\vec{q}\equiv\left(\vec{q}_1+\vec{q}_2+\vec{q}_3\right)$ in the last line consistent with Eq. (13.30).

The total phase of Eq. (13.34) suggest the following integrals:

$$\frac{ie}{\hbar c}\Phi\Longrightarrow\frac{ie}{\hbar c}\oint_S\vec{B}\cdot d\vec{s},$$
$$=\frac{ie}{\hbar c}\iiint_{V_{\text{tetrahedron}}}dv\,\vec{\nabla}\cdot\vec{B}, \tag{13.37}$$

where $d\vec{s}$ denotes the incremental surface in the r-coordinates (not the lattice coordinates). Since $\vec{\nabla}\cdot\vec{B}=0$ then $\Phi=0$. However, there is another solution for the phase, which can be written in detail as

$$\frac{ie}{\hbar c}\Phi=\frac{ie}{\hbar c}\vec{B}\cdot\left[\begin{array}{l}\left(\vec{q}_1+\vec{q}_2\right)\times\vec{q}_1+\left(\left(\vec{q}_1+\vec{q}_2+\vec{q}_3\right)\times\left(\vec{q}_1+\vec{q}_2\right)\right)\\ +\left(\vec{q}_2\times\left(\vec{q}_2+\vec{q}_3\right)\right)+\left(\vec{q}_1\times\left(\vec{q}_1+\vec{q}_2+\vec{q}_3\right)\right)\end{array}\right],$$
$$=i2\pi n, \tag{13.38}$$

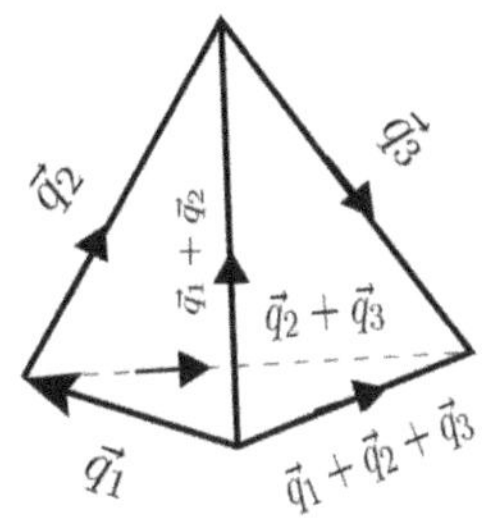

Fig. 13.1. The tetrahedron through which the emerging magnetic fluxes are determined, demanding a source of magnetic flux within the enclosed domain if the total flux do not add to zero.

which clearly exhibit an expression with all the magnetic fluxes emerging from the tetrahedron, see Fig. 13.1. Without affecting the validity of Eq. (13.36), the total phase of Eq. (13.34) for all emerging magnetic flux of Eq. (13.38) must satisfy the following expression

$$\begin{aligned} \frac{ie}{\hbar c}\Phi &= \frac{ie}{\hbar c}\oint_S \vec{B}\cdot d\vec{s} = \frac{ie}{\hbar c}\iiint_{V_{\text{tetrahedron}}} dv\vec{\nabla}\cdot\vec{B}, \\ &= i2\pi n,\ n \in \mathbb{Z},\quad n = 0, \pm 1, \pm 1, \ldots, \end{aligned} \tag{13.39}$$

if there is a magnetic charge e_m, "+" for monopoles and "−" for anti-monopoles. The magnetic field $\vec{B}$ is such that

$$\vec{B} = e_m \frac{\vec{r}}{r^3}, \tag{13.40}$$

$$\vec{\nabla}\cdot\vec{B} = 4\pi e_m \delta^3(\vec{r}) \Longrightarrow \iiint_{V_{\text{tetrahedron}}} dv\vec{\nabla}\cdot\vec{B} = 4\pi e_m, \tag{13.41}$$

$$\nabla^2 \phi_m = 4\pi e_m \delta^3(\vec{r}), \tag{13.42}$$

then it follows that the magnetic charge, e_m, is quantized,

$$\frac{ie}{\hbar c}\iiint_{V_{\text{tetrahedron}}} dv\ \vec{\nabla}\cdot\vec{B} = \frac{ie}{\hbar c}4\pi e_m = i2\pi n, \tag{13.43}$$

which yields the Dirac magnetic monopole charge,

$$e_m = n\left(\frac{\hbar c}{2e}\right) = \frac{1}{4\pi}n\phi_o. \tag{13.44}$$

Equation (13.44) was originally given by Dirac by his thought experiment of semi-infinite thin magnetic string, where $\left(\frac{\hbar c}{2e}\right) = \frac{e}{2\alpha} = \frac{137}{2}e$ (α is the fine structure constant $= \frac{e^2}{\hbar c}$) is the unit magnetic charge. Now therefore,

$$\iiint_{V_{\text{tetrahedron}}} dv\, \vec{\nabla}\cdot\vec{B} = 0, \quad \text{and} \quad \iiint_{V_{\text{tetrahedron}}} dv\, \vec{\nabla}\cdot\vec{B} \neq 0, \tag{13.45}$$

would be physically consistent if $\vec{\nabla}\cdot\vec{B} = 0$ outside a singularity and $\vec{\nabla}\cdot\vec{B} = e_m$ only at the core. This allows us to propose that

$$\vec{B} = e_m \frac{\vec{r}}{r^3}, \tag{13.46}$$

$$\vec{\nabla}\cdot\vec{B} = 4\pi e_m \delta^3(\vec{r}), \tag{13.47}$$

$$\nabla^2 \phi_m = 4\pi e_m \delta^3(\vec{r}), \tag{13.48}$$

where ϕ_m is the magnetic monopole potential. $\vec{\nabla}\cdot\vec{B} = \nabla^2\phi_m = 0$ if the singular point $\vec{r} = 0$ is excluded. We have

$$\iiint_{\delta V_{\text{core}}} dv \vec{\nabla}\cdot\vec{B} = n\left(\frac{e}{2\alpha}\right), \tag{13.49}$$

where δV_{core} is an infinitesimal domain containing the singularity at $\vec{r} = 0$ or the quantum flux is emerging through a spherical surface surrounding the point of singularity. Clearly a magnetic monopole is *infinitely localized* particle and therefore must be hugely massive. Thus, it requires very high-energy not currently readily available in experimental physics to detect it. Moreover, it may always occur in pairs, namely, monopole and anti-monopole pair (i.e., north and south pole, similar to vortex anti-vortex pairs of the X–Y model of spin systems) of infinitesimal size, which exacerbate detection, and conspire to evade experimental detection as magnetic dipoles.

13.7 The X–Y model of spin systems and BS condition

The X–Y model is a sort of a generalization of the Ising model, i.e., instead of discrete ± 1 spin value one place a spin rotor at each site which can point in any direction in a 2D plane. There is a theorem based on Landau symmetry breaking arguments that such 2D systems are not expected to exhibit long-range order due to transverse fluctuations [76]. However, this 2D system possess quasi-long-range order in finite-size systems at very low temperatures. This is the so-called Berezinskii–Kosterlitz–Thouless (BKT) transition [86, 87], marked by the occurrence of bound vortex–antivortex pairs at low temperatures to unbound or isolated vortices and antivortices above some critical temperatures.

In 1972, Kosterlitz and Thouless (KT) made a complete identification of a new type of phase transition in 2D systems where topological defects in the form of vortices and antivortices play a crucial role. For their work they were awarded the Nobel Prize in 2016. Here we are mainly concern on how the BS condition governs the nature of vortices of the X–Y model of spin systems. After the work of KT strong interest toward phase transition without symmetry breaking become mainstream. It was realized in passing that long before, the regular liquid–gas transition does not break symmetry, this was recognized without given much significance. Later, Polyakov extended the work of KT to gauge theories, for example, 2 + 1 "compact" QED has a gapped spectrum in the IR due to topological excitations [88]. The SU(N) Thirring model has a fermions condensing with finite mass in the IR without breaking the chiral symmetry of the theory [89]. It was also further extended in the context of 2D melting of crystalline solids [90] leading to a new liquid crystalline hexatic phase. Another violation of Landau symmetry breaking arguments is exemplified by the Haldane phase transition [84].

13.7.1 *The Hamiltonian for the X–Y model of spin system*

The X–Y model is a generalization of the Ising model, where Ising spins $\sigma = \pm 1$ are replaced by planar vector rotors of unit length, which can point in an arbitrary direction within the X–Y plane,

$$\vec{n} = \begin{pmatrix} \cos\varphi \\ \sin\varphi \end{pmatrix} \Longrightarrow |\vec{n}| = 1. \tag{13.50}$$

The Hamiltonian is given by

$$\begin{aligned} H &= -J\sum_{\langle i,j\rangle} \vec{n}_i \cdot \vec{n}_j, \\ &= -J\sum_{\langle i,j\rangle} \cos(\vartheta_i)\cos(\vartheta_j) + \sin(\vartheta_i)\sin(\vartheta_j), \\ &= -J\sum_{\langle i,j\rangle} \cos(\vartheta_i - \vartheta_j). \end{aligned} \tag{13.51}$$

The cosine function can be expanded in powers of $(\vartheta_i - \vartheta_j)$,

$$H = -J\sum_{\langle i,j\rangle} \left[1 - \frac{1}{2}(\vartheta_i - \vartheta_j)^2 + O\left[(\vartheta_i - \vartheta_j)^4\right]\cdots\right]. \tag{13.52}$$

In the continuum limit, Eq. (13.52) leads to the following Hamiltonian studied by BKT:

$$H \simeq -2JN + \frac{J}{2}\int d^2r\,|\nabla\theta(r)|^2, \tag{13.53}$$

where $-2JN \equiv: E_o$ is the energy of the system when all N spin rotors are aligned, and $\theta(r)$ labels the angle of the rotors at each point in the X–Y plane. The partition function in this continuum limit must account for all possible configurational function $\{\theta(r)\}$. We are thus lead to a *functional integral* of the partition function, Z,

$$Z \Longrightarrow \int D[\theta(r)]\exp\left[-\beta\left(E_o + \frac{J}{2}\int d^2r\,|\nabla\theta(r)|^2\right)\right]. \tag{13.54}$$

Remark 13.1. The X–Y model of spin system also serves as a beautiful model of Berry connection and Berry curvature in a more explicit and much simpler form. This naturally leads us to the BS quantization condition for the vortex solutions given by KBT. It is therefore expected that this model will have an impact not only in statistical physics but also in quantum field theory and subject to generalizations. For example, if we write the unit spin vector by employing the Dirac ket and bra (bra–ket) notations, we have

$$|\vec{n}\rangle = e^{-i\theta(r)},$$

then

$$\langle\vec{n}|\, i\frac{\partial}{\partial r}\,|\vec{n}\rangle = \nabla\theta(r),$$

which is the Berry connection. This site connection is indeed better appreciated by the way the sites are coupled in Eq. (13.51). This remark serves as advanced view, as it relates to the BS quantization, of the discussions that follow.

13.7.1.1 *Vortices as solutions*

To simplify the calculation of the functional integral of Eq. (13.54), we use perturbative techniques which allows us to make use of the saddle point approximation. This entails expansion of the functional in terms of small fluctuations, $\delta\theta$, around the minimum of H at θ_o. Therefore, we need

$$\left.\frac{\delta H[\theta]}{\delta\theta(r)}\right|_{\theta=\theta_o} = 0. \tag{13.55}$$

Then the field configurations are approximated, $\theta(r) \simeq \theta_o + \delta\theta(r)$. We have for the partition function,

$$Z \simeq e^{-\beta E_o} \sum_{\theta_o} \int D[\delta\theta] \times \exp\left\{-\beta\left(\begin{array}{c} H[\theta_o] \\ +\frac{1}{2}\iint dr_1 dr_2 \frac{\delta^2 H[\theta]}{\delta\theta(r_1)\,\delta\theta(r_2)} \delta\theta(r_1)\,\delta\theta(r_2) \end{array}\right)\right\}. \tag{13.56}$$

Here, it is assumed that $H[\theta_o]$ dominates and the rest are fluctuations. With $H[\theta]$ given by

$$H[\theta] \equiv \frac{J}{2}\int d^2r\, |\nabla\theta(r)|^2. \tag{13.57}$$

The minimum is determined from the solutions of Eq. (13.55), which correspond to solving the following equation:

$$\nabla^2\theta(r) = 0. \tag{13.58}$$

Although the analysis of the solutions to Eq. (13.58) proceeds classically, it has accurate analogy to quantum physics and contains the elements of Berry phase, Berry curvature and BS quantization condition. Indeed, the X–Y model has some resemblance to the potential flow in 2D hydrodynamics.[3] There, the Laplace equation, $\nabla^2\varphi(x) = 0$, is thoroughly analyzed for vortex solutions. Moreover, the X–Y model is reminiscent of the harmonic oscillator which also serves as a perfect classical forebear of the annihilation and creation operator, or ladder operator formalism, in quantum mechanics.

The two types of solutions to $\nabla^2\theta(r) = 0$,

$$\text{(i) } \theta(r) = \text{constant} \Longrightarrow \text{ ground state} \Longrightarrow E_o, \tag{13.59}$$

$$\text{(ii) } \theta(r) = \text{vortex solutions.} \tag{13.60}$$

The vortex solution is given by

$$\theta_\pm = \pm\arctan\left(\frac{y-b}{x-a}\right). \tag{13.61}$$

[3]The X–Y model has some resemblance to the potential flow in 2D hydrodynamics. There, the Laplace equation, $\nabla^2\varphi(x) = 0$, is thoroughly analyzed for vortex solutions and the boundary condition given is the so-called *circulation integral.*

Let

$$\theta_\pm = \pm \arctan w, \tag{13.62}$$

$$\theta'_\pm = \pm \frac{w'}{1+w^2}. \tag{13.63}$$

Then we have,

$$1 + w^2 = 1 + \left(\frac{y-b}{x-a}\right)^2 = \frac{1}{(x-a)^2}((x-a)^2 + (y-b)^2), \tag{13.64}$$

$$\frac{\partial w}{\partial x} = \frac{\partial}{\partial x}\left(\frac{y-b}{x-a}\right) = (y-b)\frac{\partial}{\partial x}\left(\frac{1}{x-a}\right) = -\frac{(y-b)}{(x-a)^2}, \tag{13.65}$$

$$\frac{\partial w}{\partial y} = \frac{\partial}{\partial y}\left(\frac{y-b}{x-a}\right) = \frac{1}{x-a}\frac{\partial}{\partial y}((y-b)) = \frac{1}{(x-a)}. \tag{13.66}$$

The circulation velocities are given by $v = (u, v)$

$$u = \frac{-\frac{(y-b)}{(x-a)^2}}{\frac{1}{(x-a)^2}\left((x-a)^2 + (y-b)^2\right)} = -\frac{(y-b)}{\left((x-a)^2 + (y-b)^2\right)}, \tag{13.67}$$

$$v = \frac{\frac{1}{(x-a)}}{\frac{1}{(x-a)^2}\left((x-a)^2 + (y-b)^2\right)} = \frac{x-a}{\left((x-a)^2 + (y-b)^2\right)}. \tag{13.68}$$

These results are depicted in Figs. 13.2 and 13.3. The vortex solution θ_+ corresponds to a vortex charge $+1$, and θ_- corresponds to antivortex charge -1. Both θ_+ and θ_- are singular at $(x, y) = (a, b)$.

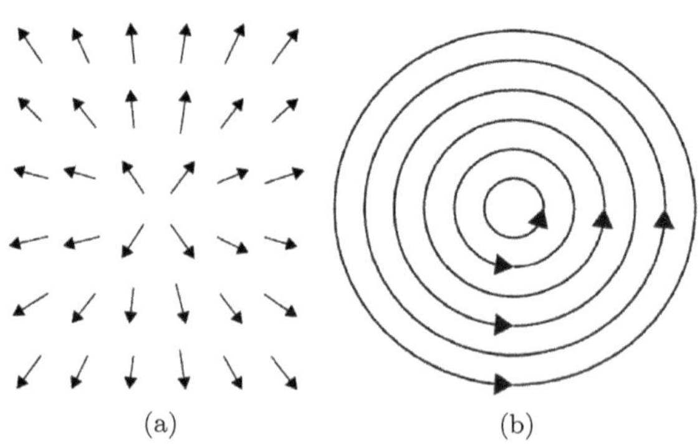

Fig. 13.2. (a) Spin configurations for the vortex θ_+. (b) The vortex's gradient, $\nabla\theta_+$. The contour integral of θ_+ is quantized by virtue of the BS quantization condition.

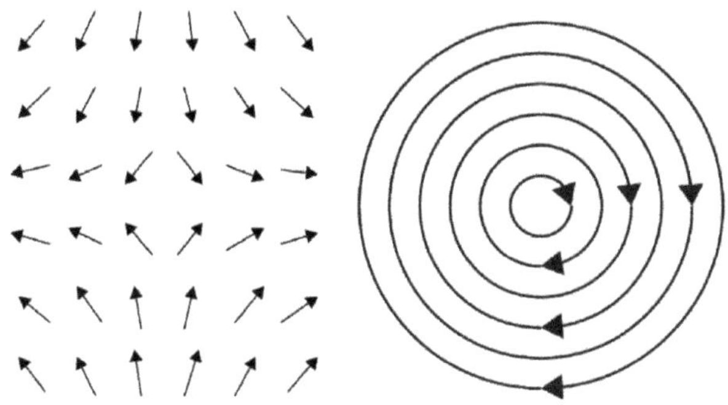

Fig. 13.3. (a) Spin configurations for the antivortex θ_-. (b) The antivortex's gradient, $\nabla\theta_-$. The contour integral of $\nabla\theta_-$ is quantized by virtue of the BS quantization condition.

13.7.1.2 *BS quantization of X–Y model vortex charges*

For the X–Y model of spin system, the quantization of vortex charge, in contrast to the magnetic monopole charge, usually proceeds classically in a form of a simple *boundary condition*, the so-called *circulation integral* (*borrowed from 2D hydrodynamics*) belonging to the domain of integers, $\mathbb{Z}$. Since the existence of a vortex is based on the existence of a singularity, the vortex charge can be singly charged, ± 1 or multiply charged, $\pm m$ reminiscent of the Dirac magnetic monopole. The boundary condition for vortex solution is given by Eq. (13.60). The condition, $\oint_C \vec{\nabla}\theta \cdot d\vec{l} = 2\pi m$ basically follows from the single valuedness of $\vec{n} = \begin{pmatrix} \cos\theta \\ \sin\theta \end{pmatrix}$. The number $m \in Z$ is also called the winding number or vortex charge (by identifying with the potential problem in electrostatics,

$$\int_V \vec{\nabla}\cdot\vec{\nabla}\theta = \int_V \vec{\nabla}\cdot\vec{E} = \int \vec{E}\cdot\vec{n}\, dS, \tag{13.69}$$

which is equal to the electric charge enclosed by the surface). Mathematically speaking, here our order parameter manifold is $S^1 \Longrightarrow \pi_1\left(S^1\right) = Z$.

Only when there is a singularity would there be a point charge, or vortex core in the present instance given by

$$\frac{1}{2\pi}\oint_{\partial S} \vec{\nabla}\theta \cdot d\vec{l} = \frac{1}{2\pi}\int_S \left(\vec{\nabla}\times\vec{\nabla}\theta\right)\cdot dS = m, \tag{13.70}$$

which is our BS quantization analogy for the X–Y model.

13.7.1.3 *Energetics of X–Y model*

The energy cost of a single vortex has been shown to be,

$$U_v = \pi J \ln \frac{L}{a}, \tag{13.71}$$

where L is the size of the system and a is the lattice constant. Whereas, the entropy of a vortex is given by

$$S_v = 2k_B \ln \frac{L}{a}. \tag{13.72}$$

Thus the free energy cost due to presence of vortex, which give the competition between order and disorder is

$$\begin{aligned} \Delta F &= U_v - TS_v, \\ &= (\pi J - 2k_B T) \ln \frac{L}{a}. \end{aligned} \tag{13.73}$$

The BKT transition occurs when $\Delta F \implies 0$. This give the transition temperature, $T_{\text{KBT}} = \frac{\pi J}{2k_B}$. For $T < T_{\text{KBT}}$, the system is unstable against the formation of vortex–antivortex pair. The energy of the vortex–antivortex pair is

$$E_{\text{pair}} = \pi J \ln \frac{r}{a}, \tag{13.74}$$

where r is the separation of the pair. This is much less than the energy of a single vortex in the limit of large system size. For $T > T_{\text{KBT}}$, the pairs become unbounded.

13.7.2 *BS quantization condition in Haldane phase*

It is worth mentioning that quantization of the internal precession frequency of spin-1 chain soliton studied by Haldane [84], i.e., soliton angular momentum, S_Z, follows that of Eq. (13.10). The gapped ground state of odd integer-spin chain is now called the Haldane phase, a new topological state of matter [91]. The Haldane phase, also called a symmetry protected topological state, is viewed as a short-range entangled which bears some similarity to the KT vortex solution of the X–Y model of spin. We will not go into the details of Haldane phase as this will take us far from the scope of this book.

13.7.2.1 *Haldane model*

A Chern insulator in a honeycomb lattice that exhibits quantum Hall effect [92] in the absence of external magnetic fields was proposed by Haldane [93] as a basic theoretical representation for the quantum anomalous Hall effect.

The calculation of the Berry connection relies on the solution of the eigenfunctions and eigenvalues of the tight-binding Hamiltonian of spinless honeycomb lattice. We refer the interested readers to some of the references listed in the bibliography.

13.8 BS quantization condition in superfluids

For superfluid the BS quantization condition is not implemented as a boundary condition but as a real BS quantization condition. Superfluid maybe viewed as a classical complex-valued matter field with emergent constant of motion, the topological order. Like the X–Y model, the complex matter field of Bose–Einstein condensate, $\psi(r)$, is given by

$$\psi(r) = |\psi(r)| e^{i\Phi(r)}. \tag{13.75}$$

Likewise the superfield velocity field is given by

$$v_s = \gamma \nabla \Phi, \tag{13.76}$$

where $\gamma = \frac{\hbar}{m}$ is some constant. Here, the circulation is effectively quantized using the BS quantization condition, i.e.,

$$\begin{aligned} \frac{m}{2\pi\hbar} \oint_C \vec{v}_s \cdot d\vec{l} &= \frac{\gamma m}{2\pi\hbar} \oint_C \nabla\Phi \cdot d\vec{l}, \\ &= 2\pi \frac{\hbar}{m} \frac{m}{2\pi\hbar} \times \text{integer}, \\ &= \text{integer} \in \mathbb{Z}, \end{aligned} \tag{13.77}$$

whereas the X–Y model is confined to 2D, superfluid in 3D behave more distinctly: In 3D the superfluid quantized vortices form a metastable closed ring or open chain ending at the surface. A vortex chain with both ends ending at the same surface appears as a bound vortex/anti-vortex pair at the surface.

13.8.1 *BS quantization of vortices in superconductors*

A well-known property of superconductors is that they expel magnetic fields, the so-called Meissner effect. In some cases, for sufficiently strong magnetic fields however, it will be energetically favorable for the superconductor to form a lattice of quantum vortices through the superconductor each which carry quantized magnetic flux. A superconductor that can support vortex lattices is called a type-II superconductor. Vortex-quantization in superconductors is general.

The BS quantization of vortices is a direct $U(1)$ theory of connection, gauge or vector potential. We have,

$$\frac{1}{2\pi\hbar}\frac{e}{c}\iint_S \vec{B}\cdot d\vec{S} = \frac{e}{hc}\oint_{\partial S} A\cdot dl = n, \tag{13.78}$$

$$\oint_{\partial S} A\cdot dl = n\phi_o, \tag{13.79}$$

where $\phi_o = \frac{hc}{e}$ is the quantum flux. This is reminiscent of Eq. (13.14) for the LL degeneracy values.

In the next section, we will discuss the BS quantization condition in coherent state (CS) formulation of quantum mechanics. This will introduce us to the Berry phase and Berry connection in quantum field theory [94], specifically non-Abelian gauge theory of elementary particles, which is still an active field of research. For example, in effective quantum field theory, the Berry phase is equivalent to Wess–Zumino–Witten action for the sigma model [95].

13.9 BS quantization in CS formulation of quantum mechanics

First we will treat the iconic harmonic oscillator in terms of ladder operators. This is the precursor to the CS formulation of quantum physics.

13.9.1 *Harmonic oscillator*

The Hamiltonian is,

$$H = \frac{p^2}{2m} + \frac{1}{2}\kappa q^2, \tag{13.80}$$

and the equations of motion are

$$\dot{q} = \frac{\partial H}{\partial p} = \frac{p}{m}, \quad \dot{p} = -\frac{\partial H}{\partial q} = -\kappa q = -m\omega^2 q. \tag{13.81}$$

We defined the ladder operators, i.e., annihilation and creation operators, a and $a^\dagger$ as

$$a = \sqrt{\frac{m\omega}{2\hbar}}\left(q + \frac{ip}{m\omega}\right), \tag{13.82}$$

$$a^\dagger = \sqrt{\frac{m\omega}{2\hbar}}\left(q - \frac{ip}{m\omega}\right). \tag{13.83}$$

In terms of this operators, we have,

$$\begin{aligned} H &= \left(\frac{1}{2}\kappa q^2 + \frac{p^2}{2m}\right), \\ &= \hbar\omega\left(a^\dagger a\right). \end{aligned} \tag{13.84}$$

13.9.1.1 *BS quantization for harmonic oscillator*

We will try to implement the BS quantization condition by using the momentum and position operators in terms of the ladder operators and their eigenvalues.

$$Q = \sqrt{\frac{\hbar}{2m\omega}}\left(a + a^\dagger\right), \tag{13.85}$$

$$P = -i\sqrt{\frac{\hbar m\omega}{2}}\left(a - a^\dagger\right), \tag{13.86}$$

$$[Q, P] = i\hbar, \tag{13.87}$$

$$\left[a, a^\dagger\right] = 1. \tag{13.88}$$

The BS quantization now simply reads

$$\begin{aligned} \frac{1}{2\pi\hbar}\oint PdQ &= \frac{-i}{2\pi\hbar}\oint\left(\sqrt{\frac{\hbar m\omega}{2}}\left(a - a^\dagger\right)\right) d\left(\sqrt{\frac{\hbar}{2m\omega}}\left(a + a^\dagger\right)\right), \\ &= \frac{-i}{2\pi\hbar}\frac{\hbar}{2}\oint\left(ada - a^\dagger da^\dagger\right) + \left(ada^\dagger - a^\dagger da\right), \\ &= n. \end{aligned} \tag{13.89}$$

Although apparently the first term contributes to zero being exact integral taken around the contour, there is actually an arbitrariness in the constants of integration besides zeros. Thus, the first term involving $ada - a^\dagger da^\dagger$ cannot be dismissed immediately as zero, since it has have arbitrary constants of integration.

But taking advantage of the arbitrariness in constants of integration, we may also conceive the following:

$$\begin{aligned}\int \left(ada - a^\dagger da^\dagger\right) &= \int \left(\left(\frac{a^2}{2} - i\gamma \right) - \left(\frac{\left(a^\dagger\right)^2}{2} + i\beta \right) \right), \\ &= \int \left(\frac{a^2}{2} - \frac{\left(a^\dagger\right)^2}{2} \right) - i\left(\gamma + \beta\right) \int d\phi \\ &= -i\left(\gamma + \beta\right) 2\pi. \end{aligned} \tag{13.90}$$

We may choose, $(\gamma + \beta) = 1$. Then

$$\int \left(ada - a^\dagger da^\dagger\right) = -i2\pi = -i\frac{4\pi}{2}. \tag{13.91}$$

Hence, we may have two quantization of the conjugate quantum-fields,

$$\frac{i}{4\pi} \left[\int \left(a^\dagger da - ada^\dagger \right) \right] = n, \tag{13.92}$$

and a possible quantization, by tailoring the constant of integration, with hindsight of the zero-point energy of quantum harmonic oscillator as

$$\frac{i}{4\pi} \left[\int \left(a^\dagger da - ada^\dagger \right) \right] = n + \frac{1}{2}. \tag{13.93}$$

Thus, the BS quantization leads us to the correct positively definite quantum energy levels of a harmonic oscillator,

$$E_n = \hbar\omega \left(n + \frac{1}{2} \right), \tag{13.94}$$

which includes the zero-point energy, without actually solving for the harmonic oscillator eigenfunctions.

13.9.2 *Coherent states and BS quantization*

CS is defined to be the *right* eigenstate of the annihilation operator $\hat{a}$,

$$\hat{a}\left|\alpha\right\rangle = \alpha\left|\alpha\right\rangle. \tag{13.95}$$

Since $\hat{a}$ is non-Hermitian, $\alpha = |\alpha| e^{i\theta}$ is complex, $|\alpha|$ and θ are the amplitude and phase of the eigenvalue, α. The conjugate state $\langle\alpha|$ is the *left* eigenstate of the creation operator $\hat{a}^\dagger$, this follows by taking the Hermitian conjugate of Eq. (13.95). The state $|\alpha\rangle$ is the so-called CS. The usefulness of CS is that they form a basis for the representation of other states. CS can never be made orthogonal, although for well-separated eigenvalues α, they can be made approximately orthogonal [12]. Moreover, the set of CS is overcomplete, in the sense that the set of CS form a basis but are not linearly independent, i.e., they are expressible in terms of each other, using transition function as explained in Part 1. The nice thing is that the complex eigenvalue α is labeled by the classical (average) values of position and momentum in the following sense:

$$\langle\alpha|\, \hat{a}\, |\alpha\rangle = \alpha = \sqrt{\frac{m\omega}{2\hbar}} \langle\alpha| \left(Q + \frac{i}{m\omega} P\right) |\alpha\rangle\,, \tag{13.96}$$

$$\langle\alpha|\, Q\, |\alpha\rangle = \sqrt{\frac{2\hbar}{m\omega}} \mathrm{Re}\,\alpha, \tag{13.97}$$

$$\langle\alpha|\, P\, |\alpha\rangle = \sqrt{2\hbar m\omega} \mathrm{Im}\,\alpha, \tag{13.98}$$

$$\mathrm{Re}\,\alpha = \frac{1}{\sqrt{2}} \langle\alpha|\, \tilde{Q}\, |\alpha\rangle\,, \tag{13.99}$$

$$\mathrm{Im}\,\alpha = \frac{1}{\sqrt{2}} \langle\alpha|\, \tilde{P}\, |\alpha\rangle\,, \tag{13.100}$$

$$\begin{aligned} 2h\mathrm{Im}\,\alpha \mathrm{Re}\,\alpha &= \frac{1}{2} \langle\alpha|\, \tilde{P}\, |\alpha\rangle \langle\alpha|\, \tilde{Q}\, |\alpha\rangle\,, \\ &= \langle\alpha|\, P\, |\alpha\rangle \langle\alpha|\, Q\, |\alpha\rangle\,, \end{aligned} \tag{13.101}$$

where $\tilde{Q}$ and $\tilde{P}$ are the scaled canonical operators given by

$$\tilde{Q} = \sqrt{\frac{m\omega}{\hbar}} Q, \tag{13.102}$$

$$\tilde{P} = \sqrt{\frac{1}{\hbar m\omega}} P. \tag{13.103}$$

Indeed, we have

$$\begin{aligned} \alpha &= \frac{1}{\sqrt{2}} (\langle\alpha|\, \tilde{Q}\, |\alpha\rangle + i \langle\alpha|\, \tilde{P}\, |\alpha\rangle), \\ &= \langle\alpha|\, \hat{a}\, |\alpha\rangle\,. \end{aligned} \tag{13.104}$$

The counting of α states can be derived from the counting of states in (p_c, q_c) phase-space, where $q_c = \langle\alpha|\, Q\, |\alpha\rangle$ and $p_c = \langle\alpha|\, P\, |\alpha\rangle$, namely,

$$\begin{aligned} \frac{1}{2\pi\hbar} dq_c dp_c &= \frac{1}{\pi}\, d\,\mathrm{Re}\,\alpha\, d\,\mathrm{Im}\,\alpha, \\ &\Rightarrow \frac{1}{\pi} d\alpha^2 \text{ (abbreviated).} \end{aligned} \tag{13.105}$$

Since $\langle\alpha|\, P\, |\alpha\rangle$ appears as the Berry connection, we proceed to apply the BS quantization as follows:

$$\frac{2\hbar}{2\pi\hbar} \oint \mathrm{Im}\,\alpha\, d\,\mathrm{Re}\alpha = n, \tag{13.106}$$

$$\frac{-i}{4\pi} \oint \left(\alpha - \alpha^\dagger\right) d\left(\alpha + \alpha^\dagger\right) = n, \tag{13.107}$$

$$\frac{-i}{4\pi} \oint \left(\alpha d\alpha - \alpha^\dagger d\alpha^\dagger\right) + \frac{i}{4\pi} \oint \left(\alpha^\dagger d\alpha - \alpha\, d\alpha^\dagger\right) = n. \tag{13.108}$$

Equation (13.108) for the CS representation exactly reproduce Eq. (13.89) for the harmonic oscillator. The same considerations for the first term of Eq. (13.108) yields the same expression as Eq. (13.93).

In agreement with published results, the second term of Eq.(13.108) appears as an approximate (semiclassical) result in the CS representation, as obtained in a BS quantization by Tochishita *et al.* [96] written as Eq. (38) in their paper,

$$\frac{i\hbar}{2} \oint_C \left(\xi^* d\xi - \xi d\xi^*\right) = 2\pi m\hbar. \tag{13.109}$$

13.10 Other areas where the BS condition have been used

There are several instances that the BS condition have been used, which are beyond the scope of this book. For example, it has been used for 1D Bogoliubov–de Gennes Hamiltonian. It may also have relevance to the so-called Fermionic oscillator in the 2-state system and its corresponding CS formulation. Its relevance and utility around non-Abelian gauge theory is indeed not yet clear and lack focused investigation.

In mathematics of geometric quantization, monodromy (and holonomy) of completely integrable Hamiltonian systems makes use of B–S quantization condition [97].

13.10.1 *BS quantization in FQHE*

Recently, the BS quantization has been employed [77, 79, 98] in the study of FQHE which generalizes the composite fermion theory of Jain [78]. In the composite fermion theory, the concept of fractional charge is central to the theory of the FQHE, thus the FQHE conductance essentially follows the formula for IQHE with renormalized electron charge. In the recent work of Jacak [77, 79, 98], the BS quantization has been employed in his holonomic approach using multi-loop orbits for incommensurate ratio of Wigner lattice constant to the magnetic length, $\frac{a}{l_B} = \frac{p}{q}$. His holonomic approach has produce a new hierarchy in FQHE which include the composite fermion hierarchy of Jain [78], as well as the IQHE. Indeed it is the commensuration and incommensuration of these two length scales that lead to the fractal spectrum of Hofstadter butterfly [99] and Wannier Diophantine equation [100] for the gaps of fractal spectrum.

The work on FQHE is still a topic of active research. Indeed, the topic of the interaction of magnetic field with matter has produced so many enigmatic and intriguing results in the history of physics, to name a few, such as the fractal spectrum of Hofstadter butterfly, IQHE, giant diamagnetism [9], and of course the ongoing active research in FQHE and fractal spectrum. All these may just be a 'tip of the iceberg' in revealing the fundamental understanding of nature which are pursued in so many diverse fronts in both geometry, physics, and the sciences.

In the following chapter, we will discuss in more details a novel quantum transport approach to the BS quantization of the IQHE in electrical conductivity. The physics of IQHE is well established, however from the point of view of BS quantization this is not fully appreciated. Moreover, a fully kinetic approach has not been given, which is in stark contrast to other approaches, such as the conventional Kubo current–current correlation perturbative approach of equilibrium systems.

Chapter 14

BS Quantization in IQHE

Here we present our new approach to integer quantum Hall effect which makes use of quantum superfield lattice Weyl transform non-equilibrium Green's function (SFLWT-NEGF) formalism [12]. This formalism includes non-equilibrium transport in superconductivity, as well as vacuum fluctuations or *zitterbewegung* in conductance measurements [101].

The calculation of integer quantum Hall effect is another good example where the IQHE conductance is directly proportional to the BS quantization of Eq. (13.8), i.e.,

$$\sigma_{xy} = \frac{e^2}{h}\sum_{\alpha}\frac{\Delta\phi_{\text{total}}}{2\pi} = \frac{e^2}{h}\sum_{\alpha} n_\alpha, \tag{14.1}$$

where BS condition enters in the following simple form,

$$\begin{aligned}\frac{\Delta\phi_{\text{total}}}{2\pi} &= \frac{1}{2\pi}\oint dk_c\ \left[\left\langle \alpha,\vec{k}\right| i\frac{\partial}{\partial k_c}\left|\alpha,\vec{k}\right\rangle\right]_{\text{contour}}, \\ &= \frac{1}{2\pi}\oint\ (\nabla_c\phi)\ dk_c = n_\alpha, \end{aligned}\tag{14.2}$$

where $\frac{e^2}{h}$ is the *quantum conductance.* Again $\Delta\phi_{\text{total}}$ is identified as the BS contour integral undergoing quantization. This will be derived Part 7 since our new quantum transport approach to IQHE is not well known. The derivation of Eq. (14.1) from a kinetic quantum transport employing phase-space quantum distribution or Wigner distribution function is detailed in Chapter 30.

It is also worth mentioning that in mesoscopic closed circuits with ballistic conducting channel and perfectly conducting leads, the conductance is simply equal to $\frac{e^2}{h}$, the *quantum conductance per electron spin, i.e.,* $n_\sigma \equiv 1$ [12].

14.1 BS quantization of orbital magnetic moment, edge states

In what follows, we will make use of the following identities:

$$\langle\alpha,\vec{p}|\,\vec{v}\,|\beta,\vec{p}\rangle = \left(E_\alpha\left(\vec{p}\right) - E_\beta\left(\vec{p}\right)\right)\langle\alpha,\vec{p}|\left(\nabla_{\vec{p}}\right)|\beta,\vec{p}\rangle\,, \tag{14.3}$$

together with

$$\hat{\mathbf{Q}}\,|\alpha,\vec{p}\rangle = -i\hbar\nabla_{\vec{p}}\,|\alpha,\vec{p}\rangle = -i\nabla_k\,|\alpha,\vec{p}\rangle, \tag{14.4}$$

and make use of the general relations,

$$\langle\alpha,\nabla_{\vec{k}}\vec{p}|\,|\beta,\vec{p}\rangle = \frac{\langle\alpha,\vec{p}|\,\vec{v}\,|\beta,\vec{p}\rangle}{\omega_{\alpha\beta}}. \tag{14.5}$$

Similarly, we have

$$\langle\beta,\vec{p}|\,\nabla_{\vec{k}}\,|\alpha,\vec{p}\rangle = \frac{\langle\beta,\vec{p}|\,\vec{v}\,|\alpha,\vec{p}\rangle}{\omega_{\beta\alpha}}. \tag{14.6}$$

The orbital magnetic moment is given by

$$\begin{aligned}\langle\vec{M}\rangle &= \operatorname{Tr}\rho\frac{e}{2m}\vec{L} = \operatorname{Tr}\rho\frac{e}{2m}\vec{Q}\times\vec{P},\\ &= -i\frac{e}{2}\sum_{\alpha,p;\beta,p} f\left(E_\alpha\left(\vec{p}\right)\right)\langle\alpha,\vec{p}|\,\nabla_{\vec{k}}\,|\beta,\vec{p}\rangle\times\langle\beta,\vec{p}|\,\vec{v}\,|\alpha,\vec{p}\rangle.\end{aligned} \tag{14.7}$$

Using Eq. (14.6), we obtain

$$\langle\vec{M}\rangle = -i\frac{e}{2}\sum_{\alpha,p;\beta,p} f\left(E_\alpha\left(\vec{p}\right)\right)\omega_{\beta\alpha}\,\langle\alpha,\vec{p}|\,\nabla_{\vec{k}}\,|\beta,\vec{p}\rangle\times\langle\beta,\vec{p}|\,\nabla_{\vec{k}}\,|\alpha,\vec{p}\rangle, \tag{14.8}$$

or

$$\langle\vec{M}\rangle = -i\frac{e}{2}\sum_{\alpha,p;\beta,p}\frac{f\left(E_\alpha\left(\vec{p}\right)\right)}{\omega_{\alpha\beta}}\langle\alpha,\vec{p}|\,\vec{v}\,|\beta,\vec{p}\rangle\times\langle\beta,\vec{p}|\,\vec{v}\,|\alpha,\vec{p}\rangle\,. \tag{14.9}$$

So Berry's curvature implies the presence of orbital magnetic moment. Using Eq. (14.6), we obtain as written in some literature

$$\langle\vec{M}\rangle = -\frac{e}{2}\left(i\right)^2\frac{\hbar^2}{m^2}\sum_{\alpha,p;\beta,p} f\left(E_\alpha\left(\vec{p}\right)\right)\frac{\left(E_\alpha\left(\vec{p}\right) - E_\beta\left(\vec{p}\right)\right)}{i\hbar}$$

$$\times \frac{\langle \alpha, \vec{p} | \, m\vec{v} \, | \beta, \vec{p} \rangle}{(E_\alpha(\vec{p}) - E_\beta(\vec{p}))} \frac{\langle \beta, \vec{p} | \, m\vec{v} \, | \alpha, \vec{p} \rangle}{(E_\alpha(\vec{p}) - E_\beta(\vec{p}))},$$

$$= -i \frac{e\hbar}{2m^2} \sum_{\alpha,p;\beta,p} f(E_\alpha(\vec{p})) \frac{\langle \alpha, \vec{p} | \, m\vec{v} \, | \beta, \vec{p} \rangle \times \langle \beta, \vec{p} | \, m\vec{v} \, | \alpha, \vec{p} \rangle}{(E_\alpha(\vec{p}) - E_\beta(\vec{p}))}. \tag{14.10}$$

In fact we can extract the Berry curvature by rewriting in the Heisenberg picture,

$$\begin{aligned}
\langle \vec{M} \rangle &= \mathrm{Tr}\, \rho \frac{e}{2m} L = \mathrm{Tr}\, \rho \frac{e}{2m} \vec{Q} \times \vec{P}, \\
&= \frac{e}{2m} \sum_{\alpha,p;\beta,p} f(E_\alpha(\vec{p})) \langle \alpha, \vec{p} | \, \vec{Q} \, | \beta, \vec{p} \rangle \times \langle \beta, \vec{p} | \, \vec{P} \, | \alpha, \vec{p} \rangle, \\
&= \frac{e}{2} \sum_{\alpha,p;\beta,p} f(E_\alpha(\vec{p})) \langle \alpha, \vec{p} | \, e^{\frac{i}{\hbar}Ht} \vec{Q}_S e^{-\frac{i}{\hbar}Ht} \, | \beta, \vec{p} \rangle \times \langle \beta, \vec{p} | \, \vec{v} \, | \alpha, \vec{p} \rangle, \\
&= \frac{e}{2} \sum_{\alpha,p;\beta,p} f(E_\alpha(\vec{p})) e^{i\omega_{\alpha\beta}t} \langle \alpha, \vec{p} | \, \vec{Q}_S \, | \beta, \vec{p} \rangle \times \langle \beta, \vec{p} | \, \vec{v} \, | \alpha, \vec{p} \rangle. \qquad (14.11)
\end{aligned}$$

From Eq. (14.4), we obtain

$$\begin{aligned}
\langle \vec{M} \rangle &= -i \frac{e}{2} \sum_{\alpha,p;\beta,p} f(E_\alpha(\vec{p})) e^{i\omega_{\alpha\beta}t} \langle \alpha, \vec{p} | \, \nabla_{\vec{k}} \, | \beta, \vec{p} \rangle \times \langle \beta, \vec{p} | \, \vec{v} \, | \alpha, \vec{p} \rangle \\
&= -i \frac{e}{2} \sum_{\alpha,p;\beta,p} f(E_\alpha(\vec{p})) e^{i\omega_{\alpha\beta}t} \langle \alpha, \vec{p} | \, \nabla_{\vec{k}} \, | \beta, \vec{p} \rangle \times \langle \beta, \vec{p} | \, \nabla_{\vec{k}} \, | \alpha, \vec{p} \rangle \, \omega_{\beta\alpha}.
\end{aligned} \tag{14.12}$$

Integrating the RHS with respect to time, the result is in dimensional units of an orbital magnetic moment multiplied by time denoted by $\langle \vec{\mathcal{M}} \rangle$,

$$\begin{aligned}
\langle \vec{\mathcal{M}} \rangle &= -i \frac{e}{2} \sum_{\alpha,p;\beta,p} f(E_\alpha(\vec{p})) \int_{-\infty}^{0} e^{i\omega_{\alpha\beta}t} dt \; \omega_{\beta\alpha} \langle \alpha, \vec{p} | \, \nabla_{\vec{k}} \, | \beta, \vec{p} \rangle \times \langle \beta, \vec{p} | \, \nabla_{\vec{k}} \, | \alpha, \vec{p} \rangle, \\
&= -i \frac{e}{2} \sum_{\alpha,p;\beta,p} f(E_\alpha(\vec{p})) \frac{\omega_{\beta\alpha}}{i\omega_{\alpha\beta}} dt \;\; \langle \alpha, \vec{p} | \, \nabla_{\vec{k}} \, | \beta, \vec{p} \rangle \times \langle \beta, \vec{p} | \, \nabla_{\vec{k}} \, | \alpha, \vec{p} \rangle, \\
&= -\frac{e}{2} \sum_{\alpha,p;\beta,p} f(E_\alpha(\vec{p})) \; \langle \alpha, \nabla_{\vec{k}} \vec{p} | \, | \beta, \vec{p} \rangle \times \langle \beta, \vec{p} | \, \nabla_{\vec{k}} \, | \alpha, \vec{p} \rangle, \\
&= -\frac{e}{2} \sum_{\alpha,p} f(E_\alpha(\vec{p})) \; \langle \alpha, \nabla_{\vec{k}} \vec{p} | \times | \alpha, \nabla_{\vec{k}} \vec{p} \rangle.
\end{aligned} \tag{14.13}$$

The last line is just

$$\begin{aligned}\langle\vec{\mathcal{M}}\rangle &= -\frac{e}{2}\frac{a^2}{(2\pi)^2}\sum_{\alpha}\int\int dk_x dk_y\, f\left(E_\alpha\left(\vec{k}\right)\right)\begin{bmatrix}\left\langle\frac{\partial}{\partial k_x}\alpha,\vec{k}\right|\frac{\partial}{\partial k_y}\left|\alpha,\vec{k}\right\rangle\\ -\left\langle\frac{\partial}{\partial k_y}\alpha,\vec{k}\right|\frac{\partial}{\partial k_x}\left|\alpha,\vec{k}\right\rangle\end{bmatrix}\\ &= -\frac{e}{2}\frac{a^2}{(2\pi)^2}\sum_{\alpha} f\left(E_\alpha\left(\vec{k}\right)\right)\int\int dk_x dk_y\left(\nabla_{\vec{k}}\times\left\langle\alpha,\vec{k}\right|\frac{\partial}{\partial\vec{k}}\left|\alpha,\vec{k}\right\rangle\right),\end{aligned} \tag{14.14}$$

explicitly revealing the Berry curvature derived from the expression for the orbital magnetic moment. Equation (14.14) has the similar dimensional units of a Bohr magneton multiplied by time t. The resulting quantization of orbital motion leads to edge states and integer quantum Hall effect under uniform magnetic fields. Analysis of edge states can be found in the works of Laughlin [71], and Halperin [72].

Chapter 15

Summary of Part Two

The thesis of this part of the book is that if we simply cast the BS quantization condition as a U(1) gauge theory, like the gauge field of the topological quantum field theory (TQFT) via the Chern–Simons gauge theory, or specifically as in topological band theory (TBT) of CMP in terms of Berry connection and curvature to make it self-consistent, then all the quantization method in all the physical phenomena treated here are unified. We have demonstrated how it permeates and pervades the whole of modern quantum physics, implicitly identifying the BS condition as the forebear of modern geometrical or topological quantum theory and even permeates the geometric quantization in mathematics [97]. We have shown its place in the theory of the Nobel prize winning discoveries of quantum Hall effect, Kosterlitz–Thouless transition in two-dimensional systems, Haldane phase, and quantized vortices in superfluids. Here, we also show how the U(1) gauge theory naturally give us the quantized magnetic charge of Dirac monopole. The BS quantization also leads to the expression of BS quantization of the quantum field theory of harmonic oscillator and in coherent state representation.

The BS quantization leads to quantum conductance, quantum flux, Landau-level degeneracies, discrete flux in multi-loop holonomic FQHE, discrete vortex charge, discrete Dirac monopole charge, energy levels of harmonic oscillators, etc., as a theory of Berry connection. The BS quantization is thus expected to continue to hold more prominent role in the advances of quantum physics and geometry.

Our new real-time SFLWT-NEGF quantum kinetic transport offered a new and entirely open system approach that leads to BS quantization of IQHE. This is treated in greater detail in Chapter 30.

It is worthwhile to mention that the real-time formalism of SFLWT-NEGF multi-spinor quantum transport equations are also able to predict various entanglements leading to different topological phases of low-dimensional and nanostructured gapped condensed matter systems. This is treated in more details in Part 6 of this book.

Part 3

Non-equilibrium Quantum Superfield Transport Theory

Chapter 16

Quantum Superfield Transport Theory

Since this book is on the application of non-equilibrium quantum transport theory to spinful and topological systems, we will give here an outline of the theory to make the book self-contained. The term "quantum superfield" indicates the use of real-time axis in contrast to the use of closed-time contour in the older non-equilibrium quantum transport formulations [102]. The theoretical foundation of computational nanoelectronics, spintronics, pseudospintronics, and most likely emerging topotronics, is based on non-equilibrium quantum transport physics coupled with the lattice Weyl transform techniques. The latter has been given a comprehensive treatment in Part 1 of this book [103]. This review will provide a unified approach for large-scale simulations of functional materials and highly nonlinear active high-speed nanodevices. This will give an alternative perspective of topological systems, providing a foundation on computational spinful and topological systems which will be discussed in detail in Parts 6–8 of this book.

16.1 Why non-equilibrium transport and lattice Weyl transform

To discover novel nanodevice functionality from fully time-dependent quantum kinetic transport simulations, namely, high switching-speed or cut-off frequency performance, self-oscillatory behavior serving as high-frequency inverter or as a THz source, as a single memory device from bistability and hysteresis behavior, as a model of reaction rates in enzymatic and reactions in biological and chemical processes, as new information processor, energy harvesting, energy storage or new battery nanodevice, etc. Active nanodevices are essentially characterized by space–time inhomogeneity, where slowly varying gradient approximations are no longer valid, as well as nonlinear quantum effects (e.g., nonlinear feedbacks and self-consistent driving fields), dynamic scatterings with excitations of surrounding environments, and openness to input and output boundaries. Nowadays, the study of non-equilibrium nanosystems is pursued in almost all fields of technology, medicine, biology, fluid dynamics, material sciences, etc., and will no doubt greatly influence the scientific and technological developments of this century. The emerging fields of quantum control [104] and quantum computing [105] are also being pursued inspired by modern fabrication techniques of nanoscale systems and advances in the study of ultrafast physical phenomena and quantum entanglements.

The invention of transistors and laser have ushered rapid computational and analytical advances in the study of non-equilibrium systems, mainly within semi-classical physics. In the past, computer-aided design of microelectronic devices and very large-scale circuit systems have been responsible in the rapid advances of integrated circuit technology, computers, data processing and the ensuing information revolution. However, the miniaturization of semiconductor transistors and lasers to nanoscale regime and the growing number of novel nanodevice concepts that are continually being pursued demanded expeditious developments in computational and theoretical applications of nonequilibrium quantum transport physics coupled with discrete phase-space Lattice Weyl transform techniques.

16.1.1 *Historical background*

The study of nonequilibrium quantum physics was initiated by Schwinger [102] in 1961. Schwinger introduced the concept of closed time contour or double time axis to differentiate forward and backward time development, in order to incorporate irreversibility, scatterings, nonlinear effects with applied fields, and interaction with environment. In 1964, Keldysh [106] employed the Feynman diagram technique to statistical system very far from thermodynamic equilibrium under the action of external fields. At about the same time as Schwinger, Kadanoff and Baym [107] discussed quantum statistical mechanics, employing similar time-contour techniques, using the properties of the so-called non-equilibrium Green's functions which are also the players in the time contour technique of Schwinger and Keldysh.

The major drawback of the time contour technique is that products of different correlation functions are determined based on the time ordering and evaluation of time integral along a time contour [108]. Another major drawback of the earlier applications of quantum transport equations is that the distribution functions in phase-space were only solved within the slowly-varying approximation [107, 109]. With the emergence of nanoscience and nanoelectronics, spintronics, pseudospintronics, topotronics, and the study of ultrafast phenomena, entanglements, and novel nanodevices [110], it became highly desirable to (a) have an algebraic manipulation for calculating quantum transport equations and many-body corrections to different correlation functions, including all joint distributions and pairings, such as Cooper pairings, and the non-equilibrium Green's functions, without the complexity of determining the pertinent ranges of time ordering and integral along the time contour.[1] It has also become crucial to (b) avoid the slowly-varying approximation paradigm in deriving kinetic quantum transport equations for the distribution functions, to enable the simulation of the high-speed performance of emerging nanodevices [11, 111–115]. These two requirements have lead us to a combined nonequilibrium quantum superfield transport and lattice Weyl transform techniques. This new formalism is expected to lead to a straightforward calculations of the needed many-body corrections [110] as well as to practicality in analyzing the nonlinear and time-dependent performance of nanodevices [14, 116]. These are achieved by using (a) a completely algebraic manipulation of Liouville space quantum dynamics [117] similar to the zero-temperature many-body techniques, and (b) lattice Weyl transform techniques [3] to do away with prevailing slowly-varying approximations. These two developments will be briefly discussed here for completeness.

16.1.2 *Discrete phase space and lattice Weyl transform techniques*

Part 1 is a comprehensive treatment of this topic covering CMP and QFT, including a third quantization scheme. Therefore the main body of this review deals with theoretical foundation of non-equilibrium many-body quantum physics based on quantum superfield or real-time approach.

16.1.3 *Super-operators and super-statevectors*

An alternative and straightforward algebraic approach to non-equilibrium quantum physics, which avoids the complex usage of time contour in older formulations [102, 106, 107], can be formulated in terms of quantum

[1] Indeed, the nonlinear effects of higher-order many-body correlations between quasiparticles have been experimentally shown to dominate the nonlinear optical response of semiconductors [110].

Liouville space (L-space) dynamics of super-operators and super-statevector [118, 119]. A real-time quantum transport approach is an idea first proposed by Schmutz [118]. Here, this is implemented by employing quantum superfield theoretical techniques [117], which accounts for all pairing correlations and dynamics. The hallmark of the quantum superfield theory (QSFT) of non-equilibrium physics is its purely algebraic and straightforward procedure for deriving the equations for the super-correlation functions that enter in the theory, a procedure exactly paralleling that of the many-body quantum field theory at zero temperature. QSFT is a real time approach and the relevant quantum transport equations comes out to be simply the consequent equations for the components of the super-correlation functions [117]. This is in contrast with the time-contour formalism where one has to trace the forward and backward time development along the time contour, in a rather awkward manner, to derive the relevant quantum transport equations [102, 106–108]. Indeed, the power of the QSFT approach for calculating the nonequilibrium electron self-energy due to Coulomb interaction, including pairing dynamics, has been given by the author to second order in the interaction potential in a recently published handbook on theoretical and computational nanotechnology [10].

Effectively, instead of the time-contour or the doubling of time axis [102, 106], together with ordinary Hilbert space (H-space) quantum-field operators in the time-contour formalism, these are replaced in the quantum superfield theoretical approach by a single real-time axis and a doubling of quantum field operators [117–119]. In QSFT, the canonically conjugate quantum field operators are elevated to super-operators, whereas density matrix operators are elevated to super-state vectors. Note that the doubling of degrees of freedom by either the time-contour approach or doubling of the quantum field operators in the QSFT approach exactly serve the same purpose, namely, these two doubling schemes differentiate the quantum dynamics of forward and backward time development, i.e., both formalisms incorporate irreversibility, scatterings, and nonlinear interactions with applied driving fields and the environment. Indeed, doubling of the quantum-field operators is quite natural in the QSFT approach, where algebraic manipulations are similar to that of zero-temperature many-body theory.

The well-known quantum mechanics of two-level atoms gives a simplest exposition of the superfield theory of nonequilibrium quantum physics. Moreover, as shown in Appendix C, the derivation of Pauli quantum mechanics of spin $\frac{1}{2}$ system also give a very enlightening illustration of the power of discrete phase space quantum mechanics treated in great details in Part 1.

16.2 Quantum Liouville space and super-Schrödinger equation

The von Neumann density-matrix evolution equation of quantum statistical dynamics in Hilbert (H)-space becomes a super-statevector dynamical equation in Liouville (L)-space. Thus, the familiar von Neumann density-matrix operator equation in H-space given by

$$i\hbar\frac{\partial}{\partial t}\rho(t) = [\mathcal{H}, \rho], \tag{16.1}$$

becomes a super-Schrödinger equation for the super-statevector, $|\rho(t)\rangle\rangle$, in L-space expressed as

$$i\hbar\frac{\partial}{\partial t}|\rho(t)\rangle\rangle = \mathcal{L}|\rho(t)\rangle\rangle. \tag{16.2}$$

Equation (16.2) is formally identical to the Schrödinger equation, and is expected to have identical formal solutions. In Eq. (16.1), $\rho(t)$ is the density-matrix operator for the whole many-body system in H-space, whereas in Eq. (16.2) $|\rho(t)\rangle\rangle$ is the corresponding super-statevector in L-space. The super-operator $\mathcal{L}$ corresponds to the commutator $[\mathcal{H}, \rho]$ of Eq. (16.1), and is referred to here as the Liouvillian. Thus, we may write the Liouvillian L as

$$\mathcal{L} = \widehat{\mathcal{H}} - \widetilde{\mathcal{H}}, \tag{16.3}$$

which defines $\widehat{\mathcal{H}}$ and $\widetilde{\mathcal{H}}$. These are super-operators with the property that $\widehat{\mathcal{H}}|\rho(t)\rangle\rangle = |\mathcal{H}\rho(t)\rangle\rangle$ and $\widetilde{\mathcal{H}}|\rho(t)\rangle\rangle = |\rho(t)\mathcal{H}^{\dagger}\rangle\rangle$. These relations are valid for fermions and bosons. For number-conserving fermion operator $\mathcal{H}$, $\widetilde{\mathcal{H}} = \widetilde{\mathcal{H}}^{\dagger}$. From Eq. (16.3), we can make the assertion that any H-space operator basically becomes the "hat" and "tilde" super-operators in L-space. This is often referred to as the doubling in the number of the degrees of freedom, to differentiate the quantum dynamics for "forward" and "backward" time development. In general, any super-operator $\tilde{A}$, which is an arbitrary product of quantum superfield operators, will be denoted as $\tilde{A}_{\epsilon}$ to take care of the commutation properties of the quantum superfield operators. This will become clear in Section 16.3.

The nature of these super-operators can be analyzed using a set of many-body basis states [118]. If a countable set $\{|n\rangle\}$ is an orthonormal basis in H-space in the number representation of many-body basis states, then the basis super-statevectors are constructed from the direct product space and is given by the set $||m\rangle\langle n|\rangle\rangle$, which represents the orthonormal basis super-statevectors in L-space. This leads to the corresponding super-statevector expansion in L-space,

$$|A\rangle\rangle = \sum_{m,n} (\langle\langle |n\rangle \langle m|| \; |A\rangle\rangle) \; ||m\rangle \langle n|\rangle\rangle, \tag{16.4}$$

where $\langle\langle |n\rangle \langle m|| \; |A\rangle\rangle = \langle m| A |n\rangle$ is the corresponding expansion coefficient, which is identical to the expansion coefficient of the operator A in H-space. In general, the super-scalar product of two super-statevectors involves "dotted" summation of all indices of $|A\rangle\rangle$ and $|B\rangle\rangle$ and is naturally given by the ordinary trace operation,

$$\langle\langle A| \; |B\rangle\rangle = \mathrm{Tr}\, A^\dagger B, \tag{16.5}$$

where the right-hand side (RHS) corresponds to taking the trace in H-space. The norm of any super-statevector $|A\rangle\rangle$ is thus defined as

$$|A| = \langle\langle A| \; |B\rangle\rangle^{\frac{1}{2}} = \left\{\mathrm{Tr}\, A^\dagger A\right\}^{\frac{1}{2}}. \tag{16.6}$$

A unit super-statevector is here defined as

$$|1\rangle\rangle = \sum_{m} ||m\rangle \langle m|\rangle\rangle. \tag{16.7}$$

The trace of any operator A in H-space can thus be written as a super dot product of the unit super-statevector with the super-statevector corresponding to A,

$$\mathrm{Tr}\; A = \langle\langle 1| \; |A\rangle\rangle\,. \tag{16.8}$$

In particular, $\mathrm{Tr}\; \rho A = \mathrm{Tr}\; A\rho = \langle\langle 1| \; |\rho A\rangle\rangle = \langle\langle 1| \; \tilde{A}_\epsilon^\dagger \; |\rho\rangle\rangle = \langle\langle 1| \; |A\rho\rangle\rangle = \langle\langle 1| \; \hat{A} \; |\rho\rangle\rangle$. Observe our definition of $\tilde{A}_\epsilon^\dagger$

$$\langle\langle 1| \; \tilde{A}_\epsilon^\dagger \; |\rho\rangle\rangle = \langle\langle 1| \; |A\rho\rangle\rangle, \tag{16.9}$$

where all ensuing (anti)commutation is canceled by appropriate factors in $\tilde{A}_\epsilon^\dagger$ for fermions, as will become clear in what follows. Since any H-space operator consists of arbitrary number of annihilation and creation quantum-field operators, we need to determine the property of the corresponding super-annihilation and super-creation operators to be able construct a QSFT of nonequilibrium systems of particles and elementary excitations. This is discussed in Section 16.3.

16.3 Doubling of H-space quantum-field operators

The Liouvillian super-operator defined above also induces the "doubling" of the H-space many-body quantum field operators, ψ and $\psi^\dagger$, to become the "hat" and "tilde", $\widehat{\psi}$ $(\widetilde{\psi})$ and $\widehat{\psi}^\dagger$ $\left(\widetilde{\psi}^\dagger\right)$, super-operators in L-space. They are easily defined for the "hat" super-operators as follows:

$$\widehat{\psi}^\dagger \, ||m\rangle \langle n|\rangle\rangle = |\psi^\dagger \, |m\rangle \langle n|\rangle\rangle, \tag{16.10}$$

$$\widehat{\psi} \, ||m\rangle \langle n|\rangle\rangle = |\widehat{\psi} \, |m\rangle \langle n|\rangle\rangle. \tag{16.11}$$

Therefore, the "hat" quantum field super-operators clearly obey the usual equal-time commutation (anticommutation) relations for bosons (fermions). The nature of the tilde quantum field super-operators can be determined by imposing the constraint implied by the Liouvillian super-operator in Eq. (16.2), and the requirement that the "hat" and "tilde" super-operators commute (anticommute) for bosons (fermions). Then, the time evolution super-operator of Eq. (16.2) factorizes into a product of the "hat" time evolution super-operator and the "tilde" time evolution super-operator, as expected from the formal solution of Eq. (16.1) in H-space.

Thus, we want the following relations to hold for the "tilde" operators

$$\left[\widehat{\psi}_1, \widehat{\psi}_2^\dagger\right]_\epsilon = \left[\widetilde{\psi}_1, \widetilde{\psi}_2^\dagger\right]_\epsilon = \delta_{1,2}. \tag{16.12}$$

Therefore,

$$\left[\widetilde{\psi}, \widetilde{\psi}^\dagger\right]_\epsilon \, ||m\rangle \langle n|\rangle\rangle = ||m\rangle \langle n|\rangle\rangle \, . \tag{16.13}$$

Upon evaluating, we come up with,

$$\begin{aligned} \widetilde{\psi}\widetilde{\psi}^\dagger \, ||m\rangle \langle n|\rangle\rangle &= \widetilde{\psi} \, ||m\rangle \langle n| \, \psi\rangle\rangle, \\ &= | \, |m\rangle \langle n| \, \psi^\dagger \psi\rangle\rangle, \\ &= \epsilon \, | \, |m\rangle \langle n| \, \psi\psi^\dagger\rangle \Big\rangle. \end{aligned} \tag{16.14}$$

Similarly

$$\widetilde{\psi}^\dagger \widetilde{\psi} \, ||m\rangle \langle n|\rangle\rangle = \epsilon \, | \, |m\rangle \langle n| \, \psi^\dagger \psi\rangle \Big\rangle, \tag{16.15}$$

yielding

$$\left[\widetilde{\psi}, \widetilde{\psi}^\dagger\right]_\epsilon = \epsilon, \tag{16.16}$$

where $\epsilon = 1$ for bosons and $\epsilon = -1$ for fermions. In order to fix this inconsistency with Eq. (16.12), let us, for the moment, write

$$\widetilde{\psi} \, ||m\rangle \langle n|\rangle\rangle = \epsilon^{\mathbb{Z}_1(m,n)} | \, |m\rangle \langle n| \, \psi^\dagger\rangle\rangle, \tag{16.17}$$

$$\widetilde{\psi}^\dagger \, ||m\rangle \langle n|\rangle\rangle = \epsilon^{\mathbb{Z}_2(m,n)} \, ||m\rangle \langle n| \, \psi\rangle\rangle, \tag{16.18}$$

where $\mathbb{Z}_j(m,n)$ is an integer determined by m and n. Combining Eqs. (16.17) and (16.18), making use of the commutation (anti-commutation) relation for the H-space quantum field operators, we must also have

$$\epsilon^{\mathbb{Z}_1(m,n)}\epsilon^{\mathbb{Z}_2(m,n)}\epsilon = 1. \tag{16.19}$$

We have another basic requirement for $\mathbb{Z}_j(m,n)$, namely, $\epsilon^{\mathbb{Z}_j(m,m)} = 1$ for compatibility with ensembles of mixed density matrix operators in the Liouville equation. Therefore, we can write

$$\epsilon^{\mathbb{Z}_j(m,n)} = \epsilon^{m-n}. \tag{16.20}$$

However, to satisfy Eq. (16.19), we have to modify one of the $\mathbb{Z}_j(m,n)$'s to include an extra ϵ. Thus, for fermions we can either have,

$$\epsilon^{\tilde{\mathbb{Z}}_1(m,n)} = \epsilon^{m-n+1}, \quad \epsilon^{\mathbb{Z}_2(m,n)} = \epsilon^{m-n}, \tag{16.21}$$

or

$$\epsilon^{\tilde{\mathbb{Z}}_2(m,n)} = \epsilon^{m-n+1}, \quad \epsilon^{\mathbb{Z}_1(m,n)} = \epsilon^{m-n}. \tag{16.22}$$

For bosons, all the $\epsilon^{\mathbb{Z}_j(m,n)}$'s can be taken equal to unity. Here, we prefer to use Eq. (16.22) for fermions for convenience in what follows, whereas Schmutz [118] use the equivalent of Eq. (16.21). Thus, we have in addition to Eqs. (16.10) and (16.11) the following relations for the "tilde" quantum field super-operators in L-space,

$$\widetilde{\psi}\, \|m\rangle\langle n|\rangle\rangle = \epsilon^{m-n}\, \left| |m\rangle\langle n|\,\psi^\dagger\rangle \right\rangle, \tag{16.23}$$

$$\widetilde{\psi^\dagger}\, \|m\rangle\langle n|\rangle\rangle = \epsilon^{m-n+1}\, \|m\rangle\langle n|\,\psi\rangle\rangle, \tag{16.24}$$

where $\left[\widehat{\psi_1}, \widehat{\psi_2^\dagger}\right]_\epsilon = \left[\widetilde{\psi_1}, \widetilde{\psi_2^\dagger}\right]_\epsilon = \delta_{1,2}$, and all other commutation or anti-commutation of "hat" and "tilde" super-operators are zero. Moreover, for "paired" tilde operators we now have

$$\widetilde{\psi}\widetilde{\psi^\dagger}\, \|m\rangle\langle n|\rangle\rangle = |\,|m\rangle\langle n|\,\psi\psi^\dagger\rangle\rangle, \tag{16.25}$$

$$\widetilde{\psi^\dagger}\widetilde{\psi}\, \|m\rangle\langle n|\rangle\rangle = |\,|m\rangle\langle n|\,\psi^\dagger\psi\rangle\rangle. \tag{16.26}$$

However, for unpaired tilde operators, there will be produced ϵ factors for all interchanges. For example,

$$\begin{aligned}
\widetilde{\psi_1^\dagger}\widetilde{\psi_2^\dagger}\, \|m\rangle\langle n|\rangle\rangle &= \widetilde{\psi_1^\dagger}\,\epsilon^{m-n+1}\, \|m\rangle\langle n|\,\psi_2\rangle\rangle, \\
&= \epsilon^{m-n+1}\epsilon^{m-n+1}\, \|m\rangle\langle n|\,\psi_1\psi_2\rangle\rangle, \\
&= \|m\rangle\langle n|\,\psi_1\psi_2\rangle\rangle, \\
&= \epsilon\, \|m\rangle\langle n|\,\psi_2\psi_1\rangle\rangle.
\end{aligned} \tag{16.27}$$

Likewise

$$\widetilde{\psi_1}\widetilde{\psi_2}\,||m\rangle\langle n|\rangle\rangle = \epsilon\,|\,|m\rangle\langle n|\,\psi_2^\dagger\psi_1^\dagger\rangle\Big\rangle. \tag{16.28}$$

Observe that by virtue of the Hermitian conjugation, the order of the ordinary operators on the RHS is reversed to the order of the tilde operators in the LHS. Thus, any super-operator $\tilde{A}_\epsilon$ consisting of product of *unpaired* quantum superfield operators will be defined by a counteracting prefactor as

$$\tilde{A}_\epsilon = e^{(\eta-1)\frac{[(\eta-1)+1]}{2}}\tilde{A}, \tag{16.29}$$

and

$$\tilde{A}_\epsilon^\dagger = e^{(\eta-1)\frac{[(\eta-1)+1]}{2}}\tilde{A}^\dagger, \tag{16.30}$$

where η is the number of unpaired $\widetilde{\psi}$ or $\widetilde{\psi}^\dagger$ in $\tilde{A}_\epsilon$. The total number of interchanges is $\eta-1$ and these are summed up from 1 up to $\eta-1$, and hence there will be $(\eta-1)$ terms in the sum. The remaining operators $\tilde{A}$ and $\tilde{A}^\dagger$ obey the property given by Eqs. (16.23)–(16.24). Note for number-conserving operator, $\eta = 0$, i.e., no unpaired $\widetilde{\psi}$ or $\widetilde{\psi}^\dagger$, hence $\tilde{\mathcal{H}}_\epsilon = \tilde{\mathcal{H}} = \tilde{\mathcal{H}}^\dagger$.

To construct QSFT we basically need two more things, (a) a starting Schwinger equation, i.e., the equation of motion of the average quantum superfield, and (b) a generating super-functional for all correlations. These will allow us to go over from super-operator theory to the statistical mechanics of $\mathbb{C}$-number fields, where the $\mathbb{C}$-number fields are elements of the Grassmann algebra for fermions [16]. With the properties of the superfield operators known, the Schwinger equation can easily be obtained. In the next two sections, we derive the generating super-functional.

16.4 Formal solutions of super-Schrödinger equation

From Eq. (16.2), we can formally write the solution for the super-statevector $|\rho(t)\rangle\rangle$ as

$$\begin{aligned} |\rho(t)\rangle\rangle &= U_{\mathcal{L}}(t,t_0)\,|\rho(t_0)\rangle\rangle \\ &= T\exp\left[\frac{-i}{\hbar}\int_{t_o}^{t}\mathcal{L}(t')\,dt'\right]|\rho(t_o)\rangle\rangle, \end{aligned} \tag{16.31}$$

where T is the usual real-time ordering operator. We take t_o as the time when the "perturbation" Liouvillian, $\mathcal{L}^{(1)} = \mathcal{L} - \mathcal{L}_o$, is turned on. Then we can also write,

$$|\rho(t_o)\rangle\rangle = T\exp\left[\frac{-i}{\hbar}\int_{\xi_o}^{t_o}\mathcal{L}_o(t')\,dt'\right]|\rho(\xi_o)\rangle\rangle,$$
$$= U_{\mathcal{L}}(t_0,\xi_o)\,|\rho(\xi_o)\rangle\rangle \tag{16.32}$$

where the system, during the time duration from ξ_o to t_o, is acted on only by the "unperturbed" Liouvillian $\mathcal{L}_o$. Therefore, we can also write

$$|\rho(t)\rangle\rangle = T\exp\left[\frac{-i}{\hbar}\int_{t_0}^{t}\mathcal{L}(t')\,dt'\right]U_{\mathcal{L}}(t_0,0)\,|\rho_0(0)\rangle\rangle. \tag{16.33}$$

where

$$|\rho_o(0)\rangle\rangle = \mathcal{U}_o(0,\xi_o)\,|\rho(\xi_o)\rangle\rangle. \tag{16.34}$$

If we let $\xi_o \Rightarrow -\infty$, and write $|\rho_o(0)\rangle\rangle = |\rho_{eq}\rangle\rangle$, then we have

$$|\rho(t)\rangle\rangle = T\exp\left[\frac{-i}{\hbar}\int_{0}^{t}\mathcal{L}(t')\,dt'\right]|\rho_{eq}\rangle\rangle. \tag{16.35}$$

It follows that

$$|\rho(0)\rangle\rangle = |\rho_{\mathcal{L}}\rangle\rangle = |\rho_I(0)\rangle\rangle, \tag{16.36}$$

which defines the super-Heisenberg representation $|\rho_{\mathcal{L}}\rangle\rangle$ of the super-statevector $|\rho(t)\rangle\rangle$. Note that $|\rho_{\mathcal{L}}\rangle\rangle$ is independent of time. Equation (16.33) also leads us to define the super-interaction representation in L-space. Thus for any super-operator O, we define,

$$O_I(t) = \mathcal{U}_o(0,t)\,\mathcal{O}(t)\,\mathcal{U}_o(t,0), \tag{16.37}$$

as the super-interaction representation of O in L-space, and

$$O_{\mathcal{L}}(t) = \mathcal{U}(0,t)\,\mathcal{O}(t)\,\mathcal{U}(t,0), \tag{16.38}$$
$$= [\mathcal{U}(0,t)\,\mathcal{U}_o(t,0)]\,O_I(t)\,[\mathcal{U}_o(0,t)\,\mathcal{U}(t,0)]$$
$$= \overline{S}(0,t)\,O_I(t)\,\overline{S}(t,0), \tag{16.39}$$

which give the relation between the super-Heisenberg representation and the super-interaction representation. Since $\mathcal{L}$, $\widetilde{\mathcal{H}}_{\epsilon}$, and $\widehat{\mathcal{H}}$ commute then we have for the "hat" and "tilde" quantum field super-operators the following relations:

$$\begin{aligned}\widehat{\psi}_{\widehat{\mathcal{H}}}(t) &= T\exp\left[\frac{i}{\hbar}\int_0^t \widehat{\mathcal{H}}(t')\,dt'\right]\widehat{\psi}\,T\exp\left[\frac{-i}{\hbar}\int_0^t \widehat{\mathcal{H}}(t')\,dt'\right],\\ &= T\exp\left[\frac{i}{\hbar}\int_0^t \mathcal{L}(t')\,dt'\right]\widehat{\psi}\,T\exp\left[\frac{-i}{\hbar}\int_0^t \mathcal{L}(t')\,dt'\right],\\ &= \widehat{\psi}_{\mathcal{L}}(t),\end{aligned} \tag{16.40}$$

and,

$$\begin{aligned}\widetilde{\psi}_{\widetilde{\mathcal{H}}_\epsilon}(t) &= T\exp\left[\frac{i}{\hbar}\int_0^t \widetilde{\mathcal{H}}_\epsilon(t')\,dt'\right]\widetilde{\psi}\,T\exp\left[\frac{-i}{\hbar}\int_0^t \widetilde{\mathcal{H}}_\epsilon(t')\,dt'\right],\\ &= T\exp\left[\frac{i}{\hbar}\int_0^t \mathcal{L}(t')\,dt'\right]\widetilde{\psi}\,T\exp\left[\frac{-i}{\hbar}\int_0^t \mathcal{L}(t')\,dt'\right],\\ &= \widetilde{\psi}_{\mathcal{L}}(t).\end{aligned} \tag{16.41}$$

Similar relations exist for the canonically conjugate quantum field operators: $\widehat{\psi}^{\dagger}_{\widehat{\mathcal{H}}}(t) = \widehat{\psi}^{\dagger}_{\mathcal{L}}(t)$, and $\widetilde{\psi}^{\dagger}_{\widetilde{\mathcal{H}}_\epsilon}(t) = \widetilde{\psi}^{\dagger}_{\mathcal{L}}(t)$.

16.5 Generating super-functional

The "super-Schrödinger equation" that we have developed in L-space and its corresponding formal solution given above are not complete. For example, it is not clear what would be the canonically conjugate counterpart of the super-statevector $|\rho(t)\rangle\rangle$ for defining generating functional, analogous to the existing canonically conjugate pair of H-space states, $|\psi_+, t\rangle$ and $|\psi_-, t\rangle$, which naturally occur in zero temperature variational quantum action principle forming the basis for deriving the time-dependent Schrödinger equation, as first given by Dirac [120].

16.5.1 *Time-dependent variational principle in Liouville space*

To complete the "super-Schrödinger quantum mechanics", we need to construct the corresponding variational principle in L-space. We follow the general construction of a variational principle [121, 122] as an optimization of the estimate for the expectation value of any operator $\mathcal{O}$ at the time t_1 from the knowledge of the initial condition for the density matrix super-statevector $|\rho(t_o)\rangle\rangle$, subject to the constraint that $|\rho(t)\rangle\rangle$ obeys the L-space super-Schrödinger equation. Therefore, we have to optimize the following super-functional [122],

$$\Phi\left(|\rho(t)\rangle\rangle, |\Lambda(t)\rangle\rangle\right) = \langle\langle\mathcal{O}(t)|\ |\rho(t_1)\rangle\rangle - \int_{t_o}^{t_1} \langle\langle\Lambda(t)|\, i\hbar\frac{\partial}{\partial t} - \mathcal{L}\,|\rho(t)\rangle\rangle, \tag{16.42}$$

where we have introduced the dual space super-statevector $\langle\langle\Lambda(t)|$ as a "Lagrange multiplier". Upon integration by parts, $\Phi\left(|\rho(t)\rangle\rangle, |\Lambda(t)\rangle\rangle\right)$ is also equal to the following super-functional

$$\begin{aligned}
&\Phi\left(|\rho(t)\rangle\rangle, |\Lambda(t)\rangle\rangle\right)\\
&\quad = \langle\langle\mathcal{O}(t)|\ |\rho(t_1)\rangle\rangle - \left[i\hbar\,\langle\langle\Lambda(t_1)|\ |\rho(t_1)\rangle\rangle - i\hbar\,\langle\langle\Lambda(t_o)|\ |\rho(t_o)\rangle\rangle\right]\\
&\qquad + \int_{t_o}^{t_1} \left\langle\left\langle i\hbar\frac{\partial\Lambda(t)}{\partial t} - \mathcal{L}\Lambda(t)\right|\ |\rho(t)\rangle\right\rangle.
\end{aligned} \tag{16.43}$$

Thus, the optimum condition for $\Phi\left(|\rho(t)\rangle\rangle, |\Lambda(t)\rangle\rangle\right)$ occurs for $i\hbar\,\langle\langle\Lambda(t_1)| = \langle\langle\mathcal{O}(t)|$, and when the following equations of motion are obeyed for $t_o < t < t_1$,

$$i\hbar\frac{\partial\,|\rho(t)\rangle\rangle}{\partial t} = \mathcal{L}\,|\rho(t)\rangle\rangle, \tag{16.44}$$

$$i\hbar\frac{\partial\,|\Lambda(t)\rangle\rangle}{\partial t} = \mathcal{L}\,|\Lambda(t)\rangle\rangle. \tag{16.45}$$

If we let $t_1 \Rightarrow \infty$ and $t_o \Rightarrow -\infty$, apply the boundary condition: $i\hbar\,\langle\langle\Lambda(t_1)| = \langle\langle\mathcal{O}(t)|$, and $|\rho(-\infty)\rangle\rangle = |\rho_{eq}\rangle\rangle$, then the solutions can be written as,

$$|\rho(t)\rangle\rangle = T\exp\left[\frac{-i}{\hbar}\int_0^t \mathcal{L}_o(t')\,dt'\right]\overline{S}(t,-\infty)\,|\rho_{eq}\rangle\rangle, \tag{16.46}$$

$$|\Lambda(t)\rangle\rangle = T^{ac}\exp\left[\frac{-i}{\hbar}\int_t^0 \mathcal{L}_o(t')\,dt'\right]\overline{S}(t,\infty)\,|\mathcal{O}\rangle\rangle, \tag{16.47}$$

where T^{ac} denotes anti-chronological time ordering. Therefore, we can write a "transition probability" in L-space as

$$\langle\langle\Lambda(t)|\ |\rho(t)\rangle\rangle = \frac{1}{i\hbar}\langle\langle\mathcal{O}|\,\overline{S}(\infty,-\infty)\,|\rho_{eq}\rangle\rangle = \frac{1}{i\hbar}\langle\langle\mathcal{O}_{\mathcal{L}}|\ |\rho_{\mathcal{L}}\rangle\rangle, \tag{16.48}$$

where $|\rho_{\mathcal{L}}\rangle\rangle$ is the super-Heisenberg representation of the super-statevector $|\rho(t)\rangle\rangle$ defined by Eq. (16.36), and similarly $\langle\langle\mathcal{O}_{\mathcal{L}}| = \langle\langle\mathcal{O}|\,\overline{S}(\infty,0)$.

Note that in terms of evaluating the expectation value of the super-operator at $t = t_1 = \infty$, we can explicitly write this as a time-ordered expectation value by rewriting Eq. (16.48) as

$$\langle\langle\Lambda(t)|\,|\rho(t)\rangle\rangle = \frac{1}{i\hbar}\langle\langle\mathcal{O}|\,\overline{S}(\infty,-\infty)\,|\rho_{eq}\rangle\rangle = \frac{1}{i\hbar}\langle\langle 1|\,\mathcal{O}\overline{S}(\infty,-\infty)\,|\rho_{eq}\rangle\rangle. \tag{16.49}$$

We are particularly interested in $\mathcal{O} = i\hbar\Im$, where $\Im$ is the identity operator. Then the canonically conjugate pair $|\Lambda(t)\rangle\rangle$ and $|\rho(t)\rangle\rangle$ is defined by the "transition probability" in L-space given by the following relation:

$$\langle\langle\Lambda(t)|\,|\rho(t)\rangle\rangle = \langle\langle 1|\,\overline{S}(\infty,-\infty)\,|\rho_{eq}\rangle\rangle = \langle\langle\Im_{\mathcal{L}}|\,|\rho_{\mathcal{L}}\rangle\rangle, \tag{16.50}$$

where $\langle\langle\Im_{\mathcal{L}}|$ is given by

$$\langle\langle\Im_{\mathcal{L}}| = \langle\langle 1|\,\overline{S}(\infty,0)\,. \tag{16.51}$$

From the time-ordered expectation value deduced by Eq. (16.49), we see that indeed $\langle\langle 1|\,\overline{S}(\infty,-\infty)\,|\rho_{eq}\rangle\rangle$ is the analog to the transition amplitude occurring in zero temperature time-dependent quantum mechanics [120].

16.5.2 *Effective action and generating super-functional: Super S-matrix theory*

In order to gain further insights into the S-matrix formalism in L-space, we formulate the constrained stationary value [123] of the "super-action" given by

$$\int dt\ \langle\langle\Lambda(t)|\,i\hbar\frac{\partial}{\partial t} - \mathcal{L}\,|\rho(t)\rangle\rangle\,, \qquad \begin{cases}\langle\langle\Lambda(t)|\,|\rho(t)\rangle\rangle = 1\\ \langle\langle\Lambda(t)|\,\Psi\,|\rho(t)\rangle\rangle = \langle\Psi\rangle\,,\end{cases} \tag{16.52}$$

where Ψ stands for an arbitrary quantum field super-operator (the extension to several species of quantum field super-operators is straightforward). Thus, we consider the stationary variation of the super-functional

$$\begin{aligned}\Omega\left(|\rho(t)\rangle\rangle\,,|\Lambda(t)\rangle\rangle\right) = &\int dt\ \langle\langle\Lambda(t)|\,i\hbar\frac{\partial}{\partial t} - \mathcal{L}\,|\rho(t)\rangle\rangle\\ &-\int dx\ \langle\langle\Lambda(t)|\,\eta(x)\cdot\Psi\,|\rho(t)\rangle\rangle\\ &-\int dt\ w(t)\,\langle\langle\Lambda(t)|\,|\rho(t)\rangle\rangle,\end{aligned} \tag{16.53}$$

where $\eta(x)$ and $w(t)$ are introduced as $\mathbb{C}$-number Lagrange multipliers. Note that $\eta(x)$ is also playing the role of the Schwinger source field.

Carrying out the variation with respect to $|\Lambda(t)\rangle\rangle$ and $|\rho(t)\rangle\rangle$, and enforcing the stationarity of $\Omega(|\rho(t)\rangle\rangle, |\Lambda(t)\rangle\rangle)$, we obtain the following equations

$$\left.\begin{aligned} i\hbar\tfrac{\partial|\rho(t)\rangle\rangle}{\partial t} - \mathcal{L}|\rho(t)\rangle\rangle - \textstyle\int dx\, \eta(x)\cdot\Psi|\rho(t)\rangle\rangle - \; w(t)\,|\rho(t)\rangle\rangle = 0, \\ i\hbar\tfrac{\partial|\Lambda(t)\rangle\rangle}{\partial t} - \mathcal{L}|\Lambda(t)\rangle\rangle - \textstyle\int dx\, \eta(x)\cdot\Psi|\Lambda(t)\rangle\rangle - \; w^*(t)\,|\Lambda(t)\rangle\rangle = 0. \end{aligned}\right\} \tag{16.54}$$

We write the solutions to these equations as

$$|\rho(t)\rangle\rangle_{\eta,w} = \exp\left[\frac{-i}{\hbar}\int_{-\infty}^{t} dt'\; w(t')\right]|\rho(t)\rangle\rangle_{\eta}, \tag{16.55}$$

$$|\Lambda(t)\rangle\rangle_{\eta,w} = \exp\left[\frac{i}{\hbar}\int_{-\infty}^{t} dt'\; w^*(t')\right]|\Lambda(t)\rangle\rangle_{\eta}, \tag{16.56}$$

where $|\rho(t)\rangle\rangle_\eta$ and $|\Lambda(t)\rangle\rangle_\eta$ obey the equations,

$$i\hbar\frac{\partial|\rho(t)\rangle\rangle_\eta}{\partial t} - \mathcal{L}|\rho(t)\rangle\rangle_\eta - \int dx\,\eta(x)\cdot\Psi|\rho(t)\rangle\rangle_\eta = 0, \tag{16.57}$$

$$i\hbar\frac{\partial|\Lambda(t)\rangle\rangle_\eta}{\partial t} - \mathcal{L}|\Lambda(t)\rangle\rangle_\eta - \int dx\,\eta(x)\cdot\Psi|\Lambda(t)\rangle\rangle_\eta = 0. \tag{16.58}$$

Therefore, we can calculate the "transition probability" in L-space from the solutions of Eq. (16.54), which is given by Eq. (16.55). We obtain the following relation:

$$\langle\langle\Lambda(t)|\;|\rho(t)\rangle\rangle_{\eta,w} = 1 = \exp\left[\frac{-i}{\hbar}\int_{-\infty}^{\infty} dt'\; w(t')\right]\langle\langle\Lambda(t)|\;|\rho(t)\rangle\rangle_\eta, \tag{16.59}$$

which leads to the "transition probability" in the presence of the Schwinger source term given by

$$\langle\langle\Lambda(t)|\;|\rho(t)\rangle\rangle_\eta = \exp\left[\frac{i}{\hbar}\int_{-\infty}^{\infty} dt'\; w(t')\right] = \exp\left[\frac{i}{\hbar}W\right]. \tag{16.60}$$

This is the same "transition probability" as that obtained in Eq. (16.50). Therefore, we obtain the equality

$$\langle\langle 1|\,\overline{S}(\infty,-\infty)\,|\rho_{eq}\rangle\rangle = \exp\left[\frac{i}{\hbar}W\right]. \tag{16.61}$$

Analogous to the zero temperature S-matrix formalism, we identify W as the generating super-functional for connected n-point super-Green's functions. This is supported by the time-ordered way of taking the average value, shown by Eq. (16.49). To have a deeper appreciation of this analogy, we take the scalar product with $\langle\langle\Lambda(t)|_{\eta,w}$ of both sides of the first line of Eq. (16.54) and integrate over time. We obtain

$$\begin{aligned} W &= \int dt' \ w(t') \\ &= \int dt \ \langle\langle\Lambda(t)| \left\{ i\hbar \frac{\partial |\rho(t)\rangle\rangle}{\partial t} - \mathcal{L} - \int dx \ \eta(x) \cdot \Psi \right\} |\rho(t)\rangle\rangle_{\eta,w}, \end{aligned} \tag{16.62}$$

from which we deduced the variational derivative

$$\frac{\delta}{\delta\eta} W = -\langle\langle\Lambda(t)| \ \Psi \ |\rho(t)\rangle\rangle_{\eta,w} = \langle\Psi\rangle, \quad \text{(at the optimum condition)}. \tag{16.63}$$

Therefore, we have the effective super-action given by

$$A_{\text{eff}} = \int_{-\infty}^{\infty} dt \ \langle\langle\Lambda(t)| \ i\hbar\frac{\partial}{\partial t} - \mathcal{L} \ |\rho(t)\rangle\rangle_{\eta,w} = W + \int dx \ \eta(x) \cdot \langle\Psi\rangle, \tag{16.64}$$

from which we also deduced the variational derivative

$$\frac{\delta}{\delta\langle\Psi\rangle} A_{\text{eff}} = \eta. \tag{16.65}$$

Hence, the effective super-action is stationary with respect to the variation of $\langle\Psi\rangle$ when $\eta(x) = 0$, which is the physical situation, i.e., $W = A_{\text{eff}}$. Thus for $\eta(x) = 0$, we have

$$W = -i\hbar \ \ln\langle\langle 1| \ \overline{S}(\infty, -\infty) \ |\rho_{eq}\rangle\rangle = \int_{-\infty}^{\infty} dt \ \langle\langle\Lambda(t)| \ i\hbar\frac{\partial}{\partial t} - \mathcal{L} \ |\rho(t)\rangle\rangle. \tag{16.66}$$

The relation derived here for nonequilibrium quantum-field theory between super $\overline{S}$-matrix and the effective super-action provides a rigorous basis supporting the functional theory of time-dependent many-body quantum mechanics discussed by by Rajagopal and Buot in a series of papers [124–126].

We will refer to $\overline{S}(\infty, -\infty)$ as the complete S-matrix super-operator for quantum-field theoretical methods of nonequilibrium systems. Equation (16.39) allows us to write the evolution equation for the $\overline{S}$-matrix operator as

$$i\hbar\frac{\partial}{\partial t}\overline{S}(t, t_o) = \mathcal{L}_I^{(1)}(t) \ \overline{S}(t, t_o), \tag{16.67}$$

where

$$\mathcal{L}_I^{(1)}(t) = \mathcal{U}_o(0, t)(\mathcal{L} - \mathcal{L}_o)\mathcal{U}_o(t, 0), \tag{16.68}$$

and

$$(\mathcal{L} - \mathcal{L}_o) = [v(1) + u(1)]\,\Psi(1) + [v(1,2) + u(1,2)]\,\Psi(1)\,\Psi(2). \quad (16.69)$$

The Schwinger external sources [127] indicated by $u(1)$ and $u(1,2)$ contain all the time-dependent part of $v(1)$ and $v(1,2)$ respectively. These Schwinger source terms are to be set equal to zero at the end of all calculations. The solution to Eq. (16.67) immediately yields,

$$\overline{S}(t, t_o) = 1 - \frac{i}{\hbar}\int_{t_o}^{t} \mathcal{L}_I^{(1)}(t')\;\overline{S}(t', t_o)\;dt', \quad (16.70)$$

which upon iteration yields,

$$\overline{S}(t, t_o) = T\exp\left[\frac{-i}{\hbar}\int_{t_o}^{t} \mathcal{L}_I^{(1)}(t')\;dt'\right]. \quad (16.71)$$

Thus, we can explicitly exhibit $\langle\langle 1|\,\overline{S}(\infty, -\infty)\,|\rho_{eq}\rangle\rangle$ as

$$\begin{aligned}\langle\langle 1|\,\overline{S}(\infty, -\infty)\,|\rho_{eq}\rangle\rangle &= \langle\langle 1|\,T\exp\left[\frac{-i}{\hbar}\int_{-\infty}^{\infty} \mathcal{L}_I^{(1)}(t')\;dt'\right]|\rho_{eq}\rangle\rangle, \\ &= \exp\left[\frac{i}{\hbar}W\right], \end{aligned} \quad (16.72)$$

where W is the effective action. Equation (16.72) shows the relation of Eq. (16.61) to $\mathcal{L}_I^{(1)}(t)$ similar to zero-temperature many-body quantum-field theory.

16.6 Generalized super-Green's functions

In the use of quantum field theoretical techniques, we need to construct the Green's function formalism in L-space which incorporates all the nonequilibrium ("hat" and "tilde") quantum field super-operators, and sufficiently general enough to be able to account for quantum transport in systems with pairing interactions. This can be conveniently done by introducing a 4-component second quantized quantum field super-operators given by

$$\Psi(1) = \begin{pmatrix} \widehat{\psi}(1) \\ \widetilde{\psi}^{\dagger}(1) \\ \widehat{\psi}^{\dagger}(1) \\ \widetilde{\psi}(1) \end{pmatrix}. \quad (16.73)$$

This multicomponent notation for the quantum field super-operator is a further generalization of the multicomponent quantum field operator introduced by De Dominicis and Martin [128] in their treatment of non-normal systems at thermal equilibrium. Their treatment is also an extension of that introduced by Nambu [129] in his gauge invariance in theory of superconductivity. Since the exact pairing mechanism is left unspecified, the equations derived here are also applicable to other exotic pairing proposed in high-T_c superconductors. Thus, the corresponding quantum transport equations represent more general quantum transport equations including all particle-pairing dynamics.

From the commutation (anti-commutation) relations of the "hat" and "tilde" super-operators, we obtain the following equal-time commutation (anticommutation) relations,

$$[\Psi(1), \Psi(2)]_\epsilon \Rightarrow \left[\Psi_\alpha(r), \Psi_\beta(r')\right]_\epsilon = \delta(r-r')\,(\tau_4^\epsilon)_{\alpha\beta}, \tag{16.74}$$

where the matrix $(\tau_4^\epsilon)_{\alpha\beta}$ is

$$\tau_4^\epsilon = \begin{pmatrix} 0 & 0 & 1 & 0 \\ 0 & 0 & 0 & -\epsilon \\ -\epsilon & 0 & 0 & 0 \\ 0 & 1 & 0 & 0 \end{pmatrix}. \tag{16.75}$$

Similarly, we have

$$[\Psi(1), \Psi^\dagger(2)]_\epsilon \Rightarrow [\Psi_\alpha(r), \Psi_\beta^\dagger(r')]_\epsilon = \delta(r-r')\,(\tau_4^\epsilon)_{\alpha\beta}\,\overline{\sigma}_x, \tag{16.76}$$

where,

$$\overline{\sigma}_x = \begin{pmatrix} 0 & \delta \\ \delta & 0 \end{pmatrix}, \qquad \delta = \begin{pmatrix} 1 & 0 \\ 0 & 1 \end{pmatrix}, \tag{16.77}$$

and

$$[\Psi^\dagger(1), \Psi^\dagger(2)]_\epsilon \Rightarrow [\Psi_\alpha^\dagger(r), \Psi_\beta^\dagger(r')]_\epsilon = \delta(r-r')\,(\tau_4^\epsilon)^{-1}_{\alpha\beta}. \tag{16.78}$$

The important point to note is that the relations between the 4-component quantum field super-operators are $\mathbb{C}$-numbers (constant matrices). This is important for the validity of the Wicks' theorem applied to the averages of products of super-operators in the diagrammatic treatment. We further note that a general Liouvillian super-operator $\mathcal{L}$ can be expressed in either of the three different ways, namely, (a) in terms of Ψ alone, (b) in terms of Ψ and $\Psi^\dagger$, analogous to the equilibrium quantum field theoretical treatment, or (c) in terms of the $\Psi^\dagger$ alone. Since there seems to be no distinct advantages for our purpose, in formulating a generalized quantum field theory in L-space using either scheme (a), (b), or (c), we are free to choose either of these three schemes. We chose (a) since the resulting L-space formalism bears a close resemblance to the theory of classical statistical systems, especially when all potentials are ϵ-symmetrized.[2]

[2]Consider the terms occurring in $i\hbar\frac{\partial}{\partial t}\Psi_{\mathcal{L}}\left(\xi''\right) = \left[\Psi_{\mathcal{L}}\left(\xi''\right), \mathcal{L}\right]$. As an example, consider the term $v\left(1234\right)\left[\Psi_{\mathcal{L}}\left(\xi''\right), \Psi_{\mathcal{L}}\left(1\right)\Psi_{\mathcal{L}}\left(2\right)\Psi_{\mathcal{L}}\left(3\right)\Psi_{\mathcal{L}}\left(4\right)\right]$. Applying the equal-time commutation relations between the fields, we have the following results:

$$\begin{aligned}
&v\left(1234\right)\left[\Psi_{\mathcal{L}}\left(\xi''\right), \Psi_{\mathcal{L}}\left(1\right)\Psi_{\mathcal{L}}\left(2\right)\Psi_{\mathcal{L}}\left(3\right)\Psi_{\mathcal{L}}\left(4\right)\right] \\
&\quad = \left(\tau_4^e\right)_{\alpha_{\xi''},\alpha_1}\delta\left(r_{\xi''}-r_1\right)\Psi_{\mathcal{L}}\left(2\right)\Psi_{\mathcal{L}}\left(3\right)\Psi_{\mathcal{L}}\left(4\right)\times v\left(1234\right) \\
&\qquad + \left(\tau_4^e\right)_{\alpha_{\xi''},\alpha_2}\delta\left(r_{\xi''}-r_2\right)\Psi_{\mathcal{L}}\left(1\right)\Psi_{\mathcal{L}}\left(3\right)\Psi_{\mathcal{L}}\left(4\right)\times v\left(1234\right) \\
&\qquad + \left(\tau_4^e\right)_{\alpha_{\xi''},\alpha_3}\delta\left(r_{\xi''}-r_3\right)\Psi_{\mathcal{L}}\left(1\right)\Psi_{\mathcal{L}}\left(2\right)\Psi_{\mathcal{L}}\left(4\right)\times v\left(1234\right) \\
&\qquad + \left(\tau_4^e\right)_{\alpha_{\xi''},\alpha_3}\delta\left(r_{\xi''}-r_4\right)\Psi_{\mathcal{L}}\left(1\right)\Psi_{\mathcal{L}}\left(2\right)\Psi_{\mathcal{L}}\left(3\right)\times v\left(1234\right).
\end{aligned} \tag{16.79}$$

It is desirable to fix the arrangement of indices in the fields. Since the 1234 indices are dummy variables and repeated indices are summed over, we can rearrange the prefactors, involving the fields, of the potential in Eq. (16.79) such that they are brought to the same form as the first prefactor. This can be done if we accompany the rearrangement of the fields by a displacement of the dummy variable 1 in $v(1234)$ by an appropriate number of places to the right and taking into account the appropriate factors due to the commutation relationship indicated in Eq. (16.79). Therefore, we can also write,

$$\begin{aligned}
&v\left(1234\right)\left[\Psi_{\mathcal{L}}\left(\xi''\right), \Psi_{\mathcal{L}}\left(1\right)\Psi_{\mathcal{L}}\left(2\right)\Psi_{\mathcal{L}}\left(3\right)\Psi_{\mathcal{L}}\left(4\right)\right] \\
&\quad = \left(\tau_4^e\right)_{\alpha_{\xi''},\alpha_1}\delta\left(r_{\xi''}-r_1\right)\left\{\sum_{n=0}^{3}\epsilon^n D_n\left(1\right)v\left(1234\right)\right\}\Psi_{\mathcal{L}}\left(2\right)\Psi_{\mathcal{L}}\left(3\right)\Psi_{\mathcal{L}}\left(4\right),
\end{aligned}$$

where $D_n\left(1\right)$ transpose the index 1 n places to the right and $\epsilon^n = \left(-1\right)^n$. We have, upon taking the time-ordered average, the following:

$$\begin{aligned}
&\left\langle\left(\tau_4^e\right)_{\alpha_{\xi''},\alpha_1}\delta\left(r_{\xi''}-r_1\right)\left\{\sum_{n=0}^{3}\epsilon^n D_n\left(1\right)v\left(1234\right)\right\}\Psi_{\mathcal{L}}\left(2\right)\Psi_{\mathcal{L}}\left(3\right)\Psi_{\mathcal{L}}\left(4\right)\right\rangle \\
&\quad = \left(\tau_4^e\right)_{\alpha_{\xi''},\alpha_1}\delta\left(r_{\xi''}-r_1\right)\frac{1}{3!}\left\{\sum_{P(1234)}\epsilon^P v\left(1234\right)\right\}\mathcal{G}\left(234\right).
\end{aligned}$$

This observation can be utilized as a trick for recasting non-equilibrium quantum field theory as a pure functional theory, similar to the classical-statistical field theory. To implement this trick, one simply rewrite the external potentials in e-symmetrized form from the very beginning, as employed by De Dominicis and Martin [128].

In this scheme, we define the one-particle super Green's function as

$$\mathcal{G}(1,2) = \frac{\langle\langle \Im_{\mathcal{L}}|\, T\, \Psi_{\mathcal{L}}(1)\, \Psi_{\mathcal{L}}(2)\, |\rho_{\mathcal{L}}\rangle\rangle}{\langle\langle \Im_{\mathcal{L}}|\, |\rho_{\mathcal{L}}\rangle\rangle}. \tag{16.80}$$

From previous developments, the averages can be written in terms of the $\bar{S}$-matrix, e.g.,

$$\mathcal{G}(1,2) = (i\hbar)^2 \frac{\frac{\delta^2 \langle\langle 1|\, \overline{S}(\infty,-\infty)\, |\rho_{eq}\rangle\rangle}{\delta u(1)\delta u(2)}}{\langle\langle 1|\, \overline{S}(\infty,-\infty)\, |\rho_{eq}\rangle\rangle}. \tag{16.81}$$

Clearly, multi time-dependent averages or moments are defined only for time-ordered quantum field super-operators. The Green's functions of the three different schemes are related to each other through the constant matrix $\overline{\sigma}_x$. We obtained the following,

$$\mathcal{G}(1,2) = i\hbar \begin{pmatrix} F(1,2) & G(1,2) \\ \epsilon G^T(1,2) & {}^{\dagger}F(1,2) \end{pmatrix}, \tag{16.82}$$

where the superscript T indicates the taking of the transpose, and,

$$F(1,2) = \begin{pmatrix} g^c_{hh}(1,2) & g^<_{hh}(1,2) \\ g^>_{hh}(1,2) & g^{ac}_{hh}(1,2) \end{pmatrix}, \tag{16.83}$$

$${}^{\dagger}F(1,2) = \begin{pmatrix} g^c_{ee}(1,2) & \epsilon g^<_{ee}(1,2) \\ \epsilon g^>_{ee}(1,2) & g^{ac}_{ee}(1,2) \end{pmatrix}, \tag{16.84}$$

$$G(1,2) = \begin{pmatrix} G^c(1,2) & \epsilon G^<(1,2) \\ G^>(1,2) & \epsilon G^{ac}(1,2) \end{pmatrix}, \tag{16.85}$$

$$\epsilon G^T(1,2) = \begin{pmatrix} \epsilon G^{cT}(1,2) & \epsilon G^{>T}(1,2) \\ G^{<T}(1,2) & G^{acT}(1,2) \end{pmatrix}. \tag{16.86}$$

The matrix elements are time-ordered averages. By virtue of the number-conserving Liouvillian and the property of quantum superfield operators, Eqs. (16.23)–(16.24), these are given, with the time-ordered averages in the super-Heisenberg representation translated into their H-space counterparts, by the following expressions:

$$\langle T\, \hat{\psi}_{\mathcal{L}}(1)\, \hat{\psi}_{\mathcal{L}}(2)\rangle = \theta(12)\, i\hbar g^>_{hh}(1,2) + \theta(21)\, i\hbar g^<_{hh}(1,2) = i\hbar g^c_{hh}(1,2), \tag{16.87}$$

$$\langle T\, \hat{\psi}_{\mathcal{L}}(1)\, \tilde{\psi}^{\dagger}_{\mathcal{L}}(2)\rangle = -\mathrm{Tr}\, \rho_{\mathcal{H}} \psi_{\mathcal{H}}(2)\, \psi_{\mathcal{H}}(1) = i\hbar g^<_{hh}(1,2), \tag{16.88}$$

$$\langle T\, \tilde{\psi}^{\dagger}_{\mathcal{L}}(1)\, \hat{\psi}_{\mathcal{L}}(2)\rangle = \mathrm{Tr}\, \rho_{\mathcal{H}} \psi_{\mathcal{H}}(1)\, \psi_{\mathcal{H}}(2) = i\hbar g^>_{hh}(1,2), \tag{16.89}$$

$$\langle T\, \tilde{\psi}^{\dagger}_{\mathcal{L}}(1)\, \tilde{\psi}^{\dagger}_{\mathcal{L}}(2)\rangle = \theta(12)\, i\hbar g^{<}_{hh}(1,2) + \theta(21)\, i\hbar g^{>}_{hh}(1,2) = i\hbar g^{ac}_{hh}(1,2), \tag{16.90}$$

$$\langle T\, \hat{\psi}^{\dagger}_{\mathcal{L}}(1)\, \hat{\psi}^{\dagger}_{\mathcal{L}}(2)\rangle = \theta(12)\, i\hbar G^{>}(1,2) + \theta(21)\, i\hbar G^{<}(1,2) = i\hbar G^{c}(12), \tag{16.91}$$

$$\langle T\, \hat{\psi}_{\mathcal{L}}(1)\, \tilde{\psi}_{\mathcal{L}}(2)\rangle = \mathrm{Tr}\, \rho_{\mathcal{H}} \psi^{\dagger}_{\mathcal{H}}(2)\, \psi_{\mathcal{H}}(1) = -i\hbar G^{<}(1,2), \tag{16.92}$$

$$\langle T\, \tilde{\psi}^{\dagger}_{\mathcal{L}}(1)\, \hat{\psi}^{\dagger}_{\mathcal{L}}(2)\rangle = \mathrm{Tr}\, \rho_{\mathcal{H}} \psi_{\mathcal{H}}(1)\, \psi^{\dagger}_{\mathcal{H}}(2) = i\hbar G^{>}(1,2), \tag{16.93}$$

$$\begin{aligned}\langle T\, \tilde{\psi}^{\dagger}_{\mathcal{L}}(1)\, \tilde{\psi}_{\mathcal{L}}(2)\rangle &= -\left\{\theta(12)\, i\hbar G^{<}(1,2) + \theta(21)\, i\hbar G^{>}(1,2)\right\} \\ &= -i\hbar G^{ac}(12),\end{aligned} \tag{16.94}$$

$$\begin{aligned}\langle T\, \hat{\psi}^{\dagger}_{\mathcal{L}}(1)\, \hat{\psi}_{\mathcal{L}}(2)\rangle &= -\left\{\theta(12)\, i\hbar G^{<}(2,1) + \theta(21)\, i\hbar G^{>}(2,1)\right\} \\ &= -i\hbar G^{cT}(12),\end{aligned} \tag{16.95}$$

$$\langle T\, \tilde{\psi}_{\mathcal{L}}(1)\, \hat{\psi}_{\mathcal{L}}(2)\rangle = -\mathrm{Tr}\, \rho_{\mathcal{H}} \psi^{\dagger}_{\mathcal{H}}(1)\, \psi_{\mathcal{H}}(2) = i\hbar G^{<T}(1,2), \tag{16.96}$$

$$\langle T\, \hat{\psi}^{\dagger}_{\mathcal{L}}(1)\, \tilde{\psi}^{\dagger}_{\mathcal{L}}(2)\rangle = -\mathrm{Tr}\, \rho_{\mathcal{H}} \psi_{\mathcal{H}}(2)\, \psi^{\dagger}_{\mathcal{H}}(1) = -i\hbar G^{>T}(1,2), \tag{16.97}$$

$$\begin{aligned}\langle T\, \tilde{\psi}_{\mathcal{L}}(1)\, \tilde{\psi}^{\dagger}_{\mathcal{L}}(2)\rangle &= \theta(12)\, i\hbar G^{>}(2,1) + \theta(21)\, i\hbar G^{<}(2,1), \\ &= i\hbar G^{acT}(12),\end{aligned} \tag{16.98}$$

$$\begin{aligned}\langle T\, \hat{\psi}^{\dagger}_{\mathcal{L}}(1)\, \hat{\psi}^{\dagger}_{\mathcal{L}}(2)\rangle &= \theta(12)\, i\hbar g^{>}_{ee}(1,2) + \theta(21)\, i\hbar g^{<}_{ee}(1,2), \\ &= i\hbar g^{c}_{ee}(1,2),\end{aligned} \tag{16.99}$$

$$\langle T\, \hat{\psi}^{\dagger}_{\mathcal{L}}(1)\, \tilde{\psi}_{\mathcal{L}}(2)\rangle = \mathrm{Tr}\, \rho_{\mathcal{H}} \psi^{\dagger}_{\mathcal{H}}(2)\, \psi^{\dagger}_{\mathcal{H}}(1) = -i\hbar g^{<}_{ee}(1,2), \tag{16.100}$$

$$\langle T\, \tilde{\psi}_{\mathcal{L}}(1)\, \hat{\psi}^{\dagger}_{\mathcal{L}}(2)\rangle = -\mathrm{Tr}\, \rho_{\mathcal{H}} \psi^{\dagger}_{\mathcal{H}}(1)\, \psi^{\dagger}_{\mathcal{H}}(2) = -i\hbar g^{>}_{ee}(1,2), \tag{16.101}$$

$$\begin{aligned}\langle T\, \tilde{\psi}_{\mathcal{L}}(1)\, \tilde{\psi}_{\mathcal{L}}(2)\rangle &= \theta(12)\, i\hbar g^{<}_{ee}(1,2) + \theta(21)\, i\hbar g^{>}_{ee}(1,2) \\ &= i\hbar g^{ac}_{ee}(1,2).\end{aligned} \tag{16.102}$$

The "retarded" and "advanced" correlations are also defined using the above components, for example, we have

$$g^{c}_{hh} - g^{<}_{hh} = g^{>}_{hh} - g^{ac}_{hh} = g^{r}_{hh}, \tag{16.103}$$

$$g^{c}_{hh} - g^{>}_{hh} = g^{<}_{hh} - g^{ac}_{hh} = g^{a}_{hh}. \tag{16.104}$$

Note that $G(1,2)$ of Eq. (16.85) corresponds to the Keldysh nonequilibrium Green's function for normal systems [11, 106].

In what follows we often refer to the super-Green's function simply as average or moment quantum distribution function. Later we define quantum super-correlation functions or quantum cumulants, analogous to the classical statistical theory of random and dissipative phenomena. This distinction is very important in treating the quantum transport of particle system which exhibit superfluidity and superconductivity. We will also refer to both classes of distribution functions as generalized quantum distribution functions (GQDF).

16.7 Self-consistent equations

In terms of the 4-component field of Eq. (16.73), let us consider a nonequilibrium system whose Liouvillian can in general be expressed as

$$\mathcal{L} = \sum_{N=1}^{M} v\,(1,2,3,\ldots,N)\;\Psi\,(1)\;\Psi\,(2)\;\Psi\,(3)\cdots\cdot\Psi\,(N) + \mathcal{L}_{\text{ext}}^{\text{Schw}}, \quad (16.105)$$

where the Schwinger source term is given by

$$\mathcal{L}_{\text{ext}}^{\text{Schw}} = \widehat{\mathcal{H}}_{\text{ext}} - \widetilde{\mathcal{H}}_{\epsilon,\text{ext}} = u\,(1)\,\Psi\,(1) + u\,(1,2)\,\Psi\,(1)\;\Psi\,(2). \quad (16.106)$$

The indices $1,2,3,\ldots,N$, each stands for all pertinent quantum labels, and repeated indices are "summed" over. For example, for a system of interacting bosons or fermions, interacting via two-body forces, we have the following expression for the potentials occurring in Eq. (16.105),

$$v\,(1) = 0, \quad (16.107)$$

$$v\,(1,2) = [\delta_{\alpha 1,3}\delta_{\alpha 2,1} - \delta_{\alpha 1,2}\delta_{\alpha 2,4}]\left\{\frac{-\hbar^2}{2m}\nabla_1^2 + v\,\{1\}\right\}\delta\,(r_1 - r_2), \quad (16.108)$$

$$v\,(1,2,3) = 0, \quad (16.109)$$

$$\begin{aligned} v\,(1,2,3,4) = {} & [\delta_{\alpha 1,3}\delta_{\alpha 2,3}\delta_{\alpha 3,1}\delta_{\alpha 4,1} - \delta_{\alpha 1,4}\delta_{\alpha 2,4}\delta_{\alpha 3,2}\delta_{\alpha 4,2}] \\ & \times \{v\,(r_1 - r_2)\}\,\delta\,(r_2 - r_3)\,\delta\,(r_1 - r_4), \end{aligned} \quad (16.110)$$

where the subscripts of the delta-function pertain to the components of Ψ in Eq. (16.73). In the super-Heisenberg representation, Ψ obeys the equation of motion as

$$i\hbar\frac{\partial}{\partial t}\Psi_{\mathcal{L}}(\xi'') = [\Psi_{\mathcal{L}}(\xi''),\mathcal{L}]. \quad (16.111)$$

Note that in the equation of motion for $\Psi_{\mathcal{L}}\left(\xi''\right)$, all times occurring in $\mathcal{L}$ are equal to $t_{\xi''}$ and therefore we can use the equal-time relations to evaluate the RHS of Eq. (16.111). The results can be written as

$$i\hbar\frac{\partial}{\partial t_{\xi''}}\Psi_{\mathcal{L}}\left(\xi''\right) = (\tau_4^{\epsilon})_{\alpha_{\xi''},\alpha_1}\,\delta\left(r_{\xi''} - r_1\right) \times \sum_{N=1}^{M}\left\{\sum_{n=0}^{N-1}\epsilon^n D_n\left(1\right) v\left(1, t_{\xi''}; 2, t_{\xi''}; \cdots\cdots N, t_{\xi''}\right)\right\} \times \Psi_{\mathcal{L}}\left(1, t_{\xi''}\right)\Psi_{\mathcal{L}}\left(2, t_{\xi''}\right)\cdots\cdots\Psi_{\mathcal{L}}\left(N, t_{\xi''}\right), \tag{16.112}$$

where $D_n\left(1\right)$ is a "transposition operator" acting only on $v(1, t_{\xi''}; 2, t_{\xi''}; \cdots ; N, t_{\xi''})$. The effect of $D_n\left(1\right)$ is to transpose the index 1, n places to the right in $v(1, t_{\xi''}; 2, t_{\xi''}; \cdots ; N, t_{\xi''})$, while the functional form of v relative to the order of its indices remains unaffected. The average of $\Psi_{\mathcal{L}}\left(\xi''\right)$, which we denote by $\langle\Psi_{\mathcal{L}}\left(\xi''\right)\rangle \equiv \mathcal{G}\left(\xi''\right)$, obeys the following equation of motion:

$$i\hbar\frac{\partial}{\partial t_{\xi''}}\mathcal{G}\left(\xi''\right) = (\tau_4^{\epsilon})_{\alpha_{\xi''},\alpha_1}\,\delta\left(r_{\xi''} - r_1\right) \times \sum_{N=1}^{M}\left\{\sum_{i=1}^{N}\delta\left(t_{\xi''} - t_i\right)\frac{v^{\epsilon}\left(1, 2, \ldots N\right)}{(N-1)!}\right\}\mathcal{G}\left(2, 3, 4, \ldots, N\right), \tag{16.113}$$

where,

$$\mathcal{G}\left(2, 3, 4, \ldots, N\right) = \langle\Psi_{\mathcal{L}}\left(2\right)\Psi_{\mathcal{L}}\left(3\right)\Psi_{\mathcal{L}}\left(4\right)\ldots\ldots\Psi_{\mathcal{L}}\left(N\right)\rangle, \tag{16.114}$$

$$v^{\epsilon}\left(1, 2, 3, \ldots, N\right) = \sum_{P}\epsilon^{P} v\left(1, 2, 3, \ldots, N\right). \tag{16.115}$$

Here $v^{\epsilon}\left(1, 2, 3, \ldots, N\right)$ is the ϵ-symmetrized potential obtained by summing all the ϵ-permutation of the indices in $v\left(1, 2, 3, \ldots, N\right)$. With Eq. (16.113), the hierarchy of equations for the GQDF, $\mathcal{G}\left(1, 2, 3, 4, \ldots, N\right)$, for any N can be generated by using the variational derivative indicated in Eq. (16.81). The variational derivative allows for an exact "closed form" equations to be written down for all the distribution functions. The major task in the calculations lies in expressing the variational derivative operations self-consistently in terms of the pertinent lower-order distribution functions and effective multi-point effective interaction potentials or vertex functions, in such a way as to decouple the hierarchy and obtain a mathematically closed set of coupled equations, the number of equations being equal to the number of unknown functions.

Analogous to classical case we also refer to $\langle\langle 1|\,\overline{S}\left(\infty, -\infty\right)|\rho_{eq}\rangle\rangle = \exp\left[\frac{i}{\hbar}W\right]$ as the characteristic functional, it is a functional of the Schwinger source fields. This can be evaluated in terms of 1-point and 2-point GQDFs, from which a fully renormalized equation of motions can be obtained. It serves as the generating functional for the quantum-field moment distributions in the presence of the Schwinger source terms. We may thus write,

$$\exp\left[\frac{i}{\hbar}W\right]\mathcal{G}(1,2,\ldots,N)=\left[i\hbar\frac{\delta}{\delta u(N)}i\hbar\frac{\delta}{\delta u(N-1)}\cdots i\hbar\frac{\delta}{\delta u(1)}\right]\exp\left[\frac{i}{\hbar}W\right], \tag{16.116}$$

where to account for anti-commuting Grassmann variables $u(i)$, in the case of fermions, we take the convention that the variational derivatives are right-hand derivatives, i.e., variational derivatives involving higher-indexed u's operates from the right of the variational derivatives involving lower-indexed u's, in a "horizontal stack" fashion. The anticommutation relationship must be taken into account if any rearrangement different from this conventional arrangement is made. This convention must be adhered to make the theory consistent. Equation (16.116) for the quantum-field moment distribution $\mathcal{G}$ yields an iterative relation between $\mathcal{G}$ of different orders,

$$\mathcal{G}(1,2,\ldots,n)=\left\{i\hbar\frac{\delta}{\delta u(n)}-\frac{\delta W}{\delta u(n)}\right\}\mathcal{G}(1,2,\ldots,n-1), \tag{16.117}$$

where

$$-\frac{\delta W}{\delta u(n)}=i\hbar\frac{\delta\ln\langle\langle 1|\,\overline{S}(\infty,-\infty)\,|\rho_{eq}\rangle\rangle}{\delta u(n)}=\mathcal{G}(n)=\mathcal{K}(n). \tag{16.118}$$

In Eq. (16.117), the operator within the curly bracket is understood to be right-handed operator in the same way as the repeated variational derivatives operators were defined.

Upon expanding the exponential in Eq. (16.72), we are also able to express the characteristic functional in terms of the generalized quantum-field moment distributions, $\mathcal{G}$'s, as

$$\exp\left[\frac{i}{\hbar}W\right]=\sum_{n=0}^{\infty}\frac{1}{n!}u(1)\,u(2)\ldots u(n)\,\mathcal{G}(1,2,\ldots,n), \tag{16.119}$$

since the $\mathcal{G}$'s are the coefficient in the Fourier series expansion of Eq. (16.72) in the limit the $u(i)$'s go to zero. We note that in the perturbative diagrammatic application of Wick's theorem (for superfluid Bose system it is assumed that $\mathcal{L}_o$ is expanded about the condensate for Wick's theorem to be applicable) to evaluate the $\mathcal{G}$'s, each term in the summation of Eq. (16.119) can be represented by a sum of diagrams or graphs in a one-to-one correspondence basis.

Connected and Irreducible Diagrams: Correlation Function $\mathcal{K}$. The set of topologically distinct diagrams corresponding to each $\mathcal{G}$ is made up of "connected" and "disconnected" subsets of graphs. An important lemma in graph theory [130] asserts that the sum of the RHS of Eq. (16.119) satisfy the following identity,

$$\sum_{n=0}^{\infty} \frac{1}{n!} u(1)\, u(2) \cdots\cdot u(n)\, \mathcal{G}(1,2,\ldots,n)$$

$$= \exp\left[\sum_{n=0}^{\infty} \frac{1}{n!} u(1)\, u(2) \cdots\cdot u(n)\, \mathcal{K}(1,2,\ldots,n)\right], \qquad (16.120)$$

where the quantum-field cumulant distribution functions or quantum correlation functions, $\mathcal{K}(1,2,\ldots,n)$, represent the topologically distinct "connected" subset of graphs occurring in the LHS of Eq. (16.120). Corresponding to the classical theory [25], these quantum correlation functions are generated by variational derivatives of $\ln \langle\langle 1| \overline{S}(\infty,-\infty) |\rho_{eq}\rangle\rangle = \frac{i}{\hbar} W$ with respect to the Schwinger external sources. Thus, we have

$$\mathcal{K}(1,2,\ldots,n) = (i\hbar)^n \frac{\delta^n \ln \langle\langle 1| \overline{S}(\infty,-\infty) |\rho_{eq}\rangle\rangle}{\delta u(n)\, \delta u(n-1) \cdots \delta u(1)}, \qquad (16.121)$$

which yields a simple iterative relation between correlation functions of different orders, namely,

$$\mathcal{K}(1,2,\ldots,n) = i\hbar \frac{\delta \mathcal{K}(1,2,\ldots,n-1)}{\delta u(n)}. \qquad (16.122)$$

Similar functional derivative relations can be obtained between GQDF with even number of indices by using the variation with respect to the external Schwinger source term $u(1,2)$. Since $u(1,2)$ is an ordinary $\mathbb{C}$-number, the order of the $u(i,j)$'s is not critical in taking the functional derivatives. We use the more elementary variation with respect to the $u(i)$'s, so as to completely resemble classical distribution function theory and to include systems which exhibit superfluidity and superconductivity. The above iterative relations are important in setting up the exact "closed" form of the equations for a GQDF of arbitrary order. For a system of bosons, one expects that the 1st, 2nd, 3rd, and 4th-order GQDF essentially determines its characteristic functional, whereas, for unconstrained fermion systems only even-ordered GQDF, 2nd and 4th order, determine its characteristic functional.

The following relations hold between the correlation quantum distribution functions (CQDFs) and the moment quantum distribution functions (MQDFs),

$$\mathcal{K}(1) = \mathcal{G}(1), \tag{16.123}$$

$$\mathcal{K}(1,2) = \mathcal{G}(1,2) - \frac{1}{2!}\sum_{P} \epsilon^{P} \mathcal{G}(1)\,\mathcal{G}(2). \tag{16.124}$$

To obtain higher-order CQDFs, we have to subtract all possible lower order CQDFs from the MQDFs difference,

$$\mathcal{K}(1,2,\ldots,n) = \mathcal{G}(1,2,\ldots,n) - \frac{1}{n!}\sum_{P} \epsilon^{P} \mathcal{G}(1)\,\mathcal{G}(2)\cdots\mathcal{G}(n). \tag{16.125}$$

This can be verified by a straightforward application of the variational derivatives defined above.

16.7.1 *Self-consistent equations for GQDF*

We multiply both sides of Eq. (16.113) by the inverse of the matrix τ_4^{ϵ} and limit ourselves to $M = 4$ in Eq. (16.105), which will cover almost all of many-body physics, and make use of Eq. (16.117) to rewrite Eq. (16.113) as

$$\begin{aligned}\mathcal{G}_o^{-1}(\xi'',2)\,\mathcal{G}(2) = \mathcal{V}(\xi'') &+ \frac{1}{(3-1)!}\mathcal{V}(\xi''23)\left\{i\hbar\frac{\delta}{\delta u(3)} + \mathcal{G}(3)\right\}\mathcal{G}(2)\\ &+ \frac{1}{(4-1)!}\mathcal{V}(\xi''234)\left\{i\hbar\frac{\delta}{\delta u(4)} + \mathcal{G}(4)\right\}\\ &\times\left\{i\hbar\frac{\delta}{\delta u(3)} + \mathcal{G}(3)\right\}\mathcal{G}(2),\end{aligned} \tag{16.126}$$

where

$$\mathcal{G}_o^{-1}(\xi'',2) = \left(\tau_4^{\epsilon}\right)^{-1} i\hbar\delta(\xi'',2)\,\tfrac{\partial}{\partial t_2} - \mathcal{V}((\xi'',2)), \tag{16.127}$$

$$\mathcal{V}(\xi'') = \delta\left(r_{\xi''} - r_1\right)\delta\left(t_{\xi''} - t_1\right)\{v(1) + u(1)\}, \tag{16.128}$$

$$\frac{1}{(n-1)!}\mathcal{V}(\xi''2\ldots n) = \delta\left(r_{\xi''} - r_1\right)\prod_{i=1}^{n}\delta\left(t_{\xi''} - t_i\right)\frac{1}{(n-1)!}v^{\epsilon}(12\ldots n). \tag{16.129}$$

We recast Eq. (16.126) as

$$\mathcal{G}_o^{-1}(1,2)\,\mathcal{G}(2) = \mathcal{V}(1) + \Xi(1), \tag{16.130}$$

where

$$\Xi(1) = \frac{1}{2!}\mathcal{V}(123)\left\{i\hbar\frac{\delta}{\delta u(3)} + \mathcal{G}(3)\right\}\mathcal{G}(2)$$
$$+\frac{1}{3!}\mathcal{V}(1234)\left\{i\hbar\frac{\delta}{\delta u(4)} + \mathcal{G}(4)\right\}\left\{i\hbar\frac{\delta}{\delta u(3)} + \mathcal{G}(3)\right\}\mathcal{G}(2). \tag{16.131}$$

Equation (16.130) for the average field is sometimes referred to as the Schwinger equation. In Eq. (16.130), $\Xi(1)$ is the effective ("internal") source function contributed by the system. We can utilize Eq. (16.130) to obtain the equation for the propagator or two-point correlation function $\mathcal{K}(23)$. We obtained

$$[\mathcal{G}_o^{-1}(1,2) - \breve{\Sigma}(12)]\frac{\mathcal{K}(23)}{i\hbar} = \delta(13), \qquad \breve{\Sigma}(12) = \frac{\delta\Xi(1)}{\delta\mathcal{G}(2)}, \tag{16.132}$$

where $\breve{\Sigma}(12)$ is the particle super self-energy. We will refer to $\mathcal{K}(23)/i\hbar$ as the super-propagator kernel or simply as the propagator. Note that the inverse of the quantum correlation function is $[\mathcal{G}_o^{-1}(1,2)-\breve{\Sigma}(12)]/i\hbar$, whereas the inverse of the propagator is $\mathcal{G}_o^{-1}(1,2) - \breve{\Sigma}(12)$.

16.7.1.1 *Schwinger equation as a many-body effective Schrödinger equation*

From Eq. (16.130) and second equation of Eq. (16.132), we may write the Schwinger equation as

$$\begin{aligned}\mathcal{G}_o^{-1}(1,2)\,\mathcal{G}(2) &= \mathcal{V}(1) + \Xi(1), \\ &= \mathcal{V}(1) + \breve{\Sigma}(12)\,\mathcal{G}(2),\end{aligned} \tag{16.133}$$

which, in the limit that the Schwinger source term $\mathcal{V}(1)$ goes to zero, yields the effective Schrödinger equation with nonlocal effective potential given by the self-energy $\breve{\Sigma}(12)$, namely, $\left\{\mathcal{G}_o^{-1}(1,2) - \breve{\Sigma}(12)\right\}\mathcal{G}(2) = 0$. A very well-known approximation scheme for approximating this self-consistent many-body effective Schrödinger equation is given by Kohn and Sham [131] in the form of a density-functional formalism. Equation (16.133) can also be deduced from the equation of motion obeyed by $\mathcal{G}(1,2)$.

Corresponding to the Kohn–Sham equation for fermions, a well-known equation for dilute many-body boson systems is the so-called Gross–Pitaevskii equation or the nonlinear Schrödinger equation for bosons.

In the high-density limit (highly nonlinear regimes) it has been shown that the Gross–Pitaevskii equation can be mapped to a Kohn–Sham scheme of density functional theory and vice versa.

16.7.1.2 *Closure problem and renormalization procedure*

Equations (16.130) and (16.132) are not closed since $\Xi(1)$ and $\breve{\Sigma}(12)$ are functionals of higher-order GQDF. In what follows we will show that the 3-point CQDF occurring in $\Xi(1)$ and $\breve{\Sigma}(12)$ can be expressed in terms of lower-order CQDFs in combination with 3-point vertex function. This 3-point vertex functions in turn obeys an iterative equation which can be solved in terms of the functional derivative of $\breve{\Sigma}(12)$ with respect to $\mathcal{G}(i)$ and $\mathcal{K}(ij)$. Similarly, the 4-point CQDF occurring in $\breve{\Sigma}(12)$ can be expressed in analogous fashion. We will see that $\breve{\Sigma}(12)$ involves 3-point and 4-point CQDF, which can be expressed in terms of 3-point and 4-point vertices in combination with 1-point and 2-point CQDF. These vertex function can in turn be expressed in terms of functional derivative of $\breve{\Sigma}(12)$ with respect to the 1-point and 2-point CQDF, thus closing the number of unknown functions and effectively decoupling the hierarchy. In the end, we basically end up with eight equations in eight unknowns, which can then be solved self-consistently.

First let us express $\Xi(1)$ and $\breve{\Sigma}(12)$ in terms of CQDF. We make use of Eqs. (16.117), (16.124), and the expansion of higher-order cumulants as function of the averages to obtain the following expressions:

$$\begin{aligned}\Xi(1) &= \frac{\mathcal{V}(123)}{2!}\{\mathcal{K}(23)+\mathcal{G}(2)\mathcal{G}(3)\}\\ &\quad+\frac{\mathcal{V}(1234)}{3!}\{\mathcal{K}(234)+3\mathcal{G}(2)\mathcal{K}(34)+\mathcal{G}(2)\mathcal{G}(3)\mathcal{G}(4)\},\end{aligned} \tag{16.134}$$

$$\begin{aligned}\breve{\Sigma}(1w) &= \frac{\mathcal{V}(123)}{2!}\left\{\begin{array}{c}\mathcal{K}(23l)\,\mathcal{K}^{-1}(lw)\\ +\epsilon\mathcal{G}(3)\,\delta(2w)+\mathcal{G}(2)\,\delta(3w)\end{array}\right\}+\frac{\mathcal{V}(1234)}{3!}\\ &\quad\times\left\{\begin{array}{c}\mathcal{K}(234l)\,\mathcal{K}^{-1}(lw)\\ +\epsilon\mathcal{K}(24)\,\delta(3w)+\mathcal{K}(34)\,\delta(2w)+\mathcal{K}(23)\,\delta(4w)\\ +\epsilon\mathcal{G}(3)\,\mathcal{K}(24l)\,\mathcal{K}^{-1}(lw)+\mathcal{G}(2)\,\mathcal{K}(34l)\,\mathcal{K}^{-1}(lw)\\ +\mathcal{G}(4)\,\mathcal{K}(23l)\,\mathcal{K}^{-1}(lw)+\mathcal{G}(2)\,\mathcal{G}(3)\,\delta(4w)\\ +\epsilon\mathcal{G}(2)\,\mathcal{G}(4)\,\delta(3w)+\mathcal{G}(3)\,\mathcal{G}(4)\,\delta(2w)\end{array}\right\}.\end{aligned} \tag{16.135}$$

We will now show that Eqs. (16.134) and (16.135) are functionals of only one-point and two-point CQDFs. In order to do this, we define the n-point 'effective interaction potentials' or vertex functions as

$$\begin{aligned}\breve{\Gamma}(123\ldots n) &= -\left\{\frac{\delta}{\delta\mathcal{G}(n)}\right\}\cdots\left\{\frac{\delta}{\delta\mathcal{G}(3)}\right\}\mathcal{K}^{-1}(12)\\ &= \frac{\delta}{\delta\mathcal{G}(n)}\breve{\Gamma}(12\ldots n-1)\\ &= \frac{\delta^n\mathcal{F}^{(1)}}{i\hbar\delta\mathcal{G}(n)\,\delta\mathcal{G}(n-1)\ldots\delta\mathcal{G}(1)},\end{aligned} \tag{16.136}$$

where $\mathcal{F}^{(1)}$ is the generating functional for the n-point super vertex function [128], obtained from the effective super-action W by Legendre transformation,

$$\mathcal{F}^{(1)} = W + \mathcal{V}(1)\,\mathcal{G}(1) + \frac{1}{2!}\mathcal{V}(12)\,\mathcal{G}(12).$$

We can relate the vertex functions to the CQDF, obtaining

$$\breve{\Gamma}(123) = \mathcal{K}^{-1}(1j)\,\mathcal{K}(jil)\,\mathcal{K}^{-1}(i2)\,\mathcal{K}^{-1}(l3), \tag{16.137}$$

which allows us to express the 3-point CQDF in terms of the 3-point vertex function in combination with 2-point CQDF,

$$\mathcal{K}(\bar{1}\bar{2}\bar{3}) = \mathcal{K}(\bar{1}1)\,\breve{\Gamma}(123)\,\mathcal{K}(3\bar{3})\,\mathcal{K}(2\bar{2}). \tag{16.138}$$

Similarly, we write

$$\breve{\Gamma}(1234) = i\hbar\frac{\delta}{\delta u(m)}\breve{\Gamma}(123)\,\frac{\delta u(m)}{i\hbar\delta\mathcal{G}(4)}. \tag{16.139}$$

Using Eqs. (16.137) and (16.138), we end up with the following expression,

$$\begin{aligned}\breve{\Gamma}(1234) &= \mathcal{K}^{-1}(1j)\,\mathcal{K}(jilm)\,\mathcal{K}^{-1}(i2)\,\mathcal{K}^{-1}(l3)\,\mathcal{K}^{-1}(m4)\\ &\quad -\breve{\Gamma}(1j4)\,\mathcal{K}(jy)\,\breve{\Gamma}(y23) - \breve{\Gamma}(1j3)\,\mathcal{K}(jy)\,\breve{\Gamma}(y24)\\ &\quad -\breve{\Gamma}(12j)\,\mathcal{K}(jy)\,\breve{\Gamma}(y34).\end{aligned} \tag{16.140}$$

The last expression enables us to express the 4-point CQDF in terms of 4-point vertex and 3-point vertex in combination with only 2-point CQDFs.

We have, corresponding to Eq. (16.138), the expression

$$\mathcal{K}(1234) = \mathcal{K}(1\bar{1})\,\mathcal{D}(\bar{1}\bar{2}\bar{3}\bar{4})\,\mathcal{K}(\bar{2}2)\,\mathcal{K}(\bar{3}3)\,\mathcal{K}(\bar{4}4), \tag{16.141}$$

where

$$\begin{aligned}\mathcal{D}(1234) = {} & \breve{\Gamma}(1234) + \breve{\Gamma}(1j4)\,\mathcal{K}(jy)\,\breve{\Gamma}(y23) \\ & + \breve{\Gamma}(1j3)\,\mathcal{K}(jy)\,\breve{\Gamma}(y24) + \breve{\Gamma}(12j)\,\mathcal{K}(jy)\,\breve{\Gamma}(y34)\,.\end{aligned} \tag{16.142}$$

Substituting the expressions given by Eqs. (16.138) and (16.141) in Eqs. (16.134) and (16.135), we obtain

$$\begin{aligned}\Xi(1) = {} & \frac{\mathcal{V}(123)}{2!}\{\mathcal{K}(23) + \mathcal{G}(2)\,\mathcal{G}(3)\} + \frac{\mathcal{V}(1234)}{3!} \\ & \times \left\{\begin{matrix} \mathcal{K}(2\bar{2})\,\breve{\Gamma}(\bar{2}\bar{3}\bar{4})\,\mathcal{K}(\bar{3}3)\,\mathcal{K}(\bar{4}4) + \epsilon\mathcal{K}(24)\,\mathcal{G}(3) \\ +\mathcal{G}(2)\,\mathcal{K}(34) + \mathcal{G}(4)\,\mathcal{K}(23) + \mathcal{G}(2)\,\mathcal{G}(3)\,\mathcal{G}(4)\end{matrix}\right\},\end{aligned} \tag{16.143}$$

$$\begin{aligned}\breve{\Sigma}(1w) = {} & \frac{\mathcal{V}(123)}{2!}\left\{\mathcal{K}(2\bar{2})\,\breve{\Gamma}(\bar{2}\bar{3}w)\,\mathcal{K}(\bar{3}3) + \epsilon\mathcal{G}(3)\,\delta(2w) + \mathcal{G}(2)\,\delta(3w)\right\} \\ & + \frac{\mathcal{V}(1234)}{3!}\left\{\begin{matrix} \mathcal{K}(2\bar{2})\,\mathcal{D}(\bar{2}\bar{3}\bar{4}w)\,\mathcal{K}(\bar{3}3)\,\mathcal{K}(\bar{4}4) \\ +\,\epsilon\mathcal{G}(3)\,\mathcal{K}(2\bar{2})\,\breve{\Gamma}(\bar{2}\bar{4}w)\,\mathcal{K}(\bar{4}4) \\ +\,\mathcal{G}(2)\,\mathcal{K}(3\bar{3})\,\breve{\Gamma}(\bar{3}\bar{4}w)\,\mathcal{K}(\bar{4}4) \\ +\,\mathcal{G}(4)\,\mathcal{K}(2\bar{2})\,\breve{\Gamma}(\bar{2}\bar{3}w)\,\mathcal{K}(\bar{3}3) \\ +\,\epsilon\mathcal{K}(24)\,\delta(3w) + \mathcal{K}(34)\,\delta(2w) \\ +\,\mathcal{K}(23)\,\delta(4w) + \mathcal{G}(2)\,\mathcal{G}(3)\,\delta(4w) \\ +\,\epsilon\mathcal{G}(2)\,\mathcal{G}(4)\,\delta(3w) + \mathcal{G}(3)\,\mathcal{G}(4)\,\delta(2w)\end{matrix}\right\}.\end{aligned} \tag{16.144}$$

16.7.2 *Iterative equations for the vertex functions*

The structure of the dependence of the vertex functions on and $\mathcal{K}(ij)$ can be exhibited by relating these vertex functions to the functional derivatives of $\breve{\Sigma}(ij)$ with respect to $\mathcal{G}(i)$ and $\mathcal{K}(ij)$. Let us first consider $\breve{\Gamma}(123)$. This can be expressed as

$$\breve{\Gamma}(123) = \left(\frac{\delta\breve{\Sigma}(12)}{i\hbar\delta\mathcal{G}(3)}\right)_{\mathcal{K}(ij)} + \left(\frac{\delta\breve{\Sigma}(12)}{i\hbar\delta\mathcal{K}(45)}\right)_{\mathcal{G}(i)} \mathcal{K}(46)\left(-\frac{\delta\mathcal{K}^{-1}(67)}{\delta\mathcal{G}(3)}\right)\mathcal{K}(75). \tag{16.145}$$

Therefore, we have the iterative equation for $\breve{\Gamma}(123)$ as

$$\breve{\Gamma}(123) = \left(\frac{\delta\breve{\Sigma}(12)}{i\hbar\delta\mathcal{G}(3)}\right)_{\mathcal{K}(ij)} + \left(\frac{\delta\breve{\Sigma}(12)}{i\hbar\delta\mathcal{K}(45)}\right)_{\mathcal{G}(i)} \mathcal{K}(46)\,\breve{\Gamma}(673)\,\mathcal{K}(75). \tag{16.146}$$

What this means is that it is indeed possible to express $\breve{\Gamma}(123)$ as functional of only $\mathcal{G}(i)$ and $\mathcal{K}(ij)$. This is also consistent with Eq. (16.144), in that $\breve{\Sigma}(12)$ is also a functional of only $\mathcal{G}(i)$ and $\mathcal{K}(ij)$ since we can also prove an analogous iterative relation for $\mathcal{D}(1234)$. Thus, Eq. (16.146) allows us to express $\breve{\Gamma}(123)$ in terms of the partial functional derivatives of the self-energy $\breve{\Sigma}(12)$ with respect to $\mathcal{G}(i)$ and $\mathcal{K}(ij)$, respectively, and exhibits the structure of the dependence of $\breve{\Gamma}(ijk)$ on $\mathcal{G}(i)$ and $\mathcal{K}(ij)$ relative to the dependence of $\breve{\Sigma}(ij)$ on these same quantities. The formal solution of Eq. (16.146) is

$$\breve{\Gamma}(123) = \Lambda(12,\bar{1}\bar{2})\left(\frac{\delta\breve{\Sigma}(\bar{1}\bar{2})}{i\hbar\delta\mathcal{G}(3)}\right)_{\mathcal{K}(ij)}, \tag{16.147}$$

where

$$\Lambda(12,\bar{1}\bar{2}) = \left[\delta(1\bar{1})\,\delta(2\bar{2}) - \left(\frac{\delta\breve{\Sigma}(12)}{i\hbar\delta\mathcal{K}(45)}\right)_{\mathcal{G}(i)} \mathcal{K}(4\bar{1})\,\mathcal{K}(\bar{2}5)\right]^{-1}. \tag{16.148}$$

Now consider the iterative equation for $\mathcal{D}(1234)$ in terms of the functional derivative of the self-energy. From Eq. (16.141), we have

$$\mathcal{D}(1234) = \mathcal{K}^{-1}(1j)\,i\hbar\frac{\delta\mathcal{K}(ji)}{\delta u(lm)}\mathcal{K}^{-1}(i2)\,\mathcal{K}^{-1}(l3)\,\mathcal{K}^{-1}(m4). \tag{16.149}$$

This can be re-expressed in terms of functional derivative of the self-energy with respect to 2-point CQDF. We obtain

$$\mathcal{D}(1234) = \left(\frac{\delta\breve{\Sigma}(12)}{i\hbar\delta\mathcal{K}(43)}\right)_{\mathcal{G}(i)} + \left(\frac{\delta\breve{\Sigma}(12)}{i\hbar\delta\mathcal{K}(56)}\right)_{\mathcal{G}(i)} \mathcal{K}(5\bar{5})\,\mathcal{D}(\bar{5}\bar{6}34)\,\mathcal{K}(\bar{6}6). \tag{16.150}$$

The formal solution for is given by

$$\mathcal{D}(1234) = \Lambda(12, \bar{3}\bar{4}) \left(\frac{\delta \breve{\Sigma}(\bar{3}\bar{4})}{i\hbar\delta\mathcal{K}(43)} \right)_{\mathcal{G}(i)}. \tag{16.151}$$

Substituting Eqs. (16.147) and (16.151) in the expression for $\Xi(1)$, Eq. (16.143) and $\breve{\Sigma}(1w)$, Eq. (16.144), we finally obtain the following expressions:

$$\Xi(1) = \frac{\mathcal{V}(123)}{2!}\{\mathcal{K}(23) + \mathcal{G}(2)\mathcal{G}(3)\} + \frac{\mathcal{V}(1234)}{3!} \times \left\{ \begin{array}{c} \mathcal{K}(2\bar{2})\Lambda(\bar{2}\bar{3}56)\left(\frac{\delta\breve{\Sigma}(56)}{i\hbar\delta\mathcal{G}(4)}\right)_{\mathcal{K}(ij)} \mathcal{K}(\bar{3}3)\mathcal{K}(\bar{4}4) \\ +\epsilon\mathcal{K}(24)\mathcal{G}(3) \\ +\mathcal{G}(2)\mathcal{K}(34) + \mathcal{G}(4)\mathcal{K}(23) + \mathcal{G}(2)\mathcal{G}(3)\mathcal{G}(4) \end{array} \right\}, \tag{16.152}$$

$$\breve{\Sigma}(1w) = \frac{\mathcal{V}(123)}{2!}\left\{ \begin{array}{c} \mathcal{K}(2\bar{2})\Lambda(\bar{2}\bar{3}56)\left(\frac{\delta\breve{\Sigma}(56)}{i\hbar\delta\mathcal{G}(w)}\right)_{\mathcal{K}(ij)} \mathcal{K}(\bar{3}3) \\ +\epsilon\mathcal{G}(3)\delta(2w) + \mathcal{G}(2)\delta(3w) \end{array} \right\} + \frac{\mathcal{V}(1234)}{3!} \times \left\{ \begin{array}{c} \mathcal{K}(2\bar{2})\Lambda(\bar{2}\bar{3},56)\left(\frac{\delta\breve{\Sigma}(56)}{i\hbar\delta\mathcal{K}(w\bar{4})}\right)_{\mathcal{G}(i)} \mathcal{K}(\bar{3}3)\mathcal{K}(\bar{4}4) \\ +\epsilon\mathcal{G}(3)\mathcal{K}(2\bar{2})\Lambda(\bar{2}\bar{4},56)\left(\frac{\delta\breve{\Sigma}(56)}{i\hbar\delta\mathcal{G}(w)}\right)_{\mathcal{K}(ij)} \mathcal{K}(\bar{4}4) \\ +\mathcal{G}(2)\mathcal{K}(3\bar{3})\Lambda(\bar{3}\bar{4},56)\left(\frac{\delta\breve{\Sigma}(56)}{i\hbar\delta\mathcal{G}(w)}\right)_{\mathcal{K}(ij)} \mathcal{K}(\bar{4}4) \\ +\mathcal{G}(4)\mathcal{K}(2\bar{2})\Lambda(\bar{2}\bar{3}56)\left(\frac{\delta\breve{\Sigma}(56)}{i\hbar\delta\mathcal{G}(w)}\right)_{\mathcal{K}(ij)} \mathcal{K}(\bar{3}3) \\ +\epsilon\mathcal{K}(24)\delta(3w) + \mathcal{K}(34)\delta(2w) \\ +\mathcal{K}(23)\delta(4w) + \mathcal{G}(2)\mathcal{G}(3)\delta(4w) \\ +\epsilon\mathcal{G}(2)\mathcal{G}(4)\delta(3w) + \mathcal{G}(3)\mathcal{G}(4)\delta(2w) \end{array} \right\}. \tag{16.153}$$

We now summarized the results of this section. We have Eqs. (16.130) and (16.132) for $\mathcal{G}(i)$ and $\mathcal{K}(ij)$, containing $\Xi(i)$ and $\breve{\Sigma}(ij)$, respectively. Equation (16.134) for $\Xi(1)$ and Eq. (16.135) for $\breve{\Sigma}(12)$ show that $\Xi(1)$ is a functional of 3-point CQDF, and $\breve{\Sigma}(12)$ is a functional of 3-point CQDFs and 4-point CQDF. This is a reflection of the hierarchical nature of the equation for the GQDFs. However, Eq. (16.138) shows that a 3-point CQDF can be expressed in terms of the lower order 2-point CQDFs in combination with 3-point vertex function. Similarly, Eq. (16.141) shows that a 4-point

CQDF can be expressed in terms of 2-point CQDFs in combination with 3-point and 4-point vertex functions. Equations (16.147) and (16.151) show that these vertex functions can be expressed in terms of 2-point CQDFs in combination with partial functional derivatives of the self-energy with respect to $\mathcal{G}(i)$ and $\mathcal{K}(ij)$, thus closing the number of unknown functions. Thus we have a closed set of coupled equations in eight unknown functions, $\mathcal{G}(i)$, $\mathcal{K}(ij)$, $\Xi(i)$, $\breve{\Sigma}(ij)$, $\mathcal{K}(ijk)$, $\mathcal{K}(ijkl)$, $\breve{\Gamma}(ijk)$, and $\breve{\Gamma}(ijkl)$, which give self-consistent solutions.

In a fully self-consistent and explicitly renormalized formulation, it is desirable to have a universal functional of the GQDFs, whereby the equations of motion describing the system are obtained as a stationary principle for this universal functional subject to constraint on pertinent conservation laws. Since W is the effective action, we expect this procedure to amount to a stationary action principle of W with respect to the GQDFs. We can construct this functional of GQDF by method of Legendre transformation introduced by De Dominicis and Martin [128] for interacting particle systems at thermodynamic equilibrium. This tedious renormalization procedure will not be given here [132]. Moreover, questions about 1-particle irreducibility, i.e., existence theorem of well-defined approximate self-energies, vertex functions, etc. will not be addressed here. These are assumed to be not a problem here, at least for nonrelativistic Fermi systems. In general, these are fundamental problems that plague the physics of many-degrees of freedom, namely, quantum field theory and statistical many-body physics. This appears in the guise of the so-called "closure" problem.

In non-equilibrium quantum physics, this problem has to do with the proper reduction of the quantum mechanics of the whole system, say $A+B$, which is governed by a unitary evolution operator, to the quantum mechanics of the subsystem A alone which is governed by a non-unitary evolution operator to express irreversibility. The "correct reduction" from the whole system unitary picture to the subsystem non-unitary picture has been the subject of intense fundamental research, and similar question like "renormalizability" in quantum field theory is expected to be a long-standing problem in physics of many-degrees of freedom. With respect to the evolution of reduced density matrix, there is a consensus that the "Lindblad criterion" [133] must be satisfied, which does hold in transport equations derived here since the trace of the right hand side of the transport equations reduce to zero, to express total probability equal to 1. We have only revealed the formal mechanism of the reduction to 1-particle, 2-particle, etc. nonequilibrium descriptions. The reduction procedure is a common theme in any treatment using quantum-field theoretical techniques.

In strongly correlated systems where higher-order correlations are significant, for example in systems containing electrons, holes, excitons, biexcitons, trions, etc. the quantum field operators representing the composite particles may have to be employed to make the problem tractable, and the formalism given here can thus be repeated for a composite-particle QSFT.

Chapter 17

Quantum Transport Equations of Particles

We derived here the general quantum transport equations valid for normal systems, Bose superfluids [134], superconductors [135], and dense electron–hole liquids. This involves the formulation of the total time evolution equations of various GQDF. From Eq. (16.132), we have the equation for the 2-point correlation function given by

$$\mathcal{K}^{-1}(1,2)\,\mathcal{K}(2,3)=\delta(1,3)\,. \tag{17.1}$$

Note that for fermions, $\mathcal{K}(1,2)=\mathcal{G}(1,2)$, however the expression for $\mathcal{K}(1,2)$ in Eq. (16.124), which may involved the condensate wavefunction, must be used for bosons. We write here the explicit expressions of the quantities occurring in Eq. (17.1). We have,

$$\mathcal{K}(1,2)=i\hbar\begin{pmatrix}\Im(1,2) & \tilde{G}(1,2)\\ \epsilon\tilde{G}^{T}(1,2) & {}^{\dagger}\Im(1,2)\end{pmatrix}, \tag{17.2}$$

with matrix elements consisting of 2×2 matrices given by

$$\left.\begin{aligned}\Im(1,2)&=F(1,2)-\wp(1,2)\,,\\ {}^{\dagger}\Im(1,2)&={}^{\dagger}F(1,2)-\wp^{*}(1,2)\,,\\ \tilde{G}(1,2)&=G(1,2)-\aleph(1,2)\,,\\ \epsilon\tilde{G}^{T}(1,2)&=\epsilon G^{T}(1,2)-\aleph^{T}(1,2)\,,\end{aligned}\right\} \tag{17.3}$$

where

$$\left.\begin{aligned} \wp(1,2) &= \tfrac{1}{i\hbar 2!}\sum_P \epsilon^P g_h(1)\, g_h(2) \begin{pmatrix} 1 & 1 \\ 1 & 1 \end{pmatrix}, \\ \aleph(1,2) &= \tfrac{1}{i\hbar 2!}\sum_P \epsilon^P g_h^*(2)\, g_h^*(1) \begin{pmatrix} 1 & 1 \\ 1 & 1 \end{pmatrix}. \end{aligned}\right\} \tag{17.4}$$

In Eq. (17.4), $i\hbar\aleph(1,2)$ is the one-particle reduced density matrix of the boson condensate, this is identically zero for fermions. For lack of appropriate terminology, we will simply refer to $i\hbar\wp(1,2)$ as the "anomalous" one-particle reduced density matrix of the condensate for bosons, this is zero for fermions. The condensate wave function is given by $g_h(i) = \langle\hat{\psi}(i)\rangle = \langle\tilde{\psi}^\dagger(i)\rangle$.

The inverse of the 2-point correlation function is determined from Eqs. (16.127) and (16.132), where $\mathcal{V}(12)$ indicates all one-body operators, such as the kinetic energy operator and the external potential. From Eqs. (16.110), (16.115), and (16.127), we have the matrix form of $\mathcal{V}(12)$, taking into account the permutation of indices, given by

$$\begin{aligned} \mathcal{V}(1,2) &= \begin{pmatrix} 0 & 0 & \epsilon\,\bar{\upsilon}(2,1) & 0 \\ 0 & 0 & 0 & -\bar{\upsilon}(1,2) \\ \bar{\upsilon}(1,2) & 0 & 0 & 0 \\ 0 & -\epsilon\,\bar{\upsilon}(2,1) & 0 & 0 \end{pmatrix}, \\ &\equiv \begin{pmatrix} 0 & \epsilon\,\vartheta^T(1,2) \\ \vartheta(1,2) & 0 \end{pmatrix}, \end{aligned} \tag{17.5}$$

where $\bar{\upsilon}(1,2) = \delta(1,2)\prod_{i=1}^{2}\delta(t_1 - t_i)\ \upsilon(1,2)$. The first term of Eq. (16.127) has the matrix form given by

$$\begin{aligned} &(\tau_4^\epsilon)^{-1}\, i\hbar\delta(1,2)\frac{\partial}{\partial t_2} \\ &= \begin{pmatrix} 0 & 0 & -\epsilon i\hbar\delta(1,2)\frac{\partial}{\partial t_2} & 0 \\ 0 & 0 & 0 & i\hbar\delta(1,2)\frac{\partial}{\partial t_2} \\ i\hbar\delta(1,2)\frac{\partial}{\partial t_2} & 0 & 0 & 0 \\ 0 & -\epsilon i\hbar\delta(1,2)\frac{\partial}{\partial t_2} & 0 & 0 \end{pmatrix}, \\ &\equiv \begin{pmatrix} 0 & \epsilon\,(G_t^o)^{-1T} \\ (G_t^o)^{-1} & 0 \end{pmatrix}. \end{aligned} \tag{17.6}$$

Equations (17.5) and (17.6) also define the 2×2 matrix elements, $\vartheta(1,2)$ and $(G_t^o)^{-1}$. Note that $(G_t^o)^{-1T}$ does not have the exact meaning of transpose since the time derivative is still acting to the right. Since the self-energy $\breve{\Sigma}(1,2)$ is a term in the expression for the inverse correlation function $\mathcal{K}^{-1}(1,2)$, we write this in the form

$$\breve{\Sigma}(1,2) = \begin{pmatrix} {}^{\dagger}\Delta(1,2) & \epsilon\Sigma^T(1,2) \\ \Sigma(1,2) & \Delta(1,2) \end{pmatrix} = \begin{pmatrix} \Delta_{ee}^c & \epsilon\Delta_{ee}^< & \epsilon\Sigma^{cT} & \epsilon\Sigma^{>T} \\ \epsilon\Delta_{ee}^> & \Delta_{ee}^{ac} & \Sigma^{<T} & \Sigma^{acT} \\ \Sigma^c & \epsilon\Sigma^< & \Delta_{hh}^c & \Delta_{hh}^< \\ \Sigma^> & \epsilon\Sigma^{ac} & \Delta_{hh}^> & \Delta_{hh}^{ac} \end{pmatrix}(1,2). \tag{17.7}$$

Here it is assumed that the "singular" part (delta function in time) of the self-energy has been absorbed in the $\mathcal{V}(1,2)$, otherwise it is understood to be the singular part of the diagonals of Δ and G, in Δ_{ee}^c, Σ^c, etc. We write the resulting equations for the 2×2 matrix elements of Eq. (17.1) as

$$G_o^{-1}\tilde{G} = \delta + \Sigma\tilde{G} + \Delta^{\dagger}\Im, \tag{17.8}$$

$$G_o^{-1T}\tilde{G}^T = \delta + \Sigma^T\tilde{G}^T + {}^{\dagger}\Delta\,\Im, \tag{17.9}$$

$$G_o^{-1T}\,{}^{\dagger}\Im = \Sigma^T\,{}^{\dagger}\Im + \epsilon\,{}^{\dagger}\Delta\,\tilde{G}, \tag{17.10}$$

$$G_o^{-1}\,\Im = \Sigma\,\Im + \epsilon\,\Delta\,\tilde{G}^T. \tag{17.11}$$

We determined from Eqs. (17.1)–(17.7) and Eqs. (16.83)–(16.86) the following relations:

$$G_o^{-1}(1,2) = \begin{pmatrix} i\hbar\delta(1,2)\frac{\partial}{\partial t_2} - \bar{v}(1,2) & 0 \\ 0 & -\epsilon\left[i\hbar\delta(1,2)\frac{\partial}{\partial t_2} - \bar{v}(2,1)\right] \end{pmatrix}, \tag{17.12}$$

$$\epsilon G_o^{-1T}(1,2) = \begin{pmatrix} -\epsilon\left[i\hbar\delta(1,2)\frac{\partial}{\partial t_2} + \bar{v}(2,1)\right] & 0 \\ 0 & i\hbar\delta(1,2)\frac{\partial}{\partial t_2} + \bar{v}(1,2) \end{pmatrix}, \tag{17.13}$$

$$\begin{aligned} \tilde{G}(1,2) &= \begin{pmatrix} G^c(1,2) & \epsilon G^<(1,2) \\ G^>(1,2) & \epsilon G^{ac}(1,2) \end{pmatrix} - \aleph(1,2), \\ &= \begin{pmatrix} \tilde{G}^c(1,2) & \epsilon\tilde{G}^<(1,2) \\ \tilde{G}^>(1,2) & \epsilon\tilde{G}^{ac}(1,2) \end{pmatrix}, \end{aligned} \tag{17.14}$$

$$\Im(1,2) = \begin{pmatrix} g^{c}_{hh}(1,2) & g^{<}_{hh}(1,2) \\ g^{>}_{hh}(1,2) & g^{ac}_{hh}(1,2) \end{pmatrix} - \wp(1,2),$$

$$= \begin{pmatrix} \tilde{g}^{c}_{hh}(1,2) & \tilde{g}^{<}_{hh}(1,2) \\ \tilde{g}^{>}_{hh}(1,2) & \tilde{g}^{ac}_{hh}(1,2) \end{pmatrix}, \tag{17.15}$$

$$^{\dagger}\Im(1,2) = \begin{pmatrix} g^{c}_{ee}(1,2) & \epsilon g^{<}_{ee}(1,2) \\ \epsilon g^{>}_{ee}(1,2) & g^{ac}_{ee}(1,2) \end{pmatrix} - \wp(1,2),$$

$$= \begin{pmatrix} \tilde{g}^{c}_{ee}(1,2) & \epsilon\tilde{g}^{<}_{ee}(1,2) \\ \epsilon\tilde{g}^{>}_{ee}(1,2) & \tilde{g}^{ac}_{ee}(1,2) \end{pmatrix}. \tag{17.16}$$

Equations (17.8) and (17.11) were also given by Aronov *et al.* [135] for superconductors, these equations formally resembles the well-known Gorkov equations [136] for superconductors at thermal equilibrium, except that now each quantity entering in Eqs. (17.8–17.11) are 2×2 matrices by virtue of the doubling of the quantum-field operators.

17.1 General quantum transport equations

We give here the quantum transport equations for fermions only. We make use of the following relations in deriving the quantum transport equations in what follows:

$$\left.\begin{aligned} G^{c}(12) &= -G^{ac}(21), \\ G^{>}(12)^{\dagger} &= -G^{>}(21), \\ G^{<}(12)^{\dagger} &= -G^{<}(21), \\ g^{c}_{ee}(12)^{\dagger} &= -g^{ac}_{hh}(21), \\ g^{>}_{ee}(12)^{\dagger} &= -g^{>}_{hh}(21), \\ g^{<}_{ee}(12)^{\dagger} &= -g^{<}_{hh}(21). \end{aligned}\right\} \tag{17.17}$$

Introducing the operator $\mathcal{XX}$, whose effect when operating on a matrix is to interchange the diagonal and off-diagonal elements, we have the total time derivative of ${}^{\dagger}\Im$ given by

$$
\begin{aligned}
i\hbar \left(\frac{\partial}{\partial t_1} + \frac{\partial}{\partial t_2} \right) {}^{\dagger}\Im(12)
&= \begin{pmatrix} -1 & 0 \\ 0 & \epsilon \end{pmatrix} [(G_t^o)^{-1}]^T(1\overline{2})^{\dagger}\Im(\overline{2}2) \\
&\quad + \mathcal{X}\mathcal{X}\left\{ \begin{pmatrix} 1 & 0 \\ 0 & \epsilon \end{pmatrix} \left[(G_t^o)^{-1}(2\overline{2})\Im(\overline{2}1) \right]^{\dagger} \begin{pmatrix} 1 & 0 \\ 0 & -1 \end{pmatrix} \right\}.
\end{aligned} \tag{17.18}
$$

Similarly, we obtained the total time derivative of $\Im$ as follows:

$$
\begin{aligned}
i\hbar \left(\frac{\partial}{\partial t_1} + \frac{\partial}{\partial t_2} \right) \Im(12)
&= \begin{pmatrix} 1 & 0 \\ 0 & -\epsilon \end{pmatrix} (G_t^o)^{-1}(1\overline{2})\Im(\overline{2}2) \\
&\quad + \mathcal{X}\mathcal{X}\left\{ \begin{pmatrix} -1 & 0 \\ 0 & -\epsilon \end{pmatrix} \left[[(G_t^o)^{-1}]^{T\dagger}\Im(\overline{2}1) \right]^{\dagger} \begin{pmatrix} 1 & 0 \\ 0 & -1 \end{pmatrix} \right\}.
\end{aligned} \tag{17.19}
$$

The total time derivative of $\tilde{G}$ is given by

$$
\begin{aligned}
i\hbar \left(\frac{\partial}{\partial t_1} + \frac{\partial}{\partial t_2} \right) \tilde{G}
&= \begin{pmatrix} 1 & 0 \\ 0 & -\epsilon \end{pmatrix} (G_t^o)^{-1}\,(1\overline{2})\,\tilde{G}\,(\overline{2}2) \\
&\quad + \mathcal{X}\mathcal{X}\left[\begin{pmatrix} \epsilon & 0 \\ 0 & 1 \end{pmatrix} \left[(G_t^o)^{-1}\,\tilde{G} \right]^{\dagger} \begin{pmatrix} 1 & 0 \\ 0 & -1 \end{pmatrix} \right].
\end{aligned} \tag{17.20a}
$$

To obtain the desired quantum transport equations, we express the chronological and anti-chronological quantities in terms of the advanced and retarded ones, as follows,

$$
F^c = F^r + F^<; \; F^{ac} = F^< - F^a, \tag{17.21}
$$

$$
F^c = F^a + F^>; \; F^{ac} = F^> - F^r, \tag{17.22}
$$

where F^r and F^a are the retarded and advanced function, respectively.

Summarizing, we have the following general quantum transport expressions for fermions [$\{A, B\}$means anticommutator of A and B]:

$$i\hbar\left(\frac{\partial}{\partial t_1}+\frac{\partial}{\partial t_2}\right)G^{<}=\left[vG^{<}-G^{<}v^{T}\right]$$
$$+\left[\Sigma^{r}G^{<}-G^{<}\Sigma^{a}\right]+\left[\Sigma^{<}G^{a}-G^{r}\Sigma^{<}\right]$$
$$+\left[\Delta^{r}_{hh}g^{<}_{ee}-g^{<}_{hh}\Delta^{a}_{ee}\right]$$
$$+\left[\Delta^{<}_{hh}g^{a}_{ee}-g^{r}_{hh}\Delta^{<}_{ee}\right], \quad (17.23)$$

$$i\hbar\left(\frac{\partial}{\partial t_1}+\frac{\partial}{\partial t_2}\right)G^{>}=\left[v^{T}G^{>}-G^{>}v\right]$$
$$+\left[\Sigma^{r}G^{>}-G^{>}\Sigma^{a}\right]+\left[\Sigma^{>}G^{a}-G^{r}\Sigma^{>}\right]$$
$$+\left[\Delta^{r}_{hh}g^{>}_{ee}-g^{>}_{hh}\Delta^{a}_{ee}\right]$$
$$+\left[\Delta^{>}_{hh}g^{a}_{ee}-g^{r}_{hh}\Delta^{>}_{ee}\right], \quad (17.24)$$

$$i\hbar\left(\frac{\partial}{\partial t_1}+\frac{\partial}{\partial t_2}\right)g^{<}_{hh}=\left\{v,g^{<}_{hh}\right\}+\left\{\Sigma^{r}g^{<}_{hh}+g^{<}_{hh}\left(\Sigma^{a}\right)^{T}\right\}$$
$$+\left\{\Sigma^{<}g^{a}_{hh}+g^{r}_{hh}\left(\Sigma^{>}\right)^{T}\right\}+g^{<}_{hh}\left(\Sigma^{>}-\Sigma^{<}\right)^{T}$$
$$-\left\{\Delta^{r}_{hh}\left(G^{>}\right)^{T}+G^{<}\Delta^{a}_{hh}\right\}$$
$$-\left\{\Delta^{<}_{hh}\left(G^{a}\right)^{T}+G^{r}\Delta^{<}_{hh}\right\}$$
$$-\Delta^{<}_{hh}\left(G^{>}-G^{<}\right)^{T}, \quad (17.25)$$

$$i\hbar\left(\frac{\partial}{\partial t_1}+\frac{\partial}{\partial t_2}\right)g^{>}_{hh}=\left\{v^{T},g^{>}_{hh}\right\}+\left\{\Sigma^{r}g^{>}_{hh}+g^{>}_{hh}\left(\Sigma^{a}\right)^{T}\right\}$$
$$+\left\{\Sigma^{>}g^{a}_{hh}+g^{r}_{hh}\left(\Sigma^{<}\right)^{T}\right\}+g^{>}_{hh}\left(\Sigma^{>}-\Sigma^{<}\right)^{T}$$
$$-\left\{\Delta^{r}_{hh}\left(G^{<}\right)^{T}+G^{>}\Delta^{a}_{hh}\right\}$$
$$-\left\{\Delta^{>}_{hh}\left(G^{a}\right)^{T}+G^{r}\Delta^{>}_{hh}\right\}$$
$$-\Delta^{>}_{hh}\left(G^{>}-G^{<}\right)^{T}, \quad (17.26)$$

$$i\hbar\left(\frac{\partial}{\partial t_1}+\frac{\partial}{\partial t_2}\right)g^{<}_{ee}=-\left\{v^{T},g^{<}_{ee}\right\}-\left\{\left(\Sigma^{r}\right)^{T}g^{<}_{ee}+g^{<}_{ee}\Sigma^{a}\right\}$$
$$-\left\{\left(\Sigma^{>}\right)^{T}g^{a}_{ee}+g^{r}_{ee}\Sigma^{<}\right\}+\left(\Sigma^{>}-\Sigma^{<}\right)^{T}g^{<}_{ee}$$

$$
\begin{aligned}
&+ \left\{\Delta^r_{ee} G^< + \left(G^>\right)^T \Delta^a_{ee}\right\} \\
&+ \left\{\Delta^<_{ee} G^a + G^{rT} \Delta^<_{ee}\right\} \\
&- \left(G^> - G^<\right)^T \Delta^<_{ee}, \qquad (17.27)
\end{aligned}
$$

$$
\begin{aligned}
i\hbar\left(\frac{\partial}{\partial t_1} + \frac{\partial}{\partial t_2}\right) g^>_{ee} = &- \left\{v, g^>_{ee}\right\} - \left\{\Sigma^{rT} g^>_{ee} + g^>_{ee}\left(\Sigma^a\right)\right\} \\
&- \left\{\left(\Sigma^<\right)^T g^a_{ee} + g^r_{ee}\Sigma^>\right\} + \left(\Sigma^> - \Sigma^<\right)^T g^>_{ee} \\
&+ \left\{\left(G^<\right)^T \Delta^a_{ee} + \Delta^r_{ee} G^>\right\} \\
&+ \left\{\Delta^>_{ee} G^a + \left(G^r\right)^T \Delta^>_{ee}\right\} \\
&- \left(G^> - G^<\right)^T \Delta^>_{ee}. \qquad (17.28)
\end{aligned}
$$

Note that the equations for the retarded and advanced quantities can be derived from the equations for the chronological Green's functions, obtained from Eqs. (17.18)–(17.20a), coupled with the quantum transport equations given above. In what follows, we will give physical meaning to the terms occurring in the above expressions by applying these general expressions in Section 17.2 to a simple two-level system which has a rich physical content very well known in quantum optics.

17.2 Lattice Weyl transform identities

To obtain the kinetic transport equations one needs to perform the lattice Weyl transformation on the above quantum transport equations. The reader is referred back to Part1 of this book for comprehensive discussions on lattice Weyl transformation. We give here the following identities for the lattice Weyl transform of the following differential operators:

$$
\left.
\begin{aligned}
i\hbar\left(\frac{\partial}{\partial t_1} + \frac{\partial}{\partial t_2}\right) F(12) &\Leftrightarrow i\hbar\frac{\partial}{\partial t} F_w(\vec{p}, \vec{q}, E, t), \\
\left(\frac{\partial^2}{\partial t_1^2} - \frac{\partial^2}{\partial t_2^2}\right) F(12) &\Leftrightarrow -\frac{2i}{\hbar} E \frac{\partial}{\partial t} F_w(\vec{p}, \vec{q}, E, t), \\
\left(\nabla_1^2 - \nabla_2^2\right) F(12) &\Leftrightarrow \frac{2i}{\hbar} p \cdot \nabla_{\vec{q}} F_w(p, q),
\end{aligned}
\right\} \qquad (17.29)
$$

where the Bloch states has the representation, $\langle q|\, p\rangle = \left(\frac{1}{N\hbar^3}\right)^{\frac{1}{2}} \exp \frac{i}{\hbar}\{\vec{p}\cdot\vec{q} - Et\}$ $[N\hbar^3 \Rightarrow h^3$ in the continuum limit], representing a traveling wave in lattice space, with group velocity in a band λ given by $\frac{dE_\lambda\{\vec{p}\}}{d\vec{p}}$.

The second is the lattice Weyl transform of product of arbitrary operators [we used here the notation: $p \equiv (\vec{p}, -E)$ and $q = (\vec{q}, t)$] in terms of "Poisson bracket operator",

$$AB\,(p,q) = \exp\left[\frac{\hbar}{i}\left(\frac{\partial^{(a)}}{\partial p}\cdot\frac{\partial^{(b)}}{\partial q} - \frac{\partial^{(a)}}{\partial q}\cdot\frac{\partial^{(b)}}{\partial p}\right)\right] a\,(p,q)\;\, b\,(p,q)\,, \tag{17.30}$$

or in terms of integral operator,

$$\begin{aligned} AB\,(p,q) &= \frac{1}{(2\pi\hbar)^8}\int dp'dq' K_A^{+}\left(p,q;p',q'\right)\; b\left(p',q'\right) \\ &= \frac{1}{(2\pi\hbar)^8}\int dp'dq' a\left(p',q'\right) K_B^{-}\left(p,q;p',q'\right), \end{aligned} \tag{17.31}$$

where integral kernels are defined by

$$\begin{aligned} K_Y^{\pm}\left(p,q;p',q'\right) = \int du\; dv \exp\left\{\frac{i}{\hbar}\left[\left(p-p'\right)\cdot v + \left(q-q'\right)\cdot u\right]\right\} \\ \times y\left(p \pm \frac{u}{2}, q \mp \frac{v}{2}\right). \end{aligned} \tag{17.32}$$

Thus, we may write the lattice Weyl transform of a commutator $[A, B]$ and an anticommutator $\{A, B\}$ as

$$\begin{aligned} [A,B]\,(p,q) = \cos\Lambda\left[a\,(p,q)\, b\,(p,q) - b\,(p,q)\, a\,(p,q)\right] \\ - i\sin\Lambda\{a\,(p,q)\, b\,(p,q) + b(p,q)a(p,q)\}, \end{aligned} \tag{17.33}$$

$$\begin{aligned} \{A,B\}\,(p,q) = \cos\Lambda\left\{a\,(p,q)\, b\,(p,q) + b\,(p,q)\, a\,(p,q)\right\} \\ -i\sin\Lambda\left[a\,(p,q)\, b\,(p,q) - b\,(p,q)\, a\,(p,q)\right], \end{aligned} \tag{17.34}$$

where $\Lambda = \frac{\hbar}{2}\left(\frac{\partial^{(a)}}{\partial p}\cdot\frac{\partial^{(b)}}{\partial q} - \frac{\partial^{(a)}}{\partial q}\cdot\frac{\partial^{(b)}}{\partial p}\right)$. In terms of integral operators, we have,

$$\begin{aligned} [A,B]\,(p,q) = \frac{1}{(2\pi\hbar)^8}\int dp'dq' K_A^{+}\left(p,q;p',q'\right)\; b\left(p',q'\right) \\ - b\left(p',q'\right) K_A^{-}\left(p,q;p',q'\right), \end{aligned} \tag{17.35}$$

$$\{A, B\}(p, q) = \frac{1}{(2\pi\hbar)^8} \int dp' dq' K_A^+ (p, q; p', q') \; b(p', q')$$
$$+ b(p', q') K_A^- (p, q; p', q') . \tag{17.36}$$

The above expressions simplify considerably when the lattice Weyl transforms are scalar functions [3, 11]. Also, for calculating currents, which involves taking the trace, similar simplifications arise. For this case, we have the "lattice" Weyl transform of a commutator and anti-commutator of two operators, $[A, B]$ and $\{A, B\}$, respectively, given by the following expressions:

$$[A, B](p, q) = \frac{1}{(2\pi\hbar)^8} \int dp' dq' K_A^s (p, q; p', q') \; b(p', q') , \tag{17.37}$$

$$\{A, B\}(p, q) = \frac{1}{(2\pi\hbar)^8} \int dp' dq' K_A^c (p, q; p', q') \; b(p', q') , \tag{17.38}$$

where

$$K_Y^{s,c} (p, q; p', q') = \int du \; dv \exp\left\{\frac{i}{\hbar} \left[(p - p') \cdot v + (q - q') \cdot u\right]\right\}$$
$$\times \left[y\left(p + \frac{u}{2}, q - \frac{v}{2}\right) \mp y\left(p - \frac{u}{2}, q + \frac{v}{2}\right)\right] . \tag{17.39}$$

17.3 Generalized Bloch equations

For multi-band quantum transport dynamics in solids, it is more convenient to consider electrons and holes. This is often the case with semiconductor physics. Thus, it is desirable to recast the general quantum transport equations, Eqs. (17.23)–(17.28) in the electron–hole picture [137]. In this sense, generalized semiconductor Bloch equations (GSBEs) refers to Eqs. (17.23)–(17.28) expressed in the so-called "defect" representation (the transformation table is given in Ref. [137] and in the handbook[10] in great detail). In GSBE, the particle kinematics within the respective bands is accounted for, as well as all relevant scattering mechanisms. Moreover, pairing interaction between electrons, between electron and hole, and between holes are also included, as well as creation of electron–hole pairs by optical excitation and impact ionization (other excitation processes is implicitly incorporated in the self-energy, as well as Auger recombination, etc.), and Zener tunneling. The first-order Hartree Coulomb pairing between an electron and a hole which leads to the formation of excitons will not be explicitly treated here as a vast literature already exist on this subject.

With the Bloch band as discrete indices, Eqs. (17.23)–(17.24) becomes

$$
\begin{aligned}
&i\hbar\left(\frac{\partial}{\partial t_1}+\frac{\partial}{\partial t_2}\right)G^{\lessgtr}_{\alpha\beta}(12)\\
&=\left[v_\alpha\delta_{\alpha\gamma}\delta_{1\bar{2}}G^{\lessgtr}_{\gamma\beta}(\bar{2}2)-G^{\lessgtr}_{\alpha\gamma}(1\bar{2})v^T_\gamma\delta_{\gamma\beta}\delta_{\bar{2}2}\right]\\
&\quad+\left[\Sigma^r_{\alpha\gamma}(1\bar{2})G^{\lessgtr}_{\gamma\beta}(\bar{2}2)-G^{\lessgtr}_{\alpha\gamma}(1\bar{2})\Sigma^a_{\gamma\beta}(\bar{2}2)\right]\\
&\quad+\left[\Sigma^{\lessgtr}_{\alpha\gamma}(1\bar{2})G^a_{\gamma\beta}(\bar{2}2)-G^r_{\alpha\gamma}(1\bar{2})\Sigma^{\lessgtr}_{\gamma\beta}(\bar{2}2)\right]\\
&\quad+\left[\Delta^r_{hh,\alpha\gamma}(1\bar{2})g^{\lessgtr}_{ee,\gamma\beta}(\bar{2}2)-g^{\lessgtr}_{hh,\alpha\gamma}(1\bar{2})\Delta^a_{ee,\gamma\beta}(\bar{2}2)\right]\\
&\quad+\left[\Delta^{\lessgtr}_{hh,\alpha\gamma}(1\bar{2})g^a_{ee,\gamma\beta}(\bar{2}2)-g^r_{hh,\alpha\gamma}(1\bar{2})\Delta^{\lessgtr}_{ee,\gamma\beta}(\bar{2}2)\right],
\end{aligned}
\tag{17.40}
$$

and similarly for Eqs. (17.25)–(17.28). The most general semiconductor Bloch equations follow from these multi-band quantum transport equations by transforming to the semiconductor defect representation using the dictionary given in the handbook[10]. The complete derivation is straightforward.

It is more instructive to give the results for the multi-band quantum transport equations for $\alpha,\beta \Rightarrow v$ and c only in the defect representation since the resulting equations serve as the generalization of the density-matrix treatment of Mandel and Wolf [138] in quantum optics, as well as the coherent-wave treatment of Stahl and Balslev [139], which also provide a closed set of equations.

17.3.1 *Generalized Bloch equations in quantum optics*

We identify the density matrix elements used by Mandel and Wolf [138] with the notations used here in Table 17.1, where the averaging account for mixed states.

Table 17.1. Density matrix and generalized notation.

Mandel and Wolf	Generalized notation in defect representation
$\rho_{vc}=\langle c_v(1)\ c^\dagger_c(2)\rangle$	$\Rightarrow \langle \phi^\dagger_v\ (1)\,\psi^\dagger_c(2)\rangle = i\hbar g^{e-h,>}_{ee,vc}(12)$
$\rho_{cv}=\langle c_c(1)\ c^\dagger_v(2)\rangle$	$\Rightarrow \langle \psi_c(1)\ \phi_v(2)\rangle = i\hbar g^{e-h,>}_{hh,cv}(12)$
$\rho_{vv}=\langle c_v(1)\ c^\dagger_v(2)\rangle$	$\Rightarrow \langle \phi^\dagger_v\ (1)\,\phi_v(2)\rangle = -i\hbar G^{e-h,<T}_{vv}(12)$
$\rho_{cc}=\langle c_c(1)\ c^\dagger_c(2)\rangle$	$\Rightarrow \langle \psi_c(1)\ \psi^\dagger_c(2)\rangle = i\hbar G^{e-h,>}_{cc}(12)$

The generalized Bloch equations in quantum optics are thus given by the following equations:

$$
\begin{aligned}
&i\hbar\left(\frac{\partial}{\partial t_1}+\frac{\partial}{\partial t_2}\right)\left(-G_{vv}^{e-h,<T}(12)\right)\\
&=\left[v_{vv}(1\bar{2})\left(-G_{vv}^{e-h,<T}(\bar{2}2)\right)-\left(-G_{vv}^{e-h,<T}(1\bar{2})\right)v_{vv}^{T}(\bar{2}2)\right]\\
&\quad+\left[\begin{array}{c}\left(-\Sigma_{vv}^{e-h,aT}(1\bar{2})\right)\left(-G_{vv}^{e-h,<T}(\bar{2}2)\right)\\ -\left(-G_{vv}^{e-h,<T}(1\bar{2})\right)\left(-\Sigma_{vv}^{e-h,rT}(\bar{2}2)\right)\end{array}\right]\\
&\quad+\left[\Delta_{ee,vc}^{e-h,r}(1\bar{2})\,g_{hh,cv}^{e-h,>}(\bar{2}2)-g_{ee,vc}^{e-h,>}(1\bar{2})\,\Delta_{hh,cv}^{e-h,a}(\bar{2}2)\right]\\
&\quad+\left[\begin{array}{c}\left(-\Sigma_{vv}^{e-h,<T}(1\bar{2})\right)\left(-G_{vv}^{e-h,rT}(\bar{2}2)\right)\\ -\left(-G_{vv}^{e-h,aT}(1\bar{2})\right)\left(-\Sigma_{vv}^{e-h,<T}(\bar{2}2)\right)\end{array}\right]\\
&\quad+\left[\Delta_{ee,vc}^{e-h,>}(1\bar{2})\,g_{hh,cv}^{e-h,a}(\bar{2}2)-g_{ee,vc}^{e-h,r}(1\bar{2})\,\Delta_{hh,cv}^{>}(\bar{2}2)\right]\\
&\quad+\left[\Delta_{ee,vv}^{e-h,r}(1\bar{2})\,g_{hh,vv}^{e-h,>}(\bar{2}2)-g_{ee,vv}^{e-h,>}(1\bar{2})\,\Delta_{hh,vv}^{e-h,a}(\bar{2}2)\right]\\
&\quad+\left[\begin{array}{c}\left(-\Sigma_{vc}^{e-h,aT}(1\bar{2})\right)\left(-G_{cv}^{e-h,<T}(\bar{2}2)\right)\\ -\left(-G_{vc}^{e-h,<T}(1\bar{2})\right)\left(-\Sigma_{ee,cv}^{rT}(\bar{2}2)\right)\end{array}\right]\\
&\quad+\left[\Delta_{ee,vv}^{e-h,>}(1\bar{2})\,g_{hh,vv}^{e-h,a}(\bar{2}2)-g_{ee,vv}^{e-h,r}(1\bar{2})\,\Delta_{hh,vv}^{e-h,>}(\bar{2}2)\right]\\
&\quad+\left[\begin{array}{c}\left(-\Sigma_{vc}^{<T}(1\bar{2})\right)\left(-G_{cv}^{e-h,rT}(\bar{2}2)\right)\\ -\left(-G_{vc}^{e-h,aT}(1\bar{2})\right)\left(-\Sigma_{cv}^{e-h,<T}(\bar{2}2)\right)\end{array}\right],
\end{aligned}
\tag{17.41}
$$

$$
\begin{aligned}
&i\hbar\left(\frac{\partial}{\partial t_1}+\frac{\partial}{\partial t_2}\right)g_{ee,vc}^{e-h,>}(12)\\
&=\left[v_{vv}(1\bar{2})\,g_{ee,vc}^{e-h,>}(\bar{2}2)-g_{ee,vc}^{e-h,>}(1\bar{2})\,v_{cc}^{T}(\bar{2}2)\right]\\
&\quad+\left[\begin{array}{c}-\Sigma_{vv}^{e-h,aT}(1\bar{2})\,g_{ee,vc}^{e-h,>}(\bar{2}2)\\ -\left(-G_{vv}^{e-h,<T}(1\bar{2})\right)\Delta_{ee,vc}^{e-h,a}(\bar{2}2)\end{array}\right]
\end{aligned}
$$

$$+\left[\Delta^{e-h,r}_{ee,vc}(1\bar{2})\,G^{e-h,>}_{cc}(\bar{2}2)-g^{e-h,>}_{ee,vc}(1\bar{2})\,\Sigma^{e-h,a}_{cc}(\bar{2}2)\right]$$

$$+\left[\begin{array}{c}-\Sigma^{e-h,<T}_{vv}(1\bar{2})\,g^{e-h,a}_{ee,vc}(\bar{2}2)\\ -\left(-G^{e-h,aT}_{vv}(1\bar{2})\right)\Delta^{e-h,>}_{ee,vc}(\bar{2}2)\end{array}\right]$$

$$+\left[\Delta^{e-h,>}_{ee,vc}(1\bar{2})\,G^{e-h,a}_{cc}(\bar{2}2)-g^{e-h,r}_{ee,vc}(1\bar{2})\,\Sigma^{e-h,>}_{cc}(\bar{2}2)\right]$$

$$+\left[\Delta^{e-h,r}_{hh,vv}(1\bar{2})\,G^{e-h,>}_{vc}(\bar{2}2)-g^{e-h,>}_{ee,vv}(1\bar{2})\,\Sigma^{e-h,a}_{vc}(\bar{2}2)\right]$$

$$+\left[\begin{array}{c}-\Sigma^{e-h,aT}_{vc}(1\bar{2})\,g^{e-h,>}_{ee,cc}(\bar{2}2)\\ -\left(-G^{e-h,<T}_{hh,vc}(1\bar{2})\right)\Delta^{e-h,a}_{ee,cc}(\bar{2}2)\end{array}\right]$$

$$+\left[\Delta^{e-h,>}_{ee,vv}(1\bar{2})\,G^{e-h,a}_{vc}(\bar{2}2)-g^{e-h,r}_{hh,vv}(1\bar{2})\,G^{e-h,>}_{vc}(\bar{2}2)\right]$$

$$+\left[\begin{array}{c}-\Sigma^{e-h,<T}_{vc}(1\bar{2})\,g^{e-h,a}_{ee,cc}(\bar{2}2)\\ -\left(-G^{e-h,aT}_{vc}(1\bar{2})\right)\Delta^{e-h,>}_{ee,cc}(\bar{2}2)\end{array}\right], \tag{17.42}$$

$$\begin{aligned}
&i\hbar\left(\frac{\partial}{\partial t_1}+\frac{\partial}{\partial t_2}\right)g^{e-h,>}_{hh,cv}(12)\\
&=\left[v_{cc}g^{e-h,>}_{hh,cv}-g^{e-h,>}_{hh,cv}v^{T}_{vv}\right]\\
&\quad+\left[\Delta^{e-h,r}_{hh,cv}\left(-G^{e-h,<T}_{vv}\right)-g^{e-h,>}_{hh,cv}\left(-\Sigma^{e-h,rT}_{vv}\right)\right]\\
&\quad+\left[\Sigma^{e-h,r}_{cc}g^{e-h,>}_{hh,cv}-G^{e-h,>}_{cc}\Delta^{a}_{hh,cv}\right]\\
&\quad+\left[\Delta^{e-h,>}_{hh,cv}\left(-G^{e-h,rT}_{vv}\right)-g^{e-h,r}_{hh,cv}\left(-\Sigma^{e-h,<T}_{vv}\right)\right]\\
&\quad+\left[\Sigma^{e-h,>}_{cc}g^{e-h,a}_{hh,cv}-G^{e-h,r}_{cc}\Delta^{e-h,>}_{hh,cv}\right]\\
&\quad+\left[\Sigma^{e-h,r}_{cv}g^{e-h,>}_{hh,vv}-G^{e-h,>}_{hh,cv}\Delta^{e-h,a}_{hh,vv}\right]\\
&\quad+\left[\Delta^{e-h,r}_{hh,cc}\left(-G^{<T}_{cv}\right)-g^{e-h,>}_{hh,cc}\left(-\Sigma^{e-h,rT}_{cv}\right)\right]\\
&\quad+\left[\Sigma^{e-h,>}_{cv}g^{e-h,a}_{hh,vv}-G^{e-h,r}_{cv}\Delta^{e-h,>}_{hh,vv}\right]\\
&\quad+\left[\Delta^{e-h,>}_{hh,cc}\left(-G^{rT}_{cv}\right)-g^{e-h,r}_{hh,cc}\left(-\Sigma^{<T}_{cv}\right)\right],
\end{aligned} \tag{17.43}$$

$$
\begin{aligned}
i\hbar&\left(\frac{\partial}{\partial t_1}+\frac{\partial}{\partial t_2}\right)G_{cc}^{e-h,>}(12)\\
&=\left[v_{cc}(1\bar{2})\,G_{cc}^{e-h,>}(\bar{2}2)-G_{cc}^{e-h,>}(1\bar{2})\,v_{cc}^{T}\right]\\
&+\left[\Delta_{hh,cv}^{e-h,r}(1\bar{2})\,g_{ee,vc}^{e-h,>}(\bar{2}2)-g_{hh,cv}^{e-h,>}(1\bar{2})\,\Delta_{ee,vc}^{e-h,a}(\bar{2}2)\right]\\
&+\left[\Sigma_{cc}^{e-h,r}(1\bar{2})\,G_{cc}^{e-h,>}(\bar{2}2)-G_{cc}^{e-h,>}(1\bar{2})\,\Sigma_{cc}^{e-h,a}(\bar{2}2)\right]\\
&+\left[\Delta_{hh,cv}^{e-h,>}(1\bar{2})\,g_{ee,vc}^{e-h,a}(\bar{2}2)-g_{hh,cv}^{e-h,r}(1\bar{2})\,\Delta_{ee,vc}^{e-h,>}(\bar{2}2)\right]\\
&+\left[\Sigma_{cc}^{e-h,>}(1\bar{2})\,G_{cc}^{e-h,a}(\bar{2}2)-G_{cc}^{e-h,r}(1\bar{2})\,\Sigma_{cc}^{e-h,>}(\bar{2}2)\right]\\
&+\left[\Sigma_{cv}^{e-h,r}(1\bar{2})\,G_{vc}^{e-h,>}(\bar{2}2)-G_{cv}^{e-h,>}(1\bar{2})\,\Sigma_{vc}^{e-h,a}(\bar{2}2)\right]\\
&+\left[\Delta_{hh,cc}^{e-h,r}(1\bar{2})\,g_{ee,cc}^{e-h,>}(\bar{2}2)-g_{hh,cc}^{e-h,>}(1\bar{2})\,\Delta_{ee,cc}^{e-h,a}(\bar{2}2)\right]\\
&+\left[\Sigma_{cv}^{e-h,>}(1\bar{2})\,G_{vc}^{e-h,a}(\bar{2}2)-G_{cv}^{e-h,r}(1\bar{2})\,\Sigma_{vc}^{e-h,>}(\bar{2}2)\right]\\
&+\left[\Delta_{hh,cc}^{e-h,>}(1\bar{2})\,g_{ee,cc}^{e-h,a}(\bar{2}2)-g_{hh,cc}^{e-h,r}(1\bar{2})\,\Delta_{ee,cc}^{e-h,>}(\bar{2}2)\right].
\end{aligned}
\tag{17.44}
$$

In applying these equations to quantum optics [138], we neglect the particle pairing terms which are contained in the last four brackets of every equations, we also neglect terms arising from the particle self-energies Σ as well as terms involving $\Delta^{e-h,\lessgtr}$, and all dissipative terms. These result in a closed set of equations. Upon substituting the corresponding quantities in Table 17.1, and identifying $\Delta_{hh,cv}^{e-h,r}(1\bar{2}) \Rightarrow \langle c|\,\mathcal{H}_I(t)\,|v\rangle\,\delta_{1\bar{2}}$, and $\Delta_{ee,vc}^{e-h,a}(\bar{2}1) \Rightarrow \langle v|\,\mathcal{H}_I(t)\,|c\rangle\,\delta_{1\bar{2}}$, with the relation $\Delta_{ee,vc}^{e-h,a}(\bar{2}1) = \Delta_{ee,vc}^{e-h,r}(\bar{2}1)$, $\Delta_{hh,cv}^{e-h,r}(1\bar{2}) = \Delta_{hh,cv}^{e-h,a}(1\bar{2})$ by virtue of the locality of time-dependence, we end up with

$$\frac{\partial}{\partial t}\rho_{vv}=\frac{1}{i\hbar}\left[\langle v|\,\mathcal{H}_I(t)\,|c\rangle\,\rho_{cv}-\rho_{vc}\,\langle c|\,\mathcal{H}_I(t)\,|v\rangle\right], \tag{17.45}$$

$$\frac{\partial}{\partial t}\rho_{vc}=\frac{1}{i\hbar}\left[-\hbar\omega_o\rho_{vc}+\langle v|\,\mathcal{H}_I(t)\,|c\rangle\,(\rho_{cc}-\rho_{vv})\right], \tag{17.46}$$

$$\frac{\partial}{\partial t}\rho_{cv}=\frac{1}{i\hbar}\left[\hbar\omega_o\rho_{cv}+\langle c|\,\mathcal{H}_I(t)\,|v\rangle\,(\rho_{vv}-\rho_{cc})\right], \tag{17.47}$$

$$\frac{\partial}{\partial t}\rho_{cc}=-\frac{1}{i\hbar}\left[\langle v|\,\mathcal{H}_I(t)\,|c\rangle\,\rho_{cv}-\rho_{vc}\,\langle c|\,\mathcal{H}_I(t)\,|v\rangle\right]. \tag{17.48}$$

Table 17.2. Coherent-wave correlation functions and generalized notations.

S-B atomic	S-B tight binding	Generalized notation
$s^{\dagger}_{\lambda,j} = c^{\dagger}_{j} d^{\dagger}_{\lambda,j}$	$Y^{\dagger}_{ij} = c^{\dagger}_{j} d^{\dagger}_{i}$	$\Rightarrow \langle \psi^{\dagger}_{c}(2)\, \phi^{\dagger}_{v}(1) \rangle = -i\hbar g^{e-h,<}_{ee,vc}(12)$
$s_{\lambda,j} = d_{\lambda,j} c_j$	$Y_{ij} = d_i c_j$	$\Rightarrow \langle \phi_v(1)\, \psi_c(2) \rangle = i\hbar g^{e-h,>}_{hh,vc}(12)$
$n_j = c^{\dagger}_{j} c_j$	$C_{ij} = c^{\dagger}_{i} c_j$	$\Rightarrow \langle \psi^{\dagger}_{c}(1)\, \psi_c(2) \rangle = -i\hbar G^{e-h,<T}_{cc}(12)$
$p_{\lambda\nu,j} = d^{\dagger}_{\lambda,j} d^{\dagger}_{\nu,j}$	$D_{ij} = d_i d_j$	$\Rightarrow \langle \phi^{\dagger}_{v}(1)\, \phi_v(2) \rangle = -i\hbar G^{e-h,<T}_{vv}(12)$

These equations are exactly the Bloch equations of a two-level atom discussed by Mandel and Wolf [138]. These equations are indeed very rich in physics and the reader is referred to Ref. [138] for further discussions on the electrodynamics of two-level atoms.

17.3.2 *Generalized coherent-wave theory*

Stahl and Balslev (SB) [139] introduced the coherent wave theory in discussing the electrodynamics of semiconductor band edge. The correlation functions they used correspond to $G^{<}_{vc}(12)$, $G^{<T}_{vc}(12)$, $G^{>}_{vv}(12)$, and $G^{<T}_{cc}(12)$ expressed in the defect representation. Table 17.2 maps the correspondence between coherent-wave correlation functions and the nonequilibrium generalization, both in the defect representation.

The generalized coherent wave theory of SB [139] is thus given by the following equations:

$$
\begin{aligned}
&i\hbar\left(\frac{\partial}{\partial t_1}+\frac{\partial}{\partial t_2}\right)\left(-G^{e-h,<T}_{vv}(12)\right)\\
&= \left[v_{vv}(1\bar{2})\left(-G^{e-h,<T}_{vv}(\bar{2}2)\right)-\left(-G^{e-h,<T}_{vv}(1\bar{2})\right)\left(v^{T}_{vv}(\bar{2}2)\right)\right]\\
&\quad+\left[\begin{array}{l}\left(-\Sigma^{e-h,aT}_{vv}(1\bar{2})\right)\left(-G^{e-h,<T}_{vv}(\bar{2}2)\right)\\ -\left(-G^{e-h,<T}_{vv}(1\bar{2})\right)\left(-\Sigma^{e-h,rT}_{vv}(\bar{2}2)\right)\end{array}\right]\\
&\quad+\left[\Delta^{e-h,r}_{ee,vc}(1\bar{2})\,g^{e-h,>}_{hh,cv}(\bar{2}2)-g^{e-h,>}_{ee,vc}(1\bar{2})\,\Delta^{e-h,a}_{hh,cv}(\bar{2}2)\right]\\
&\quad+\left[\begin{array}{l}\left(-\Sigma^{e-h,<T}_{vv}(1\bar{2})\right)\left(-G^{e-h,rT}_{vv}(\bar{2}2)\right)\\ -\left(-G^{e-h,aT}_{vv}(1\bar{2})\right)\left(-\Sigma^{e-h,<T}_{vv}(\bar{2}2)\right)\end{array}\right]
\end{aligned}
$$

$$
\begin{aligned}
&+\left[\Delta^{e-h,>}_{ee,vc}\left(1\bar{2}\right)g^{e-h,a}_{hh,cv}\left(\bar{2}2\right)-g^{e-h,r}_{ee,vc}\left(1\bar{2}\right)\Delta^{>}_{hh,cv}\left(\bar{2}2\right)\right]\\
&+\left[\Delta^{e-h,r}_{ee,vv}\left(1\bar{2}\right)g^{e-h,>}_{hh,vv}\left(\bar{2}2\right)-g^{e-h,>}_{ee,vv}\left(1\bar{2}\right)\Delta^{e-h,a}_{hh,vv}\left(\bar{2}2\right)\right]\\
&+\left[\begin{array}{l}\left(-\Sigma^{e-h,aT}_{vc}\left(1\bar{2}\right)\right)\left(-G^{e-h,<T}_{cv}\left(\bar{2}2\right)\right)\\ -\left(-G^{e-h,<T}_{vc}\left(1\bar{2}\right)\right)\left(-\Sigma^{rT}_{ee,cv}\left(\bar{2}2\right)\right)\end{array}\right]\\
&+\left[\Delta^{e-h,>}_{ee,vv}\left(1\bar{2}\right)g^{e-h,a}_{hh,vv}\left(\bar{2}2\right)-g^{e-h,r}_{ee,vv}\left(1\bar{2}\right)\Delta^{e-h,>}_{hh,vv}\left(\bar{2}2\right)\right]\\
&+\left[\begin{array}{c}\left(-\Sigma^{<T}_{vc}\left(1\bar{2}\right)\right)\left(-G^{e-h,rT}_{cv}\left(\bar{2}2\right)\right)\\ -\left(-G^{e-h,aT}_{vc}\left(1\bar{2}\right)\right)\left(-\Sigma^{e-h,<T}_{cv}\left(\bar{2}2\right)\right)\end{array}\right],
\end{aligned}
\tag{17.49}
$$

$$
\begin{aligned}
& i\hbar\left(\frac{\partial}{\partial t_1}+\frac{\partial}{\partial t_2}\right)g^{e-h,<}_{ee,vc}\left(12\right)\\
&=\left[\begin{array}{l}v_{vv}\left(1\bar{2}\right)g^{e-h,<}_{ee,vc}\left(\bar{2}2\right)\\ -g^{e-h,<}_{ee,vc}\left(1\bar{2}\right)v^{T}_{c}\left(\bar{2}2\right)\end{array}\right]\\
&\quad+\left[\begin{array}{l}-\Sigma^{e-h,aT}_{vv}\left(1\xi\right)g^{e-h,<}_{ee,vc}\left(\xi2\right)\\ +G^{e-h,>T}_{vv}\left(1\xi\right)\Delta^{e-h,a}_{ee,vc}\left(\xi2\right)\end{array}\right]\\
&\quad+\left[\begin{array}{l}\Delta^{e-h,r}_{ee,vc}\left(1\xi\right)G^{e-h,<}_{cc}\left(\xi2\right)\\ -g^{e-h,<}_{ee,vc}\left(1\xi\right)\Sigma^{e-h,a}_{cc}\left(\xi2\right)\end{array}\right]\\
&\quad+\left[\begin{array}{l}-\Sigma^{e-h,>T}_{vv}\left(1\xi\right)g^{e-h,a}_{ee,vc}\left(\xi2\right)\\ +G^{e-h,aT}_{vv}\left(1\xi\right)\Delta^{e-h,<}_{ee,vc}\left(\xi2\right)\end{array}\right]\\
&\quad+\left[\begin{array}{l}\Delta^{e-h,<}_{ee,vc}\left(1\xi\right)G^{e-h,a}_{cc}\left(\xi2\right)\\ -g^{e-h,r}_{ee,vc}\left(1\xi\right)\Sigma^{e-h,<}_{cc}\left(\xi2\right)\end{array}\right]\\
&\quad+\left[\begin{array}{l}\Delta^{e-h,r}_{ee,vv}\left(1\xi\right)G^{e-h,<}_{vc}\left(\xi2\right)\\ -g^{e-h,<}_{ee,vv}\left(1\xi\right)\Sigma^{e-h,a}_{vc}\left(\xi2\right)\end{array}\right]
\end{aligned}
$$

$$+\begin{bmatrix}-\Sigma_{vc}^{e-h,aT}(1\xi)\,g_{ee,cc}^{e-h,<}(\xi 2)\\+G_{vc}^{e-h,>T}(1\xi)\,\Delta_{ee,cc}^{e-h,a}(\xi 2)\end{bmatrix}$$

$$+\begin{bmatrix}\Delta_{ee,vv}^{e-h,<}(1\xi)\,G_{vc}^{e-h,a}(\xi 2)\\-g_{ee,vv}^{e-h,r}(1\xi)\,\Sigma_{vc}^{e-h,<}(\xi 2)\end{bmatrix}$$

$$+\begin{bmatrix}-\Sigma_{vc}^{e-h,>T}(1\xi)\,g_{ee,cc}^{e-h,a}(\xi 2)\\+G_{vc}^{e-h,aT}(1\xi)\,\Delta_{ee,cc}^{e-h,<}(\xi 2)\end{bmatrix}, \tag{17.50}$$

$$-i\hbar\left(\frac{\partial}{\partial t_1}+\frac{\partial}{\partial t_2}\right)\left(g_{hh,vc}^{e-h,>}(12)\right)$$

$$= \left[\left(g_{hh,vc}^{e-h,>}(\xi 2)\right)v_{vv}^{T}(1\xi)-v_c(\xi 2)\,g_{hh,vc}^{e-h,>}(1\xi)\right]$$

$$+\left[-\left(g_{hh,vc}^{e-h,>}(\xi 2)\right)\left(\Sigma_{vv}^{e-h,r}(1\xi)\right)-\Delta_{hh,cv}^{e-h,r}(2\xi)\left(G_{vv}^{e-h,>}(1\xi)\right)\right]$$

$$+\left[\left(-G_{cc}^{e-h,<T}(\xi 2)\right)\Delta_{hh,cv}^{e-h,a}(\xi 1)-\Sigma_{cc}^{e-h,r}(2\xi)\left(g_{hh,vc}^{e-h,>}(1\xi)\right)\right]$$

$$+\left[g_{hh,cv}^{e-h,r}(2\xi)\left(\Sigma_{vv}^{e-h,>}(1\xi)\right)+\left(\Delta_{hh,vc}^{e-h,>}(\xi 2)\right)\left(G_{vv}^{e-h,rT}(\xi 1)\right)\right]$$

$$+\left[G_{cc}^{e-h,r}(2\xi)\left(\Delta_{vc}^{e-h,>}(1\xi)\right)+\left(\Sigma_{cc}^{e-h,<T}(\xi 2)\right)g_{hh,cv}^{e-h,a}(\xi 1)\right]$$

$$+\begin{bmatrix}\left(-G_{cv}^{e-h,<}(2\xi)\right)\Delta_{hh,vv}^{e-h,a}(\xi 1)\\+\Sigma_{cv}^{e-h,r}(2\xi)\left(g_{hh,vv}^{e-h,<}(\xi 1)\right)\end{bmatrix}$$

$$+\begin{bmatrix}\left(g_{hh,cc}^{e-h,<}(2\xi)\right)\left(\Sigma_{ee,cv}^{e-h,rT}(\xi 1)\right)\\-\Delta_{hh,cc}^{e-h,r}(2\xi)\left(G_{cv}^{e-h,>T}(\xi 1)\right)\end{bmatrix}$$

$$+\begin{bmatrix}-G_{hh,cv}^{e-h,r}(2\xi)\left(\Delta_{hh,vv}^{e-h,<}(\xi 1)\right)\\+\left(\Sigma_{cv}^{e-h,<}(2\xi)\right)g_{hh,vv}^{e-h,a}(\xi 1)\end{bmatrix}$$

$$+\begin{bmatrix}g_{hh,cc}^{e-h,r}(2\xi)\left(\Sigma_{cv}^{e-h,>T}(\xi 1)\right)\\-\left(\Delta_{ee,cc}^{e-h,<}(2\xi)\right)\left(G_{cv}^{e-h,rT}(\xi 1)\right)\end{bmatrix}, \tag{17.51}$$

$$
\begin{aligned}
-i\hbar &\left(\frac{\partial}{\partial t_1}+\frac{\partial}{\partial t_2}\right)\left(-G_{cc}^{e-h,<T}(12)\right) \\
&= \left[v_{cc}(\bar{2}1)\left(-G_{cc}^{e-h,<T}(\bar{2}2)\right)-\left(-G_{cc}^{e-h,<T}(1\bar{2})\right)v_{cc}(\bar{2}2)\right] \\
&+ \left[\left(g_{hh,vc}^{>}(\bar{2}2)\right)\Delta_{ee,vc}^{e-h,a}(\bar{2}1)-\Delta_{hh,cv}^{e-h,r}(2\bar{2})\left(g_{ee,cv}^{e-h,>}(1\bar{2})\right)\right] \\
&+ \left[\left(-G_{cc}^{e-h,<T}(\bar{2}2)\right)\Sigma_{cc}^{e-h,a}(\bar{2}1)-\Sigma_{cc}^{e-h,r}(2\bar{2})\left(-G_{cc}^{e-h,<T}(1\bar{2})\right)\right] \\
&+ \left[g_{hh,cv}^{e-h,r}(2\bar{2})\left(\Delta_{ee,cv}^{e-h,>}(1\bar{2})\right)-\left(\Delta_{hh,vc}^{e-h,>}(\bar{2}2)\right)g_{ee,vc}^{e-h,a}(\bar{2}1)\right] \\
&+ \left[G_{cc}^{e-h,r}(2\bar{2})\left(-\Sigma_{cc}^{e-h,<T}(1\bar{2})\right)-\left(-\Sigma_{cc}^{e-h,<T}(\bar{2}2)\right)G_{cc}^{e-h,a}(\bar{2}1)\right] \\
&+ \left[\left(-G_{cv}^{e-h,<}(2\bar{2})\right)\Sigma_{vc}^{e-h,a}(\bar{2}1)-\Sigma_{cv}^{e-h,r}(2\bar{2})\left(-G_{vc}^{e-h,<}(\bar{2}1)\right)\right] \\
&+ \left[\left(-g_{hh,cc}^{e-h,<}(2\bar{2})\right)\Delta_{ee,cc}^{e-h,a}(\bar{2}1)-\Delta_{hh,cc}^{e-h,r}(2\bar{2})\left(-g_{ee,cc}^{e-h,<}(\bar{2}1)\right)\right] \\
&+ \left[G_{cv}^{e-h,r}(2\bar{2})\left(-\Sigma_{vc}^{e-h,<}(\bar{2}1)\right)-\left(-\Sigma_{cv}^{e-h,<}(2\bar{2})\right)G_{vc}^{e-h,a}(\bar{2}1)\right] \\
&+ \left[g_{hh,cc}^{e-h,r}(2\bar{2})\left(-\Delta_{ee,cc}^{e-h,<}(\bar{2}1)\right)-\left(-\Delta_{hh,cc}^{e-h,<}(2\bar{2})\right)g_{ee,cc}^{e-h,a}(\bar{2}1)\right].
\end{aligned}
\tag{17.52}
$$

In making contact with the work of SB [139], we dropped the particle pairing terms (the terms contained in the last four brackets of each equation), and also ignore dissipative terms involving the particle self-energies Σ and $\Delta^{\lessgtr}$. Then Eqs. (17.49)–(17.52) reduce to the following semiconductor Bloch equations corresponding to those given by SB [139].

$$
\begin{aligned}
i\hbar &\left(\frac{\partial}{\partial t_1}+\frac{\partial}{\partial t_2}\right)\left(-G_{vv}^{e-h,<T}(12)\right) \\
&= \left[\left(v_{vv}^{T}(1\bar{2})\right)\left(-G_{vv}^{e-h,<T}(\bar{2}2)\right)-\left(-G_{vv}^{e-h,<T}(1\bar{2})\right)\left(v_{vv}(\bar{2}2)\right)\right] \\
&+ \left[\Delta_{ee,vc}^{e-h,r}(1\bar{2})\,g_{hh,cv}^{e-h,>}(\bar{2}2)-g_{ee,vc}^{e-h,>}(1\bar{2})\,\Delta_{hh,cv}^{e-h,a}(\bar{2}2)\right],
\end{aligned}
\tag{17.53}
$$

$$
\begin{aligned}
i\hbar &\left(\frac{\partial}{\partial t_1}+\frac{\partial}{\partial t_2}\right)g_{ee,vc}^{e-h,<}(12) \\
&= \left[\left(v_{vv}(1\bar{2})\right)g_{ee,vc}^{e-h,<}(\bar{2}2)-g_{ee,vc}^{e-h,<}(1\bar{2})\,v_{c}^{T}(\bar{2}2)\right] \\
&+ \left[G_{vv}^{e-h,>T}(1\xi)\,\Delta_{ee,vc}^{e-h,a}(\xi 2)\right] \\
&+ \left[\Delta_{ee,vc}^{e-h,r}(1\xi)\,G_{cc}^{e-h,<}(\xi 2)\right],
\end{aligned}
\tag{17.54}
$$

$$-i\hbar\left(\frac{\partial}{\partial t_1}+\frac{\partial}{\partial t_2}\right)\left(g^{e-h,>}_{hh,vc}(12)\right)$$
$$=\left[\left(g^{e-h,>}_{hh,vc}(\xi 2)\right)v^{T}_{vv}(1\xi)-v_c(\xi 2)\,g^{e-h,>}_{hh,vc}(1\xi)\right]$$
$$-\left[\Delta^{e-h,r}_{hh,cv}(2\xi)\left(G^{e-h,>}_{vv}(1\xi)\right)\right]$$
$$-\left[G^{e-h,<T}_{cc}(\xi 2)\,\Delta^{e-h,a}_{hh,cv}(\xi 1)\right], \tag{17.55}$$

$$-i\hbar\left(\frac{\partial}{\partial t_1}+\frac{\partial}{\partial t_2}\right)\left(-G^{e-h,<T}_{cc}(12)\right)$$
$$=\left[v_{cc}(\bar{2}1)\left(-G^{e-h,<T}_{cc}(\bar{2}2)\right)-\left(-G^{e-h,<T}_{cc}(1\bar{2})\right)v_{cc}(\bar{2}2)\right]$$
$$+\left[\left(g^{>}_{hh,vc}(\bar{2}2)\right)\Delta^{e-h,a}_{ee,vc}(\bar{2}1)-\Delta^{e-h,r}_{hh,cv}(2\bar{2})\left(g^{e-h,>}_{ee,cv}(1\bar{2})\right)\right]. \tag{17.56}$$

17.3.2.1 *The tight-binding limit*

In the tight-binding limit, the energy band energies in the coordinate representation correspond to the hopping integrals between different sites. Therefore, we may write the one-body terms such as $v^{T}_{vv}(1\bar{2}) = -E_g\delta_{1\bar{2}} - T^{v}_{\bar{2}1}$, $v_{cc}(\bar{2}2) = T^{c}_{\bar{2}2}$. The externally applied light field produces interband transition which we denote by $\Delta^{e-h,r}_{hh,cv}(1\bar{2}) \Rightarrow \langle c|\mathcal{H}_I(t)|v\rangle\,\delta_{1\bar{2}} = -M_o\cdot F(1)\,\delta_{1\bar{2}}$, and $\Delta^{e-h,a}_{ee,vc}(\bar{2}1) \Rightarrow \langle v|\mathcal{H}_I(t)|c\rangle\,\delta_{1\bar{2}} = -M_o\cdot F(1)\,\delta_{1\bar{2}}$, where F is the electric field and M_o is the dipole matrix element. We also have the relation $\Delta^{e-h,a}_{ee,vc}(\bar{2}1) = \Delta^{e-h,r}_{ee,vc}(\bar{2}1)$, $\Delta^{e-h,r}_{hh,cv}(1\bar{2}) = \Delta^{e-h,a}_{hh,cv}(1\bar{2})$ by virtue of the locality of time-dependence. Upon substituting the symbols in Table 17.2, we obtain,

$$\left(\frac{\partial}{\partial t_1}+\frac{\partial}{\partial t_2}\right)D_{12}$$
$$=\frac{1}{i\hbar}\left[\left(-E_g\delta_{1\bar{2}}-T^{v}_{\bar{2}1}\right)D_{\bar{2}2}-\left(D_{1\bar{2}}\right)\left(-E_g\delta_{\bar{2}2}-T^{v}_{\bar{2}2}\right)\right]$$
$$+\frac{1}{i\hbar}\left[\left(-M_o\cdot F\delta_{1\bar{2}}\right)i\hbar g^{e-h,>}_{hh,cv}(\bar{2}2)-i\hbar g^{e-h,>}_{ee,vc}(1\bar{2})\left(-M_o\cdot F\delta_{\bar{2}2}\right)\right], \tag{17.57}$$

$$\left(\frac{\partial}{\partial t_1}+\frac{\partial}{\partial t_2}\right)Y^{\dagger}_{ij}(12) = -\frac{1}{-i\hbar}\left[\left(-E_g\delta_{1\bar{2}}-T^{v}_{\bar{2}1}\right)Y^{\dagger}_{\bar{2}2}-Y^{\dagger}_{1\bar{2}}T^{c}_{\bar{2}\bar{2}}\right]$$
$$-\frac{1}{i\hbar}\left[i\hbar G^{e-h,>T}_{vv}(1\xi)\left(-M_o\cdot F\delta_{\xi 2}\right)\right]$$
$$-\frac{1}{i\hbar}\left[\left(-M_o\cdot F\delta_{1\xi}\right)i\hbar G^{e-h,<}_{cc}(\xi 2)\right], \tag{17.58}$$

$$-\left(\frac{\partial}{\partial t_1}+\frac{\partial}{\partial t_2}\right)Y_{12}=\frac{1}{i\hbar}\left[\left(Y_{\xi 2}\right)\left(-E_g\delta_{1\xi}-T^v_{1\xi}\right)-T^c_{2\xi}Y_{1\xi}\right]$$
$$-\frac{1}{i\hbar}\left[\left(-M_o\cdot F\delta_{2\xi}\right)\left(i\hbar G^{e-h,>}_{vv}\left(1\xi\right)\right)\right]$$
$$+\frac{1}{i\hbar}\left[C_{\xi 2}\left(-M_o\cdot F\delta_{\xi 1}\right)\right], \tag{17.59}$$

$$-\left(\frac{\partial}{\partial t_1}+\frac{\partial}{\partial t_2}\right)C_{12}$$
$$=\frac{1}{i\hbar}\left[T^c_{\bar{2}1}v_{cc}C_{\bar{2}2}-C_{1\bar{2}}T^c_{\bar{2}2}v_{cc}\right]$$
$$+\frac{1}{i\hbar}\left[Y_{\bar{2}2}\left(-M_o\cdot F\delta_{\bar{2}1}\right)-\left(-M_o\cdot F\delta_{2\bar{2}}\right)\left(i\hbar g^{e-h,>}_{ee,cv}\left(1\bar{2}\right)\right)\right]. \tag{17.60}$$

Making use of the following relations (from equal-time anti-commutation relation):

$$i\hbar G^{>T}\left(12\right)=\delta_{12}+i\hbar G^{<T}\left(12\right),$$
$$i\hbar g^{e-h,>}_{hh,cv}\left(12\right)=\left\langle\psi_c\left(1\right)\psi_v^\dagger\left(2\right)\right\rangle=-\left\langle\psi_v^\dagger\left(2\right)\psi_c\left(1\right)\right\rangle=-i\hbar g^{e-h,>}_{hh,vc}\left(21\right),$$
$$i\hbar g^{e-h,>}_{ee,vc}\left(12\right)=\left\langle\psi_v\left(1\right)\psi_c^\dagger\left(2\right)\right\rangle=-\left\langle\psi_c^\dagger\left(2\right)\psi_v\left(1\right)\right\rangle=i\hbar g^{e-h,<}_{ee,vc}\left(12\right),$$
$$i\hbar g^{e-h,>}_{ee,cv}\left(12\right)=\left\langle\psi_c^\dagger\left(1\right)\psi_v\left(2\right)\right\rangle=-i\hbar g^{e-h,<}_{ee,vc}\left(21\right),$$

and completing substitution, noting that $t_1=t_2$, these equations become

$$\frac{\partial}{\partial t}D_{12}+\frac{i}{\hbar}\left[D_{1\bar{2}}T^v_{\bar{2}2}-T^v_{\bar{2}1}D_{\bar{2}2}\right]$$
$$=-\frac{iM_o\cdot}{\hbar}\left[F\left(1\right)Y_{21}-Y^\dagger_{12}F\left(2\right)\right], \tag{17.61}$$

$$\frac{\partial}{\partial t}Y^\dagger_{12}-\frac{i}{\hbar}\left[E_gY^\dagger_{12}+T^v_{\bar{2}1}Y^\dagger_{\bar{2}2}+Y^\dagger_{1\bar{2}}T^c_{\bar{2}2}\right]$$
$$=-\frac{iM_o\cdot}{\hbar}\left[\delta_{12}F\left(2\right)-C_{21}F\left(1\right)-D_{12}F\left(2\right)\right], \tag{17.62}$$

$$\frac{\partial}{\partial t}Y_{12}+\frac{i}{\hbar}\left[E_gY_{12}+T^v_{1\xi}Y_{\xi 2}+T^c_{2\xi}Y_{1\xi}\right]$$
$$=\frac{iM_o\cdot}{\hbar}\left[\left(\delta_{12}F\left(2\right)-F\left(1\right)C_{12}-F\left(2\right)D_{21}\right)\right], \tag{17.63}$$

$$\frac{\partial}{\partial t}C_{12}+\frac{i}{\hbar}\left[C_{1\bar{2}}T^c_{\bar{2}2}-T^c_{\bar{2}1}C_{\bar{2}2}\right]$$
$$=-\frac{iM_o\cdot}{\hbar}\left[F\left(1\right)Y_{12}-F\left(2\right)\left(Y^\dagger_{21}\right)\right], \tag{17.64}$$

which is exactly the coherent wave equations derived by SB [139]. Note that the hopping matrix T_{ij} is assumed to be Hermitian in Ref. [139].

17.3.2.2 *Flat band case*

For flat band case, we simply use the symbols under SB Atomic in Table 17.2, putting $i = j$ and hence taking the hopping integrals to be zero. Here we assume that the valence band consist of multiplet, and we use the indices λ and ν to distinguish two degenerate levels of the multiplet. Note that although $\mathcal{H}_I(t)$ consist of dot-product sum over all directions, the dipole matrix element picks up only certain directions. Then we obtain

$$\frac{\partial}{\partial t}p_{\lambda\nu,j} = -\frac{iM_o}{\hbar}[F_{\lambda,j}\, s_{\nu,j} - s^{\dagger}_{\lambda,j}F_{\nu,j}], \tag{17.65}$$

$$\frac{\partial}{\partial t}s^{\dagger}_{\lambda,j} - \frac{i}{\hbar}E_g s^{\dagger}_{\lambda,j} = -\frac{iM_o}{\hbar}\left[\delta_{\lambda\nu}F_{\nu,j} - n_j F_{\lambda,j} - p_{\lambda\nu,j}F_{\nu,j}\right], \tag{17.66}$$

$$\frac{\partial}{\partial t}s_{\lambda,j} + \frac{i}{\hbar}E_g s_{\lambda,j} = \frac{iM_o}{\hbar}\left[(\delta_{\lambda\nu}F_{\nu,j} - n_j F_{\lambda,j} - p_{\lambda\nu,j}F_{\nu,j})\right], \tag{17.67}$$

$$\frac{\partial}{\partial t}n_j = -\frac{iM_o}{\hbar}(s_{\lambda,j} - s^{\dagger}_{\lambda,j})F_{\lambda,j}\,, \tag{17.68}$$

which are exactly the coherent wave equations given by in Ref. [139] for flat-band case. When expressed in terms of density, we simply replace the Dirac delta $\delta_{\lambda\nu}$, which arise from the anti-commutation relation, by $\delta_{\lambda\nu}N$ where $N = \frac{1}{V}$ and replace the index j by a representative coordinate r.

We can also cast the above equations in vector-tensor notation as

$$\frac{\partial \overleftrightarrow{p}(r)}{\partial t} = -\frac{i}{\hbar}\left[\mathbf{s}(r)\otimes\mathbf{E}(r) - \mathbf{E}(r)\otimes\mathbf{s}^{\dagger}(r)\right], \tag{17.69}$$

$$\frac{\partial n(r)}{\partial t} = -\frac{i}{\hbar}\left[\mathbf{s}(r)\cdot\mathbf{E}(r) - \mathbf{E}(r)\cdot\mathbf{s}^{\dagger}(r)\right] = \mathrm{Tr}\overleftrightarrow{p}(r), \tag{17.70}$$

$$\frac{\partial \mathbf{s}^{\dagger}(r)}{\partial t} - i\omega_o\,\mathbf{s}^{\dagger}(r) = -\frac{i}{\hbar}\left\{\mathbf{E}(r)\cdot\overleftrightarrow{1}\left(N - \langle\widehat{n}(r)\rangle\right) - \overleftrightarrow{p}(r)\cdot\mathbf{E}(r)\right\}, \tag{17.71}$$

$$\frac{\partial \mathbf{s}(r)}{\partial t} + i\omega_o\mathbf{s}(r) = \frac{i}{\hbar}\left\{\left(N - \langle\widehat{n}(r)\rangle\right)\overleftrightarrow{1}\cdot\mathbf{E}(r) - \mathbf{E}(r)\cdot\overleftrightarrow{p}(r)\right\}, \tag{17.72}$$

which describe the dynamics of a coupled flat conduction "s"-band and flat valence "p"-band with arbitrary state of polarization. Here we assigned the notation $E_{\xi}(r)\,M_o^{\xi}(r) \Rightarrow \mathbf{E}$ as used in [139]. It should be noted that the coherent-wave theoretical formalism of SB [139] does not incorporate scattering as well as the dynamics of real pairing (second-order Coulomb pairing, phonon and other "excitation-induced" pairing) between electrons, between holes, and between electron and hole, which maybe important in strongly-correlated systems such as in high-T_c superconductive materials.

17.4 Quantum transport equations in phase space

The quantum distribution-function transport equations are obtained by performing the lattice Weyl transform of Eqs. (17.53)–(17.56). After integrating over the energy variable, we have

$$
\begin{aligned}
&\left(\frac{\partial}{\partial t}\right) f_w^h(\vec{p}, \vec{q}, t) \\
&= -\frac{p}{m_v} \cdot \nabla_q f_w^h(\vec{p}, \vec{q}, t) \\
&\quad + \frac{2\pi}{(2\pi\hbar^4)} \int dp'\, dv \sin\left\{\frac{i}{\hbar}\left[(\vec{p}-\vec{p}')\cdot\vec{v}\right]\right\} \begin{bmatrix} \Phi_v\left(\vec{q}+\frac{\vec{v}}{2}\right) \\ -\Phi_v\left(\vec{q}-\frac{\vec{v}}{2}\right) \end{bmatrix} f_w^h(\vec{p}', \vec{q}, t) \\
&\quad + \left(\frac{\partial}{\partial t}\right) f_w^h(\vec{p}, \vec{q}, t)\Big|_{\text{intraband collision}} \\
&\quad + \left(\frac{\partial}{\partial t}\right) f_w^h(\vec{p}, \vec{q}, t)\Big|_{\text{interband transition,}}
\end{aligned} \tag{17.73}
$$

$$
\begin{aligned}
&\frac{\partial}{\partial t} f_w^{ee}(\vec{p}, \vec{q}, t) \\
&= \frac{i}{\hbar}\left\{E_g + \frac{\left(p - \frac{i\hbar}{2}\nabla_q\right)^2}{2m^r}\right\} f_w^{ee}(\vec{p}, \vec{q}, t) \\
&\quad - \frac{i}{\hbar}\int d\vec{p}'\, \mathcal{K}(\vec{p}-\vec{p}', \vec{q})\; f_w^{ee}(\vec{p}', \vec{q}, t) \\
&\quad + \left[\frac{2\pi}{(2\pi\hbar)^4}\int d\vec{p}'\, d\vec{v} \sin\left\{\frac{i}{\hbar}\left[(\vec{p}-\vec{p}')\cdot\vec{v}\right]\right\}\right. \\
&\quad \left. \times \begin{bmatrix} e\phi\left(\vec{q}+\frac{\vec{v}}{2}\right) \\ -e\phi\left(\vec{q}-\frac{\vec{v}}{2}\right) \end{bmatrix} f_w^{ee}(\vec{p}', \vec{q}, t)\right]
\end{aligned}
$$

$$+\left[\begin{array}{c}\frac{2\pi i}{(2\pi\hbar)^4}\int d\vec{p}'\int\ d\vec{v}\exp\left\{\frac{i}{\hbar}\left[(\vec{p}-\vec{p}')\cdot\vec{v}\right]\right\}\\ \times\ \Delta_{ee,vc}^{e-h,a}\left(q+\frac{v}{2}\right)\left[1-f_w^h\left(-\vec{p}',\vec{q},t\right)\right]\\ -\frac{2\pi i}{(2\pi\hbar)^4}\int d\vec{p}'\int\ d\vec{v}\exp\left\{\frac{i}{\hbar}\left[(\vec{p}-\vec{p}')\cdot\vec{v}\right]\right\}\\ \times\ \Delta_{ee,vc}^{e-h,r}\left(q-\frac{v}{2}\right)\ f_w^c\left(\vec{p}',\vec{q},t\right)\end{array}\right], \tag{17.74}$$

$$\begin{aligned}&\frac{\partial}{\partial t}f_w^{hh}\left(\vec{p},\vec{q},t\right)\\ &=-\frac{i}{\hbar}\left\{E_g\ +\frac{\left(p+\frac{i\hbar}{2}\nabla_q\right)^2}{2m^r}\right\}f_w^{ee}\left(\vec{p},\vec{q},t\right)\\ &\quad+\frac{i}{\hbar}\int d\vec{p}'\ \mathcal{K}\left(\vec{p}-\vec{p}',\vec{q}\right)\ f_w^{ee}\left(\vec{p}',\vec{q},t\right)\\ &\quad-\left[\frac{2\pi}{(2\pi\hbar)^4}\int d\vec{p}'\ d\vec{v}\sin\left\{\frac{i}{\hbar}\left[(\vec{p}-\vec{p}')\cdot\vec{v}\right]\right\}\right.\\ &\quad\left.\times\left[\begin{array}{c}e\phi\left(\vec{q}+\frac{\vec{v}}{2}\right)\\ -e\phi\left(\vec{q}-\frac{\vec{v}}{2}\right)\end{array}\right]\ f_w^{ee}\left(\vec{p}',\vec{q},t\right)\right]\\ &\quad-\left[\begin{array}{c}\frac{2\pi i}{(2\pi\hbar)^4}\int d\vec{p}'\int\ d\vec{v}\exp\left\{\frac{i}{\hbar}\left[(\vec{p}-\vec{p}')\cdot\vec{v}\right]\right\}\\ \times\Delta_{ee,vc}^{e-h,r}\left(q-\frac{v}{2}\right)\left[1-f_w^h\left(\vec{p}',\vec{q},t\right)\right]\\ -\frac{2\pi i}{(2\pi\hbar)^4}\int d\vec{p}'\int\ d\vec{v}\exp\left\{\frac{i}{\hbar}\left[(\vec{p}-\vec{p}')\cdot\vec{v}\right]\right\}\\ \times\Delta_{ee,vc}^{e-h,a}\left(q+\frac{v}{2}\right)\ f_w^c\left(-\vec{p}',\vec{q},t\right)\end{array}\right],\end{aligned} \tag{17.75}$$

$$\begin{aligned}&\left(\frac{\partial}{\partial t}\right)f_w^c\left(\vec{p},\vec{q},t\right)\\ &=-\frac{p}{m_c}\cdot\nabla_q f_w^c\left(\vec{p},\vec{q},t\right)\\ &\quad+\frac{2\pi}{(2\pi\hbar^4)}\int dp'\ dv\sin\left\{\frac{i}{\hbar}\left[(\vec{p}-\vec{p}')\cdot\vec{v}\right]\right\}\left[\begin{array}{c}\Phi_c\left(\vec{q}+\frac{\vec{v}}{2}\right)\\ -\Phi_c\left(\vec{q}-\frac{\vec{v}}{2}\right)\end{array}\right]f_w^c\left(\vec{p},\vec{q},t\right)\\ &\quad+\left(\frac{\partial}{\partial t}\right)f_w^c\left(\vec{p},\vec{q},t\right)\Big|_{\text{intraband collision}}\\ &\quad+\left(\frac{\partial}{\partial t}\right)f_w^c\left(\vec{p},\vec{q},t\right)\Big|_{\text{interband transition,}}\end{aligned} \tag{17.76}$$

where $\phi(q)$ is the self-consistent potential, and

$$\Phi_v(q) = -e\phi(q) + \Delta E_v(q), \tag{17.77}$$

$$\Phi_c(q) = -e\phi(q) + \Delta E_c(q), \tag{17.78}$$

m^r is the reduced effective mass, and

$$f_w^h(-\vec{p}, \vec{q}, t) = \frac{1}{2\pi\hbar} \int dE \left(-i\hbar G_{vv}^{e-h,<T}(\vec{p}, E, \vec{q}, t)\right), \tag{17.79}$$

$$f_w^{ee}(\vec{p}, \vec{q}, t) = \frac{1}{2\pi\hbar} \int dE \left(-i\hbar g_{ee,vc}^{e-h,<}(\vec{p}, E, \vec{q}, t)\right), \tag{17.80}$$

$$f_w^{hh}(\vec{p}, \vec{q}, t) = \frac{1}{2\pi\hbar} \int dE \left(i\hbar g_{hh,vc}^{e-h,>}(\vec{p}, E, \vec{q}, t)\right), \tag{17.81}$$

$$f_w^c(\vec{p}, \vec{q}, t) = \frac{1}{2\pi\hbar} \int dE \left(-i\hbar G_{cc}^{e-h,<}(\vec{p}, E, \vec{q}, t)\right), \tag{17.82}$$

and,

$$\mathcal{K}(\vec{p} - \vec{p}', \vec{q}) = \frac{1}{(2\pi\hbar)^3} \int d\vec{v} \exp\left\{\frac{i}{\hbar}\left[(\vec{p} - \vec{p}') \cdot \vec{v}\right]\right\} \begin{bmatrix} \Delta E_v\left(\vec{q} - \frac{\vec{v}}{2}\right) \\ -\Delta E_c\left(\vec{q} + \frac{\vec{v}}{2}\right) \end{bmatrix}. \tag{17.83}$$

For our purpose, the complicated intraband collision term will be approximated by the relaxation time approximation, successfully used in simulating AlGaAs/GaAs RTD [11]. Thus, we have in Eqs. (17.73) and (17.76)

$$\left[\left(\frac{\partial}{\partial t}\right) f_w^h(\vec{p}, \vec{q}, t)\right]_{\text{intraband collision}} = \frac{1}{\tau_h}\left[\frac{\rho^h(q, t)}{\rho_o^h(q)} f_w^h(\vec{p}, \vec{q}, 0) - f_w^h(\vec{p}, \vec{q}, t)\right] \tag{17.84}$$

$$\left[\left(\frac{\partial}{\partial t}\right) f_w^c(\vec{p}, \vec{q}, t)\right]_{\text{intraband collision}}, = \frac{1}{\tau_c}\left[\frac{\rho^c(q, t)}{\rho_o^c(q)} f_w^c(\vec{p}, \vec{q}, 0) - f_w^c(\vec{p}, \vec{q}, t)\right], \tag{17.85}$$

where $f_w^h(\vec{p}, \vec{q}, 0)$ is the Wigner distribution function of holes at zero bias, $\rho^h(q, t) = \sum_{\vec{p}} f_w^h(\vec{p}, \vec{q}, t)$ is the hole density, and τ_h is a constant hole relaxation time calculated by considering all scattering mechanisms. Similarly,

$f_w^c(\vec{p},\vec{q},0)$ is the Wigner distribution function of conduction electrons at zero bias, $\rho^c(q,t)=\sum_{\vec{p}} f_w^c(\vec{p},\vec{q},t)$ is the electron density, and τ_c is a constant electron relaxation time calculated by considering all scattering mechanisms [11].

The interband transition terms due to Zener tunneling is the basis of the intriguing autonomous current oscillation of type II RTD. The interband transition terms are given by

$$\left(\frac{\partial}{\partial t}\right) f_w^h(\vec{p},\vec{q},t)\bigg|_{\text{interband transition}}$$

$$= \begin{bmatrix} \frac{2\pi}{(2\pi\hbar)^4}\int dp'\, dv \exp\left\{\frac{i}{\hbar}\left[(\vec{p}-\vec{p}')\cdot\vec{v}\right]\right\} \\ \times \begin{bmatrix} \Delta_{hh;cv}^{e-h,r}\left(\vec{q}-\frac{\vec{v}}{2}\right) g_{ee;vc}^{e-h,>}\left(\vec{p}',\vec{q},t\right) \\ -\Delta_{ee;vc}^{e-h,a}\left(\vec{q}+\frac{\vec{v}}{2}\right) g_{hh;cv}^{e-h,>}\left(\vec{p}',\vec{q},t\right) \end{bmatrix} \end{bmatrix}, \qquad (17.86)$$

$$\left(\frac{\partial}{\partial t}\right) f_w^c(\vec{p},\vec{q},t)\bigg|_{\text{interband transition}}$$

$$= \begin{bmatrix} \frac{2\pi}{(2\pi\hbar)^4}\int dp'\, dv \exp\left\{\frac{i}{\hbar}\left[(\vec{p}-\vec{p}')\cdot\vec{v}\right]\right\} \\ \times \begin{bmatrix} \Delta_{hh;cv}^{e-h,r}\left(\vec{q}-\frac{\vec{v}}{2}\right) g_{ee;vc}^{e-h,<}\left(\vec{p}',\vec{q},t\right) \\ -\Delta_{ee;vc}^{e-h,a}\left(\vec{q}+\frac{\vec{v}}{2}\right) g_{hh;cv}^{e-h,<}\left(\vec{p}',\vec{q},t\right) \end{bmatrix} \end{bmatrix}, \qquad (17.87)$$

with $\Delta_{hh;cv}^{e-h,r}(\vec{q})=\Delta_{hh;cv}^{e-h,r(Z)}(\vec{q})$ and $\Delta_{ee;vc}^{e-h,a}(\vec{q})=\Delta_{ee;vc}^{e-h,a(Z)}(\vec{q})$ given in Ref. [12] for the Zener electron–hole pair potential. We have used the integral instead of a gradient expansion since the Zener interband pair potential $\Delta_{hh;cv}^{e-h,r(Z)}(\vec{q})$ and $\Delta_{ee;vc}^{e-h,a(Z)}(\vec{q})$ can have a strong variation as a function of q near the drain-side of the barrier edge in type II resonant tunneling device, see Fig. 17.1.

The Zener electron–hole pair correlation function $g_{ee}^{e-h,<}$ (or pair wavefunction) can be approximated by neglecting the intraband self-energies and terms involving $\Delta^{\lessgtr}$. Equation (17.74) is a Schrödinger equation of the pair wavefunction in the mixed $\vec{p}$-$\vec{q}$ representation. The time-dependent numerical simulation of the quantum transport of electrons and holes in type-II RTD, using Eqs. (17.73)–(17.75), present a real challenge. This simulation is needed for cost-effective optimization of device structure and material properties in the design of type-II RTD devices.

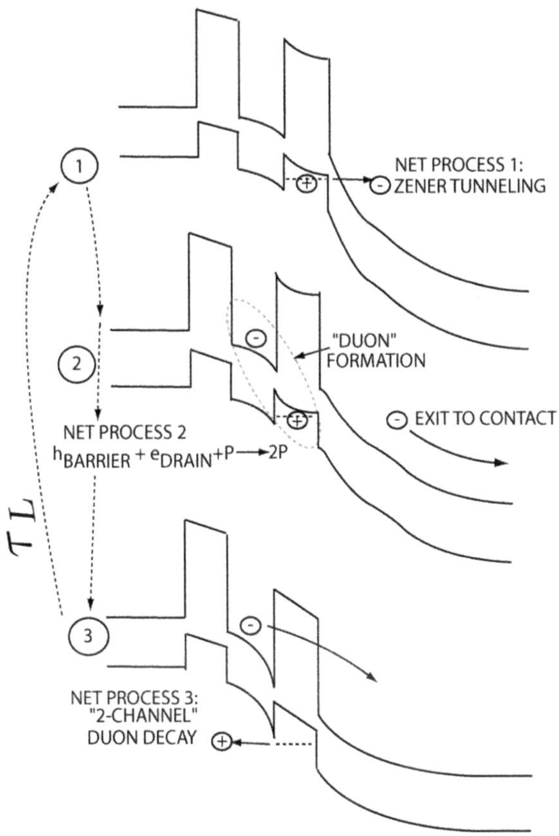

Fig. 17.1. Schematic energy-band diagram showing the mechanism of oscillation of trapped hole charge in the barrier in a solid-state THz source. Reproduced from Ref. [140].

17.4.1 *Conservation of particle in Zener tunneling*

We observe from Eq. (17.86) and Eq. (17.87) that the following equality holds:

$$\left(\frac{\partial}{\partial t}\right) f_w^h(\vec{p},\vec{q},t)\Big|_{\text{interband transition}} = \left(\frac{\partial}{\partial t}\right) f_w^c(\vec{p},\vec{q},t)\Big|_{\text{interband transition}}, \tag{17.88}$$

since $g_{ee;vc}^{e-h,<}(\vec{p}',\vec{q},t) = g_{ee;vc}^{e-h,>}(\vec{p}',\vec{q},t)$ and $g_{hh;cv}^{e-h,<}(\vec{p}',\vec{q},t) = g_{hh;cv}^{e-h,>}(\vec{p}',\vec{q},t)$. This simply states the rate of hole generation by Zener tunneling of electrons from the valence band to the conduction band is equal to the generation rate for electrons in the conduction band.

Chapter 18

Quantum Transport of Quantized Fields

Here we give the QSFT formalism for the quantized classical vibrational field in solid. Similar techniques can be followed for QSFT of other classical real fields, such as quantized electromagnetic fields and quantized plasma oscillations. In terms of the non-equilibrium superfield theoretical approach to a non-equilibrium electron-phonon system, the Liouvillian can in general be expressed as

$$\begin{aligned}\mathcal{L} &= \upsilon\left(1,2\right)\ \Psi\left(1\right)\ \Psi\left(2\right) + \upsilon\left(1,2,3,4\right)\ \Psi\left(1\right)\ \Psi\left(2\right)\ \Psi\left(3\right)\Psi\left(4\right) + \mathcal{L}_{\text{ext}}^{\text{e}-\text{Schw}} \\ &\quad + M^{o}_{i_1 i_2}\left(\xi_1\xi_2\right)\dot{U}_{i_1}\left(\xi_1\right)\dot{U}_{i_2}\left(\xi_2\right) + \Theta^{o(2)}_{i_1 i_2}\left(\xi_1\xi_2\right)U_{i_1}\left(\xi_1\right)U_{i_2}\left(\xi_2\right) \\ &\quad - g^{i}_{r}U_i\left(\xi\right) + \mathcal{L}_{\text{ext}}^{\text{ph}-\text{Schw}} \\ &\quad - X^{o}_{i}\left(12;\xi\right)\Psi\left(1\right)\ \Psi\left(2\right)U_i\left(\xi\right),\end{aligned} \tag{18.1}$$

where the electron superfield Schwinger source term is given by Eq. (16.106) and the phonon superfield Schwinger source term is

$$\mathcal{L}_{\text{ext}}^{\text{ph}-\text{Schw}} = -\varepsilon^{i}\left(\xi\right)U_i\left(\xi\right), \tag{18.2}$$

where the real phonon superfield is

$$U_i\left(\xi,t\right) = \begin{pmatrix}\sqrt{m_k}\hat{u}_i\left(\xi,t\right)\\ \sqrt{m_k}\tilde{u}_i\left(\xi,t\right)\end{pmatrix}, \tag{18.3}$$

$$\varepsilon^{i}\left(\xi\right) = \begin{pmatrix}\phi^{i}_{\text{Schw}}\\ -\phi^{i}_{\text{Schw}}\end{pmatrix}. \tag{18.4}$$

The $\hat{u}(\xi,t)$ and $\tilde{u}(\xi,t)$ are the "hat" and "tilde" displacement operator, respectively, of atoms in site ξ. The indices $1,2,3,4$ each stands for all pertinent quantum labels, and repeated indices are "summed" over. The index ξ includes the index k which labels the k_{th} atom in the unit cell. Thus m_k refers to the mass of the k_{th} atom. Note that the boson field source terms are purposely written with negative sign while the fermion field source term takes a positive sign, following the convention analogous to the minimal coupling interaction of charge particle and electromagnetic vector potential. This leads to the equation for the average boson field with a positive source term. This will result in a different sign of the factor $\frac{1}{i\hbar}$ in the definition of both super-propagators for phonons, plasmons, and photons compared to those of fermion super-propagators, Eq. (16.132).

In the Liouvillian for the real phonon superfield, Eq. (18.1) we have

$$\Theta^{o(2)}_{i_1 i_2}(\xi_1\xi_2) = [\delta_{\alpha_1,1}\delta_{\alpha_2,1} - \delta_{\alpha_1,2}\delta_{\alpha_2,2}] \frac{\phi^{(2)}_{i_1 i_2}(\xi_1\xi_2)}{2\sqrt{m_{k_1}m_{k_2}}} \prod_{j=\xi_1}^{j=\xi_2} \delta(t_j - t), \quad (18.5)$$

$$\phi^{(2)}_{i_1 i_2}(\xi_1\xi_2) = \frac{\partial^2 V(\xi_1\xi_2....\xi_N..)}{\partial u_{i_1}(\xi_1)\,\partial u_{i_2}(\xi_2)}, \quad (i_1, i_2 = \vec{x}, \vec{y}, \vec{z}), \quad (18.6)$$

where $V(\xi_1\xi_2\cdots\xi_N\cdots)$ is the total effective potential energy function for the nuclear motion, and

$$M^o_{i_1 i_2}(\xi_1\xi_2) = [\delta_{\alpha_1,1}\delta_{\alpha_2,1} - \delta_{\alpha_1,2}\delta_{\alpha_2,2}] \frac{1}{2}\delta_{i_1 i_2}\delta(\xi_1 - \xi_2) \prod_{j=\xi_1}^{j=\xi_2} \delta(t_j - t), \quad (18.7)$$

where the subscripts of the delta-function pertain to the components of the phonon superfield $U_i(\xi,t)$. The last term in the Liouvillian $\mathcal{L}$ represents the electron–phonon interaction, where

$$X^o_i(12;\xi) = \sum_{P(12)} \epsilon^P \left[\delta_{\alpha_1,3}\delta_{\alpha_2,1}\delta_{\alpha_\xi,1} - \delta_{\alpha_1,2}\delta_{\alpha_2,2}\delta_{\alpha_\xi,2}\right] \frac{\chi^i(12;\xi)}{\sqrt{m_{k_\xi}}} \times \prod_{j=1}^{j=\xi} \delta(t_j - t), \quad (18.8)$$

$$\chi^i_{n,m}(12;\xi) = \langle r_1, n|\, z^* e \nabla^i_{r_\xi} \frac{1}{|x - r_\xi|}\, |r_2, m\rangle, \quad (18.9)$$

where z^* is the effective charge of the nucleus located at ξ. The energy-band index n, m included in indices $1, 2$ are explicitly written in the definition of the matrix element for $\chi^i(12;\xi) \equiv \chi^i_{n,m}(12;\xi)$. In bulk materials, $\chi^i(12;\xi)$ is often determined using knowledge of the deformation potential [141], diagonal in band index, $n = m$. Various approximation schemes also treat $r_1 = r_\xi = r_2$.

In time-dependent perturbation theory leading to the Fermi golden rule, $\chi^i(12;\xi)$ enters the Fermi golden rule for the transition probability

$$\mathcal{P}_i^f = \frac{2\pi}{\hbar} \left| \langle i | \mathcal{H}_{e-ph} | f \rangle \right|^2 \delta(E_f - E_i), \tag{18.10}$$

where

$$\langle i | \mathcal{H}_{e-ph} | f \rangle = \langle i | \chi^i(12;\xi) \;\; u_i(\xi, t) | f \rangle . \tag{18.11}$$

Expressing the ion displacement $u_i(\xi, t)$ at lattice point ξ in terms of the normal modes of the dynamical matrix expressed in terms phonon annihilation and creation operators and polarization vectors in bulk materials, then one finds that only longitudinal phonons interact with electrons if Umklapp processes are neglected. In semiconductors with the diamond or zinc-blende structure, the two non-equivalent atoms per unit cell can allow electron interaction with optical vibrational modes corresponding to movement of the two atoms relative to each other. The optical electron–phonon interaction is also determined using knowledge of optical deformation potential. Electrons interact with polar modes, which is important in ionic crystals.

18.1 Liouvillian-space phonon dynamics

The free phonon quantum superfield equation in the super-Heisenberg or Liouvillian representation can be written as

$$\frac{\partial^2}{\partial t^2} U_i(\xi, t) = -\sum_{\xi', j} \mathcal{D}_{ij}(\xi, \xi') \, U_j(\xi', t), \tag{18.12}$$

where summation is explicitly indicated for the repeated indices and the super-dynamical matrix $\mathcal{D}_{ij}(\xi, \xi')$ is defined as

$$\mathcal{D}_{ij}(\xi, \xi') = \frac{\Theta^{(2)}_{i_1 i_2}(\xi\xi')}{\sqrt{m_k m_{k'}}}. \tag{18.13}$$

We write the solution of Eq.(18.12) as $U_i(\xi,t) = u_i(k)\exp\left\{\frac{i}{\hbar}\left(p\cdot X_l^o - Et\right)\right\}$. Then we obtain[1]

$$\left(\frac{E}{\hbar}\right)^2 u_i(k) = \sum_{\xi',j} \mathcal{D}_{ij}(\xi,\xi')\exp\left\{\frac{i}{\hbar}p\cdot\left(X_{l'}^o - X_l^o\right)\right\}u_j(k'). \quad (18.14)$$

This resulting eigenvalue problem can be solved for the eigenvalues and corresponding eigenvector. Denote the eigenvalues as $\omega_\lambda^o(p)^2$ and the corresponding eigenvector as $e_\lambda^{k,j}(p)$, where λ is the index for the vibrational branch and j labels the independent spatial directions. Then we have

$$\omega_\lambda^o(p)^2 e_\lambda^{k,i}(p)\exp\left\{\frac{i}{\hbar}p\cdot X_l^o\right\} = \sum_{\xi',j}\mathcal{D}_{ij}(\xi,\xi')\exp\left\{\frac{i}{\hbar}p\cdot X_{l'}^o\right\}e_\lambda^{k',j}(p). \quad (18.15)$$

Instead of the eigenvector $e_\lambda^{k,j}(p)$, it is more convenient to use as basis function the following expression:

$$B_\lambda^i(p;\xi) \equiv B_\lambda^i(p;l.k) = e_\lambda^{k,i}(p)\exp\left\{\frac{i}{\hbar}p\cdot X_l^o\right\}. \quad (18.16)$$

The following orthonormality and completeness relation hold for $e_\lambda^{k,i}(p)$ and $B_\lambda^i(p;l.k)$

$$\sum_{k,j} e_\lambda^{k,j}(p)\ e_{\lambda'}^{k,j}(p)^* = \delta_{\lambda\lambda'}, \quad (18.17)$$

$$\sum_{k,j} e_\lambda^{k,j}(p)\ e_\lambda^{k',j'}(p)^* = \delta_{kk''}\delta_{jj'}, \quad (18.18)$$

and

$$\sum_{k,j} B_\lambda^j(p;l.k)\ B_{\lambda'}^j(p;l.k)^* = \delta_{\lambda\lambda'}, \quad (18.19)$$

$$\sum_{k,j} B_\lambda^j(p;l.k)\ B_\lambda^{j'}(p;l.k')^* = \delta_{kk''}\delta_{jj'}. \quad (18.20)$$

[1]The use of i as dummy label, either subscript or superscript, should not be confused with its use as multiplicative factor as $i=\sqrt{-1}$.

We also have

$$e_{\lambda}^{k,j}(p)^* = e_{\lambda}^{k,j}(-p), \tag{18.21}$$

$$B_{\lambda}^{j}(p;l.k)^* = B_{\lambda}^{j}(-p;l.k). \tag{18.22}$$

We now expand the ion super-displacement operator in terms of the eigenfunction of the super-dynamical matrix $\mathcal{D}_{ij}(\xi,\xi')$,

$$U_i(\xi,t) = \frac{1}{\sqrt{N}}\sum_{\lambda,p} Q_{\lambda}(p,t)\ B_{\lambda}^{i}(p;\xi), \tag{18.23}$$

where N is the number of crystal lattice points, $Q_{\lambda}(p,t)$ is the normal mode coordinate super-operator in the super-Heisenberg representation. Since the displacement super-operator is real, then $Q_{\lambda}(p,t)^{\dagger} = Q_{\lambda}(-p,t)$. From the canonical commutation relation,

$$\left[\dot{U}_i(\xi,t), U_j(\xi',t)\right] = \frac{\hbar}{i}\delta_{ij}\delta_{\xi\xi'}\begin{pmatrix}1 & 0\\ 0 & 1\end{pmatrix}, \tag{18.24}$$

we also have

$$\left[\dot{Q}_{\lambda}(p,t), Q_{\lambda'}(p',t)\right] = \frac{\hbar}{i}\delta_{\lambda\lambda'}\delta_{p,-p'}\begin{pmatrix}1 & 0\\ 0 & 1\end{pmatrix}. \tag{18.25}$$

Moreover, from the super-Heisenberg equation of motion for the displacement operator, we have

$$\frac{\partial^2}{\partial t^2}Q_{\lambda}(p,t) = -\omega_{\lambda}^{o}(p)^2\ Q_{\lambda}(p,t), \tag{18.26}$$

which demonstrates the independent harmonic oscillator character of the normal modes.

The phonon super-annihilation and super-creation operators $\alpha_{\lambda}(p)$ and $\alpha_{\lambda}(p)^{\dagger}$, respectively, are defined in the usual fashion in terms of the independent oscillator coordinate by the following relations:

$$Q_{\lambda}(p,t) = \sqrt{\frac{\hbar}{2\omega_{\lambda}^{o}(p)}}(\alpha_{\lambda}(p,t) + \alpha_{\lambda}^{\dagger}(-p,t)), \tag{18.27}$$

$$\dot{Q}_{\lambda}(p,t) = \sqrt{\frac{\hbar\omega_{\lambda}^{o}(p)}{2}}(\alpha_{\lambda}(p,t) - \alpha_{\lambda}^{\dagger}(-p,t)), \tag{18.28}$$

which satisfy the usual commutation relations

$$\left[\alpha_\lambda(p,t),\alpha_{\lambda'}(p',t)\right]=\left[\alpha_\lambda^\dagger(p,t),\alpha_{\lambda'}^\dagger(p',t)\right]=0, \tag{18.29}$$

$$\left[\alpha_\lambda^\dagger(p,t),\ \alpha_{\lambda'}(p',t)\right]=\delta_{\lambda\lambda'}\delta_{p,p'}\begin{pmatrix}1&0\\0&1\end{pmatrix}. \tag{18.30}$$

In the super-annihilation and super-creation super-operator representation, the Liouvillian for free phonon system becomes a sum of independent harmonic oscillators,

$$\mathcal{L}=\sum_{\lambda,p}\left\{\alpha_\lambda^\dagger(p)\ \alpha_\lambda(p)+\frac{1}{2}\right\}\hbar\,\Omega_\lambda(p), \tag{18.31}$$

where

$$\Omega_\lambda(p)=\begin{pmatrix}\omega_\lambda^o(p)&0\\0&-\omega_\lambda^o(p)\end{pmatrix}, \tag{18.32}$$

and similarly, we have

$$\alpha_\lambda(p)=\begin{pmatrix}\hat{a}_\lambda(p)\\\tilde{a}_\lambda(p)\end{pmatrix},\quad \alpha_\lambda^\dagger(p)=\begin{pmatrix}\hat{a}_\lambda^\dagger(p)\\\tilde{a}_\lambda^\dagger(p)\end{pmatrix}. \tag{18.33}$$

Corresponding to Eq. (18.27), we define the corresponding Hilbert space operators

$$q_\lambda(p,t)=\sqrt{\frac{\hbar}{2\omega_\lambda^o(p)}}\left(a_\lambda(p,t)+a_\lambda^\dagger(-p,t)\right), \tag{18.34}$$

$$q_\lambda^\dagger(p,t)=\sqrt{\frac{\hbar}{2\omega_\lambda^o(p)}}\left(\alpha_\lambda(-p,t)+\alpha_\lambda^\dagger(p,t)\right). \tag{18.35}$$

18.2 The phonon super-Green's function

The phonon super-correlation function $S_{ij}(\xi\xi';tt')$ is defined by the functional derivative of the average super-displacement operator with respect to the Schwinger source for phonon displacement field

$$S_{ij}(\xi\xi';tt')=\frac{\delta}{\delta\varepsilon^j(\xi')}\langle U_i(\xi,t)\rangle$$

$$=\begin{pmatrix}S_{ij}^c(\xi\xi';tt')&-S_{ij}^<(\xi\xi';tt')\\S_{ij}^>(\xi\xi';tt')&-S_{ij}^{ac}(\xi\xi';tt')\end{pmatrix}. \tag{18.36}$$

The "reduced density matrix" for phonons is defined by $S_{ij}^{<}(\xi\xi';tt')$ which is given by

$$
\begin{aligned}
S_{ij}^{<}(\xi\xi';tt') &= \frac{i}{\hbar}\langle T\ \hat{u}_i(\xi,t)\ \tilde{u}_j(\xi',t')\rangle \\
&= \frac{1}{N}\sum_{\{\lambda,p\}} B_\lambda^i(p,\xi)\,B_{\lambda'}^j(p',\xi')\,\frac{i}{\hbar}\langle T\ \hat{q}_\lambda(p,t)\ \tilde{q}_{\lambda'}(p',t')\rangle \\
&= \frac{1}{N}\sum_{\{\lambda,p\}} B_\lambda^i(p,\xi)\,B_{\lambda'}^{j\dagger}(p',\xi')\,\frac{i}{\hbar}Tr\ \rho\ q_{\lambda'}^\dagger(p',t')\,q_\lambda(p,t),
\end{aligned}
\tag{18.37}
$$

where the last line is written in Hilbert space representation. Substituting the relation between $q_\lambda(p,t)$ and $q_\lambda^\dagger(p,t)$ in terms of Hilbert space annihilation, $a_\lambda(p,t)$, and creation operators, $a_\lambda^\dagger(p,t)$, using Eqs. (18.34)–(18.35), we obtain

$$
S_{ij}^{<}(\xi\xi';tt') = \frac{1}{N}\sum_{\{\lambda,p\}} B_\lambda^i(p,\xi)\,B_{\lambda'}^{j\dagger}(p',\xi')\,\frac{i}{2\sqrt{\omega_\lambda^o(p)\,\omega_{\lambda'}^o(p')}}
\times\mathrm{Tr}\rho\left\{\begin{array}{c} a_{\lambda'}(-p',t')\,a_\lambda(p,t) \\ +a_{\lambda'}(-p',t')\,a_\lambda^\dagger(-p,t) \\ +a_{\lambda'}^\dagger(p',t')\,a_\lambda(p,t) \\ +a_{\lambda'}^\dagger(p',t')\,a_\lambda^\dagger(-p,t) \end{array}\right\}.
\tag{18.38}
$$

For free phonons at thermal equilibrium, only the phonon conserving combinations, with $\lambda=\lambda'$ and $p=p'$, survived giving the result

$$
-iS_{ij}^{<}(\xi\xi';tt') = \frac{1}{N}\sum_{\lambda,p} e_\lambda^{k,i}(p)\,e_\lambda^{k',j}(p)^\dagger\exp\left\{\frac{i}{\hbar}p\cdot(X_{l'}^o - X_l^o)\right\}
\times\frac{1}{2\omega_\lambda^o(p)}\left\{\begin{array}{c}(1+n(\omega_\lambda^o(p)))\,e^{i\omega_\lambda^o(p)(t-t')} \\ +n(\omega_\lambda^o(p))\,e^{-i\omega_\lambda^o(p)(t-t')}\end{array}\right\},
\tag{18.39}
$$

where

$$
n(\omega_\lambda^o(p)) = \frac{1}{e^{\beta\hbar\omega_\lambda^o(p)}-1},
\tag{18.40}
$$

is the equilibrium distribution function for phonons. By virtue of translational invariance with respect to lattice coordinates and time, the Weyl transformation to phase space can immediately be written as

$$-iS_{ij}^{<}\left(kk'p;\omega\right) = \sum_{\lambda} e_{\lambda}^{k,i}\left(p\right) e_{\lambda}^{k',j}\left(p\right)^{\dagger} \times \frac{2\pi}{2\omega_{\lambda}^{o}\left(p\right)} \left\{ \begin{array}{c} \left(1+n\left(\omega_{\lambda}^{o}\left(p\right)\right)\right)\delta\left(\omega+\omega_{\lambda}^{o}\left(p\right)\right) \\ +n\left(\omega_{\lambda}^{o}\left(p\right)\right)\delta\left(\omega-\omega_{\lambda}^{o}\left(p\right)\right) \end{array} \right\}. \quad (18.41)$$

The terms in the curly bracket can be combined by noting that $(1+n(-\omega_{\lambda}^{o}(p))) = -n(\omega_{\lambda}^{o}(p))$, and representing the Dirac-delta functions as limit of their respective analytical expressions. The result is

$$\begin{aligned} -iS_{ij}^{<}\left(kk'p;\omega\right) &= 2\pi \text{ sgn } \omega \sum_{\lambda} e_{\lambda}^{k,i}\left(p\right) e_{\lambda}^{k',j}\left(p\right)^{\dagger} \\ &\quad \times \text{Lim}_{\Gamma\Rightarrow 0} \frac{\Gamma}{\left(\omega^{2}-\omega_{\lambda}^{o}\left(p\right)^{2}\right)^{2}+\left(\frac{\Gamma}{2}\right)^{2}} n\left(\omega_{\lambda}^{o}\left(p\right)\right), \\ &= A_{ij}\left(kk'p;\omega\right) n\left(\omega_{\lambda}^{o}\left(p\right)\right), \end{aligned} \quad (18.42)$$

where the spectral function $A_{ij}(kk'p;\omega)$ for thermal phonons appears expanded in terms of the eigenvectors (polarization vector) of the dynamical matrix

$$A_{ij}\left(kk'p;\omega\right) = \sum_{\lambda} e_{\lambda}^{k,i}\left(p\right) e_{\lambda}^{k',j}\left(p\right)^{\dagger} \; \{2\pi \; sgn \; \omega \; \delta(\omega^{2}-\omega_{\lambda}^{o}\left(p\right)^{2})\}. \quad (18.43)$$

Using similar procedure, we also have

$$-iS_{ij}^{>}\left(kk'p;\omega\right) = A_{ij}\left(kk'p;\omega\right)\left(1+n\left(\omega_{\lambda}^{o}\left(p\right)\right)\right). \quad (18.44)$$

The expressions for $-iS_{ij}^{<}(kk'p;\omega)$ and $-iS_{ij}^{>}(kk'p;\omega)$ leads to the relation

$$-i\left(S_{ij}^{>}\left(kk'p;\omega\right) - S_{ij}^{<}\left(kk'p;\omega\right)\right) = A_{ij}\left(kk'p;\omega\right). \quad (18.45)$$

The sum rule obeyed by the spectral function $A_{ij}(kk'p;\omega)$ reads

$$\frac{1}{2\pi}\int_{-\infty}^{\infty} \omega \; A_{ij}\left(kk'p;\omega\right) d\omega = \delta_{kk'}\delta_{ij}, \quad (18.46)$$

which also follows from the canonical commutation relation. The phonon density of states in the vibration branch index λ is defined by

$$\sigma(\omega) = \sum_p N \frac{\omega}{\pi} 2\pi \ \mathrm{sgn}\ \omega \ \delta(\omega^2 - \omega_\lambda^o(p)^2),$$

$$= \frac{N}{(2\pi)^3} \int dp^3 \frac{\omega}{\pi} 2\pi \ \frac{\omega}{|\omega|} \ \delta(\omega^2 - \omega_\lambda^o(p)^2),$$

$$= \frac{N}{8\pi^3} \oint_{\omega_\lambda^o(p)=\omega} \frac{dS_p}{\left|\nabla_p \omega_\lambda^o(p)\right|}, \tag{18.47}$$

where the momentum integral is over the constant energy surface, S_p, at $\omega = \omega_\lambda^o(p)$.

The expressions for $-iS_{ij}^{<}(kk'p;\omega)$ and $-iS_{ij}^{>}(kk'p;\omega)$ has so far only given for a phonon system at thermal equilibrium. When the system is not in equilibrium, the reduced density matrices not only depend on the relative coordinates but also depend on the center coordinates. For non-equilibrium system, in analogy with Eq. (18.42) for thermal equilibrium, the following expression:

$$-iS_{ij}^{<}(k,k';p,l_c;\omega,t_c) = A_{ij}(k,k';p,l_c;\omega,t_c)\, n(p,l_c;\omega,t_c), \tag{18.48}$$

is obtained after lattice Weyl transformation and expressing the result as expanded in terms of the eigenvectors of the dynamical matrix. By similar token, we can also write the phonon "hole" distribution as

$$-iS_{ij}^{>}(k,k';p,l_c;\omega,t_c) = A_{ij}(k,k';p,l_c;\omega,t_c)\,[1 + n(p,l_c;\omega,t_c)]. \tag{18.49}$$

18.3 Equation for the phonon super-correlation function

The phonon super-correlation function or super-propagator obeys

$$S_{ij}^o(\xi\bar{\xi})^{-1} \frac{S_{jm}(\bar{\xi}\xi')}{-i\hbar} = \delta_{im}(\xi - \xi') + \Pi_{ij}(\xi\bar{\xi}) \frac{S_{jm}(\bar{\xi}\xi')}{-i\hbar}, \tag{18.50}$$

where

$$S_{ij}^o(\xi\bar{\xi})^{-1} = 2M_{ij}^o(\xi\bar{\xi}) \frac{\partial^2}{\partial t^2} + 2\frac{\Theta_{ij}^{o(2)}(\xi\bar{\xi})}{\sqrt{m_{k_\xi} m_{k_{\bar{\xi}}}}}. \tag{18.51}$$

Taking into account Eqs. (18.7)–(18.5), we may write Eq. (18.50) as

$$\frac{\partial^2}{\partial t^2}S_{ij}\left(\xi\xi';tt'\right)+\frac{\bar{\Theta}^{o(2)}_{iy}\left(\xi\xi''\right)}{\sqrt{m_k m_{k''}}}S_{yj}\left(\xi''\xi';tt'\right)-\Pi_{iy}\left(\xi\xi'';tt''\right)S_{yj}\left(\xi''\xi';t''t'\right)$$
$$=\delta_{ij}\delta_{\xi\xi'}\delta\left(t-t'\right), \tag{18.52}$$

where diagonal matrix in the "hat" and "tilde" indices, $\bar{\Theta}^{o(2)}_{iy}\left(\xi\xi''\right)$, defined by

$$\bar{\Theta}^{o(2)}_{iy}\left(\xi\xi''\right)=\begin{pmatrix}\phi^{(2)}_{iy}\left(\xi\xi''\right)/\sqrt{m_k m_{k''}} & 0\\ 0 & \phi^{(2)}_{iy}\left(\xi\xi''\right)/\sqrt{m_k m_{k''}}\end{pmatrix}$$
$$=\bar{\mathcal{D}}_{iy}\left(\xi\xi''\right), \tag{18.53}$$

in contrast to the use of $\Theta^{o(2)}_{iy}\left(\xi\xi''\right)$ of Eq. (18.51), which is defined by Eq. (18.5). The phonon transport equation is determined from Eq. (18.52) and its adjoint, and is given by

$$\left(\frac{\partial^2}{\partial t^2}-\frac{\partial^2}{\partial t'^2}\right)S_{ij}\left(\xi\xi';tt'\right)$$
$$=-\bar{\mathcal{D}}_{iy}\left(\xi\xi''\right)S_{yj}\left(\xi''\xi';tt'\right)$$
$$+\mathcal{X}\mathcal{X}\left\{\begin{pmatrix}-1 & 0\\ 0 & 1\end{pmatrix}\left[\bar{\mathcal{D}}_{jy}\left(\xi\xi''\right)S_{yj}\left(\xi''\xi';tt'\right)\right]^{\dagger}\begin{pmatrix}-1 & 0\\ 0 & 1\end{pmatrix}\right\}$$
$$+\Pi_{iy}\left(\xi\xi'';tt''\right)S_{yj}\left(\xi''\xi';t''t'\right)$$
$$-\mathcal{X}\mathcal{X}\left\{\begin{pmatrix}-1 & 0\\ 0 & 1\end{pmatrix}\left[\diamond_{jy}\left(\xi\xi''\right)S_{yj}\left(\xi''\xi';tt'\right)\right]^{\dagger}\begin{pmatrix}-1 & 0\\ 0 & 1\end{pmatrix}\right\}, \tag{18.54}$$

where $\mathcal{X}\mathcal{X}$ indicates the operation of interchanging the diagonal as well as the diagonal elements of the resulting matrix. With the "super quantities" as matrices in the "hat" and "tilde" indices, we can write Eq. (18.54) as coupled equations of the matrix elements. The consistency of the resulting coupled transport equations are verified by checking for the equation of the different expressions for retarded propagator, as well as for the equation of the advanced propagator.

18.4 Kinetic phonon transport equations in phase space

As usual in deriving the quantum transport equations, we are interested in the equation for which defines the statistics of phonons and phonon "holes", as well as the equation of which defines the energetics of their particle behavior. From the coupled equations for the matrix elements, we obtain the following expressions:

$$\left(\frac{\partial^2}{\partial t^2} - \frac{\partial^2}{\partial t'^2}\right) S^{>,<} = \left[\text{Re}\Pi^r - d^{(2)}, S^{>,<}\right] + \left[\Pi^{>,<}, \text{Re}S^r\right]$$
$$+ i\{\text{Im}\Pi^r, S^{>,<}\} - i\{\Pi^{>,<}, \text{Im}S^r\}$$
$$= \left[\text{Re}\Pi^r - d^{(2)}, S^{>,<}\right] + \left[\Pi^{>,<}, \text{Re}S^r\right]$$
$$\pm\frac{1}{2}\left\{\Pi^{>,<}, S^{<,>}\right\} \mp \frac{1}{2}\left\{\Pi^{<,>}, S^{>,<}\right\}, \qquad (18.55)$$

$$\left(\frac{\partial^2}{\partial t^2} - \frac{\partial^2}{\partial t'^2}\right) S^{r,a} = -[d^{(2)}, S^{r,a}] + [\Pi^{r,a}, S^{r,a}], \qquad (18.56)$$

where we have used the definition,

$$d_{ij}^{(2)}\left(\xi\xi'\right) = \frac{\phi_{ij}^{(2)}\left(\xi\xi'\right)}{\sqrt{m_k m_{k'}}}, \qquad (18.57)$$

and the following identities:

$$i\text{Im}S^r = \frac{S^> - S^<}{2}, \quad i\text{Im}\Pi^r = \frac{\Pi^> - \Pi^<}{2}. \qquad (18.58)$$

What remains to be done is to take the lattice Weyl transform (LWT) of Eqs. (18.55) and (18.56) to obtain the transport equations in $(\vec{p}, \vec{X}_l, \omega, t) \equiv (\vec{p}, \vec{q}, E, t)$ phase-space. First we note that from Eqs. (18.48) and (18.49), and by virtue of the orthonormality and completeness of the eigenvectors of the dynamical matrix, the lattice Weyl transforms are matrices in the indices, λ, k, i where k refers to the k_{th} nucleus in a unit cell, λ refers to the vibration branch index, and i refers to the Cartesian directions. The lattice Weyl transform of $d_{ij}^{(2)}(\xi, \xi')$ is diagonal in the vibration branch index with the diagonal elements as matrices in the remaining indices, i.e.,

$$d_{ij}^{(2)}\left(kk'; \vec{p}, \vec{q}, E, t\right) = \sum_{\lambda} \left[\omega_\lambda^o\left(\vec{p}\right)\right]^2 e_\lambda^{k,i}\left(\vec{p}\right)\ e_\lambda^{k',j}\left(\vec{p}\right)^\dagger, \qquad (18.59)$$

which is diagonal in the vibration index and independent of $(\vec{q}, E, t)$ set of variables. In view of the trivial dependence on indices of the resulting matrix of all the lattice Weyl transforms we make further simplification by simply considering the vibration branch index only. Making use of the relations given in Eqs. (17.33)–(17.34), we obtain the following phonon transport equation in phase space:

$$\begin{aligned}
&\frac{-2i}{\hbar}E\frac{\partial}{\partial t}f^{\gtrless}_{\lambda\lambda'}(p,q)+\cos\Lambda\begin{bmatrix}\omega^{o}_{\lambda}(p)^{2}\,f^{\gtrless}_{\lambda\lambda'}(p,q)\\ -f^{\gtrless}_{\lambda\lambda'}(p,q)\,\omega^{o}_{\lambda'}(p)^{2}\end{bmatrix}\\
&\qquad -i\sin\Lambda\begin{Bmatrix}\omega^{o}_{\lambda}(p)^{2}\,f^{\gtrless}_{\lambda\lambda'}(p,q)\\ +f^{\gtrless}_{\lambda\lambda'}(p,q)\,\omega^{o}_{\lambda'}(p)^{2}\end{Bmatrix}\\
&\qquad\qquad =\cos\Lambda\left[\mathrm{Re}\Pi^{r}(p,q),f^{\gtrless}(p,q)\right]_{\lambda\lambda'}\\
&\qquad\qquad\quad -i\sin\Lambda\left\{\mathrm{Re}\Pi^{r}(p,q),f^{\gtrless}(p,q)\right\}_{\lambda\lambda'}\\
&\qquad\qquad\quad +\cos\Lambda\left[\mathrm{Re}\Pi^{\gtrless}(p,q),\mathrm{Re}f^{r}_{r}(p,q)\right]_{\lambda\lambda'}\\
&\qquad\qquad\quad -i\sin\Lambda\left\{\mathrm{Re}\Pi^{\gtrless}(p,q),\mathrm{Re}f^{r}(p,q)\right\}_{\lambda\lambda'}\\
&\qquad\qquad\quad +i\cos\Lambda\left\{\mathrm{Im}\Pi^{r}(p,q),f^{\gtrless}(p,q)\right\}_{\lambda\lambda'}\\
&\qquad\qquad\quad +\sin\Lambda\left[\mathrm{Im}\Pi^{r}(p,q),f^{\gtrless}(p,q)\right]_{\lambda\lambda'}\\
&\qquad\qquad\quad -i\cos\Lambda\left\{\Pi^{\gtrless}(p,q),\mathrm{Im}f^{r}(p,q)\right\}_{\lambda\lambda'}\\
&\qquad\qquad\quad -\sin\Lambda\left[\Pi^{\gtrless}(p,q),\mathrm{Im}f^{r}(p,q)\right]_{\lambda\lambda'},
\end{aligned}\tag{18.60}$$

where the Λ's are the appropriate differential "Poisson bracket" operators, and

$$f^{\gtrless}_{\lambda\lambda'}(p,q)=-iS^{\gtrless}_{\lambda\lambda'}(\vec{p},\vec{q},-E,t).$$

We can also express the transport equation as integro-differential equations by using Eqs. (17.35) and (17.36), as

$$\begin{aligned}
\frac{\partial}{\partial t}f^{\gtrless}(p,q)=&-\frac{\partial\hbar\omega^{o}_{\lambda}(p)}{\partial\vec{p}}\cdot\frac{\partial f^{\gtrless}_{\lambda\lambda}(\vec{p},\vec{q},E,t)}{\partial\vec{q}}\\
&-\frac{1}{2i}\frac{1}{(h^{4})^{2}}\int dp'dq'\int du\;dv\\
&\times\exp\left\{\frac{i}{\hbar}\left[(p-p')\cdot v+(q-q')\cdot u\right]\right\}
\end{aligned}$$

$$\times \left[\begin{array}{c} \left[\begin{array}{c} \frac{\mathrm{Re}\Pi^r\left(p+\frac{u}{2},q-\frac{v}{2}\right)}{\omega} f^{\gtrless}\left(p',q'\right) \\ -f^{\gtrless}\left(p',q'\right)\frac{\mathrm{Re}\Pi^r\left(p-\frac{u}{2},q+\frac{v}{2}\right)}{\omega} \end{array} \right] \\ + \left[\begin{array}{c} \frac{\Pi^{\gtrless}\left(p+\frac{u}{2},q-\frac{v}{2}\right)}{\omega} f^{r}\left(p',q'\right) \\ -f^{r}\left(p',q'\right)\frac{\Pi^{\gtrless}\left(p-\frac{u}{2},q+\frac{v}{2}\right)}{\omega} \end{array} \right] \\ \pm \left\{ \begin{array}{c} \frac{\Pi^{\gtrless}\left(p+\frac{u}{2},q-\frac{v}{2}\right)}{\omega} f^{\lessgtr}\left(p',q'\right) \\ + f^{\lessgtr}\left(p',q'\right)\frac{\Pi^{\gtrless}\left(p-\frac{u}{2},q+\frac{v}{2}\right)}{\omega} \end{array} \right\} \\ \mp \left\{ \begin{array}{c} \frac{\Pi^{\lessgtr}\left(p+\frac{u}{2},q-\frac{v}{2}\right)}{\omega} f^{\gtrless}\left(p',q'\right) \\ + f^{\gtrless}\left(p',q'\right)\frac{\Pi^{\lessgtr}\left(p-\frac{u}{2},q+\frac{v}{2}\right)}{\omega} \end{array} \right\} \end{array} \right], \tag{18.61}$$

where the quantities without subscripts are matrices in branch index λ and do not necessarily commute. Equation (18.61) will serve as a basis for simulating the transport of phonon across interfaces and boundaries of intervening different materials. The LWT of the retarded correlation function equation is given by the following differential form:

$$\begin{aligned} &\frac{-2i}{\hbar}E\frac{\partial}{\partial t}f^{r,a}_{\lambda\lambda'}(p,q) + \cos\Lambda \left[\begin{array}{c} \omega^o_\lambda(p)^2 f^{r,a}_{\lambda\lambda'}(p,q) \\ -f^{r,a}_{\lambda\lambda'}(p,q)\,\omega^o_{\lambda'}(p)^2 \end{array} \right] \\ &\quad -i\sin\Lambda \left\{ \begin{array}{c} \omega^o_\lambda(p)^2 f^{r,a}_{\lambda\lambda'}(p,q) \\ +f^{r,a}_{\lambda\lambda'}(p,q)\,\omega^o_{\lambda'}(p)^2 \end{array} \right\} \\ &= \cos\Lambda\left[\Pi^{r,a}(p,q), f^{r,a}(p,q)\right]_{\lambda\lambda'} \\ &\quad -i\sin\Lambda\left\{\Pi^{r,a}(p,q), f^{r,a}(p,q)\right\}_{\lambda\lambda'}, \end{aligned} \tag{18.62}$$

or by its corresponding integral form through the use of Eqs. (17.35) and (17.36).

18.5 The phonon Boltzmann equation

More revealing equations can be seen by neglecting off-diagonal or "interbranch" terms, and expanding the equations in terms of the gradients. Using the differential "Poisson bracket" operator in Eqs .(17.33)–(17.34) to implement the gradient expansion to the leading order, we obtain

$$\begin{aligned}
&\frac{-E}{\hbar}\frac{\partial}{\partial t}f^{\gtrless}_{\lambda\lambda}(p,q)-\left\{\frac{\hbar}{2}\nabla_p\Omega^2_{\lambda\lambda}(p,q)\cdot\nabla_q f^{\gtrless}_{\lambda\lambda}(p,q)\right\}\\
&=\left\{\frac{\hbar}{2}\frac{\partial}{\partial E}\mathrm{Re}\Pi^r_{\lambda\lambda}(p,q)\frac{\partial}{\partial t}f^{\gtrless}_{\lambda\lambda}(p,q)\right\}-\left\{\frac{\hbar}{2}\frac{\partial}{\partial t}\mathrm{Re}\Pi^r_{\lambda\lambda}(p,q)\frac{\partial}{\partial E}f^{\gtrless}_{\lambda\lambda}(p,q)\right\}\\
&\quad-\left[\left\{\frac{\hbar}{2}\nabla_p\Pi^{\gtrless}_{\lambda\lambda}(p,q)\cdot\nabla_q\mathrm{Re}f^r_{\lambda\lambda}(p,q)\right\}\right.\\
&\quad\left.-\left\{\frac{\hbar}{2}\frac{\partial}{\partial E}\Pi^{\gtrless}_{\lambda\lambda}(p,q)\frac{\partial}{\partial t}\mathrm{Re}f^r_{\lambda\lambda}(p,q)\right\}\right]\\
&\quad+\left[\left\{\frac{\hbar}{2}\nabla_q\Pi^{\gtrless}_{\lambda\lambda}(p,q)\cdot\nabla_p\mathrm{Re}f^r_{\lambda\lambda}(p,q)\right\}\right.\\
&\quad\left.-\left\{\frac{\hbar}{2}\frac{\partial}{\partial t}\Pi^{\gtrless}_{\lambda\lambda}(p,q)\frac{\partial}{\partial E}\mathrm{Re}f^r_{\lambda\lambda}(p,q)\right\}\right]\\
&\quad+\frac{i}{2}\Pi^{>}_{\lambda\lambda}(p,q)f^{<}_{\lambda\lambda}(p,q)-\frac{i}{2}\Pi^{<}_{\lambda\lambda}(p,q)f^{>}_{\lambda\lambda}(p,q),
\end{aligned}\tag{18.63}$$

where we have defined the renormalized kinematic frequency by $\Omega^2_{\lambda\lambda}(p,q)=\omega^o_\lambda(p)^2-\mathrm{Re}\Pi^r_{\lambda\lambda}(p,q)$. The leading terms in the right-hand side of Eq. (18.63) containing the Planck's constant $\hbar$ are bonafide quantum corrections to the phonon Boltzmann equation. The Boltzmann equation for the distribution function of phonons, $n_{\lambda\lambda}(p,q)$, from vibration branch λ readily follows by neglecting these leading quantum corrections. The result upon substituting $\frac{E}{\hbar}=\omega$ is given by the following expression:

$$\begin{aligned}
&\frac{\partial}{\partial t}n_{\lambda\lambda}(p,q)+\frac{\Omega_{\lambda\lambda}(p,q)}{\omega}\nabla_p\Omega_{\lambda\lambda}(p,q)\cdot\nabla_q n_{\lambda\lambda}(p,q)\\
&\quad=-\frac{i}{2}\left\{\frac{\Pi^{>}_{\lambda\lambda}(p,q)}{\omega}\right\}n_{\lambda\lambda}(p,q)+\frac{i}{2}\left\{\frac{\Pi^{<}_{\lambda\lambda}(p,q)}{\omega}\right\}[1+n_{\lambda\lambda}(p,q)],
\end{aligned}\tag{18.64}$$

with the spectral function satisfying the equation

$$\frac{\partial}{\partial t}A_{\lambda\lambda}(p,q)=\frac{\Omega_{\lambda\lambda}(p,q)}{\omega}\nabla_p\hbar\Omega_{\lambda\lambda}(p,q)\cdot\nabla_q A_{\lambda\lambda}(p,q).\tag{18.65}$$

We thus obtain the familiar interpretation [108] relating as the scattering-out rate and as the scattering-in rate. The simplest solution to the second line of Eq. (18.65) is to assume that the spectral function, $A_{\lambda\lambda}(p,q)$, to be independent of crystal lattice coordinates and time. Further by taking the self-consistent and renormalized frequency to be given by the solution of the equality $\Omega_{\lambda\lambda}(p,q) = \omega$, with this solution denoted by $\omega_\lambda(\mathbf{p},t)$, we finally obtain,

$$\begin{aligned}
\frac{\partial}{\partial t} n_\lambda(\vec{p},\vec{q},\omega_\lambda,t) &= -\nabla_{\vec{p}}\hbar\omega_\lambda(p)\cdot\nabla_{\vec{q}} n_\lambda(\vec{p},\vec{q},\omega_\lambda,t) \\
&\quad + \left(\frac{\partial}{\partial t} n_\lambda(\vec{p},\vec{q},\omega_\lambda,t)\right)_c, \\
\left(\frac{\partial}{\partial t} n_\lambda(\vec{p},\vec{q},\omega_\lambda,t)\right)_c &= i\frac{\Pi_\lambda^{<}(\vec{p},\vec{q},\omega_\lambda,t)}{2\omega_\lambda(p)}\left[1+n_\lambda(\vec{p},\vec{q},\omega_\lambda,t)\right] \quad \text{“scattering in”} \\
&\quad -i\frac{\Pi_\lambda^{>}(\vec{p},\vec{q},\omega_\lambda,t)}{2\omega_\lambda(p)} n_\lambda(\vec{p},\vec{q},\omega_\lambda,t) \quad \text{“scattering out”}.
\end{aligned} \tag{18.66}$$

The calculation of $\Pi_{ij}(\xi,\bar{\xi})$ and corresponding Feynman diagrams are given in Appendix B of Ref. [10]. Equation (18.66) is similar to the frequently used form of the Boltzmann equation for phonons characterized by the use of self-consistently renormalized phonon frequency and phonon group velocity [142] arising from the real part of the phonon self-energy.

A study of the collision term of Eq. (18.66) for anharmonic phonons, with phonon–phonon interaction arising from the cubic terms in the ionic displacements, can be found in the work of the author [143] and in the study of phonon hydrodynamics and second sound [144].

Chapter 19

Summary of Part Three

It is worthwhile to mention at the outset that in applying our quantum transport formalism to an alternative perspective of spinful and topological systems, the following equations are of crucial importance. These are Eqs. (17.23), (17.33), and (17.34). Equations (17.37) and (17.38) also play a role. The discussions on discrete phase space of Part 1, especially Section 5.1 of Chapter 5, are also of utmost importance.

QSFT is a straightforward algebraic approach to non-equilibrium quantum many-body physics exactly paralleling the quantum-field theoretical techniques at zero-temperature. The lattice Weyl transformation technique avoids the slowly-varying approximation usually employed in pre-1990 formulation of quantum transport equations in lattice phase space.

In this review, the electron quantum field operators are considered as the relevant quantum field operators. Clearly continued advances in ultrafast optical experimental techniques would yield more information on the appropriate quantum field operators to use in constructing the non-equilibrium quantum physics using the procedure reviewed here. For example, in the non-equilibrium theory of more complex system involving excitons, biexcitons, trions, etc., one may employ the approximate creation and annihilation operators for these quasiparticles to be the relevant quantum fields to derived the coupled quantum transport equations of composite particles similar to those derived here. Remarkably, in the third quantization of QFT of Part 1, Chapter 11, third quantization functional fields are creation and annihilation functional fields operators.

We have generalized the derivation of quantum distribution-function transport equations using L-space QSFT and lattice Weyl–Wigner formulation of the quantum theory of solids [3]. The QSFT technique allows a straightforward "algebraic" approach to the statistical quantum field-theoretical treatment of a time-dependent system of interacting quantized fields. The lattice Weyl transform technique provides the rigorous vehicle for transforming all time-evolution equations of the Green's function technique into kinetic transport equations in phase-space needed for large-scale time-dependent computer simulations, in contrast to the use of gradient expansion theories [107, 108, 145, 146] whose range of validity has no relevance to the high-speed performance analyses of quantum nanodevices. The QSFT results when applied to Fermi systems yield the coherent wave theory of Stahl and Balslev [139], Bloch equations of a two-level system in quantum optics [138] and have also been used in our large-scale simulation of the high-frequency time-dependent behavior of RTDs [14]. The power and simplicity of the regional renormalization scheme for modeling the time-dependent behavior of type I and type II RTDs under biasing condition for limit-cycle autonomous current-oscillation operation, serving as novel solid-state THz sources is discussed in Ref. [12]. The generalized semiconductor Bloch equation (GSBE) will no doubt become an important vehicle for investigating the high-speed performance of other multi-band tunneling devices. It may also have relevance to non-equilibrium superconductivity of high-T_c materials.

The present self-consistent dynamical approach is needed for ultra-fast open systems where dissipation, decay, as well as creation of excitations in a highly changing environment effectively renders model mean-field Hamiltonians invalid. Theories of relaxation going beyond mean-field approximation and able to describe very-short-time dynamics, including boundary conditions, is urgently needed in ultra-fast optics of confined nanostructures [110]. This demands that one considers the relevant quantum fields as the only invariant entities with all correlations between these fields highly changing in space and time.

Indeed, a major challenge in analyzing the high-speed performance of semiconductor gain material for investigating microcavity lasers is the fact that the electron–hole system properties are strongly affected by many-particle Coulomb interactions, and strong coupling to the light and crystal lattice-phonon fields [110, 147]. In highly excited semiconductor systems, the exciton, e–h Cooper pairs, and electron–hole plasma energetics and their accompanying distributions greatly affect the polaritons, phonons, biexcitons, trions, and higher-order "pairing" (excitonic molecule) distributions, their wavefunctions and energies.

19.1 New computational challenge

We have also discussed QSFT for quantized real vibrational fields. Similar formalism can be carried out for electromagnetic fields. Coupled quantum transport equations for electrons and phonons is urgently needed in the fundamental investigation and computer simulation of Kapitza resistance in energy transport across solid-solid interfaces where electron–phonon coupling dominates, such as in Au-diamond and Pb-diamond interfaces. The understanding of Kapitza resistance to heat transport, exhibited by a temperature difference across interfaces between metals and liquid helium and between solid–solid interfaces, is still not well understood even after six decades of its discovery [148–153]. In fact by assuming that heat is entirely carried by phonons, the lattice dynamics heat transport simulation for Pb-diamond gives result which differs from the experimental value by a factor of 100! This problem may involved non-equilibrium system of highly-coupled phonons and surface-plasmons at the metal-insulator interface [154]. Furthermore, the emerging field of integrated nanooptics presents another computational challenge as a non-equilibrium system of highly-coupled photons and surface-plasmon–polaritons [155, 156].

With the exploding surge of interest on spintronics and nanomagnetics [157], there is an urgent need for a fully quantum transport extension of the classical Bloch equation to guide the numerical simulation of the speed-power switching performance of realistic spin nanostructures/transistors. This may involved highly-coupled quantum transport equations for spin magnetization and charge, and in addition in the case of multi-band spin, coupling to pseudo-spin magnetization equations. Moreover, in spin/pseudospin caloritronics and spin batteries coupling of spin system, thermal excitations and phonons needs more powerful numerical simulations. More complex heterostructures with new functionality and multidimensional aspects of the relevant quantum distributions in phase space clearly present highly-demanding computational challenges [158].

With the increasing capability of fast and efficient computer resources together with modern experimental techniques to probe interfaces, nanosystems, and ultrafast dynamics, there is a critical need for quantum transport equations that is not based on model mean-field Hamiltonians, to investigate the complexities of highly non-equilibrium ultrafast quantum phenomena and to define the range of applicability of various approximation schemes. This brief review and other cited works by the author provide the theoretical quantum transport basis for meeting these challenges and provide a unified approach for large-scale simulations of functional materials and highly nonlinear active nanodevices in nanoelectronics, spintronics, pseudospintronics, caloritronics, and more recently in giving foundation of computational topotronics discussed in greater details in Parts 6–8 of this book.

Part 4

Quantum Transport of Spin and Pseudospin Systems

Chapter 20

Spin Magnetization Quantum Transport Equations: Spin and Charge Correlations

20.1 Introduction

The ultrafast-switching-speed and power-dissipation performance analyses of nanoelectronic devices has revealed the inadequacy of the classical Boltzmann transport equation (BTE), and ushered its extension to fully time-dependent and highly-nonlinear non-equilibrium quantum distribution function (QDF) transport equations for charge carriers. This has been achieved through the use of non-equilibrium Green's function, obtained either by the time-contour quantum field formulation of Schwinger [102], Keldysh [106], and Kadanoff and Baym [107], or by the real-time quantum superfield formulation of Buot coupled with his lattice Weyl (LW) transformation technique [10, 12] for realistic nanodevice transport-physics simulations. The extension of BTE to QDF transport equation has proved to be highly crucial in discovering autonomous current oscillations in resonant tunneling devices through numerical simulations, and in resolving controversial issues concerning highly-nonlinear and bistable current-voltage characteristics found in the experiments [11, 14]. Indeed, the QDF approach in phase space has so far been the most successful technique in the time-dependent analyses of open and active nanosystem and nanodevices [14].

To the authors' knowledge, the extension of the well-known classical Bloch equation for spin transport, which is the analogue of the classical BTE for particle transport, to fully time-dependent and highly-nonlinear non-equilibrium QDF transport equations for the magnetization has not been reported by other authors. With the exploding surge of interest on spintronics and nanomagnetics [157], there is an urgent need for this fully quantum transport extension of the classical Bloch equation to guide the numerical simulation of the speed-power switching performance of functional spin-nanostructures and realistic spin-transistors.

The well-known classical Bloch equations for spin systems is usually given in the diffusive regime [159] as

$$\frac{\partial M_x}{\partial t} = \left[D\nabla^2 M_x - \frac{M_x}{T_2}\right] + \gamma(\vec{M} \times \vec{B})_x, \tag{20.1}$$

$$\frac{\partial M_y}{\partial t} = \left[D\nabla^2 M_y - \frac{M_y}{T_2}\right] + \gamma(\vec{M} \times \vec{B})_y, \tag{20.2}$$

$$\frac{\partial M_z}{\partial t} = \left[D\nabla^2 M_z - \frac{M_z - M_o}{T_1}\right] + \gamma(\vec{M} \times \vec{B})_z, \tag{20.3}$$

where γ is the gyromagnetic ratio, T_1 is the spin-relaxation characteristic time, and T_2 is the spin-dephasing characteristic time. In analyzing the current-induced spin orientation of electrons in semiconductors, with spin-orbit coupling, D'yakonov–Perel' [160] used the following transport equation for the spin spatial-density vector, $\vec{S}$, as

$$\frac{\partial}{\partial t}\vec{S} = \left[-\vec{\nabla}\cdot\vec{Q} + \frac{\vec{S}}{\tau_s}\right] + \vec{\Omega} \times \vec{S}, \tag{20.4}$$

where $\vec{Q}$ is the spin flux density,

$$\vec{\Omega} = \frac{1}{\hbar}\mu_B g \vec{B}, \tag{20.5}$$

μ_B is the Bohr magneton, g is the g-factor for electrons, and $\vec{B}$ is the effective magnetic field.

What these earlier classical treatments have shown is that spin transport consist of terms similar to particle transport, i.e., with spin-independent transport coefficients or parameters, such as the diffusion coefficient, D, and relaxation times in Eq. (20.3), and terms describing the torques in the system. In micromagnetics, dealing mostly with micro-domain wall dynamics, the so-called classical Landau–Lifshitz–Gilbert equation [161],

$$\frac{\partial \vec{M}}{\partial t} = \gamma\vec{M} \times \vec{H} - \frac{\alpha}{|M|}\vec{M} \times \frac{\partial \vec{M}}{\partial t}, \tag{20.6}$$

has become the governing time-dependent equation for the magnetization. Indeed, its various modifications and phenomenological damping terms, has become the major player in the analyses of the dynamics of the magnetization field [161–163]. What Gilbert has shown is that the damping terms can contribute to the torque in the system. In Eq. (20.6) for example, the Rayleigh damping field, $\eta\frac{\partial \vec{M}}{\partial t}$, introduces a small torque, $\frac{\alpha}{|M|}\vec{M} \times \frac{\partial \vec{M}}{\partial t}$, proportional to the damping parameter α [161]. In quantum field theoretical language, this means that the spin-dependent self-energies will contribute to the torque in the system as will be shown more rigorously in what follows.

The purpose here is to extend this classical Bloch equations to a fully time-dependent and highly-nonlinear non-equilibrium SMQDFT equations on the same level as in the extension of the classical BTE to full QDF transport equation based on non-equilibrium Green's function technique [102, 106, 107]. In this chapter, we will first treat spin transport within a single energy band. The multiband spin quantum transport, where pseudo-spin also plays a role, will follow.

20.2 Real-time Green's function and LW transform

Our starting point is the general non-equilibrium quantum transport expressions for fermions, as obtained from the real-time quantum superfield theoretical technique of Part 3. Upon dropping the Cooper pairing terms in those expressions which do not concern us here (their corresponding transport equations are important in non-equilibrium superconductivity), the non-equilibrium Green's function transport equations reduce to the following known expressions [102, 106, 107]:

$$i\hbar\left(\frac{\partial}{\partial t_1}+\frac{\partial}{\partial t_2}\right)G^{\gtrless} = \left[vG^{\gtrless}-G^{\gtrless}v^T\right]+\left[\Sigma^r G^{\gtrless}-G^{\gtrless}\Sigma^a\right] + \left[\Sigma^{\gtrless}G^a - G^r\Sigma^{\gtrless}\right]. \tag{20.7}$$

The quantum distribution-function transport equations in $(p,q)=(\vec{p},\vec{q},E,t)$-space are obtained by applying the LW transformation discussed in Part 1 [3] (although continuum approximation is employed in this chapter, this is not essential and we use the word "lattice" when referring to solid-state problems). The LW transform $a(p,q)$ of any operator $\hat{A}$ is defined by the following relation:

$$a(p,q) = \int dv \exp\left(\frac{ip\cdot v}{\hbar}\right)\left\langle q-\frac{v}{2}\right|\hat{A}\left|q+\frac{v}{2}\right\rangle \equiv \mathcal{W}A\left(\vec{v},\tau,\vec{q},t\right), \tag{20.8}$$

where the matrix element in the integrand is evaluated between two Wannier functions, and $\mathcal{W}$ indicates the operation of taking the LW transform. This transformation enables the numerical simulation of real time-dependent dynamical open systems by transforming from *two-point* correlation functions to numerically manageable and physically meaningful *local* functions in phase space.[1] The well-known electron Wigner distribution function (WDF), $f_w(\vec{p}, \vec{q}, t)$, is obtained from the correlation function $-i\hbar G^<$ by performing integration over the energy variable

$$f_w(\vec{p}, \vec{q}, t) = \frac{1}{2\pi\hbar} \int dE \left(-i\hbar G^< (\vec{p}, E, \vec{q}, t)\right). \tag{20.9}$$

In the absence of particle pairing of superconductivity to obtain Eq. (20.7), we may also write the general transport equation, written for "<"-quantities as [here $\{A, B\}$ and $[A, B]$ means anticommutator and commutator, respectively, of A and B],

$$i\hbar\left(\frac{\partial}{\partial t_1} + \frac{\partial}{\partial t_2}\right) G^< = \left[\tilde{\mathcal{H}}, G^<\right] + \left[\Sigma^<, \operatorname{Re} G^r\right] - \frac{i}{2}\left\{\Gamma, G^<\right\} + \frac{i}{2}\left\{\Sigma^<, A\right\}, \tag{20.10}$$

where the single-particle Hamiltonian in phase space is given by

$$\tilde{\mathcal{H}} = \mathcal{H}_o + \Sigma^\delta + \operatorname{Re}\Sigma^r, \tag{20.11}$$

with Σ^δ being the "singular part" (delta function in time) of the self-energy.

$$\mathcal{H}_o = E(\vec{p}) + V(\vec{q}), \tag{20.12}$$

where $E(\vec{p})$ is the electron dispersion relation or energy-band function, $\vec{p}$ is the crystal momentum $V(\vec{q})$ is the self-consistent single-particle potential function and $\vec{q}$ is the lattice coordinates. In obtaining Eq. (20.10) from Eq. (20.7), we made use of the relations, $(G^a)^\dagger = G^r$, and $\Sigma^{<\dagger}(1,2) = -\Sigma^<(2,1)$, and the following relations:

$$iA(1,2) = -2i\operatorname{Im} G^r = -\left(G^>(1,2) - G^<(1,2)\right), \tag{20.13}$$

$$i\Gamma(1,2) = -2i\operatorname{Im}\Sigma^r = -\left(\Sigma^>(1,2) - \Sigma^<(1,2)\right). \tag{20.14}$$

The second term in Eq. (20.10) is a commutator and also describes evolution in phase-space although only serving to complicate, by virtue of the scattering term, $\Sigma^<$, the kinetics of particle motion described by the first term, a sort of complex interference phenomena similar to *zitterbewegung* in Dirac quantum mechanics [5]. Although often neglected in considering quantum transport of particles as this is expected to make a small corrections to the particle current, we will include the $[\Sigma^<, \operatorname{Re} G^r]$ term in our

---

[1]In contrast to Fourier transformation, the matrix element in the integrand of Eq. (20.8) which is diagonal in either lattice position or crystal momentum is invariant under the lattice Weyl transformation. Clearly, the crystal momentum and lattice position themselves, considered as matrix element of momentum and position operator, respectively, are invariant under lattice Weyl transformation.

present considerations for completeness. Note that in the gradient expansion of Eq. (20.10), with the $[\Sigma^<, \mathrm{Re}\, G^r]$ term neglected, the leading terms immediately give the classical BTE for spinless systems.[2] However, even this is no longer quite true for spin system as we shall see later, since the spinor scattering terms contribute to the torque in the system.

The third term of Eq. (20.10), which is proportional to $G^<$, is the *scattering-out* term, and the last term which is proportional to the spectral density is the *scattering-in* term, in analogy with the collision terms in the classical Boltzmann equation. The collision terms can also be rewritten using the identity,

$$-\frac{i}{2}\left\{\Gamma, G^<\right\} + \frac{i}{2}\left\{\Sigma^<, A\right\} = -\frac{1}{2}\left\{\Sigma^<, G^>\right\} + \frac{1}{2}\left\{\Sigma^>, G^<\right\}, \qquad (20.15)$$

which also leads to $i\Sigma^>$ and $-i\Sigma^<$ terms, upon performing the $\mathcal{W}$ operation, as scattering-out and scattering-in collision terms, respectively [107].

The equation for G^r and $G^a = G^{r\dagger}$ are simpler and are given as, $\hat{G}^{-1}G^r = 0$, $G^a\left(\hat{G}^{-1}\right)^\dagger = 0$, where

$$\hat{G}^{-1}G^r(1,2) = i\hbar\left(\frac{\partial}{\partial t_1} + \frac{\partial}{\partial t_2}\right)G^r(1,2) - \left[\tilde{\mathcal{H}} + \Sigma^r, G^r\right](1,2). \qquad (20.16)$$

The above equations are then cast in the language of phase-space QDF transport equations, similar to that of the classical Boltzmann equation by the use of Buot lattice Weyl–Wigner formulation of the band dynamics of electrons in a solid Part 1 [3]. In the *continuum approximation*, the first thing to do is to recast the *two-point* space and time arguments in all the transport equations as follows: $t_1 = t - \frac{\tau}{2}$, $t_2 = t + \frac{\tau}{2}$, $\vec{q}_1 = \vec{q} - \frac{\tilde{\mathrm{v}}}{2}$, $\vec{q}_2 = \vec{q} + \frac{\tilde{\mathrm{v}}}{2}$. We also need to define a $(3+1)$-dimensional canonical variables, namely, $p = (\vec{p}, -E)$, $q = (\vec{q}, t)$, which allows us to write $v = (\vec{v}, \tau)$. The transformation of Eq. (20.8) is then applied.

For simplicity in spin quantum transport equations, we consider only a single energy band having spin indices, $\downarrow$ and $\uparrow$. Incorporating the $\downarrow$ and $\uparrow$ variables in Eqs. (20.7) and (20.6), we obtain exactly four transport equations for $G^<_{\uparrow\uparrow}(1,2)$, $G^<_{\uparrow\downarrow}(1,2)$, $G^<_{\downarrow\uparrow}(1,2)$, and $G^<_{\downarrow\downarrow}(1,2)$. In spintronics, we are interested in the time-dependent evolution of the magnetic polarization, $S_z = i\hbar\left(G^<_{\uparrow\uparrow} - G^<_{\downarrow\downarrow}\right)$, as this is transported across the device. In order to achieve this, we need to perform *linearly-independent* combinations of the

[2]That the gradient expansion of Eq. (20.10), with the $[\Sigma^<, \mathrm{Re}\, G^r]$ term neglected, yields the classical BTE for spinless system is often viewed as justifying the neglect of $[\Sigma^<, \mathrm{Re}\, G^r]$ in the quantum transport equations [107].

components of the spinor non-equilibrium Green's function consisting of the set: $\left\{G^{<}_{\uparrow\uparrow}(1,2), G^{<}_{\uparrow\downarrow}(1,2), G^{<}_{\downarrow\uparrow}(1,2), G^{<}_{\downarrow\downarrow}(1,2)\right\}$. The resulting new set will become the relevant and independent correlation functions more pertinent to the spin-transport problem, which consists of the separation into scalar charge (particle number) distribution and spin magnetization vector.[3] Here, the terms correlation function and distribution function are interchangeably used since correlation functions used here has one-to-one correspondence with their distribution functions in phase space upon applying the LW transformation.

20.3 Charge and spin correlation functions

The separation in terms of a scalar and vector representing the total charge and spin-vector correlation functions, naturally occurs as expansion coefficient in terms of the Pauli matrices,

$$\begin{pmatrix} G^{<}_{\uparrow\uparrow} & G^{<}_{\uparrow\downarrow} \\ G^{<}_{\downarrow\uparrow} & G^{<}_{\downarrow\downarrow} \end{pmatrix} = \frac{1}{2}(S_o\hat{I} + \vec{S}\cdot\vec{\sigma}), \tag{20.17}$$

where $\hat{I}$ is the 2×2 identity matrix, and

$$\left.\begin{aligned} S_x &= \left(G^{<}_{\downarrow\uparrow} + G^{<}_{\uparrow\downarrow}\right), \\ iS_y &= \left(G^{<}_{\downarrow\uparrow} - G^{<}_{\uparrow\downarrow}\right), \\ S_z &= \left(G^{<}_{\uparrow\uparrow} - G^{<}_{\downarrow\downarrow}\right), \\ S_o &= \left(G^{<}_{\uparrow\uparrow} + G^{<}_{\downarrow\downarrow}\right), \end{aligned}\right\} \tag{20.18}$$

[3]It is worth mentioning that the transformation from four independent operators or quantum states to their four linearly-independent combinations has yielded some of the deepest concepts of quantum mechanics, namely, spin angular momentum and quantum teleportation. A two-state eigenvector space, denoted by basis $|\uparrow\rangle$ and $|\downarrow\rangle$, has the following complete set of four basis operators, $\hat{b}_{\downarrow\uparrow} = |\downarrow\rangle\langle\uparrow|$, $\hat{b}^{\dagger}_{\downarrow\uparrow} = |\uparrow\rangle\langle\downarrow|$, $\hat{d}_{\uparrow\uparrow} = |\uparrow\rangle\langle\uparrow|$, $\hat{d}_{\downarrow\downarrow} = |\downarrow\rangle\langle\downarrow|$. The spin operators are obtained as simple linear combinations of the above basis operators as, $\hat{S}_o = \frac{\hbar}{2}\left(\hat{d}_{\uparrow\uparrow} + \hat{d}_{\downarrow\downarrow}\right)$, $\hat{S}_x = \frac{\hbar}{2}\left(\hat{b}^{\dagger}_{\downarrow\uparrow} + \hat{b}_{\downarrow\uparrow}\right)$, $\hat{S}_y = \frac{\hbar}{2i}\left(\hat{b}^{\dagger}_{\downarrow\uparrow} - \hat{b}_{\downarrow\uparrow}\right)$, $\hat{S}_z = \frac{\hbar}{2}\left(\hat{d}_{\uparrow\uparrow} - \hat{d}_{\downarrow\downarrow}\right)$, obeying the quantum-mechanical spin angular-momentum commutation relations, namely, $\left[\hat{S}_j, \hat{S}_k\right] = i\hbar\epsilon_{jkl}\hat{S}_l$, $i, j, k \neq o$. In matrix form, these are represented by the well-known Pauli matrix, $\vec{\sigma}$, namely, $\hat{S}_x = \frac{\hbar}{2}\sigma_x$, $\hat{S}_y = \frac{\hbar}{2}\sigma_y$, $\hat{S}_z = \frac{\hbar}{2}\sigma_z$. Note that operator $\hat{S}_o$ represent the identity operator (by virtue of the completeness relation for the eigenvectors; statistically referred to as a pure state) multiplied by $\frac{\hbar}{2}$. Any 2×2 matrix in "spin" space can be expanded in terms of the Pauli and identity operators.

where we drop the "<" superscript in the spin correlation functions, S_j. In the frame of reference where the z direction is fixed by the magnetic field, the S_x and S_y evolution equation describe *dephasing* mechanisms in the X–Y plane of the "Bloch sphere". The total charge of the system, $\rho(p,q,t)$ is represented by the scalar S_o i.e., from Eq. (20.9) we have,

$$\rho(\vec{q},t) = \frac{e}{(2\pi\hbar)^2} \int d\vec{p}dE \left(-i\hbar S_o(\vec{p},E,\vec{q},t)\right), \tag{20.19}$$

where e is the electron charge and $S_o(\vec{p},E,\vec{q},t)$ is the LW transform of correlation S_o. Thus Eq. (20.17) performs the separation of the charge and spin correlation functions as the relevant variables of any fermion system with spin degree of freedom.

For slowly varying system where the total charge vary slowly with space and time, perhaps in some spin transfer torque systems, we can reduce the number of important spin correlation functions in the spin quantum transport problem by making an approximation to S_o (for two-level atomic system, $S_o = 1$). Assuming the conservation of particles possessing spin degree of freedom within semiconductor channel of, say, a spintronic transfer torque transistor, we must have the following relation for the "$\uparrow\uparrow$" and "$\downarrow\downarrow$" rates of change, $\frac{\partial\rho^<_{\uparrow\uparrow}(p,q,t)}{\partial t} = -\frac{\partial\rho^<_{\downarrow\downarrow}(p,q,t)}{\partial t}$, where $\rho_{\uparrow\uparrow}(p,q,t)$ and $\rho_{\downarrow\downarrow}(p,q,t)$ are the LW transforms of $i\hbar G^<_{\uparrow\uparrow}$ and $i\hbar G^<_{\downarrow\downarrow}$, respectively.

Generally however, S_o will be varying in both space and time, i.e., a two-point non-equilibrium correlation function in space and time, describing the varying charge within the spintronic device due to depletion and accumulation in various regions within the device, such as in spin transfer torque devices employing thin barriers between the channel and ferromagnetic leads.

20.3.1 *Spin canonical forms of the Hamiltonian and self-energies*

We can also write the spinor self-energies used in Eq. (20.7) as independent linear expansion into spin-independent scalar [denoted by overscript-bar symbol] and vector quantities,

$$\Sigma^{a,r,<}_{\alpha\beta} = \frac{1}{2}(\bar{\Sigma}^{a,r,<}\hat{I} + \vec{\Xi}^{a,r,<} \cdot \vec{\sigma}), \tag{20.20}$$

which we write explicitly as

$$\Sigma^r_{\alpha\beta} = \frac{1}{2}\begin{pmatrix} \bar{\Sigma}^r + \Xi^r_z & \Xi^r_x - i\Xi^r_y \\ \Xi^r_x + i\Xi^r_y & \bar{\Sigma}^r - \Xi^r_z \end{pmatrix}, \tag{20.21}$$

$$\Sigma^a_{\alpha\beta} = \frac{1}{2}\begin{pmatrix} \bar{\Sigma}^a + \Xi^a_z & \Xi^a_x - i\Xi^a_y \\ \Xi^a_x + i\Xi^a_y & \bar{\Sigma}^a - \Xi^a_z \end{pmatrix}, \tag{20.22}$$

$$\Sigma^<_{\alpha\beta} = \frac{1}{2}\begin{pmatrix} \bar{\Sigma}^< + \Xi^<_z & \Xi^<_x - i\Xi^<_y \\ \Xi^<_x + i\Xi^<_y & \bar{\Sigma}^< - \Xi^<_z \end{pmatrix}. \tag{20.23}$$

Likewise, the spinor quantities used in Eq. (20.7), namely, the spin-dependent *scattering-out* Γ term is cast in the form,

$$\begin{aligned}\frac{\Gamma_{\alpha\beta}}{2} &= \frac{1}{2}(\bar{\Gamma}\hat{I} + \vec{\gamma}\cdot\vec{\sigma}) = -\mathrm{Im}\,\Sigma^r_{\alpha\beta} \\ &= -\frac{1}{2}(\mathrm{Im}\,\bar{\Sigma}^r\hat{I} + \mathrm{Im}\,\vec{\Xi}^r\cdot\vec{\sigma}),\end{aligned} \tag{20.24}$$

so that

$$\bar{\Gamma} = -\mathrm{Im}\,\bar{\Sigma}^r, \tag{20.25}$$

$$\vec{\gamma} = -\mathrm{Im}\,\vec{\Xi}^r. \tag{20.26}$$

Similarly, we have for the spin-dependent spectral function, A, in the *scattering-in* term as

$$\begin{aligned}\frac{A_{\alpha\beta}}{2} &= \frac{1}{2}(\bar{A}\hat{I} + \vec{\mathcal{A}}\cdot\vec{\sigma}) = -\mathrm{Im}\,S^r_{\alpha\beta} \\ &= -\frac{1}{2}(\mathrm{Im}\,\bar{S}^r\hat{I} + \mathrm{Im}\,\vec{S}^r\cdot\vec{\sigma}),\end{aligned} \tag{20.27}$$

so that

$$\bar{A} = -\mathrm{Im}\,\bar{S}^r, \tag{20.28}$$

$$\vec{\mathcal{A}} = -\mathrm{Im}\,\vec{S}^r. \tag{20.29}$$

In the effective-mass approximation, we may take the corresponding single-particle Hamiltonian $\tilde{\mathcal{H}}_s$ for spin systems in a magnetic field with spin-orbit coupling as

$$\begin{aligned}\tilde{\mathcal{H}}_s = &\left(\frac{\hat{\mathcal{P}}^2}{2m^*} - eV(\hat{Q}) + \frac{1}{2}\mathrm{Re}\,\bar{\Sigma}^r\right)\hat{I} + H_{so}(\hat{\mathcal{P}},\hat{Q}) \\ &+ \frac{1}{2}(\mathrm{Re}\,\vec{\Sigma}^r\cdot\vec{\sigma} - g_{\text{eff}}\mu_B\vec{B}\cdot\vec{\sigma}),\end{aligned} \tag{20.30}$$

where $\hat{\mathcal{P}} = \hat{P} + \frac{e}{c}A(\hat{Q})$, electric charge, $e = |e|$, $V(\hat{Q})$ is the self-consistent applied potential, and $\vec{B} = \nabla_{\vec{q}} \times A(q)$. The second term, $H_{so}(\hat{\mathcal{P}},\hat{Q})$, is the spin-orbit coupling which may consists of Dresselhaus and/or Rashba term[4] as well as transistor-gate and strain-induced spin-orbit coupling. The term $\mathrm{Re}\,\vec{\Sigma}^r \cdot \vec{\sigma}$ comes from the spin vector of the spin-dependent part of $\mathrm{Re}\,\Sigma^r_{\alpha\beta}$ in the single particle Hamiltonian. Its presence basically modifies the effective magnetic field due to external fields and spin-orbit coupling.

We have in general,

$$H_{so}(\hat{\mathcal{P}}, \hat{Q}) = \mu_B \frac{\hat{Q} \times \hat{\mathcal{P}}}{2\hbar m^* c^2} \left| \frac{\vec{E}}{q} \right| \cdot \vec{\sigma}, \tag{20.31}$$

which may acquire definite form in lower dimensional system based on symmetry considerations. In what follows, we write the most general form of the single-particle Hamiltonian, $\tilde{\mathcal{H}}_s$, as

$$\tilde{\mathcal{H}}_s = \frac{1}{2}(\bar{H}(\hat{P}, \hat{Q})\hat{I} - \vec{h}(\hat{P}, \hat{Q}) \cdot \vec{\sigma}), \tag{20.32}$$

where $\vec{h}(\hat{P}, \hat{Q}) = \left(g_s \frac{e}{2mc} \vec{B}_{\text{eff}} - \mathrm{Re}\,\vec{\Sigma}^r\right)$. Therefore

$$\tilde{\mathcal{H}}_s = \frac{1}{2}(\bar{H}\hat{I} - \vec{\sigma} \cdot \vec{\mathcal{B}}), \tag{20.33}$$

where

$$\vec{\mathcal{B}} = g_s \frac{e}{2mc} \vec{B}_{\text{eff}}, \tag{20.34}$$

and for electron-spin angular momentum, the *g-factor*, $g_s \simeq 2$. Here the effective $\vec{B}_{\text{eff}}(\hat{P}, \hat{Q})$ includes the effects of transistor-gate or strain induced spin-orbit coupling, and applied magnetic fields. $H(\hat{P}, \hat{Q})$ consists of the energy-band function and spin-independent confining potential due to the device heterostructure coupled with the applied bias at the terminals. The spin-orbit coupling in $\vec{B}_{\text{eff}}(\hat{P}, \hat{Q})$ also depends very much on the electric field of the self-consistent potential inside the device. This in turn depends on the total electric charge, represented by S_o, through the Poisson equation. Although we could include $\mathrm{Re}\,\vec{\Sigma}^r$ in the definition of $\vec{\mathcal{B}}$ above, we treat this separately to emphasize the possible role of scattering terms in contributing to the torque in the system.

---

[4]For example, for symmetric quantum well, the Rashba spin-orbit coupling can be expressed as $H_R = \frac{\alpha}{\hbar}\left(\hat{e}_z \times \vec{p}\right) \cdot \vec{\sigma}$, where α is the Rashba spin-orbit parameter.

20.4 Spin and charge correlation transport equations

The resulting equations obeyed by the spin correlation functions defined by Eq. (20.18) in accordance with Eq. (20.7) can be written in vector notation as

$$\begin{aligned}
&i\hbar\left(\frac{\partial}{\partial t_1}+\frac{\partial}{\partial t_2}\right)\vec{S}^<\\
&=\frac{1}{2}[\bar{H},\vec{S}^<]+\frac{1}{2}[\bar{\Sigma}^r\vec{S}^<-\vec{S}^<\bar{\Sigma}^a]+\frac{1}{2}[\bar{\Sigma}^<\vec{S}^a-\vec{S}^r\bar{\Sigma}^<]\\
&+\frac{i}{2}[\vec{\mathcal{B}}\times\vec{S}^<-\vec{S}^<\times\vec{\mathcal{B}}]\\
&+i\frac{1}{2}[\vec{\Xi}^r\times\vec{S}^<-\vec{S}^<\times\vec{\Xi}^a]+i\frac{1}{2}[\vec{\Xi}^<\times\vec{S}^a-\vec{S}^r\times\vec{\Xi}^<]\\
&+\frac{1}{2}[\vec{\mathcal{B}},S_o^<]+\frac{1}{2}[\vec{\Xi}^rS_o^<-S_o^<\vec{\Xi}^a]+\frac{1}{2}[\vec{\Xi}^<S_o^a-S_o^r\vec{\Xi}^<],
\end{aligned}\tag{20.35}$$

and for the *scalar* correlation function representing the total charge,

$$\begin{aligned}
&i\hbar\left(\frac{\partial}{\partial t_1}+\frac{\partial}{\partial t_2}\right)S_o^<\\
&=\frac{1}{2}[\bar{H},S_o^<]+\frac{1}{2}[\bar{\Sigma}^rS_o^<-S_o^<\bar{\Sigma}^a]+\frac{1}{2}[\bar{\Sigma}^<S_o^a-S_o^r\bar{\Sigma}^<]\\
&+\frac{1}{2}[\vec{\mathcal{B}}\cdot\vec{S}^<-\vec{S}^<\cdot\vec{\mathcal{B}}]+\frac{1}{2}[\vec{\Xi}^r\cdot\vec{S}^<-\vec{S}^<\cdot\vec{\Xi}^a]\\
&+\frac{1}{2}[\vec{\Xi}^<\cdot\vec{S}^a-\vec{S}^r\cdot\vec{\Xi}^<].
\end{aligned}\tag{20.36}$$

For self-consistency, one must also be solved for potential, Φ, or electric field, $\vec{E}$, using the Poisson equation,

$$\nabla^2\Phi=e\frac{\rho(\vec{q},t)-\rho_o}{\varepsilon_o},\tag{20.37}$$

where $\Phi=V$ in Eq. (20.30), ρ_o is the positive background charge and $\rho(\vec{q},t)$ is given by Eq. (20.19).

We also have the equivalent equation for $\vec{S}^<$,

$$\begin{aligned}
&i\hbar\left(\frac{\partial}{\partial t_1}+\frac{\partial}{\partial t_2}\right)\vec{S}^<\\
&=\frac{1}{2}[\bar{H}+\mathrm{Re}\,\bar{\Sigma}^r,\vec{S}^<]+\frac{i}{2}\{\mathrm{Im}\,\bar{\Sigma}^r,\vec{S}^<\}-\frac{i}{2}\{\bar{\Sigma}^<,\mathrm{Im}\,\vec{S}^r\}+[\bar{\Sigma}^<,\mathrm{Re}\,\vec{S}^r]\\
&+\frac{i}{2}[(\vec{\mathcal{B}}+\mathrm{Re}\,\vec{\Xi}^r)\times\vec{S}^<-\vec{S}^<\times(\vec{\mathcal{B}}+\mathrm{Re}\,\vec{\Xi}^r)]\\
&-\frac{1}{2}\{\mathrm{Im}\,\vec{\Xi}^r\times\vec{S}^<+\vec{S}^<\times\mathrm{Im}\,\vec{\Xi}^r\}+\frac{1}{2}\{\vec{\Xi}^<\times\mathrm{Im}\,\vec{S}^r+\mathrm{Im}\,\vec{S}^r\times\vec{\Xi}^<\}
\end{aligned}$$

$$+ \frac{i}{2}[\vec{\Xi}^< \times \mathrm{Re}\,\vec{S}^r - \mathrm{Re}\,\vec{S}^r \times \vec{\Xi}^<] + \frac{1}{2}[\vec{\mathcal{B}}, S_o^<] + \frac{1}{2}[\mathrm{Re}\,\vec{\Xi}^r, S_o^<]$$
$$+ \frac{i}{2}\{\mathrm{Im}\,\vec{\Xi}^r, S_o^<\} + [\vec{\Xi}^<, \mathrm{Re}\,S_o^r] - \frac{i}{2}\{\vec{\Xi}^<, \mathrm{Im}\,S_o^r\}, \tag{20.38}$$

which can be reduced to that which one can derive from Eq. (20.6), by using the relations in Eqs. (20.24) and (20.26). The result is

$$i\hbar\left(\frac{\partial}{\partial t_1} + \frac{\partial}{\partial t_2}\right)\vec{S}^<$$
$$= \frac{1}{2}[\bar{H} + \mathrm{Re}\,\bar{\Sigma}^r, \vec{S}^<] - \frac{i}{2}\{\bar{\Gamma}, \vec{S}^<\} + \frac{i}{2}\{\bar{\Sigma}^<, \vec{A}\} + [\bar{\Sigma}^<, \mathrm{Re}\,\vec{S}^r]$$
$$+ \frac{i}{2}[(\vec{\mathcal{B}} + \mathrm{Re}\,\vec{\Xi}^r) \times \vec{S}^< - \vec{S}^< \times (\vec{\mathcal{B}} + \mathrm{Re}\,\vec{\Xi}^r)]$$
$$+ \frac{1}{2}\{\vec{\gamma} \times \vec{S}^< + \vec{S}^< \times \vec{\gamma}\} - \frac{1}{2}\{\vec{\Xi}^< \times \vec{A} + \vec{A} \times \vec{\Xi}^<\}$$
$$+ \frac{i}{2}[\vec{\Xi}^< \times \mathrm{Re}\,\vec{S}^r - \mathrm{Re}\,\vec{S}^r \times \vec{\Xi}^<]$$
$$+ \frac{1}{2}[(\vec{\mathcal{B}} + \mathrm{Re}\,\vec{\Xi}^r), S_o^<] + [\vec{\Xi}^<, \mathrm{Re}\,S_o^r] - \frac{i}{2}\{\vec{\gamma}, S_o^<\} - \frac{i}{2}\{\vec{\Xi}^<, \mathrm{Im}\,S_o^r\}. \tag{20.39}$$

Note that our definition of $\vec{\mathcal{B}}$, Eq. (20.34), has the unit of energy, just like the self-energies.

20.4.1 *Spin magnetization quantum transport equations*

The equivalent expression derived from Eq. (20.6) in terms of the magnetization vector, $\vec{\mathcal{M}} = g_s \frac{e\hbar}{2m^* c}\vec{\mathcal{S}} = g_s \mu_B$ (the g-factor for spin, $g_s \simeq 2$) is

$$i\hbar\left(\frac{\partial}{\partial t_1} + \frac{\partial}{\partial t_2}\right)\vec{\mathcal{M}}$$
$$= \frac{1}{2}[\bar{H} + \mathrm{Re}\,\bar{\Sigma}^r, \vec{\mathcal{M}}] - \frac{i}{2}\{\bar{\Gamma}, \vec{\mathcal{M}}\} + \frac{i}{2}g_s\mu_B\{\bar{\Sigma}^<, \vec{A}\} + \frac{1}{2}[\bar{\Sigma}^<, \mathrm{Re}\,\vec{\mathcal{M}}^r]$$
$$+ \frac{i}{2}[(\vec{\mathcal{B}} + \mathrm{Re}\,\vec{\Xi}^r) \times \vec{\mathcal{M}} - \vec{\mathcal{M}} \times (\vec{\mathcal{B}} + \mathrm{Re}\,\vec{\Xi}^r)]$$
$$+ \frac{1}{2}\{\vec{\gamma} \times \vec{\mathcal{M}} + \vec{\mathcal{M}} \times \vec{\gamma}\} - \frac{1}{2}g_s\mu_B\{\vec{\Xi}^< \times \vec{A} + \vec{A} \times \vec{\Xi}^<\}$$
$$+ \frac{i}{2}[\vec{\Xi}^< \times \mathrm{Re}\,\vec{\mathcal{M}}^r - \mathrm{Re}\,\vec{\mathcal{M}}^r \times \vec{\Xi}^<]$$
$$+ \frac{1}{2}[(\vec{\mathcal{B}} + \mathrm{Re}\,\vec{\Xi}^r), M_o^<] + \frac{1}{2}[\vec{\Xi}^<, \mathrm{Re} M_o^r] - \frac{i}{2}\{\vec{\gamma}, M_o^<\} - \frac{i}{2}\{\vec{\Xi}^<, \mathrm{Im}\,M_o^r\}, \tag{20.40}$$

where we defined,

$$\vec{\mathcal{M}}^r = g_s\mu_B\vec{S}^r, \tag{20.41}$$

$$M_o^< = g_s\mu_B S_o^<. \tag{20.42}$$

First, let us evaluate Eq. (20.40) by assuming constant magnetic field $\mathcal{B}$ and in the absence of scatterings. We obtain

$$\frac{\partial}{\partial t}\vec{\mathcal{M}} = \mathcal{W}\left\{\frac{1}{i\hbar}\left[\bar{H},\vec{\mathcal{M}}\right] + \frac{1}{\hbar}\vec{\mathcal{B}}\times\vec{\mathcal{M}}\right\}, \tag{20.43}$$

where $\mathcal{W}$ indicates the operation of taking the LW transform. Equation (20.43) is the ballistic version of the expression given by Eq. (20.3).

In carrying out the LW transform of Eqs. (20.35), (20.39) and (20.40), the following relations between vector cross products and commutation or anti-commutation relations are useful,

$$\vec{A}\times\vec{B} - \vec{B}\times\vec{A} = \hat{I}_i\hat{\epsilon}_{ijk}\left\{A_j, B_k\right\}, \tag{20.44}$$

$$\vec{A}\times\vec{B} + \vec{B}\times\vec{A} = \hat{I}_i\hat{\epsilon}_{ijk}\left[A_j, B_k\right], \tag{20.45}$$

where $\hat{I}_i$ is the unit dyadic symmetric tensor or *idemfactor*, and $\hat{\epsilon}_{ijk}$ is the anti-symmetric unit tensor. For the purpose of making gradient expansion to make contact with classical kinetics, we give here for convenience the following LW transform of a commutator $[A, B]$ and an anticommutator $\{A, B\}$ in terms of Poisson bracket differential operator, Λ, as

$$\begin{aligned}[A,B](p,q) &= \cos\Lambda\left[a(p,q)\,b(p,q) - b(p,q)\,a(p,q)\right]\\ &\quad - i\sin\Lambda\left\{a(p,q)\,b(p,q) + b(p,q)\,a(p,q)\right\},\end{aligned} \tag{20.46}$$

$$\begin{aligned}\{A,B\}(p,q) &= \cos\Lambda\left\{a(p,q)\,b(p,q) + b(p,q)\,a(p,q)\right\}\\ &\quad - i\sin\Lambda\left[a(p,q)\,b(p,q) - b(p,q)\,a(p,q)\right],\end{aligned} \tag{20.47}$$

where $\Lambda = \frac{\hbar}{2}\left(\frac{\partial^{(a)}}{\partial p}\cdot\frac{\partial^{(b)}}{\partial q} - \frac{\partial^{(a)}}{\partial q}\cdot\frac{\partial^{(b)}}{\partial p}\right)$. Thus, if the LW transforms are not matrices, then to lowest order we have

$$[A,B](p,q) = -i\frac{\hbar}{2}\left(\frac{\partial a(p,q)}{\partial p}\cdot\frac{\partial b(p,q)}{\partial q} - \frac{\partial a(p,q)}{\partial q}\cdot\frac{\partial b(p,q)}{\partial p}\right), \tag{20.48}$$

$$\{A,B\}(p,q) = \left\{a(p,q)\,b(p,q) + b(p,q)\,a(p,q)\right\} = 2a(p,q)\,b(p,q). \tag{20.49}$$

If we compare Eqs. (20.35)–(20.39) with the particle quantum transport equation of Eq. (20.10), without the *zitterbewegung* term, $[\Sigma^<, \mathrm{Re}\, G^r]$, we see that the leading terms of the first line of the RHS of Eq. (20.39), corresponds to the classical BTE. The next three lines give the torque, due to spin-orbit coupling (included in the effective magnetic field), and terms coming from the spin-dependent relaxation and dephasing scattering mechanisms. The last line explicitly involves coupling to the total particle density (representing total charge).

20.4.2 *Leading terms of gradient expansion in phase space*

Carrying out the $\mathcal{W}$ operation in Eq. (20.40), we find that the leading gradient terms in the RHS reads

$$
\begin{aligned}
\frac{\partial}{\partial t}\vec{\mathcal{M}} = &-\frac{1}{m}\left(\vec{p}+\frac{e}{c}\vec{A}(\vec{q})\right)\cdot\nabla_q\vec{\mathcal{M}} - \frac{1}{\hbar}\bar{\Gamma}\vec{\mathcal{M}} + \frac{e}{mc}\bar{\Sigma}^{<}\vec{\mathcal{A}} - e\vec{E}\cdot\nabla_p\vec{\mathcal{M}} \\
&- \frac{e}{2c}(\vec{v}\times\vec{B})\cdot\nabla_p\vec{\mathcal{M}} + \frac{1}{\hbar}\vec{\mathcal{B}}\times\vec{\mathcal{M}} - \frac{1}{2}\left[\frac{\partial \mathrm{Re}\,\bar{\Sigma}^{r}}{\partial p}\frac{\partial\vec{\mathcal{M}}}{\partial q} - \frac{\partial \mathrm{Re}\,\bar{\Sigma}^{r}}{\partial q}\frac{\partial\vec{\mathcal{M}}}{\partial p}\right] \\
&- \frac{1}{2}\left[\frac{\partial\bar{\Sigma}^{<}}{\partial p}\frac{\partial \mathrm{Re}\,\vec{\mathcal{M}}^{r}}{\partial q} - \frac{\partial\bar{\Sigma}^{<}}{\partial q}\frac{\partial \mathrm{Re}\,\vec{\mathcal{M}}^{r}}{\partial p}\right] + \frac{1}{\hbar}\mathrm{Re}\,\vec{\Xi}^{r} \\
&\times\vec{\mathcal{M}} + \frac{1}{\hbar}\vec{\Xi}^{<}\times\mathrm{Re}\,\vec{\mathcal{M}}^{r} - \frac{1}{2}\hat{I}_i\hat{\epsilon}_{ijk}\left[\frac{\partial\gamma_j}{\partial p}\frac{\partial\mathcal{M}_k}{\partial q} - \frac{\partial\gamma_j}{\partial q}\frac{\partial\mathcal{M}_k}{\partial p}\right] \\
&+ \frac{1}{2}\frac{e}{2mc}\hat{I}_i\hat{\epsilon}_{ijk}\left[\frac{\partial\Xi_j}{\partial p}\frac{\partial\mathcal{A}_k}{\partial q} - \frac{\partial\Xi_j}{\partial q}\frac{\partial\mathcal{A}_k}{\partial p}\right] - \frac{1}{2}\left[\frac{\partial\vec{\mathcal{B}}}{\partial p}\frac{\partial M_o^{<}}{\partial q} - \frac{\partial\vec{\mathcal{B}}}{\partial q}\frac{\partial M_o^{<}}{\partial p}\right] \\
&- \frac{1}{\hbar}\vec{\gamma}M_o^{<} - \frac{1}{\hbar}\vec{\Xi}^{<}\mathrm{Im}\,M_o^{r} - \frac{1}{2}\left[\frac{\partial \mathrm{Re}\,\vec{\Xi}^{r}}{\partial p}\frac{\partial M_o^{<}}{\partial q} - \frac{\partial \mathrm{Re}\,\vec{\Xi}^{r}}{\partial q}\frac{\partial M_o^{<}}{\partial p}\right] \\
&- \frac{1}{2}\left[\frac{\partial\vec{\Xi}^{<}}{\partial p}\frac{\partial \mathrm{Re}\,M_o^{r}}{\partial q} - \frac{\partial\vec{\Xi}^{<}}{\partial q}\frac{\partial \mathrm{Re}\,M_o^{r}}{\partial p}\right], \qquad (20.50)
\end{aligned}
$$

where all quantities above are LW transforms, i.e., they are functions defined in phase space $(p, q, t) = (\vec{p}, -E, \vec{q}, t)$. In the first line of Eq. (20.50) and the first term of the second line arise from the relation,

$$
\begin{aligned}
&-\frac{1}{2}\left[\frac{\partial\left(\bar{H}+\mathrm{Re}\,\bar{\Sigma}^{r}\right)}{\partial p}\frac{\partial\vec{\mathcal{M}}}{\partial q} - \frac{\partial\left(\bar{H}+\mathrm{Re}\,\bar{\Sigma}^{r}\right)}{\partial q}\frac{\partial\vec{\mathcal{M}}}{\partial p}\right] \\
&= -\frac{1}{m}\left(\vec{p}+\frac{e}{c}\vec{A}(\vec{q})\right)\cdot\nabla_{\vec{q}}\vec{\mathcal{M}} - e\left(\vec{E}+\frac{1}{2c}(\vec{v}\times\vec{B})\right)\cdot\nabla_{\vec{p}}\vec{\mathcal{M}} \\
&\quad - \frac{1}{2}\left[\frac{\partial \mathrm{Re}\,\bar{\Sigma}^{r}}{\partial p}\frac{\partial\vec{\mathcal{M}}}{\partial q} - \frac{\partial \mathrm{Re}\,\bar{\Sigma}^{r}}{\partial q}\frac{\partial\vec{\mathcal{M}}}{\partial p}\right], \qquad (20.51)
\end{aligned}
$$

where the RHS arise from the expression given by the first terms of Eq. (20.30). We have used the gauge: $\vec{A}(\vec{q}) = (\frac{1}{2}\vec{B} \times \vec{q})$, $\vec{B}$ is the uniform external magnetic field.[5] The appearance of the vector potential in Eq. (20.50) seems to suggest that the equation is not gauge invariant. However, since we have already chosen the symmetric gauge of the vector potential, $\vec{A}(q) = \left(\frac{1}{2}\vec{B} \times \vec{q}\right)$, the term $\frac{e}{mc}\vec{A}(\vec{q}) \cdot \nabla_{\vec{q}}\vec{\mathcal{M}}$ does not contribute by virtue of the fact that $\vec{A}(\vec{q})$ is orthogonal to the direction of $\nabla_{\vec{q}}$. This means that only the physical electromagnetic fields contribute in Eq. (20.50). Hence we can replace

$$-\frac{1}{m}\left(\vec{p} + \frac{e}{c}\vec{A}(\vec{q})\right) \cdot \nabla_q\vec{\mathcal{M}} \Rightarrow -\vec{v} \cdot \nabla_{\vec{q}}\vec{\mathcal{M}}, \tag{20.52}$$

in Eq. (20.50) to cast in gauge-invariant form.

The first line in the RHS of Eq. (20.50) is the Boltzmann equation for the magnetization vector distribution. The second line incorporates the ordinary Hall effect and extra-ordinary Hall effect [164] due to the external magnetic field and effects of spin-orbit coupling. This describes the Lorentz force or ordinary Hall effect as well as the extra-ordinary Hall effect due to spin-orbit coupling [164–166]. Note that $\vec{\mathcal{B}}$ includes the effects of external magnetic field and effective internal magnetic field due to spin-orbit coupling. The third line is the contribution to the Boltzmann equation if the effects of $\mathrm{Re}\,\bar{\Sigma}^r$ and $\mathrm{Re}\,\vec{\mathcal{M}}^r$ are included. The rest of this equation incorporates spin-dependent scattering, $\mathrm{Re}\,\vec{\Xi}^r$, which directly contribute to the torque. The precession of the magnetic moment, $\vec{\mathcal{M}}$, is also correlated with the torque exerted by $\vec{\gamma}$, and similarly for *scattering-in* precession (fluctuation) of $\vec{\mathcal{A}}$ due to torque exerted by $\vec{\Xi}$. The last two lines describe the coupling to the total charge. Thus, spin-dependent scattering terms are capable of describing the *spatio–temporal-dependent* scattering dynamics of various mechanisms of spin relaxations and precession, namely, Elliott–Yafet (spin-dependent impurities and

---

[5]In taking the derivative of $\frac{\partial \bar{H}}{\partial q}$, we make the expansion of kinetic energy terms in $\bar{H} \Rightarrow \frac{\vec{p}^2}{2m} + \frac{e}{mc}\vec{p} \cdot \vec{A} + \left(\frac{e}{mc}\right)^2 \vec{A} \cdot \vec{A}$ using the symmetric Landau gauge: $\vec{A} = \frac{1}{2}(\vec{B} \times \vec{q})$. Then in taking the gradient with respect to $\vec{q}$, the $\vec{A} \cdot \vec{A}$ term does not contribute and we are left with,

$$\begin{aligned}\frac{e}{mc}\vec{p} \cdot \frac{\partial}{\partial \vec{q}}\vec{A} &= \frac{e}{mc}\vec{p} \cdot \frac{\partial}{\partial \vec{q}}\frac{1}{2}(\vec{B} \times \vec{q}) \\ &= -\frac{e}{mc}\frac{1}{2}\vec{p} \cdot \frac{\partial}{\partial \vec{q}}(\vec{q} \times \vec{B}) = -\frac{e}{mc}\frac{1}{2}\vec{p} \cdot (\hat{I} \times \vec{B}) \\ &= -\frac{e}{mc}\frac{1}{2}(\vec{p} \times \vec{B}) = -\frac{e}{2c}(\vec{v} \times \vec{B})\end{aligned}$$

phonon scatterings), D'yakonov–Perel' (due to electron–electron scatterings, with electrons executing random-walk in momentum space, resulting in the fluctuations of magnitude and direction of spin precession axis). Moreover, it incorporates ordinary and extra-ordinary Hall effect [164] and as well all non-classical terms for describing magnetic systems of interest in spintronics and micromagnetics. The Bir–Aronov–Pikus scattering (due to the interaction with holes in valence band, serving as a spin sink), and the hyperfine interactions are not relevant in our present treatment.

20.5 Concluding remarks

Until now the treatments of spin transport are mainly focused on scattering mechanisms, based on quasi-classical steady-state conditions and/or where non-uniformity in real space is often not considered, common in micromagnetics. On the other hand for spintronic transistor devices, strong inhomogeneity in real space is of prime considerations. The calculations of the self energy Σ in the literature often have the goal of obtaining the relaxation times to be used in classical kinetic equations.

A self-energy spinor calculation of the many-body effects in spin-orbit coupling has been given by Rajagopal [167]. A first principle *ab initio* calculation of the self-energies has been given by Zhukov and Chulkov [168] for the Elliot-Yafet mechanism in Al, Cu, Au, Nb, and Ta, and by Mower, Vignale, and Tokatly [169] for D'yakonov–Perel' spin relaxation mechanism in photo-excited electron–hole liquid in intrinsic semiconductors exhibiting spin-split band structure.

Our derivation of the non-equilibrium spin magnetization QDF transport (SMQDFT) equations is based on the spinor non-equilibrium Green's function which is decomposed into charge (particle density) and spin magnetization correlations that are highly coupled as SMQDFT equations in phase space, for adapting to numerical simulations. The resulting SMQDFT equations should provide the fundamental framework for carrying out the numerical simulation of the switching or time-dependent performance analyses of spintronic devices, such as those that make use of thin insulating layers between conducting metal layers (this heterostructure configuration solves the mismatch problem between the conductivities of metal layers which has hindered achieving high spin-injection efficiency). Potential barriers of up to 1 eV exist in promising spintronic devices, exemplified by the spin valve and magnetic tunnel transistors [170].

Chapter 21

Multi-Band Quantum Transport: Spin, Pseudo-Spin, and Charge Coupling

21.1 Introduction

The need to analyze the ultrafast-switching-speed and power-dissipation (i.e., speed-power product) performance of nanoelectronic devices has ushered the extension of the classical Boltzmann transport equation to a fully time-dependent and highly-nonlinear non-equilibrium quantum distribution function (QDF) transport equations for charge carriers. This has been achieved through the use of non-equilibrium Green's function, obtained either by the time-contour quantum field formulation of Schwinger, [102] Keldysh, [106] and Kadanoff and Baym, [107] or by the real-time quantum superfield formulation of Buot, coupled with his lattice Weyl (LW) transformation technique in Part 3 of this book (see also [171, 172]). This extension to QDF transport equation has proved to be highly crucial in discovering autonomous THz current oscillations in resonant tunneling devices through numerical simulations, and in resolving controversial issues concerning the highly-nonlinear and bistable current–voltage characteristics found in the experiments [11, 14]. Indeed, the phase space QDF kinetic approach has so far been the most successful technique in the time-dependent or transient-switching analyses of open and active nanosystem and nanodevices, as evidenced by the numerical work of Jensen and Buot on resonant tunneling heterostructures [11]. Recently, Sakurai and Tanimura [173] employ a more detailed model of phonon scattering using a Caldeira–Leggett Hamiltonian to perform the Wigner distribution function transport simulation of resonant tunneling diode (RTD). Their results confirm the essential role of emitter quantization in the occurrence of autonomous current oscillation in RTD, as originally proposed by Buot *et al.* [112] and in analysis using a generalized semiconductor Bloch equations by Buot [174] more than a decade ago. The presence of double plateaus found in Ref. [173] is very likely due to the presence of two discrete levels within the quantum well, although this aspect was not elucidated in their paper. It is worth mentioning that the quantum phase-space distribution or Wigner function transport approach can also determine the number of quantum levels within the quantum well, which can be probed by allowing a wider range of biasing condition at the drain contact [11].

With the exploding surge of interest on spintronics and nanomagnetics [157], there is an equally urgent need for a fully quantum transport extension of the classical Bloch equation to guide the time-dependent numerical simulation of the speed-power switching performance and reliability analyses of realistic functional spin nanostructures and transistors. The extension of the classical spin density equation or the diffusive Bloch equation for spin transport, [160] which is the analogue of the classical Boltzmann equation for charged-particle transport, to fully space-time dependent and highly-nonlinear non-equilibrium QDF transport equations for the magnetization has been given for a single energy band in Chapter 20 (see also [158]).

The present chapter is a generalization of the treatment used in Chapter 20 to derive the spin and multi-band quantum transport equations (SMQTEs). The multi-band results obtained here, in the electron–hole representation, maybe of considerable interest in exploring the physical characteristics and properties of quantum plasmas in various nanoscale devices. Indeed, quantum plasmas have been found in modern sciences and technologies, such as nanowires, quantum dots, quantum-wells, and semiconductor devices [175]. Moreover, the non-equilibrium processes in quantum plasmas have been extensively investigated by using the linearized quantum hydrodynamic equation encompassing the Bohm potential term. There is as well a strong and emerging interest in the spin magnetization of two particle-specie systems, such as in plasmas [176]. In fact, general interest for the effect of spin magnetization in transport physics of two spinning particle-specie systems, such as electron–positron plasma, is just recently developing [177]. There is also a strong interest the properties of electron–hole plasma in semiconductors of nanoelectronics including the spin magnetization quantum transport properties. All these developments require a multi-band approach to spin magnetization quantum transport equations of two particle-specie systems.

21.1.1 *Background and objective*

An immense activity with time-dependent spin magnetization equation, driven by technological applications, occurs in the field of micromagnetics dealing mostly with micro-domain wall dynamics. The so-called classical Landau–Lifshitz–Gilbert equation, with its various modifications and phenomenological damping terms, has become the major player in the analyses of the dynamics of the magnetization field [161, 163, 165]. In magnetoelectronics and spintronics, the mutual dependence of electron transport properties and magnetic properties have resulted in various phenomena referred to as giant magnetoresistance [178], spin modulator [179] based on spin precession with or without magnetic field (i.e., based on spin-orbit coupling), and spin transfer torque [180, 181] which have become springboards of several novel spin-based device concepts.

In analyzing the current-induced spin orientation of electrons in semiconductors, with spin-orbit coupling, D'yakonov–Perel' [174] used the following transport equation for the spin spatial-density vector, $\vec{S}$, as

$$\frac{\partial}{\partial t}\vec{S} = \left[-\vec{\nabla}\cdot\vec{Q} - \frac{\vec{S}}{\tau_s}\right] + \vec{\Omega}\times\vec{S}, \tag{21.1}$$

where $\vec{Q}$ is the spin flux density,

$$\vec{\Omega} = \frac{1}{\hbar}\mu_B g\vec{B}, \tag{21.2}$$

μ_B is the Bohr magneton, g is the g-factor for electrons, and $\vec{B}$ is the effective magnetic field. Earlier, in the 1950s Torrey [159] employed the classical Bloch equations for spin systems in the diffusive regime as

$$\frac{\partial\vec{M}}{\partial t} = [D\nabla^2\vec{M} - \vec{M}_\tau] + \gamma\vec{M}\times\vec{B}, \text{ where } \vec{M}_\tau = \left\{\frac{M_x}{\tau_2}, \frac{M_y}{\tau_2}, \frac{M_z}{\tau_1}\right\}, \tag{21.3}$$

where D is the diffusion coefficient, τ_1 is the spin-magnetization relaxation characteristic time, and τ_2 is the spin-dephasing characteristic time for the M_x and M_y components. Most of the works that follows on spin transport also make use of classical spin transport equations. These classical techniques are no longer valid for the analyses of the ultrafast switching speed and power-dissipation performance of the emerging spintronic devices.

What these classical treatments have earlier shown is that spin transport consist of terms similar to particle transport, i.e., with spin-independent transport parameters, such as the diffusion coefficient, D, and relaxation times in Eq. (21.3), and terms describing the torques in the system.[1] However, the spin-charge coupling for a single band transport [182], and

[1]Gilbert [161] has shown that the damping terms can contribute to the torque in the system. In quantum field theoretical language this means that the spin-dependent self-energies will contribute to the torque in the system.

spin/pseudo-spin/charge interaction [183] in multi-band transport are lacking in the equations or at best not fully treated self-consistently within these classical and semiclassical treatments [184]. Moreover, quantum transport aspects needed for the performance analyses of spin-based nanodevices are absent in these earlier treatments.

As in Chapter 20, the purpose of this chapter is to extend the above classical magnetization transport equations to a fully time-dependent and highly-nonlinear non-equilibrium spin magnetization QDF transport equations in phase space, based on non-equilibrium Green's function technique, on the same level as in the extension of the classical Boltzmann kinetic equation to full QDF transport equation for electrons. In the single-band spin magnetization QDF transport equations of Chapter 20, the spin-charge coupling is explicitly incorporated.

In deriving the multi-band SMQTEs, the following procedure is employed. (1) Firstly, we cast quantities with Pauli–Dirac or real spin degrees of freedom into their canonical expressions in terms of Pauli matrices and identity matrix. The coefficient of the Pauli matrices define the real-spin magnetization density, whereas the coefficient of the identity matrix defines a "real-spin scalar", i.e., independent of the real-spin degrees of freedom, but still dependent on the band indices. (2) Secondly, the real-spin scalar quantities are expressed in terms of Pauli matrices and identity matrix. The coefficient of the Pauli matrices define the pseudo-spin magnetization density, whereas the coefficient of the identity matrix defines an absolute scalar, independent of both the real-spin degrees of freedom and energy band indices, which therefore represents the total number of particle density and determines the spatio-temporal electronic charge of the system. The general idea is to transform quantities characterized by two discrete indices into a vector and scalar quantities. This is done for both the spin and band indices in this chapter.

In our present multi-band generalization, we found that besides the highly-coupled transport equations for the Pauli–Dirac or real-spin density, nonlinear coupling to the pseudo-spin density is also present in all of the spin magnetization quantum transport equations for coupled electrons and holes. Conversely, the pseudo-spin magnetization quantum transport equations incorporate the pseudo-spin/real-spin coupling, as well as the pseudo-spin coupling with the total charge in the system. The resulting quantum transport equations are summarily cast in more physically transparent forms in Section 21.5.

21.2 Non-equilibrium quantum superfield theory

Our starting point is the general quantum transport expressions for fermions as obtained from the real-time quantum superfield theoretical formulation of Part 3, namely,

$$\begin{aligned} i\hbar\left(\frac{\partial}{\partial t_1}+\frac{\partial}{\partial t_2}\right)G^{\gtrless} &= \left[vG^{\gtrless}-G^{\gtrless}v^T\right] \\ &\quad+\left[\Sigma^r G^{\gtrless}-G^{\gtrless}\Sigma^a\right]+\left[\Sigma^{\gtrless}G^a-G^r\Sigma^{\gtrless}\right] \\ &\quad+\left[\Delta^r_{hh}g^{\gtrless}_{ee}-g^{\gtrless}_{hh}\Delta^a_{ee}\right] \\ &\quad+\left[\Delta^{\gtrless}_{hh}g^a_{ee}-g^r_{hh}\Delta^{\gtrless}_{ee}\right]. \end{aligned} \tag{21.4}$$

The last two brackets account for the Cooper pairings between Fermions of the same specie. These do not concern us here (their corresponding transport equations are important in non-equilibrium superconductivity). In what follows we will drop these last two brackets of the RHS of Eq. (21.4).[2]

21.3 Multi-band quantum transport equations

In the absence of pairing between fermions of the same-specie Eq. (21.4) becomes, by separately writing the discrete energy-band quantum-label arguments, as

$$\begin{aligned} &i\hbar\left(\frac{\partial}{\partial t_1}+\frac{\partial}{\partial t_2}\right)G^{\lessgtr}_{\alpha\beta}(12) \\ &\quad=\left[v_\alpha\delta_{\alpha\gamma}\delta_{1\bar{2}}G^{\lessgtr}_{\gamma\beta}(\bar{2}2)-G^{\lessgtr}_{\alpha\gamma}(1\bar{2})\,v^T_\gamma\delta_{\gamma\beta}\delta_{\bar{2}2}\right] \\ &\qquad+\left[\Sigma^r_{\alpha\gamma}(1\bar{2})\,G^{\lessgtr}_{\gamma\beta}(\bar{2}2)-G^{\lessgtr}_{\alpha\gamma}(1\bar{2})\,\Sigma^a_{\gamma\beta}(\bar{2}2)\right] \\ &\qquad+\left[\Sigma^{\lessgtr}_{\alpha\gamma}(1\bar{2})\,G^a_{\gamma\beta}(\bar{2}2)-G^r_{\alpha\gamma}(1\bar{2})\,\Sigma^{\lessgtr}_{\gamma\beta}(\bar{2}2)\right], \end{aligned} \tag{21.5}$$

[2]Here, we consider only pairing between fermions of different species, namely, electrons and holes. The pairing potential is derived from the interband self-energies, $\Sigma_{\alpha\beta}\,(\alpha\neq\beta)$, and hence the electron–hole coupling is denoted by $\Delta^{e-h}_{\alpha\beta}$ in the electron–hole picture or "defect" representation. Remarkably, interband pairing has been treated in the context of quantum chromodynamics QCD and cold atoms, heavy fermions, cuprates, and BCS superconductivity [185–187]. In heavy fermions, the interband pairing occurs if two Fermi surfaces arising from different bands are very close to one other. In QCD and cold atoms the pairing is between particles with very different masses.

where the Greek subscript indices correspond to discrete energy-band indices, and the numeral indices correspond to the two-point space–time arguments. In what follows we will treat the two-band model of a semiconductor and replace Greek subscripts by v and c, the valence and conduction energy-band quantum labels, respectively.

21.3.1 *Electron–hole picture*

In Bloch electron dynamics of multi-band systems, it is more convenient to employ the electron–hole picture or defect representation. The transport equation for the conduction-band electrons in the electron–hole or "defect" representation reduces to

$$
\begin{aligned}
i\hbar &\left(\frac{\partial}{\partial t_1}+\frac{\partial}{\partial t_2}\right) G_{cc}^{e-h,<}(12) \\
&= \left[v_c(1\xi)\, G_{cc}^{<}(\xi 2) - G_{cc}^{<}(1\xi)\, v_c^{T}(\xi 2)\right] \\
&\quad + \left[\Sigma_{cc}^{r}(1\xi)\, G_{cc}^{<}(\xi 2) - G_{cc}^{<}(1\xi)\, \Sigma_{cc}^{a}(\xi 2)\right] \\
&\quad + \left[\Sigma_{cc}^{<}(1\xi)\, G_{cc}^{a}(\xi 2) - G_{cc}^{r}(1\xi)\, \Sigma_{cc}^{<}(\xi 2)\right] \\
&\quad + \left[\Delta_{ee,cv}^{e-h,r}(1\xi)\, g_{hh,vc}^{e-h,<}(\xi 2) - g_{ee,cv}^{e-h,<}(1\xi)\, \Delta_{hh,vc}^{e-h,a}(\xi 2)\right] \\
&\quad + \left[\Delta_{ee,cv}^{e-h,<}(1\xi)\, g_{hh,vc}^{e-h,a}(\xi 2) - g_{ee,cv}^{e-h,r}(1\xi)\, \Delta_{hh,vc}^{e-h,<}(\xi 2)\right],
\end{aligned}
\tag{21.6}
$$

where the off-diagonal self-energies, $\Sigma_{\alpha\beta}$, and Green's functions, $G_{\alpha\beta}$ are denoted by Δ^{e-h} and g^{e-h}, respectively, in analogy to the Δ-function and anomalous Green's function, $\mathcal{F}$, i.e., $g^{e-h} \Rightarrow \mathcal{F}$ in the theory of superconductivity [185–187]. In Eq. (21.6), there is neither Cooper pairing between electrons nor between holes. Moreover, the electron and electron–hole pictures coincides for the conduction band transport equation.

We have the mapping between electron picture and electron–hole picture using Table 21.1 [12].

We also have Table 21.2, which can also be similarly applied to the self-energies.

From the first table, we see that the hole density, $-i\hbar G_{vv}^{e-h,<}(12)$ correspond to $i\hbar G_{vv}^{>T}(12)$ in the electron energy-band representation. Thus, to obtain the equation for the hole density given by $-i\hbar G_{vv}^{e-h,<}(12)$, one may look for the equation for $i\hbar G_{vv}^{>T}(12)$ in the general non-equilibrium formulation of electrons. The equation for $i\hbar G_{vv}^{>T}(12)$ can be obtained from the equation for $G_{vv}^{>}$, by taking the complex conjugate of the above equation

Table 21.1. Electron to electron–hole picture.

electron picture	$\langle e-\text{field}\rangle$	$\langle e-h\text{ field}\rangle$	$\mathbf{e}-\mathbf{h}$ picture
$-i\hbar G^{<}_{vv}(12)$	$\langle\psi^{\dagger}_{v}(2)\,\psi_{v}(1)\rangle$	$\langle\phi_{v}(2)\,\phi^{\dagger}_{v}(1)\rangle$	$i\hbar G^{h,>T}_{vv}(12)$
$-i\hbar G^{<}_{vc}(12)$	$\langle\psi^{\dagger}_{c}(2)\,\psi_{v}(1)\rangle$	$\langle\psi^{\dagger}_{c}(2)\,\phi^{\dagger}_{v}(1)\rangle$	$-i\hbar g^{e-h,<}_{ee,vc}(12)$
$-i\hbar G^{<}_{cv}(12)$	$\langle\psi^{\dagger}_{v}(2)\,\psi_{c}(1)\rangle$	$\langle\phi_{v}(2)\,\psi_{c}(1)\rangle$	$-i\hbar g^{e-h,<}_{hh,cv}(12)$
$-i\hbar G^{<}_{cc}(12)$	$\langle\psi^{\dagger}_{c}(2)\,\psi_{c}(1)\rangle$	$\langle\psi^{\dagger}_{c}(2)\,\psi_{c}(1)\rangle$	$-i\hbar G^{e-h,<}_{cc}(12)$
$i\hbar G^{>T}_{vv}(12)$	$\langle\psi_{v}(2)\,\psi^{\dagger}_{v}(1)\rangle$	$\langle\phi^{\dagger}_{v}(2)\,\phi_{v}(1)\rangle$	$-i\hbar G^{h,<}_{vv}(12)$
$i\hbar G^{>}_{vv}(12)$	$\langle\psi_{v}(1)\,\psi^{\dagger}_{v}(2)\rangle$	$\langle\phi^{\dagger}_{v}(1)\,\phi_{v}(2)\rangle$	$-i\hbar G^{h,<T}_{vv}(12)$

Table 21.2. Similar relations for self-energies.

(a)		(b)	
electron picture	$\mathbf{e}-\mathbf{h}$ picture	**electron** picture	$\mathbf{e}-\mathbf{h}$ picture
$G^{r}_{vv}(12)$	$-G^{e-h,aT}_{vv}(12)$	$G^{a}_{vv}(12)$	$-G^{e-h,rT}_{vv}(12)$
$G^{r}_{vc}(12)$	$g^{e-h,r}_{ee,vc}(12)$	$G^{a}_{vc}(12)$	$g^{e-h,a}_{ee,vc}(12)$
$G^{r}_{cv}(12)$	$g^{e-h,r}_{hh,cv}(12)$	$G^{a}_{cv}(12)$	$g^{e-h,a}_{hh,cv}(12)$
$G^{r}_{cc}(12)$	$G^{e-h,r}_{cc}(12)$	$G^{a}_{cc}(12)$	$G^{e-h,a}_{cc}(12)$

and making use of the relation

$$G^{>\dagger}(12) = -G^{>T}(12)\,. \tag{21.7}$$

We have the transport equation for $G^{>}(12)$ given by

$$\begin{aligned} i\hbar\left(\frac{\partial}{\partial t_1}+\frac{\partial}{\partial t_2}\right)&G^{>}_{vv}(12)\\ =&\left[v^{T}_{vv}(1\xi)\,G^{>}_{vv}(\xi 2)-G^{>}_{vv}(1\xi)\,v_{vv}(\xi 2)\right]\\ &+\left[\Sigma^{r}_{vv}(1\xi)\,G^{>}_{vv}(\xi 2)-G^{>}_{vv}(1\xi)\,\Sigma^{a}_{vv}(\xi 2)\right]\\ &+\left[\Sigma^{r}_{vc}(1\xi)\,G^{>}_{cv}(\xi 2)-G^{>}_{vc}(1\xi)\,\Sigma^{a}_{cv}(\xi 2)\right]\\ &+\left[\Sigma^{>}_{vv}(1\xi)\,G^{a}_{vv}(\xi 2)-G^{r}_{vv}(1\xi)\,\Sigma^{>}_{vv}(\xi 2)\right]\\ &+\left[\Sigma^{>}_{vc}(1\xi)\,G^{a}_{cv}(\xi 2)-G^{r}_{vc}(1\xi)\,\Sigma^{>}_{cv}(\xi 2)\right] \end{aligned} \tag{21.8}$$

Taking the complex conjugate of Eq. (21.8), we obtain

$$\begin{aligned} i\hbar\left(\frac{\partial}{\partial t_1}+\frac{\partial}{\partial t_2}\right)&G^{>T}_{vv}(12)\\ =&\left[-G^{>}_{vv}(2\xi)\,v^{T}_{vv}(\xi 1)+v^{T}_{vv}(\xi 2)\,G^{>T}_{vv}(1\xi)\right] \end{aligned}$$

$$
\begin{aligned}
&+\left[-G^{>}_{vv}(2\xi)\,\Sigma^{a}_{vv}(\xi 1)+\Sigma^{r}_{vv}(2\xi)\,G^{>}_{vv}(\xi 1)\right]\\
&+\left[-G^{>}_{cv}(2\xi)\,\Sigma^{a}_{vc}(\xi 1)+\Sigma^{r}_{cv}(2\xi)\,G^{>}_{vc}(\xi 1)\right]\\
&+\left[-G^{r}_{vv}(2\xi)\,\Sigma^{>}_{vv}(\xi 1)+\Sigma^{>}_{vv}(2\xi)\,G^{a}_{vv}(\xi 1)\right]\\
&+\left[-G^{r}_{cv}(2\xi)\,\Sigma^{>}_{vc}(\xi 1)+\Sigma^{>}_{cv}(2\xi)\,G^{a}_{vc}(\xi 1)\right].
\end{aligned}
\tag{21.9}
$$

Applying the relation

$$
-G^{>T}_{vv}(12)=-G^{>}_{vv}(21)=G^{e-h,<}_{vv}(12),
$$

and going entirely to the defect representation for the rest of the terms, we obtain for the time evolution of the hole density without Cooper pairing between holes.

$$
\begin{aligned}
&-i\hbar\left(\frac{\partial}{\partial t_1}+\frac{\partial}{\partial t_2}\right)G^{e-h,<}_{vv}(12)\\
&=-\left[-v_{vv}(1\xi)\,G^{e-h,<}_{vv}(\xi 2)+G^{e-h,<}_{vv}(1\xi)\,v^{T}_{vv}(\xi 2)\right]\\
&\quad-\left[-\Sigma^{e-h,r}_{vv}(1\xi)\,G^{e-h,<}_{vv}(\xi 2)+G^{e-h,<}_{vv}(1\xi)\,\Sigma^{e-h,a}_{vv}(\xi 2)\right]\\
&\quad-\left[-\Sigma^{e-h,r}_{vc}(1\xi)\,G^{e-h,<}_{cv}(\xi 2)+G^{e-h,<}_{vc}(1\xi)\,\Sigma^{e-h,a}_{cv}(\xi 2)\right]\\
&\quad-\left[-\Sigma^{e-h,<}_{vv}(1\xi)\,G^{e-h,a}_{vv}(\xi 2)+G^{e-h,r}_{vv}(1\xi)\,\Sigma^{e-h,<}_{vv}(\xi 2)\right]\\
&\quad-\left[-\Sigma^{e-h,<}_{vc}(1\xi)\,G^{e-h,a}_{cv}(\xi 2)+G^{e-h,r}_{vc}(1\xi)\,\Sigma^{e-h,<}_{cv}(\xi 2)\right].
\end{aligned}
\tag{21.10}
$$

Note that in Eq. (21.10), we have effectively rearranged the numerical indices in Eq. (21.9). The single-particle Hamiltonian v_{vv} is a delta-function in its two arguments, unlike the correlation functions and self-energies. Replacing the off-diagonal self-energies, $\Sigma_{\alpha\beta}$, and Green's functions, $G_{\alpha\beta}$ by Δ^{e-h} and g^{e-h}, respectively, as was done in Eq. (21.6), allow us to rewrite the equation as

$$
\begin{aligned}
&i\hbar\left(\frac{\partial}{\partial t_1}+\frac{\partial}{\partial t_2}\right)G^{e-h,<}_{vv}(12)\\
&=-\left[v_{vv}(1\xi)\,G^{e-h,<}_{vv}(\xi 2)-G^{e-h,<}_{vv}(1\xi)\,v^{T}_{vv}(\xi 2)\right]\\
&\quad-\left[\Sigma^{e-h,r}_{vv}(1\xi)\,G^{e-h,<}_{vv}(\xi 2)-G^{e-h,<}_{vv}(1\xi)\,\Sigma^{e-h,a}_{vv}(\xi 2)\right]
\end{aligned}
$$

$$
\begin{aligned}
&- \left[\Delta^{e-h,r}_{ee,vc}(1\xi)\, g^{e-h,<}_{hh,cv}(\xi 2) - g^{e-h,<}_{ee,vc}(1\xi)\, \Delta^{e-h,a}_{hh,cv}(\xi 2)\right] \\
&- \left[\Sigma^{e-h,<}_{vv}(1\xi)\, G^{e-h,a}_{vv}(\xi 2) - G^{e-h,r}_{vv}(1\xi)\, \Sigma^{e-h,<}_{vv}(\xi 2)\right] \\
&- \left[\Delta^{e-h,<}_{ee,vc}(1\xi)\, g^{e-h,a}_{hh,cv}(\xi 2) - g^{e-h,r}_{ee,vc}(1\xi)\, \Delta^{e-h,<}_{hh,cv}(\xi 2)\right]. \qquad (21.11)
\end{aligned}
$$

Compared with Eq. (21.6) for the conduction band, one can readily see that for flat bands, or atomic limit, the remaining terms in these two equations are equal, i.e.,

$$
\begin{aligned}
i\hbar\left(\frac{\partial}{\partial t_1}+\frac{\partial}{\partial t_2}\right) G^{e-h,<}_{vv}(12) &\Rightarrow -\left[\Delta^{e-h,r}_{ee,vc}\, g^{e-h,<}_{hh,cv} - g^{e-h,<}_{ee,vc}\, \Delta^{e-h,a}_{hh,cv}\right] \\
&= \left[\Delta^{e-h,r}_{hh,cv}\, g^{e-h,<}_{ee,vc} - g^{e-h,<}_{hh,cv}\, \Delta^{e-h,a}_{ee,vc}\right], \\
i\hbar\left(\frac{\partial}{\partial t_1}+\frac{\partial}{\partial t_2}\right) G^{e-h,<}_{cc}(12) &\Rightarrow \left[\Delta^{e-h,r}_{hh,cv}\, g^{e-h,<}_{ee,vc} - g^{e-h,<}_{hh,cv}\, \Delta^{e-h,a}_{ee,vc}\right], \qquad (21.12)
\end{aligned}
$$

since $\Delta^{e-h,r}_{ee,vc} = \Delta^{e-h,a}_{ee,vc}$, and $\Delta e^{e-h,a}_{hh,cv} = \Delta^{e-h,r}_{hh,cv}$. This equality simply states that the rate of change of the holes in the valence band is the same as the rate of change of the electrons in the conduction band, or the rate of electron creation in conduction band equals the rate of hole creation in the valence band, and *vice versa*.

If we use the electron picture to write the transport equation for the valence band, the resulting equation would be similar to Eq. (21.11) without the superscript $^{e-h}$ and with the right-hand side of that equation multiplied by -1. The above equation reduces to

$$
\begin{aligned}
&i\hbar\left(\frac{\partial}{\partial t_1}+\frac{\partial}{\partial t_2}\right) G^{<}_{vv}(12) \\
&= \left[v_{vv}(1\xi)\, G^{<}_{vv}(\xi 2) - G^{<}_{vv}(1\xi)\, v^{T}_{vv}(\xi 2)\right] \\
&\quad + \left[\Sigma^{r}_{vv}(1\xi)\, G^{<}_{vv}(\xi 2) - G^{<}_{vv}(1\xi)\, \Sigma^{a}_{vv}(\xi 2)\right] \\
&\quad + \left[\Sigma^{<}_{vv}(1\xi)\, G^{a}_{vv}(\xi 2) - G^{r}_{vv}(1\xi)\, \Sigma^{<}_{vv}(\xi 2)\right] \\
&\quad + \left[g^{>}_{hh,vc}(2\xi)\, \Delta^{a}_{ee,cv}(\xi 1) - \Delta^{r}_{hh,vc}(2\xi)\, g^{>}_{ee,cv}(\xi 1)\right] \\
&\quad + \left[g^{r}_{hh,vc}(2\xi)\, \Delta^{>}_{ee,cv}(\xi 1) - \Delta^{>}_{hh,vc}(2\xi)\, g^{a}_{ee,cv}(\xi 1)\right]. \qquad (21.13)
\end{aligned}
$$

Then one can also easily see from Eq. (21.13) that for flat-band case, the rate of change of electrons in the conduction band and that of the valence band would have opposite sign, i.e., the rate of creation of electrons in the conduction band occurs at the expense of equal rate of loss of electrons in the valence band.

Later, we will also use the electron picture to investigate the pseudo-spin transport equations. There, the total charge determines the total net charge to be used in the Poisson equation. Of course the total net charge at each point of the phase-space is the charge represented by the total number of electrons in the conduction and valence band minus the total positive background charge of the crystal lattice.

The equations for the interband or polarization terms, $g_{hh}^{e-h,<}$, $g_{ee}^{e-h,<}$, etc. can be obtained from the equations for $G_{\alpha\beta}^{<}(12)$, where $\alpha \neq \beta$. In the electron–hole or defect representation, the equation for $g_{ee,cv}^{e-h,<}$ is determined from the general superfield formalism given by the equation for $G_{cv}^{<}(12)$. Upon transforming to the electron–hole picture, the change of polarization due to the destruction of electron–hole pairs brought about by Auger recombination, optical de-excitation, and other recombination processes is given in the defect representation as

$$\begin{aligned}
i\hbar&\left(\frac{\partial}{\partial t_1}+\frac{\partial}{\partial t_2}\right)g_{hh,cv}^{e-h,<}(12)\\
&=\left[v_{cc}(\xi 2)\,g_{hh,cv}^{e-h,<}(1\xi)-g_{hh,cv}^{e-h,<}(\xi 2)\,v_{vv}^{T}(1\xi)\right]\\
&\quad+\left[\Sigma_{cc}^{e-h,r}(1\xi)\,g_{hh,cv}^{e-h,<}(\xi 2)-G_{cc}^{e-h,<}(1\xi)\,\Delta_{hh,cv}^{e-h,a}(\xi 2)\right]\\
&\quad+\left[\Sigma_{cc}^{e-h,<}(1\xi)\,g_{hh,cv}^{e-h,a}(\xi 2)-G_{cc}^{e-h,r}(1\xi)\,\Delta_{hh,cv}^{e-h,<}(\xi 2)\right]\\
&\quad+\left[-\Delta_{hh,cv}^{e-h,r}(1\xi)\,G_{vv}^{e-h,>T}(\xi 2)+g_{hh,cv}^{e-h,<}(1\xi)\,\Sigma_{vv}^{e-h,rT}(\xi 2)\right]\\
&\quad+\left[-\Delta_{hh,cv}^{e-h,<}(1\xi)\,G_{vv}^{e-h,rT}(\xi 2)+g_{hh,cv}^{e-h,r}(1\xi)\,\Sigma_{vv}^{e-h,>T}(\xi 2)\right]. \qquad (21.14)
\end{aligned}$$

Similarly, upon going over to the defect representation of the reverse process, i.e., the change of polarization due to the creation of electron–hole pairs by impact ionization, optical excitation, Zener tunneling, and other excitation processes is therefore given by

$$\begin{aligned}
i\hbar&\left(\frac{\partial}{\partial t_1}+\frac{\partial}{\partial t_2}\right)g_{ee,vc}^{e-h,<}(12)\\
&=\left[v_{vv}(1\xi)\,g_{ee,vc}^{e-h,<}(\xi 2)-g_{ee,vc}^{e-h,<}(1\xi)\,v_{c}^{T}(\xi 2)\right]\\
&\quad+\left[-\Sigma_{vv}^{e-h,aT}(1\xi)\,g_{ee,vc}^{e-h,<}(\xi 2)+G_{vv}^{e-h,>T}(1\xi)\,\Delta_{ee,vc}^{e-h,a}(\xi 2)\right]\\
&\quad+\left[-\Sigma_{vv}^{e-h,>T}(1\xi)\,g_{ee,vc}^{e-h,a}(\xi 2)+G_{vv}^{e-h,aT}(1\xi)\,\Delta_{ee,vc}^{e-h,<}(\xi 2)\right]
\end{aligned}$$

$$+\left[\Delta_{ee,vc}^{e-h,r}(1\xi)\,G_{cc}^{e-h,<}(\xi 2)-g_{ee,vc}^{e-h,<}(1\xi)\,\Sigma_{cc}^{e-h,a}(\xi 2)\right]$$
$$+\left[\Delta_{ee,vc}^{e-h,<}(1\xi)\,G_{cc}^{e-h,a}(\xi 2)-g_{ee,vc}^{e-h,r}(1\xi)\,\Sigma_{cc}^{e-h,<}(\xi 2)\right]. \qquad (21.15)$$

Note the presence of the "transposed" terms, namely, $G_{vv}^{e-h,>T}(\xi 2)$, $\Sigma_{vv}^{e-h,rT}(\xi 2)$, $G_{vv}^{e-h,rT}(\xi 2)$, and $\Sigma_{vv}^{e-h,>T}(\xi 2)$ in Eq. (21.14) and $\Sigma_{vv}^{e-h,aT}(1\xi)$, $G_{vv}^{e-h,aT}(1\xi)$, $G_{vv}^{e-h,>T}(1\xi)$, and $\Sigma_{vv}^{e-h,>T}(1\xi)$ in Eq. (21.15) of the interband "pairing" correlation functions. More will said about this later.

In the calculation of the *pseudo-spin* transport equations the transposed quantities, referred to above, in Eqs. (21.14) and (21.15) do not enter since we use the electron picture to provide the two states for the electrons, namely the two quantum labels c and v in order to derive the pseudo-spin quantum transport equations. Since there is a one-to-one mapping between the *electron picture* and *electron–hole picture* except for the valence band as shown in Tables 21.1–21.2, all one does to go from electron–hole picture to the electron picture is to change the sign of the right-hand side of Eq. (21.13) and substitute for the transposed quantities their equivalent expressions in the electron picture.

21.4 Spinor correlations for Bloch electrons with spin

In the presence of Pauli–Dirac spin degree of freedom, Eqs. (21.6), (21.13), (21.14), and (21.15) become matrix equations. These are given in Appendices D and D.1.

21.4.1 *Spin canonical* 2 × 2 *matrix forms*

In spintronics, we are interested in the time-dependent evolution of the multi-band polarization and magnetization densities, $S_{\alpha\beta,z} = i\hbar\left(G_{\alpha\beta,\uparrow\uparrow}^{<}-G_{\alpha\beta,\downarrow\downarrow}^{<}\right)$, as these are transported across the device. This leads us to transform all 2×2 matrices into their spin canonical forms, defined here as expansion in terms of the Pauli spin matrices and identity.

First let us partition the *total* non-equilibrium Green's functions (TNEGFs) into 2×2 submatrix components, by virtue of the spin indices, as

$$\begin{pmatrix} G_{cc,\uparrow\uparrow}^{e-h,<}(12) & G_{cc,\uparrow\downarrow}^{e-h,<}(12) & g_{hh,cv,\uparrow\uparrow}^{e-h,<}(12) & g_{hh,cv,\uparrow\downarrow}^{e-h,<}(12) \\ G_{cc,\downarrow\uparrow}^{e-h,<}(12) & G_{cc,\downarrow\downarrow}^{e-h,<}(12) & g_{hh,cv,\downarrow\uparrow}^{e-h,<}(12) & g_{hh,cv,\downarrow\downarrow}^{e-h,<}(12) \\ g_{ee,vc,\downarrow\downarrow}^{e-h,<}(12) & g_{ee,vc,\downarrow\uparrow}^{e-h,<}(12) & G_{vv,\uparrow\uparrow}^{e-h,<}(12) & G_{vv,\uparrow\downarrow}^{e-h,<}(12) \\ g_{ee,vc,\uparrow\downarrow}^{e-h,<}(12) & g_{ee,vc,\uparrow\uparrow}^{e-h,<}(12) & G_{vv,\downarrow\uparrow}^{e-h,<}(12) & G_{vv,\downarrow\downarrow}^{e-h,<}(12) \end{pmatrix} = \begin{pmatrix} (CC) & (CV) \\ (VC) & (VV) \end{pmatrix}, \qquad (21.16)$$

where (CC), (CV), (VC), and (VV) are 2×2 submatrices. We can form four coupled spin magnetization quantum transport equations for each submatrix, giving us 16 coupled magnetization quantum transport equations.

Thus, we first transform the 2×2 matrix variables in Eq. (21.16) into their spin-canonical forms using the Pauli-matrices,

$$(CC) = \frac{1}{2}(S_{cc,o}I + \vec{S}_{cc} \cdot \vec{\sigma}), \tag{21.17}$$

$$(CV) = \frac{1}{2}(S_{cv,o}I + \vec{S}_{cv} \cdot \vec{\sigma}), \tag{21.18}$$

$$(VC) = \frac{1}{2}(S_{vc,o}I + \vec{S}_{vc} \cdot \vec{\sigma}), \tag{21.19}$$

$$(VV) = \frac{1}{2}(S_{vv,o}I + \vec{S}_{vv} \cdot \vec{\sigma}), \tag{21.20}$$

where we dropped the superscript $^<$ in the above spin-canonical form, this is to be understood in what follows unless otherwise specifically specified. The spin canonical expansion in terms of the Pauli matrices basically separates the spin-independent terms from the spin-dependent terms.

In the case of the conduction band, Eq. (21.17), and valence band, Eq. (21.20), the coefficients of the identity matrix, namely, $S_{cc,o}$ and $S_{vv,o}$ represent the charge density of electrons in the conduction band and charge density of holes in the valence band, respectively. Similarly, $S_{cv,o}$ and $S_{vc,o}$ represent the probability density for the annihilation and creation of electron–hole pairs, respectively. The coefficients of the vector, $\vec{\sigma}$, represent the corresponding spin magnetization vectors. Thus the canonical form basically performs the essential function of separating particle charge and spin. We will later show that the coefficients of the identity matrix, dependent only on band indices, given above can also be expressed in terms of the pseudo-spin vector and the total charge in our multi-band system.

To avoid confusion, a few words about the convention adopted for different symbols used in what follows. Basically, the logical scheme for transforming the 2×2 discrete indices into a vector and a scalar goes as follows. Let $F_{\alpha\beta,\sigma\sigma'}$ be a quantity labeled by discrete band and spin indices. First we transform the spin indices into a vector and spin-scalar (i.e., independent of spin degree of freedom), which we denote by $\vec{F}_{\alpha\beta}$ and $F_{\alpha\beta,o}$, respectively. Then we transform the band indices in spin-scalar $F_{\alpha\beta,o}$ into a pseudo-spin vector and absolute scalar, which we denote as $\vec{F}_o$ and $F_{o,o}$, respectively. We sometimes indicate the scalars by a bar, such as $\bar{\Delta}$ for the coupling interaction between bands. We can also transform the i-th components of spin vector with band indices, namely, $(\vec{F}_{\alpha\beta})_i = F_{\alpha\beta,i}$ into a pseudo-spin vector plus pseudo-spin scalar, which we denote by $\vec{F}_{o,i}$ and $F_{o,o,i} \equiv \bar{F}_{o,i} \equiv \bar{F}_i$, respectively, where the index i is the real spin-vector component index. Observe that $\vec{F}_{o,i}$ is a dyadic. Alternatively, we can also transform the band indices in the real-spin vector $\vec{F}_{\alpha\beta}$ into a dyadic $I_i\ \vec{F}_{o,i}$ and a vector, $I_i F_{o,o,i} \equiv I_i \bar{F}_i$, where I_i is a component of *idemfactor* or unit dyadic.

21.4.2 *Intraband non-equilibrium spin magnetization correlation functions*

To establish the notations used here, we give the spin canonical forms of the single particle Hamiltonian, of various correlation functions and self-energies. For the intraband canonical terms, we have for the intraband spin-*subtraces* or spin-*scalar* correlation function for each band,

$$S^{r,a,<}_{cc,o} = (G^{e-h,r,a,<}_{cc,\uparrow\uparrow} + G^{e-h,r,a,<}_{cc,\downarrow\downarrow}), \tag{21.21}$$

$$S^{r,a,<}_{vv,o} = (G^{e-h,r,a,<}_{vv,\uparrow\uparrow} + G^{e-h,r,a,<}_{vv,\downarrow\downarrow}). \tag{21.22}$$

The intraband spin-correlation vector components for electrons are,

$$S^{r,a,<}_{cc,x} = (G^{e-h,r,a,<}_{cc,\downarrow\uparrow} + G^{e-h,r,a,<}_{cc,\uparrow\downarrow}), \tag{21.23}$$

$$iS^{r,a,<}_{cc,y} = (G^{e-h,r,a,<}_{cc,\downarrow\uparrow} - G^{e-h,r,a,<}_{cc,\uparrow\downarrow}), \tag{21.24}$$

$$S^{r,a,<}_{cc,z} = (G^{e-h,r,a,<}_{cc,\uparrow\uparrow} - G^{e-h,r,a,<}_{cc,\downarrow\downarrow}), \tag{21.25}$$

and for holes,

$$S^{r,a,<}_{vv,x} = (G^{e-h,r,a,<}_{vv,\downarrow\uparrow} + G^{e-h,r,a,<}_{vv,\uparrow\downarrow}), \tag{21.26}$$

$$iS^{r,a,<}_{vv,y} = (G^{e-h,r,a,<}_{vv,\downarrow\uparrow} - G^{e-h,r,a,<}_{vv,\uparrow\downarrow}), \tag{21.27}$$

$$S^{r,a,<}_{vv,z} = (G^{e-h,r,a,<}_{vv,\uparrow\uparrow} - G^{e-h,r,a,<}_{vv,\downarrow\downarrow}). \tag{21.28}$$

21.4.3 *Interband non-equilibrium spin correlation functions*

For the interband canonical terms, we have interband spin-*subtraces* and spin-*scalar* correlation functions,

$$S^{r,a,<}_{cv,o} = (g^{e-h,r,a,<}_{hh,cv,\uparrow\uparrow} + g^{e-h,r,a,<}_{hh,cv,\downarrow\downarrow}), \tag{21.29}$$

$$S^{r,a,<}_{vc,o} = (g^{e-h,r,a,<}_{ee,vc,\downarrow\downarrow} + g^{e-h,r,a,<}_{ee,vc,\uparrow\uparrow}), \tag{21.30}$$

and for the interband spin-vector correlation components,

$$S^{r,a,<}_{cv,x} = (g^{e-h,r,a,<}_{hh,cv,\downarrow\uparrow} + g^{e-h,r,a,<}_{hh,cv,\uparrow\downarrow}), \tag{21.31}$$

$$iS^{r,a,<}_{cv,y} = (g^{e-h,r,a,<}_{hh,cv,\downarrow\uparrow} - g^{e-h,r,a,<}_{hh,cv,\uparrow\downarrow}), \tag{21.32}$$

$$S^{r,a,<}_{cv,z} = (g^{e-h,r,a,<}_{hh,cv,\uparrow\uparrow} - g^{e-h,r,a,<}_{hh,cv,\downarrow\downarrow}). \tag{21.33}$$

The corresponding interband reverse process "conjugate" correlations are,

$$S^{r,a,<}_{vc,x} = (g^{e-h,r,a,<}_{ee,vc,\uparrow\downarrow} + g^{e-h,r,a,<}_{ee,vc,\downarrow\uparrow}), \tag{21.34}$$

$$iS^{r,a,<}_{vc,y} = (g^{e-h,r,a,<}_{ee,vc,\uparrow\downarrow} - g^{e-h,r,a,<}_{ee,vc,\downarrow\uparrow}), \tag{21.35}$$

$$S^{r,a,<}_{vc,z} = (g^{e-h,r,a,<}_{ee,vc,\downarrow\downarrow} - g^{e-h,r,a,<}_{ee,vc,\uparrow\uparrow}). \tag{21.36}$$

The relations given by Eqs. (21.21)–(21.36) can of course be inverted to express each spin-dependent correlation function for the two-band model in terms of the real-spin magnetization correlation functions, defined on the left side of these relations.

21.4.4 *Single-particle Hamiltonian spin canonical forms*

We have for the single-particle Hamiltonian for the electrons expressed in canonical forms, assuming the presence of external magnetic field and/or spin-orbit coupling,

$$v_{c,\sigma\sigma'} = \frac{1}{2}(\bar{H}_c\hat{I} + \vec{\mathcal{B}}_c \cdot \vec{\sigma}), \tag{21.37}$$

where the upper bar in $\bar{H}_c$ indicates the subtrace of the 2×2 spin matrix for the conduction band. Similarly, for the holes we have,

$$v_{v,\sigma\sigma'} = \frac{1}{2}(\bar{H}_v\hat{I} + \vec{\mathcal{B}}_v \cdot \vec{\sigma}). \tag{21.38}$$

These lead to the matrix expression for electrons,

$$v_{c,\sigma\sigma'} = \frac{1}{2}\begin{pmatrix} \bar{H}_c + \mathcal{B}_{c,z} & \mathcal{B}_{c,x} - i\mathcal{B}_{c,y} \\ \mathcal{B}_{c,x} + i\mathcal{B}_{c,y} & \bar{H}_c - \mathcal{B}_{c,z} \end{pmatrix}, \tag{21.39}$$

and for holes,

$$v_{v,\sigma\sigma'} = \frac{1}{2}\begin{pmatrix} \bar{H}_v + \mathcal{B}_{v,z} & \mathcal{B}_{v,x} - i\mathcal{B}_{v,y} \\ \mathcal{B}_{v,x} + i\mathcal{B}_{v,y} & \bar{H}_v - \mathcal{B}_{v,z} \end{pmatrix}. \tag{21.40}$$

The components of $\vec{\mathcal{B}}_c$ and $\vec{\mathcal{B}}_v$ are real valued, since the single-particle Hamiltonian, v, is Hermitian. These are the effective magnetic field components, multiplied by $\mu_B g$, which accounts for the external magnetic fields as well as the effects of spin-orbit coupling.

21.4.5 *Canonical forms of spin-dependent electrons self-energy*

Similarly, we have for the spin-canonical forms for the electron self-energies,

$$\Sigma^r_{cc,\sigma\sigma'} = \frac{1}{2}\begin{pmatrix} \bar{\Sigma}^r_{cc} + \Xi^r_{cc,z} & \Xi^r_{cc,x} - i\Xi^r_{cc,y} \\ \Xi^r_{cc,x} + i\Xi^r_{cc,y} & \bar{\Sigma}^r_{cc} - \Xi^r_{cc,z} \end{pmatrix}, \tag{21.41}$$

$$\Sigma^a_{cc,\sigma\sigma'} = \frac{1}{2}\begin{pmatrix} \bar{\Sigma}^a_{cc} + \Xi^a_{cc,z} & \Xi^a_{cc,x} - i\Xi^a_{cc,y} \\ \Xi^a_{cc,x} + i\Xi^a_{cc,y} & \bar{\Sigma}^a_{cc} - \Xi^a_{cc,z} \end{pmatrix}, \tag{21.42}$$

$$\Sigma^<_{cc,\sigma\sigma'} = \frac{1}{2}\begin{pmatrix} \bar{\Sigma}^<_{cc} + \Xi^<_{cc,z} & \Xi^<_{cc,x} - i\Xi^<_{cc,y} \\ \Xi^<_{cc,x} + i\Xi^<_{cc,y} & \bar{\Sigma}^<_{cc} - \Xi^<_{cc,z} \end{pmatrix}. \tag{21.43}$$

21.4.6 *Spin-canonical forms for the holes self-energy*

For the holes, we

$$\Sigma^r_{vv,\sigma\sigma'} = \frac{1}{2}\begin{pmatrix} \bar{\Sigma}^r_{vv} + \Xi^r_{vv,z} & \Xi^r_{vv,x} - i\Xi^r_{vv,y} \\ \Xi^r_{vv,x} + i\Xi^r_{vv,y} & \bar{\Sigma}^r_{vv} - \Xi^r_{vv,z} \end{pmatrix}, \tag{21.44}$$

$$\Sigma^a_{vv,\sigma\sigma'} = \frac{1}{2}\begin{pmatrix} \bar{\Sigma}^a_{vv} + \Xi^a_{vv,z} & \Xi^a_{vv,x} - i\Xi^a_{vv,y} \\ \Xi^a_{vv,x} + i\Xi^a_{vv,y} & \bar{\Sigma}^a_{vv} - \Xi^a_{vv,z} \end{pmatrix}, \tag{21.45}$$

$$\Sigma^<_{vv,\sigma\sigma'} = \frac{1}{2}\begin{pmatrix} \bar{\Sigma}^<_{vv} + \Xi^<_{vv,z} & \Xi^<_{vv,x} - i\Xi^<_{vv,y} \\ \Xi^<_{vv,x} + i\Xi^<_{vv,y} & \bar{\Sigma}^<_{vv} - \Xi^<_{vv,z} \end{pmatrix}. \tag{21.46}$$

21.4.7 *Canonical forms of electron–hole pairing self-energy*

For the electron–hole recombination "self-energy", we have

$$\Delta^{e-h,r}_{hh,cv,\sigma\sigma'} = \frac{1}{2}(\bar{\Delta}^r_{hh,cv} + \vec{\delta}^r_{hh,cv}\cdot\vec{\sigma}), \tag{21.47}$$

$$\Delta^{e-h,<}_{hh,cv,\sigma\sigma'} = \frac{1}{2}(\bar{\Delta}^<_{hh,cv} + \vec{\delta}^<_{hh,cv}\cdot\vec{\sigma}), \tag{21.48}$$

$$\Delta^{e-h,r}_{hh,cv,\sigma\sigma'} = \frac{1}{2}\begin{pmatrix} \bar{\Delta}^r_{hh,cv} + \delta^r_{hh,cv,z} & \delta^r_{hh,cv,x} - i\delta^r_{hh,cv,y} \\ \delta^r_{hh,cv,x} + i\delta^r_{hh,cv,y} & \bar{\Delta}^r_{hh,cv} - \delta^r_{hh,cv,z} \end{pmatrix}, \tag{21.49}$$

$$\Delta^{e-h,<}_{hh,cv,\sigma\sigma'} = \frac{1}{2}\begin{pmatrix} \bar{\Delta}^<_{hh,cv} + \delta^<_{hh,cv,z} & \delta^<_{hh,cv,x} - i\delta^<_{hh,cv,y} \\ \delta^<_{hh,cv,x} + i\delta^<_{hh,cv,y} & \bar{\Delta}^<_{hh,cv} - \delta^<_{hh,cv,z} \end{pmatrix}, \tag{21.50}$$

and for the electron–hole pair generation,

$$\Delta^{e-h,a}_{ee,vc,\sigma\sigma'} = \frac{1}{2}(\bar{\Delta}^a_{ee,vc} + \vec{\delta}^a_{ee,vc}\cdot\vec{\sigma}), \tag{21.51}$$

$$\Delta^{e-h,<}_{ee,vc,\sigma\sigma'} = \frac{1}{2}(\bar{\Delta}^<_{ee,vc} + \vec{\delta}^<_{ee,vc}\cdot\vec{\sigma}), \tag{21.52}$$

$$\Delta^{e-h,a}_{ee,vc,\sigma\sigma'} = \frac{1}{2}\begin{pmatrix} \bar{\Delta}^a_{ee,vc} + \delta^a_{ee,vc,z} & \delta^a_{ee,vc,x} - i\delta^a_{ee,vc,y} \\ \delta^a_{ee,vc,x} + i\delta^a_{ee,vc,y} & \bar{\Delta}^a_{ee,vc} - \delta^a_{ee,vc,z} \end{pmatrix}, \tag{21.53}$$

$$\Delta^{e-h,<}_{ee,vc,\sigma\sigma'} = \frac{1}{2}\begin{pmatrix} \bar{\Delta}^<_{ee,vc} + \delta^<_{ee,vc,z} & \delta^<_{ee,vc,x} - i\delta^<_{ee,vc,y} \\ \delta^<_{ee,vc,x} + i\delta^<_{ee,vc,y} & \bar{\Delta}^<_{ee,vc} - \delta^<_{ee,vc,z} \end{pmatrix}. \tag{21.54}$$

21.5 Multi-band vector and scalar transport equations

We summarize the main results of this chapter by recasting the pertinent equations into more meaningful expressions. We refer the readers to the appendices for the calculations of main results.

21.5.1 *Semiconductor Bloch equations for magnetization vector*

We start by rewriting the RHS of semiconductor Bloch equations for the spin-magnetization vector (Appendix D.2), i.e., the Pauli–Dirac MSMVQTEs into three groups, given in the electron–hole picture, as

$$\begin{aligned} i\hbar\left(\frac{\partial}{\partial t_1}+\frac{\partial}{\partial t_2}\right)\vec{S}^<_{cc} &= +\frac{1}{2}\left[\begin{array}{c} \left[\bar{H}_c, \vec{S}^<_{cc}\right] \\ +\left[\left(\bar{\Sigma}^r_{cc}\right)\vec{S}^<_{cc} - \vec{S}^<_{cc}\left(\bar{\Sigma}^a_{cc}\right)\right] + \left[\bar{\Sigma}^<_{cc}\left(\vec{S}^a_{cc}\right) - \left(\vec{S}^r_{cc}\right)\bar{\Sigma}^<_{cc}\right] \\ +\left[\left(\bar{\Delta}^r_{hh,cv}\right)\vec{S}^<_{vc} - \vec{S}^<_{cv}\left(\bar{\Delta}^a_{ee,vc}\right)\right] + \left[\bar{\Delta}^<_{hh,cv}\,\vec{S}^a_{vc} - \vec{S}^r_{cv}\bar{\Delta}^<_{ee,vc}\right] \end{array}\right] \\ &+\frac{1}{2}\left[\begin{array}{c} +i\left[\vec{\mathcal{B}}_{cc}\times\vec{S}^<_{cc} - \vec{S}^<_{cc}\times\vec{\mathcal{B}}_{cc}\right] \\ +i\left[\left(\vec{\Xi}^r_{cc}\right)\times\vec{S}^<_{cc} - \vec{S}^<_{cc}\times\left(\vec{\Xi}^a_{cc}\right)\right] + i\left[\vec{\Xi}^<_{cc}\times\left(\vec{S}^a_{cc}\right) - \left(\vec{S}^r_{cc}\right)\times\vec{\Xi}^<_{cc}\right] \\ +i\left[\vec{\delta}^r_{hh,cv}\times\vec{S}^<_{vc} - \vec{S}^<_{cv}\times\vec{\delta}^a_{ee,vc}\right] + i\left[\vec{\delta}^<_{hh,cv}\times\vec{S}^a_{vc} - \vec{S}^r_{cv}\times\vec{\delta}^<_{ee,vc}\right] \end{array}\right] \\ &+\frac{1}{2}\left[\begin{array}{c} +\left[\vec{\mathcal{B}}_{cc}, S^<_{cc,o}\right] \\ +\left[\left(\vec{\Xi}^r_{cc}\right)S^<_{cc,o} - S^<_{cc,o}\left(\vec{\Xi}^a_{cc}\right)\right] + \left[\vec{\Xi}^<_{cc}\left(S^a_{cc,o}\right) - \left(S^r_{cc,o}\right)\vec{\Xi}^<_{cc}\right] \\ +\left[\left(\vec{\delta}^r_{hh,cv}\right)S^<_{vc,o} - S^<_{cv,o}\left(\vec{\delta}^a_{ee,vc}\right)\right] + \left[\vec{\delta}^<_{hh,cv}S^a_{vc,o} - S^r_{cv,o}\vec{\delta}^<_{ee,vc}\right] \end{array}\right], \end{aligned} \tag{21.55}$$

$$
\begin{aligned}
&i\hbar\left(\frac{\partial}{\partial t_1}+\frac{\partial}{\partial t_2}\right)\vec{S}^{<}_{cv} \\
&=\frac{1}{2}\left[\begin{array}{c}
\left[\bar{H}_c\vec{S}^{<}_{cv}-\vec{S}^{<}_{cv}\bar{H}_v\right] \\
+\left[\bar{\Sigma}^{r}_{cc}\vec{S}^{<}_{cv}-\vec{S}^{<}_{cc}\bar{\Delta}^{\,a}_{hh,cv}\right]+\left[\bar{\Sigma}^{<}_{cc}\vec{S}^{a}_{cv}-\vec{S}^{r}_{cc}\bar{\Delta}^{<}_{hh,cv}\right] \\
-\left[\bar{\Delta}^{r}_{hh,cv}\vec{S}^{>T}_{vv}-\vec{S}^{<}_{cv}\bar{\Sigma}^{rT}_{vv}\right]-\left[\bar{\Delta}^{<}_{hh,cv}\vec{S}^{rT}_{vv}-\vec{S}^{r}_{cv}\bar{\Sigma}^{>T}_{vv}\right]
\end{array}\right] \\
&+\frac{1}{2}\left[\begin{array}{c}
+i\left[\vec{\mathcal{B}}_{cc}\times\vec{S}^{<}_{cv}-\vec{S}^{<}_{cv}\times\vec{\mathcal{B}}_{vv}\right] \\
+i\left[\vec{\Xi}^{r}_{cc}\times\vec{S}^{<}_{cv}-\vec{S}^{<}_{cc}\times\vec{\delta}^{\,a}_{hh,cv}\right]+i\left[\vec{\Xi}^{<}_{cc}\times\vec{S}^{a}_{cv}-\vec{S}^{r}_{cc}\times\vec{\delta}^{<}_{hh,cv}\right] \\
-i\left[\vec{\delta}^{r}_{hh,cv}\times\vec{S}^{,>T}_{vv}-\vec{S}^{<}_{cv}\times\vec{\Xi}^{rT}_{vv}\right]-i\left[\vec{\delta}^{<}_{hh,cv}\times\vec{S}^{rT}_{vv}-\vec{S}^{r}_{cv}\times\vec{\Xi}^{>T}_{vv}\right]
\end{array}\right] \\
&+\frac{1}{2}\left[\begin{array}{c}
+\left[\vec{\mathcal{B}}_{cc}S^{<}_{cv,o}-S^{<}_{cv,o}\vec{\mathcal{B}}_{vv}\right] \\
+\left[\vec{\Xi}^{r}_{cc}S^{<}_{cv,o}-S^{<}_{cc,o}\vec{\delta}^{\,a}_{hh,cv}\right]+\left[\vec{\Xi}^{<}_{cc}S^{a}_{cv,o}-S^{r}_{cc,o}\vec{\delta}^{<}_{hh,cv}\right] \\
-\left[\vec{\delta}^{r}_{hh,cv}S^{>T}_{vv,o}-S^{<}_{cv,o}\vec{\Xi}^{rT}_{vv}\right]-\left[\vec{\delta}^{<}_{hh,cv}S^{rT}_{vv,o}-S^{r}_{cv,o}\vec{\Xi}^{>T}_{vv}\right]
\end{array}\right],
\end{aligned}
\tag{21.56}
$$

$$
\begin{aligned}
&i\hbar\left(\frac{\partial}{\partial t_1}+\frac{\partial}{\partial t_2}\right)\vec{S}^{<}_{vc} \\
&=\frac{1}{2}\left[\begin{array}{c}
\left[\bar{H}_v\vec{S}^{<}_{vc}-\vec{S}^{<}_{vc}\bar{H}_c\right] \\
-\left[\bar{\Sigma}^{a,T}_{vv}\vec{S}^{<}_{vc}-\vec{S}^{>T}_{vv}\bar{\Delta}^{a}_{ee,vc}\right]-\left[\bar{\Sigma}^{>T}_{vv}\vec{S}^{a}_{vc}-\vec{S}^{aT}_{vv}\bar{\Delta}^{<}_{ee,vc}\right] \\
+\left[\bar{\Delta}^{r}_{ee,vc}\vec{S}^{<}_{cc}-\vec{S}^{<}_{vc}\bar{\Sigma}^{a}_{cc}\right]+\left[\bar{\Delta}^{<}_{ee,vc}\vec{S}^{a}_{cc}-\vec{S}^{r}_{vc}\bar{\Sigma}^{<}_{cc}\right]
\end{array}\right] \\
&+\frac{1}{2}\left[\begin{array}{c}
+i\left[\vec{\mathcal{B}}_{vv}\times\vec{S}^{<}_{vc}-\vec{S}^{<}_{vc}\times\vec{\mathcal{B}}_{cc}\right] \\
+i\left[\vec{S}^{aT}_{vv}\times\vec{\delta}^{<}_{ee,vc}-\vec{\Xi}^{>T}_{vv}\times\vec{S}^{a}_{vc}\right]+i\left[\vec{S}^{>T}_{vv}\times\vec{\delta}^{a}_{ee,vc}-\vec{\Xi}^{a,T}_{vv}\times\vec{S}^{<}_{vc}\right] \\
+i\left[\vec{\delta}^{r}_{ee,vc}\times\vec{S}^{<}_{cc}-\vec{S}^{<}_{vc}\times\vec{\Xi}^{a}_{cc}\right]+i\left[\vec{\delta}^{<}_{ee,vc}\times\vec{S}^{a}_{cc}-\vec{S}^{r}_{vc}\times\vec{\Xi}^{<}_{cc}\right]
\end{array}\right] \\
&+\frac{1}{2}\left[\begin{array}{c}
+\left[\vec{\mathcal{B}}_{vv}S^{<}_{vc,o}-S^{<}_{vc,o}\vec{\mathcal{B}}_{cc}\right] \\
-\left[\vec{\Xi}^{a,T}_{vv}S^{<}_{vc,o}-S^{>T}_{vv,o}\vec{\delta}^{a}_{ee,vc}\right]-\left[\vec{\Xi}^{>T}_{vv}S^{a}_{vc,o}-S^{aT}_{vv,o}\vec{\delta}^{<}_{ee,vc}\right] \\
+\left[\vec{\delta}^{r}_{ee,vc}S^{<}_{cc,o}-S^{<}_{vc,o}\vec{\Xi}^{a}_{cc}\right]+\left[\vec{\delta}^{<}_{ee,vc}S^{a}_{cc,o}-S^{r}_{vc,o}\vec{\Xi}^{<}_{cc}\right]
\end{array}\right],
\end{aligned}
\tag{21.57}
$$

$$
\begin{aligned}
&i\hbar\left(\frac{\partial}{\partial t_1}+\frac{\partial}{\partial t_2}\right)\vec{S}^{<}_{vv} \\
&=\frac{1}{2}\left\{\begin{array}{c}
\left[\bar{H}_v,\vec{S}^{<}_{vv}\right] \\
+\left[\bar{\Sigma}^{r}_{vv}\vec{S}^{<}_{vv}-\vec{S}^{<}_{vv}\bar{\Sigma}^{a}_{vv}\right]+\left[\bar{\Sigma}^{<}_{vv}\vec{S}^{a}_{vv}-\vec{S}^{r}_{vv}\bar{\Sigma}^{<}_{vv}\right] \\
+\left[\bar{\Delta}^{r}_{ee,vc}\vec{S}^{<}_{cv}-\vec{S}^{<}_{vc}\bar{\Delta}^{a}_{hh,cv}\right]+\left[\bar{\Delta}^{<}_{ee,vc}\vec{S}^{a}_{cv}-\vec{S}^{r}_{vc}\bar{\Delta}^{<}_{hh,cv}\right]
\end{array}\right\} \\
&+\frac{1}{2}\left[\begin{array}{c}
+i\left[\vec{\mathcal{B}}_{vv}\times\vec{S}^{<}_{vv}-\vec{S}^{<}_{vv}\times\vec{\mathcal{B}}_{vv}\right] \\
+i\left[\vec{\Xi}^{r}_{vv}\times\vec{S}^{<}_{vv}-\vec{S}^{<}_{vv}\times\vec{\Xi}^{a}_{vv}\right]+i\left[\vec{\Xi}^{<}_{vv}\times\vec{S}^{a}_{vv}-\vec{S}^{r}_{vv}\times\vec{\Xi}^{<}_{vv}\right] \\
+i\left[\vec{\delta}^{<}_{ee,vc}\times\vec{S}^{a}_{cv}-\vec{S}^{r}_{vc}\times\vec{\delta}^{<}_{hh,cv}\right]+i\left[\vec{\delta}^{r}_{ee,vc}\times\vec{S}^{<}_{cv}-\vec{S}^{<}_{vc}\times\vec{\delta}^{a}_{hh,cv}\right]
\end{array}\right] \\
&+\frac{1}{2}\left[\begin{array}{c}
+\left[\vec{\mathcal{B}}_{vv},S^{<}_{vv,o}\right] \\
+\left[\vec{\Xi}^{r}_{vv}S^{<}_{vv,o}-S^{<}_{vv,o}\vec{\Xi}^{a}_{vv}\right]+\left[\vec{\Xi}^{<}_{vv}S^{a}_{vv,o}-S^{r}_{vv,o}\vec{\Xi}^{<}_{vv}\right] \\
+\left[\vec{\delta}^{<}_{ee,vc}S^{a}_{cv,o}-S^{r}_{vc,o}\vec{\delta}^{<}_{hh,cv}\right]+\left[\vec{\delta}^{r}_{ee,vc}S^{<}_{cv,o}-S^{<}_{vc,o}\vec{\delta}^{a}_{hh,cv}\right]
\end{array}\right]. \qquad (21.58)
\end{aligned}
$$

The right-hand-side of each SMVQTEs, Eqs. (21.55)–(21.58), essentially consist of three group of bracketed terms contributing to the rate of change, namely, (1) a group of terms describing the transport of spin magnetization vector with spin-independent transport parameters similar to the equation for spinless particle quantum transport, (2) a group of terms describing various sources of torques influencing the spin orientation and directional flow of spin magnetization correlations or phase-space magnetization density, and (3) a group of terms describing the transport the spin-scalar or charge for each band index with spin-dependent transport parameters or expressing the coupling of the real spin magnetization with the pseudo-spin magnetization. In principle the above semiconductor Bloch equations for the spin-magnetization vector can be reduced to single equation for the spin/pseudo-spin magnetization dyadic/tensor. We demonstrate this process in Section 22.

21.5.2 *Pseudo-spin magnetization vector transport equation*

We also write the RHS of the pseudo-spin transport equation into a similar groups of terms, first group independent of the real spin plus terms involving real spin, given in the electron picture, as

$$
\begin{aligned}
i\hbar\left(\frac{\partial}{\partial t_1}+\frac{\partial}{\partial t_2}\right)\vec{S}_o
&= \begin{bmatrix} \frac{1}{4}\left[\bar{H},\vec{S}_o\right]+\frac{1}{4}\left[\bar{\zeta}^<,\mathrm{Re}\,\vec{S}^r_o\right] \\ +\frac{i}{4}\left\{\bar{\zeta},\vec{S}_o\right\}-\frac{i}{4}\left\{\bar{\zeta}^<,\mathrm{Im}\,\vec{S}^r_o\right\} \end{bmatrix} \\
&\quad + \begin{bmatrix} \frac{i}{4}\left[\vec{\beta}\times\vec{S}_o-\vec{S}_o\times\vec{\beta}\right]+\frac{i}{4}\left[\vec{\zeta}^<\times\mathrm{Re}\,\vec{S}^r_o-\mathrm{Re}\,\vec{S}^r_o\times\vec{\zeta}^<\right] \\ -\frac{1}{4}\left\{\vec{\zeta}\times\vec{S}_o+\vec{S}_o\times\vec{\zeta}\right\}+\frac{1}{4}\left\{\vec{\zeta}^<\times\mathrm{Im}\,\vec{S}^r_o+\mathrm{Im}\,\vec{S}^r_o\times\vec{\zeta}^<\right\} \end{bmatrix} \\
&\quad + \begin{bmatrix} \frac{i}{4}\left[\vec{\mathcal{B}}_i\times\vec{S}_{o,i}-\vec{S}_{o,i}\times\vec{\mathcal{B}}_i\right] \\ +\frac{i}{4}\left[\vec{\zeta}^<_i\times\mathrm{Re}\,\vec{S}^r_{o,i}-\mathrm{Re}\,\vec{S}^r_{o,i}\times\vec{\zeta}^<_i\right] \\ -\frac{1}{4}\left\{\vec{\zeta}_i\times\vec{S}_{o,i}+\vec{S}_{o,i}\times\vec{\zeta}_i\right\} \\ +\frac{1}{4}\left\{\vec{\zeta}^<_i\times\mathrm{Im}\,\vec{S}^r_{o,i}+\mathrm{Im}\,\vec{S}^r_{o,i}\times\vec{\zeta}^<_i\right\} \end{bmatrix} \\
&\quad + \begin{bmatrix} \frac{1}{4}\left[\bar{\mathcal{B}}_i,\vec{S}_{o,i}\right]+\frac{1}{4}\left[\vec{\mathcal{B}}_i,\bar{S}_{o,i}\right] \\ +\frac{i}{4}\left\{\bar{\zeta}_i,\vec{S}_{o,i}\right\}+\frac{i}{4}\left\{\vec{\zeta}_i,\bar{S}_{o,i}\right\} \\ +\frac{1}{4}\left[\bar{\zeta}^<_i,\mathrm{Re}\,\vec{S}^r_{o,i}\right]+\frac{1}{4}\left[\vec{\zeta}^<_i,\mathrm{Re}\,\bar{S}^r_{o,i}\right] \\ -\frac{i}{4}\left\{\bar{\zeta}^<_i,\mathrm{Im}\,\vec{S}^r_{o,i}\right\}-\frac{i}{4}\left\{\vec{\zeta}^<_i,\mathrm{Im}\,\bar{S}^r_{o,i}\right\} \end{bmatrix} \\
&\quad + \begin{bmatrix} +\frac{1}{4}\left[\vec{\beta},S_{o,o}\right]+\frac{1}{4}\left[\vec{\zeta}^<,\mathrm{Re}\,S^r_{o,o}\right] \\ +\frac{i}{4}\left\{\vec{\zeta},S_{o,o}\right\}-\frac{i}{4}\left\{\vec{\zeta}^<,\mathrm{Im}\,S^r_{o,o}\right\} \end{bmatrix}.
\end{aligned}
\tag{21.59}
$$

where we employ the Einstein summation convention for the Pauli–Dirac spin vector "dot product" indicated by the repeated index i (virtually eliminating the spin-degree of freedom).

The equation for the pseudo-spin magnetization has five groups of bracketed terms contributing to the rate of change, namely, (1) terms describing transport of pseudo-spin magnetization with transport parameters independent of both real spin and pseudo-spin degrees of freedom, (2) a bracketed terms describing various sources of pseudo-torque for the pseudo-spin magnetization, (3) another bracketed terms containing the pseudo-torque contribution from the pseudo-spin/real spin dyadic, with the real-spin dependence integrated out (sum over), (4) a bracketed terms containing two types of contribution: (a) contribution from the transport pseudo-spin/real spin dyadic with pseudo-spin independent transport parameters, with the real-spin dependence integrated out (sum over) and (b) contribution from the transport of pseudo-spin scalar of spin magnetization vector, e.g., $S_{o,o,i} \equiv \bar{S}_{o,i} \equiv \bar{S}_i$, with pseudo-spin dependent transport parameters and with the real-spin dependence integrated out (sum over), and (5) a bracket terms containing the transport of total charge (absolute scalar) with pseudo-spin dependent transport parameters. Thus, the pseudo-spin magnetization equations incorporate the pseudo-spin/real-spin coupling, as well as the pseudo-spin coupling to the total charge.

21.5.3 *The absolute scalar transport equation*

The corresponding absolute scalar[3] equation is calculated in Appendix D.4, and is given as

$$
\begin{aligned}
i\hbar\left(\frac{\partial}{\partial t_1}+\frac{\partial}{\partial t_2}\right) S_{o,o}
&= \frac{1}{4}\,[\bar{H}, S_{o,o}] + \frac{1}{4}\,[\bar{\zeta}^{<}, \mathrm{Re}\, S^{r}_{o,o}] + \frac{i}{4}\,[\{\bar{\zeta}, S_{o,o}\}] - \frac{i}{4}\,\{\bar{\zeta}^{<}, \mathrm{Im}\, S^{r}_{o,o}\} \\
&\quad + \frac{1}{4}\left[\vec{\beta}^{\cdot psp}, \vec{S}_o\right] + \frac{1}{4}\left[\vec{\zeta}^{<\cdot psp}, \mathrm{Re}\, \vec{S}^{r}_o\right] + \frac{i}{4}\left\{\vec{\zeta}^{\cdot psp}, \vec{S}_o\right\} - \frac{i}{4}\left\{\vec{\zeta}^{<\cdot psp}, \mathrm{Im}\, \vec{S}^{r}_o\right\} \\
&\quad + \frac{1}{4}\,[\bar{\mathcal{B}}_i, \bar{S}_i] + \frac{1}{4}\,[\bar{\zeta}^{<}_i, \mathrm{Re}\, \bar{S}^{r}_i] + \frac{i}{4}\,\{\bar{\zeta}_i, \bar{S}_i\} - \frac{i}{4}\,\{\bar{\zeta}^{<}_i, \mathrm{Im}\, \bar{S}^{r}_i\} \\
&\quad + \frac{1}{4}\left[\vec{\mathcal{B}}^{\cdot psp}_i, \vec{S}_i\right] + \frac{1}{4}\left[\vec{\zeta}^{<\cdot psp}_i, \mathrm{Re}\, \vec{S}^{r}_i\right] + \frac{i}{4}\left\{\vec{\zeta}^{\cdot psp}_i, \vec{S}_i\right\} - \frac{i}{4}\left\{\vec{\zeta}^{<\cdot psp}_i, \mathrm{Im}\, \vec{S}^{r}_i\right\}.
\end{aligned}
\tag{21.60}
$$

where some of the symbols are defined in Eqs. (D.41)–(D.43) of Appendix D.3.2. In Eq. (21.60), the superscripts $^{\cdot psp}$ indicates the process of "integrating out" the pseudo-spin degree of freedoms, aside from the dot product of the Pauli–Dirac spin vector indicated by the Einstein summation convention for the repeated i index. The first line of Eq. (21.60) describes the familiar quantum transport equation of particles or charges. The second line

[3] $S_{o,o}$ represents the trace of the 4×4 matrix of Eq. (21.16) and hence independent of the Pauli–Dirac and pseudo-spin degrees of freedom. The time variation does not in general vanish by virtue of the non-locality of the space and time arguments of the two-point correlations involved in its equation.

gives the contribution of pseudo-spin magnetization with pseudo-spin dependent transport parameters but with the pseudo-spin degrees of freedom integrated out to produce absolute scalars. The third line gives the contribution from the transport of pseudo-spin scalar of spin magnetization vector, e.g. $S_{o,o,i} \equiv \bar{S}_{o,i} \equiv \bar{S}_i$, with pseudo-spin independent transport parameters and with the real-spin dependence integrated out (sum over). The fourth line gives the contribution from pseudo-spin/spin dyadic with pseudo-spin/spin dependent transport parameters, but with pseudo-spin and spin degrees of freedom integrated out to produce absolute scalars.

Equations (D.5)–(D.8) with (D.58) and (21.60) constitute the non-equilibrium multi-band magnetization quantum transport equations. These 16 equations for the components of the spin vectors are the main results of this chapter. Equations (21.55)–(21.59) and (21.60) constitute the non-equilibrium spin magnetization quantum transport equations of this chapter, written into at least three groups of more meaningful terms. One note that the first group of terms of all equations is similar to that of the non-equilibrium spinless particle-correlation transport equations by virtue of the spin-scalar or pseudo-spin scalar, as the case maybe, transport coefficients, second group usually contains all the torque terms, and the last group generally contains coupling between spin, pseudo-spin, and charge. Note that the equation for the pseudo-spin magnetization vector involves coupling to the total charge represented by $S_{o,o}$.

21.5.3.1 *Single conduction band limit*

In the single conduction band limit for the electrons, Eq. (D.5) reduces to

$$
\begin{aligned}
&i\hbar\left(\frac{\partial}{\partial t_1}+\frac{\partial}{\partial t_2}\right)\vec{S}^<_{cc}\\
&=\frac{1}{2}\left[\bar{H}_c,\vec{S}^<_{cc}\right]+\frac{1}{2}\left[\vec{\mathcal{B}}_{cc},S^<_{cc,o}\right]+\frac{i}{2}\left[\vec{\mathcal{B}}_{cc}\times\vec{S}^<_{cc}-\vec{S}^<_{cc}\times\vec{\mathcal{B}}_{cc}\right]\\
&\quad+\frac{1}{2}\begin{bmatrix}+\left[\bar{\Sigma}^r_{cc}\vec{S}^<_{cc}-\vec{S}^<_{cc}\bar{\Sigma}^a_{cc}\right]+\left[\vec{\Xi}^r_{cc}S^<_{cc,o}-S^<_{cc,o}\vec{\Xi}^a_{cc}\right]\\+i\left[\vec{\Xi}^r_{cc}\times\vec{S}^<_{cc}-\vec{S}^<_{cc}\times\vec{\Xi}^a_{cc}\right]\end{bmatrix}\\
&\quad+\frac{1}{2}\begin{bmatrix}\left[\bar{\Sigma}^<_{cc}\vec{S}^a_{cc}-\vec{S}^r_{cc}\bar{\Sigma}^<_{cc}\right]+\left[\vec{\Xi}^<_{cc}S^a_{cc,o}-S^r_{cc,o}\vec{\Xi}^<_{cc}\right]\\+i\left[\vec{\Xi}^<_{cc}\times\vec{S}^a_{cc}-\vec{S}^r_{cc}\times\vec{\Xi}^<_{cc}\right]\end{bmatrix}.
\end{aligned}
\tag{21.61}
$$

Upon ignoring the effect of $S^{<}_{cc,o}$,[4] the above further reduces to,

$$
\begin{aligned}
& i\hbar\left(\frac{\partial}{\partial t_1}+\frac{\partial}{\partial t_2}\right)\vec{S}^{<}_{cc} \\
&= \frac{1}{2}\left[\bar{H}_c+\mathrm{Re}\,\bar{\Sigma}^{r}_{cc},\vec{S}^{<}_{cc}\right]+\frac{i}{2}\left\{\mathrm{Im}\,\bar{\Sigma}^{r}_{cc},\vec{S}^{<}_{cc}\right\} \\
&\quad -\frac{i}{2}\left\{\bar{\Sigma}^{<}_{cc},\mathrm{Im}\,\vec{S}^{r}_{cc}\right\}+\left[\bar{\Sigma}^{<}_{cc},\mathrm{Re}\,\vec{S}^{r}_{cc}\right] \\
&\quad +\frac{i}{2}\left[\vec{\mathcal{B}}_{cc}\times\vec{S}^{<}_{cc}-\vec{S}^{<}_{cc}\times\vec{\mathcal{B}}_{cc}\right]+\frac{i}{2}\left[\mathrm{Re}\,\vec{\Xi}^{r}_{cc}\times\vec{S}^{<}_{cc}-\vec{S}^{<}_{cc}\times\mathrm{Re}\,\vec{\Xi}^{r}_{cc}\right] \\
&\quad -\frac{1}{2}\left\{\mathrm{Im}\,\vec{\Xi}^{r}_{cc}\times\vec{S}^{<}_{cc}+\vec{S}^{<}_{cc}\times\mathrm{Im}\,\vec{\Xi}^{r}_{cc}\right\}+\frac{i}{2}\left[\vec{\Xi}^{<}_{cc}\times\mathrm{Re}\,\vec{S}^{r}_{cc}-\mathrm{Re}\,\vec{S}^{r}_{cc}\times\vec{\Xi}^{<}_{cc}\right] \\
&\quad +\frac{1}{2}\left\{\vec{\Xi}^{<}_{cc}\times\mathrm{Im}\,\vec{S}^{r}_{cc}+\mathrm{Im}\,\vec{S}^{r}_{cc}\times\vec{\Xi}^{<}_{cc}\right\}. \qquad (21.62)
\end{aligned}
$$

Further upon neglecting the effects of $\mathrm{Re}\,\vec{S}^{r}_{cc}$,[5] as was done in Chapter 20 (see also Ref. [158]), we finally obtain

$$
\begin{aligned}
& i\hbar\left(\frac{\partial}{\partial t_1}+\frac{\partial}{\partial t_2}\right)\vec{S}^{<}_{cc} \\
&= \frac{1}{2}\left[\bar{H}_c+\mathrm{Re}\,\bar{\Sigma}^{r}_{cc},\vec{S}^{<}_{cc}\right]+\frac{i}{2}\left\{\mathrm{Im}\,\bar{\Sigma}^{r}_{cc},\vec{S}^{<}_{cc}\right\}-\frac{i}{2}\left\{\bar{\Sigma}^{<}_{cc},\mathrm{Im}\,\vec{S}^{r}_{cc}\right\} \\
&\quad +\frac{i}{2}\left[\left(\vec{\mathcal{B}}_{cc}+\mathrm{Re}\,\vec{\Xi}^{r}_{cc}\right)\times\vec{S}^{<}_{cc}-\vec{S}^{<}_{cc}\times\left(\vec{\mathcal{B}}_{cc}+\mathrm{Re}\,\vec{\Xi}^{r}_{cc}\right)\right] \\
&\quad -\frac{1}{2}\left\{\mathrm{Im}\,\vec{\Xi}^{r}_{cc}\times\vec{S}^{<}_{cc}+\vec{S}^{<}_{cc}\times\mathrm{Im}\,\vec{\Xi}^{r}_{cc}\right\} \\
&\quad +\frac{1}{2}\left\{\vec{\Xi}^{<}_{cc}\times\mathrm{Im}\,\vec{S}^{r}_{cc}+\mathrm{Im}\,\vec{S}^{r}_{cc}\times\vec{\Xi}^{<}_{cc}\right\}. \qquad (21.63)
\end{aligned}
$$

---

[4] $S^{<}_{cc,o}$ represents the trace of 2×2 spin matrices, which can be equated to unity in pure stationary quantum state. However, in non-equilibrium quantum transport physics, $S^{<}_{cc,o}$ represent the spin-independent charge of the conduction electrons, which could vary in space and time. Therefore the transport equation for $S^{<}_{cc,o}$ must also be solved for $S^{<}_{cc,o}$ to be used in calculating the charge to be fed to the Poisson equation for self-consistency [158]. For now, we postpone this treatment as this will be incorporated later in the pseudo-spin formulation, where it is the total charge in our multi-band system that is the relevant variable to be fed to the Poisson equation for self-consistency in the potential distribution.

[5] This term is often attributed to electron *zitterbewegung* or fluctuating motion and is traditionally been neglected in the treatment of non-equilibrium quantum transport employed in practical applications.

Using the following relations:

$$-\mathrm{Im}\,\vec{\Xi}^{r}_{cc} = \vec{\gamma}, \tag{21.64}$$

$$-\mathrm{Im}\,\bar{\Sigma}^{r}_{cc} = \bar{\Gamma}, \tag{21.65}$$

$$-\mathrm{Im}\,\vec{S}^{r}_{cc} = \vec{A} \tag{21.66}$$

$$-\mathrm{Im}\,S^{r}_{cc,o} = A \tag{21.67}$$

$$\vec{\mathcal{B}}_{cc} \Rightarrow -\vec{\mathcal{B}}_{cc} = -g_s \frac{e\hbar}{2m^* c}\vec{B}_{\mathrm{eff}} \ \text{ for electrons.} \tag{21.68}$$

Then, we obtain

$$\begin{aligned} i\hbar &\left(\frac{\partial}{\partial t_1} + \frac{\partial}{\partial t_2}\right)\vec{S}^{<}_{cc} \\ &= \frac{1}{2}\left[\bar{H}_c + \mathrm{Re}\,\bar{\Sigma}^{r}_{cc}, \vec{S}^{<}_{cc}\right] - \frac{i}{2}\left\{\bar{\Gamma}, \vec{S}^{<}_{cc}\right\} + \frac{i}{2}\left\{\bar{\Sigma}^{<}_{cc}, \vec{A}\right\} \\ &+ \frac{i}{2}\left[\vec{S}^{<}_{cc} \times \left(\vec{\mathcal{B}}_{cc} - \mathrm{Re}\,\vec{\Xi}^{r}_{cc}\right) - \left(\vec{\mathcal{B}}_{cc} - \mathrm{Re}\,\vec{\Xi}^{r}_{cc}\right) \times \vec{S}^{<}_{cc}\right] \\ &+ \frac{1}{2}\left\{\vec{\gamma} \times \vec{S}^{<}_{cc} + \vec{S}^{<}_{cc} \times \vec{\gamma}\right\} - \frac{1}{2}\left\{\vec{\Xi}^{<}_{cc} \times \vec{A} + \vec{A} \times \vec{\Xi}^{<}_{cc}\right\}, \end{aligned} \tag{21.69}$$

which agrees with the single-band spin-vector magnetization transport equation given in Chapter 20, within the approximation used and using Eqs. (21.64)-(21.68) except for the presence of the extra term in Eq. (21.69) given by

$$\frac{i}{2}\left[\mathrm{Re}\,\vec{\Xi}^{r}_{cc} \times \vec{S}^{<}_{cc} - \vec{S}^{<}_{cc} \times \mathrm{Re}\,\vec{\Xi}^{r}_{cc}\right].$$

These terms was overlooked in casting the $\mathrm{Re}\,\Sigma^r$ single-particle Hamiltonian $\tilde{\mathcal{H}}$ in Ref. [188] into its 2×2 spin canonical form. Equation (21.61) is the more exact expression for the single-band spin magnetization vector transport equation.

However, to be quite exact, we should include the effect of $S^{<}_{cc,o}$. The result will be adding the following terms to the RHS of Eq. (21.69).

$$\begin{aligned} &\left[\mathrm{Re}\,\vec{\Xi}^{r}_{cc}, S^{<}_{cc,o}\right] + i\left\{\mathrm{Im}\,\vec{\Xi}^{r}_{cc}, S^{<}_{cc,o}\right\} + \left[\vec{\Xi}^{<}_{cc}, \mathrm{Re}\,S^{r}_{cc,o}\right] - i\left\{\vec{\Xi}^{<}_{cc}, \mathrm{Im}\,S^{r}_{cc,o}\right\} \\ &= \left[\mathrm{Re}\,\vec{\Xi}^{r}_{cc}, S^{<}_{cc,o}\right] - i\left\{\vec{\gamma}, S^{<}_{cc,o}\right\} + \left[\vec{\Xi}^{<}_{cc}, \mathrm{Re}\,S^{r}_{cc,o}\right] + i\left\{\vec{\Xi}^{<}_{cc}, A_o\right\} \end{aligned}$$

yielding the formally exact result,

$$
\begin{aligned}
i\hbar\left(\frac{\partial}{\partial t_1}+\frac{\partial}{\partial t_2}\right)\vec{S}^<_{cc}
&=\frac{1}{2}\left[\bar{H}_c+\mathrm{Re}\,\bar{\Sigma}^r_{cc},\vec{S}^<_{cc}\right]-\frac{i}{2}\left\{\bar{\Gamma},\vec{S}^<_{cc}\right\}+\frac{i}{2}\left\{\bar{\Sigma}^<_{cc},\vec{\mathcal{A}}\right\}+\left[\bar{\Sigma}^<_{cc},\mathrm{Re}\,\vec{S}^r_{cc}\right]\\
&+\frac{1}{2}\left[\mathrm{Re}\,\vec{\Xi}^r_{cc},S^<_{cc,o}\right]-\frac{i}{2}\left\{\vec{\gamma},S^<_{cc,o}\right\}+\frac{i}{2}\left\{\vec{\Xi}^<_{cc},A_o\right\}+\frac{1}{2}\left[\vec{\Xi}^<_{cc},\mathrm{Re}\,S^r_{cc,o}\right]\\
&+\frac{i}{2}\left[\vec{S}^<_{cc}\times\left(\vec{\mathcal{B}}_{cc}-\mathrm{Re}\,\vec{\Xi}^r_{cc}\right)-\left(\vec{\mathcal{B}}_{cc}-\mathrm{Re}\,\vec{\Xi}^r_{cc}\right)\times\vec{S}^<_{cc}\right]\\
&+\frac{1}{2}\left\{\vec{\gamma}\times\vec{S}^<_{cc}+\vec{S}^<_{cc}\times\vec{\gamma}\right\}-\frac{1}{2}\left\{\vec{\Xi}^<_{cc}\times\vec{\mathcal{A}}+\vec{\mathcal{A}}\times\vec{\Xi}^<_{cc}\right\},\\
&+\frac{i}{2}\left[\vec{\Xi}^<_{cc}\times\mathrm{Re}\,\vec{S}^r_{cc}-\mathrm{Re}\,\vec{S}^r_{cc}\times\vec{\Xi}^<_{cc}\right],
\end{aligned}
\tag{21.70}
$$

where we have also restored the effect of $\mathrm{Re}\,\vec{S}^r_{cc}$, ignored in Eq. (21.69). We observe that by including the particle-density correlation (spin-scalar), $S^<_{cc,o}$, the transport of total electron density in the conduction band is governed by the spin-vector transport coefficients. The equation for $S^<_{cc,o}$ itself is given by (see Appendix D.3)

$$
\begin{aligned}
i\hbar\left(\frac{\partial}{\partial t_1}+\frac{\partial}{\partial t_2}\right)S^<_{cc,o}
&=\frac{1}{2}\left[\bar{H}_c,S^<_{cc,o}\right]+\frac{1}{2}\left[\bar{\Sigma}^r_{cc}S^<_{cc,o}-\ S^<_{cc,o}\bar{\Sigma}^a_{cc}\right]+\frac{1}{2}\left[\bar{\Sigma}^<_{cc}S^a_{cc,o}-\ S^r_{cc,o}\bar{\Sigma}^<_{cc}\right]\\
&+\frac{1}{2}\left[\vec{\mathcal{B}}_{cc}\cdot\vec{S}^<_{cc}-\vec{S}^<_{cc}\cdot\vec{\mathcal{B}}_{cc}\right]+\frac{1}{2}\left[\vec{\Xi}^r_{cc}\cdot\vec{S}^<_{cc}-\vec{S}^<_{cc}\cdot\vec{\Xi}^a_{cc}\right]\\
&+\frac{1}{2}\left[\vec{\Xi}^<_{cc}\cdot\vec{S}^a_{cc}-\vec{S}^r_{cc}\cdot\vec{\Xi}^<_{cc}\right],
\end{aligned}
\tag{21.71}
$$

giving the ordinary quantum transport of particle density (first line) plus coupling to spin-vector, $\vec{S}^<_{cc}$, through the dot product terms (second line). The first line, after the lattice Weyl transformation, has in fact been the workhorse in the simulation of resonant tunneling devices (RTDs) where the spin degrees of freedom are not included.

Chapter 22

Two Tight-Binding Energy Bands

22.1 Two-level atom with spin degrees of freedom

Let us examine the ramifications of the semiconductor Bloch equations for the spin magnetization vector, Eqs. (21.55)–(21.58), in a two flat-band limit or two-level atomic system possessing real-spin degrees of freedom. To the author's knowledge, this has not been rigorously treated in the literature. Here, we expect the physics to be simpler for understanding a particle–antiparticle description versus a description as single particle specie which occupies two energy states, of basically the same physical process. We will demonstrate that the resulting equations can be reduced to an equation for a spin magnetization dyadic tensor, incorporating both the spin and pseudo-spin degrees of freedom.

For flat band limit or two-level atomic system, we will ignore terms arising from the particle self-energies, Σ, including its off-diagonal terms involving $\bar{\Delta}^{\lessgtr}$. We retained $\bar{\Delta}^r$ and $\bar{\Delta}^a$ as interband coupling in response to any external influence. The above four equations for the spin magnetization in the electron–hole or *defect* representation reduce to the following shorter expressions:

$$\begin{aligned} i\hbar\left(\frac{\partial}{\partial t_1}+\frac{\partial}{\partial t_2}\right)\vec{S}^{<}_{cc} &= \frac{1}{2}\left\{\left[\bar{H}_c,\vec{S}^{<}_{cc}\right]+\left[\vec{\mathcal{B}}_{cc},S^{<}_{cc,o}\right]+i\left[\vec{\mathcal{B}}_{cc}\times\vec{S}^{<}_{cc}-\vec{S}^{<}_{cc}\times\vec{\mathcal{B}}_{cc}\right]\right\} \\ &\quad+\frac{1}{2}\left[\begin{array}{c}\frac{1}{i\hbar}\left[\bar{\Delta}^r_{hh,cv}\vec{S}^{<}_{vc}-\vec{S}^{<}_{cv}\bar{\Delta}^a_{ee,vc}\right]+\frac{1}{i\hbar}\left[\vec{\delta}^r_{hh,cv}S^{<}_{vc,o}-S^{<}_{cv,o}\vec{\delta}^a_{ee,vc}\right] \\ +\frac{1}{\hbar}\left[\vec{\delta}^r_{hh,cv}\times\vec{S}^{<}_{vc}-\vec{S}^{<}_{cv}\times\vec{\delta}^a_{ee,vc}\right]\end{array}\right], \end{aligned} \tag{22.1}$$

$$
\begin{aligned}
&i\hbar\left(\frac{\partial}{\partial t_1}+\frac{\partial}{\partial t_2}\right)\vec{S}^<_{cv}\\
&=\frac{1}{2}\left[\begin{matrix}\left[\bar{H}_c\vec{S}^<_{cv}-\vec{S}^<_{cv}\bar{H}_v\right]+\left[\vec{\mathcal{B}}_{cc}S^<_{cv,o}-S^<_{cv,o}\vec{\mathcal{B}}_{vv}\right]\\ +i\left[\vec{\mathcal{B}}_{cc}\times\vec{S}^<_{cv}-\vec{S}^<_{cv}\times\vec{\mathcal{B}}_{vv}\right]\end{matrix}\right]\\
&\quad\times\frac{1}{2}\left[\left[-\vec{S}^<_{cc}\bar{\Delta}^a_{hh,cv}\right]+\left[-S^<_{cc,o}\vec{\delta}^a_{hh,cv}\right]+i\left[-\vec{S}^<_{cc}\times\vec{\delta}^a_{hh,cv}\right]\right]\\
&\quad\times\frac{1}{2}\left[-\left[\bar{\Delta}^r_{hh,cv}\vec{S}^{>T}_{vv}-\right]-\left[\vec{\delta}^r_{hh,cv}S^{>T}_{vv,o}-\right]+i\left[-\vec{\delta}^r_{hh,cv}\times\vec{S}^{>T}_{vv}\right]\right],
\end{aligned}
\tag{22.2}
$$

$$
\begin{aligned}
&i\hbar\left(\frac{\partial}{\partial t_1}+\frac{\partial}{\partial t_2}\right)\vec{S}^<_{vc}\\
&=\frac{1}{2}\left[\begin{matrix}\left[\bar{H}_v\vec{S}^<_{vc}-\vec{S}^<_{vc}\bar{H}_c\right]+\left[\vec{\mathcal{B}}_{vv}S^<_{vc,o}-S^<_{vc,o}\vec{\mathcal{B}}_{cc}\right]\\ +i\left[\vec{\mathcal{B}}_{vv}\times\vec{S}^<_{vc}-\vec{S}^<_{vc}\times\vec{\mathcal{B}}_{cc}\right]\end{matrix}\right]\\
&\quad+\frac{1}{2}\left[-\left[-\vec{S}^{>T}_{vv}\bar{\Delta}^a_{ee,vc}\right]-\left[-S^{>T}_{vv,o}\vec{\delta}^a_{ee,vc}\right]+i\left[\vec{S}^{>T}_{vv}\times\vec{\delta}^a_{ee,vc}-\right]\right]\\
&\quad+\frac{1}{2}\left[\left[\bar{\Delta}^r_{ee,vc}\vec{S}^<_{cc}-\right]+\left[\vec{\delta}^r_{ee,vc}S^<_{cc,o}-\right]+i\left[\vec{\delta}^r_{ee,vc}\times\vec{S}^<_{cc}-\right]\right],
\end{aligned}
\tag{22.3}
$$

$$
\begin{aligned}
&i\hbar\left(\frac{\partial}{\partial t_1}+\frac{\partial}{\partial t_2}\right)\vec{S}^<_{vv}\\
&=-\frac{1}{2}\left\{\left[\bar{H}_v,\vec{S}^<_{vv}\right]+\left[\vec{\mathcal{B}}_{vv},S^<_{vv,o}\right]+i\left[\vec{\mathcal{B}}_{vv}\times\vec{S}^<_{vv}-\vec{S}^<_{vv}\times\vec{\mathcal{B}}_{vv}\right]\right\}\\
&\quad-\frac{1}{2}\left[\begin{matrix}\frac{1}{i\hbar}\left[\bar{\Delta}^r_{ee,vc}\vec{S}^<_{cv}-\vec{S}^<_{vc}\bar{\Delta}^a_{hh,cv}\right]+\frac{1}{i\hbar}\left[\vec{\delta}^r_{ee,vc}S^<_{cv,o}-S^<_{vc,o}\vec{\delta}^a_{hh,cv}\right]\\ +\frac{1}{\hbar}\left[\vec{\delta}^r_{ee,vc}\times\vec{S}^<_{cv}-\vec{S}^<_{vc}\times\vec{\delta}^a_{hh,cv}\right]\end{matrix}\right].
\end{aligned}
\tag{22.4}
$$

Let the zero of energy lies in the middle of the energy gap. Thus $\bar{H}_c=-\bar{H}_v=\frac{\hbar\omega_o}{2}$. We identify the interband matrix elements as $\bar{\Delta}^r_{ee,vc}=\langle v|\,\mathcal{H}_I\,|c\rangle$, and $\bar{\Delta}^a_{hh,cv}=\langle c|\,\mathcal{H}_I\,|v\rangle\equiv\bar{\Delta}_{cv}$, so that $\bar{\Delta}^a_{hh,vc}=\bar{\Delta}^r_{ee,vc}\equiv\bar{\Delta}_{vc}$, by virtue of the locality of space-time dependence. We have the spin vector, $\vec{\mathcal{B}}_{cc}$, determined by

$$
H_{cc,\sigma\sigma'}=\frac{1}{2}\begin{pmatrix}\bar{H}_{cc}+\mathcal{B}_{cc,z} & \mathcal{B}_{cc,x}-i\mathcal{B}_{cc,y}\\ \mathcal{B}_{cc,x}+i\mathcal{B}_{cc,y} & \bar{H}_{cc}-\mathcal{B}_{cc,z}\end{pmatrix},
\tag{22.5}
$$

generally indicating a spin-split conduction band $H_{cc,\sigma\sigma'}$ into coupled spin-up level and spin-down level, with $\bar{H}_{cc}$ in the middle of the two split levels. The *external* magnetic which is spin-independent basically *contributes* to the z-component of $\vec{\mathcal{B}}_{cc}$ (shift of the band), whereas the rest of the components arise from the spin-orbit interaction. Similarly for holes, we have the spin vector, $\vec{\mathcal{B}}_{vv}$, determined from the relation,

$$H_{vv,\sigma\sigma'} = \frac{1}{2}\begin{pmatrix} \bar{H}_{vv} + \mathcal{B}_{vv,z} & \mathcal{B}_{vv,x} - i\mathcal{B}_{vv,y} \\ \mathcal{B}_{vv,x} + i\mathcal{B}_{vv,y} & \bar{H}_{vv} - \mathcal{B}_{vv,z} \end{pmatrix}. \tag{22.6}$$

The components of $\vec{\mathcal{B}}_{cc}$ and $\vec{\mathcal{B}}_{vv}$ are real valued, since $H_{vv,\sigma\sigma'}$, the *single-particle* Hamiltonian, is Hermitian. These are the effective magnetic field components multiplied by μ_B, which accounts for the external magnetic fields as well as the effects of spin-orbit coupling. Thus, these quantities in Eq. (22.6) have the units of energy.

The most general form of the single-particle Hamiltonian for each band, $\tilde{\mathcal{H}}_{\alpha\alpha}$, is of the form as

$$\tilde{\mathcal{H}}_{\alpha\alpha} = \frac{1}{2}(\bar{H}_{\alpha\alpha}(\hat{P},\hat{Q})\hat{I} - \vec{h}(\hat{P},\hat{Q})\cdot\vec{\sigma}_s), \tag{22.7}$$

where the spin vector, $\vec{h}(\hat{P},\hat{Q}) = (g_s\frac{e\frac{\hbar}{2}}{2mc}\vec{B}_{\text{eff}} - \text{Re}\,\vec{\Sigma}^r)$. Therefore

$$\tilde{\mathcal{H}}_{\alpha\alpha} = \frac{1}{2}(\bar{H}_{\alpha\alpha}\hat{I} - \vec{\sigma}_s\cdot\vec{\mathcal{B}}_{\alpha\alpha}), \tag{22.8}$$

where

$$\vec{\mathcal{B}}_{\alpha\alpha} = \left(\frac{e\hbar}{2mc}\vec{B}_{\text{eff}} - \text{Re}\,\vec{\Sigma}^r\right), \tag{22.9}$$

and for electron-spin angular momentum, the g_s-*factor*, $g_s \simeq 2$. Here the effective $\vec{B}_{\text{eff},\alpha\alpha}(\hat{P},\hat{Q})$ includes the effects of transistor-gate or strain induced spin-orbit coupling, and applied magnetic fields. $H(\hat{P},\hat{Q})$ consists of the energy-band structure and spin-independent confining potential due to the device heterostructure coupled with the applied bias at the terminals. The spin-orbit coupling in $\vec{B}_{\text{eff}}(\hat{P},\hat{Q})$ also depends very much on the electric field of the self-consistent potential inside the device.

From Eqs. (22.1)–(22.4), the magnetization correlation functions with the band indices in the electron–hole representation are: $\vec{S}^<_{cc}$, $\vec{S}^<_{cv}$, $\vec{S}^<_{vc}$, $\vec{S}^<_{vv}$, and $\vec{S}^{>T}_{vv}$. The Pauli-blocking terms, $\vec{S}^{>T}_{vv}$ and $S^{>T}_{vv,o}$, is related to $\vec{S}^<_{vv}$ and $S^{<T}_{vv,o}$, respectively, through the spectral-function vector, $\vec{\mathcal{A}}$, and spin-scalar, A, respectively,

$$\vec{S}^{>T}_{vv} = -i\vec{\mathcal{A}} + \vec{S}^{<T}_{vv}, \tag{22.10}$$

$$S^{>T}_{vv,o} = -iA + S^{<T}_{vv,o}, \tag{22.11}$$

where $\vec{A}$ is the spin-dependent spectral-function vector whereas A is the spin-independent spectral function for holes. The occurrence of the vector, $\vec{A}$, and scalar, A, spectral functions in the equation for the recombination and electron–hole pair creation rates may have to do with having two different systems consisting of electrons and holes, or a two particle-specie problem, generating or annihilating electron–hole pairs between the two species. Thus, only the spectral functions for the holes are needed in the above equations. For a two-level atom, assuming locality in space–time variables, we can simply write $\vec{S}^{<T}_{vv} = \vec{S}^{<}_{vv}$, and $S^{<T}_{vv,o} = S^{<}_{vv,o}$. Basically, the electron picture can be considered a *single-particle* representation of the two tight-binding energy bands or two-level atomic system, whereas the electron–hole picture is inherently a *many-body* representation of the same process.

Now define the following transformation from the 2×2 matrix in the energy-band indices into a *spin/pseudo-spin* canonical *dyadic* as

$$\begin{pmatrix} \vec{S}^{<}_{cc} & \vec{S}^{<}_{cv} \\ \vec{S}^{<}_{vc} & \vec{S}^{<}_{vv} \end{pmatrix} = \frac{1}{2}\begin{pmatrix} \vec{\Im} + \vec{S}_z & \vec{S}_x - i\vec{S}_y \\ \vec{S}_x + i\vec{S}_y & \vec{\Im} - \vec{S}_z \end{pmatrix}, \tag{22.12}$$

where,

$$\vec{S}_x = \left(\vec{S}^{<}_{vc} + \vec{S}^{<}_{cv}\right), \tag{22.13}$$

$$i\vec{S}_y = \left(\vec{S}^{<}_{vc} - \vec{S}^{<}_{cv}\right), \tag{22.14}$$

$$\vec{S}_z = \left(\vec{S}^{<}_{cc} + \vec{S}^{<}_{vv}\right), \tag{22.15}$$

$$\vec{\Im} = \left(\vec{S}^{<}_{cc} - \vec{S}^{<}_{vv}\right), \tag{22.16}$$

where we drop the "<" superscript in the *tensor/dyadic* correlation functions, $\vec{S}_j$. Note that the spin magnetic moment of holes is directed opposite to that of electrons. Here, $\vec{S}_z$ represents the *net-spin* magnetization vector of the system which may vary in space and time. We shall see later that $\vec{\Im}$ has only a *vector* character of *spin* magnetization, hence not a component of a dyadic.

22.1.1 *Spin-pseudospin Hamiltonian for a two-level system*

We have for the spin-independent term for the 2-band Hamiltonian,

$$\bar{H}_{\alpha\beta} = \begin{pmatrix} \frac{1}{2}\hbar\omega_o & 0 \\ 0 & -\frac{1}{2}\hbar\omega_o \end{pmatrix}. \tag{22.17}$$

Likewise, we make the following canonical transformation of the band indices for the spin-independent, $\bar{\Delta}$'s. Noting that these are off-diagonal in band indices, we have

$$\bar{\Delta}^{e-h,r}_{hh,\alpha\beta} = \frac{1}{2}\begin{pmatrix} 0 & \Lambda^r_{hh,x} - i\Lambda^r_{hh,y} \\ \Lambda^r_{hh,x} + i\Lambda^r_{hh,y} & 0 \end{pmatrix}, \tag{22.18}$$

We can combine Eqs. (22.17) and (22.18) whose entry components form a vector which we shall designate as $\vec{\Lambda}$

$$\bar{\Delta}^{e-h,r}_{ee,\alpha\beta} + \bar{H}_{\alpha\beta} = \frac{1}{2}\begin{pmatrix} \hbar\omega_o = \Lambda^a_{ee,z} & (\Lambda^a_{ee,x} - i\Lambda^a_{ee,y}) \\ (\Lambda^a_{ee,x} + i\Lambda^a_{ee,y}) & -\hbar\omega_o = (-\Lambda^a_{ee,z}) \end{pmatrix}, \tag{22.19}$$

where $\bar{\Delta}^{e-h,r}_{ee,\alpha\beta} + \bar{H}_{\alpha\beta}$ designates the spin-independent quantity, dependent only on band indices. We also have for the spin-vector, $\vec{\delta}$'s, off-diagonal in band indices, as

$$\vec{\delta}^{e-h,r}_{hh,\alpha\beta} = \frac{1}{2}\begin{pmatrix} 0 & \vec{\delta}^r_{hh,x} - i\vec{\delta}^r_{hh,y} \\ \vec{\delta}^r_{hh,x} + i\vec{\delta}^r_{hh,y} & 0 \end{pmatrix}. \tag{22.20}$$

We can also write $\vec{\mathcal{B}}_{\alpha\beta}$ which is diagonal in band indices, as

$$\vec{\mathcal{B}}_{\alpha\alpha} = \frac{1}{2}\begin{pmatrix} \vec{\mathcal{B}}_z & 0 \\ 0 & -\vec{\mathcal{B}}_z \end{pmatrix} \equiv^{\text{define}} \frac{1}{2}\begin{pmatrix} \vec{\delta}^r_{hh,z} & 0 \\ 0 & -\vec{\delta}^r_{hh,z} \end{pmatrix}. \tag{22.21}$$

We see from Eqs. (22.20) and (22.21) that we can define a tensor determined by the entries of the sum of the RHS of these two equations, namely,

$$\begin{aligned} &\frac{1}{2}\begin{pmatrix} 0 & \vec{\delta}^r_{hh,x} - i\vec{\delta}^r_{hh,y} \\ \vec{\delta}^r_{hh,x} + i\vec{\delta}^r_{hh,y} & 0 \end{pmatrix} + \frac{1}{2}\begin{pmatrix} \vec{\delta}^r_{hh,z} & 0 \\ 0 & -\vec{\delta}^r_{hh,z} \end{pmatrix} \\ &\quad = \frac{1}{2}\begin{pmatrix} \vec{\delta}^r_{hh,z} & \vec{\delta}^r_{hh,x} - i\vec{\delta}^r_{hh,y} \\ \vec{\delta}^r_{hh,x} + i\vec{\delta}^r_{hh,y} & -\vec{\delta}^r_{hh,z} \end{pmatrix}. \end{aligned} \tag{22.22}$$

We denote the corresponding dyadic, formed from the components of Eq. (22.22), as Π. We shall see later that the 2×2 (in the band indices) spin-vector Hamiltonian, where each entry is a 2×2 in the spin indices, can now be written as

$$\tilde{\mathcal{H}}_{\alpha\beta,\sigma\sigma'} \Longrightarrow^{\text{transformed}}_{\text{canonical}} \vec{\Lambda} \cdot \vec{\sigma}_{ps} + \vec{\sigma}_s \cdot \Pi \cdot \vec{\sigma}_{ps} + \vec{\delta} \cdot (\vec{S} + \vec{S}_o), \tag{22.23}$$

where $\vec{\Lambda}$ is the vector formed from the components given in Eq. (22.19), $\vec{\delta}$ includes the external magnetic fields, $\vec{S}$ is the total Pauli–Dirac spin of the system, $\vec{S}_o$ is the pseudo-spin, and Π is the tensor defined by

$$\Pi = \vec{\delta}_j \hat{I}_j,$$

where $\hat{I}_j$ is unit *dyadic* symmetric tensor or *idemfactor*. Here $\vec{\sigma}_s$ indicates Pauli–spin matrix due to spin indices, whereas $\vec{\sigma}_{ps}$ indicates the Pauli pseudo-spin matrix due to energy-band indices. The first term in Eq. (22.23) indicates that the *spin-independent* off-diagonal interaction between the conduction and valence band creates a pseudo-magnetic field, $\vec{\mathcal{B}}_o$, interacting with pseudo-spin, $\vec{S}_o$. The term $\vec{\sigma}_s \cdot \Pi \cdot \vec{\sigma}_{ps}$ clearly shows the coupling of real spin, S_i, and pseudo-spin, $S_{j,o}$, through the *interaction dyadic*, Ω_{ij}. Equation (22.23) can be recast as

$$\tilde{\mathcal{H}}_{\alpha\beta,\sigma\sigma'} \Longrightarrow^{\text{transformed}}_{\text{canonical}} = \vec{\mathcal{B}}_o \cdot \vec{S}_o + S_i \Omega_{ij} S_{j,o} + \vec{\delta} \cdot (\vec{S} + \vec{S}_o) \tag{22.24}$$

The last term seems to suggest to the use of g-factor $= 2$ for the spin interaction with effective magnetic fields, corresponding to the Dirac relativistic quantum mechanics. But clearly this seemingly is not the whole story since there are other terms in the Hamiltonian which may also give rise to the so-called anomalous g-factor, *akin* to Dirac relativistic quantum dynamics.

22.2 Spin/pseudo-spin magnetization dyadic: "Pseudo components"

Here we calculate the transport equations obeyed by the pseudo-components of the spin-magnetization dyadic tensor.

22.2.1 *Transport equation for* $\vec{S}_x = \left(\vec{S}^<_{vc} + \vec{S}^<_{cv}\right)$

We have

$$\begin{aligned}
&i\hbar\left(\frac{\partial}{\partial t_1} + \frac{\partial}{\partial t_2}\right)\vec{S}_x \\
&= \frac{1}{2}\left[\left[\bar{H}_c\vec{S}^<_{cv} - \vec{S}^<_{cv}\bar{H}_v\right] + \left[\vec{\mathcal{B}}_{cc}S^<_{cv,o} - S^<_{cv,o}\vec{\mathcal{B}}_{vv}\right]\right. \\
&\quad \left. +i\left[\vec{\mathcal{B}}_{cc} \times \vec{S}^<_{cv} - \vec{S}^<_{cv} \times \vec{\mathcal{B}}_{vv}\right]\right] + \frac{1}{2}\left[\left[-\vec{S}^<_{cc}\bar{\Delta}^a_{hh,cv}\right]\right. \\
&\quad \left. + \left[-S^<_{cc,o}\vec{\delta}^a_{hh,cv}\right] + i\left[-\vec{S}^<_{cc} \times \vec{\delta}^a_{hh,cv}\right]\right] \\
&\quad + \frac{1}{2}\left[-\left[\bar{\Delta}^r_{hh,cv}\vec{S}^{>T}_{vv}-\right] - \left[\vec{\delta}^r_{hh,cv}S^{>T}_{vv,o}-\right] + i\left[-\vec{\delta}^r_{hh,cv} \times \vec{S}^{>T}_{vv}\right]\right] \\
&\quad + \frac{1}{2}\left[\left[\bar{H}_v\vec{S}^<_{vc} - \vec{S}^<_{vc}\bar{H}_c\right] + \left[\vec{\mathcal{B}}_{vv}S^<_{vc,o} - S^<_{vc,o}\vec{\mathcal{B}}_{cc}\right]\right.
\end{aligned}$$

$$+i\left[\vec{\mathcal{B}}_{vv}\times\vec{S}^{<}_{vc}-\vec{S}^{<}_{vc}\times\vec{\mathcal{B}}_{cc}\right]\Big]+\frac{1}{2}\Big[-\left[-\vec{S}^{>T}_{vv}\bar{\Delta}^{a}_{ee,vc}\right]$$
$$-\left[-S^{>T}_{vv,o}\vec{\delta}^{a}_{ee,vc}\right]+i\left[\vec{S}^{>T}_{vv}\times\vec{\delta}^{a}_{ee,vc}-\right]\Big]$$
$$+\frac{1}{2}\Big[\left[\bar{\Delta}^{r}_{ee,vc}\vec{S}^{<}_{cc}-\right]+\left[\vec{\delta}^{r}_{ee,vc}S^{<}_{cc,o}-\right]+i\left[\vec{\delta}^{r}_{ee,vc}\times\vec{S}^{<}_{cc}-\right]\Big], \tag{22.25}$$

$$i\hbar\left(\frac{\partial}{\partial t_1}+\frac{\partial}{\partial t_2}\right)\vec{S}_x$$
$$=\frac{1}{2}\begin{bmatrix}-i\hbar\omega_o\vec{S}^{<}_{y}-2i\vec{\mathcal{B}}_{cc}\vec{S}^{<}_{y.o}\\ -i\vec{\mathcal{B}}_{cc}\times\left(\vec{S}^{<}_{vc}+\vec{S}^{<}_{cv}\right)+i\vec{\mathcal{B}}_{cc}\times\left(\vec{S}^{<}_{vc}+\vec{S}^{<}_{cv}\right)\Longrightarrow 0\end{bmatrix}$$
$$+\frac{1}{2}\left[\left(\bar{\Delta}^{r}_{ee,vc}-\bar{\Delta}^{a}_{hh,cv}\right)\left(\vec{S}^{<}_{cc}+\vec{S}^{>T}_{vv}\right)+\left(\vec{\delta}^{r}_{ee,vc}-\vec{\delta}^{a}_{hh,cv}\right)\left(S^{<}_{cc,o}+S^{>T}_{vv,o}\right)\right]$$
$$-\frac{i}{2}\left[\left(\vec{S}^{<}_{cc}-\vec{S}^{>T}_{vv}\right)\times\left(\vec{\delta}^{a}_{ee,vc}+\vec{\delta}^{r}_{hh,cv}\right)\right]. \tag{22.26}$$

Substituting for the Pauli-blocking term $\vec{S}^{>T}_{vv}$ the relation given by Eq. (22.11), we have,

$$i\hbar\left(\frac{\partial}{\partial t_1}+\frac{\partial}{\partial t_2}\right)\vec{S}_x$$
$$=\frac{1}{2}\begin{bmatrix}-i\hbar\omega_o\vec{S}^{<}_{y}-2i\vec{\mathcal{B}}_{cc}S^{<}_{y.o}\\ -i\vec{\mathcal{B}}_{cc}\times\left(\vec{S}^{<}_{vc}+\vec{S}^{<}_{cv}\right)+i\vec{\mathcal{B}}_{cc}\times\left(\vec{S}^{<}_{vc}+\vec{S}^{<}_{cv}\right)\Longrightarrow 0\end{bmatrix}$$
$$+\frac{1}{2}\Big[\left(\bar{\Delta}^{r}_{ee,vc}-\bar{\Delta}^{a}_{hh,cv}\right)\left(\vec{S}^{<}_{cc}+\left(-i\vec{\mathcal{A}}+\vec{S}^{<T}_{vv}\right)\right)$$
$$+\left(\vec{\delta}^{r}_{ee,vc}-\vec{\delta}^{a}_{hh,cv}\right)\left(S^{<}_{cc,o}+\left(-iA+S^{<T}_{vv,o}\right)\right)\Big]$$
$$-\frac{i}{2}\left[\left(\vec{S}^{<}_{cc}-\left(-i\vec{\mathcal{A}}+\vec{S}^{<T}_{vv}\right)\right)\times\left(\vec{\delta}^{a}_{ee,vc}+\vec{\delta}^{r}_{hh,cv}\right)\right], \tag{22.27}$$

$$i\hbar\left(\frac{\partial}{\partial t_1}+\frac{\partial}{\partial t_2}\right)\vec{S}_x$$
$$=\frac{1}{2}\left[-i\hbar\omega_o\vec{S}^{<}_{y}-2i\vec{\mathcal{B}}_{cc}S^{<}_{y,o}\right]$$
$$+\frac{1}{2}\left[i\Lambda_y\left[\vec{S}^{<}_{z}-i\vec{\mathcal{A}}\right]+i\vec{\delta}_y\left[S^{<}_{z,o}-iA\right]\right]$$
$$-\frac{i}{2}\left[\left[\vec{\Im}+i\vec{\mathcal{A}}\right]\times\vec{\delta}_x\right]. \tag{22.28}$$

We finally end with,[1]

$$i\hbar\left(\frac{\partial}{\partial t_1}+\frac{\partial}{\partial t_2}\right)\vec{S}_x=\frac{1}{2}\left[i\left(\Lambda_y\vec{S}_z^<-\Lambda_z\vec{S}_y^<\right)+i\left(\vec{\delta}_yS_{z,o}^<-\vec{\delta}_zS_{y,o}^<\right)+i\vec{\delta}_x\times\vec{\Im}\right]$$
$$+\vec{\mathcal{A}}\times\vec{\delta}_x+\vec{\delta}_yA+\Lambda_y\vec{\mathcal{A}}. \tag{22.29}$$

22.2.2 *Transport equation for* $i\vec{S}_y=\left(\vec{S}_{vc}^<-\vec{S}_{cv}^<\right)$

We write

$$\begin{aligned}
&i\hbar\left(\frac{\partial}{\partial t_1}+\frac{\partial}{\partial t_2}\right)\left(\vec{S}_{vc}^<-\vec{S}_{cv}^<\right)\\
&=\frac{1}{2}\left[\begin{matrix}\left[\bar{H}_v\vec{S}_{vc}^<-\vec{S}_{vc}^<\bar{H}_c\right]+\left[\vec{\mathcal{B}}_{vv}S_{vc,o}^<-S_{vc,o}^<\vec{\mathcal{B}}_{cc}\right]\\+i\left[\vec{\mathcal{B}}_{vv}\times\vec{S}_{vc}^<-\vec{S}_{vc}^<\times\vec{\mathcal{B}}_{cc}\right]\end{matrix}\right]\\
&\quad+\frac{1}{2}\left[-\left[-\vec{S}_{vv}^{>T}\bar{\Delta}_{ee,vc}^a\right]-\left[-S_{vv,o}^{>T}\vec{\delta}_{ee,vc}^a\right]+i\left[\vec{S}_{vv}^{>T}\times\vec{\delta}_{ee,vc}^a-\right]\right]\\
&\quad+\frac{1}{2}\left[\left[\bar{\Delta}_{ee,vc}^r\vec{S}_{cc}^<-\right]+\left[\vec{\delta}_{ee,vc}^rS_{cc,o}^<-\right]+i\left[\vec{\delta}_{ee,vc}^r\times\vec{S}_{cc}^<-\right]\right],\\
&\quad-\frac{1}{2}\left[\begin{matrix}\left[\bar{H}_c\vec{S}_{cv}^<-\vec{S}_{cv}^<\bar{H}_v\right]+\left[\vec{\mathcal{B}}_{cc}S_{cv,o}^<-S_{cv,o}^<\vec{\mathcal{B}}_{vv}\right]\\+i\left[\vec{\mathcal{B}}_{cc}\times\vec{S}_{cv}^<-\vec{S}_{cv}^<\times\vec{\mathcal{B}}_{vv}\right]\end{matrix}\right]\\
&\quad-\frac{1}{2}\left[\left[-\vec{S}_{cc}^<\bar{\Delta}_{hh,cv}^a\right]+\left[-S_{cc,o}^<\vec{\delta}_{hh,cv}^a\right]+i\left[-\vec{S}_{cc}^<\times\vec{\delta}_{hh,cv}^a\right]\right]\\
&\quad-\frac{1}{2}\left[-\left[\bar{\Delta}_{hh,cv}^r\vec{S}_{vv}^{>T}-\right]-\left[\vec{\delta}_{hh,cv}^rS_{vv,o}^{>T}-\right]+i\left[-\vec{\delta}_{hh,cv}^r\times\vec{S}_{vv}^{,>T}\right]\right],
\end{aligned} \tag{22.30}$$

[1]We note that the cross product of vector and tensor/dyadic, and tensor with tensor, are also tensors/dyadics, i.e.,

$$\begin{aligned}
(\vec{v}\times T)_{i\beta}&=\varepsilon_{ijk}v_jT_{k\beta}=\mathcal{F}_{i\beta}\\
(T\times\vec{v})_{i\beta}&=\varepsilon_{ijk}T_{\beta j}v_k=\mathcal{F}_{\beta i}\\
\vec{T}_j\times\vec{T}_j'&=\epsilon_{\beta ik}T_{ij}T_{kj}'\\
&=\Gamma_{\beta j}\text{ another tensor.}
\end{aligned}$$

$$i\hbar\left(\frac{\partial}{\partial t_1}+\frac{\partial}{\partial t_2}\right)i\vec{S}_y$$
$$=\frac{1}{2}\begin{bmatrix}\left[\bar{H}_v\left(\vec{S}^<_{vc}+\vec{S}^<_{cv}\right)-\left(\vec{S}^<_{vc}+\vec{S}^<_{cv}\right)\bar{H}_c\right]\\+\left[\vec{\mathcal{B}}_{vv}\left(S^<_{vc,o}+S^<_{cv,o}\right)-\left(S^<_{vc,o}+S^<_{cv,o}\right)\vec{\mathcal{B}}_{cc}\right]\\+i\left[\vec{\mathcal{B}}_{vv}\times\left(\vec{S}^<_{vc}-\vec{S}^<_{cv}\right)-\left(\vec{S}^<_{vc}-\vec{S}^<_{cv}\right)\times\vec{\mathcal{B}}_{cc}\right]\end{bmatrix}$$
$$+\frac{1}{2}\vec{S}^{>T}_{vv}\bar{\Delta}^a_{ee,vc}+\frac{1}{2}\vec{S}^<_{cc}\bar{\Delta}^a_{hh,cv}+\frac{1}{2}S^{>T}_{vv,o}\vec{\delta}^a_{ee,vc}+\frac{1}{2}S^<_{cc,o}\vec{\delta}^a_{hh,cv}$$
$$+\frac{1}{2}i\vec{S}^<_{cc}\times\vec{\delta}^a_{hh,cv}+\frac{1}{2}i\vec{S}^{>T}_{vv}\times\vec{\delta}^a_{ee,vc}+\frac{1}{2}\bar{\Delta}^r_{ee,vc}\vec{S}^<_{cc}$$
$$+\frac{1}{2}\bar{\Delta}^r_{hh,cv}\vec{S}^{>T}_{vv}+\frac{1}{2}\vec{\delta}^r_{ee,vc}S^<_{cc,o}+\frac{1}{2}\vec{\delta}^r_{hh,cv}S^{>T}_{vv,o}+\frac{1}{2}i\vec{\delta}^r_{ee,vc}\times\vec{S}^<_{cc}$$
$$+\frac{1}{2}i\vec{\delta}^r_{hh,cv}\times\vec{S}^{>T}_{vv},\tag{22.31}$$

$$i\hbar\left(\frac{\partial}{\partial t_1}+\frac{\partial}{\partial t_2}\right)i\vec{S}_y$$
$$=\frac{1}{2}\left[-2\bar{H}_{cc}\left(\vec{S}^<_{vc}+\vec{S}^<_{cv}\right)-2\vec{\mathcal{B}}_{cc}\left(S^<_{vc,o}+S^<_{cv,o}\right)\right]$$
$$+\frac{1}{2}\vec{S}^{>T}_{vv}\left(\bar{\Delta}^a_{ee,vc}+\bar{\Delta}^r_{hh,cv}\right)+\frac{1}{2}S^{>T}_{vv,o}\left(\vec{\delta}^a_{ee,vc}+\vec{\delta}^r_{hh,cv}\right)$$
$$+\frac{1}{2}i\vec{S}^{>T}_{vv}\times\left(\vec{\delta}^a_{ee,vc}-\vec{\delta}^r_{hh,cv}\right)+\frac{1}{2}\left(\bar{\Delta}^r_{ee,vc}+\bar{\Delta}^a_{hh,cv}\right)\vec{S}^<_{cc}$$
$$+\frac{1}{2}\left(\vec{\delta}^r_{ee,vc}+\vec{\delta}^a_{hh,cv}\right)S^<_{cc,o}+\frac{1}{2}i\left(\vec{\delta}^r_{ee,vc}-\vec{\delta}^a_{hh,cv}\right)\times\vec{S}^<_{cc},\tag{22.32}$$

$$i\hbar\left(\frac{\partial}{\partial t_1}+\frac{\partial}{\partial t_2}\right)i\vec{S}_y$$
$$=\frac{1}{2}\left[-2\bar{H}_{cc}\vec{S}_x-2\vec{\mathcal{B}}_{cc}S_{x,o}\right]+\frac{1}{2}\Lambda_x\left(\vec{S}^<_{cc}+\vec{S}^{>T}_{vv}\right)$$
$$+\frac{1}{2}\vec{\delta}_x\left(S^<_{cc,o}+S^{>T}_{vv,o}\right)+\frac{1}{2}i\left(i\vec{\delta}_y\right)\times\left(\vec{S}^<_{cc}-\vec{S}^{>T}_{vv}\right),\tag{22.33}$$

$$i\hbar\left(\frac{\partial}{\partial t_1}+\frac{\partial}{\partial t_2}\right)\vec{S}_y$$
$$=\frac{i}{2}\left(\Lambda_z\vec{S}_x-\Lambda_x\vec{S}^<_z\right)+\frac{i}{2}\left(\vec{\delta}_zS_{x,o}-\vec{\delta}_xS^<_{z,o}\right)+i\frac{1}{2}\vec{\delta}_y\times\vec{\Im}$$
$$-\frac{1}{2}\vec{\delta}_xA-\frac{1}{2}\Lambda_x\vec{\mathcal{A}}-\frac{1}{2}\vec{\delta}_y\times\vec{\mathcal{A}}.\tag{22.34}$$

22.2.3 *Transport equation for* $\vec{\mathfrak{S}} = \left(\vec{S}^<_{cc} - \vec{S}^<_{vv}\right)$

$$
\begin{aligned}
& i\hbar\left(\frac{\partial}{\partial t_1} + \frac{\partial}{\partial t_2}\right)\left(\vec{S}^<_{cc} - \vec{S}^<_{vv}\right) \\
&\quad = \frac{1}{2}\left\{\left[\bar{H}_c, \vec{S}^<_{cc}\right] + \left[\vec{\mathcal{B}}_{cc}, S^<_{cc,o}\right] + i\left[\vec{\mathcal{B}}_{cc} \times \vec{S}^<_{cc} - \vec{S}^<_{cc} \times \vec{\mathcal{B}}_{cc}\right]\right\} \\
&\qquad + \frac{1}{2}\left\{\left[\bar{H}_v, \vec{S}^<_{vv}\right] + \left[\vec{\mathcal{B}}_{vv}, S^<_{vv,o}\right] + i\left[\vec{\mathcal{B}}_{vv} \times \vec{S}^<_{vv} - \vec{S}^<_{vv} \times \vec{\mathcal{B}}_{vv}\right]\right\} \\
&\qquad + \frac{1}{2}\begin{bmatrix} \frac{1}{i\hbar}\left[\bar{\Delta}^r_{hh,cv}\vec{S}^<_{vc} - \vec{S}^<_{cv}\bar{\Delta}^a_{ee,vc}\right] + \frac{1}{i\hbar}\left[\vec{\delta}^r_{hh,cv}S^<_{vc,o} - S^<_{cv,o}\vec{\delta}^a_{ee,vc}\right] \\ + \frac{1}{\hbar}\left[\vec{\delta}^r_{hh,cv} \times \vec{S}^<_{vc} - \vec{S}^<_{cv} \times \vec{\delta}^a_{ee,vc}\right] \end{bmatrix} \\
&\qquad + \frac{1}{2}\begin{bmatrix} \frac{1}{i\hbar}\left[\bar{\Delta}^r_{ee,vc}\vec{S}^<_{cv} - \vec{S}^<_{vc}\bar{\Delta}^a_{hh,cv}\right] + \frac{1}{i\hbar}\left[\vec{\delta}^r_{ee,vc}S^<_{cv,o} - S^<_{vc,o}\vec{\delta}^a_{hh,cv}\right] \\ + \frac{1}{\hbar}\left[\vec{\delta}^r_{ee,vc} \times \vec{S}^<_{cv} - \vec{S}^<_{vc} \times \vec{\delta}^a_{hh,cv}\right] \end{bmatrix}.
\end{aligned} \tag{22.35}
$$

Write

$$
\begin{aligned}
& \bar{\Delta}^r_{hh,cv}\vec{S}^<_{vc} - \vec{S}^<_{cv}\bar{\Delta}^a_{ee,vc} + \bar{\Delta}^r_{ee,vc}\vec{S}^<_{cv} - \vec{S}^<_{vc}\bar{\Delta}^a_{hh,cv} \\
&\quad = \left(\bar{\Delta}^r_{hh,cv} + \bar{\Delta}^a_{ee,vc}\right)\vec{S}^<_{vc} - \bar{\Delta}^a_{ee,vc}\vec{S}^<_{vc} - \vec{S}^<_{cv}\left(\bar{\Delta}^a_{ee,vc} + \bar{\Delta}^r_{hh,cv}\right) \\
&\qquad + \vec{S}^<_{cv}\bar{\Delta}^r_{hh,cv} + \bar{\Delta}^r_{ee,vc}\vec{S}^<_{cv} - \vec{S}^<_{vc}\bar{\Delta}^a_{hh,cv} \\
&\quad = \left(\bar{\Delta}^a_{ee,vc} + \bar{\Delta}^r_{hh,cv}\right)\left(\vec{S}^<_{vc} - \vec{S}^<_{cv}\right) \\
&\qquad - \left(\bar{\Delta}^a_{ee,vc} + \bar{\Delta}^a_{hh,cv}\right)\vec{S}^<_{vc} + \left(\bar{\Delta}^r_{ee,vc} + \bar{\Delta}^r_{hh,cv}\right)\vec{S}^<_{cv} \\
&\quad = \left(\bar{\Delta}^a_{ee,vc} + \bar{\Delta}^r_{hh,cv}\right)\left(\vec{S}^<_{vc} - \vec{S}^<_{cv}\right) - \left(\bar{\Delta}^a_{ee,vc} + \bar{\Delta}^a_{hh,cv}\right)\left(\vec{S}^<_{vc} - \vec{S}^<_{cv}\right) \\
&\quad = 0.
\end{aligned} \tag{22.36}
$$

Also write

$$
\begin{aligned}
& \left(\vec{\delta}^a_{ee,vc} + \vec{\delta}^r_{hh,cv}\right)S^<_{vc,o} - \vec{\delta}^a_{ee,vc}S^<_{vc,o} - S^<_{cv,o}\left(\vec{\delta}^a_{ee,vc} + \vec{\delta}^r_{hh,cv}\right) \\
&\quad + S^<_{cv,o}\vec{\delta}^r_{hh,cv} + \vec{\delta}^r_{ee,vc}S^<_{cv,o} - S^<_{vc,o}\vec{\delta}^a_{hh,cv} \\
& = \left(\vec{\delta}^a_{ee,vc} + \vec{\delta}^r_{hh,cv}\right)\left(S^<_{vc,o} - S^<_{cv,o}\right) - \left(\vec{\delta}^a_{ee,vc} + \vec{\delta}^a_{hh,cv}\right)\left(S^<_{vc,o} - S^<_{cv,o}\right) = 0.
\end{aligned} \tag{22.37}
$$

We now write

$$\begin{aligned}
&\vec{\delta}^r_{hh,cv} \times \vec{S}^<_{vc} - \vec{S}^<_{cv} \times \vec{\delta}^a_{ee,vc} + \vec{\delta}^r_{ee,vc} \times \vec{S}^<_{cv} - \vec{S}^<_{vc} \times \vec{\delta}^a_{hh,cv} \\
&\quad = \left(\vec{\delta}^r_{hh,cv} + \vec{\delta}^a_{ee,vc}\right) \times \vec{S}^<_{vc} - \vec{\delta}^a_{ee,vc} \times \vec{S}^<_{vc} - \vec{S}^<_{cv} \times \left(\vec{\delta}^a_{ee,vc} + \vec{\delta}^r_{hh,cv}\right) \\
&\qquad + \vec{S}^<_{cv} \times \vec{\delta}^r_{hh,cv} + \vec{\delta}^r_{ee,vc} \times \vec{S}^<_{cv} \ - \vec{S}^<_{vc} \times \vec{\delta}^a_{hh,cv} \\
&\quad = \left(\vec{\delta}^r_{hh,cv} + \vec{\delta}^r_{ee,vc}\right) \times \left(\vec{S}^<_{vc} + \vec{S}^<_{cv}\right) - \left(\vec{\delta}^a_{ee,vc} - \vec{\delta}^a_{hh,cv}\right) \\
&\qquad \times \vec{S}^<_{vc} + \left(\vec{\delta}^r_{ee,vc} - \vec{\delta}^r_{hh,cv}\right) \times \vec{S}^<_{cv} \\
&\quad = \left(\vec{\delta}_{cv} + \vec{\delta}_{vc}\right) \times \left(\vec{S}^<_{vc} + \vec{S}^<_{cv}\right) - \left(\vec{\delta}_{vc} - \vec{\delta}_{cv}\right) \times \left(\vec{S}^<_{vc} - \vec{S}^<_{cv}\right) \\
&\quad = \vec{\delta}_x \times \vec{S}^<_x - i\vec{\delta}_y \times i\vec{S}^<_y \\
&\quad = \vec{\delta}_x \times \vec{S}^<_x + \vec{\delta}_y \times \vec{S}^<_y.
\end{aligned} \tag{22.38}$$

Now also write

$$\begin{aligned}
&\left[\vec{\mathcal{B}}_{cc} \times \vec{S}^<_{cc} - \vec{S}^<_{cc} \times \vec{\mathcal{B}}_{cc}\right] + \left[\vec{\mathcal{B}}_{vv} \times \vec{S}^<_{vv} - \vec{S}^<_{vv} \times \vec{\mathcal{B}}_{vv}\right] \\
&\quad = 2\vec{\mathcal{B}}_{cc} \times \vec{S}^<_{cc} + 2\vec{\mathcal{B}}_{vv} \times \vec{S}^<_{vv} \\
&\quad = 2\frac{1}{2}\vec{\delta}_z \times \vec{S}^<_{cc} - 2\frac{1}{2}\vec{\delta}_z \times \vec{S}^<_{vv} \\
&\quad = \vec{\delta}_z \times \left(\vec{S}^<_{cc} - \vec{S}^<_{vv}\right) \\
&\quad = \vec{\delta}_z \times \vec{S}^<_z.
\end{aligned} \tag{22.39}$$

Therefore

$$\begin{aligned}
i\hbar\left(\frac{\partial}{\partial t_1} + \frac{\partial}{\partial t_2}\right)\vec{\Im} &= \vec{\delta}_x \times \vec{S}^<_x + \vec{\delta}_y \times \vec{S}^<_y + \vec{\delta}_z \times \vec{S}^<_z \\
&= \vec{\delta} \times \vec{S}^<\Big|_{.ps},
\end{aligned} \tag{22.40}$$

which is just equal to a *spin vector*, not a dyadic! The subscript in $|_{.ps}$ indicates contraction/dot product with respect to *pseudo-spin components*.

22.2.4 *Transport equation for* $\vec{S}_z = \left(\vec{S}^<_{cc} + \vec{S}^<_{vv}\right)$

In the electron–hole picture, $\vec{S}_z = \left(\vec{S}^<_{cc} + \vec{S}^<_{vv}\right)$, represent the *net spin* in the system since for holes the spin is directed opposite to that of electrons.

Then, we have,

$$
\begin{aligned}
& i\hbar\left(\frac{\partial}{\partial t_1}+\frac{\partial}{\partial t_2}\right)\left(\vec{S}^<_{cc}+\vec{S}^<_{vv}\right) \\
&= \frac{1}{2}\left\{\left[\bar{H}_c,\vec{S}^<_{cc}\right]+\left[\vec{\mathcal{B}}_{cc},S^<_{cc,o}\right]+i\left[\vec{\mathcal{B}}_{cc}\times\vec{S}^<_{cc}-\vec{S}^<_{cc}\times\vec{\mathcal{B}}_{cc}\right]\right\} \\
&\quad -\frac{1}{2}\left\{\left[\bar{H}_v,\vec{S}^<_{vv}\right]+\left[\vec{\mathcal{B}}_{vv},S^<_{vv,o}\right]+i\left[\vec{\mathcal{B}}_{vv}\times\vec{S}^<_{vv}-\vec{S}^<_{vv}\times\vec{\mathcal{B}}_{vv}\right]\right\} \\
&\quad +\frac{1}{2}\left[\begin{array}{c}\frac{1}{i\hbar}\left[\bar{\Delta}^r_{hh,cv}\vec{S}^<_{vc}-\vec{S}^<_{cv}\bar{\Delta}^a_{ee,vc}\right]+\frac{1}{i\hbar}\left[\vec{\delta}^r_{hh,cv}S^<_{vc,o}-S^<_{cv,o}\vec{\delta}^a_{ee,vc}\right] \\ +\frac{1}{\hbar}\left[\vec{\delta}^r_{hh,cv}\times\vec{S}^<_{vc}-\vec{S}^<_{cv}\times\vec{\delta}^a_{ee,vc}\right]\end{array}\right] \\
&\quad -\frac{1}{2}\left[\begin{array}{c}\frac{1}{i\hbar}\left[\bar{\Delta}^r_{ee,vc}\vec{S}^<_{cv}-\vec{S}^<_{vc}\bar{\Delta}^a_{hh,cv}\right]+\frac{1}{i\hbar}\left[\vec{\delta}^r_{ee,vc}S^<_{cv,o}-S^<_{vc,o}\vec{\delta}^a_{hh,cv}\right] \\ +\frac{1}{\hbar}\left[\vec{\delta}^r_{ee,vc}\times\vec{S}^<_{cv}-\vec{S}^<_{vc}\times\vec{\delta}^a_{hh,cv}\right]\end{array}\right].
\end{aligned}
\tag{22.41}
$$

Write

$$
\begin{aligned}
& i\left(\vec{\mathcal{B}}_{cc}\times\vec{S}^<_{cc}-\vec{S}^<_{cc}\times\vec{\mathcal{B}}_{cc}-\vec{\mathcal{B}}_{vv}\times\vec{S}^<_{vv}+\vec{S}^<_{vv}\times\vec{\mathcal{B}}_{vv}\right) \\
&= i\left(2\vec{\mathcal{B}}_{cc}\times\vec{S}^<_{cc}-2\vec{\mathcal{B}}_{vv}\times\vec{S}^<_{vv}\right) \\
&= i\left(2\frac{1}{2}\vec{\delta}_z\times\vec{S}^<_{cc}+2\frac{1}{2}\vec{\delta}_z\times\vec{S}^<_{vv}\right) \\
&= i\vec{\delta}_z\times\left(\vec{S}^<_{cc}+\vec{S}^<_{vv}\right) \\
&= i\vec{\delta}_z\times\vec{S}^<_z.
\end{aligned}
\tag{22.42}
$$

Write

$$
\begin{aligned}
& \bar{\Delta}^r_{hh,cv}\vec{S}^<_{vc}-\vec{S}^<_{cv}\bar{\Delta}^a_{ee,vc}-\bar{\Delta}^r_{ee,vc}\vec{S}^<_{cv}+\vec{S}^<_{vc}\bar{\Delta}^a_{hh,cv} \\
&= \left(\bar{\Delta}^r_{hh,cv}+\bar{\Delta}^a_{ee,vc}\right)\vec{S}^<_{vc}-\bar{\Delta}^a_{ee,vc}\vec{S}^<_{vc}-\vec{S}^<_{cv}\left(\bar{\Delta}^a_{ee,vc}+\bar{\Delta}^r_{hh,cv}\right) \\
&\quad +\vec{S}^<_{cv}\bar{\Delta}^r_{hh,cv}-\bar{\Delta}^r_{ee,vc}\vec{S}^<_{cv}+\vec{S}^<_{vc}\bar{\Delta}^a_{hh,cv} \\
&= \left(\bar{\Delta}^r_{hh,cv}+\bar{\Delta}^a_{ee,vc}\right)\left(\vec{S}^<_{vc}-\vec{S}^<_{cv}\right) \\
&\quad -\left(\bar{\Delta}^a_{ee,vc}-\bar{\Delta}^a_{hh,cv}\right)\vec{S}^<_{vc}-\left(\bar{\Delta}^r_{ee,vc}-\bar{\Delta}^r_{hh,cv}\right)\vec{S}^<_{cv} \\
&= \left(\bar{\Delta}^r_{hh,cv}+\bar{\Delta}^a_{ee,vc}\right)\left(\vec{S}^<_{vc}-\vec{S}^<_{cv}\right)-\left(\bar{\Delta}^a_{ee,vc}-\bar{\Delta}^a_{hh,cv}\right)\left(\vec{S}^<_{vc}+\vec{S}^<_{cv}\right)
\end{aligned}
$$

$$
\begin{aligned}
&= \bar{\Delta}_x i\vec{S}^{<}_{y} - i\bar{\Delta}_y \vec{S}^{<}_{x} \\
&= i\left(\bar{\Delta}_x \vec{S}^{<}_{y} - \bar{\Delta}_y \vec{S}^{<}_{x}\right).
\end{aligned} \tag{22.43}
$$

Similarly, write

$$
\begin{aligned}
&\vec{\delta}^{r}_{hh,cv} S^{<}_{vc,o} - S^{<}_{cv,o}\vec{\delta}^{a}_{ee,vc} - \vec{\delta}^{r}_{ee,vc} S^{<}_{cv,o} + S^{<}_{vc,o}\vec{\delta}^{a}_{hh,cv} \\
&\quad = \left(\vec{\delta}^{r}_{hh,cv} + \vec{\delta}^{a}_{ee,vc}\right) S^{<}_{vc,o} - \vec{\delta}^{a}_{ee,vc} S^{<}_{vc,o} - S^{<}_{cv,o}\left(\vec{\delta}^{a}_{ee,vc} + \vec{\delta}^{r}_{hh,cv}\right) \\
&\qquad + S^{<}_{cv,o}\vec{\delta}^{r}_{hh,cv} - \vec{\delta}^{r}_{ee,vc} S^{<}_{cv,o} + S^{<}_{vc,o}\vec{\delta}^{a}_{hh,cv} \\
&\quad = \left(\vec{\delta}^{r}_{hh,cv} + \vec{\delta}^{a}_{ee,vc}\right)\left(S^{<}_{vc,o} - S^{<}_{cv,o}\right) \\
&\qquad - \left(\vec{\delta}^{a}_{ee,vc} - \vec{\delta}^{a}_{hh,cv}\right) S^{<}_{vc,o} - \left(\vec{\delta}^{r}_{ee,vc} - \vec{\delta}^{r}_{hh,cv}\right) S^{<}_{cv,o} \\
&\quad = \left(\vec{\delta}_{cv} + \vec{\delta}_{vc}\right)\left(S^{<}_{vc,o} - S^{<}_{cv,o}\right) - \left(\vec{\delta}_{vc} - \vec{\delta}_{cv}\right)\left(S^{<}_{vc,o} + S^{<}_{cv,o}\right) \\
&\quad = i\vec{\delta}_x S^{<}_{y,o} - i\vec{\delta}_y S^{<}_{x,o}.
\end{aligned} \tag{22.44}
$$

Now also write

$$
\begin{aligned}
&\vec{\delta}^{r}_{hh,cv} \times \vec{S}^{<}_{vc} - \vec{S}^{<}_{cv} \times \vec{\delta}^{a}_{ee,vc} - \vec{\delta}^{r}_{ee,vc} \times \vec{S}^{<}_{cv} + \vec{S}^{<}_{vc} \times \vec{\delta}^{a}_{hh,cv} \\
&\quad = \left(\vec{\delta}^{r}_{hh,cv} + \vec{\delta}^{a}_{ee,vc}\right) \times \vec{S}^{<}_{vc} - \vec{\delta}^{a}_{ee,vc} \times \vec{S}^{<}_{vc} - \vec{S}^{<}_{cv} \times \left(\vec{\delta}^{a}_{ee,vc} + \vec{\delta}^{r}_{hh,cv}\right) \\
&\qquad + \vec{S}^{<}_{cv} \times \vec{\delta}^{r}_{hh,cv} - \vec{\delta}^{r}_{ee,vc} \times \vec{S}^{<}_{cv} + \vec{S}^{<}_{vc} \times \vec{\delta}^{a}_{hh,cv} \\
&\quad = \left(\vec{\delta}^{r}_{hh,cv} + \vec{\delta}^{a}_{ee,vc}\right) \times \left(\vec{S}^{<}_{vc} + \vec{S}^{<}_{cv}\right) - \left(\vec{\delta}^{a}_{ee,vc} + \vec{\delta}^{a}_{hh,cv}\right) \\
&\qquad \times \vec{S}^{<}_{vc} - \left(\vec{\delta}^{r}_{ee,vc} + \vec{\delta}^{r}_{hh,cv}\right) \times \vec{S}^{<}_{cv} \\
&\quad = \left(\vec{\delta}^{r}_{hh,cv} + \vec{\delta}^{a}_{ee,vc}\right) \times \left(\vec{S}^{<}_{vc} + \vec{S}^{<}_{cv}\right) - \left(\vec{\delta}^{a}_{ee,vc} + \vec{\delta}^{a}_{hh,cv}\right) \times \left(\vec{S}^{<}_{vc} + \vec{S}^{<}_{cv}\right) \\
&\quad = 0.
\end{aligned} \tag{22.45}
$$

Therefore,

$$
i\hbar\left(\frac{\partial}{\partial t_1} + \frac{\partial}{\partial t_2}\right)\vec{S}^{<}_{z} = i\vec{\delta}_z \times \vec{S}^{<}_{z} + i\left(\bar{\Delta}_x \vec{S}^{<}_{y} - \bar{\Delta}_y \vec{S}^{<}_{x}\right) + i\left(\vec{\delta}_x S^{<}_{y,o} - \vec{\delta}_y S^{<}_{x,o}\right). \tag{22.46}
$$

Thus, in the electron–hole picture it is more convenient to define appropriately the *net spin* as $\vec{S}^{<}_{z} = \left(\vec{S}^{<}_{cc} + \vec{S}^{<}_{vv}\right)$, a dyadic component, and instead use a scalar in band indices (only a spin vector) as $\vec{\Im} = \left(\vec{S}^<_{cc} - \vec{S}^<_{vv}\right)$. Then we have

$$i\hbar\left(\frac{\partial}{\partial t_1}+\frac{\partial}{\partial t_2}\right)\vec{S}_x = \frac{1}{2}\left[i\left(\Lambda_y\vec{S}^<_z - \Lambda_z\vec{S}^<_y\right) + i\left(\vec{\delta}_y S^<_{z,o} - \vec{\delta}_z S^<_{y,o}\right) + i\,\vec{\delta}_x \times \vec{\Im}\right] + \vec{\mathcal{A}}\times\vec{\delta}_x + \vec{\delta}_y A + \Lambda_y\vec{\mathcal{A}}, \tag{22.47}$$

$$i\hbar\left(\frac{\partial}{\partial t_1}+\frac{\partial}{\partial t_2}\right)\vec{S}_y = \frac{1}{2}\left[i\left(\Lambda_z\vec{S}_x - \Lambda_x\vec{S}^<_z\right) + i\left(\vec{\delta}_z S_{x,o} - \vec{\delta}_x S^<_{z,o}\right) + i\vec{\delta}_y \times \vec{\Im}\right] - \frac{1}{2}\vec{\delta}_y\times\vec{\mathcal{A}} - \frac{1}{2}\vec{\delta}_x A - \frac{1}{2}\Lambda_x\vec{\mathcal{A}}, \tag{22.48}$$

$$i\hbar\left(\frac{\partial}{\partial t_1}+\frac{\partial}{\partial t_2}\right)\vec{S}^<_z = i\left(\bar{\Delta}_x\vec{S}^<_y - \bar{\Delta}_y\vec{S}^<_x\right) + i\left(\vec{\delta}_x S^<_{y,o} - \vec{\delta}_y S^<_{x,o}\right) + i\vec{\delta}_z\times\vec{S}^<_z, \tag{22.49}$$

$$i\hbar\left(\frac{\partial}{\partial t_1}+\frac{\partial}{\partial t_2}\right)\vec{\Im} = \vec{\delta}_x\times\vec{S}^<_x + \vec{\delta}_y\times\vec{S}^<_y + \vec{\delta}_z\times\vec{S}^<_z, \text{ a spin vector only.} \tag{22.50}$$

22.3 Electron picture

In the electron picture which is appropriate for two level atom, the equations acquire a much simpler form. In Eqs. (22.1)–(22.4), we have to change the sign of the RHS of Eq. (22.4), and change $\vec{S}^{>T}_{vv} \Longrightarrow -\vec{S}^<_{vv}$, using the conversion Table 21.1, wherever this occur in all of these equations. Now evaluate

$$\begin{aligned} &i\hbar\left(\frac{\partial}{\partial t_1}+\frac{\partial}{\partial t_2}\right)\left(\vec{S}^<_{cc} - \vec{S}^<_{vv}\right) \\ &= \frac{1}{2}\left\{\left[\bar{H}_c, \vec{S}^<_{cc}\right] + \left[\vec{\mathcal{B}}_{cc}, S^<_{cc,o}\right] + i\left[\vec{\mathcal{B}}_{cc}\times\vec{S}^<_{cc} - \vec{S}^<_{cc}\times\vec{\mathcal{B}}_{cc}\right]\right\} \\ &\quad - \frac{1}{2}\left\{\left[\bar{H}_v, \vec{S}^<_{vv}\right] + \left[\vec{\mathcal{B}}_{vv}, S^<_{vv,o}\right] + i\left[\vec{\mathcal{B}}_{vv}\times\vec{S}^<_{vv} - \vec{S}^<_{vv}\times\vec{\mathcal{B}}_{vv}\right]\right\} \\ &\quad + \frac{1}{2}\left[\begin{array}{c} \frac{1}{i\hbar}\left[\bar{\Delta}^r_{hh,cv}\vec{S}^<_{vc} - \vec{S}^<_{cv}\bar{\Delta}^a_{ee,vc}\right] + \frac{1}{i\hbar}\left[\vec{\delta}^r_{hh,cv}S^<_{vc,o} - S^<_{cv,o}\vec{\delta}^a_{ee,vc}\right] \\ +\frac{1}{\hbar}\left[\vec{\delta}^r_{hh,cv}\times\vec{S}^<_{vc} - \vec{S}^<_{cv}\times\vec{\delta}^a_{ee,vc}\right] \end{array}\right] \\ &\quad - \frac{1}{2}\left[\begin{array}{c} \frac{1}{i\hbar}\left[\bar{\Delta}^r_{ee,vc}\vec{S}^<_{cv} - \vec{S}^<_{vc}\bar{\Delta}^a_{hh,cv}\right] + \frac{1}{i\hbar}\left[\vec{\delta}^r_{ee,vc}S^<_{cv,o} - S^<_{vc,o}\vec{\delta}^a_{hh,cv}\right] \\ +\frac{1}{\hbar}\left[\vec{\delta}^r_{ee,vc}\times\vec{S}^<_{cv} - \vec{S}^<_{vc}\times\vec{\delta}^a_{hh,cv}\right] \end{array}\right], \end{aligned} \tag{22.51}$$

$$
\begin{aligned}
&i\hbar\left(\frac{\partial}{\partial t_1}+\frac{\partial}{\partial t_2}\right)\left(\vec{S}^<_{cc}-\vec{S}^<_{vv}\right)\\
&\quad=\frac{1}{2}\left\{\begin{array}{c}\left(\left[\frac{1}{2}\hbar\omega_o,\vec{S}^<_{cc}\right]+\left[\frac{1}{2}\hbar\omega_o,\vec{S}^<_{vv}\right]\right)+\left(\left[\frac{1}{2}\vec{\delta}_z,S^<_{cc,o}\right]+\left[\frac{1}{2}\vec{\delta}_z,S^<_{vv,o}\right]\right)\\ +i\left(\vec{\delta}_z\times\left(\vec{S}^<_{cc}+\vec{S}^<_{vv}\right)\right)\end{array}\right\}\\
&\qquad+\frac{1}{2}\left[\begin{array}{c}\left[\bar{\Delta}^r_{hh,cv}\vec{S}^<_{vc}-\vec{S}^<_{cv}\bar{\Delta}^a_{ee,vc}\right]-\left[\bar{\Delta}^r_{ee,vc}\vec{S}^<_{cv}-\vec{S}^<_{vc}\bar{\Delta}^a_{hh,cv}\right]\\ +\left[\vec{\delta}^r_{hh,cv}S^<_{vc,o}-S^<_{cv,o}\vec{\delta}^a_{ee,vc}\right]-\left[\vec{\delta}^r_{ee,vc}S^<_{cv,o}-S^<_{vc,o}\vec{\delta}^a_{hh,cv}\right]\\ +i\left[\vec{\delta}^r_{hh,cv}\times\vec{S}^<_{vc}-\vec{S}^<_{cv}\times\vec{\delta}^a_{ee,vc}\right]-i\left[\vec{\delta}^r_{ee,vc}\times\vec{S}^<_{cv}-\vec{S}^<_{vc}\times\vec{\delta}^a_{hh,cv}\right]\end{array}\right].
\end{aligned}
\tag{22.52}
$$

Consider the terms,

$$
\begin{aligned}
&\left[\bar{\Delta}^r_{hh,cv}\vec{S}^<_{vc}-\vec{S}^<_{cv}\bar{\Delta}^a_{ee,vc}\right]-\left[\bar{\Delta}^r_{ee,vc}\vec{S}^<_{cv}-\vec{S}^<_{vc}\bar{\Delta}^a_{hh,cv}\right]\\
&\quad=\bar{\Delta}^r_{hh,cv}\vec{S}^<_{vc}-\vec{S}^<_{cv}\bar{\Delta}^a_{ee,vc}-\bar{\Delta}^r_{ee,vc}\vec{S}^<_{cv}+\vec{S}^<_{vc}\bar{\Delta}^a_{hh,cv}\\
&\quad=\left(\bar{\Delta}^r_{hh,cv}+\bar{\Delta}^a_{ee,vc}\right)\vec{S}^<_{vc}-\bar{\Delta}^a_{ee,vc}\vec{S}^<_{vc}-\vec{S}^<_{cv}\left(\bar{\Delta}^a_{ee,vc}+\bar{\Delta}^r_{hh,cv}\right)\\
&\qquad+\vec{S}^<_{cv}\bar{\Delta}^r_{hh,cv}-\bar{\Delta}^r_{ee,vc}\vec{S}^<_{cv}+\vec{S}^<_{vc}\bar{\Delta}^a_{hh,cv}\\
&\quad=\left(\bar{\Delta}^r_{hh,cv}+\bar{\Delta}^a_{ee,vc}\right)\left(\vec{S}^<_{vc}-\vec{S}^<_{cv}\right)-\left(\bar{\Delta}^a_{ee,vc}-\bar{\Delta}^a_{hh,cv}\right)\vec{S}^<_{vc}\\
&\qquad-\left(\bar{\Delta}^r_{ee,vc}-\bar{\Delta}^r_{hh,cv}\right)\vec{S}^<_{cv}\\
&\quad=\left(\bar{\Delta}^r_{hh,cv}+\bar{\Delta}^a_{ee,vc}\right)\left(\vec{S}^<_{vc}-\vec{S}^<_{cv}\right)-\left(\bar{\Delta}^a_{ee,vc}-\bar{\Delta}^a_{hh,cv}\right)\left(\vec{S}^<_{vc}+\vec{S}^<_{cv}\right)\\
&\quad=\bar{\Delta}_x i\vec{S}^<_y-i\bar{\Delta}_y\vec{S}^<_x\\
&\quad=i\left(\bar{\Delta}_x\vec{S}^<_y-\bar{\Delta}_y\vec{S}^<_x\right).
\end{aligned}
\tag{22.53}
$$

Also

$$
\begin{aligned}
&\left[\vec{\delta}^r_{hh,cv}S^<_{vc,o}-S^<_{cv,o}\vec{\delta}^a_{ee,vc}\right]-\left[\vec{\delta}^r_{ee,vc}S^<_{cv,o}-S^<_{vc,o}\vec{\delta}^a_{hh,cv}\right]\\
&\quad=\vec{\delta}^r_{hh,cv}S^<_{vc,o}-S^<_{cv,o}\vec{\delta}^a_{ee,vc}-\vec{\delta}^r_{ee,vc}S^<_{cv,o}+S^<_{vc,o}\vec{\delta}^a_{hh,cv}\\
&\quad=\left(\vec{\delta}^r_{hh,cv}+\vec{\delta}^a_{ee,vc}\right)S^<_{vc,o}-\vec{\delta}^a_{ee,vc}S^<_{vc,o}-S^<_{cv,o}\left(\vec{\delta}^a_{ee,vc}+\vec{\delta}^r_{hh,cv}\right)
\end{aligned}
$$

$$
\begin{aligned}
&+ S^{<}_{cv,o}\vec{\delta}^{r}_{hh,cv} - \vec{\delta}^{r}_{ee,vc}S^{<}_{cv,o} + S^{<}_{vc,o}\vec{\delta}^{a}_{hh,cv} \\
&= \left(\vec{\delta}^{r}_{hh,cv} + \vec{\delta}^{a}_{ee,vc}\right)\left(S^{<}_{vc,o} - S^{<}_{cv,o}\right) - \left(\vec{\delta}^{a}_{ee,vc} - \vec{\delta}^{a}_{hh,cv}\right) S^{<}_{vc,o} \\
&\quad - \left(\vec{\delta}^{r}_{ee,vc} - \vec{\delta}^{r}_{hh,cv}\right) S^{<}_{cv,o} \\
&= \left(\vec{\delta}_{cv} + \vec{\delta}_{vc}\right)\left(S^{<}_{vc,o} - S^{<}_{cv,o}\right) - \left(\vec{\delta}_{vc} - \vec{\delta}_{cv}\right)\left(S^{<}_{vc,o} + S^{<}_{cv,o}\right) \\
&= i\vec{\delta}_x S^{<}_{y,o} - i\vec{\delta}_y S^{<}_{x,o},
\end{aligned} \tag{22.54}
$$

and further, we have

$$
\begin{aligned}
&\left[\vec{\delta}^{r}_{hh,cv} \times \vec{S}^{<}_{vc} - \vec{S}^{<}_{cv} \times \vec{\delta}^{a}_{ee,vc}\right] - \left[\vec{\delta}^{r}_{ee,vc} \times \vec{S}^{<}_{cv} - \vec{S}^{<}_{vc} \times \vec{\delta}^{a}_{hh,cv}\right] \\
&= \vec{\delta}^{r}_{hh,cv} \times \vec{S}^{<}_{vc} - \vec{S}^{<}_{cv} \times \vec{\delta}^{a}_{ee,vc} - \vec{\delta}^{r}_{ee,vc} \times \vec{S}^{<}_{cv} + \vec{S}^{<}_{vc} \times \vec{\delta}^{a}_{hh,cv} \\
&= \left(\vec{\delta}^{r}_{hh,cv} + \vec{\delta}^{a}_{ee,vc}\right) \times \vec{S}^{<}_{vc} - \vec{\delta}^{a}_{ee,vc} \times \vec{S}^{<}_{vc} - \vec{S}^{<}_{cv} \times \left(\vec{\delta}^{a}_{ee,vc} + \vec{\delta}^{r}_{hh,cv}\right) \\
&\quad + \vec{S}^{<}_{cv} \times \vec{\delta}^{r}_{hh,cv} - \vec{\delta}^{r}_{ee,vc} \times \vec{S}^{<}_{cv} + \vec{S}^{<}_{vc} \times \vec{\delta}^{a}_{hh,cv} \\
&= \left(\vec{\delta}^{r}_{hh,cv} + \vec{\delta}^{a}_{ee,vc}\right) \times \left(\vec{S}^{<}_{vc} + \vec{S}^{<}_{cv}\right) - \left(\vec{\delta}^{a}_{ee,vc} + \vec{\delta}^{a}_{hh,cv}\right) \\
&\quad \times \vec{S}^{<}_{vc} - \left(\vec{\delta}^{r}_{ee,vc} + \vec{\delta}^{r}_{hh,cv}\right) \times \vec{S}^{<}_{cv} \\
&= \left(\vec{\delta}^{r}_{hh,cv} + \vec{\delta}^{a}_{ee,vc}\right) \times \left(\vec{S}^{<}_{vc} + \vec{S}^{<}_{cv}\right) - \left(\vec{\delta}^{a}_{ee,vc} + \vec{\delta}^{a}_{hh,cv}\right) \times \left(\vec{S}^{<}_{vc} + \vec{S}^{<}_{cv}\right) \\
&= 0.
\end{aligned} \tag{22.55}
$$

Therefore, we have

$$
\begin{aligned}
&i\hbar\left(\frac{\partial}{\partial t_1} + \frac{\partial}{\partial t_2}\right)\left(\vec{S}^{<}_{cc} - \vec{S}^{<}_{vv}\right) \\
&= \frac{1}{2}\left[i\vec{\delta}_z \times \vec{S}^{<}_{T} + i\left(\bar{\Delta}_x \vec{S}^{<}_{y} - \bar{\Delta}_y \vec{S}^{<}_{x}\right) + i\left(\vec{\delta}_x S^{<}_{y,o} - \vec{\delta}_y S^{<}_{x,o}\right)\right],
\end{aligned} \tag{22.56}
$$

where $\vec{S}^{<}_{T} = \vec{S}^{<}_{cc} + \vec{S}^{<}_{vv}$ indicates the total electron spin, a vector quantity not a dyadic, and $\vec{\delta}_z$ contains the effect of external magnetic field and spin-orbit coupling. Thus this term represent the precession of the vector $\vec{S}^{<}_{T}$ about the vector, $\vec{\delta}_z$. We end up with

$$
\begin{aligned}
i\hbar\left(\frac{\partial}{\partial t_1} + \frac{\partial}{\partial t_2}\right)\vec{S}^{<}_{z} &= \frac{i}{2}\left(\bar{\Delta}_x \vec{S}^{<}_{y} - \bar{\Delta}_y \vec{S}^{<}_{x}\right) \\
&\quad + \frac{i}{2}\left(\vec{\delta}_x S^{<}_{y,o} - \vec{\delta}_y S^{<}_{x,o}\right) + \frac{i}{2}\vec{\delta}_z \times \vec{S}^{<}_{T}.
\end{aligned} \tag{22.57}
$$

Note that the equation for $\left(\vec{S}_{cc}^{<}-\vec{S}_{vv}^{<}\right)$ in the *electron representation* for the two-level atom is formally identical to the equation for $\left(\vec{S}_{cc}^{<}+\vec{S}_{vv}^{<}\right)$ in the *electron–hole representation.*

Now we evaluate the terms involving the off-diagonal in band indices. We have

$$
\begin{aligned}
&i\hbar\left(\frac{\partial}{\partial t_1}+\frac{\partial}{\partial t_2}\right)\left(\vec{S}_{vc}^{<}-\vec{S}_{cv}^{<}\right)\\
&\quad=\frac{1}{2}\left[\begin{array}{c}\left[\bar{H}_v\vec{S}_{vc}^{<}-\vec{S}_{vc}^{<}\bar{H}_c\right]+\left[\vec{\mathcal{B}}_{vv}S_{vc,o}^{<}-S_{vc,o}^{<}\vec{\mathcal{B}}_{cc}\right]\\ +i\left[\vec{\mathcal{B}}_{vv}\times\vec{S}_{vc}^{<}-\vec{S}_{vc}^{<}\times\vec{\mathcal{B}}_{cc}\right]\end{array}\right]\\
&\quad\quad-\frac{1}{2}\left[\begin{array}{c}\left[\bar{H}_c\vec{S}_{cv}^{<}-\vec{S}_{cv}^{<}\bar{H}_v\right]+\left[\vec{\mathcal{B}}_{cc}S_{cv,o}^{<}-S_{cv,o}^{<}\vec{\mathcal{B}}_{vv}\right]\\ +i\left[\vec{\mathcal{B}}_{cc}\times\vec{S}_{cv}^{<}-\vec{S}_{cv}^{<}\times\vec{\mathcal{B}}_{vv}\right]\end{array}\right]\\
&\quad\quad-\frac{1}{2}\left[\left[\vec{S}_{vv}^{<}\bar{\Delta}_{ee,vc}^{a}\right]+\left[\vec{S}_{vv,o}^{<}\vec{\delta}_{ee,vc}^{a}\right]+i\left[\vec{S}_{vv}^{<}\times\vec{\delta}_{ee,vc}^{a}\right]\right]\\
&\quad\quad+\frac{1}{2}\left[\vec{S}_{cc}^{<}\bar{\Delta}_{hh,cv}^{a}+S_{cc,o}^{<}\vec{\delta}_{hh,cv}^{a}+i\ \vec{S}_{cc}^{<}\times\vec{\delta}_{hh,cv}^{a}\right]\\
&\quad\quad+\frac{1}{2}\left[\bar{\Delta}_{ee,vc}^{r}\vec{S}_{cc}^{<}+\vec{\delta}_{ee,vc}^{r}S_{cc,o}^{<}+i\vec{\delta}_{ee,vc}^{r}\times\vec{S}_{cc}^{<}\right]\\
&\quad\quad-\frac{1}{2}\left[\bar{\Delta}_{hh,cv}^{r}\vec{S}_{vv}^{<}+\vec{\delta}_{hh,cv}^{r}S_{vv,o}^{<}+i\vec{\delta}_{hh,cv}^{r}\times\vec{S}_{vv}^{<}\right],
\end{aligned}
\tag{22.58}
$$

$$
\begin{aligned}
&i\hbar\left(\frac{\partial}{\partial t_1}+\frac{\partial}{\partial t_2}\right)\left(\vec{S}_{vc}^{<}-\vec{S}_{cv}^{<}\right)\\
&\quad=\frac{1}{2}\left[\begin{array}{c}\left[-2\frac{1}{2}\hbar\omega_o\left(\vec{S}_{vc}^{<}+\vec{S}_{cv}^{<}\right)\right]+\left[-2\frac{1}{2}\vec{\delta}_z\left(S_{vc,o}^{<}+S_{cv,o}^{<}\right)\right]\\ +i\left[-\frac{1}{2}\vec{\delta}_z\times\left(\vec{S}_{vc}^{<}-\vec{S}_{cv}^{<}\right)-\left(\vec{S}_{vc}^{<}-\vec{S}_{cv}^{<}\right)\times\frac{1}{2}\vec{\delta}_z\right]\end{array}\right]\\
&\quad\quad-\frac{1}{2}\left[\begin{array}{l}-\left(\bar{\Delta}_{ee,vc}^{r}+\bar{\Delta}_{hh,cv}^{a}\right)\left(\vec{S}_{cc}^{<}-\vec{S}_{vv}^{<}\right)\\ -\left(\vec{\delta}_{ee,vc}^{r}+\vec{\delta}_{hh,cv}^{a}\right)\left(S_{cc,o}^{<}-\vec{S}_{vv,o}^{<}\right)\\ -i\left(\vec{\delta}_{ee,vc}^{r}-\vec{\delta}_{hh,cv}^{a}\right)\times\left(\vec{S}_{cc}^{<}+\vec{S}_{vv}^{<}\right)\end{array}\right],
\end{aligned}
\tag{22.59}
$$

$$
\begin{aligned}
&i\hbar\left(\frac{\partial}{\partial t_1}+\frac{\partial}{\partial t_2}\right)\left(\vec{S}_{vc}^{<}-\vec{S}_{cv}^{<}\right)\\
&\quad=\frac{1}{2}\left[\begin{array}{c}\bar{\Delta}_x^r\vec{S}_z^{<}-\bar{\Delta}_z^r\vec{S}_x^{<}-\vec{\delta}_zS_{x,o}^{<}+\vec{\delta}_x^rS_{z,o}^{<}\\ -\left[-\frac{1}{2}\vec{\delta}_z\times\vec{S}_y^{<}+\frac{1}{2}\vec{\delta}_z\times\vec{S}_y^{<}\right]\end{array}\right]-\frac{1}{2}\vec{\delta}_y^r\times\vec{S}_T^{<},
\end{aligned}
\tag{22.60}
$$

$$i\hbar\left(\frac{\partial}{\partial t_1}+\frac{\partial}{\partial t_2}\right)\vec{S}_y^<$$
$$=-\frac{i}{2}\left[\bar{\Delta}_x^r\vec{S}_z^<-\bar{\Delta}_z^r\vec{S}_x^<-\vec{\delta}_z S_{x,o}^<+\vec{\delta}_x^r S_{z,o}^<\right]+\frac{i}{2}\vec{\delta}_y^r\times\vec{S}_T^<$$
$$=+\frac{i}{2}\left[-\bar{\Delta}_x^r\vec{S}_z^<+\bar{\Delta}_z^r\vec{S}_x^<+\vec{\delta}_z S_{x,o}^<-\vec{\delta}_x^r S_{z,o}^<\right]+\frac{i}{2}\vec{\delta}_y^r\times\vec{S}_T^<$$
$$=\frac{i}{2}\left[\left(\bar{\Delta}_z^r\vec{S}_x^<-\bar{\Delta}_x^r\vec{S}_z^<\right)+\left(\vec{\delta}_z S_{x,o}^<-\vec{\delta}_x^r S_{z,o}^<\right)\right]+\frac{i}{2}\vec{\delta}_y^r\times\vec{S}_T^<. \quad (22.61)$$

Next we have,

$$i\hbar\left(\frac{\partial}{\partial t_1}+\frac{\partial}{\partial t_2}\right)\left(\vec{S}_{vc}^<+\vec{S}_{cv}^<\right)$$
$$=\frac{1}{2}\begin{bmatrix}\left[\bar{H}_v\vec{S}_{vc}^<-\vec{S}_{vc}^<\bar{H}_c\right]+\left[\vec{\mathcal{B}}_{vv}S_{vc,o}^<-S_{vc,o}^<\vec{\mathcal{B}}_{cc}\right]\\ +i\left[\vec{\mathcal{B}}_{vv}\times\vec{S}_{vc}^<-\vec{S}_{vc}^<\times\vec{\mathcal{B}}_{cc}\right]\end{bmatrix}$$
$$+\frac{1}{2}\begin{bmatrix}\left[\bar{H}_c\vec{S}_{cv}^<-\vec{S}_{cv}^<\bar{H}_v\right]+\left[\vec{\mathcal{B}}_{cc}S_{cv,o}^<-S_{cv,o}^<\vec{\mathcal{B}}_{vv}\right]\\ +i\left[\vec{\mathcal{B}}_{cc}\times\vec{S}_{cv}^<-\vec{S}_{cv}^<\times\vec{\mathcal{B}}_{vv}\right]\end{bmatrix}$$
$$-\frac{1}{2}\left[\left[\vec{S}_{vv}^<\left(\bar{\Delta}_{ee,vc}^a-\bar{\Delta}_{hh,cv}^r\right)\right]+\left[S_{vv,o}^<\left(\vec{\delta}_{ee,vc}^a-\vec{\delta}_{hh,cv}^r\right)\right]\right.$$
$$\left.+i\vec{S}_{vv}^<\times\left(\vec{\delta}_{ee,vc}^a+\vec{\delta}_{hh,cv}^r\right)\right]+\frac{1}{2}\left[\left(\bar{\Delta}_{ee,vc}^r-\bar{\Delta}_{hh,cv}^a\right)\vec{S}_{cc}^<\right.$$
$$\left.+\left(\vec{\delta}_{ee,vc}^r-\vec{\delta}_{hh,cv}^a\right)S_{cc,o}^<+i\left(\vec{\delta}_{ee,vc}^r+\vec{\delta}_{hh,cv}^a\right)\times\vec{S}_{cc}^<\right], \quad (22.62)$$

$$i\hbar\left(\frac{\partial}{\partial t_1}+\frac{\partial}{\partial t_2}\right)\left(\vec{S}_{vc}^<+\vec{S}_{cv}^<\right)$$
$$=\frac{1}{2}\begin{bmatrix}\left[-2\tfrac{1}{2}\hbar\omega_o\vec{S}_{vc}^<\right]+\left[-2\tfrac{1}{2}\vec{\delta}_z S_{vc,o}^<\right]\\ +i\left[-\tfrac{1}{2}\vec{\delta}_z\times\vec{S}_{vc}^<-\vec{S}_{vc}^<\times\tfrac{1}{2}\vec{\delta}_z\right]\end{bmatrix}$$
$$+\frac{1}{2}\begin{bmatrix}2\tfrac{1}{2}\hbar\omega_o\vec{S}_{cv}^<+2\tfrac{1}{2}\vec{\delta}_z S_{cv,o}^<\\ +i\left[\tfrac{1}{2}\vec{\delta}_z\times\vec{S}_{cv}^<-\vec{S}_{cv}^<\times\left(-\tfrac{1}{2}\vec{\delta}_z\right)\right]\end{bmatrix}$$
$$-\frac{1}{2}\left[+\vec{S}_{vv,o}^<\vec{\delta}_{ee,vc}^a+i\vec{S}_{vv}^<\times\vec{\delta}_{ee,vc}^a\right]$$

$$+\frac{1}{2}\left[\bar{\Delta}^{r}_{ee,vc}\left(\vec{S}^{<}_{cc}-\vec{S}^{<}_{vv}\right)+\vec{\delta}^{r}_{ee,vc}\left(S^{<}_{cc,o}-\right)+i\vec{\delta}^{r}_{ee,vc}\times\vec{S}^{<}_{cc}\right]$$

$$-\frac{1}{2}\left[\vec{S}^{<}_{cc}\bar{\Delta}^{a}_{hh,cv}+S^{<}_{cc,o}\vec{\delta}^{a}_{hh,cv}+i\vec{S}^{<}_{cc}\times\vec{\delta}^{a}_{hh,cv}\right]$$

$$+\frac{1}{2}\left[\bar{\Delta}^{r}_{hh,cv}\vec{S}^{<}_{vv}+\vec{\delta}^{r}_{hh,cv}S^{<}_{vv,o}+i\vec{\delta}^{r}_{hh,cv}\times\vec{S}^{<}_{vv}\right], \tag{22.63}$$

$$i\hbar\left(\frac{\partial}{\partial t_1}+\frac{\partial}{\partial t_2}\right)\vec{S}^{<}_{x}$$

$$=\frac{i}{2}\left[\left(\bar{\Delta}_{y}\vec{S}^{<}_{z}-\bar{\Delta}_{z}\vec{S}^{<}_{y}\right)+\left(\vec{\delta}_{y}\left(S^{<}_{z,o}\right)-\vec{\delta}_{z}S^{<}_{y,o}\right)+\vec{\delta}^{r}_{x}\times\vec{S}^{<}_{T}\right]. \tag{22.64}$$

Summarizing the results in the electron picture, we have

$$i\hbar\frac{\partial}{\partial t}\vec{S}^{<}_{x}=\frac{i}{2}\left[\left(\bar{\Delta}_{y}\vec{S}^{<}_{z}-\bar{\Delta}_{z}\vec{S}^{<}_{y}\right)+\left(\vec{\delta}_{y}\left(S^{<}_{z,o}\right)-\vec{\delta}_{z}S^{<}_{y,o}\right)+\vec{\delta}^{r}_{x}\times\vec{S}^{<}_{T}\right], \tag{22.65}$$

$$i\hbar\frac{\partial}{\partial t}\vec{S}^{<}_{y}=\frac{i}{2}\left[\left(\bar{\Delta}^{r}_{z}\vec{S}^{<}_{x}-\bar{\Delta}^{r}_{x}\vec{S}^{<}_{z}\right)+\left(\vec{\delta}_{z}S^{<}_{x,o}-\vec{\delta}^{r}_{x}S^{<}_{z,o}\right)+\vec{\delta}^{r}_{y}\times\vec{S}^{<}_{T}\right], \tag{22.66}$$

$$i\hbar\frac{\partial}{\partial t}\vec{S}^{<}_{z}=\frac{i}{2}\left[\left(\bar{\Delta}_{x}\vec{S}^{<}_{y}-\bar{\Delta}_{y}\vec{S}^{<}_{x}\right)+\left(\vec{\delta}_{x}S^{<}_{y,o}-\vec{\delta}_{y}S^{<}_{x,o}\right)+\vec{\delta}_{z}\times\vec{S}^{<}_{T}\right], \tag{22.67}$$

$$i\hbar\frac{\partial}{\partial t}\vec{S}^{<}_{T}=\frac{i}{2}\left(\vec{\delta}^{r}_{x}\times\vec{S}^{<}_{x}+\vec{\delta}^{r}_{y}\times\vec{S}^{<}_{y}+\vec{\delta}_{z}\times\vec{S}^{<}_{z}\right). \tag{22.68}$$

We can cast the above three equations, *schematically*, as a single dyadic equation as follows:

$$i\hbar\frac{\partial}{\partial t}\vec{S}^{<}_{\text{spin}}\vec{S}^{<}_{\text{pseudospin}}=\frac{i}{2}\begin{bmatrix}\vec{S}^{<}_{\text{spin}}\left(\vec{\Delta}_{\text{pseudospin}}\times\vec{S}^{<}_{\text{pseudospin}}\right)\\ +\vec{\delta}_{\text{spin}}\left(\vec{\delta}_{\text{pseudospin}}\times\vec{S}^{<}_{\text{pseudospin},o}\right)\\ +\left(\vec{\delta}_{\text{spin}}\times\vec{S}^{<}_{T}\right)\vec{\delta}_{\text{pseudospin}}\end{bmatrix}, \tag{22.69}$$

and

$$i\hbar\frac{\partial}{\partial t}\vec{S}^{<}_{T}=\frac{i}{2}\left(\vec{\delta}^{r}_{x}\times\vec{S}^{<}_{x}+\vec{\delta}^{r}_{y}\times\vec{S}^{<}_{y}+\vec{\delta}_{z}\times\vec{S}^{<}_{z}\right), \tag{22.70}$$

where $\left(\vec{\delta}_{\text{spin}}\times\vec{S}^{<}_{T}\right)\vec{\delta}_{\text{pseudospin}}$ gives a dyadic of the form, $\vec{\Omega}_{\text{spin}}\vec{\delta}_{\text{pseudospin}}$. Later, we will give the exact mathematical expression [i.e., given in Eq. (22.84)] for the single equation for the spin-magnetization dyadic. This dyadic equation is supplemented by the *vector* equation for $\vec{S}_T^<$ derived as follows:

$$
\begin{aligned}
&i\hbar\left(\frac{\partial}{\partial t_1}+\frac{\partial}{\partial t_2}\right)\left(\vec{S}_{cc}^<+\vec{S}_{vv}^<\right)\\
&\quad=\frac{1}{2}\left\{\left[\frac{1}{2}\hbar\omega_o,\left(\vec{S}_{cc}^<-\vec{S}_{vv}^<\right)\right]+\left[\frac{1}{2}\vec{\delta}_z,\left(S_{cc,o}^<-S_{vv,o}^<\right)\right]\right.\\
&\qquad\left.+i\left[\vec{\delta}_z\times\left(\vec{S}_{cc}^<-\vec{S}_{vv}^<\right)\right]\right\}\\
&\qquad+\frac{1}{2}\left[\bar{\Delta}_{hh,cv}^r\vec{S}_{vc}^<-\vec{S}_{cv}^<\bar{\Delta}_{ee,vc}^a+\bar{\Delta}_{ee,vc}^r\vec{S}_{cv}^<-\vec{S}_{vc}^<\bar{\Delta}_{hh,cv}^a\right]\\
&\qquad+\frac{1}{2}\left[\vec{\delta}_{hh,cv}^r S_{vc,o}^<-S_{cv,o}^<\vec{\delta}_{ee,vc}^a+\vec{\delta}_{ee,vc}^r S_{cv,o}^<-S_{vc,o}^<\vec{\delta}_{hh,cv}^a\right]\\
&\qquad+\frac{1}{2}\left[i\left[\vec{\delta}_{hh,cv}^r\times\vec{S}_{vc}^<-\vec{S}_{cv}^<\times\vec{\delta}_{ee,vc}^a\right]+i\left[\vec{\delta}_{ee,vc}^r\times\vec{S}_{cv}^<-\vec{S}_{vc}^<\times\vec{\delta}_{hh,cv}^a\right]\right],
\end{aligned}
\tag{22.71}
$$

$$
\begin{aligned}
&i\hbar\left(\frac{\partial}{\partial t_1}+\frac{\partial}{\partial t_2}\right)\left(\vec{S}_{cc}^<+\vec{S}_{vv}^<\right)\\
&\quad=\frac{1}{2}\left\{\left[\frac{1}{2}\hbar\omega_o,\vec{S}_z^<\right]+\left[\frac{1}{2}\vec{\delta}_z,S_{z,o}^<\right]+i\left[\vec{\delta}_z\times\vec{S}_z^<\right]\right\}\\
&\qquad+\frac{1}{2}\left[\left(\bar{\Delta}_{hh,cv}^r+\bar{\Delta}_{ee,vc}^a\right)\left(\vec{S}_{vc}^<-\vec{S}_{cv}^<\right)-\left(\bar{\Delta}_{ee,vc}^a+\bar{\Delta}_{hh,cv}^a\right)\left(\vec{S}_{vc}^<-\vec{S}_{cv}^<\right)\right]\\
&\qquad+\frac{1}{2}\left[\left(\vec{\delta}_{hh,cv}^r+\vec{\delta}_{ee,vc}^a\right)\left(S_{vc,o}^<-S_{cv,o}^<\right)-\left(\vec{\delta}_{ee,vc}^a+\vec{\delta}_{hh,cv}^a\right)\left(S_{vc,o}^<-S_{cv,o}^<\right)\right]\\
&\qquad+\frac{i}{2}\left[2\vec{\delta}_{hh,cv}^r\times\vec{S}_{vc}^<+2\vec{\delta}_{ee,vc}^r\times\vec{S}_{cv}^<\right],
\end{aligned}
\tag{22.72}
$$

$$
\begin{aligned}
&i\hbar\left(\frac{\partial}{\partial t_1}+\frac{\partial}{\partial t_2}\right)\left(\vec{S}_{cc}^<+\vec{S}_{vv}^<\right)\\
&\quad=\frac{i}{2}\vec{\delta}_z\times\vec{S}_z^<\\
&\qquad+i\left[\frac{1}{4}\left(\vec{\delta}_x^r-i\vec{\delta}_y^r\right)\times\left(\vec{S}_x^<+i\vec{S}_y^<\right)+\frac{1}{4}\left(\vec{\delta}_x^r+i\vec{\delta}_y^r\right)\times\left(\vec{S}_x^<-i\vec{S}_y^<\right)\right].
\end{aligned}
\tag{22.73}
$$

Consider

$$
\begin{aligned}
&\left(\vec{\delta}_x^r-i\vec{\delta}_y^r\right)\times\left(\vec{S}_x^<+i\vec{S}_y^<\right)+\left(\vec{\delta}_x^r+i\vec{\delta}_y^r\right)\times\left(\vec{S}_x^<-i\vec{S}_y^<\right)\\
&\quad=\vec{\delta}_x^r\times\vec{S}_x^<+\vec{\delta}_x^r\times\vec{S}_x^<
\end{aligned}
$$

$$+\vec{\delta}_y^r \times \vec{S}_y^< + \vec{\delta}_y^r \times \vec{S}_y^<$$
$$+\vec{\delta}_x^r \times i\vec{S}_y^< + \vec{\delta}_x^r \times \left(-i\vec{S}_y^<\right)$$
$$\left(-i\vec{\delta}_y^r\right) \times \vec{S}_x^< + i\vec{\delta}_y^r \times \left(\vec{S}_x^<\right)$$
$$= 2\vec{\delta}_x^r \times \vec{S}_x^< + 2\vec{\delta}_y^r \times \vec{S}_y^<. \tag{22.74}$$

Thus

$$i\hbar\frac{\partial}{\partial t}\vec{S}_T^< = \frac{i}{2}\left(\vec{\delta}_x^r \times \vec{S}_x^< + \vec{\delta}_y^r \times \vec{S}_y^< + \vec{\delta}_z \times \vec{S}_z^<\right). \tag{22.75}$$

The last result is a vector since it shows a contraction with respect to the pseudospin vector components.

22.4 Results for defect and electron representations

We have for the defect/electron–hole representations the following equations, a two particle-specie problem,

$$i\hbar\left(\frac{\partial}{\partial t_1}+\frac{\partial}{\partial t_2}\right)\vec{S}_x = \frac{1}{2}\left[i\left(\Lambda_y\vec{S}_z^< - \Lambda_z\vec{S}_y^<\right) + i\left(\vec{\delta}_y S_{z,o}^< - \vec{\delta}_z S_{y,o}^<\right) + i\vec{\delta}_x \times \vec{\Im}\right]$$
$$+\vec{\mathcal{A}} \times \vec{\delta}_x + \vec{\delta}_y A + \Lambda_y\vec{\mathcal{A}}, \tag{22.76}$$

$$i\hbar\left(\frac{\partial}{\partial t_1}+\frac{\partial}{\partial t_2}\right)\vec{S}_y = \frac{1}{2}\left[i\left(\Lambda_z\vec{S}_x - \Lambda_x\vec{S}_z^<\right) + i\left(\vec{\delta}_z S_{x,o} - \vec{\delta}_x S_{z,o}^<\right) + i\vec{\delta}_y \times \vec{\Im}\right]$$
$$-\frac{1}{2}\vec{\delta}_y \times \vec{\mathcal{A}} - \frac{1}{2}\vec{\delta}_x A - \frac{1}{2}\Lambda_x\vec{\mathcal{A}}, \tag{22.77}$$

$$i\hbar\left(\frac{\partial}{\partial t_1}+\frac{\partial}{\partial t_2}\right)\vec{S}_z^< = i\left(\bar{\Delta}_x\vec{S}_y^< - \bar{\Delta}_y\vec{S}_x^<\right) + i\left(\vec{\delta}_x S_{y,o}^< - \vec{\delta}_y S_{x,o}^<\right) + i\vec{\delta}_z \times \vec{S}_z^<, \tag{22.78}$$

$$i\hbar\left(\frac{\partial}{\partial t_1}+\frac{\partial}{\partial t_2}\right)\vec{\Im} = \vec{\delta}_x \times \vec{S}_x^< + \vec{\delta}_y \times \vec{S}_y^< + \vec{\delta}_z \times \vec{S}_z^<, \quad \text{a spin vector only.} \tag{22.79}$$

The results for the electron picture is given by, a single particle-specie problem,

$$i\hbar\frac{\partial}{\partial t}\vec{S}_x^< = \frac{i}{2}\left[\left(\bar{\Delta}_y\vec{S}_z^< - \bar{\Delta}_z\vec{S}_y^<\right) + \left(\vec{\delta}_y\left(S_{z,o}^<\right) - \vec{\delta}_z S_{y,o}^<\right) + \vec{\delta}_x^r \times \vec{S}_T^<\right],$$

$$i\hbar\frac{\partial}{\partial t}\vec{S}_y^< = \frac{i}{2}\left[\left(\bar{\Delta}_z^r\vec{S}_x^< - \bar{\Delta}_x^r\vec{S}_z^<\right) + \left(\vec{\delta}_z S_{x,o}^< - \vec{\delta}_x^r S_{z,o}^<\right) + \vec{\delta}_y^r \times \vec{S}_T^<\right],$$
$$i\hbar\frac{\partial}{\partial t}\vec{S}_z^< = \frac{i}{2}\left[\left(\bar{\Delta}_x\vec{S}_y^< - \bar{\Delta}_y\vec{S}_x^<\right) + \left(\vec{\delta}_x S_{y,o}^< - \vec{\delta}_y S_{x,o}^<\right) + \vec{\delta}_z \times \vec{S}_T^<\right],$$
$$i\hbar\frac{\partial}{\partial t}\vec{S}_T^< = \frac{i}{2}\left(\vec{\delta}_x^r \times \vec{S}_x^< + \vec{\delta}_y^r \times \vec{S}_y^< + \vec{\delta}_z \times \vec{S}_z^<\right).$$

22.4.1 *Generalized Bloch equation for two-level atomic system*

Gathering the above results, we have in the electron picture,

$$\begin{aligned} i\hbar\frac{\partial}{\partial t}\vec{S}_x^< &= \frac{i}{2}\left[\left(\bar{\Delta}_y\vec{S}_z^< - \bar{\Delta}_z\vec{S}_y^<\right) + \left(\vec{\delta}_y\left(S_{z,o}^<\right) - \vec{\delta}_z S_{y,o}^<\right) + \vec{\delta}_x^r \times \vec{S}_T^<\right] \\ &= \frac{i}{2}I_\alpha\varepsilon_{xjk}\mathcal{B}_j S_{\alpha k} + \frac{i}{2}I_\alpha\varepsilon_{xjk}\delta_{\alpha j}S_{k,o} + \frac{i}{2}I_\alpha\varepsilon_{\alpha jk}\delta_{jx}\mathcal{S}_k, \end{aligned} \tag{22.80}$$

$$\begin{aligned} i\hbar\frac{\partial}{\partial t}\vec{S}_y^< &= \frac{i}{2}\left[\left(\bar{\Delta}_z^r\vec{S}_x^< - \bar{\Delta}_x^r\vec{S}_z^<\right) + \left(\vec{\delta}_z S_{x,o}^< - \vec{\delta}_x^r S_{z,o}^<\right) + \vec{\delta}_y^r \times \vec{S}_T^<\right] \\ &= \frac{i}{2}\varepsilon_{yjk}\mathcal{B}_j S_{\alpha k} + \frac{i}{2}I_\alpha\varepsilon_{yjk}\delta_{\alpha j}S_{k,o} + \frac{i}{2}I_\alpha\varepsilon_{\alpha jk}\delta_{jy}\mathcal{S}_k, \end{aligned} \tag{22.81}$$

$$\begin{aligned} i\hbar\frac{\partial}{\partial t}\vec{S}_z^< &= \frac{i}{2}\left[\left(\bar{\Delta}_x\vec{S}_y^< - \bar{\Delta}_y\vec{S}_x^<\right) + \left(\vec{\delta}_x S_{y,o}^< - \vec{\delta}_y S_{x,o}^<\right) + \vec{\delta}_z \times \vec{S}_T^<\right] \\ &= \frac{i}{2}I_\alpha\varepsilon_{zjk}\mathcal{B}_j S_{\alpha k} + \frac{i}{2}I_\alpha\varepsilon_{zjk}\delta_{\alpha j}S_{k,o} + \frac{i}{2}I_\alpha\varepsilon_{\alpha jk}\delta_{jz}\mathcal{S}_k, \end{aligned} \tag{22.82}$$

$$i\hbar\frac{\partial}{\partial t}\vec{S}_T^< = \frac{i}{2}\left(\vec{\delta}_x^r \times \vec{S}_x^< + \vec{\delta}_y^r \times \vec{S}_y^< + \vec{\delta}_z \times \vec{S}_z^<\right). \tag{22.83}$$

We can reduced these equations for the pseudo-components of the dyadic into a *single* dyadic equation as

$$i\hbar\frac{\partial}{\partial t}I_l\vec{S}_l^< = \frac{i}{2}I_l\varepsilon_{ljk}\mathcal{B}_j S_{\alpha k} + \frac{i}{2}I_l I_\alpha\varepsilon_{ljk}\delta_{\alpha j}S_{k,o} + \frac{i}{2}I_l I_\alpha\varepsilon_{\alpha jk}\delta_{jl}\mathcal{S}_k. \tag{22.84}$$

We observe that aside from the interaction with the pseudo-spin, $\vec{S}_o$, and total Pauli–Dirac spin magnetization, $\vec{\mathcal{S}}$, the terms included in the parenthesis for the spin-magnetization tensor is formally the same as the Bloch equation without spin. In the absence of Pauli–Dirac spin, Eq. (22.83) reduces to the familiar vector Bloch equation in quantum optics, namely,

$$\hbar\left(\frac{\partial}{\partial t}\right)\vec{S}_o = \frac{1}{2}\left[\vec{\mathcal{B}}\times\vec{S}_o\right]. \tag{22.85}$$

22.5 Retarded Green's function and self-consistency

Clearly, some important observables, described by the spectral correlation function, $\mathrm{Im}\, G^r$, and occupation number correlation function, $i\hbar G^<$, are the governing dynamical variables in quantum transport physics. Thus, strictly speaking one must also solve for the transport equation of the nonequilibrium retarded Green's function and self-energy in the presence of spin. The nonequilibrium quantum superfield formalism yields the equation for $G^r(1,2)$ as (e.g., see [189]),

$$i\hbar\left(\frac{\partial}{\partial t_1}+\frac{\partial}{\partial t_2}\right)G^r(1,2)=\left[\left(\bar{v}+\Sigma^r\right),G^r\right]_{1,2}, \tag{22.86}$$

where $\bar{v}$ is the single-particle Hamiltonian, and all quantum labels are absorbed in the two-point arguments denoted by the numeral $1, 2$. Then similar procedure can be followed in bringing the above equation into a multi-band matrix form. Then the submatrices can be expressed in the spin-canonical form as in Eq. (21.16), and one solves for the retarded spin correlation functions. This process will yield another 16 nonequilibrium transport equations.

However, usually one is focused mainly in solving the spin magnetization transport equations, and treat Eq. (22.86) as separate calculations.[2] In most cases, the solution to Eq. (22.86) are obtained by some sort of approximation and the results plugged into the spin magnetization equations as was done in the simulation of quantum transport equations for charge carriers in nanoelectronic devices. This is coupled with some further simplification by ignoring the effect of $\mathrm{Re}\, S^r$ in the spin magnetization transport equations.[3] For these reasons, we will no longer spend time in rigorously including Eq. (22.86) in our spin magnetization transport equations.

22.5.1 *Self-consistent electric field and potential*

There is also an important ingredient in the nonlinearity of the SMQTEs introduced through the need for the self-consistency of the potential

[2]The retarded Green's function, G^r, has to do with the energy structure of the system. We are separating the calculation of G^r in the same sense where we usually treat the band structure as a separate calculations, already done before we construct our two-band model of the semiconductors. Indeed, in molecular nanodevices, the density-functional numerical approach is more efficient in calculating the energetics of the nanosystem through G^r.

[3]That the gradient expansion of SMQTEs, with the $\left[\Sigma^<, \mathrm{Re}\, S^r\right]$ term neglected, yields the classical Boltzmann transport equation for spinless system is often viewed as justifying the neglect of $\left[\Sigma^<, \mathrm{Re}\, S^r\right]$ in the quantum transport equations [107].

distribution [162, 190]. This in turn affects the spin-orbit coupling terms in the single particle Hamiltonian and even affects the corresponding many-body aspects of spin-orbit coupling [167]. Thus one also need to solve the Poisson equation for the self-consistent potential, similar to what was done in the numerical simulation of resonant tunneling devices [11, 14, 162]. Clearly then the total charge, represented by $S_{o,o}$ obeying Eq. (21.60) is also coupled, after correcting for the background positive charge, to the potential through the Poisson equation affecting the single particle Hamiltonian and many-body effects in a highly nonlinear self-consistent loop to be accounted for in real device numerical simulations, where clever approximations are eventually needed to bring the problem to a manageable proportion.

22.5.1.1 *Selfconsistency of induced vector fields*

If needed one may use the following relations for selfconsistency of the magnetic fields. To calculate the magnetic field *produced* by a system containing magnetized materials we have to use the following form of Ampere's law,

$$\frac{1}{\mu_o}(\nabla \times B) = J_{\text{free}} + J_{\text{bound}} = J_{\text{free}} + \nabla \times \left[\mu_B\left(\vec{S}_o + \vec{\mathcal{S}}\right)\right], \tag{22.87}$$

$$\nabla \times \left(\frac{B}{\mu_o} - \mu_B\left(\vec{S}_o + \vec{\mathcal{S}}\right)\right) = J_{\text{free}}. \tag{22.88}$$

We define

$$\vec{H} = \frac{\vec{B}}{\mu_o} - \mu_B\left(\vec{S}_o + \vec{\mathcal{S}}\right). \tag{22.89}$$

Then

$$\begin{aligned}\nabla \cdot \vec{H} &= \nabla \cdot \left(\frac{\vec{B}}{\mu_o} - \mu_B\left(\vec{S}_o + \vec{\mathcal{S}}\right)\right)\\ &= \left(\frac{\nabla \cdot \vec{B}}{\mu_o} - \mu_B \nabla \cdot \left(\vec{S}_o + \vec{\mathcal{S}}\right)\right) = -\mu_B \nabla \cdot \left(\vec{S}_o + \vec{\mathcal{S}}\right),\end{aligned} \tag{22.90}$$

$$\nabla \cdot \vec{H} = -\mu_B \nabla \cdot \left(\vec{S}_o + \vec{\mathcal{S}}\right). \tag{22.91}$$

Thus, after calculating $\vec{H}$, we can obtained $\vec{B}_{\text{induced}}$ from Eq. (22.89) which has to be multiplied by μ_B and added to $\vec{\mathcal{B}}_{cc}$ and $\vec{\mathcal{B}}_{vv}$ for selfconsistency of the effective magnetic field at each time-step of a numerical simulation.

22.6 Phase-space spin quantum transport equations

We will first give formal prescriptions on how to transform all the *nonlocal two space–time-points* correlation-function quantum transport equations, given in Eqs. (D.5)–(D.8) with (D.58) and (21.60) into kinetic QDF transport equations defined in *local phase-space points*, i.e., in $(p, q; E, t)$ phase space. This is achieved through the use of Buot's discrete phase-space transformation for condensed matter [3], or his discrete formulation of quantum mechanics leading to the lattice version of the Weyl transformation. Then we will examine what has so far been attempted in the literature towards the treatment of kinetic QDF transport in spintronic devices.

We observe that all of the terms entering in Eqs. (D.5)–(D.8) wit (D.58) and (21.60) involve commutators and anticommutators of two scalar functions and of a scalar and a vector functions, as well as sum and difference of cross products of vector functions. It is more convenient to cast everything in terms of commutators and anticommutators before making the necessary transformation to kinetic equations in phase space. Thus, we need to change the expressions involving cross product of vectors to commutator or anticommutator, as the case maybe. We make use of the following identities for two vectors $\vec{A}$ and $\vec{B}$:

$$\vec{A} \times \vec{B} - \vec{B} \times \vec{A} = \hat{I}_i \hat{\epsilon}_{ijk} \{A_j, B_k\}, \tag{22.92}$$

$$\vec{A} \times \vec{B} + \vec{B} \times \vec{A} = \hat{I}_i \hat{\epsilon}_{ijk} [A_j, B_k], \tag{22.93}$$

where $\hat{I}_i$ is the unit dyadic symmetric tensor or *idemfactor*, and $\hat{\epsilon}_{ijk}$ is the anti-symmetric unit tensor. As used before, the square bracket stands for the commutator and the curly bracket stands for anti-commutator of the two vector components separated by a comma. We also have terms that goes like, $\vec{A} \times \vec{B} - \vec{C} \times \vec{D}$ occurring in Eqs. (D.6) and (D.7), for example the term $\vec{\Xi}^{<}_{cc} \times \vec{S}^{a}_{cv} - \vec{S}^{r}_{cc} \times \vec{\delta}^{<}_{hh,cv}$ in Eq. (D.6). For the typical cross products involving four spin vectors in Eq. (22.94), we will simply expand this as

$$\vec{A} \times \vec{B} - \vec{C} \times \vec{D} = \hat{I}_i \hat{\epsilon}_{ijk} A_j B_k - \hat{I}_i \hat{\epsilon}_{ijk} C_j D_k \tag{22.94}$$

22.6.1 *Transformation to* (p, q, E, t) *phase-space*

The QDF kinetic transport equations in (p, q, E, t)-space are obtained by applying the "lattice" Weyl transformation of the correlation function equations by using the following set of identities [see also Ref. [12]] (although continuum approximation is interchangeably used, this is not essential and we adapt the word "lattice" when referring to solid-state problems). For convenience, we give these here the LW transformation of the following two space–time-point correlation functions:

$$i\hbar\left(\frac{\partial}{\partial t_1}+\frac{\partial}{\partial t_2}\right)F(12)\Leftrightarrow i\hbar\frac{\partial}{\partial t}F_w(\vec{p},\vec{q},E,t), \tag{22.95}$$

where the Bloch states has the representation, $\langle q|\,p\rangle = \left(\frac{1}{N\hbar^3}\right)^{\frac{1}{2}}\exp\frac{i}{\hbar}\{\vec{p}\cdot\vec{q}-Et\}$ [$N\hbar^3 \Rightarrow h^3$ in the continuum limit], representing a traveling wave in lattice space, with group velocity in a band λ given by $\frac{dE_\lambda\{\vec{p}\}}{d\vec{p}}$. The above identities readily follow from the definition of the LW transform, which in the continuum approximation can simply be written as

$$\begin{aligned}i\hbar\left(\frac{\partial}{\partial t_1}+\frac{\partial}{\partial t_2}\right)F(12) &= i\hbar\frac{\partial}{\partial t}\int dv\; d\tau e^{\frac{i}{\hbar}(p\cdot v-E\tau)}\\ &\quad\times F\left(q-\frac{v}{2},t-\frac{\tau}{2};q+\frac{v}{2},t+\frac{\tau}{2}\right)\\ &\Leftrightarrow i\hbar\frac{\partial}{\partial t}F_w(\vec{p},\vec{q},E,t).\end{aligned} \tag{22.96}$$

The second set is the LW transform of a product of two-point correlation functions [$p\equiv(\vec{p},-E)$ and $q=(\vec{q},t)$] in terms of "Poisson bracket operator",

$$AB(p,q)=\exp\left[\frac{\hbar}{2i}\left(\frac{\partial^{(a)}}{\partial p}\cdot\frac{\partial^{(b)}}{\partial q}-\frac{\partial^{(a)}}{\partial q}\cdot\frac{\partial^{(b)}}{\partial p}\right)\right]a(p,q)\,b(p,q), \tag{22.97}$$

or in terms of integral operator,

$$\begin{aligned}AB(p,q) &= \frac{1}{(2\pi\hbar)^8}\int dp'dq'K_A^+(p,q;p',q')\,b(p',q')\\ &= \frac{1}{(2\pi\hbar)^8}\int dp'dq'a(p',q')\,K_B^-(p,q;p',q'),\end{aligned} \tag{22.98}$$

where the factor $\frac{1}{(2\pi\hbar)^8}$ accounts for the proper normalization of the integration (counting of states in terms of unit action) in (p,q,E,t)-space, and the integral kernels are defined by

$$\begin{aligned}K_Y^{\pm}(p,q;p',q') &= \int du\; dv\exp\left\{\frac{i}{\hbar}\left[(p-p')\cdot v+(q-q')\cdot u\right]\right\}\\ &\quad\times y\left(p\pm\frac{u}{2},q\mp\frac{v}{2}\right).\end{aligned} \tag{22.99}$$

For numerical purposes using discrete lattice points, the following expressions of $K_Y^{\pm}(p,q;p',q')$ is more preferable

$$\begin{aligned}&K_Y^{\pm}(p,q;p',q')\\ &\quad=\int du\,dv\exp\left\{\frac{2i}{\hbar}\left[(p-p')\cdot v+(q-q')\cdot u\right]\right\}y(p\pm u,q\mp v).\end{aligned} \tag{22.100}$$

Thus, we may write the LW transform of a commutator $[A,B]$ and an anticommutator $\{A,B\}$ in terms of Poisson bracket differential operator, Λ, as

$$[A,B](p,q) = \cos\Lambda\,[a(p,q)\,b(p,q) - b(p,q)\,a(p,q)] - i\sin\Lambda\,\{a(p,q)\,b(p,q) + b(p,q)\,a(p,q)\}, \tag{22.101}$$

$$\{A,B\}(p,q) = \cos\Lambda\,\{a(p,q)\,b(p,q) + b(p,q)\,a(p,q)\} - i\sin\Lambda\,[a(p,q)\,b(p,q) - b(p,q)\,a(p,q)], \tag{22.102}$$

where $\Lambda = \frac{\hbar}{2}\left(\frac{\partial^{(a)}}{\partial p}\cdot\frac{\partial^{(b)}}{\partial q} - \frac{\partial^{(a)}}{\partial q}\cdot\frac{\partial^{(b)}}{\partial p}\right)$. In terms of integral operators, we have,

$$[A,B](p,q) = \frac{1}{(2\pi\hbar)^8}\int dp'dq' K_A^{+}(p,q;p',q')\,b(p',q') - b(p',q')\,K_A^{-}(p,q;p',q'), \tag{22.103}$$

$$\{A,B\}(p,q) = \frac{1}{(2\pi\hbar)^8}\int dp'dq' K_A^{+}(p,q;p',q')\;b(p',q') + b(p',q')\,K_A^{-}(p,q;p',q'). \tag{22.104}$$

The above expressions simplify considerably when the LW transforms are scalar functions. For this case, we have the "lattice" Weyl transform of a commutator and anti-commutator of two operators, $[A,B]$ and $\{A,B\}$, respectively, given by the following expressions:

$$[A,B](p,q) = \frac{1}{(2\pi\hbar)^8}\int dp'dq' K_A^{s}(p,q;p',q')\,b(p',q'), \tag{22.105}$$

$$\{A,B\}(p,q) = \frac{1}{(2\pi\hbar)^8}\int dp'dq' K_A^{c}(p,q;p',q')\,b(p',q'), \tag{22.106}$$

where

$$K_Y^{s}(p,q;p',q') = \int du\,dv \exp\left\{\frac{2i}{\hbar}\left[(p-p')\cdot v + (q-q')\cdot u\right]\right\} \times [y(p+u,q-v) - y(p-u,q+v)]. \tag{22.107}$$

and

$$K_Y^{c}(p,q;p',q') = \int du\;dv \exp\left\{\frac{2i}{\hbar}\left[(p-p')\cdot v + (q-q')\cdot u\right]\right\} \times [y(p+u,q-v) + y(p-u,q+v)]. \tag{22.108}$$

22.7 Concluding remarks

What we have accomplished is the demonstration of the formal mathematical structure of nonequilibrium multi-band spin-correlation transport equations. In essence, the four equations for the nonequilibrium Green functions given by Eqs. (5), (12), (13), and (14) become vector equations for the spin magnetization yielding 12 coupled equations for the vector components. The last four equations are provided by the vector equation for the pseudo-spin magnetization and the quantum transport equation for the absolute scalar independent of real spin and band indices.

It has been demonstrated that spin-dependent self-energies due to many-body effects give rise to torques in the system. For example, the many-body effects in spin-orbit coupling has lead to the separation of self-energy into spin-independent part and the corresponding spin vector [167–169]. The multi-band aspect give rise to pseudo-spin degree of freedom. Thus, it is expected that the results here will serve as a fundamental basis for constructing realistic transport equations, embodying various approximation schemes, for engineering expediency.

The range of their validity of all simpler and manageable approximate equations can be assessed in the light of the present mathematical structure given in this part. This must be accurate enough for treating the space and time dependent spin relaxation and dephasing scattering mechanisms between conduction electrons, between valence holes, between electrons and holes, their coupling to pseudo-spin, and to the total charges. The total electron charges represented by $S_{o,o}$ minus the positive background charge will be fed to the Poisson equation for self-consistency in the potential. It is expected that the accompanying physical interpretations will acquire deeper insights when applied to highly nonequilibrium situations for spintronic device-performance applications, based on the fundamental structure of multi-band SMQTEs presented in this chapter.

In particular, we have shown that the pseudo-spin is highly coupled to Pauli–Dirac spin transport equations and to the particle charge density. It is not clear how this pseudo-spin aspect of spin transport is described by conventional methods of Pauli–Dirac spin scattering physics so far employed in the literature on the theory of spin relaxation and dephasing mechanisms in solids. Perhaps, this can just be incorporated through the use of "g -factor" *akin* to the Dirac relativistic quantum mechanics.

The two-band model employed here is reminiscent of Dirac relativistic quantum mechanics, where the positive (negative) energy states are doubly degenerate and by time reversal invariance (Kramers' theorem), the quantum states are orthogonal and hence have two degrees of freedom yielding the Pauli–Dirac spin, whereas the two "discrete" bands, of infinite widths, yield some sort of pseudo-spin, whose effect could render the simplest explanation for the source of the g-factor $= 2$ for Dirac electrons. Indeed, the decoupling of the positive from the negative energy states by means of the Foldy–Wouthuysen transformation[4] results in the g-factor for the electron spin. Clearly the g-factor depends on the presence of the two energy bands, which give rise to pseudo-spin in our present treatment of multi-band dynamics in solids. But this is not the whole story about pseudospin dynamics as we shown here, the additional effects may point to the anomalous g-factor for Dirac electrons coming from Dirac relativistic quantum mechanics.

[4]Reduction of the Dirac equation for an electron in a magnetic field to its non-relativistic limit yields the Schrödinger equation with a correction term which takes account of the interaction of the electron's intrinsic magnetic moment with the magnetic field giving the correct energy.

Part 5

A Perspective on Quantum Entanglements

Chapter 23

A Perspective on Entanglement Characterization and Measures

In this chapter, we give a perspective of the basic foundation of the concepts of entanglement characterization and quantitative measures of entanglement. These concepts are demonstrated by giving concrete examples. In the Concluding Remarks section, we give some comments on the famous Bell's inequality, the EPR paradox, as well as the contrasting and envisage geometric nature of Einstein spooky action at a distance of entanglement. We shed light on quantum-mechanical joint distribution as having inherent characteristics of quantum entanglement. Joint distributions are ubiquitous in quantum transport physics of Chapter 28. Furthermore, we proposed a natural measure of entanglements, namely, concurrence, entanglement of formation, and entropic distance from maximally entangled qubits.

23.1 Diagrammatic calculation of entangled basis states

The principal characteristics of quantum mechanics, which are alien to classical physics, are as follows. First is the notion of *uncertainty*, i.e., not all the classical physical observable properties of a system can be simultaneously determined with exact precision. For example, position and momentum cannot both be known at the same time. There may other sets of observable properties that share the uncertainty properties. Second, the notion of *superposition*, for example, a cat is a superposition of alive and dead cat. In other words an arbitrary quantum state is a superposition of several quantum sates that describe measurable events. Third, the notion of *entanglement*, for example, if two photons or two particle spins become entangled, i.e., they are allowed to interact initially so that they will subsequently be defined by a single quantum state, then once they are separated, they will still share the same quantum state. So measuring one will determine the state of the other: for example, with a spin-zero entangled state, if one particle is measured to be in a spin-up state, the other is instantly forced to be in a spin-down state. We would like to address this weirdest aspect of quantum mechanics, i.e., the notion of quantum entanglement.

Today, quantum entanglement forms the basis of several quantum technologies. In quantum teleportation and quantum cryptography, entangled particles are used to transmit signals that cannot be intercepted by an eavesdropper. The preliminary viable quantum cryptography systems are now being used by banks. And the exploding field of quantum computation uses superposition and entangled quantum states to perform computational calculations in parallel and at ultra-speed, so that some types of classically impractical calculations can be done by quantum computers in reasonable length of time.

Historically, in a 1935 paper, Einstein, Boris Podolsky and Nathan Rosen argue that quantum mechanics was not a complete physical theory. Known today as the "EPR paradox", the thought experiment was meant to demonstrate the inherent conceptual, in terms of classical analogies, of quantum theory. Entanglement claims that the result of a measurement on one particle of an entangled quantum system can have an instantaneous effect on another particle, regardless of their distance. But the EPR paradox did help deepen our understanding of quantum mechanics by exposing the fundamentally non-classical characteristics of the measurement process. Before that paper, most physicists viewed a measurement as a physical disturbance inflicted directly on the measured system: one shines light onto an electron to determine its position, but this disturbs the electron and produces uncertainties. The EPR paradox shows that a "measurement" can be performed on a particle without disturbing it directly, by performing a measurement on a distant entangled particle.

Einstein postulated the existence of hidden variables, yet unknown local properties of the system which should account for the discrepancy, so that no instantaneous spooky action would be necessary. Bohr disagreed vehemently with this view and defended the far stricter Copenhagen interpretation of quantum mechanics. The two men were passionate about the subject, especially at the Solvay Conferences of 1927 and 1930, neither Einstein nor Bohr conceded defeat.

Here, we try to propose a simple mechanical model of a classical analogy to the "spooky at a distance" property of quantum entanglement. This is a simple mechanical model, an ideal classical analogy for Einstein arguments. For what it is worth, it gives a simple classical/mechanical implementation and pedagogical value to the concept of quantum entanglement, which may revive interest in some intermediary medium (not hidden variables, but part of the whole "system", i.e., nonlocality and hence does violate Bell's theorem) in quantum entanglement.

23.1.1 *Mechanical implementation of quantum entanglements*

Here, we now give a simple and pedagogical implementation of quantum entanglement using a chain of mechanical inverters, see-saws or swings, or an envisage reversible tightly-coupled quantum-inverter circuits. This mechanical model has been previously used by the author in investigating the fundamental physical limits of computational processes [191]. We employ the mechanical inverter-chain model for its intuitive simplicity.

23.1.2 *Mechanical model for entangled two qubits*

A schematic picture of an entangled two qubits is shown in Figs. 23.1 and 23.2. The frictionless guides and fulcrum would make the inverter chain reversible. It does not really matters, for our purpose, whether there is small swing angle or large swing angle. For large swing angle one can arrange to have extensible arms through the use of sliding female tubing. What is more important though, from the point of view of entanglement, is the presence of rigid coupling. Rigid coupling allows for simultaneous events to take place irrespective of thc length of the intermediary chain, thus capturing the essence of quantum entanglement.

No communication is involved since the whole process is a reconfiguration of the whole chain system. No hidden variables and the deterministic reconfiguration process does violate Bell's inequality. In other words, this mechanical model is conceptually an ideal model of quantum entanglement. One is tempted to speculate whether spacetime provides rigid coupling in realistic situations, indeed due to spacetime entanglement [192].

Figure 23.1 shows a configuration for a triplet entanglement.

Another independent configuration of the chain embodies the spin singlet configuration shown in Fig. 23.2

23.2 A mechanical derivation of entanglement

Here, we can now give a *purely* mechanical characterization of quantum entanglement as deduced from figures, Figs. 23.1 and 23.2, respectively.

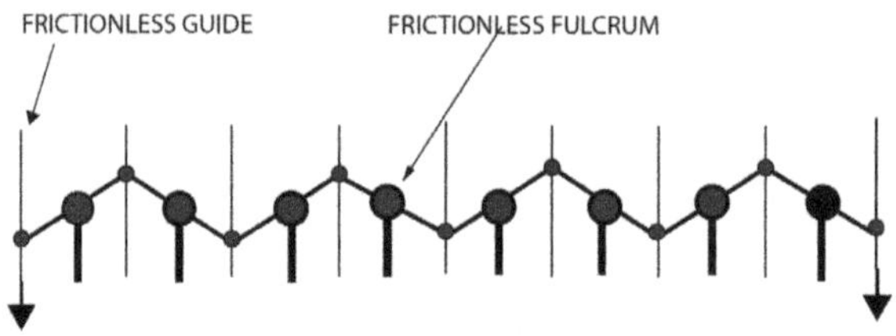

Fig. 23.1. An entangled spin triplet. We assume rigid-coupling between inverters to effect simultaneity of events at both ends, independent of distance.

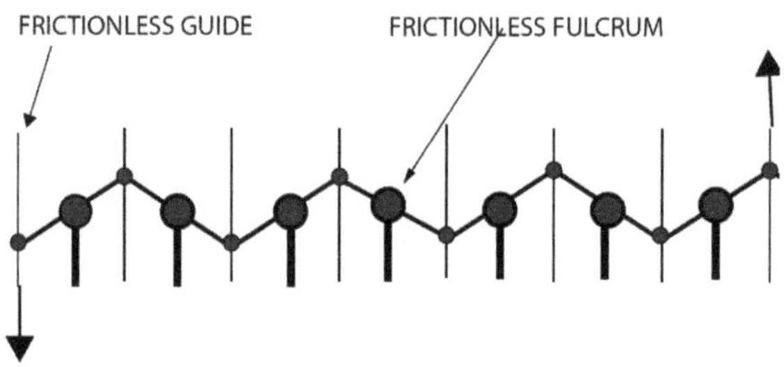

Fig. 23.2. An entangled spin singlet. We assume rigid-coupling between inverters to effect simultaneity of events at both ends, independent of distance.

These figures are clearly showing that a "flip" of one spin simultaneously flips the other spin. Thus, from the mechanical viewpoint, an entangled quantum states must be a superposition of one configuration and its flip configuration. This is crucial in conceptually bringing in the "non-local rigid connection (*independent of separation distance*)" of the two spins in the diagrams. By treating the two configurations, one and its flip partner, as two sites, then the unitary superposition can be implemented by a Hadamard transform, a discrete Fourier transform. The resulting superposed state is a especial case of a Bloch states or crystal momentum states [12]. This allows us to broaden the concept of *momentum states* as joint distribution or entangled superposition of position states or Wannier states at each lattice sites.

Indeed, the self-consistent momentum-based Bohr–Sommerfeld quantization of Chapter 13 may also be viewed as originally bringing in the topological aspect of entanglement [81], for example, standard model of particle physics, gauge theories, Berry phases and curvatures. Hence, it is not surprising that speculations are being advance in some corners that geometric spacetime of general relativity as emergent from entanglement [192].

This is also extended to the *many-body quasiparticle distribution* in quantum transport physics, namely, $G^{<}(q_1, q_2, t_1, t_2, \downarrow, \uparrow)$, which generalizes to joint distribution in position, time, and spin degrees of freedom, important in discussing topological spin and pseudospin systems in Chapter 28 of this book. Effective and relevant superposition is basically implemented by the accompanying complex quantum transport equations and their solutions. This also applies to site coupling in the use of hopping terms in tight binding Hamiltonians. Hence, linear superpositions and entanglement seems ubiquitous in quantum mechanics. Remarkably, the search for efficient measure of entanglement in quantum transport processes is still an ongoing and active field of research. It is, for example, very much desired to understand and characterize the entanglement in exciton transport in photosynthesis [193].

For a simple two degrees of freedom, this in turn allows us to readily characterize entangled states as having *concurrency* equal to unity, corresponding to the normalization of effective Bloch states via the Hadamard transformation. Here, concurrency is a measure of entanglement and is defined as the *absolute value* of the dot product of Bloch states and its Hermitian conjugate, i.e., the flip states of the Bloch states or entangled states. On the other hand, the "lattice site" states, such as $|0,1\rangle$ and $|1,0\rangle$, are orthogonal as given, just like the Wannier states at different sites. This momentum viewpoint of entanglement, however, does not apply to classical mechanics because there is no concept of superposition of states, or off-diagonal density matrix elements in the classical sense.

23.2.1 *Diagrammatic techniques*

In the following diagrammatic analyses, the inverter chains of Figs. 23.1 and 23.2 are abstracted into simple line segment representations (more faithfully as a *sawtooth* segment).

23.2.2 *Two-level diagram representations of entangled two qubits*

The line diagrams below is an abstractions of the inverter chain of Figs. 23.1 and 23.2, respectively. The first line of Fig. 23.3 represents a two-state diagram, with entangled basis $|0\rangle\,|0\rangle$ and $|1\rangle\,|1\rangle$. The second line represents a two-state diagram, with entangled basis $|0\rangle\,|1\rangle$ and $|1\rangle\,|0\rangle$.

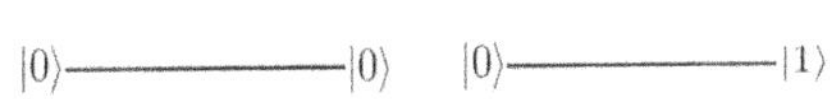

Fig. 23.3. Two two-state diagrams for an entangled two qubits. The calculational entangled basis is obtained using Hadamard transformation or discrete Fourier transform for two-state systems.

23.2.3 *Hadamard transformation for calculational entangled basis states*

The Hadamard transform of the entangled basis states of two qubits given above is what we referred to as the calculational entangled basis states. These are the Bell entangled basis states derived formally before. For the first diagram of Fig. 23.3, we have,

$$\begin{pmatrix} |\Phi^+\rangle \\ |\Phi^-\rangle \end{pmatrix} = \frac{1}{\sqrt{2}} \begin{pmatrix} 1 & 1 \\ 1 & -1 \end{pmatrix} \begin{pmatrix} |0\rangle\,|0\rangle \\ |1\rangle\,|1\rangle \end{pmatrix}, \tag{23.1}$$

and for the second diagram,

$$\begin{pmatrix} |\Psi^+\rangle \\ |\Psi^-\rangle \end{pmatrix} = \frac{1}{\sqrt{2}} \begin{pmatrix} 1 & 1 \\ 1 & -1 \end{pmatrix} \begin{pmatrix} |0\rangle\,|1\rangle \\ |1\rangle\,|0\rangle \end{pmatrix}. \tag{23.2}$$

23.2.4 *Mechanical modeling of entangled three qubits*

For entangled n qubits, there are $\frac{2^n}{2}$ integer number of two-state entangled basis diagrams. Thus, for three qubits the four diagrams shown in Fig. 23.4.

From the diagrams of Fig. 23.4, and using the Hadamard transformation, we can immediately write down the calculational entangled basis states for

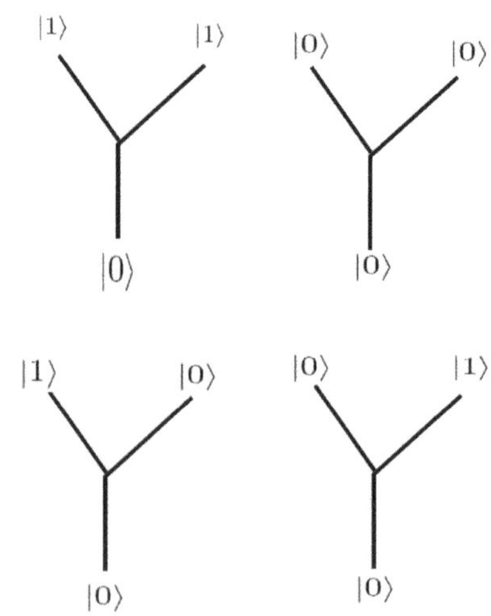

Fig. 23.4. Two-state four diagrams for entangled three qubits.

three qubits.

$$\begin{pmatrix} \Xi^+ \\ \Xi^- \end{pmatrix} = \frac{1}{\sqrt{2}} \begin{pmatrix} 1 & 1 \\ 1 & -1 \end{pmatrix} \begin{pmatrix} |0\rangle\,|1\rangle\,|1\rangle \\ |1\rangle\,|0\rangle\,|0\rangle \end{pmatrix}, \tag{23.3}$$

$$\begin{pmatrix} \Theta_3^+ \\ \Theta_3^{--} \end{pmatrix} = \frac{1}{\sqrt{2}} \begin{pmatrix} 1 & 1 \\ 1 & -1 \end{pmatrix} \begin{pmatrix} |0\rangle\,|0\rangle\,|0\rangle \\ |1\rangle\,|1\rangle\,|1\rangle \end{pmatrix}, \tag{23.4}$$

$$\begin{pmatrix} \Omega^+ \\ \Omega^- \end{pmatrix} = \frac{1}{\sqrt{2}} \begin{pmatrix} 1 & 1 \\ 1 & -1 \end{pmatrix} \begin{pmatrix} |0\rangle\,|1\rangle\,|0\rangle \\ |1\rangle\,|0\rangle\,|1\rangle \end{pmatrix}, \tag{23.5}$$

$$\begin{pmatrix} \Gamma^+ \\ \Gamma^- \end{pmatrix} = \frac{1}{\sqrt{2}} \begin{pmatrix} 1 & 1 \\ 1 & -1 \end{pmatrix} \begin{pmatrix} |0\rangle\,|0\rangle\,|1\rangle \\ |1\rangle\,|1\rangle\,|0\rangle \end{pmatrix}. \tag{23.6}$$

23.2.5 *Mechanical modeling of entangled four qubits*

We can consider eight two-entangled-state diagrams for entangled four qubits. These are shown in Fig. 23.5.

Following the same procedure using Hadamard Fourier transformation, we can also immediately write down the calculational entangled basis states. Here, we use our own notations, $\Phi_j^{\pm}$, for the transformed calculational basis states. Following the order of diagrams in Fig. 23.5, we have

$$\begin{pmatrix} \Phi_1^+ \\ \Phi_1^- \end{pmatrix} = \frac{1}{\sqrt{2}} \begin{pmatrix} 1 & 1 \\ 1 & -1 \end{pmatrix} \begin{pmatrix} |0\rangle\,|0\rangle\,|0\rangle\,|0\rangle \\ |1\rangle\,|1\rangle\,|1\rangle\,|1\rangle \end{pmatrix}, \tag{23.7}$$

$$\begin{pmatrix} \Phi_2^+ \\ \Phi_2^- \end{pmatrix} = \frac{1}{\sqrt{2}} \begin{pmatrix} 1 & 1 \\ 1 & -1 \end{pmatrix} \begin{pmatrix} |0\rangle\,|0\rangle\,|0\rangle\,|1\rangle \\ |1\rangle\,|1\rangle\,|1\rangle\,|0\rangle \end{pmatrix}, \tag{23.8}$$

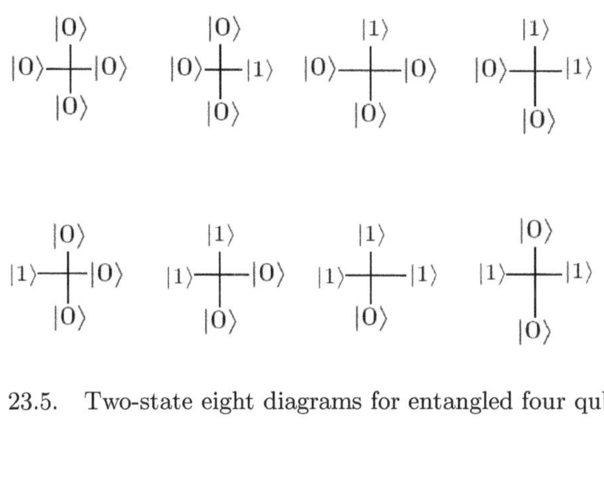

Fig. 23.5. Two-state eight diagrams for entangled four qubits.

$$\begin{pmatrix} \Phi_3^+ \\ \Phi_3^- \end{pmatrix} = \frac{1}{\sqrt{2}} \begin{pmatrix} 1 & 1 \\ 1 & -1 \end{pmatrix} \begin{pmatrix} |0\rangle\,|0\rangle\,|1\rangle\,|0\rangle \\ |1\rangle\,|1\rangle\,|0\rangle\,|1\rangle \end{pmatrix}, \tag{23.9}$$

$$\begin{pmatrix} \Phi_4^+ \\ \Phi_4^- \end{pmatrix} = \frac{1}{\sqrt{2}} \begin{pmatrix} 1 & 1 \\ 1 & -1 \end{pmatrix} \begin{pmatrix} |0\rangle\,|0\rangle\,|1\rangle\,|1\rangle \\ |1\rangle\,|1\rangle\,|0\rangle\,|0\rangle \end{pmatrix}, \tag{23.10}$$

$$\begin{pmatrix} \Phi_5^+ \\ \Phi_5^- \end{pmatrix} = \frac{1}{\sqrt{2}} \begin{pmatrix} 1 & 1 \\ 1 & -1 \end{pmatrix} \begin{pmatrix} |0\rangle\,|1\rangle\,|0\rangle\,|0\rangle \\ |1\rangle\,|0\rangle\,|1\rangle\,|1\rangle \end{pmatrix}, \tag{23.11}$$

$$\begin{pmatrix} \Phi_6^+ \\ \Phi_6^- \end{pmatrix} = \frac{1}{\sqrt{2}} \begin{pmatrix} 1 & 1 \\ 1 & -1 \end{pmatrix} \begin{pmatrix} |0\rangle\,|1\rangle\,|1\rangle\,|0\rangle \\ |1\rangle\,|0\rangle\,|0\rangle\,|1\rangle \end{pmatrix}, \tag{23.12}$$

$$\begin{pmatrix} \Phi_7^+ \\ \Phi_7^- \end{pmatrix} = \frac{1}{\sqrt{2}} \begin{pmatrix} 1 & 1 \\ 1 & -1 \end{pmatrix} \begin{pmatrix} |0\rangle\,|1\rangle\,|1\rangle\,|1\rangle \\ |1\rangle\,|0\rangle\,|0\rangle\,|0\rangle \end{pmatrix}, \tag{23.13}$$

$$\begin{pmatrix} \Phi_8^+ \\ \Phi_8^- \end{pmatrix} = \frac{1}{\sqrt{2}} \begin{pmatrix} 1 & 1 \\ 1 & -1 \end{pmatrix} \begin{pmatrix} |0\rangle\,|1\rangle\,|0\rangle\,|1\rangle \\ |1\rangle\,|0\rangle\,|1\rangle\,|0\rangle \end{pmatrix}. \tag{23.14}$$

In general, one can systematically write down the calculational entangled basis states for any number, n, of qubits by using the physically meaningful diagrammatic techniques prescribed here. One does not even need to draw the diagrams themselves, but simply list the $\frac{2^n}{2}$ entangled basis basis to deduce the calculational entangled basis states by Hadamard transformation of the two configurations of imagined in rigidly coupled spins. It is the superposition of coupled spin states by Hadamard transformation that give rise to the emergent entanglement, as fully depicted by the schematic diagrams.

Chapter 24

Entanglement Measures

Quantum entanglement has developed from a mere intellectual curiosity [194] of the fundamental structure of quantum mechanics[1] to become an important and practical resource for quantum information processing in the evolving theory of quantum information and ultra-fast computing. Thus, the quantitative measure of entanglement has developed into one of the most active fields of theoretical and experimental research. Here we will try to shed more light on some of the important concepts in the quantification of quantum entanglement by using a concrete simple mechanical model of a bipartite system of qubits or chain of qubits. This treatment will be in contrast with mostly abstract and statistical treatment of entanglement measure in the literature.

For the two-qubit entanglement, we will focus on the so-called entanglement of formation and concurrence, two of the most important concepts to characterize entanglement resource. Here we consider a qubit as a two-state system. Moreover, we also consider an entangled qubit as effectively a two-state system, an *emergent qubit* as depicted in our physical inverter-chain link model. A two-state system has a unity entropy. Thus, a maximally entangled state has entropy equal to 1. From the discrete phase space point of view, any two-state system can be considered to possess two lattice-site states. Discrete Fourier transform or Hadamard transform implements unitary superposition of the two *lattice-site states* (also referred to as *Wannier functions*) to yield a sort of *crystal-momentum states* (also referred to as

[1]Note that although Bell's theorem asserts the nonlocality of quantum mechanics, the EPR inquiry is still not resolved, i.e., what is is still left unanswered is the mysterious "link" between qubits corresponding to our "see-saw" or mechanical inverter-chain link.

(*Note*: Bell's inequality theorem is widely discussed in the literature and websites, we prefer not to cite specific reference).

Bloch functions). For example, take the $|00\rangle$ and $|11\rangle$ *lattice-site states*, then the Hadamard bijective discrete transformation gives the *crystal-momentum states*, Φ^+ and Φ^-, which are two of the Bell basis states (or *Bloch function* states).

24.1 Bell basis deduced from inverter-chain link model

We sketch here the derivation of the Bell basis states from our physical model, as depicted in Fig. 24.1. The derivation is based in considering each entangled diagram as a two-state (binary system) of *emergent* qubit. Clearly, the entanglement of two *bare* qubits is divided into two orthogonal spaces of triplet[2] and singlet entanglement states.

We also have the inverse relationship

$$\begin{pmatrix} |00\rangle \\ |11\rangle \end{pmatrix} = \frac{1}{\sqrt{2}} \begin{pmatrix} 1 & 1 \\ 1 & -1 \end{pmatrix} \begin{pmatrix} |\Phi^+\rangle \\ |\Phi^-\rangle \end{pmatrix}, \tag{24.1}$$

$$\begin{pmatrix} |01\rangle \\ |10\rangle \end{pmatrix} = \frac{1}{\sqrt{2}} \begin{pmatrix} 1 & 1 \\ 1 & -1 \end{pmatrix} \begin{pmatrix} |\Psi^+\rangle \\ |\Psi^-\rangle \end{pmatrix}. \tag{24.2}$$

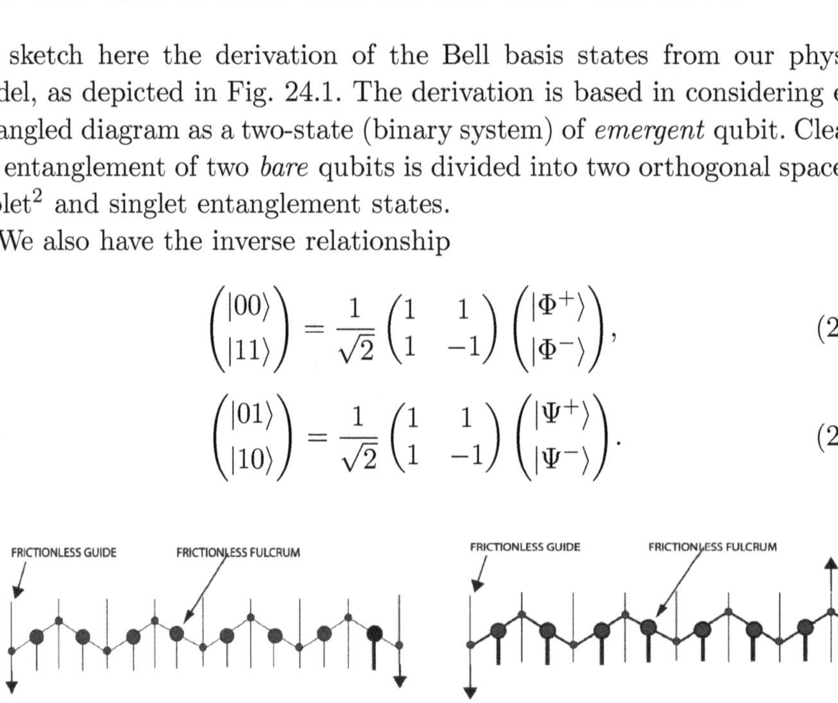

$$\begin{pmatrix} |\Phi^+\rangle \\ |\Phi^-\rangle \end{pmatrix} = \frac{1}{\sqrt{2}} \begin{pmatrix} 1 & 1 \\ 1 & -1 \end{pmatrix} \begin{pmatrix} |0\rangle\,|0\rangle \\ |1\rangle\,|1\rangle \end{pmatrix} \qquad \begin{pmatrix} |\Psi^+\rangle \\ |\Psi^-\rangle \end{pmatrix} = \frac{1}{\sqrt{2}} \begin{pmatrix} 1 & 1 \\ 1 & -1 \end{pmatrix} \begin{pmatrix} |0\rangle\,|1\rangle \\ |1\rangle\,|0\rangle \end{pmatrix}$$

$$|\Phi^+\rangle = \frac{1}{\sqrt{2}}|00\rangle + \frac{1}{\sqrt{2}}|11\rangle \qquad |\Psi^+\rangle = \frac{1}{\sqrt{2}}|01\rangle + \frac{1}{\sqrt{2}}|10\rangle$$

$$|\Phi^-\rangle = \frac{1}{\sqrt{2}}|00\rangle - \frac{1}{\sqrt{2}}|11\rangle \qquad |\Psi^-\rangle = \frac{1}{\sqrt{2}}|01\rangle - \frac{1}{\sqrt{2}}|10\rangle$$

Fig. 24.1. Physical diagrammatic model of "triplet" (left) and singlet (right) entanglement. By construction, each diagram is viewed as a two-state system, respectively. The Bell basis are readily derived below each diagram, using the Hadamard transformation. The actual physical implementation of the chain of inverters may need frictionless male/female sliding tube coupling for large-angle swing, but this is beside the point. We assume a rigid coupling between inverters to implement a simultaneity of events of the qubits on both ends of the link.

[2]The use of the term "triplet" is actually a misnomer here since the entangled system is not free to assume a singlet or zero spin state. Thus, this term is used here only as a label.

We refer to $\{|00\rangle, |11\rangle\}$ as the *Wannier functions* space and the $\{|\Phi^+\rangle, |\Phi^-\rangle\}$ as the corresponding *Bloch functions* space of a two-state *triplet* entanglement system. Likewise, $\{|01\rangle, |10\rangle\}$ as the *Wannier functions* space and the $\{|\Psi^+\rangle, |\Psi^-\rangle\}$ as the corresponding *Bloch functions* space of a two-state singlet entanglement system.

By virtue of this bijective relationship, any function of Wannier functions will have a corresponding function of Bloch functions. For example, a maximally *mixed* Wannier function states will generate a maximally mixed Bloch function states or the so-called maximally *mixed* entanglement states. Mixed and pure states will be further discussed below.

24.2 Entangled qubits as an emergent qubit

The virtue of our inverter-chain link model is that the emergent two-state property of entangled qubits is very transparent, since changing the state of one of the qubits also *immediately* change the states of the rest of the entangled partner(s).[3] Here, we will rigorously justify our claim that entangled qubits behave, as a whole, as an *emergent* qubit.

Here, we employ the matrix representation of states and operators. We now represent $|\Phi^+\rangle$ and as

$$|\Phi^+\rangle = \frac{1}{\sqrt{2}}\begin{pmatrix}1\\1\end{pmatrix}, \tag{24.3}$$

$$|\Phi^-\rangle = \frac{1}{\sqrt{2}}\begin{pmatrix}1\\-1\end{pmatrix}. \tag{24.4}$$

Then we have,

$$\langle\Phi^+|\,\sigma_x\,|\Phi^+\rangle = \frac{1}{2}\begin{pmatrix}1 & 1\end{pmatrix}\begin{pmatrix}0 & 1\\1 & 0\end{pmatrix}\begin{pmatrix}1\\1\end{pmatrix} = 1, \tag{24.5}$$

$$\langle\Phi^-|\,\sigma_x\,|\Phi^-\rangle = \frac{1}{2}\begin{pmatrix}1 & -1\end{pmatrix}\begin{pmatrix}0 & 1\\1 & 0\end{pmatrix}\begin{pmatrix}1\\-1\end{pmatrix} = -1. \tag{24.6}$$

proving the two eigenvalues, like a qubit property of the triplet system.

[3]The idea that entanglement is due to conservation of momentum does not hold for triplet entanglement since its two states give opposing spin-angular momentum. However, one may interpret that quantum superposition of the two-opposing angular momentum states conserves the overall zero net spin-angular momentum, as supported by their eigenvalues similar to a single qubit. On the other hand, for singlet entanglement, the zero angular momentum is conserve in both two states. This seemingly apparent physical difference of the triplet and singlet entanglements underscores the importance of resolving the "mysterious link" in the EPR inquiry [194], in order to further advance theoretical physics.

For the singlet, it is more convenient to change the phase of the second term by i so that we have

$$\left|\Psi^{+}\right\rangle = \frac{1}{\sqrt{2}}\begin{pmatrix}1\\ i\end{pmatrix}, \tag{24.7}$$

$$\left|\Psi^{-}\right\rangle = \frac{1}{\sqrt{2}}\begin{pmatrix}1\\ -i\end{pmatrix}. \tag{24.8}$$

Then, we have,

$$\left\langle\Psi^{+}\right|\sigma_y\left|\Psi^{+}\right\rangle = \frac{1}{2}\begin{pmatrix}1 & -i\end{pmatrix}\begin{pmatrix}0 & -i\\ i & 0\end{pmatrix}\begin{pmatrix}1\\ i\end{pmatrix} = 1, \tag{24.9}$$

$$\left\langle\Psi^{-}\right|\sigma_y\left|\Psi^{-}\right\rangle = \frac{1}{2}\begin{pmatrix}1 & i\end{pmatrix}\begin{pmatrix}0 & -i\\ i & 0\end{pmatrix}\begin{pmatrix}1\\ -i\end{pmatrix} = -1. \tag{24.10}$$

again proving the qubit property of the singlet system, where the two-state aspect which is quite apparent in Fig. 24.1.

The above manipulation for the singlet is a bit round-about trick for the singlet entanglement. The most straightforward manipulation is to recognize that the singlet system as an independent system with its own two states which is orthogonal to the triplet states. In this sense we can also represent the singlet states just like that of the triplet states, namely,

$$\left|\Psi^{+}\right\rangle = \frac{1}{\sqrt{2}}\begin{pmatrix}1\\ 1\end{pmatrix}, \tag{24.11}$$

$$\left|\Psi^{-}\right\rangle = \frac{1}{\sqrt{2}}\begin{pmatrix}1\\ -1\end{pmatrix}. \tag{24.12}$$

Note that in the "Bloch-state" space, σ_z in the "Wannier-state" space is transform to σ_x in entangled states,

$$\begin{aligned}\sigma_x &= H\sigma_z H,\\ &= \frac{1}{\sqrt{2}}\begin{pmatrix}1 & 1\\ 1 & -1\end{pmatrix}\begin{pmatrix}1 & 0\\ 0 & -1\end{pmatrix}\frac{1}{\sqrt{2}}\begin{pmatrix}1 & 1\\ 1 & -1\end{pmatrix},\\ &= \begin{pmatrix}0 & 1\\ 1 & 0\end{pmatrix}.\end{aligned} \tag{24.13}$$

Thus, we also have for the singlet states,

$$\left\langle\Psi^{+}\right|\sigma_x\left|\Psi^{+}\right\rangle = \frac{1}{2}\begin{pmatrix}1 & 1\end{pmatrix}\begin{pmatrix}0 & 1\\ 1 & 0\end{pmatrix}\begin{pmatrix}1\\ 1\end{pmatrix} = 1, \tag{24.14}$$

$$\left\langle\Psi^{-}\right|\sigma_x\left|\Psi^{-}\right\rangle = \frac{1}{2}\begin{pmatrix}1 & -1\end{pmatrix}\begin{pmatrix}0 & 1\\ 1 & 0\end{pmatrix}\begin{pmatrix}1\\ -1\end{pmatrix} = -1. \tag{24.15}$$

Following this notion, Eqs. (24.14) and (24.15) can easily be extended to all entangled qubits, bipartite or multi-partite qubit systems to yield an emergent qubit. It is far more simpler to analyze the inverter-chain link diagrams or employ diagrammatic analyses. We also refer the readers to Section 25.1 where maximally entangled qubits have entropy equal to 1, affirming as an emergent qubit.

24.3 Translational/shift invariance of Bell basis

The translational invariance of maximally entangled Bell basis states demonstrates a unique characteristic of entangled qubits. This unique property has been used to detect or measure how much entanglement is present in arbitrary pure and mixed states. This is implemented in terms of the notion of concurrence, to be elaborated below.

The translational or shift operation (also referred to as the *flip* operation in the literature) is demonstrated here for the Bell Basis states and their superpositions. First, let us consider all the Bell basis states. We have

$$\Phi^{+} = \frac{1}{\sqrt{2}} \left(|0\rangle\,|0\rangle + |1\rangle\,|1\rangle \right). \tag{24.16}$$

Upon applying the translation (shift by $+1 \,(\mathrm{mod}\, 2)$) operator, $T\,(+1)$, we obtain,

$$\begin{aligned} T\,(+1)\,\Phi^{+} &= \frac{1}{\sqrt{2}} \left(|0+1\rangle\,|0+1\rangle + |1+1\rangle\,|1+1\rangle \right), \\ &= \frac{1}{\sqrt{2}} \left(|1\rangle\,|1\rangle + |0\rangle\,|0\rangle \right), \\ &= \Phi^{+}, \end{aligned} \tag{24.17}$$

where addition obeys modular arithmetic (*mod* 2). Similarly, we have

$$T\,(+1)\,\Phi^{-} = -\Phi^{-}, \tag{24.18}$$

$$T\,(+1)\,\Psi^{-} = -\Psi^{-}, \tag{24.19}$$

$$T\,(+1)\,\Psi^{+} = \Psi^{+}. \tag{24.20}$$

We still consider the $-\Phi^{-}$ and $-\Psi^{-}$ as invariant since it differs from the unshifted Bell states by a global phase factor. In the literature this translation operation is the so-called flipping operation first used by Wooters *et al.*, [195, 196].

24.3.1 *Translational property of superposition of Bell basis*

Now let us consider the superposition of maximally entangled Bell basis states. We can readily see that applying the Hadamard transformation to two *Bloch* states will yield the *Wannier* states. We have

$$\frac{1}{\sqrt{2}}\begin{pmatrix}1 & 1\\ 1 & -1\end{pmatrix}\begin{pmatrix}\Phi^+\\ \Phi^-\end{pmatrix}=\begin{pmatrix}|0\rangle\,|0\rangle\\ |1\rangle\,|1\rangle\end{pmatrix}. \tag{24.21}$$

Thus,

$$\frac{1}{\sqrt{2}}\left(\Phi^+ + \Phi^-\right) = |0\rangle\,|0\rangle\,, \tag{24.22}$$

$$\frac{1}{\sqrt{2}}\left(\Phi^+ - \Phi^-\right) = |1\rangle\,|1\rangle\,. \tag{24.23}$$

The thing to notice is that although the superposition is made up of two maximally entangled Bell basis, the results are not shift invariant (i.e., generation of another *Wannier* state yields *Wannier* state located at another *site, which is* orthogonal to the unshifted one). This means that the superposition given above yields unentangled states (product states) in Eq. (24.23). In what follows, we will see that only the following combination of entangled basis states yields another entangled basis states as long as they both belong to either "even" or "odd" spaces, i.e., by diagrammatic construction, we have,

$$triplet^{\pm} \otimes triplet^{\pm} = triplet^{\pm} \text{ or } singlet^{\pm}, \tag{24.24}$$

$$triplet^{\pm} \otimes singlet^{\pm} = singlet^{\pm} \text{ or } triplet^{\pm}, \tag{24.25}$$

$$singlet^{\pm} \otimes singlet^{\pm} = triplet^{\pm} \text{ or } singlet^{\pm}. \tag{24.26}$$

Equation (24.25) is quite interesting because it holds on a complete expansion of a direct product of two qubit states. These relations can easily be deduced from the physical diagrammatic model, see Fig. 24.1. For example, if we define the operation as a superposition such as

$$\begin{aligned}\frac{1}{\sqrt{2}}\left(\Phi^+ + \Psi^+\right) &= \frac{1}{\sqrt{2}}\left(\frac{1}{\sqrt{2}}\left(|1\rangle\,|1\rangle + |0\rangle\,|0\rangle\right) + \frac{1}{\sqrt{2}}\left(|0\rangle\,|1\rangle + |1\rangle\,|0\rangle\right)\right),\\ &= \frac{1}{2}\left[\left(|1\rangle\,|1\rangle + |0\rangle\,|0\rangle\right) + \left(|0\rangle\,|1\rangle + |1\rangle\,|0\rangle\right)\right],\\ &= \left[\frac{1}{\sqrt{2}}\left(|1\rangle + |0\rangle\right) \otimes \frac{1}{\sqrt{2}}\left(|1\rangle + |0\rangle\right)\right].\end{aligned} \tag{24.27}$$

We see that Eq. (24.27) is a direct product state of two qubits. We will see in what follows that this is an entangled state and corresponds to Eq. (24.25) of the *singlet*$^+$ or *triplet*$^+$ of the model diagrams depending on the actual linkage. Similarly, we have,

$$\begin{aligned}\frac{1}{\sqrt{2}}\left(\Phi^- + \Psi^-\right) &= \frac{1}{\sqrt{2}}\left(\frac{1}{\sqrt{2}}\left(|1\rangle|1\rangle - |0\rangle|0\rangle\right) + \frac{1}{\sqrt{2}}\left(|0\rangle|1\rangle - |1\rangle|0\rangle\right)\right),\\ &= \frac{1}{2}\left[\left(|1\rangle|1\rangle - |0\rangle|0\rangle\right) + \left(|0\rangle|1\rangle - |1\rangle|0\rangle\right)\right],\\ &= \left[\frac{1}{\sqrt{2}}\left(|1\rangle + |0\rangle\right) \otimes \frac{1}{\sqrt{2}}\left(|1\rangle - |0\rangle\right)\right],\end{aligned} \tag{24.28}$$

corresponds to Eq. (24.25) of the *singlet* or *triplet* of the model diagrams, depending on the actual linking.

However, the following combinations of "even" and "odd" entangled states result in *unentangled* states, namely,

$$\begin{aligned}\frac{1}{\sqrt{2}}\left(\Phi^+ + \Psi^-\right) &= \frac{1}{\sqrt{2}}\left(\frac{1}{\sqrt{2}}\left(|1\rangle|1\rangle + |0\rangle|0\rangle\right) + \frac{1}{\sqrt{2}}\left(|0\rangle|1\rangle - |1\rangle|0\rangle\right)\right),\\ &= \frac{1}{2}\left[\left(|1\rangle|1\rangle + |0\rangle|0\rangle\right) + \left(|0\rangle|1\rangle - |1\rangle|0\rangle\right)\right],\end{aligned} \tag{24.29}$$

and

$$\begin{aligned}\frac{1}{\sqrt{2}}\left(\Phi^- + \Psi^+\right) &= \frac{1}{\sqrt{2}}\left(\frac{1}{\sqrt{2}}\left(|1\rangle|1\rangle - |0\rangle|0\rangle\right) + \frac{1}{\sqrt{2}}\left(|0\rangle|1\rangle + |1\rangle|0\rangle\right)\right),\\ &= \frac{1}{2}\left[\left(|1\rangle|1\rangle - |0\rangle|0\rangle\right) + \left(|0\rangle|1\rangle + |1\rangle|0\rangle\right)\right],\end{aligned} \tag{24.30}$$

by virtue of the failure to have global sign factors. All these claims are justified through the concept of concurrence, an inherent property of entangled qubits, to be discussed in what follows. Moreover, this feature of failing to have global sign factor is also reflected in the failure to represent by our inverter-chain link diagrams.

24.3.2 *Bipartite system*

Let the Hilbert spaces of a bipartite system consisting of A and B be denoted by $\mathcal{H}_A$ and $\mathcal{H}_B$, respectively. A bipartite system is a system with Hilbert space equal to the direct product of $\mathcal{H}_A$ and $\mathcal{H}_B$, i.e.,

$$\mathcal{H}_{AB} = \mathcal{H}_A \otimes \mathcal{H}_B. \tag{24.31}$$

Let the density matrix for the whole system be denoted by ρ. Then the reduced density matrix of a subsystem A is given by the partial trace,

$$\rho_A = \mathrm{Tr}_B\ \rho. \tag{24.32}$$

The entanglement entropy, S_A, is the *von Neumann entropy* of the reduced density matrix, ρ_A,

$$S_A = -\mathrm{Tr}\ \rho_A \log \rho_A. \tag{24.33}$$

Example 24.1. Let Ω be the number of distinct states in subsystem A. Assume a uniform distribution among states, hence ρ_A has eigenvalues $\frac{1}{\Omega}$, i.e., ρ_A can be represented by a diagonal $\Omega \times \Omega$ matrix with identical matrix elements given by $\frac{1}{\Omega}$. Thus, in taking the trace we can use the eigenvalues of the reduced density matrix operator, ρ_A. Therefore, we have

$$\begin{aligned} S_A &= -\mathrm{Tr}\,\frac{1}{\Omega}\log\frac{1}{\Omega}, \\ &= -\Omega\left(\frac{1}{\Omega}\log\frac{1}{\Omega}\right), \\ &= -\log\frac{1}{\Omega}, \\ &= \log\ \Omega. \end{aligned} \tag{24.34}$$

Upon multiplying by the Boltzmann constant, k_B, we obtain,

$$k_B S_A = k_B \log\ \Omega, \tag{24.35}$$

which is the Boltzmann thermodynamic entropy, based on ergodic theorem.

Example 24.2. Two qubit system.

A qubit is simply a quantum bit whose number of distinct eigenstates is 2. We denote these eigenstates as $|0\rangle$ and $|1\rangle$, i.e., a two-state system. If each subsystem A or B is a single qubit, then the Hilbert space of the whole system is span by the following 4 direct product states, namely,

$$|00\rangle\,,\ |01\rangle\,,\ |10\rangle\,,\ |11\rangle\,. \tag{24.36}$$

Here, the first bit refers to subsystem A and the second bit refers to subsystem B, where we write

$$|ij\rangle = |i\rangle_A\ |j\rangle_B\ \epsilon\ \mathcal{H}_A \otimes \mathcal{H}_B. \tag{24.37}$$

Now, the density matrix of a pure state is,

$$\rho = |\psi\rangle \langle\psi|, \tag{24.38}$$

then

$$\begin{aligned}\rho^2 &= |\psi\rangle |\psi\rangle \langle\psi| |\psi\rangle \langle\psi|,\\ &= |\psi\rangle \langle\psi|,\\ &= \rho.\end{aligned} \tag{24.39}$$

An operator whose square is equal to itself must have an eigenvalue equal to unity. Let us write for pure state of the two qubits as

$$|\psi\rangle = \frac{1}{\sqrt{2}}\left(|00\rangle + |11\rangle\right), \tag{24.40}$$

$$\langle\psi| = \frac{1}{\sqrt{2}}\left(\langle 00| + \langle 11|\right), \tag{24.41}$$

$$|\psi\rangle \langle\psi| = \frac{1}{2}\left(|00\rangle + |11\rangle\right)\left(\langle 00| + \langle 11|\right). \tag{24.42}$$

We have

$$\begin{aligned}\langle\psi| |\psi\rangle &= \frac{1}{2}\left(\langle 00| + \langle 11|\right)\left(|00\rangle + |11\rangle\right),\\ &= \frac{1}{2}\left(\langle 00| |00\rangle + \langle 11| |11\rangle\right),\\ &= 1.\end{aligned} \tag{24.43}$$

Thus, indeed,

$$\begin{aligned}\rho^2 &= |\psi\rangle \langle\psi| |\psi\rangle \langle\psi|,\\ &= |\psi\rangle \langle\psi|,\\ &= \rho.\end{aligned} \tag{24.44}$$

We refer to $|\psi\rangle$ as a maximally entangled state in the sense that the first qubit is exactly "mapped" to the second qubit. The reduced density matrix for subsystem A is

$$\begin{aligned}\rho_A &= \mathrm{Tr}_B\ \rho,\\ &= \frac{1}{2}\langle 0_B| \left(|00\rangle + |11\rangle\right)\left(\langle 00| + \langle 11|\right) |0_B\rangle\end{aligned}$$

$$
\begin{aligned}
&+\frac{1}{2}\langle 1_B|(|00\rangle+|11\rangle)(\langle 00|+\langle 11|)|1_B\rangle, \\
&=\frac{1}{2}(|0_A\rangle\langle 0_A|+|1_A\rangle\langle 1_A|). \qquad (24.45)
\end{aligned}
$$

Now clearly

$$
\begin{aligned}
\rho_A^2 &= \frac{1}{4}(|0_A\rangle\langle 0_A|+|1_A\rangle\langle 1_A|)(|0_A\rangle\langle 0_A|+|1_A\rangle\langle 1_A|), \\
&= \frac{1}{4}(|0_A\rangle\langle 0_A|+|1_A\rangle\langle 1_A|), \\
&= \frac{1}{2}\rho_A, \qquad (24.46)
\end{aligned}
$$

so that ρ_A is not a pure state but mixed, i.e., a mixture of two pure states, $|0_A\rangle$ and $|1_A\rangle$. Note that the mixed states density matrix do not possess *off-diagonal elements*. From the last identity, the eigenvalues of ρ_A is $\frac{1}{2}$ and since the subsystem A is a single qubit with two distinct states, then eigenvalues of ρ_A corresponds to $\frac{1}{\Omega}$ of our first example above. Thus we refer to ρ_A as "uniformly"" mixed often referred to as "*maximally mixed*" with the initial state $|\psi\rangle$ as "*maximally entangled*".

24.4 Entanglement of formation of multi-partite qubit systems

The entanglement entropy of subsystem A can be calculated using the eigenvalues of the 2×2 matrix of ρ_A, which is $\lambda_A=\frac{1}{2}$. Therefore we have for the entanglement entropy or entanglement of formation given by

$$
\begin{aligned}
S_A &= -\mathrm{Tr}\,\rho_A\log\rho_A, \\
&= -\mathrm{Tr}\,\frac{1}{2}\log\frac{1}{2}, \\
&= -2\left(\frac{1}{2}\log\frac{1}{2}\right), \\
&= \log 2 \equiv \log 2^1, \qquad (24.47)
\end{aligned}
$$

where exponent base 2, here 1, is the number of qubits that is entangled with system B. This will be made clear in the next example.

Example 24.3. A four qubit system:

If each subsystem A or B has two qubits, then the Hilbert space of the whole system is span by the following 16 direct product states, i.e.,

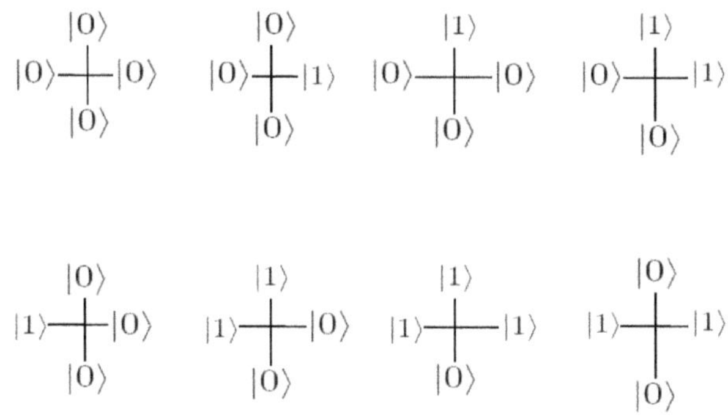

Fig. 24.2. Two-state eight diagrams for entangled four qubits. The so-called flip operation yields the second state for each of the above diagrams. The entangled basis is constructed by the superposition, via the Hadamard transformation, of each diagram and its corresponding flipped diagram.

$2^4 = 16$ states. The maximally entangled pure state is determined by the following eight diagrams [23], see Fig. 24.2, and their flipped or translated states.

Any combination or superposition of both "even" triplet and singlet states comprised a maximally entangled state, i.e.,

$$\begin{pmatrix} \Phi_1^+ \\ \Phi_1^- \end{pmatrix} = \frac{1}{\sqrt{2}} \begin{pmatrix} 1 & 1 \\ 1 & -1 \end{pmatrix} \begin{pmatrix} |0\rangle\,|0\rangle\,|0\rangle\,|0\rangle \\ |1\rangle\,|1\rangle\,|1\rangle\,|1\rangle \end{pmatrix}, \tag{24.48}$$

$$\begin{pmatrix} \Phi_8^+ \\ \Phi_8^- \end{pmatrix} = \frac{1}{\sqrt{2}} \begin{pmatrix} 1 & 1 \\ 1 & -1 \end{pmatrix} \begin{pmatrix} |0\rangle\,|1\rangle\,|0\rangle\,|1\rangle \\ |1\rangle\,|0\rangle\,|1\rangle\,|0\rangle \end{pmatrix}. \tag{24.49}$$

Thus, from the states given above, we have for example the maximally entangled state,

$$\begin{aligned} |\psi\rangle &= \frac{1}{\sqrt{4}} \left(|00,00\rangle + |01,01\rangle + |10,10\rangle + |11,11\rangle\right), \\ &= \frac{1}{\sqrt{2}} \left(\Phi_1^+ + \Phi_8^+\right), \end{aligned} \tag{24.50}$$

and similarly, we can also form another entangled state,

$$\begin{aligned} |\psi\rangle &= \frac{1}{\sqrt{4}} \left(|00,00\rangle + |01,01\rangle - |10,10\rangle - |11,11\rangle\right), \\ &= \frac{1}{\sqrt{2}} \left(\Phi_1^- + \Phi_8^-\right), \end{aligned} \tag{24.51}$$

where the first two bits belongs to subsystem A and the second two bits belong to subsystem B. We note that

$$\begin{aligned}\langle\psi|\,|\psi\rangle &= \frac{1}{4}\left(\langle 00,00| + \langle 01,01| + \langle 10,10| + \langle 11,11|\right) \\ &\quad \times \left(|00,00\rangle + |01,01\rangle + |10,10\rangle + |11,11\rangle\right), \\ &= 1. \end{aligned} \tag{24.52}$$

Therefore

$$\rho^2 = \left(|\psi\rangle\,\langle\psi|\right)^2, \tag{24.53}$$

$$= |\psi\rangle\,\langle\psi| = \rho. \tag{24.54}$$

So $|\psi\rangle$ is a pure state.

In general, the following maximally entangled state corresponds to a chain of entangled basis states, namely,

$$|\psi^+\rangle = \frac{1}{\sqrt{8}}\left(\Phi_1^+ + \Phi_2^+ + \Phi_3^+ + \Phi_4^+ + \Phi_5^+ + \Phi_6^+ + \Phi_7^+ + \Phi_8^+\right), \tag{24.55}$$

as well as

$$|\psi^-\rangle = \frac{1}{\sqrt{8}}\left(\Phi_1^- + \Phi_2^- + \Phi_3^- + \Phi_4^- + \Phi_5^- + \Phi_6^- + \Phi_7^- + \Phi_8^-\right), \tag{24.56}$$

of a four qubit system.

The density matrix operator for the whole 4-qubit system can be written as

$$\rho = \frac{1}{\sqrt{2}}\left(|\Phi_1^- + \Phi_8^-\rangle\right)\frac{1}{\sqrt{2}}\left(\langle\Phi_1^- + \Phi_8^-|\right), \tag{24.57}$$

$$\begin{aligned}\rho = \frac{1}{4}&\left[|00,00\rangle + |01,01\rangle + |10,10\rangle + |11,11\rangle\right]\left[\langle 00,00| + \langle 01,01|\right. \\ &\left. + \langle 10,10| + \langle 11,11|\right]. \end{aligned} \tag{24.58}$$

The reduced density matrix operator for subsystem A is again obtained by taking partial trace with respect to subsystem B.

$$\begin{aligned}\rho_A &= \mathrm{Tr}_B\ \rho, \\ &= \langle 00_B|\,\rho\,|00_B\rangle + \langle 01_B|\,\rho\,|01_B\rangle + \langle 10_B|\,\rho\,|10_B\rangle + \langle 11_B|\,\rho\,|11_B\rangle, \\ &= \frac{1}{4}\,|00_A\rangle\,\langle 00_A| + |01_A\rangle\,\langle 01_A| + |10_A\rangle\,\langle 10_A| + |11_A\rangle\,\langle 11_A|. \end{aligned} \tag{24.59}$$

To determine the eigenvalues for ρ_A, we take its square,

$$\begin{aligned}\rho_A^2 &= \frac{1}{16}\left(|00_A\rangle\langle 00_A| + |01_A\rangle\langle 01_A| + |10_A\rangle\langle 10_A| + |11_A\rangle\langle 11_A|\right)\\ &\quad\times\left(|00_A\rangle\langle 00_A| + |01_A\rangle\langle 01_A| + |10_A\rangle\langle 10_A| + |11_A\rangle\langle 11_A|\right),\\ &= \frac{1}{16}\left(|00_A\rangle\langle 00_A| + |01_A\rangle\langle 01_A| + |10_A\rangle\langle 10_A| + |11_A\rangle\langle 11_A|\right),\\ &= \frac{1}{4}\rho_A. \end{aligned} \tag{24.60}$$

So we have

$$\rho_A^2 = \frac{1}{4}\rho_A, \tag{24.61}$$

and the eigenvalues of ρ_A is $\frac{1}{4}$. We can now calculate the entanglement entropy of subsystem A. We have,

$$\begin{aligned} S_A &= -\mathrm{Tr}\,\rho_A \log \rho_A,\\ &= -4\left[\frac{1}{4}\log\frac{1}{4}\right],\\ &= -\log\frac{1}{4},\\ &= \log 2^2. \end{aligned} \tag{24.62}$$

The exponent 2 correspond to the number of qubits that is entangled with subsystem B. In general, for maximally entangled bipartite system A and B, each having k number of qubits S_A is given by

$$S_A = \log 2^k. \tag{24.63}$$

Now of course, for this bipartite system,

$$S_A = S_B, \tag{24.64}$$

which simply means a complete matching of configurations of each system A and B, respectively, otherwise some degrees of freedom will be hanging, not matched or cannot be entangled.

Example 24.4. Tripartite system of three qubits:

The following diagrams in Fig. 24.3 represent the entangled tripartite system of qubits. The entangled basis states are as follows:

$$\begin{pmatrix}\Xi^+\\ \Xi^-\end{pmatrix} = \frac{1}{\sqrt{2}}\begin{pmatrix}1 & 1\\ 1 & -1\end{pmatrix}\begin{pmatrix}|0\rangle\,|1\rangle\,|1\rangle\\ |1\rangle\,|0\rangle\,|0\rangle\end{pmatrix}, \tag{24.65}$$

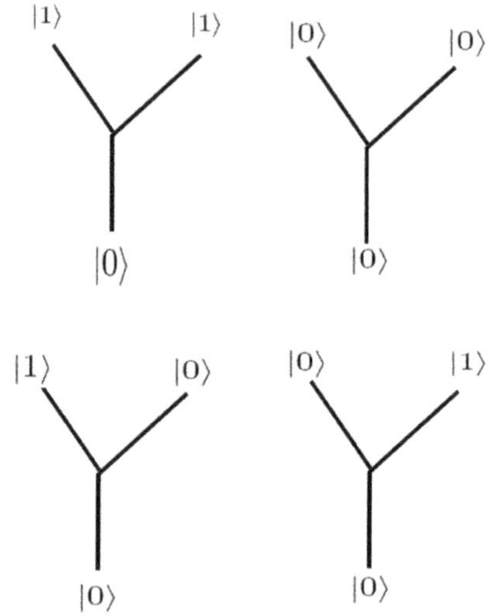

Fig. 24.3. Two-state four diagrams for entangled three qubits of a tripartite system. Flip operations yield the respective second states.

$$\begin{pmatrix} \Theta_3^+ \\ \Theta_3^{--} \end{pmatrix} = \frac{1}{\sqrt{2}} \begin{pmatrix} 1 & 1 \\ 1 & -1 \end{pmatrix} \begin{pmatrix} |0\rangle\,|0\rangle\,|0\rangle \\ |1\rangle\,|1\rangle\,|1\rangle \end{pmatrix}, \tag{24.66}$$

$$\begin{pmatrix} \Omega^+ \\ \Omega^- \end{pmatrix} = \frac{1}{\sqrt{2}} \begin{pmatrix} 1 & 1 \\ 1 & -1 \end{pmatrix} \begin{pmatrix} |0\rangle\,|1\rangle\,|0\rangle \\ |1\rangle\,|0\rangle\,|1\rangle \end{pmatrix}, \tag{24.67}$$

$$\begin{pmatrix} \Gamma^+ \\ \Gamma^- \end{pmatrix} = \frac{1}{\sqrt{2}} \begin{pmatrix} 1 & 1 \\ 1 & -1 \end{pmatrix} \begin{pmatrix} |0\rangle\,|0\rangle\,|1\rangle \\ |1\rangle\,|1\rangle\,|0\rangle \end{pmatrix}. \tag{24.68}$$

A chain or superposition of any choice of entangled *even* basis states also form a maximally entangled state, e.g.,

$$|\Psi^4\rangle = \frac{1}{\sqrt{4}} \left[\Xi^+ + \Theta_3^+ + \Omega^+ + \Gamma^+\right], \tag{24.69}$$

$$|\Psi^3\rangle = \frac{1}{\sqrt{3}} \left[\Xi^+ + \Theta_3^+ + \Omega^+\right], \tag{24.70}$$

$$|\Psi^2\rangle = \frac{1}{\sqrt{2}} \left[\Xi^+ + \Theta_3^+\right], \tag{24.71}$$

$$|\Psi^1\rangle = \Xi^+, \tag{24.72}$$

are all, respectively, maximally entangled states with emergent two-state or qubit properties.

Example 24.5. Here, we need to define what we mean by entanglement of formation as the entanglement entropy of unentangling the three-qubit tripartite system into three separate one-qubit systems. It is important to point out that any entangled qubits, irrespective of their number, identically behave like a single qubit, i.e., behaving exactly like a two-state system. Thus, the natural order of disentangling is as follows: First, one entangles one qubit from the remaining two entangled qubits. The entanglement of formation is equivalent to one qubit. Next, one disentangled the remaining entangled two qubits. This further give an entanglement of formation of one qubit. Thus, the total entanglement of formation is $1 + 1 = 2$ qubits.

24.4.1 *Monogamy inequality in multi-partite entanglement*

The reasoning we have given above yields exact equality of the monogamy relations, usually given as *inequality* in the literature as deduced from statistical analysis, e.g., one given by Kim [197], The exact equality relation for the entanglement of formation is diagrammatically shown in Fig. 24.4.

The exact equality comes about by construction, since on both sides of Fig. 24.4, two two-qubit entangled states are being unentangled to obtain the entanglement of formation of this multi-partite system. This comes about through the knowledge that any multi-partite entangled qubits behave as an emergent qubit.

Similarly, for an entangled 4-partite system, the entanglement of formation is determined schematically by the diagrams of Fig. 24.5, where the unentangling operations to be done on the left side are itemized on the unentangling operations of *three* entangled two-qubits on the right side of the equality sign.

This sort of diagrammatic analysis of the entropy of entanglement of formation can straightforwardly be employed to all multi-partite qubit systems

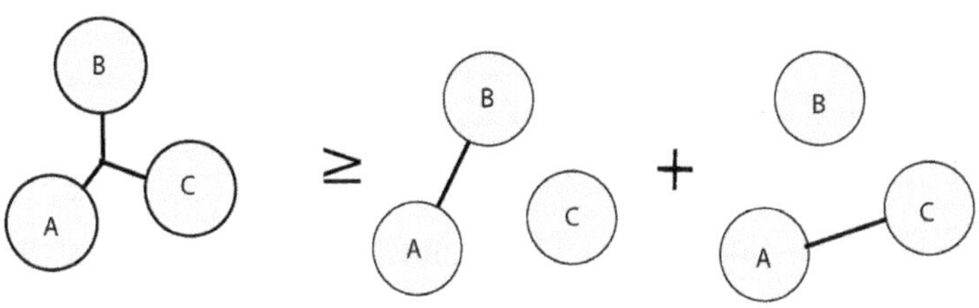

Fig. 24.4. In our diagrammatic analysis the right-hand side of the figure yields $1+1 = 2$ as the entanglement of formation, being the sum of two two-qubit entanglements. Thus, in our diagrammatic analysis, we obtained exact monogamy equality not inequality. Reproduced from Ref. [197].

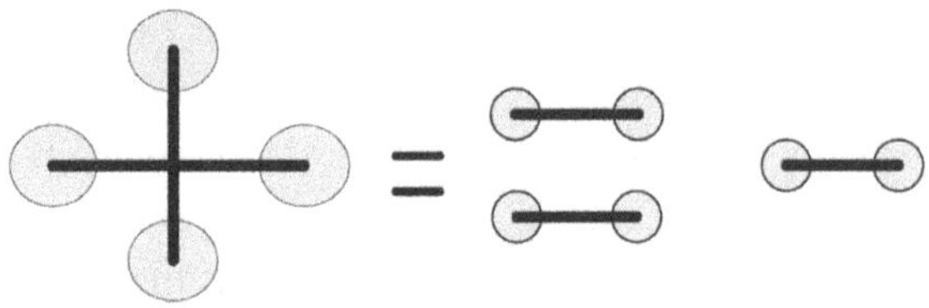

Fig. 24.5. In our diagrammatic analysis the right-hand side of the figure yields $1 + 1 + 1 = 3$ as the entanglement of formation, being the sum of three two-qubit entanglements. Thus, in our diagrammatic analysis, we obtained exact equality not inequality.

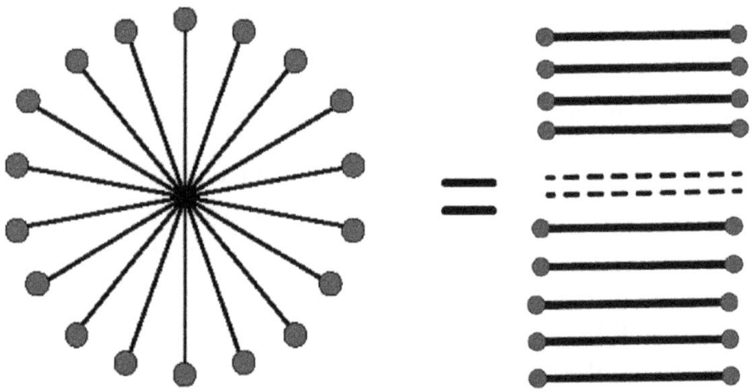

Fig. 24.6. In our diagrammatic analysis the right-hand side of the figure yields $1 + 1 + 1, \ldots\ldots, 1 = 17$ qubits as the entanglement of formation, being the sum of 17 two-qubit entanglements of formation. There are 18 multi-party entangled qubits in the left-hand side.

by a simple counting argument as is done here. Although our diagrammatic analysis is based on maximally entangled qubits, we believe the diagrammatic construction holds also for any arbitrarily entangled multipartite qubit systems. For example, for 18 multi-partite entangled qubits in Fig. 24.6, we have the entanglement of formation schematically depicted by the equality where there is 17 number of monogamy in the right-hand side yielding 17 qubits of entropy of entanglement of formation.

24.5 Expansion of product states in Bell basis

Let us consider $|00\rangle$, $|01\rangle$, $|10\rangle$, $|11\rangle$ as our direct product basis in expanding $|\psi\rangle$,

$$\begin{aligned} |\psi\rangle &= \alpha_0 |00\rangle + \alpha_1 |01\rangle + \alpha_2 |10\rangle + \alpha_3 |11\rangle, \\ &= \sum_i \beta_i e_i, \end{aligned} \tag{24.73}$$

where e_i denotes the Bell basis. We have the Bell basis given by

$$e_1 = \left|\Phi^+\right\rangle = \frac{1}{\sqrt{2}}\left(|0\rangle|0\rangle + |1\rangle|1\rangle\right), \tag{24.74}$$

$$e_2 = \left|\Phi^-\right\rangle = \frac{1}{\sqrt{2}}\left(|0\rangle|0\rangle - |1\rangle|1\rangle\right), \tag{24.75}$$

$$e_3 = \left|\Psi^+\right\rangle = \frac{1}{\sqrt{2}}\left(|0\rangle|1\rangle + |1\rangle|0\rangle\right), \tag{24.76}$$

$$e_4 = \left|\Psi^-\right\rangle = \frac{1}{\sqrt{2}}\left(|0\rangle|1\rangle - |1\rangle|0\rangle\right). \tag{24.77}$$

Standard basis can be express in terms of Bell basis, a re-statement of Eqs. (24.1) and (24.2), namely,

$$|0\rangle|0\rangle = \frac{1}{\sqrt{2}}\left(e_1 + e_2\right), \tag{24.78}$$

$$|1\rangle|1\rangle = \frac{1}{\sqrt{2}}\left(e_1 - e_2\right), \tag{24.79}$$

$$|0\rangle|1\rangle = \frac{1}{\sqrt{2}}\left(e_3 + e_4\right), \tag{24.80}$$

$$|1\rangle|0\rangle = \frac{1}{\sqrt{2}}\left(e_3 - e_4\right), \tag{24.81}$$

so we can also expand $|\psi\rangle$ as

$$\begin{aligned}|\psi\rangle &= \beta_1 e_1 + \beta_2 e_2 + \beta_3 e_3 + \beta_4 e_4,\\ &= \beta_1\left(\frac{1}{\sqrt{2}}\left(|0\rangle|0\rangle + |1\rangle|1\rangle\right)\right) + \beta_2\left(\frac{1}{\sqrt{2}}\left(|0\rangle|0\rangle - |1\rangle|1\rangle\right)\right)\\ &\quad + \beta_3\left(\frac{1}{\sqrt{2}}\left(|0\rangle|1\rangle + |1\rangle|0\rangle\right)\right) + \beta_4\left(\frac{1}{\sqrt{2}}\left(|0\rangle|1\rangle - |1\rangle|0\rangle\right)\right),\end{aligned} \tag{24.82}$$

$$\begin{aligned}|\psi\rangle &= \frac{1}{\sqrt{2}}\left(\beta_1 + \beta_2\right)|0\rangle|0\rangle + \frac{1}{\sqrt{2}}\left(\beta_1 - \beta_2\right)|1\rangle|1\rangle\\ &\quad + \frac{1}{\sqrt{2}}\left(\beta_3 + \beta_4\right)|0\rangle|1\rangle + \frac{1}{\sqrt{2}}\left(\beta_3 - \beta_4\right)|1\rangle|0\rangle.\end{aligned} \tag{24.83}$$

Therefore, equating coefficients, we have the equality,

$$\begin{aligned}|\psi\rangle &= \alpha_0|00\rangle + \alpha_1|01\rangle + \alpha_2|10\rangle + \alpha_3|11\rangle,\\ &= \frac{1}{\sqrt{2}}\left(\beta_1 + \beta_2\right)|0\rangle|0\rangle + \frac{1}{\sqrt{2}}\left(\beta_1 - \beta_2\right)|1\rangle|1\rangle\\ &\quad + \frac{1}{\sqrt{2}}\left(\beta_3 + \beta_4\right)|0\rangle|1\rangle + \frac{1}{\sqrt{2}}\left(\beta_3 - \beta_4\right)|1\rangle|0\rangle,\end{aligned} \tag{24.84}$$

yielding

$$\alpha_0 = \frac{1}{\sqrt{2}}(\beta_1 + \beta_2), \tag{24.85}$$

$$\alpha_1 = \frac{1}{\sqrt{2}}(\beta_3 + \beta_4), \tag{24.86}$$

$$\alpha_2 = \frac{1}{\sqrt{2}}(\beta_3 - \beta_4), \tag{24.87}$$

$$\alpha_3 = \frac{1}{\sqrt{2}}(\beta_1 - \beta_2). \tag{24.88}$$

Then

$$\begin{aligned}\sum_0^3 |\alpha_i|^2 &= \sum_1^4 |\beta_i|^2 = 1,\\ &= \left|\frac{1}{\sqrt{2}}(\beta_1 + \beta_2)\right|^2 + \left|\frac{1}{\sqrt{2}}(\beta_1 - \beta_2)\right|^2 + \left|\frac{1}{\sqrt{2}}(\beta_3 + \beta_4)\right|^2\\ &\quad + \left|\frac{1}{\sqrt{2}}(\beta_3 - \beta_4)\right|^2,\\ &= \frac{1}{2}\left(2\left(\beta_1^2 + \beta_2^2\right)\right) + \frac{1}{2}\left(2\left(\beta_3^2 + \beta_4^2\right)\right) = \sum_1^4 |\beta_i|^2 = 1. \end{aligned} \tag{24.89}$$

Summarizing, the bijective relation between the Bell basis states and the product states given in Eq. (24.77) reflects the discrete phase-space physics of these two-state systems.

24.6 Chiral degrees of freedom and entangled qubits

Observe that the pseudo-spin variables in semiconductor Bloch equation is defined by the following expressions,

$$S_0 = (\rho_{cc} + \rho_{vv}), \tag{24.90}$$

$$S_x = \rho_{vc} + \rho_{cv}, \tag{24.91}$$

$$S_y = -i(\rho_{vc} - \rho_{cv}), \tag{24.92}$$

$$S_z = (\rho_{cc} - \rho_{vv}). \tag{24.93}$$

This maps to

$$e_0 = \left|\Phi^+\right\rangle = \frac{1}{\sqrt{2}}(|1\rangle|1\rangle + |0\rangle|0\rangle), \tag{24.94}$$

$$e_x = |\Psi^+\rangle = \frac{1}{\sqrt{2}} (|0\rangle |1\rangle + |1\rangle |0\rangle), \tag{24.95}$$

$$ie_y = i |\Psi^-\rangle = \frac{1}{\sqrt{2}} (|0\rangle |1\rangle - |1\rangle |0\rangle), \tag{24.96}$$

$$e_z = |\Phi^-\rangle = \frac{1}{\sqrt{2}} (|1\rangle |1\rangle - |0\rangle |0\rangle). \tag{24.97}$$

We refer to the basis (e_0, e_x, e_y, e_z) as the pseudo-spin component basis, which differ from the Bell basis in $|\Psi^-\rangle$ by the factor i in $|\Psi^-\rangle$. An important observation that follows from this is that the entangling of chiral degrees of freedom creates another chiral degrees of freedom, e.g., Eq. (24.51).

This is a very important observation. This will aid in understanding the *entanglement-induced delocalization,* for example, in topological insulators where entanglement occurs at the edge of the sample. This is treated by the quantum transport approach given in Part 6 of this book [198].

Chapter 25

The Concurrence Concept and Emergent Qubit

By virtue of our diagrammatic construction, coupled with discrete phase space Hadamard transformation, in deriving the entangled basis states, the concurrence concept defined by Wooters [195, 196], as well as the two-state (qubit) properties of entangled multi-partite qubits, naturally coincides with the mathematical description of our physical model of qubit entanglement. In other words, the essence of the quantum description of an entangled qubits in Fig. 24.1 consists of the superposition of product "site states" and their corresponding translated "site states" of all qubits. This means a superposition of the two states of the entangled two qubits of Fig. 24.1. One observe that irrespective of how many qubits are entangled, the resulting entangled state is an *emergent* two-state system and therefore behave just like one qubit with two-states, namely, the first state and its *translated* state [23]. Therefore its associated entropy is just one, Eq. (24.47).[1]

The concept of concurrence is basically contained by construction of our physical model of qubit entanglement. It is defined by

$$\left|\langle\Psi|\left|\tilde{\Psi}\right\rangle\right| = C, \tag{25.1}$$

where C is the quantitative value of concurrence. This number lies between 0 and 1, $0 \leq C \leq 1$. Here $\tilde{\Psi}$ is the corresponding translated qubits of Ψ.

25.1 A natural measure of entanglement

The two properties of any entangled qubits, namely, concurrence and its *emergent* qubit behavior, lead Wooters [195, 196] to introduce a measure of

[1] This is also deduced when concurrence $C = 1$ [196].

entanglement, as incorporated in the two formulas,

$$E(C) = H\left(\frac{1}{2} + \frac{1}{2}\sqrt{1 - C^2}\right), \tag{25.2}$$

$$H(x) = -x \ln x - (1 - x) \ln (1 - x). \tag{25.3}$$

Equation (25.2) basically say that if there is complete concurrence, i.e., $C = 1$, then Eq. (25.3) says that the system behave as an emergent qubit or two-state system, as depicted clearly in our diagrams. Thus, for maximally entangled multi-partite qubits, we have $C = 1$,

$$E(C) = H\left(\frac{1}{2}\right), \tag{25.4}$$

$$\begin{aligned} H\left(\frac{1}{2}\right) &= \frac{1}{2}\ln 2 + \frac{1}{2}\ln 2, \\ &= 1, \end{aligned} \tag{25.5}$$

affirming that entangled qubits behave as a two-state system or as an *emergent qubit*, yielding entropy equals one.

25.2 Entropic distance as entanglement measure

From the above developments, one can introduce an entanglement entropic distance by the formula

$$\mathcal{E} = |\mathrm{Tr}(\rho \ln \rho) - 1|, \tag{25.6}$$

where $\mathrm{Tr}(\rho \ln \rho)$ is evaluated using Eqs. (25.2) and (25.3). This distance has a range: $0 \leq \mathcal{E} \leq 1$. When the concurrence, $C = 1$, $\mathrm{Tr}(\rho \ln \rho) = 1 = H\left(\frac{1}{2}\right)$, then we have the entropy-distance from maximally entangled state, $\mathcal{E} = 0$. When $C = 0$, $Tr(\rho \ln \rho) = 0 = H(1)$, then we have the entropy-distance from maximally entangled state, $\mathcal{E} = 1$.

An example where the $C < 1$ may occur in the following singlet state, with less entanglement,

$$|\psi\rangle = \alpha|01\rangle + \beta|10\rangle, \tag{25.7}$$

where

$$\alpha^2 + \beta^2 = 1. \tag{25.8}$$

Then

$$C = \left|\langle\psi| \left|\tilde{\psi}\right\rangle\right| = (\alpha\langle 01| + \beta\langle 10|)(\alpha|10\rangle + \beta|01\rangle),$$
$$= \alpha\beta + \beta\alpha \leq 1. \tag{25.9}$$

25.3 Mixed states and pure states

In discussing mixed and pure states, one makes use of density-matrix operators. This is the domain of abstract statistical treatment usually found in the literature, perhaps following the statistical tradition of Bell's theorem. To elucidate the basic physics, we will here avoid abstract statistical treatment and only discuss specific situations and examples.

25.3.1 *Mixed states from mixed entanglements*

Consider the maximally mixed state of a triplet system,

$$\hat{\rho}_W = \frac{1}{2}(|00\rangle\langle 00| + |11\rangle\langle 11|). \tag{25.10}$$

From Eqs. (24.1), we obtain the mixed entanglements given by

$$\hat{\rho}_B = \frac{1}{2}\left(\left|\Phi^+\Phi^+\right\rangle\left\langle\Phi^+\Phi^+\right| + \left|\Phi^-\Phi^-\right\rangle\left\langle\Phi^-\Phi^-\right|\right). \tag{25.11}$$

Similarly, consider the maximally mixed state of a singlet system,

$$\hat{\rho}_W = \frac{1}{2}(|01\rangle\langle 01| + |10\rangle\langle 10|). \tag{25.12}$$

From Eqs. (24.2), we obtain the mixed entanglements given by

$$\hat{\rho}_B = \frac{1}{2}\left(\left|\Psi^+\Psi^+\right\rangle\left\langle\Psi^+\Psi^+\right| + \left|\Psi^-\Psi^-\right\rangle\left\langle\Psi^-\Psi^-\right|\right). \tag{25.13}$$

25.3.2 *Mixed state from pure state (entangled state)*

Consider the example of a bipartite of two qubit system. Consider a pure entangled state,

$$|\psi\rangle = \frac{1}{\sqrt{2}}(|00\rangle + |11\rangle). \tag{25.14}$$

Then the density matrix is,

$$\rho = \frac{1}{2}(|00\rangle + |11\rangle)(\langle 00| + \langle 11|), \tag{25.15}$$

where the first qubit belongs to party A and the second qubit belongs to party B.[2] We have,

$$\rho^2 = \rho, \tag{25.16}$$

so that Eq. (25.15) is a pure state. Again, we have, by tracing the party B we obtain,

$$\begin{aligned}\rho_A &= \mathrm{Tr}_B\ \rho,\\ &= \frac{1}{2}\langle 0_B|\left(|00\rangle + |11\rangle\right)\left(\langle 00| + \langle 11|\right)|0_B\rangle\\ &\quad + \frac{1}{2}\langle 1_B|\left(|00\rangle + |11\rangle\right)\left(\langle 00| + \langle 11|\right)|1_B\rangle,\\ &= \frac{1}{2}\left(|0_A\rangle\langle 0_A| + |1_A\rangle\langle 1_A|\right).\end{aligned} \tag{25.17}$$

Now clearly

$$\begin{aligned}\rho_A^2 &= \frac{1}{4}\left(|0_A\rangle\langle 0_A| + |1_A\rangle\langle 1_A|\right)\left(|0_A\rangle\langle 0_A| + |1_A\rangle\langle 1_A|\right),\\ &= \frac{1}{4}\left(|0_A\rangle\langle 0_A| + |1_A\rangle\langle 1_A|\right),\\ &= \frac{1}{2}\rho_A.\end{aligned} \tag{25.18}$$

so that ρ_A is not a pure state but mixed, i.e., a mixture of two pure states, $|0_A\rangle$ and $|1_A\rangle$.[3] Note that the mixed states density matrix do not possess *off-diagonal elements*. The tracing operation basically eliminates the contribution of the "*off-diagonal*" terms. From the last identity, the eigenvalues of ρ_A is $\frac{1}{2}$ and since the subsystem A is an emergent qubit, in general with two distinct states, then eigenvalues of ρ_A corresponds to $\frac{1}{\Omega}$ of our first example above. Thus we refer to ρ_A as "uniformly"' mixed often referred to as *maximally mixed* with the initial state $|\psi\rangle$ as *maximally entangled*.

[2]What we mean by party A and B is in the general sense since any entangled number of qubits behave as an emergent qubit.

[3]If the parties A and B are entangled states, then what we have obtain are also mixed entanglement through the inverse relationships, Eqs. (24.1) and (24.2).

Chapter 26

Joint Distribution as a Form of Entanglement

In this section, we will explicitly show the discrete phase-space physics of qubit entanglements. We will demonstrate that a *joint Bloch-function* distributions in *momentum space* transforms into entangled qubits (*Wannier states*) in l*attice-site space*.[1] To the author's knowledge, this demonstrates for the first time the entanglement properties of joint distributions in quantum mechanics. This is crucial to the entanglement-induced delocalization and localization phase of an alternative quantum transport theory of topological insulators, discussed in Chapter 28.

Consider the generalized Fourier transformation between two Hilbert spaces,

$$|p\rangle = \sum_q \langle q|\, p\rangle \;\; |q\rangle, \tag{26.1}$$

where the $\langle q|\, p\rangle$ is the transformation function. This is particularly given by the discrete Fourier transform function,

$$\langle q|\, p\rangle = \frac{1}{\sqrt{N}} \exp\left(-\frac{i}{\hbar} p \cdot q\right), \tag{26.2}$$

where N is the number of discrete points in bijective transformations (this number N is conveniently taken as prime integers).

[1] We maintained the *Bloch function/Wannier function* analogy for convenience, and to stress the wide-ranging impact of the discrete phase-space physics.

26.1 Two-point quantum mechanical joint distributions

We have a joint distribution in p-space as

$$\left|p'\right\rangle\left|p''\right\rangle = \frac{1}{\sqrt{N}}\sum_{q'}\exp\left(-\frac{i}{\hbar}p'\cdot q'\right)\left|q'\right\rangle\frac{1}{\sqrt{N}}\sum_{q''}\exp\left(-\frac{i}{\hbar}p''\cdot q''\right)\left|q''\right\rangle. \tag{26.3}$$

Therefore, for a qubit or a two-state system, we write,

$$|p\rangle = \frac{1}{\sqrt{2}}\sum_{q=0}^{1}\exp\left(-\frac{i}{\hbar}p\cdot q\right)|q\rangle. \tag{26.4}$$

Consider the *joint distribution* of two qubits in *Bloch-function* space. We can form four joint distributions of *Bloch-function* in what follows. To derive these, first, we write the following matrix-representation and Hadamard-transformation identity,

$$\begin{aligned}
&\pi\hbar\begin{pmatrix}|0\rangle\\|1\rangle\end{pmatrix}\otimes\pi\hbar\begin{pmatrix}|0\rangle\\|1\rangle\end{pmatrix}\\
&\quad=\frac{1}{\sqrt{2}}\begin{pmatrix}1&1\\1&-1\end{pmatrix}\begin{pmatrix}|0\rangle\\|1\rangle\end{pmatrix}\otimes\frac{1}{\sqrt{2}}\begin{pmatrix}1&1\\1&-1\end{pmatrix}\begin{pmatrix}|0\rangle\\|1\rangle\end{pmatrix},\\
&\quad=\frac{1}{\sqrt{2}}\frac{1}{\sqrt{2}}\begin{pmatrix}(|0\rangle+|1\rangle)\\(|0\rangle-|1\rangle)\end{pmatrix}\otimes\begin{pmatrix}(|0\rangle+|1\rangle)\\(|0\rangle-|1\rangle)\end{pmatrix}.
\end{aligned} \tag{26.5}$$

We will denote the joint distribution in p-space as $\mathcal{D}(p_1, p_2, \ldots, p_n)$. We will denote these four *Bloch-function* distributions by the standard notations for the Bell basis in the literature, namely, Φ^+, Φ^-, Ψ^+ and Ψ^-. Then, we have

$$\begin{aligned}
(\pi\hbar)^2|0\rangle|0\rangle &= \frac{1}{\sqrt{2}}\frac{1}{\sqrt{2}}(|0\rangle+|1\rangle)(|0\rangle+|1\rangle),\\
&= \frac{1}{\sqrt{2}}\frac{1}{\sqrt{2}}\{|0\rangle|0\rangle+|1\rangle|1\rangle+|0\rangle|1\rangle+|1\rangle|0\rangle\},\\
\mathcal{D}(0,0) &= \frac{1}{\sqrt{2}}\{\Phi^+ + \Psi^+\},
\end{aligned} \tag{26.6}$$

$$\begin{aligned}
(\pi\hbar)^2|1\rangle|1\rangle &= \frac{1}{\sqrt{2}}\frac{1}{\sqrt{2}}\{(|0\rangle-|1\rangle)(|0\rangle-|1\rangle)\},\\
&= \frac{1}{\sqrt{2}}\frac{1}{\sqrt{2}}\{|0\rangle|0\rangle+|1\rangle|1\rangle-|0\rangle|1\rangle-|1\rangle|0\rangle\},\\
\mathcal{D}(1,1) &= \frac{1}{\sqrt{2}}(\Phi^+ - \Psi^+),
\end{aligned} \tag{26.7}$$

$$\begin{aligned}
(\pi\hbar)^2 |0\rangle |1\rangle &= \frac{1}{\sqrt{2}}\frac{1}{\sqrt{2}}\{(|0\rangle + |1\rangle)\}(|0\rangle - |1\rangle), \\
&= \frac{1}{\sqrt{2}}\frac{1}{\sqrt{2}}\{|0\rangle|0\rangle - |1\rangle|1\rangle - |0\rangle|1\rangle + |1\rangle|0\rangle\} \\
\mathcal{D}(0,1) &= \frac{1}{\sqrt{2}}\left(\Phi^- - \Psi^-\right), \qquad (26.8) \\
(\pi\hbar)^2 |1\rangle |0\rangle &= \frac{1}{\sqrt{2}}\frac{1}{\sqrt{2}}\{(|0\rangle - |1\rangle)(|0\rangle + |1\rangle)\} \\
&= \frac{1}{\sqrt{2}}\frac{1}{\sqrt{2}}\{|0\rangle|0\rangle - |1\rangle|1\rangle + |0\rangle|1\rangle - |1\rangle|0\rangle\}, \\
\mathcal{D}(1,0) &= \frac{1}{\sqrt{2}}\left(\Phi^- + \Psi^-\right). \qquad (26.9)
\end{aligned}$$

We observe the following relations for the Bell basis states, namely,

$$\begin{pmatrix}\Phi^+\\ \Phi^-\end{pmatrix} = \frac{1}{\sqrt{2}}\begin{pmatrix}1 & 1\\ 1 & -1\end{pmatrix}\begin{pmatrix}|0\rangle|0\rangle\\ |1\rangle|1\rangle\end{pmatrix}, \qquad (26.10)$$

$$\begin{pmatrix}\Psi^+\\ \Psi^-\end{pmatrix} = \frac{1}{\sqrt{2}}\begin{pmatrix}1 & 1\\ 1 & -1\end{pmatrix}\begin{pmatrix}|0\rangle|1\rangle\\ |1\rangle|0\rangle\end{pmatrix}. \qquad (26.11)$$

As it turns out, Eqs. (26.10) and (26.11) are relations for emergent two-state or qubit systems, respectively. This becomes very transparent in our diagrammatic approaches.

26.2 Three-point quantum mechanical joint distributions

Similarly, for three-fold joint distribution in p-space, we have,

$$\begin{aligned}
|p_1\rangle|p_2\rangle|p_3\rangle = \frac{1}{\sqrt{N}}\sum_{q_1}\exp\left(-\frac{i}{\hbar}p_1\cdot q_1\right)|q_1\rangle \\
\times\frac{1}{\sqrt{N}}\sum_{q_2}\exp\left(-\frac{i}{\hbar}p_2\cdot q_2\right)|q_2\rangle \\
\times\frac{1}{\sqrt{N}}\sum_{q_3}\exp\left(-\frac{i}{\hbar}p_3\cdot q_3\right)|q_3\rangle. \qquad (26.12)
\end{aligned}$$

Again, in matrix form, we express the above equation as

$$\pi\hbar\begin{pmatrix}|0\rangle\\|1\rangle\end{pmatrix}\otimes\pi\hbar\begin{pmatrix}|0\rangle\\|1\rangle\end{pmatrix}\otimes\pi\hbar\begin{pmatrix}|0\rangle\\|1\rangle\end{pmatrix}$$

$$=\frac{1}{\sqrt{2}}\begin{pmatrix}1&1\\1&-1\end{pmatrix}\begin{pmatrix}|0\rangle\\|1\rangle\end{pmatrix}\otimes\frac{1}{\sqrt{2}}\begin{pmatrix}1&1\\1&-1\end{pmatrix}\begin{pmatrix}|0\rangle\\|1\rangle\end{pmatrix}\otimes\frac{1}{\sqrt{2}}\begin{pmatrix}1&1\\1&-1\end{pmatrix}\begin{pmatrix}|0\rangle\\|1\rangle\end{pmatrix},$$

$$=\frac{1}{\sqrt{2}}\frac{1}{\sqrt{2}}\frac{1}{\sqrt{2}}\begin{pmatrix}(|0\rangle+|1\rangle)\\(|0\rangle-|1\rangle)\end{pmatrix}\otimes\begin{pmatrix}(|0\rangle+|1\rangle)\\(|0\rangle-|1\rangle)\end{pmatrix}\otimes\begin{pmatrix}(|0\rangle+|1\rangle)\\(|0\rangle-|1\rangle)\end{pmatrix}. \quad (26.13)$$

From Eq. (26.13), we can define eight joint distributions in p-space, namely, $\mathcal{D}(0,0,0)$, $\mathcal{D}(0,0,1)$, $\mathcal{D}(0,1,0)$, $\mathcal{D}(0,1,1)$, $\mathcal{D}(1,0,0)$, $\mathcal{D}(1,0,1)$, $\mathcal{D}(1,1,0)$, and $\mathcal{D}(1,1,1)$. Therefore, we have

$$\mathcal{D}(0,0,0)=\frac{1}{\sqrt{2}}\frac{1}{\sqrt{2}}\frac{1}{\sqrt{2}}\left\{(|0\rangle+|1\rangle)^3\right\},$$

$$=\frac{1}{\sqrt{2}}\frac{1}{\sqrt{2}}\frac{1}{\sqrt{2}}\left\{\begin{pmatrix}|0\rangle|0\rangle|0\rangle+|1\rangle|1\rangle|0\rangle+|0\rangle|1\rangle|0\rangle+|1\rangle|0\rangle|0\rangle\\|0\rangle|0\rangle|1\rangle+|1\rangle|1\rangle|1\rangle+|0\rangle|1\rangle|1\rangle+|1\rangle|0\rangle|1\rangle\end{pmatrix}\right\}. \quad (26.14)$$

We make the following definitions of entangled basis states for three qubits, which have concurrency[2] equals 1,

$$\frac{1}{\sqrt{2}}(|0\rangle|0\rangle|0\rangle+|1\rangle|1\rangle|1\rangle)=\Theta_3^+, \quad (26.15)$$

$$\frac{1}{\sqrt{2}}(|0\rangle|0\rangle|1\rangle+|1\rangle|1\rangle|0\rangle)=\Gamma^+, \quad (26.16)$$

$$\frac{1}{\sqrt{2}}(|0\rangle|1\rangle|0\rangle+|1\rangle|0\rangle|1\rangle)=\Omega^+, \quad (26.17)$$

$$\frac{1}{\sqrt{2}}(|0\rangle|1\rangle|1\rangle+|1\rangle|0\rangle|0\rangle)=\Xi^+. \quad (26.18)$$

Thus,

$$\mathcal{D}(0,0,0)=\frac{1}{\sqrt{2}}\frac{1}{\sqrt{2}}\left\{\left(\Theta_3^++\Gamma^++\Xi^++\Omega^+\right)\right\}, \quad (26.19)$$

[2]Concurrency is one of the most important characterization of entangled qubits.

is expanded in terms of entangled qubits. We also have

$$\mathcal{D}(0,0,1) = \frac{1}{\sqrt{2}}\frac{1}{\sqrt{2}}\frac{1}{\sqrt{2}}\left\{(|0\rangle + |1\rangle)^2\right\}(|0\rangle - |1\rangle),$$
$$= \frac{1}{\sqrt{2}}\frac{1}{\sqrt{2}}\frac{1}{\sqrt{2}}\{(|0\rangle|0\rangle + |1\rangle|1\rangle + |0\rangle|1\rangle + |1\rangle|0\rangle)\}(|0\rangle - |1\rangle),$$

$$\mathcal{D}(0,0,1) = \frac{1}{\sqrt{2}}\frac{1}{\sqrt{2}}\frac{1}{\sqrt{2}}\left\{\begin{pmatrix} |0\rangle|0\rangle|0\rangle - |1\rangle|1\rangle|1\rangle \\ +|1\rangle|1\rangle|0\rangle - |0\rangle|0\rangle|1\rangle \\ +|0\rangle|1\rangle|0\rangle - |1\rangle|0\rangle|1\rangle \\ +|1\rangle|0\rangle|0\rangle - |0\rangle|1\rangle|1\rangle \end{pmatrix}\right\}. \tag{26.20}$$

We further define

$$\frac{1}{\sqrt{2}}(|0\rangle|0\rangle|0\rangle - |1\rangle|1\rangle|1\rangle) = \Theta_3^-, \tag{26.21}$$

$$\frac{1}{\sqrt{2}}(|0\rangle|0\rangle|1\rangle - |1\rangle|1\rangle|0\rangle) = \Gamma^-, \tag{26.22}$$

$$\frac{1}{\sqrt{2}}(|0\rangle|1\rangle|0\rangle - |1\rangle|0\rangle|1\rangle) = \Omega^-, \tag{26.23}$$

$$\frac{1}{\sqrt{2}}(|0\rangle|1\rangle|1\rangle - |1\rangle|0\rangle|0\rangle) = \Xi^-. \tag{26.24}$$

Then, we have

$$\mathcal{D}(0,0,1) = \frac{1}{\sqrt{2}}\frac{1}{\sqrt{2}}\{(\Theta_3^- + \Gamma^- + \Xi^- + \Omega^-)\}. \tag{26.25}$$

Similarly,

$$\mathcal{D}(0,1,0) = \frac{1}{\sqrt{2}}\frac{1}{\sqrt{2}}\frac{1}{\sqrt{2}}\begin{pmatrix}(|0\rangle + |1\rangle) \\ (|0\rangle - |1\rangle)\end{pmatrix} \otimes \begin{pmatrix}(|0\rangle + |1\rangle) \\ (|0\rangle - |1\rangle)\end{pmatrix} \otimes \begin{pmatrix}(|0\rangle + |1\rangle) \\ (|0\rangle - |1\rangle)\end{pmatrix},$$
$$= \frac{1}{\sqrt{2}}\frac{1}{\sqrt{2}}\frac{1}{\sqrt{2}}\{(|0\rangle + |1\rangle)(|0\rangle - |1\rangle)(|0\rangle + |1\rangle)\},$$
$$\mathcal{D}(0,1,0) = \frac{1}{\sqrt{2}}\frac{1}{\sqrt{2}}\left\{\Theta_3^- - \Xi^- + \Gamma^- - \Omega^-\right\}. \tag{26.26}$$

$$\mathcal{D}(0,1,1) = \frac{1}{\sqrt{2}}\frac{1}{\sqrt{2}}\frac{1}{\sqrt{2}}\{(|0\rangle + |1\rangle)(|0\rangle - |1\rangle)(|0\rangle - |1\rangle)\},$$

$$= \frac{1}{\sqrt{2}}\frac{1}{\sqrt{2}}\frac{1}{\sqrt{2}}\left\{\begin{array}{c} |0\rangle|0\rangle|0\rangle + |1\rangle|1\rangle|1\rangle - |0\rangle|0\rangle|1\rangle + |1\rangle|1\rangle|0\rangle \\ - |0\rangle|1\rangle|0\rangle + |1\rangle|0\rangle|1\rangle \\ |1\rangle|0\rangle|0\rangle + |0\rangle|1\rangle|1\rangle \end{array}\right\},$$

$$\mathcal{D}(0,1,1) = \frac{1}{\sqrt{2}}\frac{1}{\sqrt{2}}\{\Theta_3^+ - \Gamma^+ - \Omega^+ + \Xi^+\}, \tag{26.27}$$

$$\mathcal{D}(1,0,0) = \frac{1}{\sqrt{2}}\frac{1}{\sqrt{2}}\frac{1}{\sqrt{2}}\{(|0\rangle - |1\rangle)(|0\rangle + |1\rangle)(|0\rangle + |1\rangle)\},$$

$$= \frac{1}{\sqrt{2}}\frac{1}{\sqrt{2}}\frac{1}{\sqrt{2}}\{(|0\rangle - |1\rangle)(|0\rangle|0\rangle + |1\rangle|0\rangle + |0\rangle|1\rangle + |1\rangle|1\rangle)\},$$

$$\mathcal{D}(1,0,0) = \frac{1}{\sqrt{2}}\frac{1}{\sqrt{2}}\{\Theta_3^- + \Omega^- + \Gamma^- + \Xi^-\}, \tag{26.28}$$

$$\mathcal{D}(1,0,1) = \frac{1}{\sqrt{2}}\frac{1}{\sqrt{2}}\frac{1}{\sqrt{2}}\{(|0\rangle - |1\rangle)(|0\rangle + |1\rangle)(|0\rangle - |1\rangle)\},$$

$$\mathcal{D}(1,0,1) = \frac{1}{\sqrt{2}}\frac{1}{\sqrt{2}}\{\Theta_3^+ - \Xi^+ + \Omega^+ - \Gamma^+\}, \tag{26.29}$$

$$\mathcal{D}(1,1,0) = \frac{1}{\sqrt{2}}\frac{1}{\sqrt{2}}\frac{1}{\sqrt{2}}\{(|0\rangle - |1\rangle)(|0\rangle - |1\rangle)(|0\rangle + |1\rangle)\},$$

$$\mathcal{D}(1,1,0) = \frac{1}{\sqrt{2}}\frac{1}{\sqrt{2}}\{\Theta_3^+ - \Xi^+ - \Omega^+ + \Gamma^+\}, \tag{26.30}$$

$$\mathcal{D}(1,1,1) = \frac{1}{\sqrt{2}}\frac{1}{\sqrt{2}}\frac{1}{\sqrt{2}}\{(|0\rangle - |1\rangle)(|0\rangle - |1\rangle)(|0\rangle - |1\rangle)\},$$

$$\mathcal{D}(1,1,1) = \frac{1}{\sqrt{2}}\frac{1}{\sqrt{2}}\{\Theta_3^- + \Xi^- - \Omega^- - \Gamma^-\}. \tag{26.31}$$

Summarizing, we have the joint distributions in p-space expressed as entangled qubits in q-space,

$$\mathcal{D}(0,0,0) = \frac{1}{\sqrt{2}}\frac{1}{\sqrt{2}}\{(\Theta_3^+ + \Gamma^+ + \Xi^+ + \Omega^+)\}, \tag{26.32}$$

$$\mathcal{D}(0,0,1) = \frac{1}{\sqrt{2}}\frac{1}{\sqrt{2}}\{(\Theta_3^- + \Gamma^- + \Xi^- + \Omega^-)\}, \tag{26.33}$$

$$\mathcal{D}(0,1,0) = \frac{1}{\sqrt{2}}\frac{1}{\sqrt{2}}\{\Theta_3^- - \Xi^- + \Gamma^- - \Omega^-\}, \tag{26.34}$$

$$\mathcal{D}(0,1,1) = \frac{1}{\sqrt{2}}\frac{1}{\sqrt{2}}\{\Theta_3^+ - \Gamma^+ - \Omega^+ + \Xi^+\}, \tag{26.35}$$

$$\mathcal{D}(1,0,0) = \frac{1}{\sqrt{4}}\{\Theta_3^- + \Omega^- + \Gamma^- + \Xi^-\}, \tag{26.36}$$

$$\mathcal{D}(1,0,1) = \frac{1}{\sqrt{4}}\{\Theta_3^+ - \Xi^+ + \Omega^+ - \Gamma^+\}, \tag{26.37}$$

$$\mathcal{D}(1,1,0) = \frac{1}{\sqrt{4}}\{\Theta_3^+ - \Xi^+ - \Omega^+ + \Gamma^+\}, \tag{26.38}$$

$$\mathcal{D}(1,1,1) = \frac{1}{\sqrt{4}}\{\Theta_3^- + \Xi^- - \Omega^- - \Gamma^-\}. \tag{26.39}$$

A crucial observation is that only even (+) combinations of entangled basis states correspond to joint distributions and entanglements, and similarly only odd (−) combinations of entangled basis states correspond to joint distributions and entanglements as well. One can continue the above matrix approach to find all the entangled basis for any number of more than three qubits. However, a powerful, straightforward and simpler procedure can be carried out using our diagrammatic approach of Section 23.1 [23].

Chapter 27

Part 5: Concluding Remarks and Perspective

The apparently perfect scheme of the mechanical model in describing quantum entanglement seems to advocate the existence of entangled spacetime medium to take the place of the chain of inverters in some unknown forms. An interesting classical analogy is the easily generated bound/entangled pair of vortices on the surface of fluids (swimming pools) which are really the ends of a vortex tube forming a U-shape vortex chain underneath the surface. This corresponds to the triplet-spin pairing of Fig. 24.1. If a vortex chain underneath the surface of the fluid undergoes a soliton transition (e.g., an inversion point or line of travel, by some ingenious fluid excitation) in its middle portion, this would also bind a vortex anti-vortex pair as it opens up at the surface. This will correspond to the singlet-spin pairing of Fig. 24.1. The fact that soliton plays a crucial role in entanglement of vortex anti-vortex pair is found in works where soliton-like perturbation is found to decay into more stable vortex anti-vortex pair [199]. Vortex anti-vortex "inverter" chain [200] can also be excited which would physically implements the inverter-chain entanglement-link of our model, Fig. 24.1.

Thus, it seems extra dimensions are needed in spacetime to have a theory of entanglement. The work of Ooguri [201] and collaborators shows that this quantum entanglement generates the extra dimensions of the gravitational theory. "It appears that it seems possible to generate a geometric connection between entangled qubits, even though there is no direct interaction between the two systems [202]. Furthermore, the solid and reliable structure of spacetime is due to the ghostly features of entanglement".

The question is thus raised: could it be that besides the geometrical spacetime aspects of gravity, there is a purely quantum mechanical aspect of spacetime geometry with extra dimensions that give rise to entanglement?

A hint along this idea is also given by Maldacena [202] when he stated that "One can consider, therefore, a pair of black holes where all the microstates are "entangled". Namely, if we observe one of the black holes in one particular microstate, then the other has to be in exactly the same microstate. A pair of black holes in this particular EPR entangled state would develop a wormhole, or Einstein–Rosen bridge, connecting them through the underneath. The geometry of this wormhole is given by the fully extended Schwarzschild geometry. It is interesting that both wormholes and entanglement naively appear to lead to a propagation of signals faster than light." "It was known that quantum entanglement is related to deep issues in the unification of general relativity and quantum mechanics, such as the black hole information paradox and the firewall paradox", says Hirosi Ooguri [201]. Remarkably, all this can very simply be viewed as bound/entangled pair of vortices on the surface of fluids, a commonplace occurrence, which are really the ends of a vortex chain (vortex tube or "Einstein-Rosen bridge") forming a U-shape underneath the surface or as a meeting two vertical funnels ("wormholes") if the surface is theoretically bowed into a U-shaped itself. There is in fact some bold idea already advanced that the wormholes of Einstein–Rosen bridge is a relativistic version of the mysterious link in quantum entanglement, this idea is now popularized by the equation ER = EPR.

In any case, from the computational point of view, deriving the computational entangled basis states have now been given for the first-time a systematic and physically meaningful procedure using a diagrammatic foundation. One can also define a group theory of entanglement diagrams.

27.1 On Bell's inequality and geometric nature of entanglement

Bell's famous inequality [203] is derived from purely classical statistical considerations using *added parameters*, i.e., *local* hidden variables, subject to rigorous probabilistic considerations. The conclusion is that realism in quantum mechanics, if nonlocal, is expected to violate the Bell's inequality, which strictly defines domain of locality. Some experiments have been interpreted this way, as violation of Bell's inequality, thus confirming that quantum mechanics is indeed *non-local,* although one expects this from the beginning of quantum mechanics due to the superposition principle. However, Bell's inequality does not seem to have anything to do with Einstein's spooky action at a distance paradox, raised in the EPR thought experiment [194]. In other words, incompleteness in quantum mechanics raised by the EPR still cannot be dismissed by Bell's inequality. Hence incompleteness of quantum mechanics is still open. Maybe Einstein's incompleteness assertion to understand quantum entanglement indicates a much deeper inquiry and carries a serious validity. In fact, entanglement maybe entirely *geometric* in nature, not entirely subject to probabilistic arguments, if quantum mechanics has to be unified with Einstein's general relativity theory.

In our discussion above, a familiar phenomenon of a bound vortex pair in a surface of a fluid hides the vortex chain that lie underneath the surface that bind the two vortices. Clearly, a 2D surface is insufficient to understand the entanglement of the two vertices. In short, one needs to go beyond two dimension to see the vortex chain under the surface that entangles the vortex/anti-vortex pair. Hence, there is an implication that generally higher dimensions are needed for a complete description of the Einstein spooky action at a distance of entanglement.

Another example, is a chain of inverters, implemented by a chain of swings or see-saws, discussed above. Indeed, the vortex chain in fluid dynamics above maybe thought as corresponding to the chain of inverters for coupled spins in our diagrammatic scheme of entanglement. This mechanical arrangement amply illustrates the entanglement of singlet and triplet spin systems. Clearly, quantum mechanics cannot provide what corresponds to the "invisible" inverter-chain of swings, and seems to suggest a purely geometric nature of entanglement. Again, Einstein assertion's that quantum mechanics is incomplete holds.

Spacetime entanglement discussed by Ooguri [201] may have something to say on all this for the sought after completeness. This remains to be seen, for a resolution of the Einstein spooky action at a distance paradox. Recently, Maldacena suggested the *geometric nature of entanglement*, i.e., the *connection* (quantum version of our vortex chain and chain of inverters) that provides the spooky action at a distance paradox is realized or envisage by the so-called wormhole, or Einstein–Rosen bridge. Maldacena considered a pair of black holes where all the microstates are "entangled". If we observe one of the black holes in one particular microstate, then the other has to be in exactly corresponding microstate. A pair of black holes in this particular EPR entangled state would develop a wormhole, or Einstein–Rosen bridge, *connecting* them through the inside. The geometry of this wormhole is given by the fully extended Schwarzschild geometry. Indeed, bound/entangled pair of vortices on the surface of fluids, a commonplace occurrence, which are really the ends of a vortex chain (vortex tube or "Einstein–Rosen bridge") forming a U-shape vortex chain underneath the surface, or as a meeting two vertical funnels ("wormholes") if the surface is theoretically bowed into U-shaped itself, seems to give a simple view of entangled black holes.

27.2 On entanglement measures

The discrete phase-space viewpoint for qubit entanglement has been very fruitful in analytically deriving the entanglement basis states for any number of entangled qubits as demonstrated by the author in his book [12], as well as in Ref. [23]. Moreover, the inverter-chain mechanical model of entanglement link has been demonstrated to faithfully implement the discrete phase space viewpoint [23]. The crucial observation that arise from this inverter-chain link model is that any multi-partite qubit entangled system has the property of an *emergent* qubit, which has been rigorously justified. This aspect of the model readily lead us to the equality relations of the entropy of entanglement formation in the so-called monogamy inequality of entanglement formation, using statistical arguments [197] discussed in the literature. The mixed states and mixed entanglements are related by the Hadamard transformations. In addition, we show that mixed state can be extracted from pure entangled state, where the party A or B need not be a single qubit themselves but can each be an entangled multi-partite qubit system, respectively, by virtue of *emergent* qubit property of entangled qubit systems.

The natural measure of entanglement which have pure geometrical significance should be based on the following: (a) entropy of entanglement formation, (b) concurrence, and (c) entropic distance from maximally entangled reference.

Part 6

Entanglement Delocalization of Topological Insulators

Chapter 28

Chiral Entanglement and Topological Phases

The recent history of condensed matter physics has shown that the study of vortices, cyclotron orbits, spinors, Berry connections (in older form as Peierls phase factor), Berry curvatures, chiral degrees of freedom, and Chern numbers have ushered the incarnation of topology in quantum physics [204, 205]. Indeed, the importance of several discrete-two degrees of freedom, i.e., spinors and their entanglements has emerge as ubiquitous in the physics of new and low-dimensional materials, such as graphene, TMDs, silicene, and topological systems [205, 206]. For example, Pauli–Dirac spin, valley spin (isospin), pseudospin due to low-energy electron–hole symmetry at Dirac points, and bilayer pseudospin have gained importance in consideration of the spin quantum transport of 2D materials.

The effect of atomic-layer pseudospin degrees of freedom in bilayer graphene and TMD materials has been vigorously pursued both theoretically and experimentally due to exotic properties, namely, exciton condensation and a new mechanism for the onset of superconductivity [206, 207]. Entangled electron–hole pairs have also been proposed for monolayer graphene [208]. Indeed, in bilayer materials a new mechanism for the onset of superconductivity has been an intriguing discovery. This is related to the onset of exciton condensate due to the entanglement of layer pseudospin with $\nu = 1$ filling of the lowest Landau levels (LLLs), i.e., lowest Landau orbit (LLO) in each of the layers [209, 210], i.e., with intralayer quantum Hall effect at $\nu = 1$ of LL0. In topological insulators, the Anderson type of localization of metallic edge states due to interaction with magnetic impurities causing successive spin flips has also been proposed [211] (in view of our recent findings, this kind of edge-states localization maybe attributed to the entanglement with impurity spin). Spin Kondo effect at the metallic edge states has also been treated [212]. In other words, the edge states have been studied and subjected to the usual treatment of 1D conductors, mostly based on the usual Hamiltonian and exchange interaction context.

Here, we propose a rounded picture and an entirely new perspective on the physics of topological insulators, QHE-TI, QSHE-TI, and Kondo-TI based on quantum non-locality as a result of the entanglement of torques induced by the various spin and chiral degrees of freedom. This is inferred from our results of the generalized non-equilibrium quantum transport equations of spin and pseudospins, and their entanglements. The present proposal implies a new mechanism for localization and delocalization based on the series of spin and pseudospin torque entanglements. The sort of localization being referred to here is typified by cyclotron-orbit current localized around orbit center due to either external magnetic field or intrinsic Berry curvature in strongly spin-orbit coupled materials. On the other hand, the sort of delocalization referred to is represented by the metallic edge states in topological insulators [211], as a result of the *quantum non-locality* brought about by spin and pseudospin entanglements, the type of delocalization responsible for the *resonant quantum tunneling transport* phenomena in resonant tunneling diodes [11, 14].

Here, the point of view in all of these is that of torque entanglement in materials with multi-spin degrees of freedom. Localization and delocalization of vortices have also found some treatments in the literature that could be interpreted as due to torque interactions [213–215]. In fact, these sort of localization and *quantum-non-locality* delocalization effects also shed light on the experimentally observed and ubiquitous phenomena in spin-dependent quantum transport physics. These are the so-called *weak localization* (WL) [216] and *weak anti-localization* (WAL), [217, 218] often referred to in the literature. By virtue of their competing effects in non-equilibrium quantum transport, the crossover from WL to WAL has also been experimentally observed [219].

The analysis presented here is based on the use of the phase-space formulation of quantum mechanics [3, 5]. The use of position and momentum variables should not be construed as employing a semi-classical analysis. This is founded on a basic quantum superposition principle, i.e., on the expansion of any quantum mechanical operator in terms of a complete set of orthogonal basis operators or *projectors,* discussed in detail in Part 1. In the phase-space formalism, either discrete or continuum, the orthogonal and complete basis operators are the *phase-space point projectors*, $\hat{\Delta}(p, q)$. Thus, in *operator Hilbert-space* method [12] of quantum theory, any operator $\hat{A}$ can be expanded as the sum or integral of the basis operator, $\hat{\Delta}(p, q)$. As an integral expansion we have,

$$\hat{A} = \left(\frac{1}{2\pi\hbar}\right)^d \int d^d p' d^d q' A\left(p', q'\right) \hat{\Delta}\left(p', q'\right), \tag{28.1}$$

where d is the dimensionality and the expansion coefficients are the $A\,(p,q)$'s, where each coefficient is given by the trace of the operator with the *phase-space point projectors*, $\hat{\Delta}\,(p,q)$,

$$A\left(p, q\right) = \mathrm{Tr}[\hat{\Delta}\left(p, q\right) \hat{A}]. \tag{28.2}$$

The *c*-function $A\,(p,q)$ is known as the *lattice Weyl transform* of the operator $\hat{A}$ [12]. Notice that the *lattice Weyl transform* operates on matrices. If $\hat{A} = \hat{\rho}$ (the density matrix operator), then $\rho\,(p,q) \equiv f_w\,(p,q)$, which is the Wigner distribution function.

It follows that any commutator or anti-commutator of $\hat{A}$ and $\hat{B}$ can be treated similarly in the operator Hilbert-space method [12],

$$[\hat{A}, \hat{B}] = \left(\frac{1}{2\pi\hbar}\right)^d \int d^d p' d^d q' [\hat{A}, \hat{B}]\left(p', q'\right) \hat{\Delta}\left(p', q'\right), \tag{28.3}$$

where the coefficient of expansion is given by the trace of the operator with $\hat{\Delta}\,(p,q)$,

$$[\hat{A}, \hat{B}]\left(p, q\right) = \mathrm{Tr}(\hat{\Delta}\left(p, q\right) [\hat{A}, \hat{B}]), \tag{28.4}$$

and

$$\{\hat{A}, \hat{B}\} = \left(\frac{1}{2\pi\hbar}\right)^d \int d^d p' d^d q' \{\hat{A}, \hat{B}\}\left(p', q'\right) \hat{\Delta}\left(p', q'\right), \tag{28.5}$$

where

$$\{\hat{A}, \hat{B}\}\left(p, q\right) = \mathrm{Tr}(\hat{\Delta}(p, q)\{\hat{A}, \hat{B}\}), \tag{28.6}$$

respectively. Note that no semi-classical approximation is involved or needed whatsoever in the above phase-space formalism of operator Hilbert-space method. The exact expression for $[\hat{A}, \hat{B}]\,(p,q)$ and $\{\hat{A}, \hat{B}\}\,(p,q)$ are given in terms of powers of Poisson bracket (equivalent to expansion to all powers of $\hbar$) or in terms of integral kernels as shown in Section E.7 of the Appendix. It is only when the first power of the Poisson bracket is used that we have a semi-classical approximation. Indeed, the exact quantum non-locality property of the commutator has been rigorously demonstrated in resonant tunneling diodes, where the theoretical numerical simulation resolves the controversial experimental quantum transport measurements on these nanodevices [14]. Moreover, the simulation was able to resolve the discrete energy levels of the quantum wells [11], and yields the THz current oscillations responsible for the current plateau in the dc-current measurements [14]. Even in the semi-classical approximation, the quantum non-locality property of the commutator is already apparent. However, no semiclassical approximation is capable of simulating the current–voltage characteristics of resonant tunneling diodes.

In the context of operator Hilbert-space method, the von Neumann density-matrix evolution equation of quantum statistical dynamics in H-space becomes a super-statevector quantum dynamical equation in L-space or operator Hilbert-space. The ensuing quantum transport formalism was discussed in detail in Part 3. To recap, the familiar von Neumann density-matrix operator equation in H-space given by

$$i\hbar\frac{\partial}{\partial t}\rho\left(t\right)=\left[\mathcal{H},\rho\right], \tag{28.7}$$

becomes a super-Schrödinger equation for the super-statevector in L-space expressed as

$$i\hbar\frac{\partial}{\partial t}\left|\rho\left(t\right)\right\rangle\rangle=\mathcal{L}\left|\rho\left(t\right)\right\rangle\rangle. \tag{28.8}$$

In Eq. (28.7), $\rho(t)$ is the density-matrix operator for the whole many-body system in H-space, whereas in Eq. (28.8) $|\rho(t)\rangle\rangle$ is the corresponding super-statevector in L-space. The superoperator $\mathcal{L}$ corresponds to the commutator $[\mathcal{H},\rho]$ of Eq. (28.7), and is referred to as the Liouvillian. The corresponding many-body quantum-field operators becomes quantum superfield operators in L-space. Similar to the zero temperature Green's function technique, the super-non-equilibrium Green's functions are calculated in the L-space theory. The details in deriving the non-equilibrium quantum transport equations including Cooper pairings are summarized in Part 3 of this book. Readers are also referred to the author's book and references therein [12].

On the other hand, without treating the Cooper pairing it is not clear what information, if any, concerning gapless Majorana edge states in p_x+ip_y-superconductors, the so-called spin-triplet superconductors. The reason why superconductors are the mining field for Majoranas is that in general Cooper pairings, i.e., the presence of anomalous Green's functions, correspond trivially to paired Majorana fermions in non-topological superconductors, but a non-trivial pairing of $m_i = (\Psi_i^\dagger + \Psi_i)/2$ and $m_j = (\Psi_j^\dagger - \Psi_j)/2i$, with $\langle m_i m_j\rangle \neq 0$ as condensed pair of Majorana fermions at neighboring sites, i.e., $i \neq j$, describes a spinless or ferromagnetic 1D topological superconductor. This combination is non-trivial precisely because when $\langle m_i \rangle = 0$, then $\langle m_j \rangle \neq 0$ and *vice versa*, which indicates the presence of unpaired Majorana at site j or i as the case maybe. This can be induced in a Kitaev wire [220] by proximity effect with s-type superconductor. This calls for use of anomalous Green's function and ordinary Green's function, say $\frac{1}{4i}[\langle \Psi_i^\dagger \Psi_j^\dagger \rangle - \langle \Psi_i \Psi_j \rangle - 2\langle \Psi_i^\dagger \Psi_j \rangle]$, to simulate $\langle m_i m_j \rangle$ in our non-equilibrium quantum transport equations, However, Cooper pairings are all beyond the scope of this book.

The general outline is that first we give all the non-equilibrium quantum transport equations for various number of spin degrees of freedom in a system. This is followed by a discussion and concluding remarks in Section 29. Some of the detailed calculations are given in Appendix E.

28.1 Quantum transport equations

Our starting point is the general quantum transport expressions for fermions, where non-equilibrium quantum superfields and their correlations are the basic variables. These are obtained from the real-time non-equilibrium quantum superfield theoretical transport formulation of Part 3 [12, 117, 158]:

$$\begin{aligned} i\hbar\left(\frac{\partial}{\partial t_1} + \frac{\partial}{\partial t_2}\right) G^{\gtrless} &= [\mathcal{H}G^{\gtrless} - G^{\gtrless}\mathcal{H}] \\ &\quad + [\Sigma^r G^{\gtrless} - G^{\gtrless}\Sigma^a] + [\Sigma^{\gtrless} G^a - G^r \Sigma^{\gtrless}] \\ &\quad + [\Delta^r_{hh} g^{\gtrless}_{ee} - g^{\gtrless}_{hh}\Delta^a_{ee}] \\ &\quad + [\Delta^{\gtrless}_{hh} g^a_{ee} - g^r_{hh}\Delta^{\gtrless}_{ee}]. \end{aligned} \tag{28.9}$$

The last two brackets account for the Cooper pairings between fermions of the same specie. These do not concern us here (their corresponding transport equations [12, 117] are important in non-equilibrium superconductivity, where we also need to solve for the non-equilibrium anomalous Green functions). In what follows, we will drop these last two square brackets of the RHS of Eq. (28.9).

28.2 'Cube' matrix quantum transport equations

In the absence of Cooper pairing between fermions of the same-specie Eq. (28.9) becomes, by separately writing the discrete spin quantum-label arguments, essentially as 8×8 matrix equations on the spin and pseudospin indices,

$$i\hbar\left(\frac{\partial}{\partial t_1}+\frac{\partial}{\partial t_2}\right)G^{\lessgtr}_{kk'll'mm'}(12)$$
$$=[\mathcal{H}_{kjl\gamma m\alpha}G^{\lessgtr}_{jk'\gamma l'\alpha m'}(\bar{2}2)-G^{\lessgtr}_{kjl\gamma m\alpha}(1\bar{2})\mathcal{H}_{jk'\gamma l'\alpha m'}\delta_{\bar{2}2}]$$
$$+[\Sigma^{r}_{kjl\gamma m\alpha}(1\bar{2})G^{\lessgtr}_{jk'\gamma l'\alpha m'}(\bar{2}2)-G^{\lessgtr}_{kjl\gamma m\alpha}(1\bar{2})\Sigma^{a}_{jk'\gamma l'\alpha m'}(\bar{2}2)]$$
$$+[\Sigma^{\lessgtr}_{kjl\gamma m\alpha}(1\bar{2})G^{a}_{jk'\gamma l'\alpha m'}(\bar{2}2)-G^{r}_{kjl\gamma m\alpha}(1\bar{2})\Sigma^{\lessgtr}_{jk'\gamma l'\alpha m'}(\bar{2}2)], \tag{28.10}$$

where the Greek subscript indices correspond to discrete degrees of freedom, namely, Pauli–Dirac spin ($k, k' = \{\downarrow, \uparrow\}$), valley indices ($l, l' = \{K, K'\}$), and layer or band indices as the case maybe (e.g., $m, m' = \{t, b\}$). The numeral indices correspond to the two-point space–time arguments. In what follows, we will treat either the two-layer model of chiral Dirac fermions (m, m'), e.g., the bottom and top layer of graphene and TMDs in strong magnetic field or other 2D materials with strong spin-orbit coupling but without magnetic fields [221].

In the absence of Cooper pairing of superconductivity, we may also write the general transport equation for '$\gtrless$'-quantities as [here $\{A, B\}$ and $[A, B]$ means anti-commutator and commutator, respectively, of operators A and B], by dropping all arguments and discrete indices, as

$$i\hbar\left(\frac{\partial}{\partial t_1}+\frac{\partial}{\partial t_2}\right)G^{\lessgtr}$$
$$=\left[\tilde{\mathcal{H}}, G^{\lessgtr}\right]+\left[\Sigma^{\lessgtr}, \mathrm{Re}\, G^{r}\right]-\frac{i}{2}\left\{\Gamma, G^{\lessgtr}\right\}+\frac{i}{2}\left\{\Sigma^{\lessgtr}, A\right\}, \tag{28.11}$$

where $\Gamma = -\frac{1}{2}\mathrm{Im}\,\Sigma^r$, is the scattering-out matrix, $A = -\frac{1}{2}\mathrm{Im}\, G^r$ is the spectral function, and the single-particle Hamiltonian is given by

$$\tilde{\mathcal{H}} = \mathcal{H}_o + \Sigma^{\delta} + \mathrm{Re}\,\Sigma^r, \tag{28.12}$$

where Σ^{δ} and $\mathrm{Re}\,\Sigma^r$ correspond to the renormalization of the bands or mass terms induced by the self-energies [12, 117]. The appearance of the imaginary i in Eq. (28.11) in the anti-commutator is commensurate when viewed in their gradient expansions in terms of Poisson bracket differential operator wherein a commutator of Eq. (28.11) has a factor i whereas the anti-commutator does not have. As is well-known, in the classical limit, the gradient expansion leads to the Boltzmann kinetic transport equation [158].

28.3 Canonical spinor form of a 'cube' matrix

Consider $G^{<}_{k,k'l,l'm,m'}$, where the indices denote the elements of the 8×8 matrix. Here k,k' denote the spin indices, l,l' denote the isospin or valley indices, and m,m' denote the pseudospin of either band or bilayer indices. We will now demonstrate how to reduce the 64 variables of the 8×8 matrix to just 8 tensor variables, which include pure scalars or total charge. This is reminiscent to a decomposition, in the absence of dissipation, of the $SU(8)$ into a direct product of $SU(2)\otimes SU(2)\otimes SU(2)$, i.e., in order to keep track of the spin-like degrees of freedom. In what follows, we make constant use of the theorem of Appendix E.

A *First Stage*: Reduction of Pauli–Dirac spin indices to scalar and vector or Pauli–Dirac spin components, we have, where x,y,z denote components of vector quantities,

$$G^{<}_{k,k'l,l'm,m'} = \frac{1}{2}\begin{pmatrix} G^{<}_{o,l,l'm,m'} + G^{<}_{z,l,l'm,m'} & G^{<}_{x,l,l'm,m'} - iG^{<}_{y,l,l'm,m'} \\ G^{<}_{x,l,l'm,m'} + iG^{<}_{y,l,l'm,m'} & G^{<}_{o,l,l'm,m'} - G^{<}_{z,l,l'm,m'} \end{pmatrix}. \tag{28.13}$$

B *Second stage*: Reduction of the remaining valley indices to scalar and vector or isospin components.

$$\frac{1}{2}G^{<}_{o,l,l'm,m'} = \frac{1}{4}\begin{pmatrix} G^{<}_{o,o,m,m'} + G^{<}_{o,z,m,m'} & G^{<}_{o,x,m,m'} - iG^{<}_{o,y,m,m'} \\ G^{<}_{o,x,m,m'} + iG^{<}_{o,y,m,m'} & G^{<}_{o,o,m,m'} - G^{<}_{o,z,m,m'} \end{pmatrix}, \tag{28.14}$$

$$\frac{1}{2}G^{<}_{k,l.l',m,m'} = \frac{1}{4}\begin{pmatrix} G^{<}_{k,o,m,m'} + G^{<}_{k,z,m,m'} & G^{<}_{k,x,m,m'} - iG^{<}_{k,y,m,m'} \\ G^{<}_{k,x,m,m'} + iG^{<}_{o,y,m,m'} & G^{<}_{k,o,m,m'} - G^{<}_{o,z,m,m'} \end{pmatrix}. \tag{28.15}$$

C *Third and final stage*: Reduction of the last band or layer indices to scalar and vector or pseudospin components.

$$\frac{1}{4}G^{<}_{o,o,m,m'} = \left(\frac{1}{2}\right)^3\begin{pmatrix} G^{<}_{o,o,o} + G^{<}_{o,o,z} & G^{<}_{o,o,x} - iG^{<}_{o,o,y} \\ G^{<}_{o,o,x} + iG^{<}_{o,o,y} & G^{<}_{o,o,o} - G^{<}_{o,o,z} \end{pmatrix}, \tag{28.16}$$

$$\frac{1}{4}G^{<}_{o,l,m,m'} = \left(\frac{1}{2}\right)^3\begin{pmatrix} G^{<}_{o,l,o} + G^{<}_{o,l,z} & G^{<}_{o,l,x} - iG^{<}_{o,l,y} \\ G^{<}_{o,l,x} + iG^{<}_{o,l,y} & G^{<}_{o,l,o} - G^{<}_{o,l,z} \end{pmatrix}, \tag{28.17}$$

$$\frac{1}{4}G^{<}_{k,o,m,m'} = \left(\frac{1}{2}\right)^3 \begin{pmatrix} G^{<}_{k,o,o} + G^{<}_{k,o,z} & G^{<}_{k,o,x} - iG^{<}_{k,o,y} \\ G^{<}_{k,o,x} + iG^{<}_{k,o,y} & G^{<}_{k,o,o} - G^{<}_{k,o,z} \end{pmatrix}, \tag{28.18}$$

$$\frac{1}{4}G^{<}_{k,l',m,m'} = \left(\frac{1}{2}\right)^3 \begin{pmatrix} G^{<}_{k,l',o} + G^{<}_{k,l',z} & G^{<}_{k,l',x} - iG^{<}_{k,l',y} \\ G^{<}_{k,l',x} + iG^{<}_{k,l',y} & G^{<}_{k,l',o} - G^{<}_{k,l',z} \end{pmatrix}. \tag{28.19}$$

Collecting results of the third and final reduction stage, we finally transformed $G^{<}_{k,k'l,l'm,m'}$ in to its spinor form as

$$(2)^3 G^{<}_{k,k'l,l'm,m'} \xrightarrow{\text{spinor}} \left\{ \begin{aligned} &\begin{pmatrix} G^{<}_{o,o,o} + G^{<}_{o,o,z} & G^{<}_{o,o,x} - iG^{<}_{o,o,y} \\ G^{<}_{o,o,x} + iG^{<}_{o,o,y} & G^{<}_{o,o,o} - G^{<}_{o,o,z} \end{pmatrix} \\ &+ \begin{pmatrix} G^{<}_{o,l,o} + G^{<}_{o,l,z} & G^{<}_{o,l,x} - iG^{<}_{o,l,y} \\ G^{<}_{o,l,x} + iG^{<}_{o,l,y} & G^{<}_{o,l,o} - G^{<}_{o,l,z} \end{pmatrix} \\ &+ \begin{pmatrix} G^{<}_{k,o,o} + G^{<}_{k,o,z} & G^{<}_{k,o,x} - iG^{<}_{k,o,y} \\ G^{<}_{k,o,x} + iG^{<}_{k,o,y} & G^{<}_{k,o,o} - G^{<}_{k,o,z} \end{pmatrix} \\ &+ \begin{pmatrix} G^{<}_{k,l',o} + G^{<}_{k,l',z} & G^{<}_{k,l',x} - iG^{<}_{k,l',y} \\ G^{<}_{k,l',x} + iG^{<}_{k,l',y} & G^{<}_{k,l',o} - G^{<}_{k,l',z} \end{pmatrix} \end{aligned} \right\}. \tag{28.20}$$

Thus, from Eq. (28.20), we end up with the eight tensor transport variables, similar to the number of distinct configurations of three spin-qubits, which include joint distributions or spin-torque entanglements, [222] defined in the Table 28.1,

Note that in Table 28.1 the eight tensor variables, including the total charge, are mapped to the configurations of three qubits as exhibited by the subscripts of the non-equilibrium tensorial Green's function, $G^{<}$.

Table 28.1. The eight transport variables.

$G^{<}_{0,0,0}$	Complete scalar, i.e., particle number density or total charge density
$G^{<}_{k,0,0}$	Pauli–Dirac spin magnetization density
$G^{<}_{0,l,0}$	Isospin magnetization density
$G^{<}_{0,0,m}$	Pseudospin magnetization density
$G^{<}_{k,l,0}$	Entangled or joint Pauli–Dirac spin and isospin magnetization density
$G^{<}_{k,0,m}$	Entangled or joint Pauli–Dirac spin and pseudospin magnetization density
$G^{<}_{0,l,m}$	Entangled or joint isospin-pseudospin magnetization density
$G^{<}_{k,l,m}$	Entangled or joint Pauli–Dirac spin, isospin, pseudospin magnetization density

28.4 Quantum transport equations for $G^{\gtrless}$, $N_s = 3$

By treating all $SU(2)$ indices on equal footing, we evaluate the canonical form of every matrix involved in all matrix products as a series of product of 2×2 matrices: $(kj)(jk'), (l\gamma)(\gamma l'), (m\alpha)(\alpha m')$, as was done in Section 28.3. We finally end up with the eight coupled quantum transport equations for $G^{<}_{o,o,o}$, $G^{<}_{k,o,o}$, $G^{<}_{o,l,o}$, $G^{<}_{o,o,m}$, $G^{<}_{k,l,o}$, $G^{<}_{k,o,m}$, $G^{<}_{o,l,m}$, and $G^{<}_{k,l,m}$ defined in Table 28.1. The technique followed in the derivation is explained in Appendix E, Section E.3.8. Note that in line with the four terms in the right-hand-side (RHS) of Eq. (28.11), the RHS of each of the eight coupled transport equations correspondingly consist of four groups. We have generalized the Hamiltonian spinor to account for either of the magnetic field and/or Dresselhaus and/or Rashba spin-orbit coupling [223].

Guided by the results of Section E.3, we obtain the non-equilibrium quantum transport equations of a system with three spin degrees of freedom. For example, we have for the total scalar,

$$2^{N_s} i\hbar \left(\frac{\partial}{\partial t_1} + \frac{\partial}{\partial t_2} \right) G^{\gtrless}_{o,o,o}$$

$$= [\tilde{\mathcal{H}}, G^{\lessgtr}]_{o,o,o} + [\Sigma^{\lessgtr}, \operatorname{Re} G^r]_{o,o,o} - \frac{i}{2}\{\Gamma, G^{\lessgtr}\}_{o,o,o} + \frac{i}{2}\{\Sigma^{\lessgtr}, A\}_{o,o,o}, \tag{28.21}$$

where

$$[\tilde{\mathcal{H}}, G^{\lessgtr}]_{o,o,o}$$

$$= \begin{matrix} [H_{o,o,o}, G^{\gtrless}_{o,o,o}] + [H_{o,o,m}, G^{\gtrless}_{o,o,m}] + [H_{o,l,o}, G^{\gtrless}_{o,l,o}] + [H_{o,l,m}, G^{\gtrless}_{o,l,m}] \\ + [H_{k,o,o}, G^{\gtrless}_{k,o,o}] + [H_{k,o,m}, G^{\lessgtr}_{k,o,m}] + [H_{k,l,o}, G^{\gtrless}_{k,l,o}] + [H_{k,l,m}, G^{\gtrless}_{k,l,m}] \end{matrix}, \tag{28.22}$$

$$[\Sigma^{\lessgtr}, \operatorname{Re} G^r]_{o,o,o}$$

$$= \begin{matrix} [\Sigma^{\gtrless}_{o,o,o}, \operatorname{Re} G^r_{o,o,o}] + [\Sigma^{\lessgtr}_{o,o,m}, \operatorname{Re} G^r_{o,o,m}] + [\Sigma^{\gtrless}_{o,l,o}, \operatorname{Re} G^r_{o,l,o}] + [\Sigma^{\gtrless}_{o,l,m}, \operatorname{Re} G^r_{o,l,m}] \\ +[\Sigma^{\lessgtr}_{k,o,o}, \operatorname{Re} G^r_{k,o,o}] + [\Sigma^{\gtrless}_{k,o,m}, \operatorname{Re} G^r_{k,o,m}] + [\Sigma^{\lessgtr}_{k,l,o}, \operatorname{Re} G^r_{k,l,o}] \\ +[\Sigma^{\lessgtr}_{k,l,m}, \operatorname{Re} G^r_{k,l,m}] \end{matrix}, \tag{28.23}$$

$$\{\Gamma, G^{\lessgtr}\}_{o,o,o}$$

$$= \begin{matrix} \{\Gamma_{o,o,o},\ G^{\gtrless}_{o,o,o}\} + \{\Gamma_{o,o,m}, G^{\gtrless}_{o,o,m}\} + \{\Gamma_{o,l,o}, G^{\gtrless}_{o,l,o}\} + \{\Gamma_{o,l,m},\ G^{\gtrless}_{o,l,m}\} \\ +\{\Gamma_{k,o,o},\ G^{\gtrless}_{k,o,o}\} + \{\Gamma_{k,o,m}, G^{\lessgtr}_{k,o,m}\} + \{\Gamma_{k,l,o}, G^{\gtrless}_{k,l,o}\} + \{\Gamma_{k,l,m}, G^{\gtrless}_{k,l,m}\} \end{matrix}, \tag{28.24}$$

$$
\begin{aligned}
&\{\Sigma^{\lessgtr}, A\}_{o,o,o} \\
&\quad = \{\Sigma^{\gtrless}_{o,o,o},\ A_{o,o,o}\} + \{\Sigma^{\lessgtr}_{o,o,m}, A_{o,o,m}\} + \{\Sigma^{\gtrless}_{o,l,o}, A_{o,l,o}\} + \{\Sigma^{\gtrless}_{o,l,m}, A_{o,l,m}\} \\
&\qquad + \{\Sigma^{\lessgtr}_{k,o,o},\ A_{k,o,o}\} + \{\Sigma^{\gtrless}_{k,o,m},\ A_{k,o,m}\} + \{\Sigma^{\lessgtr}_{k,l,o}, A_{k,l,o}\} + \{\Sigma^{\lessgtr}_{k,l,m}, A_{k,l,m}\},
\end{aligned} \tag{28.25}
$$

where repeated vector-component indices, corresponding to dot products, are summed over following the Einstein summation convention, and N_s is the integer number of spin-like degrees of freedom. We have Pauli–Dirac spin, valley spin (or isospin) and pseudospin degrees of freedom acting on the systems. Here in Eqs. (28.21)–(28.25), we have $N_s = 3$. The rest of the equations for $G^{<}_{k,o,o}$, $G^{\lessgtr}_{o,l,o}$, $G^{\lessgtr}_{o,o,m}$, $G^{\lessgtr}_{k,l,o}$, $G^{\lessgtr}_{k,o,m}$, $G^{\lessgtr}_{o,l,m}$, and $G^{\lessgtr}_{k,l,m}$ are given in Section E.4 of the Appendix.

28.5 Limiting case of two spin degrees of freedom, $N_s = 2$

To obtain the non-equilibrium quantum transport equations for $N_s = 2$ from the above quantum transport equations for $N_s = 3$, one simply deletes one of the three indices. For example, for the case of valley spin and pseudospin degrees of freedom, we obtain

$$
\begin{aligned}
&2^{N_s} i\hbar \left(\frac{\partial}{\partial t_1} + \frac{\partial}{\partial t_2}\right) G^{\gtrless}_{o,o} \\
&\quad = [\tilde{\mathcal{H}}, G^{\lessgtr}]_{o,o} + [\Sigma^{\lessgtr}, \operatorname{Re} G^r]_{o,o} - \frac{i}{2}\{\Gamma, G^{\lessgtr}\}_{o,o} + \frac{i}{2}\{\Sigma^{\lessgtr}, A\}_{o,o},
\end{aligned} \tag{28.26}
$$

where

$$
\begin{aligned}
&[\tilde{\mathcal{H}}, G^{\lessgtr}]_{o,o} \\
&\quad = [H_{o,o}, G^{\gtrless}_{o,o}] + [H_{o,m}, G^{\gtrless}_{o,m}] + [H_{l,o}, G^{\gtrless}_{l,o}] + [H_{l,m}, G^{\gtrless}_{l,m}],
\end{aligned} \tag{28.27}
$$

$$
\begin{aligned}
&[\Sigma^{\lessgtr}, \operatorname{Re} G^r]_{o,o} \\
&\quad = [\Sigma^{\gtrless}_{o,o}, \operatorname{Re} G^r_{o,o}] + [\Sigma^{\lessgtr}_{o,m}, \operatorname{Re} G^r_{o,m}] + [\Sigma^{\gtrless}_{l,o}, \operatorname{Re} G^r_{l,o}] + [\Sigma^{\gtrless}_{l,m}, \operatorname{Re} G^r_{l,m}],
\end{aligned} \tag{28.28}
$$

$$
\begin{aligned}
&\{\Gamma, G^{\lessgtr}\}_{o,o} \\
&\quad = \{\Gamma_{o,o}, G^{\gtrless}_{o,o}\} + \{\Gamma_{o,m}, G^{\gtrless}_{o,m}\} + \{\Gamma_{l,o}, G^{\gtrless}_{l,o}\} + \{\Gamma_{l,m}, G^{\gtrless}_{l,m}\},
\end{aligned} \tag{28.29}
$$

$$
\begin{aligned}
&\{\Sigma^{\lessgtr}, A\}_{o,o} \\
&\quad = \{\Sigma^{\gtrless}_{o,o},\ A_{o,o}\} + \{\Sigma^{\lessgtr}_{o,m}, A_{o,m}\} + \{\Sigma^{\gtrless}_{l,o}, A_{l,o}\} + \{\Sigma^{\gtrless}_{l,m}, A_{l,m}\},
\end{aligned} \tag{28.30}
$$

$$2^{N_s} i\hbar \left(\frac{\partial}{\partial t_1} + \frac{\partial}{\partial t_2}\right) G^{<}_{l,o}$$
$$= [\tilde{\mathcal{H}}, G^{\lessgtr}]_{l,o} + [\Sigma^{\lessgtr}, \mathrm{Re}\, G^r]_{l,o} - \frac{i}{2}\left\{\Gamma, G^{\lessgtr}\right\}_{l,o} + \frac{i}{2}\{\Sigma^{\lessgtr}, A\}_{l,o}, \tag{28.31}$$

where

$$[\tilde{\mathcal{H}}, G^{\lessgtr}]_{l,o}$$
$$= \begin{array}{l} [H_{o,o}, G^{\gtrless}_{l,o}] + [H_{o,m}, G^{\gtrless}_{l,m}] + i\epsilon_{ll_1l_2}\{H_{l_1,o}, G^{\gtrless}_{l_2,o}\} \\ +[H_{l,o}, G^{\lessgtr}_{o,o}] + [H_{l,m}, G^{\gtrless}_{o,m}] + i\epsilon_{ll_1l_2}\{H_{l_1,m}, G^{\gtrless}_{l_2,m}\} \end{array}, \tag{28.32}$$

$$[\Sigma^{\lessgtr}, \mathrm{Re}\, G^r]_{l,o}$$
$$= \begin{array}{l} +[\Sigma^{\gtrless}_{o,o}, \mathrm{Re}\, G^r_{l,o}] + [\Sigma^{\gtrless}_{o,m}, \mathrm{Re}\, G^r_{l,m}] + i\epsilon_{ll_1l_2}\{\Sigma^{\gtrless}_{l_1,o}, \mathrm{Re}\, G^r_{l_2,o}\} \\ +[\Sigma^{\gtrless}_{l,o}, \mathrm{Re}\, G^r_{o,o}] + [\Sigma^{\gtrless}_{l,m}, \mathrm{Re}\, G^r_{o,m}] + i\epsilon_{ll_1l_2}\{\Sigma^{\gtrless}_{l_1,m}, \mathrm{Re}\, G^r_{l_2,m}\} \end{array}, \tag{28.33}$$

$$\{\Gamma, G^{\lessgtr}\}_{l,o}$$
$$= \begin{array}{l} \{\Gamma_{o,o}, G^{\gtrless}_{l,o}\} + \{\Gamma_{o,m}, G^{\gtrless}_{l,m}\} + i\epsilon_{ll_1l_2}[\Gamma_{l_1,o}, G^{\gtrless}_{l_2,o}] \\ +\{\Gamma_{l,o}, G^{\lessgtr}_{o,o}\} + \{\Gamma_{l,m}, G^{\gtrless}_{o,m}\} + i\epsilon_{ll_1l_2}[\Gamma_{l_1,m}, G^{\gtrless}_{l_2,m}] \end{array}, \tag{28.34}$$

$$\left\{\Sigma^{\lessgtr}, A\right\}_{l,o}$$
$$= \begin{array}{l} \left\{\Sigma^{\gtrless}_{o,o}, A_{l,o}\right\} + \left\{\Sigma^{\gtrless}_{o,m}, A_{l,m}\right\} + i\epsilon_{ll_1l_2}\left[\Sigma^{\gtrless}_{l_1,o}, A_{l_2,o}\right] \\ +\left\{\Sigma^{\gtrless}_{l,o}, A_{o,o}\right\} + \left\{\Sigma^{\gtrless}_{l,m}, A_{o,m}\right\} + i\epsilon_{ll_1l_2}\left[\Sigma^{\gtrless}_{l_1,m}, A_{l_2,m}\right] \end{array}, \tag{28.35}$$

$$2^{N_s} i\hbar \left(\frac{\partial}{\partial t_1} + \frac{\partial}{\partial t_2}\right) G^{\gtrless}_{o,m}$$
$$= \left[\tilde{\mathcal{H}}, G^{\lessgtr}\right]_{o,m} + [\Sigma^{\lessgtr}, \mathrm{Re}\, G^r]_{o,m} - \frac{i}{2}\{\Gamma, G^{\lessgtr}\}_{o,m} + \frac{i}{2}\{\Sigma^{\lessgtr}, A\}_{o,m}, \tag{28.36}$$

where

$$[\tilde{\mathcal{H}}, G^{\lessgtr}]_{o,m}$$
$$= \begin{array}{l} [H_{o,o}, G^{\gtrless}_{o,m}] + [H_{o,m}, G^{\gtrless}_{o,o}] + i\epsilon_{mm_1m_2}\{H_{o,m_1}, G^{\gtrless}_{o,m_2}\} \\ +[H_{l,o}, G^{\gtrless}_{l,m}] + [H_{l,m}, G^{\gtrless}_{l,o}] + i\epsilon_{mm_1m_2}\{H_{l,m_1}, G^{\gtrless}_{l,m_2}\} \end{array}, \tag{28.37}$$

$$[\Sigma^{\lessgtr}, \mathrm{Re}\, G^r]_{o,m}$$
$$= \begin{array}{l} +[\Sigma^{\lessgtr}_{o,o}, \mathrm{Re}\, G^r_{o,m}] + [\Sigma^{\lessgtr}_{o,m}, \mathrm{Re}\, G^r_{o,o}] + i\epsilon_{mm_1m_2}\{\Sigma^{\lessgtr}_{o,m_1}, \mathrm{Re}\, G^r_{o,m_2}\} \\ +[\Sigma^{\lessgtr}_{l,o}, \mathrm{Re}\, G^r_{l,m}] + [\Sigma^{\lessgtr}_{l,m}, \mathrm{Re}\, G^r_{l,o}] + i\epsilon_{mm_1m_2}\{\Sigma^{\lessgtr}_{l,m_1}, \mathrm{Re}\, G^r_{l,m_2}\}, \end{array} \tag{28.38}$$

$$\{\Gamma, G^{\lessgtr}\}_{o,m}$$

$$= \left\{ \begin{array}{l} \{\Gamma_{o,o}, G^{\gtrless}_{o,o,m}\} + \{\Gamma_{o,m}, G^{\gtrless}_{o,o,o}\} + i\epsilon_{mm_1m_2}\left[\Gamma_{o,m_1}, G^{\gtrless}_{o,m_2}\right] \\ +\{\Gamma_{l,o}, G^{\gtrless}_{,l,m}\} + \{\Gamma_{l,m}, G^{\gtrless}_{l,o}\} + i\epsilon_{mm_1m_2}\left[\Gamma_{l,m_1}, G^{\gtrless}_{l,m_2}\right] \end{array} \right\}, \quad (28.39)$$

$$\left\{\Sigma^{\lessgtr}, A\right\}_{o,m}$$

$$= \left\{ \begin{array}{l} \left\{\Sigma^{\lessgtr}_{o,o}, A_{o,o,m}\right\} + \left\{\Sigma^{\lessgtr}_{o,m}, A_{o,o}\right\} + i\epsilon_{mm_1m_2}\left[\Sigma^{\lessgtr}_{o,m_1}, A_{o,m_2}\right] \\ +\left\{\Sigma^{\lessgtr}_{l,o}, A_{l,m}\right\} + \left\{\Sigma^{\lessgtr}_{l,m}, A_{l,o}\right\} + i\epsilon_{mm_1m_2}\left[\Sigma^{\lessgtr}_{l,m_1}, A_{l,m_2}\right] \end{array} \right\}, \quad (28.40)$$

$$2^{N_s} i\hbar \left(\frac{\partial}{\partial t_1} + \frac{\partial}{\partial t_2}\right) G^{<}_{l,m}$$

$$= \left[\tilde{\mathcal{H}}, G^{\lessgtr}\right]_{l,m} + \left[\Sigma^{\lessgtr}, \mathrm{Re}\, G^{r}\right]_{l,m} - \frac{i}{2}\left\{\Gamma, G^{\lessgtr}\right\}_{l,m} + \frac{i}{2}\{\Sigma^{\lessgtr}, A\}_{l,m}, \quad (28.41)$$

where

$$[\tilde{\mathcal{H}}, G^{\lessgtr}]_{l,m}$$

$$= \left\{ \begin{array}{c} [H_{o,o}, G^{\lessgtr}_{l,m}] + [H_{o,m}, G^{\lessgtr}_{l,o}] + [H_{l,o}, G^{\lessgtr}_{o,m}] + [H_{l,m}, G^{\lessgtr}_{o,o}] \\ +i\epsilon_{mm_1m_2}\{H_{o,m_1}, G^{\lessgtr}_{l,m_2}\} + i\epsilon_{mm_1m_2}\{H_{l,m_1}, G^{\lessgtr}_{o,m_2}\} \\ +i\epsilon_{ll_1l_2}\{H_{l_1,o}, G^{\lessgtr}_{l_2,m}\} + i\epsilon_{ll_1l_2}\{H_{l_1,m}, G^{\lessgtr}_{l_2,o}\} \\ +\frac{1}{2}i\epsilon_{ll_1l_2}i\epsilon_{mm_1m_2}[H_{l_1,m_1}, G^{\lessgtr}_{l_2,m_2}] \\ +\frac{1}{2}i\epsilon_{ll_1l_2}i\epsilon_{mm_1m_2}[H_{l_1,m_1}, G^{\lessgtr}_{l_2,m_2}] \end{array} \right\}, \quad (28.42)$$

$$\left[\Sigma^{\lessgtr}, \mathrm{Re}\, G^{r}\right]_{l,m}$$

$$= \left\{ \begin{array}{c} \left[\Sigma^{\lessgtr}_{o,o}, \mathrm{Re}\, G^{r}_{l,m}\right] + \left[\Sigma^{\lessgtr}_{o,m}, \mathrm{Re}\, G^{r}_{l,o}\right] + \left[\Sigma^{\lessgtr}_{l,o}, \mathrm{Re}\, G^{r}_{o,m}\right] + \left[\Sigma^{\lessgtr}_{l,m}, \mathrm{Re}\, G^{r}_{o,o}\right] \\ +i\epsilon_{mm_1m_2}\{\Sigma^{\lessgtr}_{o,m_1}, \mathrm{Re}\, G^{r}_{l,m_2}\} + i\epsilon_{mm_1m_2}\{\Sigma^{\lessgtr}_{l,m_1}, \mathrm{Re}\, G^{r}_{o,m_2}\} \\ +i\epsilon_{ll_1l_2}\{\Sigma^{\lessgtr}_{l_1,o}, \mathrm{Re}\, G^{r}_{l_2,m}\} + i\epsilon_{ll_1l_2}\{\Sigma^{\lessgtr}_{l_1,m}, \mathrm{Re}\, G^{r}_{l_2,o}\} \\ +\frac{1}{2}i\epsilon_{ll_1l_2}i\epsilon_{mm_1m_2}\left[\Sigma^{\lessgtr}_{l_1,m_1}, \mathrm{Re}\, G^{r}_{l_2,m_2}\right] \\ +\frac{1}{2}i\epsilon_{ll_1l_2}i\epsilon_{mm_1m_2}\left[\Sigma^{\lessgtr}_{l_1,m_1}, \mathrm{Re}\, G^{r}_{l_2,m_2}\right] \end{array} \right\},$$

$$(28.43)$$

$$\left\{\Gamma, G^{\lessgtr}\right\}_{l,m} = \left\{\begin{array}{c} \{\Gamma_{o,o}, G^{\lessgtr}_{l,m}\} + \{\Gamma_{o,m}, G^{\lessgtr}_{l,o}\}\{\Gamma_{l,o}, G^{\lessgtr}_{o,m}\} + \{\Gamma_{l,m}, G^{\lessgtr}_{o,o}\} \\ +i\epsilon_{mm_1m_2}\left[\Gamma_{o,m_1}, G^{\lessgtr}_{l,m_2}\right] + i\epsilon_{mm_1m_2}\left[\Gamma_{l,m_1}, G^{\lessgtr}_{o,m_2}\right] \\ +i\epsilon_{ll_1l_2}\left[\Gamma_{l_1,o}, G^{\lessgtr}_{l_2,m}\right] + i\epsilon_{ll_1l_2}\left[\Gamma_{l_1,m}, G^{\lessgtr}_{l_2,o}\right] \\ +\frac{1}{2}i\epsilon_{ll_1l_2}i\epsilon_{mm_1m_2}\{\Gamma_{l_1,m_1}, G^{\lessgtr}_{l_2,m_2}\} + \frac{1}{2}i\epsilon_{ll_1l_2}i\epsilon_{mm_1m_2}\{\Gamma_{l_1,m_1}, G^{\lessgtr}_{l_2,m_2}\} \end{array}\right\}, \tag{28.44}$$

$$\left\{\Sigma^{\lessgtr}, A\right\}_{l,m} = \left\{\begin{array}{c} \{\Sigma^{\lessgtr}_{o,o}, A_{l,m}\} + \{\Sigma^{\lessgtr}_{o,m}, A_{l,o}\} + \{\Sigma^{\lessgtr}_{l,o}, A_{o,m}\} + \{\Sigma^{\lessgtr}_{l,m}, A_{o,o}\} \\ +i\epsilon_{mm_1m_2}\{\Sigma^{\lessgtr}_{o,m_1}, A_{l,m_2}\} + i\epsilon_{mm_1m_2}\{\Sigma^{\lessgtr}_{l,m_1}, A_{o,m_2}\} \\ +i\epsilon_{ll_1l_2}\{\Sigma^{\lessgtr}_{l_1,o}, A_{l_2,m}\} + i\epsilon_{ll_1l_2}\{\Sigma^{\lessgtr}_{l_1,m}, A_{l_2,o}\} \\ +\frac{1}{2}i\epsilon_{ll_1l_2}i\epsilon_{mm_1m_2}\{\Sigma^{\lessgtr}_{l_1,m_1}, A_{l_2,m_2}\} + \frac{1}{2}i\epsilon_{ll_1l_2}i\epsilon_{mm_1m_2}\{\Sigma^{\lessgtr}_{l_1,m_1}, A_{l_2,m_2}\} \end{array}\right\}. \tag{28.45}$$

The limiting case of Pauli–Dirac spin and pseudospin, $N_s = 2$ has been treated in Part 4. There the Dirac spin semiconductor Bloch equations (DSSBEs) are treated in the electron–hole picture (two-particle species). When we cast the DSSBEs in the electron picture (single particle specie), either by virtue of the symmetry at low energies of the conduction-valence bands of Dirac points, or on account of layer pseudospin in bilayer systems, the resulting equations for Pauli–Dirac spin and pseudospin, by the method of using appropriate combinations of the DSSBEs employed there, [224] exactly reproduce the results as given in the present limiting case. The explicit form of the expressions containing the product of two Levi-Civita symbols can be obtained by the method of Ref. [224] using the DSSBEs derived there but cast in the electron picture. The typical result containing two Levi-Civita symbols, for example, the x-component of pseudospin in our present notation, is

$$2^{N_s} i\hbar\left(\frac{\partial}{\partial t_1} + \frac{\partial}{\partial t_2}\right)\vec{G}^{<}_x \Longrightarrow \left[\begin{array}{c} \frac{i}{2}\left\{\left(\vec{\Gamma}_y \times \vec{G}^{<}_z - \vec{\Gamma}_z \times \vec{G}^{<}_y\right) + \left(\vec{G}^{<}_y \times \vec{\Gamma}_z - \vec{G}^{<}_z \times \vec{\Gamma}_y\right)\right\} \\ -\frac{i}{2}\left\{\left(\vec{\Sigma}^{<}_y \times \vec{A}_z - \vec{\Sigma}^{<}_z \times \vec{A}_y\right) + \left(\vec{A}_y \times \vec{\Sigma}^{<}_z - \vec{A}_z \times \vec{\Sigma}^{<}_y\right)\right\} \end{array}\right], \tag{28.46}$$

where the vector denotes the Pauli–Dirac spin. This translates to our result in term of tensor components, for arbitrary vector components for both Dirac spin and pseudospin, as

$$2^{N_s} i\hbar \left(\frac{\partial}{\partial t_1} + \frac{\partial}{\partial t_2} \right) G^{<}_{k,m}$$

$$\Longrightarrow + \begin{bmatrix} \frac{i}{2} \epsilon_{kk_1k_2} \epsilon_{mm_1m_2} \{\Gamma_{k_1,m_1}, G^{<}_{k_2,m_2}\} \\ -\frac{i}{2} \epsilon_{kk_1k_2} \epsilon_{mm_1m_2} \{\Sigma_{k_1,m_1}, A_{k_2,m_2}\} \end{bmatrix}. \tag{28.47}$$

Equations (28.46)–(28.47) give a clear meaning of the terms containing *two* Levi-Civita symbols in the transport equations. Indeed for the scattering terms, it restores the locality character as contained in the original equations, Eq. (28.11) [refer to the discussion in Section 29.1.2].

28.5.1 *Comparison with non-equilibrium multi-band spin quantum transport*

In the light of the present notations used, the corresponding quantum transport equation for Pauli–Dirac spin and pseudospin with different notations employed in Section 21.3, as rewritten in Section D.2 of the appendix, is shown to be identical to the present results.

28.6 Limiting case of a single spin, $N_s = 1$

To obtain the non-equilibrium quantum transport equations for $N_s = 1$ from the above quantum transport equations for $N_s = 3$, one simply deletes two of the three indices. For example, for the case of pseudospin degrees of freedom, we obtain,

$$2^{N_s} i\hbar \left(\frac{\partial}{\partial t_1} + \frac{\partial}{\partial t_2} \right) G^{\gtrless}_{o}$$

$$\begin{aligned} &= [H_o, G^{\gtrless}_o] + [H_m, G^{\gtrless}_m] \\ &\quad + [\Sigma^{\gtrless}_o, \mathrm{Re}\, G^r_o] + [\Sigma^{\lessgtr}_m, \mathrm{Re}\, G^r_m] \\ &\quad -\frac{i}{2} \{\{\Gamma_o, G^{\gtrless}_o\} + \{\Gamma_m, G^{\gtrless}_m\}\} \\ &\quad +\frac{i}{2} \{\{\Sigma^{\gtrless}_o, A_o\} + \{\Sigma^{\lessgtr}_m, A_m\}\}, \end{aligned} \tag{28.48}$$

where $N_s = 1$. The pseudospin magnetization density is given by

$$\begin{aligned}
&2^{N_s} i\hbar \left(\frac{\partial}{\partial t_1} + \frac{\partial}{\partial t_2}\right) G_m^{\gtrless} \\
&= \left[H_o, G_m^{\gtrless}\right] + \left[H_m, G_o^{\gtrless}\right] + i\epsilon_{mm_1m_2}\left\{H_{m_1}, G_{m_2}^{\gtrless}\right\} \\
&\quad + \left[\Sigma_o^{\lessgtr}, \mathrm{Re}\, G_m^r\right] + \left[\Sigma_m^{\lessgtr}, \mathrm{Re}\, G_o^r\right] + i\epsilon_{mm_1m_2}\left\{\Sigma_{m_1}^{\lessgtr}, \mathrm{Re}\, G_{m_2}^r\right\} \\
&\quad - \frac{i}{2}\left\{\left\{\Gamma_o, G_m^{\gtrless}\right\} + \left\{\Gamma_m, G_o^{\gtrless}\right\} + i\epsilon_{mm_1m_2}\left[\Gamma_{m_1}, G_{m_2}^{\gtrless}\right]\right\} \\
&\quad + \frac{i}{2}\left\{\left\{\Sigma_o^{\lessgtr}, A_m\right\} + \left\{\Sigma_m^{\lessgtr}, A_o\right\} + i\epsilon_{mm_1m_2}\left[\Sigma_{m_1}^{\lessgtr}, A_{m_2}\right]\right\}. \qquad (28.49)
\end{aligned}$$

28.6.1 *Comparison with spin magnetization quantum transport equations*

The equation we obtain for Dirac spin is identical to that obtained in Section 20.4 of Part 4 which is rearranged as follows,

$$\begin{aligned}
&2i\hbar\left(\frac{\partial}{\partial t_1} + \frac{\partial}{\partial t_2}\right)\vec{S}^{<} \\
&= \left\{\begin{matrix} [\bar{H} + \mathrm{Re}\,\bar{\Sigma}^r, \vec{S}^{<}] \\ +[(\vec{\mathcal{B}} + \mathrm{Re}\,\vec{\Xi}^r), S_o^{<}] + i[(\vec{\mathcal{B}} + \mathrm{Re}\,\vec{\Xi}^r) \times \vec{S}^{<} - \vec{S}^{<} \times (\vec{\mathcal{B}} + \mathrm{Re}\,\vec{\Xi}^r)] \end{matrix}\right\} \\
&\quad + [\bar{\Sigma}^{<}, \mathrm{Re}\,\vec{S}^r] + [\vec{\Xi}^{<}, \mathrm{Re}\,S_o^r] + i[\vec{\Xi}^{<} \times \mathrm{Re}\,\vec{S}^r - \mathrm{Re}\,\vec{S}^r \times \vec{\Xi}^{<}] \\
&\quad - \frac{i}{2}\{\{\bar{\Gamma}, \vec{S}^{<}\} - \frac{i}{2}\{\vec{\gamma}, S_o^{<}\} + \frac{i}{2}\{\vec{\gamma} \times \vec{S}^{<} + \vec{S}^{<} \times \vec{\gamma}\}\} \\
&\quad + \frac{i}{2}\{\{\bar{\Sigma}^{<}, \vec{\mathcal{A}}\} + \frac{i}{2}\{\vec{\Xi}^{<}, \mathcal{A}_o\} - \frac{i}{2}\{\vec{\Xi}^{<} \times \vec{\mathcal{A}} + \vec{\mathcal{A}} \times \vec{\Xi}^{<}\}\}, \qquad (28.50)
\end{aligned}$$

where presently we used the following corresponding notations as follows:

$$\vec{S}^{<} \Longrightarrow G_k^{<},\ S_o^{<} \Longrightarrow G_o^{<},\ \bar{H} \Longrightarrow H_o,\ \vec{\mathcal{B}} + \mathrm{Re}\,\vec{\Xi}^r \Longrightarrow H_k \qquad (28.51)$$

$$\bar{\Gamma} \Longrightarrow \Gamma_o,\ \vec{\gamma} \Longrightarrow \Gamma_k,\ \mathrm{Re}\,\bar{\Sigma}^r \Longrightarrow \text{included in } H_o, \qquad (28.52)$$

$$\bar{\Sigma}^{<} \Longrightarrow \Sigma_o^{\lessgtr},\ \vec{\Xi}^{<} \Longrightarrow \Sigma_k^{\lessgtr} \qquad (28.53)$$

$$\mathrm{Re}\,\vec{S}^r \Longrightarrow \mathrm{Re}\,G_k^r,\ \mathrm{Im}\,S_o^r \Longrightarrow -\frac{A_o}{2},\ \vec{\mathcal{A}} \Longrightarrow A_k. \qquad (28.54)$$

Note that there were typos in Ref. [158], (Eq. (20), where some terms have missing extra factor $\frac{1}{2}$ and has been corrected in Eq. (28.50).

Chapter 29

Part 6: Discussion and Concluding Remarks

Here we give a more detailed discussion on the physical meaning of the analytical results as it pertains to an entirely new perspective of localization and delocalization for systems with entangled chiral degrees of freedom.

29.1 Entanglement-induced localization and delocalization

One of the revealing aspects of entanglement in non-equilibrium superfield spinor quantum-transport physics coupled with lattice Weyl transform techniques [11] is the axial/chiral localization and delocalization through the presence of Levi-Civita chiral/directional tensors. The delocalization implies diffusive processes in phase space or drift-diffusion and quantum tunneling transport in position space. The chiral localization and delocalization revolve around the the lattice Weyl transform of terms, either commutator or anti-commutator, in the transport equations as a function of the number of chiral degrees of freedom, multiplied by the Levi-Civita symbols giving the chiral nature of the resulting dynamics.

Indeed, the diffusive property of the commutator is exhibited even in the leading order in gradient expansion. The commutator represents the inherent non-locality in quantum mechanics and describes diffusive and tunneling motion in phase space of transport kinetics, whereas the anti-commutator describes *local or non-diffusive* events in $(\vec{p}, \vec{q})$ space typified either by cyclotron-orbit current tied to orbit center or by decay and growth of a phase-space distribution function tied to a point $(\vec{p}, \vec{q})$ in phase space. For example, the non-diffusive character of the scattering anti-commutator has been simply approximated by a relaxation-time approximation in Refs. [11, 14], whereas the commutator is entirely responsible for bringing in the non-locality of quantum transport physics, specifically the quantum tunneling transport in resonant tunneling nanostructures [11, 14] which for the first time resolves the controversial aspects of the current–voltage characteristics of resonant tunneling diodes [14].

For a simple demonstration (see Appendix Section E.7 for more details), we give here the following lattice-Weyl (LW) transform to phase space of a commutator $[A, B]$ and an anti-commutator $\{A, B\}$ in terms of Poisson bracket differential operator, Λ, as expansion to all powers of $\hbar$,

$$[A,B](p,q) = \cos\Lambda\ [a(p,q)\,b(p,q) - b(p,q)\,a(p,q)] \\ - i\sin\Lambda\,\{a(p,q)\,b(p,q) + b(p,q)\,a(p,q)\}, \quad (29.1)$$

$$\{A,B\}(p,q) = \cos\Lambda\ \{a(p,q)\,b(p,q) + b(p,q)\,a(p,q)\} \\ - i\sin\Lambda\ [a(p,q)\,b(p,q) - b(p,q)\,a(p,q)], \quad (29.2)$$

where $\Lambda = \frac{\hbar}{2}\left(\frac{\partial^{(a)}}{\partial p}\cdot\frac{\partial^{(b)}}{\partial q} - \frac{\partial^{(a)}}{\partial q}\cdot\frac{\partial^{(b)}}{\partial p}\right)$, the Poisson bracket operator. Thus if the LW transforms $a(p,q)$ and $b(p,q)$ are not matrices, then to lowest order, we have

$$[A,B](p,q) = -i\frac{\hbar}{2}\left(\frac{\partial a(p,q)}{\partial p}\cdot\frac{\partial b(p,q)}{\partial q} - \frac{\partial a(p,q)}{\partial q}\cdot\frac{\partial b(p,q)}{\partial p}\right), \quad (29.3)$$

$$\{A,B\}(p,q) = \{a(p,q)\,b(p,q) + b(p,q)\,a(p,q)\} = 2\ a(p,q)\,b(p,q), \quad (29.4)$$

clearly showing that even to the lowest order the local and diffusive properties in phase space of anti-commutator and commutator, respectively, are apparent. Likewise, in its full integral representations, Eqs. (29.18) and (29.19), which is suitable for carrying numerical simulations, Eq. (29.1) has a non-local kernel, whereas Eq. (29.2) has a local kernel of integration. It is precisely the non-local character of the kernel given in Eq. (29.20) that is responsible for quantum tunneling in resonant tunneling diodes [11, 14], whose unique current–voltage characteristics cannot be explained by any degree of semi-classical approximations [225].

29.1.1 *Hierarchy of entanglements: Alternating sequence of commutators and anti-commutators as function of N_s*

Let us for example start with Pauli–Dirac spinor form of the commutator of A and B which can be written in tensorial form as

$$[A,B] = \frac{1}{4}([a_o,\vec{b}] + [\vec{a},b_o] + i[\vec{a}\times\vec{b} - \vec{b}\times\vec{a}]), \quad (29.5)$$

$$[A,B]_k = \frac{1}{4}([a_o,b_k] + [a_k,b_o] + i\epsilon_{kk_1k_2}\{a_{k_1},b_{k_2}\}), \quad (29.6)$$

where the *subscript o* indicates total independence with respect to the spin degrees of freedom, $\{a_{k_1}, b_{k_2}\}$ is the anti-commutation of a_{k_1} and b_{k_2}. Note the change from original commutator to anti-commutator for the term containing one Levi-Civita symbol. In including dependence on other spin degrees of freedom, in all instances the theorem on binary product of 2×2 matrices, Eq. (29.6) is successively applied. Thus, the fully entangled 'torque' [coming from the last term of Eq. (29.6)] now becomes,

$$i\epsilon_{ll_1l_2} i\epsilon_{kk_1k_2} [a_{k_1,l_1}, b_{k_2,l_2}] = -\epsilon_{ll_1l_2}\epsilon_{kk_1k_2} [a_{k_1,l_1}, b_{k_2,l_2}]. \tag{29.7}$$

Upon adding dependence on one more pseudospin degrees of freedom, such the pseudospin arising either from the symmetry at low energies of the band indices at Dirac points in graphene or from layer pseudospin in bilayer graphene, the entanglement of the torques becomes

$$-i\epsilon_{mm_1m_2}\epsilon_{ll_1l_2}\epsilon_{kk_1k_2} \{a_{k_1,l_1,m_1}, b_{k_2,l_2,m_2}\}, \tag{29.8}$$

which is a third rank tensor. Assuming we have considered only the band pseudospin in Eq. (29.8), if one further consider the pseudospin from layer degrees of freedom in bilayer graphene, for example, then the fully entangled "torque" becomes

$$\epsilon_{nn_1n_2}\epsilon_{mm_1m_2}\epsilon_{ll_1l_2}\epsilon_{kk_1k_2} [a_{k_1,l_1,m_1,n_1}, b_{k_2,l_2,m_2,n_2}]. \tag{29.9}$$

Thus starting from the commutation of $[A, B]$, the dependence in spin degrees of freedom results in the alternating commutator and anti-commutator of the spin tensor components in the expression for entangled torques, with commutator for even number of $i\epsilon_{ijk}$ (*Levi-Civita* symbols) and anti-commutator for odd number of $i\epsilon_{ijk}$. Had we started with $\{A, B\}$, then there will be alternating anti-commutator and commutator instead, as the number of spin degrees of freedom grows.

29.1.2 *Local and non-local terms*

The commutator rigorously incorporates the quantum non-locality in phase space or quantum-diffusive motion of transport kinetics in phase space [this is evident even in the semi-classical approximation leading to Boltzmann transport equation, i.e., within the first power of Poisson bracket] and exhibits quantum tunneling across abrupt barriers when all powers of $\hbar$ are taken into account by using the integral kernel [14], whereas the anti-commutator describes local or non-diffusive events in $(\vec{p}, \vec{q})$ space as typified by cyclotron-orbit current tied to orbit center or decay and growth of a phase-space distribution function tied to a point $(\vec{p}, \vec{q})$ in phase space. Moreover, the exact integral form of the commutator and anti-commutator exhibits non-local kernel and local kernel of integration, respectively. See also the discussion in Section E.7 on transport equations in phase space.

The transport equations in phase space are obtained by applying the "lattice" Weyl transformation (although continuum approximation is often employed in this chapter, this is not essential and we use the word "lattice" when referring to solid-state problems [3, 11, 12, 14]) of the propagator equations for the respective excitations by using the following identities.

29.1.3 *"Lattice" Weyl transform of differential operators*

$$i\hbar\left(\frac{\partial}{\partial t_1}+\frac{\partial}{\partial t_2}\right)Y\left(t_1,t_2\right)\Leftrightarrow i\hbar\frac{\partial}{\partial t}Y\left(E,t\right), \tag{29.10}$$

$$\left(\frac{\partial^2}{\partial t_1^2}-\frac{\partial^2}{\partial t_2^2}\right)Y\left(t_1,t_2\right)\Leftrightarrow \frac{2i}{\hbar}E\frac{\partial}{\partial t}Y\left(E,t\right), \tag{29.11}$$

$$\left(\nabla_1^2-\nabla_2^2\right)Y\left(r_1,r_2\right)\Leftrightarrow \frac{2i}{\hbar}p\cdot\nabla_q Y\left(p,q\right). \tag{29.12}$$

29.1.4 *Lattice Weyl transform of product of arbitrary operators*

29.1.4.1 *Poisson bracket expansion*

In terms of differential "Poisson bracket" operator,

$$AB\left(p,q\right)=\exp\frac{\hbar}{i}\left(\frac{\partial^{(a)}}{\partial p}\cdot\frac{\partial^{(b)}}{\partial q}-\frac{\partial^{(a)}}{\partial q}\cdot\frac{\partial^{(b)}}{\partial p}\right)a\left(p,q\right)\ b\left(p,q\right). \tag{29.13}$$

Thus, we are lead to the lattice Weyl transform of a commutator, $[A,B]\left(p,q\right)$, and an anti-commutator, $\{A,B\}\left(p,q\right)$ as

$$\begin{aligned}[A,B]\left(p,q\right)&=cos\Lambda\left[a\left(p,q\right)\ b\left(p,q\right)-b\left(p,q\right)a\left(p,q\right)\right]\\&\quad-i\sin\Lambda\left[a\left(p,q\right)\ b\left(p,q\right)+b\left(p,q\right)a\left(p,q\right)\right],\end{aligned} \tag{29.14}$$

$$\begin{aligned}\{A,B\}\left(p,q\right)&=cos\Lambda\left[a\left(p,q\right)\ b\left(p,q\right)+b\left(p,q\right)a\left(p,q\right)\right]\\&\quad-i\sin\Lambda\left[a\left(p,q\right)\ b\left(p,q\right)-b\left(p,q\right)a\left(p,q\right)\right],\end{aligned} \tag{29.15}$$

where $\Lambda=\frac{\hbar}{2}\left(\frac{\partial^{(a)}}{\partial p}\cdot\frac{\partial^{(b)}}{\partial q}-\frac{\partial^{(a)}}{\partial q}\cdot\frac{\partial^{(b)}}{\partial p}\right)$. Note that so far no semi-classical approximation is involved in the derivation.

29.1.4.2 *Integral representation*

In terms of integral operators, we have,

$$AB\,(p,q) = \frac{1}{(\hbar^4)^2}\int dp'\ dq'\ K_A^+\,(p,q;p'q')\ b\,(p',q'),$$
$$= \frac{1}{(\hbar^4)^2}\int dp'\ dq'\ a\,(p',q')\ K_B^-\,(p,q;p'q'), \tag{29.16}$$

where integral kernels are defined by

$$K_Y^{\pm}\,(p,q;p'q') = \int dudv \exp\left\{\frac{i}{\hbar}\left[(p-p')\cdot v + (q-q')\cdot u\right]\right\} y$$
$$\times\left(p \pm \frac{u}{2}, q \mp \frac{v}{2}\right). \tag{29.17}$$

Therefore, in terms of integral operators,

$$[A,B]\,(p,q) = \frac{1}{(\hbar^4)^2}\int dp'\ dq'\ [K_A^+\,(p,q;p'q')\ b\,(p',q') - b\,(p',q')$$
$$\times K_A^-\,(p,q;p'q')], \tag{29.18}$$

$$\{A,B\}\,(p,q) = \frac{1}{(\hbar^4)^2}\int dp'\ dq'$$
$$\times [K_A^+\,(p,q;p'q')\ b\,(p',q') + b\,(p',q')\ K_A^-\,(p,q;p'q')]. \tag{29.19}$$

29.1.4.3 *Local and non-local integral kernels*

The above expressions simplify considerably when the lattice Weyl transform $a\,(p,q)$ and $b\,(p,q)$ are scalar functions. We have for the integral representations,

$$[A,B]\,(p,q) = \frac{1}{(\hbar^4)^2}\int dp'\ dq' dudv \exp\left\{\frac{i}{\hbar}\left[(p-p')\cdot v + (q-q')\cdot u\right]\right\}$$
$$\times\begin{bmatrix} a\left(p+\frac{u}{2}, q-\frac{v}{2}\right) \\ -a\left(p-\frac{u}{2}, q+\frac{v}{2}\right)\end{bmatrix} b\,(p',q'), \tag{29.20}$$

$$\{A,B\}\,(p,q) = \frac{1}{(\hbar^4)^2}\int dp'\ dq' dudv \exp\left\{\frac{i}{\hbar}\left[(p-p')\cdot v + (q-q')\cdot u\right]\right\}$$
$$\times\begin{bmatrix} a\left(p+\frac{u}{2}, q-\frac{v}{2}\right) \\ +a\left(p-\frac{u}{2}, q+\frac{v}{2}\right)\end{bmatrix} b\,(p',q'). \tag{29.21}$$

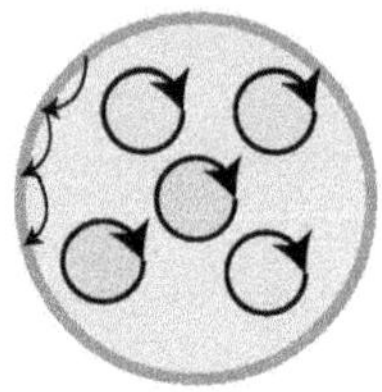

Fig. 29.1. Classical skipping orbits in space, showing the *holographic* nature of topological entanglement, i.e., *bulk-boundary correspondence or* entanglement of ↑-chirality of the boundary with ↓-chirality of the bulk. Reproduced from Ref. [205].

Thus, in its integral forms, which are better suited for carrying numerical simulations, Eq. (29.20) exhibits a *non-local* kernel for the commutator, $[A, B](p, q)$, whereas Eq. (29.21) shows an *averaging* type of kernel tied at point (p, q) or *local* kernel. This is already shown for the anti-commutator $\{A, B\}(p, q)$ to a leading order. Indeed, it is precisely the non-local character of the kernel in Eq. (29.20) that is entirely responsible for quantum-tunneling transport in resonant tunneling diodes [11, 14] which exhibit an entirely quantum non-locality characteristics.

One observes that by virtue of Eqs. (29.6)–(29.7), the entangling with a second pseudospin can transform local terms in the transport equation into diffusive or mobile terms, a sort of *delocalization*. This type of delocalization is not surprising as it is exhibited even in the classical case, often used to explain the absence of diamagnetism in classical free electron gas [205]. In the classical case, when localized currents or cyclotron orbits interacts with a constriction or opposite boundaries (providing torque interaction), delocalized currents in opposite directions or chirality reside in opposite boundaries, respectively. Thus, the quantum edge states in integer quantum Hall effect (precursor to topological insulators) are often portrayed semiclassically as counterpropagating "skipping orbits" that propagate (delocalize) along the boundary of the system in the opposite sense to that of the Landau orbits, when a particle in a Landau orbit interacts with the boundary, and specularly bounces off it (see Fig. 29.1) [205].

Thus, consider for instance a term in the quantum transport equation for $N_s = 2$ of the form, $i\epsilon_{ll_1l_2}\{H_{l_1,o}, G^{\gtrless}_{l_2,o}\}$, which is the term coming from the Hamiltonian [our spinor Hamiltonian is generalized to account for the spinor Re Σ^r [167, 226] (which is included in H, Eq. (28.12) as a spinor mass term) and/or magnetic field and/or Dresselhaus and/or Rashba spin-orbit coupling] that generates the Landau orbit and is thus a local term by virtue of the anti-commutator. However, when the dependence on pseudospin, m, comes into play this local term is transformed into a non-local term, specifically through the term, $i\epsilon_{ll_1l_2}i\epsilon_{mm_1m_2}[H_{l_1,m_1}G^{\lessgtr}_{l_2,m_2}]$ which is now a non-local term, i.e., mobile term in terms of transport point of view, as expressed by the commutator.

29.1.5 *Integer quantum Hall effect topological insulator*

Quantum mechanically speaking, in the specular reflection at the boundary, the incoming and outgoing states present the boundary pseudospin resulting in the boundary-pseudospin dependent interaction of the Landau orbits. This boundary-pseudospin dependent interaction results in the delocalization of the electron current along the boundary. Since the interaction is elastic and does not cost energy the delocalized current excitations do not have mass or are zero-excitation modes resulting in the chiral/helical dispersion relations or Dirac point of the excitations as in the case of integer quantum Hall effect. This is the mechanism leading to QHE-TI. In Fig. 29.1, one may consider the situation as an entanglement of "+" chirality of the boundary with the "−" chirality of the bulk, a highly localized version of the familiar highly-separated EPR entangled pair of spins in a singlet state [194].

Therefore, starting from the commutator $[A, B]$ of Eq. (28.11), the dependence in spinor pseudospin degrees of freedom results in a series of alternating commutator and anti-commutator of the spin tensor components as the number of entangled spin degrees of freedom, N_s, grows, with commutator for even number of $i\epsilon_{ijk}$ and anti-commutator for odd number of $i\epsilon_{ijk}$. Had we started with $\{A, B\}$, then there will be a series of alternating anti-commutator and commutator instead. *In other words, entanglement of spins will result in transformations from local to diffusive motion in phase space, and vice versa.* This novel delocalization mechanism seems to have direct relevance to topological insulators where torques, due either to spin-orbit coupling (Berry curvature) [221, 227] or magnetic fields result in localization or a gapped spectrum in the bulk [*leading* to integer quantum Hall effect in electrical conductivity], are then entangled with pseudospin torque at the boundary consistent with Dirac points, *akin* to classical specular reflection of cyclotron orbits at boundaries. This result in delocalization at the edges (conductive edge quantum states) of 2D systems.

We therefore expect that localization of the edge states can occur if further entanglement with spin-dependent (spinor) impurities (resulting in spin flips) [211], or when layer pseudospin becomes important in bilayer topological insulators, i.e., when three spin degrees of freedom become entangled at the edge. There are actually some evidence on this aspect of edge-state localization in bilayer topological quantum Hall 2D systems where edge states are not even mentioned in the discussion of transport [228]. On the other hand, the intralayer gapped bulk states of the LLO in bilayer systems become mobile due to interaction with layer pseudospin, in fact in pseudospin ferromagnetic state this becomes superconducting mediated by the condensation of bilayer bosons of excitons [210].

The inherent non-locality property of commutators is fundamentally rooted in the uncertainty principle of quantum mechanics where the smallest non-locality in phase space is measured by the Planck constant $\hbar$. Indeed, the quantum uncertainty principle naturally follows from the formalism discussed in Part 1.

29.1.6 *Metallic edge states and bilayer superconductivity*

If the pseudospin m is the boundary pseudospin, then the delocalization will result in the so-called edge metallic states. Similarly, in bilayer system with each layer in their LLO localized state, the entangling with the layer pseudospin will delocalized the LLO states into non-local conducting states. In fact in the layer-pseudospin ferromagnetic state, experiments has shown that the electron–hole boson condensate at non-zero momentum leads to a new mechanism for the onset of superconductivity.

29.1.7 *Localization and delocalization*

Thus, we expect, in the absence of magnetic field, localization from Dresselhaus and Rashba spin-orbit coupling as incorporated in the Hamiltonian, H, yielding the *orbit-term* $i\epsilon_{mm_1m_2}\{H_{m_1}, G^{\gtrless}_{m_2}\}$, whereas delocalization contribution is expected from the D'yakonov–Perel' and Elliott–Yafet mechanism as incorporated in the *scattering-out* Γ-term, and *scattering-in* A-term. These are the commutator terms containing one Levi-Civita symbol, which describe a complex motion similar to the motion and deformation of strong magnetic field-induced Landau-orbits, due to the torque exerted by the electric field [229] resulting in the build-up of Hall voltage [230]. There are indeed theoretical and experimental works which support this assertion, the effect is related to what is often referred to in the literature as weak localization due to Dresselhaus and Rashba spin-orbit couplings [216] and weak anti-localization or suppression of scattering rates due to the D'yakonov–Perel' and Elliott–Yafet scattering mechanisms [217, 218] or entanglement of spin–orbit couplings with another torque (spinor) degrees of freedom. Delocalization or non-local scattering terms may also occur where the scattering centers have a pseudo-spin character, such as the two-level dependence of the scattering in Wigner-function quantum transport formalism recently studied by Rossi *et al.* [231, 232].

29.1.7.1 *Crossover from weak localization to weak anti-localization*

In the literature, the term weak localization (WL) and weak anti-localization (WAL) often specifically refers to the the presence of both spin-orbit coupling and magnetic field in a 2D system. The effects of the magnetic field is to induce weak localization where carriers execute Wilson loops (i.e., attributed to $i\epsilon_{kk_1k_2}\{H_{k_1,o}, G^{\gtrless}_{k_2,o}\}$ anti-commutator term in transport equation) along its path, but the effect of spin-orbit coupling is to induce weak anti-localization (WAL) by trying to modify the Wilson loops into a quantum diffusive term (i.e., from the $i\epsilon_{kk_1k_2}i\epsilon_{mm_1m_2}[H_{k_1,m_1}G^{\lessgtr}_{k_2,m_2}]$ commutator term). These contrasting and competing effects as well as their respective dependence on magnetic-field strength for Wilson localization loops, and gated electric-field strength or thickness of a 2D layer system for the spin-orbit delocalization effects, can result in an observed crossover between WL and WAL in the magnetoconductivity measurements [219] as a function of gate voltage or magnetic field. The observation of weak anti-localization implies that a strong spin-orbit coupling is the dominant effects [233].

In experimental conditions, we can assume that statistically an Ω number of particles are executing Wilson loops in their transport, i.e., $i\epsilon_{kk_1k_2}\{H_{k_1,o}, G^{\gtrless}_{k_2,o}\}$ due to the effects of applied magnetic fields, whereas $(N - \Omega)$ number of particles already felt as well the applied gate-electric field resulting in a stronger spin-orbit coupling leading to $i\epsilon_{kk_1k_2}i\epsilon_{mm_1m_2}[H_{k_1,m_1}G^{\lessgtr}_{k_2,m_2}]$ torque-entanglement modification. At constant gate voltage, the magnetoconductivity is a decreasing function of the magnetic field. However by increasing the gate voltage, the spin-orbit coupling start to become dominant in counteracting the effects of the Wilson weak-localization loops and therefore a crossover point in the magnetoconductivity measurement is expected to occur at $\Omega = 0.5N$ after which the magnetoconductivity will start to increase with the applied gate voltage. Indeed, the crossover point or minimum point of the magnetoconductivity has been observed in the experiments. The terminology *weak* as used in the experimental observations is due to the fact that there are other important terms that contribute to quantum non-locality and hence magnetoconductivity transport measurements, e.g., non-locality of spin-dependent scattering terms [230].

29.1.8 *Quantum spin Hall effect topological insulator*

We remark on quantum spin Hall effect topological insulator (QSHE-TI) as it relates to the above discussion on QHE-TI. In time-reversal symmetric systems with strong spin-orbit coupling, Kramer's degenerate states play an important role. Since spin-orbit coupling does not break time reversal symmetry, the Kramer's degenerate pair remains intact and moves in opposite directions with opposite spins. Thus, the effective spin-orbit magnetic field rotates the Kramer's degenerate pair in opposite sense, say clockwise and counterclockwise rotating pair. When this pair interacts with the boundary pseudospin, in the same sense as discussed above, this oppositely moving Kramer's Landau-orbit pair forms metallic edge states of oppositely moving currents having opposite spins, i.e., helical edge states. This is the mechanism leading to QSHE-TI.

29.1.9 *Topological Kondo insulator*

It is worthwhile to add comments on topological Kondo insulator [234], which represents the intersection of topology and strongly correlated and heavy fermion systems. Within the quantum transport perspective, the hybridization of localized f and mobile d-band, which gives rise to a hybridization gap insulator, is within the domain of quantum transport localization term expressed by the *orbit-term* $i\epsilon_{kk_1k_2}\{H_{k_1}, G^{\gtrless}_{k_2}\}$ driven by strong spin-orbit interaction of the f-electrons [234] as discussed above. Likewise, the metallic edge states are described by the delocalization term $i\epsilon_{kk_1k_2}i\epsilon_{mm_1m_2}[H_{k_1,m_1}G^{\lessgtr}_{k_2,m_2}]$, similar to the mechanism leading to QSHE-TI with helical edge or surface states. It should be pointed out that since the orbit centers are the atoms themselves the delocalization at the surface could results in the restructuring of the atomic surface atoms resulting in twisting or warping, as exhibited by some Kondo insulator materials, such as *CeNiSn* possessing non-symmorphic symmetries, i.e., containing three glide reflections, three screw rotations and an inversion symmetry [235]. There, non-symmorphic symmetries give rise to a momentum-dependent twist that enables the surface states to be detached from the bulk on the glide plane leading to the so-called Möbius-twisted surface states [235].

In summary, it is worth noting that although topological phases are often discussed in the Hamiltonian context, it has also been shown that associated topological protection and phenomena can also emerge in open quantum systems with engineered dissipation [236]. In this chapter, we have shown that a rounded picture and intuitive aspects of topological phases and bilayer superconductivity arise from the entanglement of different pseudospins in highly nonlinear and non-equilibrium quantum transport equations of spin and pseudospins. These mechanisms also shed light in the so-called weak localization (WL) and delocalization (WAL) as observed experimentally [219]. Two entangled torques from two spin degrees of freedom, one in the bulk extending to the boundary and one at the boundary, are needed to characterize topological insulators. Moreover in most experiments on WL and WAL, two entangled torques, one from the applied magnetic field and one from spin-orbit coupling is also needed to to characterize the results of experiments on WL versus WAL and their crossover point.

The non-equilibrium quantum transport results of this chapter also serve as springboard to applications in the emerging field of spin-caloritronics and pseudo-spin-caloritronics [237, 238]. The formalism employed here can be extended to the inclusion of the spinor form of the electron–phonon/plasmon and spinon/magnon self-energies, in accounting for the coupling of the equations obtained here to the spin-dependent non-equilibrium quantum transport of phonons and plasmons [117]. This will have relevance concerning heat flow in spin-caloritronics and pseudo-spin-caloritronics.

Part 7

Quantum Transport Theory of IQHE and FQHE

Chapter 30

Quantum Superfield Transport Theory of IQHE

Here, we consider the Hall conductance in a 2D periodic potential, a non-equilibrium system driven by an electromotive force. The quantization of Hall conductance in a 2D periodic potential was first explained by Thouless, Kohmoto, Nightingale, and den Nijs (TKNN) [69] using the Kubo current–current correlation formula. Similar approach was employed by Streda [70]. Earlier, Laughlin [71], and later Halperin [72], study the effects produced by changes in the vector potential on the states at the edges of a finite system, where quantization of the conductance is made explicit, but it was not obvious that the result is insensitive to boundary conditions. In contrast, the use of Kubo formula by TKNN is for bulk 2D conductors.

These theoretical studies were motivated by the Nobel Prize winning experimental discovery of von Klitzing, Dorda, and Pepper [73] on the quantization of the Hall conductance of a 2D electron gas in a strong magnetic field. The strong magnetic field basically provides the gapped energy structure for the experiments. In the TKNN approach, periodic potential in crystalline solid is being treated. A strong magnetic field is not needed to provide the gapped energy structure in their theory, only peculiar, gapped energy-band structures. In principle, in the presence of strong electric field in pure crystalline solid, the discrete Landau levels in magnetic field is replaced by the *unstable* discrete Stark ladder-energy levels [24, 239].

In this chapter, we simply make use of the gapped energy-band structure of solids under external electric field. We employ the lattice Weyl transformation technique to go from the bare Hamiltonian to effective or renormalized crystalline solid Hamiltonian. We then employ the real-time superfield non-equilibrium Green's function (SFLWT-NEGF) [12] quantum transport formalism coupled with lattice Weyl transform technique [3, 11]. This is then expanded to first-order gradient expansion for deriving the topological Chern number of the integer quantum Hall effect (IQHE) in 2D systems. Remarkably, the IQHE is an integral multiple of quantum conductance, also known as contact conductance in mesoscopic physics [12].

With the SFLWT-NEGF approach, the quantization of Hall effect strictly occurs not to first order in the electric field *per se* but rather to first-order gradient expansion in the quantum transport equation. The Berry connection and Berry curvature is the fundamental physics [13] behind the demonstration of the exact quantization of Hall conductance in units of $\frac{e^2}{h}$, which also happens to coincide with the source and drain *contact* conductance per spin in a closed circuit of mesoscopic quantum transport [12]. Indeed the quantization is also reminiscent of the self-consistent Bohr–Sommerfeld quantization of the old quantum mechanics era. This Bohr–Sommerfeld quantization is discussed in Chapter 13.

The new method employed here does not employ the conventional use of Kubo formula originally employed by TKNN [69]. Instead, we have identified the topological invariant in $(p.q; E.t)$-space kinetic quantum transport, in the absence of magnetic field, to be given by

$$\frac{1}{(2\pi\hbar)}\int\int\int d\vec{\mathcal{K}}_x d\vec{\mathcal{K}}_y dt \left[\frac{\partial^{(a)}}{\partial\vec{\mathcal{K}}_x}\frac{\partial^{(b)}}{\partial\vec{\mathcal{K}}_y} - \frac{\partial^{(a)}}{\partial\vec{\mathcal{K}}_y}\frac{\partial^{(b)}}{\partial\vec{\mathcal{K}}_x}\right]$$
$$\times\sum_{\alpha,\beta} H^{(a)}_{\alpha\beta}\left(\vec{\mathcal{K}},\mathcal{E}\right)(-iG^{<(b)}_{\beta\alpha}(\vec{\mathcal{K}},\mathcal{E})). \tag{30.1}$$

Here, the uniform electric field, $\vec{F}$, is in the x-direction and $\vec{\mathcal{K}}=\vec{p}+e\vec{F}t$. Moreover, the formula is also applicable to gapped Landau-level structure of a free electron gas in intense magnetic field [73] since the variable $\vec{\mathcal{K}}$ can incorporate the external vector potential if present. Note that the change of variables from $\vec{\mathcal{K}}$ to $\vec{p}$ in the integration over the whole Brillouin zone has a Jacobian unity. The above formula in $\vec{\mathcal{K}}$-space is given as a trace over discrete band indices. The diagonal Hall current for each band α is obtained after the summation over β index is done in the treatment that follows.

This approach is an entirely new alternative viewpoint to derive topological invariant in Chern condensed matter system. Clearly, a non-equilibrium quantum transport approach is called for since any system under a external electric field is a non-equilibrium system where current flows. In our approach, we have derived the linear response theory of Kubo current–current formula strictly from real-time non-equilibrium quantum superfield theory of transport physics adapted to time-dependent electric fields. The derivation of the Kubo current–current correlation does not involve any "pullback" process, to be discussed below, i.e., it is entirely done in phase space. The Kubo current–current approach approximates a non-equilibrium situation as a disturbed equilibrium system under an electric field. Similarly, the orbital magnetic moment, and its related Berry curvature are also calculated.

Indeed, we can readily derive the FQHE by straightforward extension of our new approach to IQHE. This is given in the following Chapter 32. Moreover, this new approach allows us to give a new characterization of the Hofstadter butterfly magnetic spectrum in terms of the IQHE and FQHE regions of the spectrum. This is taken up in Part 8. In both IQHE and FQHE, we make use of the trace formula of Chapter 5 to convert the quantum kinetic transport equation to a matrix equation, which we termed as a "pullback" procedure in what follows.

30.1 The Wigner distribution function and density matrix

In the non-equilibrium many-body Green's function technique, the principal quantities of interest are the "reduced" or single-particle correlation functions defined as

$$-iG^{<}(1,2) = \mathrm{Tr}\,[\rho_H(\psi_H^{\dagger}(2))\psi_H(1)] = \langle(\psi_H^{\dagger}(2))\psi_H(1)\rangle, \qquad (30.2)$$

$$iG^{>}(1,2) = \mathrm{Tr}\,[\rho_H(\psi_H(1)\,\psi_H^{\dagger}(2))] = \langle(\psi_H(1)\,\psi_H^{\dagger}(2))\rangle, \qquad (30.3)$$

where $\psi_H(1)$ and $\psi_H^{\dagger}(2)$ are the particle annihilation and creation operators in the Heisenberg representation, respectively. The indices 1 and 2 subsume all space-time indices and other quantum-label indices. As we shall see in what follows, the equation for $G^{<}(1,2)$ plays a principal role in particle quantum transport since it stands for particle distribution in phase space.

If we write the second quantization operator, $\hat{f}$, for the one-particle $(\vec{p},\vec{q};E,t)$-phase space distribution function[1] as

$$\hat{f}_{\lambda\lambda'\sigma\sigma'}\left(\vec{p},\vec{q};E,t\right)=\sum_{\vec{v}}e^{\frac{i}{\hbar}(2\vec{p}\cdot\vec{v}+E\tau)}\psi^{\dagger}_{\lambda\sigma}\left(\vec{q}+\vec{v},t+\frac{\tau}{2}\right)\psi_{\lambda'\sigma'}\left(\vec{q}-\vec{v},t-\frac{\tau}{2}\right), \tag{30.4}$$

where λ label the band index and σ the spin index [here we drop the Heisenberg representation subscripts H for economy of indices], then upon taking the average [see also Section 11.16],

$$\langle\hat{f}_{\lambda\lambda'\sigma\sigma'}\left(\vec{p},\vec{q};E,t\right)\rangle=\sum_{\vec{v}}e^{\frac{i}{\hbar}(2\vec{p}\cdot\vec{v}+E\tau)}\left\langle\psi^{\dagger}_{\lambda\sigma}\left(\vec{q}+\vec{v},t+\frac{\tau}{2}\right)\psi_{\lambda'\sigma'}\right.$$
$$\left.\times\left(\vec{q}-\vec{v},t-\frac{\tau}{2}\right)\right\rangle. \tag{30.5}$$

Upon employing a 4D notation: $p=(\vec{p},E)$ and $q=(\vec{q},t)$, we obtain particle distribution function $\rho_{\lambda\lambda'\sigma\sigma'}\left(p,q\right)$,

$$\rho_{\lambda\lambda'\sigma\sigma'}\left(p,q\right)=\langle\hat{f}_{\lambda\lambda'\sigma\sigma'}\left(p,q\right)\rangle, \tag{30.6}$$

Equation (30.5) is indeed the lattice Weyl transform of the density matrix operator $\hat{\rho}$ as

$$\begin{aligned}\rho_{\lambda\lambda'\sigma\sigma'}\left(p,q\right)&=\langle\hat{f}_{\lambda\lambda'\sigma\sigma'}\left(p,q\right)\rangle\\&=\sum_{v}\exp\left(\frac{2i}{\hbar}p\cdot v\right)\langle q-v;\lambda',\sigma'|\,\hat{\rho}\,|q+v;\lambda,\sigma\rangle,\end{aligned} \tag{30.7}$$

where the RHS is the lattice Weyl transform (LWT) of the density matrix operator, which is identical to the LWT of $-iG^{<}(1,2)$. For convenience in our 4D notation, we have also change variable of continuous time t from $\frac{t}{2}$ to t resulting in an overall factor 2 in the exponential, so that $v=(\vec{v},\tau)$ in the discrete summation over $\vec{v}$ and continuous integration over τ. For convenience, we just used the summation symbol in Eq. (30.7) to mean summation and integration over the respective discrete and continuous variables.

[1]The Buot's formulation of discrete lattice Weyl transform is generally based on a finite field represented by a finite prime modulus number, in all periodic directions of lattice points on a "torus" obeying modular arithmetic, closed under addition, multiplication and division, where division by all non-zero numbers is possible if and only if the modulus is prime. In the terminology of abstract algebra, the ability to perform division means that modular arithmetic modulo a prime number forms a finite field. This specific aspect of Buot's theory has been given a firm theoretical foundation, discussed in detail in the Concluding section.

Thus, the expectation value of one-particle operator $\hat{A}$ can be calculated *in phase-space* like the classical averages using a distribution function,

$$\mathrm{Tr}\,(\hat{\rho}\,\hat{A}) := \langle \hat{A} \rangle = \sum_{p,q;\lambda\lambda'\sigma\sigma'} A_{\lambda\lambda'\sigma\sigma'}\,(p,q)\,\rho_{\lambda'\lambda\sigma'\sigma}\,(p,q)\,, \tag{30.8}$$

clearly exhibiting the trace of binary operator product as a trace of the product of their respective LWT's. This general observation is *crucial* in most of the calculations that follows. Equation (30.8) is a special case of the general trace formula discussed in Chapter 5.

The Wigner distribution function $f_W(\vec{p},\vec{q},t)$ maybe given by

$$\begin{aligned} f_W(\vec{p},\vec{q},t) &= \frac{1}{2\pi\hbar}\int dE\,\left(-iG^<(\vec{p},\vec{q};E,t)\right) \\ &= \frac{1}{2\pi\hbar}\int dE\,\left(-iG^<(\vec{p},\vec{q};E,t)\right). \end{aligned} \tag{30.9}$$

We further note that,

$$\begin{aligned} \hat{\rho}\,(t) &= e^{-\frac{i}{\hbar}\hat{H}t}\hat{\rho}\,(0)\,e^{\frac{i}{\hbar}\hat{H}t} \\ &= \hat{U}\,(t)\,\hat{\rho}\,(0)\,\hat{U}^{\dagger}\,(t), \end{aligned} \tag{30.10}$$

provides the major time dependence in the transport equation that follows.

Chapter 2 discussed the renormalized dynamical variables appropriate for solid-state problems.

30.1.1 *Phase factors in space and time translation operators*

Here, we give a cursory derivation of the *phase factors* entering in the time and space translation operators. Chapter 31 give a more detailed derivation. Here our derivation make use of the action principle, namely,

$$\hat{T}\,(\vec{q})\,\hat{T}\,(t) = \exp\frac{i}{\hbar}(\hat{P}\cdot\vec{q} - \mathcal{H}\;t), \tag{30.11}$$

where $\hat{P}$ is the momentum operator, $-i\hbar\nabla_{\vec{\mathbf{q}}}\psi = \hat{P}\;\psi$, and $\mathcal{H}$ is the Hamiltonian operator, explicitly given by $\mathcal{H} = i\hbar\frac{\partial}{\partial t}$ since in the Schrödinger equation, $i\hbar\frac{\partial}{\partial t}\psi = \mathcal{H}\;\psi$. Equation (30.11) is showing $\hat{T}\,(t)$ and $\hat{T}\,(\vec{q})$, respectively, as a product on the left side since by virtue of the Baker–Campbell–Hausdorff formula, $\hat{T}\,(\vec{q})$ and $\hat{T}\,(t)$ commute and therefore equal to the right-hand side of the equation. We obtain the following relations:

$$\frac{\partial\hat{T}\,(-\vec{q})}{\partial t} = \frac{i}{\hbar}[\mathcal{H},\hat{T}\,(-\vec{q})] = \frac{i}{\hbar}e\vec{F}\cdot(-\vec{q})\;\hat{T}(-\vec{q}), \tag{30.12}$$

where $\vec{F}$ is the electric field and e is the electric charge. Similarly, we have

$$\frac{\partial \hat{T}(t)}{\partial \vec{q}} = \frac{i}{\hbar}[\vec{\mathcal{K}}, \hat{T}(t)] = \left(-\frac{i}{\hbar}e\vec{F}t\right) \hat{T}(t). \tag{30.13}$$

Equations (30.12) and (30.13) suggest that in the presence of electric field, gauge invariant quantities that are *displaced* in space and time acquires a generalized Peierls phase factors [11]. As an example, for non-equilibrium translational symmetric and steady-state condition,

$$\langle \vec{q}_1, t_1 | \hat{H}^{(1)} | \vec{q}_2, t_2 \rangle \Longrightarrow e^{-i\frac{e}{\hbar}\vec{F}t\cdot(\vec{q}_1-\vec{q}_2)} e^{-i\frac{e}{\hbar}\vec{F}\cdot\vec{q}(t_1-t_2)} H^{(1)}(\vec{q}_1 - \vec{q}_2, t_1 - t_2), \tag{30.14}$$

where

$$\vec{q} = \frac{1}{2}(\vec{q}_1 + \vec{q}_2), \tag{30.15}$$

$$t = \frac{1}{2}(t_1 + t_2). \tag{30.16}$$

Equation (30.14) is a generalization of the well-known Peierls phase factor in solid-state physics. It is critical in getting the correct lattice Weyl transform of similar quantities, which are non-local functions of space and time, to phase space.

Using the four dimensional notation: $p = (\vec{p}, E)$ and $q = (\vec{q}, t)$, the Weyl transform $A(p, q)$ of any operator $\hat{\mathbf{A}}$ is defined by

$$A_{\lambda\lambda'}(p.q) = \sum_v e^{\left(\frac{2i}{\hbar}\right)p\cdot v} \langle q - v, \lambda | \hat{\mathbf{A}} | q + v, \lambda' \rangle, \tag{30.17}$$

$$= \sum_u e^{\left(\frac{2i}{\hbar}\right)q\cdot u} \langle p + u, \lambda | \hat{\mathbf{A}} | p - u, \lambda' \rangle, \tag{30.18}$$

where λ and λ' stands for other discrete quantum numbers, e.g., energy band and spin indices. Viewed as a transformation of a matrix, we see that the Weyl transform of the matrix $\langle q', \lambda' | \hat{\mathbf{A}} | q'', \lambda'' \rangle$ is given by Eq. (30.17) and the lattice Weyl transform of $\langle p', \lambda' | \hat{\mathbf{A}} | p'', \lambda'' \rangle$ is given by Eq. (30.18). Denoting the operation of taking the lattice Weyl transform by the symbol $\mathcal{W}$ then it is easy to see that the lattice Weyl transform of total partial derivatives,

$$\mathcal{W}\left(\frac{\partial}{\partial q'} + \frac{\partial}{\partial q''}\right) \langle q', \lambda' | \hat{\mathbf{A}} | q'', \lambda'' \rangle = \frac{\partial}{\partial q} \sum_v e^{\left(\frac{2i}{\hbar}\right)p\cdot v} \langle q - v, \lambda' | \hat{\mathbf{A}} | q + v, \lambda'' \rangle$$

$$= \frac{\partial}{\partial q} A_{\lambda'\lambda''}(p.q). \tag{30.19}$$

Similarly

$$\mathcal{W}\left(\frac{\partial}{\partial p'}+\frac{\partial}{\partial p''}\right)\langle p',\lambda'|\,\hat{\mathbf{A}}\,|p'',\lambda''\rangle = \frac{\partial}{\partial p}\sum_{u} e^{\left(\frac{2i}{\hbar}\right)q\cdot u}\langle p+u,\lambda|\,\hat{\mathbf{A}}\,|p-u,\lambda'\rangle$$

$$= \frac{\partial}{\partial p}A_{\lambda'\lambda''}(p.q). \tag{30.20}$$

Note that the derivatives on the LHS of Eqs. (30.19) and (30.20) obviously operate only on the wavefunctions or state vectors not on the operator.

Writing Eq. (30.17) explicitly, we have

$$A_{\lambda\lambda'}(\vec{p}.\vec{q};E,t) = \sum_{\vec{v};\tau} e^{\left(\frac{2i}{\hbar}\right)\vec{p}\cdot\vec{v}} e^{\left(\frac{i}{\hbar}\right)E\tau}\left\langle \vec{q}-\vec{v};t-\frac{\tau}{2},\lambda\right|\hat{\mathbf{A}}\left|\vec{q}+\vec{v};t+\frac{\tau}{2},\lambda'\right\rangle. \tag{30.21}$$

Using the form of matrix elements in Eq. (30.14), we have

$$\begin{aligned}&\left\langle \vec{q}-\vec{v};t-\frac{\tau}{2},\lambda\right|\hat{\mathbf{A}}\left|\vec{q}+\vec{v};t+\frac{\tau}{2},\lambda'\right\rangle\\ &\quad= e^{-i\frac{e}{\hbar}\vec{F}t\cdot(\vec{q}_1-\vec{q}_2)}e^{-i\frac{e}{\hbar}\vec{F}\cdot\vec{q}(t_1-t_2)}A(\vec{q}_1-\vec{q}_2,t_1-t_2),\\ &\quad= e^{i\frac{e}{\hbar}\vec{F}t\cdot(2\vec{v})}e^{i\frac{e}{\hbar}\vec{F}\cdot\vec{q}\tau}A_{\lambda\lambda'}(\vec{q}_1-\vec{q}_2,t_1-t_2).\end{aligned} \tag{30.22}$$

Thus

$$\begin{aligned}A_{\lambda\lambda'}(\vec{p}.\vec{q};E,t) &= \sum_{\vec{v};\tau} e^{\left(\frac{2i}{\hbar}\right)\vec{p}\cdot\vec{v}}e^{\left(\frac{i}{\hbar}\right)E\tau}e^{i\frac{e}{\hbar}\vec{F}t\cdot(2\vec{v})}e^{i\frac{e}{\hbar}\vec{F}\cdot\vec{q}\tau}A_{\lambda\lambda'}(\vec{q}_1-\vec{q}_2,t_1-t_2),\\ &= \sum_{\vec{v};\tau} e^{\left(\frac{2i}{\hbar}\right)(\vec{p}+e\vec{F}t)\cdot\vec{v}}e^{\left(\frac{i}{\hbar}\right)(E+e\vec{F}\cdot\vec{q})\tau}A_{\lambda\lambda'}(\vec{q}_1-\vec{q}_2,t_1-t_2),\\ &= A_{\lambda\lambda'}((\vec{p}+e\vec{F}t);(E+e\vec{F}\cdot q)),\\ &= A_{\lambda\lambda'}\left(\vec{\mathcal{K}};\mathcal{E}\right).\end{aligned} \tag{30.23}$$

Hence the expected dynamical variables in the phase space including the time variable occurs in particular combinations of $\vec{\mathcal{K}}$ and $\mathcal{E}$. Therefore, besides the crystal momentum varying in time as

$$\vec{\mathcal{K}} = \vec{p}+e\vec{F}t, \tag{30.24}$$

the energy variable vary with $\vec{q}$ as

$$\mathcal{E} = E + e\vec{F} \cdot \vec{q}. \tag{30.25}$$

Equation (30.25) accounts for the gain of charged-electron kinetic energy under the electric field. Thus, partial derivatives with respect to position and momentum that enter in the Poisson operator are given by

$$\frac{\partial}{\partial \vec{q}} = \frac{\partial \mathcal{E}}{\partial \vec{q}} \frac{\partial}{\partial \mathcal{E}} = e\vec{F} \frac{\partial}{\partial \mathcal{E}}, \tag{30.26}$$

$$\frac{\partial}{\partial \vec{p}} = \frac{\partial}{\partial \vec{\mathcal{K}}} = \frac{\partial \mathcal{E}}{\partial \vec{\mathcal{K}}} \frac{\partial}{\partial \mathcal{E}} = \vec{v}_g \frac{\partial}{\partial \mathcal{E}}, \tag{30.27}$$

where v_g is the group velocity. The LWT of the effective or renormalized lattice Hamiltonian $\mathcal{H}_{\text{eff}} \leftrightarrows H(\vec{p}, \vec{q}; E, t)$ can therefore be analyzed on $(\vec{\mathcal{K}}, \mathcal{E})$-space as

$$H\left(\vec{p}, \vec{q}; E, t\right) \Longrightarrow H(\vec{\mathcal{K}}, \mathcal{E}). \tag{30.28}$$

Equations (30.26) and (30.27) follows from Eqs. (30.24)–(30.25). Of course in the absence of the electric field, the dependence in phase space becomes the familiar $H\left(\vec{p}, \omega\right)$ for translationally symmetric and steady-state system. But with $\vec{F} \neq 0$ all gauge invariant quantities are functions of $(\vec{\mathcal{K}}, \mathcal{E})$ such as the *electric Bloch function* [11] or Houston wavefunction [239] and *electric Wannier function*, i.e., the electric-field dependent generalization of Wannier function. In particular, the Weyl transform of a commutator,

$$\begin{aligned} \mathcal{W}[H, G^<] &= -i \sin \Lambda \left(H\ G^< + G^<\ H\right), \\ &= -i \sin \Lambda \left\{H,\ G^<\right\}, \end{aligned} \tag{30.29}$$

where H and $G^<$ are Weyl-transformed quantities, and Λ is the Poisson bracket operator,

$$\begin{aligned} \Lambda &= \frac{\hbar}{2} \left(\frac{\partial^{(a)}}{\partial p} \cdot \frac{\partial^{(b)}}{\partial q} - \frac{\partial^{(a)}}{\partial q} \cdot \frac{\partial^{(b)}}{\partial p} \right), \\ &= \frac{\hbar}{2} \left(\frac{\partial^{(a)}}{\partial p} \cdot e\vec{F} \frac{\partial^{(b)}}{\partial \mathcal{E}} - e\vec{F} \frac{\partial^{(a)}}{\partial \mathcal{E}} \cdot \frac{\partial^{(b)}}{\partial p} \right), \\ &= \frac{\hbar}{2} e\vec{F}_x \left(\frac{\partial^{(a)}}{\partial \mathcal{K}_x} \frac{\partial^{(b)}}{\partial \mathcal{E}} - \frac{\partial^{(a)}}{\partial \mathcal{E}} \frac{\partial^{(b)}}{\partial \mathcal{K}_x} \right), \end{aligned} \tag{30.30}$$

in $(\vec{\mathcal{K}}, \mathcal{E})$-phase space.

30.2 SFLWT-NEGF transport equation

We now make use of the SFLWT-NEGF quantum transport formalism [12]. To recapitulate, in the absence of Cooper pairings and superconducting transport mechanism, the SFLWT-NEGF phase-space transport equation reduces to,

$$\begin{aligned}&\frac{\partial}{\partial t}\operatorname{Tr}G^{<}(\vec{p},\vec{q};E,t)\\&\quad=-\frac{1}{\hbar}\sin\hat{\Lambda}\left\{\operatorname{Tr}\left\{H(p,q),G^{<}(p,q)\right\}+\operatorname{Tr}\left\{\Sigma^{<}(p,q),\operatorname{Re}G^{r}(p,q)\right\}\right\}\\&\qquad+\frac{1}{2\hbar}\cos\hat{\Lambda}\left\{\operatorname{Tr}\left\{\Sigma^{<}(p,q),A(p,q)\right\}-\operatorname{Tr}\left\{\Gamma(p,q),G^{<}(p,q)\right\}\right\}.\end{aligned} \tag{30.31}$$

By cyclic invariance under trace operation, we have,

$$\begin{aligned}&\frac{\partial}{\partial t}\operatorname{Tr}\,G^{<}(\vec{p},\vec{q};E,t)\\&\quad=-\frac{2}{\hbar}\sin\hat{\Lambda}\left\{\operatorname{Tr}\left\{H(p,q)\,G^{<}(p,q)\right\}+\operatorname{Tr}\left\{\Sigma^{<}(p,q)\ \operatorname{Re}G^{r}(p,q)\right\}\right\}\\&\qquad+\frac{1}{\hbar}\cos\hat{\Lambda}\left\{\operatorname{Tr}\left\{\Sigma^{<}(p,q)\,A(p,q)\right\}-\operatorname{Tr}\left\{\Gamma(p,q)\,G^{<}(p,q)\right\}\right\}.\end{aligned} \tag{30.32}$$

Considering the contribution for each band index α, we have the diagonal element,

$$\begin{aligned}&\frac{\partial}{\partial t}\sum_{\alpha}G^{<}_{\alpha\alpha}(\vec{p},\vec{q};E,t)\\&\quad=-\frac{2}{\hbar}\sin\hat{\Lambda}\left\{\operatorname{Tr}_{\beta}\left\{H(p,q)\,G^{<}(p,q)\right\}+\operatorname{Tr}_{\beta}\left\{\Sigma^{<}(p,q)\ \operatorname{Re}G^{r}(p,q)\right\}\right\}\\&\qquad+\frac{1}{\hbar}\cos\hat{\Lambda}\left\{\operatorname{Tr}_{\beta}\left\{\Sigma^{<}(p,q)\,A(p,q)\right\}-\operatorname{Tr}_{\beta}\left\{\Gamma(p,q)\,G^{<}(p,q)\right\}\right\},\end{aligned} \tag{30.33}$$

where $\operatorname{Tr}_{\beta}$ is a partial trace over β index. In the right-hand side of Eq. (30.31) the 4D notation of phase space is employed. If we expand Eq. (30.31) to first-order in the gradient, i.e., $\sin\Lambda\simeq\Lambda$, and $\cos\hat{\Lambda}\simeq 1$. We also make use of the new variables typified by Eq. (30.28). We have, using the Einstein summation convention,

$$\frac{\partial}{\partial t}G^{<}_{\alpha\alpha}(\vec{\mathcal{K}},\mathcal{E})$$
$$= -\frac{2}{\hbar}\hat{\Lambda}\{\{H_{\alpha\beta}(\vec{\mathcal{K}},\mathcal{E})G^{<}_{\beta\alpha}(\vec{\mathcal{K}},\mathcal{E})\} + \{\Sigma^{<}_{\alpha\beta}(\vec{\mathcal{K}},\mathcal{E})\mathrm{Re}\, G^{r}_{\beta\alpha}(\vec{\mathcal{K}},\mathcal{E})\}\}$$
$$+ \frac{1}{\hbar}\{\{\Sigma^{<}_{\alpha\beta}(\vec{\mathcal{K}},\mathcal{E})A_{\beta\alpha}(\vec{\mathcal{K}},\mathcal{E})\} - \{\Gamma_{\alpha\beta}(\vec{\mathcal{K}},\mathcal{E})G^{<}_{\beta\alpha}(\vec{\mathcal{K}},\mathcal{E})\}\}, \quad (30.34)$$

where $G^{r}(\vec{\mathcal{K}},\mathcal{E})$ is the LWT of the retarded Green's function, $A(\vec{\mathcal{K}},\mathcal{E})$ is the spectral function describing 'scattering-in' rate, and $\Gamma(\vec{\mathcal{K}},\mathcal{E})$ is the corresponding 'scattering-out' rate.

30.2.1 *Ballistic transport and diffusion*

We will simplify Eq. (30.34) by neglecting the self-energies, i.e., we limit to non-interacting particles. Then we have the following reduced quantum transport equation,

$$\frac{\partial}{\partial t}G^{<}_{\alpha\alpha}(\vec{\mathcal{K}},\mathcal{E}) = -\frac{2}{\hbar}\hat{\Lambda}\{H_{\alpha\beta}(\vec{\mathcal{K}},\mathcal{E})G^{<}_{\beta\alpha}(\vec{\mathcal{K}},\mathcal{E})\}, \quad (30.35)$$

which can be written in terms of the Poisson bracket of Eq. (30.30) as

$$\frac{\partial}{\partial t}G^{<}_{\alpha\alpha}(\vec{\mathcal{K}},\mathcal{E}) = -e\vec{F}\cdot\left(\frac{\partial^{(h)}}{\partial\vec{\mathcal{K}}}\frac{\partial^{(g)}}{\partial\mathcal{E}} - \frac{\partial^{(h)}}{\partial\mathcal{E}}\frac{\partial^{(g)}}{\partial\vec{\mathcal{K}}}\right)\{H_{\alpha\beta}(\vec{\mathcal{K}},\mathcal{E})G^{<}_{\beta\alpha}(\vec{\mathcal{K}},\mathcal{E})\}. \quad (30.36)$$

Therefore, integrating with respect to time,

$$G^{<}_{\alpha\alpha}(\vec{\mathcal{K}},\mathcal{E}) = -e\vec{F}\cdot\int dt\left[\frac{\partial^{(h)}}{\partial\vec{\mathcal{K}}}\frac{\partial^{(g)}}{\partial\mathcal{E}} - \frac{\partial^{(h)}}{\partial\mathcal{E}}\frac{\partial^{(g)}}{\partial\vec{\mathcal{K}}}\right]\{H_{\alpha\beta}(\vec{\mathcal{K}},\mathcal{E})G^{<}_{\beta\alpha}(\vec{\mathcal{K}},\mathcal{E})\}. \quad (30.37)$$

With the electric field in the x-direction, then we have

$$G^{<}_{\alpha\alpha}(\vec{\mathcal{K}},\mathcal{E}) = -e|\vec{F}|\int dt\left[\frac{\partial^{(h)}}{\partial\vec{\mathcal{K}}_x}\frac{\partial^{(g)}}{\partial\mathcal{E}} - \frac{\partial^{(h)}}{\partial\mathcal{E}}\frac{\partial^{(g)}}{\partial\vec{\mathcal{K}}_x}\right]\{H_{\alpha\beta}(\vec{\mathcal{K}},\mathcal{E})G^{<}_{\beta\alpha}(\vec{\mathcal{K}},\mathcal{E})\}. \quad (30.38)$$

The Hall current density in the y-direction for each band α is thus given by the following equation,

$$\begin{aligned}
J_y &= \frac{a^2}{(2\pi\hbar)^2}\int\int d\vec{\mathcal{K}}_x d\vec{\mathcal{K}}_y \left(\frac{e}{a^2}\frac{\partial\mathcal{E}}{\partial\vec{\mathcal{K}}_y}\right)(-iG^<_{\alpha\alpha}(\vec{\mathcal{K}},\mathcal{E})),\\
&= -e^2|\vec{F}|\frac{1}{(2\pi\hbar)^2}\int\int\int d\vec{\mathcal{K}}_x d\vec{\mathcal{K}}_y\, dt\\
&\quad\times\frac{\partial\mathcal{E}}{\partial\vec{\mathcal{K}}_y}\left[\frac{\partial^{(h)}}{\partial\vec{\mathcal{K}}_x}\frac{\partial^{(g)}}{\partial\mathcal{E}}-\frac{\partial^{(h)}}{\partial\mathcal{E}}\frac{\partial^{(g)}}{\partial\vec{\mathcal{K}}_x}\right]\{H_{\alpha\beta}(\vec{\mathcal{K}},\mathcal{E})(-iG^<_{\beta\alpha}(\vec{\mathcal{K}},\mathcal{E}))\},\\
&= -e^2|\vec{F}|\frac{1}{(2\pi\hbar)^2}\int\int\int d\vec{\mathcal{K}}_x d\vec{\mathcal{K}}_y dt\\
&\quad\times\left[\frac{\partial^{(h)}}{\partial\vec{\mathcal{K}}_x}\frac{\partial^{(g)}}{\partial\vec{\mathcal{K}}_y}-\frac{\partial^{(h)}}{\partial\vec{\mathcal{K}}_y}\frac{\partial^{(g)}}{\partial\vec{\mathcal{K}}_x}\right]\{H_{\alpha\beta}(\vec{\mathcal{K}},\mathcal{E})(-iG^<_{\beta\alpha}(\vec{\mathcal{K}},\mathcal{E}))\}. \quad (30.39)
\end{aligned}$$

If we are only interested in linear response we may consider all the quantities in the integrand to be of zero-order in the electric field, although this is not necessary if we allow very weak electric field leading to time dependence being dominated by the time dependence of the density matrix, as we shall see in what follows.

30.2.2 *Quantization of Hall conductance*

From Eq. (30.39), we claim that the quantized Hall conductivity is given by

$$\begin{aligned}
&(\sigma_{yx})_{\alpha\alpha}\\
&\quad= \left(\frac{e^2}{h}\right)\frac{1}{(2\pi\hbar)}\int\int\int d\vec{\mathcal{K}}_x d\vec{\mathcal{K}}_y dt\\
&\qquad\times\left[\frac{\partial^{(h)}}{\partial\vec{\mathcal{K}}_x}\frac{\partial^{(g)}}{\partial\vec{\mathcal{K}}_y}-\frac{\partial^{(h)}}{\partial\vec{\mathcal{K}}_y}\frac{\partial^{(g)}}{\partial\vec{\mathcal{K}}_x}\right]\{H_{\alpha\beta}(\vec{\mathcal{K}},\mathcal{E})(-iG^<_{\beta\alpha}(\vec{\mathcal{K}},\mathcal{E}))\},
\end{aligned} \quad (30.40)$$

and is quantized in units of $\frac{e^2}{h}$, i.e., $\sigma_{yx} = \frac{e^2}{h}\mathbb{Z}$, where $\mathbb{Z}$ is in the domain of integers or the first Chern numbers. In doing the integration with respect to time, t, we need to examine the implicit time-dependence of the matrix element of $-iG^<$ in the "pull back" representation defined below, which is a process of reverting to matrix element representation, the domain of the lattice Weyl transformation.

First, we digress in the following chapter a more rigorous treatment of the phase factors in spatio-temporal translation operation, which is only given a cursory treatment in Section 30.1.1 above.

Chapter 31

Covariant Derivatives, Wilson Lines, and Chern Numbers

In Section 30.1.1, we have shown that the $(p.q; E.t)$ phase-space is renormalized to that of $\left(\vec{\mathcal{K}}, \mathcal{E}\right)$ phase-space, where

$$\vec{\mathcal{K}} = \vec{p} + e\vec{F}t, \tag{31.1}$$

and

$$\mathcal{E} = E_0 + e\vec{F} \cdot \vec{q}. \tag{31.2}$$

Here, the uniform electric field, $\vec{F}$, is in the x-direction, and the Hall current is in the y-direction.

We have identified the topological invariant in $(\vec{\mathcal{K}}, \mathcal{E})$ phase-space non-equilibrium quantum transport equation leading to IQHE. Moreover, the formula is also applicable to gapped Landau-level structure of a free electron gas in intense magnetic field, experimentally done in Ref. [73], since the variable $\vec{\mathcal{K}}$ can incorporate the external vector potential, and its corresponding parallel transport, if present. Note that the change of variables from $\vec{\mathcal{K}}$ to $\vec{p}$ in the integration over the whole Brillouin zone has a Jacobian unity.

In the present chapter, we show that Eqs. (31.1) and (31.2) directly arise from the *local* gauge symmetry under a uniform electric field of localized Wannier functions centered on lattice sites. These follow from covariant derivatives leading to the dynamical spatio-temporal finite translation operators that is coupled with the Wilson line parallel-transport operators, although Wilson lines are only non-trivial in the presence of magnetic fields, or connection in topological terminology.

31.1 Translation operators in uniform electric fields

In gauge theory, the minimally-coupled gauge covariant derivative is defined as

$$D_\mu = \partial_\mu + i\alpha^\mu A_\mu, \tag{31.3}$$

where A_μ is the electromagnetic four "vector" potential, and α^μ is the serve as a coupling "charge". In electrodynamics, $\alpha^{x,y,z} = \frac{e}{\hbar c}$, whereas $\alpha^0 = \frac{e}{\hbar}$. Therefore a finite translation operator $T(q_\mu)$ by the four coordinates q_μ can be written as

$$T(q_\mu) = \exp\left[q_\mu (D_\mu)\right] = \exp\left[q_\mu (\partial_\mu + i\alpha^\mu A_\mu)\right], \tag{31.4}$$

where we have used the Einstein summation convention.

31.2 Wilson lines

The part of $T(q_\mu)$ given by $\exp\left[q_\mu (i\alpha^\mu A_\mu)\right]$ represent the Wilson lines. In general, for non-Abelian gauge theories, the Wilson line is often written as

$$W_C = \mathcal{P}\exp\left[i\int_{q=0}^{q_\mu} A_\mu dx^\mu\right], \tag{31.5}$$

where α's are absorbed in A_μ, $\mathcal{P}$ is the path ordering operator since the A_μ's are generally non-Abelian and do not commute, e.g., like the Pauli matrices.

The Wilson loop is defined as the average of the trace of Wilson lines around a closed loop, where the average is taken using the Chern–Simons Lagrangian which we will not go into. Moreover, we will not be dealing with non-Abelian A_μ's and our Wilson lines are explicitly gauge invariant, being expressed in terms of the gauge-field strength.

The importance of the Wilson line as a parallel transport operator in the gauge independent formulation of gauge theories has been emphasized by Mandelstam [240] and further developed by Wu and Yang [241]. The term given by $\int_L A_\mu dx^\mu$ is the electromagnetic Berry phase. Precisely, the definition of the Berry phase ϕ reads

$$\begin{aligned}\exp i\phi\ |\psi(x+q_x, t+\tau)\rangle &= T(q_\mu)\,|\psi(x,t)\rangle = \exp\left[q_\mu(\partial_\mu + i\alpha^\mu A_\mu)\right]|\psi(x,t)\rangle, \\ &= \exp\left\{[q_x\cdot\partial_x + \tau\partial_t] + i\phi\right\}|\psi(x,t)\rangle, \\ &= \exp\left\{\frac{i}{\hbar}[\hat{P}\cdot q_x - \tau\mathcal{H}] + i\phi\right\}|\psi(x,t)\rangle,\end{aligned} \tag{31.6}$$

where $\hat{P} = -i\hbar\frac{\partial}{\partial\vec{x}}$ is the momentum operator, $\mathcal{H} = i\hbar\frac{\partial}{\partial t}$, is the time displacement operator, A_μ is the electromagnetic Berry connection. The last line serves as a generalization in terms of quantum mechanical operators, $\hat{P}$ and $\mathcal{H}$. The extra phase acquired in translation given by

$$\phi = \int \alpha^\mu A_\mu dx^\mu, \tag{31.7}$$

is the geometrical phase. It is identified with the generalized *Peierls phase factor* in energy band dynamics of solid state physics. In quantum physics, the adiabatic contour integral of Eq. (31.7) is now generally known as the Berry phase and A_μ is now generally known mathematically as the *connection*, i.e., *magnetic vector potential.* In the absence of magnetic fields, ϕ turns out to be zero. Thus clearly, the magnetic vector potential or *connection*, $\vec{A}$, is the real geometrical (topological) factor.

31.2.1 *Peierls phase factor*

The factor $\exp i\phi$ on the left-hand side of Eq. (31.6) is the Peierls-phase factor well-known in energy band dynamics of solids [24].

31.3 Gauge potential in uniform electric fields

In this section, we determined the gauge potential,

$$A_\mu = (A_0, A_x, A_y, A_z) = (A_0, \vec{A}), \tag{31.8}$$

where A_0 is the time component. We have, in the presence of a uniform electric field,

$$\vec{A}_e = \vec{F}ct, \tag{31.9}$$

where $\vec{F}$ is the uniform electric field. Clearly, $\alpha^\mu A_\mu = \frac{e}{\hbar c}\vec{A}_e$, has the same units as $\vec{\partial}$ whereas $\frac{e}{c}\vec{A}_e$ has the same dimensional units as the momentum, i.e., by combining we have,

$$\hat{P} + \frac{e}{c}\vec{A}_e, \tag{31.10}$$

represents the minimal coupling in quantum electrodynamics.

For A_0, we have

$$\alpha^0 A_0 = -\frac{1}{\hbar}e\vec{F}\cdot\vec{q}, \tag{31.11}$$

has the same units as ∂_t and hence from Eq. (31.6), we have the combination,

$$\frac{1}{\hbar}(E_o + e\vec{F}\cdot\vec{q}), \tag{31.12}$$

represents the electron kinetic energy gain from the zero-field energy, E_o, at arbitrary lattice site $\vec{q}$, due to the electric field acting on the electron charge. Note that the negative sign of Eq. (31.11) is dictated by the equation for the force as the negative gradient of the potential, namely,

$$e\vec{F} = -\vec{\nabla}(-e\vec{F}\cdot\vec{q}). \tag{31.13}$$

Therefore under uniform electric fields, we have the Wilson lines for our Abelian gauge potential given by

$$W_C = \exp\left[i\int \alpha^\mu A_\mu dx^\mu\right], \tag{31.14}$$

$$= \exp\left[i(\alpha^0 A_0 t + \alpha\vec{A}_e\cdot q)\right],$$

$$= \exp\left[i\left(\left\{-\frac{1}{\hbar}e\vec{F}\cdot\vec{q}\right\}t + \left\{\frac{e}{\hbar c}\vec{F}ct\right\}\cdot q\right)\right], \tag{31.15}$$

$$= \mathbf{1}. \tag{31.16}$$

This means that in the absence of the magnetic field the Wilson line is trivial, an identity. This is the same as saying the that in the absence of *connection*, the Wilson line is an identity. In the presence of a uniform magnetic field, there is an added terms in Eq. (31.14) precisely given by $\frac{e}{\hbar c}A_m$, where A_m is the Maxwell's vector potential. We have in the presence of magnetic field,

$$W_C = \exp\left[i\int \alpha^\mu A_\mu dx^\mu\right],$$

$$= \exp\left[i\frac{e}{\hbar c}\int_0^q \vec{A}_m\cdot dx^\mu\right], \tag{31.17}$$

where, in symmetric gauge, $\vec{A}_m = \frac{1}{2}\vec{B}\times\vec{x}$, is the vector potential or Berry connection. Thus, the Wilson lines is the generalization of *magnetic* Peierls phase factor (geometrical factor) in solid state physics.

31.4 Alternative derivation of gauge potentials and Peierls phase factor

In contrast to the derivation given above, which directly use the covariant derivative in U(1) gauge theory, here we will use a *self-consistent* Heisenberg equation for the spatio-temporal displacement operators to determine the gauge potentials and hence the generalized Peierls phase factor. Remarkably, the self-consistent Heisenberg equation of motion automatically fixes the negative sign of the scalar gauge of Eq. (31.11).

Note that *finite* displacement operators are characteristically exponential operators amenable to Fourier series expansion. Remarkably, *for self-consistency reasons*, phase factors are generally acquired due to displacement or motion in parameter space in the presence of electromagnetic fields. This is ubiquitous in solid-state physics (e.g., Peierls phase factor in magnetic fields) before the Berry connection become mainstream and fashionable. To begin, we write our bare Hamiltonian of electrons as

$$H = H_o - e\vec{F} \cdot \vec{r}, \tag{31.18}$$

where H_o is the periodic Hamiltonian in the absence of the electric field, $\vec{F}$, and e is the charge of the electron.

In this section, we want to show that the displacement operator,

$$\hat{T}(q) = \exp\left[\frac{i}{\hbar}((\vec{q} \cdot -i\hbar\nabla_{\vec{r}}))\right], \tag{31.19}$$

results in a displaced function acquiring a phase factor in the presence of electric field. This is incorporated in $\hat{T}(q)$ by

$$\begin{aligned}\tilde{T}(\vec{q}) &= \exp\left[\frac{i}{\hbar}\left(\vec{q} \cdot \left(\frac{e}{c}\vec{F}ct\right) + (\vec{q} \cdot -i\hbar\nabla_{\vec{r}})\right)\right],\\ &= \exp\left[\frac{i}{\hbar}\left(\vec{q} \cdot \left(\frac{e}{c}\vec{F}ct\right) + (\vec{q} \cdot \hat{P})\right)\right],\end{aligned} \tag{31.20}$$

where we indicate by $\tilde{T}$ the translation operator $\hat{T}$ with a phase factor.

Similarly, for the time displacement operator, we will show that

$$\hat{T}(t) = \exp\left[-\frac{i}{\hbar}\left(\left(t \cdot \; i\hbar\frac{\partial}{\partial t}\right)\right)\right], \tag{31.21}$$

also acquires a phase factor and is given by

$$\tilde{T}(t) = \exp\left[-\frac{i}{\hbar}(t(e\vec{F} \cdot \vec{q}) + (t\;\mathcal{H}))\right], \tag{31.22}$$

where $\tilde{T}(t)$ differs from $\hat{T}(t)$ by a phase factor. Here, $i\hbar\frac{\partial}{\partial t} \; : = \mathcal{H}$ is the energy operator of the system.

In other words, displacement in space by lattice vector q acquires phase factors equal to $\exp(\frac{i}{\hbar}e\vec{F}t \cdot \vec{q})$. Similarly, displacement in time by t acquires phase factor equal to $\exp(-\frac{i}{\hbar}(e\vec{F} \cdot \vec{q})t)$. In the absence of the electric field these phase factors give unity, e.g., $\tilde{T}(t) \Longrightarrow_{\vec{F}\to 0} \hat{T}(t)$, reduces to ordinary translation operator.

The physics behind these phase factors is dictated by a self-consistent translation of *local* functions in space and time under a uniform electric field, $\vec{F}$. These follow from the *self-consistent* Heisenberg equation of motion for quantum operators. For efficient bookkeeping and for ease in taking lattice Weyl transform, it is usually more convenient to attach these phase factors to displaced local functions themselves, Eq. (31.6), rather than on the operator. This yields a generalization of Peierls phase factor, well-known for solid-state problems in magnetic fields. The derivation goes as follows.

31.4.1 *Non-locality in coordinates*

The nature of the derivatives of exponential displacement operator is determined, e.g., by the following operation:

$$i\hbar\frac{\partial}{\partial t}\tilde{T}\left(q\right)W_{\lambda}\left(r-0\right)=i\hbar\frac{\partial\phi}{\partial t}\tilde{T}\left(q\right)W_{\lambda}\left(r-0\right), \tag{31.23}$$

$$\frac{\partial}{\partial t}\tilde{T}\left(q\right)=\frac{\partial\phi}{\partial t}\tilde{T}\left(q\right), \tag{31.24}$$

$$\frac{d\tilde{T}\left(q\right)}{\tilde{T}\left(q\right)}=d\phi, \tag{31.25}$$

where specifically $\hat{T}\left(-q\right)W_{\lambda}\left(r-0\right)$ is a translation of the center coordinate of a localized Wannier function centered in the origin to another lattice point q, yielding $W_{\lambda}\left(r-q\right)$. Therefore $\hat{T}\left(-q\right)$ resembles the physical process of transferring a localized function centered at the origin to a localized function centered at another lattice point q. Here we define total ϕ as undetermined for the moment as

$$\phi=-\frac{i}{\hbar}(f(q,t)+(\vec{q}\cdot\hat{P})), \tag{31.26}$$

where $f\left(q,t\right)$ is to be determined. Note that the presence of $f\left(q,t\right)$ is needed for selfconsistency in the presence of electric field. Now consider the second term in the exponent, namely,

$$\begin{aligned}\phi_2&=-\frac{i}{\hbar}\vec{q}\cdot\hat{P}\\&=-\frac{i}{\hbar}\vec{q}\cdot\left(-i\hbar\frac{\partial}{\partial\vec{r}}\right),\\&=(-q)\cdot\frac{\partial}{\partial\vec{r}},\end{aligned} \tag{31.27}$$

leading to Fourier series expansion of the translation, $\hat{T}(-q)$. Then, we obtain

$$-i\hbar\frac{\partial\phi_2}{\partial t} = [H, \phi_2] = \frac{i}{\hbar}[H, ((-\vec{q}) \cdot \hat{P})] = e\vec{F} \cdot (-\vec{q}), \tag{31.28}$$

since $-\frac{i}{\hbar}f(q,t)$ commutes with the Hamiltonian. Therefore, we have,

$$\frac{d\tilde{T}(-q)}{\tilde{T}(-q)} = d\phi = \frac{i}{\hbar}e\vec{F} \cdot (-\vec{q})\ dt. \tag{31.29}$$

We may thus write

$$-i\hbar\frac{\partial}{\partial t}\tilde{T}(\vec{q},t) = [H, \tilde{T}(\vec{q},t)] = e\vec{F} \cdot (\vec{q})\ \tilde{T}(\vec{q},t). \tag{31.30}$$

Therefore

$$\begin{aligned} \tilde{T}(q) &= \exp\frac{i}{\hbar}[(e\vec{F}t) \cdot \vec{q}\ + \hat{P} \cdot \vec{q}], \\ &= \exp\frac{i}{\hbar}[\hat{P} + e\vec{F}t] \cdot \vec{q}. \end{aligned} \tag{31.31}$$

This means that a displacement by $\vec{q}$ of localized function acquires a phase factor, PF, given by

$$PF = \exp\left[\frac{i}{\hbar}(e\vec{F}t \cdot \vec{q})\right]. \tag{31.32}$$

We can also deduce from Eq. (31.28) the relation for the momentum operator,

$$\begin{aligned} \frac{i}{\hbar}[H, [(\vec{q}) \cdot -i\hbar\nabla_{\vec{r}}]] &= e\vec{F} \cdot (\vec{q}) = \frac{\partial\hat{P}}{\partial t} \cdot (q), \Longrightarrow \frac{\partial\hat{P}}{\partial t} = e\vec{F}, \\ &\Longrightarrow \hat{P} = \hat{P}_o + e\vec{F}t, \end{aligned} \tag{31.33}$$

which indicates a covariant derivative in spatial coordinates,

$$\vec{D}_\mu = \vec{\partial}_\mu + i\frac{e}{\hbar c}\vec{A}_e, \tag{31.34}$$

with

$$\vec{A}_\mu = \vec{F}ct, \tag{31.35}$$

as obtained before.

31.4.2 *Simultaneous eigenvalues for* $\hat{H}$ *and* $\tilde{T}(\vec{q}, t)$

One very important conclusion is implied in Eq. (F.25), which we rewrite here for convenience

$$[\hat{H}, \tilde{T}(\vec{q}, t)] = e\vec{F} \cdot \vec{q}\, \tilde{T}(\vec{q}, t)\,. \tag{31.36}$$

What the above relation means is that if $\tilde{T}(\vec{q}, t)$ is diagonal then $[H, \tilde{T}(\vec{q}, t)]$ is also diagonal. But if $\tilde{T}(\vec{q}, t)$ is diagonal, then $\hat{H}$ is also diagonal with the same eigenvalues. The eigenfunction of $\tilde{T}(\vec{q}, t)$ is labeled by the quantum label $\vec{\mathcal{K}} = \vec{p}_o + e\vec{F}t$. This implies that $\hat{H}$ is also diagonal in $\vec{\mathcal{K}}$. The electric Bloch function labeled by $B\left(k_o + \frac{e}{\hbar}Ft, \ldots\right)$ is the eigenfunction of $\tilde{T}(\vec{q}, t)$ as well as that of the renormalized $\hat{H} \Longrightarrow H_{\text{renor}}$,

$$H_{\text{renor}}(\hat{K}, \hat{Q}) \Longleftrightarrow W_n(\vec{\mathcal{K}}, \mathcal{E}), \tag{31.37}$$

where the double pointed arrow denotes lattice Weyl correspondence, see Eq. (30.23) below, using the localized electric Wannier function.

We now show the the energy variable $\mathcal{E}$ in Eq. (31.37) does incorporate the coordinates through the gauge potential A_0.

31.4.3 *Non-locality in time*

Because non-local arguments in time acquires phase factor also, the energy variable of the theory now varies with $e\vec{F} \cdot \vec{q}$, as discussed next. From Eq. (31.21) for a displacement in time,

$$\hat{T}(t) = \exp \frac{i}{\hbar}(-\mathcal{H}t)\,, \tag{31.38}$$

or

$$\tilde{T}(t) \equiv \mathcal{F} \exp\left(\frac{-i}{\hbar}\left(i\hbar t \frac{\partial}{\partial t}\right)\right) = \mathcal{F} \exp\left(t\frac{\partial}{\partial t}\right), \tag{31.39}$$

where $\mathcal{F}$ incorporates the scalar gauge potential, A_0. Let

$$\begin{aligned} \mathcal{F} &= \exp\left[i\alpha^0 A_0 t\right], \\ &= \exp\left[-\frac{i}{\hbar}\left(f_t(q, t)\right)\right], \end{aligned} \tag{31.40}$$

is a phase factor to be determined. The total phase is

$$\phi_t(q) = -\frac{i}{\hbar}\left(f_t(q, t) + t\left(i\hbar\frac{\partial}{\partial t'}\right)\right). \tag{31.41}$$

From the equation of motion,

$$-i\hbar\frac{\partial\tilde{T}(t)}{\partial\vec{q}} = -i\hbar\frac{\partial\phi_t(q)}{\partial\vec{q}}\tilde{T}(t),$$

$$-i\hbar\frac{\partial\phi_t(q)}{\partial\vec{q}} = -\frac{i}{\hbar}\left[\vec{\mathcal{K}}(t'), t\left(i\hbar\frac{\partial}{\partial t'}\right)\right] = \left[\vec{p}+\frac{e}{c}\vec{A}, t\nabla_{t'}\right],$$

$$= [\vec{p}+e\vec{F}t', t\nabla_{t'}] = -e\vec{F}t. \tag{31.42}$$

Therefore

$$-i\hbar\frac{\partial\phi_t(q)}{\partial\vec{q}} = -e\vec{F}t,$$

$$\frac{\partial\phi_t(q)}{\partial\vec{q}} = -\frac{i}{\hbar}e\vec{F}t. \tag{31.43}$$

Thus, we obtained,

$$\frac{\partial\tilde{T}(t)}{\partial\vec{q}} = \left(-\frac{i}{\hbar}e\vec{F}t\right)\tilde{T}(t), \tag{31.44}$$

and hence,

$$d\ln\tilde{T}(t) = \left(-\frac{i}{\hbar}e\vec{F}t\right)\cdot d\vec{q},$$

$$\ln\tilde{T}(t) = \left(-\frac{i}{\hbar}e\vec{F}t\right)\cdot\Delta\vec{q}. \tag{31.45}$$

Hence a displacement in time carries a phase factor given by $\exp\left(-\frac{i}{\hbar}\left(e\vec{F}t\right)\cdot\Delta\vec{q}\right)$ and

$$\tilde{T}(t) = \exp\left\{-\frac{i}{\hbar}\left[t(e\vec{F}\cdot\vec{q})+t\left(i\hbar\frac{\partial}{\partial t'}\right)\right]\right\},$$

$$= \exp\left\{-\frac{i}{\hbar}\left[t(e\vec{F}\cdot\vec{q})+t\left(i\hbar\frac{\partial}{\partial t'}\right)\right]\right\}. \tag{31.46}$$

Once more this gives support to the covariant derivative,

$$D_0 = \partial_t + i\alpha^0 A_0, \tag{31.47}$$

where

$$i\alpha^0 A_0 = -i\frac{e\vec{F}\cdot\vec{q}}{\hbar}. \tag{31.48}$$

Therefore,

$$-i\hbar\frac{\partial}{\partial\vec{q}}\tilde{T}\left(\vec{q},t\right)=\left[\vec{\mathcal{K}}\left(t\right),\tilde{T}\left(\vec{q},t\right)\right]=\left(-e\vec{F}t\right)\tilde{T}\left(\vec{q},t\right). \tag{31.49}$$

Now from second line of Eq. (31.42), we have

$$\begin{aligned}-\frac{i}{\hbar}\left[\vec{\mathcal{K}}\left(t'\right),\left(t\right)i\hbar\frac{\partial}{\partial t'}\right]&=\left[\vec{p}+e\vec{F}t',\left(t\right)\frac{\partial}{\partial t'}\right]=-e\vec{F}t,\\ \left[\vec{\mathcal{K}}\left(t'\right),\mathcal{H}\right]&=-i\hbar e\vec{F}=-i\hbar\frac{\partial\mathcal{E}}{\partial\vec{q}},\end{aligned} \tag{31.50}$$

which leads to to the expression for in Eq. (31.40) for f_t in the phase factor as

$$\frac{\partial\mathcal{E}}{\partial\vec{q}}=e\vec{F}\Longrightarrow f_t=e\vec{F}\cdot\vec{q}. \tag{31.51}$$

All these results, Eqs. (31.31) and (31.46), lead to the time-dependent wave vector,

$$\hbar\vec{k}=\hbar\vec{k}_o+\hbar e\vec{F}t, \tag{31.52}$$

and to the position-dependent energy,

$$\mathcal{E}=E_o+e\vec{F}\cdot\vec{q}, \tag{31.53}$$

respectively. Again, for for taking lattice Weyl transform, it is more convenient to attach these phase factor to the displaced local functions themselves. This allows us to generalize the Peierls phase factor to space and time displacements, originally well-known for solid-state problems for magnetic fields.

As an example, for non-equilibrium translational symmetric and steady-state condition.

$$\left\langle q_1,t_1\right|\mathcal{O}\left|q_2,t_2\right\rangle\Longrightarrow e^{i\frac{e}{\hbar}\vec{F}t\cdot\left(\vec{q}_2-\vec{q}_1\right)}e^{-i\frac{e}{\hbar}\vec{F}\cdot\vec{q}\left(t_2-t_1\right)}O\left(\vec{q}_2-\vec{q}_1,t_2-t_1\right), \tag{31.54}$$

where,

$$\vec{q}=\frac{1}{2}\left(\vec{q}_1+\vec{q}_2\right), \tag{31.55}$$

$$t=\frac{1}{2}\left(t_1+t_2\right). \tag{31.56}$$

31.5 Lattice Weyl transforms

Using the four dimensional notation: $p = (\vec{p}, E)$ and $q = (q, t)$, the lattice Weyl transform (LWT), $A(p,q)$ of any operator $\hat{A}$ is defined by

$$\begin{aligned} A_{\lambda\lambda'}(p,q) &= \sum_v e^{\left(\frac{2i}{\hbar}\right)p.v} \langle q-v,\lambda|\,\hat{A}\,|q+v,\lambda'\rangle, \\ &= \sum_u e^{\left(\frac{2i}{\hbar}\right)q.u} \langle p+u,\lambda|\,\hat{A}\,|p-u,\lambda'\rangle. \end{aligned} \tag{31.57}$$

Writing Eq. (31.57) explicitly, we have

$$A_{\lambda\lambda'}(\vec{p}.\vec{q};E,t) = \sum_{\vec{v};\tau} e^{\left(\frac{2i}{\hbar}\right)\vec{p}\cdot\vec{v}} e^{\left(-\frac{i}{\hbar}\right)E\tau} \left\langle \vec{q}-\vec{v};t-\frac{\tau}{2},\lambda\right| \hat{\mathbf{A}} \left|\vec{q}+\vec{v};t+\frac{\tau}{2},\lambda'\right\rangle. \tag{31.58}$$

Using the form of matrix elements in Eq. (31.54), we have

$$\begin{aligned} &\left\langle \vec{q}-\vec{v};t-\frac{\tau}{2},\lambda\right| \hat{\mathbf{A}} \left|\vec{q}+\vec{v};t+\frac{\tau}{2},\lambda'\right\rangle \\ &\quad = e^{i\frac{e}{\hbar}\vec{F}t\cdot(\vec{q}_1-\vec{q}_2)} e^{-i\frac{e}{\hbar}\vec{F}\cdot\vec{q}(t_1-t_2)} A(\vec{q}_1-\vec{q}_2,t_1-t_2), \\ &\quad = e^{i\frac{e}{\hbar}\vec{F}t\cdot(2\vec{v})} e^{-i\frac{e}{\hbar}\vec{F}\cdot\vec{q}\tau} A_{\lambda\lambda'}(\vec{q}_1-\vec{q}_2,t_1-t_2). \end{aligned} \tag{31.59}$$

Thus,

$$\begin{aligned} A_{\lambda\lambda'}(\vec{p}.\vec{q};E,t) &= \sum_{\vec{v};\tau} e^{\left(\frac{i}{\hbar}\right)\vec{p}\cdot 2\vec{v}} e^{i\frac{e}{\hbar}\vec{F}t\cdot(2\vec{v})} e^{\left(-\frac{i}{\hbar}\right)E\tau} e^{-i\frac{e}{\hbar}\vec{F}\cdot\vec{q}\tau} A_{\lambda\lambda'}(\vec{q}_1-\vec{q}_2,t_1-t_2), \\ &= \sum_{\vec{v};\tau} e^{\left(\frac{2i}{\hbar}\right)(\vec{p}+e\vec{F}t)\cdot\vec{v}} e^{\left(-\frac{i}{\hbar}\right)(E+e\vec{F}\cdot\vec{q})\tau} A_{\lambda\lambda'}(2v,\tau), \\ &= A_{\lambda\lambda'}[(\vec{p}+e\vec{F}t),(E+e\vec{F}\cdot\vec{q})], \\ &= A_{\lambda\lambda'}(\vec{\mathcal{K}};\mathcal{E}). \end{aligned} \tag{31.60}$$

Hence the expected dynamical variables in the phase space including the time variable occurs in particular combinations of $\vec{\mathcal{K}}$ and $\mathcal{E}$. Therefore, besides the crystal momentum varying in time as

$$\vec{\mathcal{K}} = \vec{p}_o + e\vec{F}t, \tag{31.61}$$

the energy variable vary with $\vec{q}$ as

$$\mathcal{E} = E_o + e\vec{F}\cdot\vec{q}. \tag{31.62}$$

physically representing the gain of *electron* kinetic due to the electric field. This is the result of covariant derivatives for energy band dynamics in the presence of uniform electric fields. Thus, differentiation with respect to momentum and position variables are now relegated to differentiation with respect to energy $\vec{\mathcal{K}}$ and $\mathcal{E}$, respectively,

$$\frac{\partial}{\partial \vec{q}} = \frac{\partial \mathcal{E}}{\partial \vec{q}} \frac{\partial}{\partial \mathcal{E}} = e\vec{F} \frac{\partial}{\partial \mathcal{E}}, \tag{31.63}$$

$$\frac{\partial}{\partial \vec{p}} = \frac{\partial}{\partial \vec{\mathcal{K}}} = \left(\frac{\partial \mathcal{E}}{\partial \vec{\mathcal{K}}}\right) \frac{\partial}{\partial \mathcal{E}} = \vec{v}_g \frac{\partial}{\partial \mathcal{E}}, \tag{31.64}$$

$$\frac{\partial}{\partial t} = \frac{\partial \vec{\mathcal{K}}}{\partial t} \cdot \frac{\partial}{\partial \vec{\mathcal{K}}} = e\vec{F} \cdot \frac{\partial}{\partial \vec{\mathcal{K}}}, \tag{31.65}$$

where v_g is the group velocity. The LWT of the effective or renormalized lattice Hamiltonian $\mathcal{H}_{\text{eff}} \leftrightarrows H\left(\vec{p}, \vec{q}; E_o, t\right)$ can therefore be analyzed on $(\vec{\mathcal{K}}, E)$-space as

$$H\left(\vec{p}, \vec{q}; E_o, t\right) \Longrightarrow H(\vec{\mathcal{K}}, \mathcal{E}). \tag{31.66}$$

The last line is by virtue of Eqs. (30.24)–(30.25). Of course in the absence of the electric field, the dependence in phase space becomes the familiar $H\left(\vec{p}, \omega\right)$ for translationally symmetric and steady-state system. But with $\vec{F} \neq 0$ all gauge invariant quantities are functions of $(\vec{\mathcal{K}}, \mathcal{E})$ such as the *electric Bloch function* [11, 24] or Houston wavefunction [239]] and *electric Wannier function*, i.e., the electric-field dependent generalization of Wannier function.

31.5.1 *Weyl transform of commutator*

The Weyl transform of commutator is derived from the expression for the Weyl transform of a binary product of operators, [3], which we rewrite from Eq. (17.33),

$$\begin{aligned} AB\left(p,q\right) &= \sum \exp\left[-i\frac{\hbar}{2}\left(\frac{\partial}{\partial p}\cdot\frac{\partial}{\partial q} - \frac{\partial}{\partial q}\cdot\frac{\partial}{\partial p}\right)\right] A\left(p,q\right) B\left(p,q\right), \\ &= \cos\frac{\hbar}{2}\left(\frac{\partial}{\partial p}\cdot\frac{\partial}{\partial q} - \frac{\partial}{\partial q}\cdot\frac{\partial}{\partial p}\right) A\left(p,q\right) B\left(p,q\right) \\ &\quad - i\sin\frac{\hbar}{2}\left(\frac{\partial}{\partial p}\cdot\frac{\partial}{\partial q} - \frac{\partial}{\partial q}\cdot\frac{\partial}{\partial p}\right) A\left(p,q\right) B\left(p,q\right). \end{aligned} \tag{31.67}$$

Then,

$$
\begin{aligned}
[A,B](p,q) = \cos\frac{\hbar}{2}&\left(\frac{\partial^{(a)}}{\partial p}\cdot\frac{\partial^{(b)}}{\partial q}-\frac{\partial^{(a)}}{\partial q}\cdot\frac{\partial^{(b)}}{\partial p}\right)A(p,q)\,B(p,q)\\
-\cos\frac{\hbar}{2}&\left(\frac{\partial^{(b)}}{\partial p}\cdot\frac{\partial^{(a)}}{\partial q}-\frac{\partial^{(b)}}{\partial q}\cdot\frac{\partial^{(a)}}{\partial p}\right)B(p,q)\,A(p,q)\\
-i\sin\frac{\hbar}{2}&\left(\frac{\partial^{(a)}}{\partial p}\cdot\frac{\partial^{(b)}}{\partial q}-\frac{\partial^{(a)}}{\partial q}\cdot\frac{\partial^{(b)}}{\partial p}\right)A(p,q)\,B(p,q)\\
+i\sin\frac{\hbar}{2}&\left(\frac{\partial^{(b)}}{\partial p}\cdot\frac{\partial^{(a)}}{\partial q}-\frac{\partial^{(b)}}{\partial q}\cdot\frac{\partial^{(a)}}{\partial p}\right)B(p,q)\,A(p,q)\,.
\end{aligned}
\tag{31.68}
$$

Thus,

$$
\begin{aligned}
&[A,B](p,q)\\
&= \cos\frac{\hbar}{2}\left(\frac{\partial^{(a)}}{\partial p}\cdot\frac{\partial^{(b)}}{\partial q}-\frac{\partial^{(a)}}{\partial q}\cdot\frac{\partial^{(b)}}{\partial p}\right)\begin{bmatrix}A(p,q)\,B(p,q)\\ -B(p,q)\,A(p,q)\end{bmatrix}\\
&\quad -i\sin\frac{\hbar}{2}\left(\frac{\partial^{(a)}}{\partial p}\cdot\frac{\partial^{(b)}}{\partial q}-\frac{\partial^{(a)}}{\partial q}\cdot\frac{\partial^{(b)}}{\partial p}\right)\begin{bmatrix}A(p,q)\,B(p,q)\\ +B(p,q)\,A(p,q)\end{bmatrix}.
\end{aligned}
\tag{31.69}
$$

In particular, using the argument of Eq. (31.66), we can just consider the second term of Eq. (31.69) since under the trace operation, expected in applying to quantum transport formalism that follows, the contribution of the first term is zero.

$$
\mathcal{W}\left[H,G^{<}\right] = -i\sin\Lambda\,\{H(\vec{\mathcal{K}},\mathcal{E})G^{<}(\vec{\mathcal{K}},\mathcal{E})+G^{<}(\vec{\mathcal{K}},\mathcal{E})H(\vec{\mathcal{K}},\mathcal{E})\},
\tag{31.70}
$$

where Λ is the Poisson bracket operator. For the same reason, under the same trace operation that occurs in the transport equation we can just add the two terms in Eq. (31.70), acting only on $H(\vec{\mathcal{K}},\mathcal{E})G^{<}(\vec{\mathcal{K}},\mathcal{E})$. Using Eq. (31.64), we can therefore write the Poisson bracket operator Λ, as

$$
\begin{aligned}
\Lambda &= \frac{\hbar}{2}\left[\frac{\partial^{(h)}}{\partial p}\cdot\frac{\partial^{(g)}}{\partial q}-\frac{\partial^{(h)}}{\partial q}\cdot\frac{\partial^{(g)}}{\partial p}\right],\\
&= \frac{\hbar}{2}\left[\frac{\partial^{(h)}}{\partial \vec{\mathcal{K}}}\cdot\frac{\partial^{(g)}}{\partial q}-\frac{\partial^{(h)}}{\partial q}\cdot\frac{\partial^{(g)}}{\partial \vec{\mathcal{K}}}\right],
\end{aligned}
$$

$$= \frac{\hbar}{2}\left[\frac{\partial^{(h)}}{\partial\vec{\mathcal{K}}}\cdot e\vec{F}\frac{\partial^{(g)}}{\partial\mathcal{E}} - e\vec{F}\frac{\partial^{(h)}}{\partial\mathcal{E}}\cdot\frac{\partial^{(g)}}{\partial\vec{\mathcal{K}}}\right],$$

$$= \frac{\hbar}{2}e\vec{F}_x\left[\frac{\partial^{(h)}}{\partial\vec{\mathcal{K}}_x}\frac{\partial^{(g)}}{\partial\mathcal{E}} - \frac{\partial^{(h)}}{\partial\mathcal{E}}\frac{\partial^{(g)}}{\partial\vec{\mathcal{K}}_x}\right], \tag{31.71}$$

with the electric field in the x-direction on $(\vec{\mathcal{K}},\mathcal{E})$-phase space.

31.6 Application to non-equilibrium quantum transport equation

From Eq. (17.23), and neglecting the Cooper pairing terms [$v \equiv H$, a single particle non-interacting term], we have,

$$i\hbar\left(\frac{\partial}{\partial t_1}+\frac{\partial}{\partial t_2}\right)G^< = \left[H\ G^< - G^<\ H\right] + \left[\Sigma^r G^< - G^<\Sigma^a\right] + \left[\Sigma^< G^a - G^r\Sigma^<\right]. \tag{31.72}$$

The ballistic phase-space quantum transport equation, we neglect the second line of Eq. (31.72). We have the lattice Weyl transform in terms of Poisson bracket expansion,

$$\frac{\partial}{\partial t}\operatorname{Tr}G^<\left(\vec{p},\vec{q};E,t\right) = -\frac{1}{\hbar}\sin\hat{\Lambda}\ \operatorname{Tr}\left\{H\left(p,q\right)G^<\left(p,q\right)+G^<\left(p,q\right)H\left(p,q\right)\right\} \tag{31.73}$$

where in the right side the 4D notation of phase space is employed. Under a uniform electric field, this simplifies to

$$\frac{\partial}{\partial t}\operatorname{Tr}G^<(\vec{\mathcal{K}},\mathcal{E}) = -\frac{1}{\hbar}\sin\Lambda\ \operatorname{Tr}\{H(\vec{\mathcal{K}},\mathcal{E})G^<(\vec{\mathcal{K}},\mathcal{E})+G^<(\vec{\mathcal{K}},\mathcal{E})H(\vec{\mathcal{K}},\mathcal{E})\}, \tag{31.74}$$

where, from Eq. (31.71),

$$\Lambda = \frac{\hbar}{2}e\vec{F}\cdot\left[\frac{\partial^{(h)}}{\partial\vec{\mathcal{K}}}\frac{\partial^{(g)}}{\partial\mathcal{E}} - \frac{\partial^{(h)}}{\partial\mathcal{E}}\frac{\partial^{(g)}}{\partial\vec{\mathcal{K}}}\right]. \tag{31.75}$$

If we expand Eq. (31.74) to first-order in the gradient, i.e., $\sin\Lambda \simeq \Lambda$, and using Eq. (31.71), the phase-space transport equation can be written in a compact form as

$$\begin{aligned}
&\frac{\partial}{\partial t}\mathrm{Tr}\ G^{<}\left(\vec{\mathcal{K}},\mathcal{E}\right)\\
&\quad = -\frac{1}{\hbar}\frac{\hbar}{2}e\vec{F}_x\ \mathrm{Tr}\left[\frac{\partial^{(h)}}{\partial\vec{\mathcal{K}}_x}\frac{\partial^{(g)}}{\partial\mathcal{E}} - \frac{\partial^{(h)}}{\partial\mathcal{E}}\frac{\partial^{(g)}}{\partial\vec{\mathcal{K}}_x}\right]\\
&\qquad \times\left\{\begin{array}{c} H^{(a)}(\vec{\mathcal{K}},\mathcal{E})G^{<(b)}(\vec{\mathcal{K}},\mathcal{E}) \\ +G^{<(b)}(\vec{\mathcal{K}},\mathcal{E})H^{(a)}(\vec{\mathcal{K}},\mathcal{E}) \end{array}\right\},
\end{aligned} \tag{31.76}$$

with the electric field in the x-direction. By the cyclic invariance of trace [see also Chapter 5], we can write,

$$\mathrm{Tr}\,G^{<}\left(\vec{\mathcal{K}},\mathcal{E}\right) = -e\vec{F}\cdot\int dt\ \left[\frac{\partial^{(h)}}{\partial\vec{\mathcal{K}}}\frac{\partial^{(g)}}{\partial\mathcal{E}} - \frac{\partial^{(h)}}{\partial\mathcal{E}}\frac{\partial^{(g)}}{\partial\vec{\mathcal{K}}}\right]\mathrm{Tr}\left\{H^{(a)}(\vec{\mathcal{K}},\mathcal{E})G^{<(b)}(\vec{\mathcal{K}},\mathcal{E})\right\}. \tag{31.77}$$

With the electric field in the x-direction, the Hall current in the y-direction for each band α is given by the following equation, applied to the left-hand-side of Eq. (31.77)[1]

$$J_y = \frac{a^2}{(2\pi\hbar)^2}\int\int d\vec{\mathcal{K}}_x d\vec{\mathcal{K}}_y\left(\frac{e}{a^2}\frac{\partial\mathcal{E}}{\partial\vec{\mathcal{K}}_y}\right)(-iG^{<}_{\alpha\alpha}(\vec{\mathcal{K}},\mathcal{E})), \tag{31.78}$$

which leads to $J_y = \sigma_{yx}\vec{F}_x$. Then, we have (using the Einstein summation convention),

$$\begin{aligned}
(\sigma_{yx})_{\alpha\alpha} &= \left(\frac{e^2}{h}\right)\frac{1}{(2\pi\hbar)}\int\int\int d\vec{\mathcal{K}}_x d\vec{\mathcal{K}}_y dt\\
&\quad\times\left[\frac{\partial^{(a)}}{\partial\vec{\mathcal{K}}_x}\frac{\partial^{(b)}}{\partial\vec{\mathcal{K}}_y} - \frac{\partial^{(a)}}{\partial\vec{\mathcal{K}}_y}\frac{\partial^{(b)}}{\partial\vec{\mathcal{K}}_x}\right]\{H_{\alpha\beta}(\vec{\mathcal{K}},\mathcal{E})(iG^{<}_{\beta\alpha}(\vec{\mathcal{K}},\mathcal{E}))\}.
\end{aligned} \tag{31.79}$$

31.7 Reverting to matrix elements

Now to show that the Eq. (31.79) give $\sigma_{yx} = \left(\frac{e^2}{h}\right)n$, where $n \in \mathbb{Z}$, we need to transform the integral of Eq. (30.40) to the integral of the curvature of Berry connection in a closed loop. This necessitates a "pull back" (i.e., undoing) the lattice transformation of Eq. (30.40), i.e., we revert to corresponding matrix elements before taking the lattice Weyl transform.

[1]The same expression can be derived for current through a discrete lattice, as derived in Appendix G.

31.7.1 "Pull back" of the lattice Weyl transformation

The pull-back process means we have to undo the lattice transformation of the SFLWT-NEGF transport equation. Consider the integrand in Eq. (30.40) given by the partial derivatives of lattice Weyl transformed quantities.

$$\left[\frac{\partial^{(h)}}{\partial\vec{\mathcal{K}}_x}\frac{\partial^{(g)}}{\partial\vec{\mathcal{K}}_y} - \frac{\partial^{(h)}}{\partial\vec{\mathcal{K}}_y}\frac{\partial^{(g)}}{\partial\vec{\mathcal{K}}_x}\right]\{H_{\alpha\beta}(\vec{\mathcal{K}},\mathcal{E})(-iG^{<}_{\beta\alpha}(\vec{\mathcal{K}},\mathcal{E}))\}$$
$$= \left[\frac{\partial H_{\alpha\beta}(\vec{\mathcal{K}},\mathcal{E})}{\partial\vec{\mathcal{K}}_x}\frac{(-iG^{<}_{\beta\alpha}(\vec{\mathcal{K}},\mathcal{E}))}{\partial\vec{\mathcal{K}}_y} - \frac{\partial H_{\alpha\beta}(\vec{\mathcal{K}},\mathcal{E})}{\partial\vec{\mathcal{K}}_y}\frac{(-iG^{<}_{\beta\alpha}(\vec{\mathcal{K}},\mathcal{E}))}{\partial\vec{\mathcal{K}}_x}\right]. \tag{31.80}$$

The trick is to "pull back" (undo) the lattice Weyl transformation to touch base with Berry connection and Berry curvature. Take first the term of Eq. (31.80), where,

$$\frac{\partial H^{(a)}(\vec{\mathcal{K}},\mathcal{E})}{\partial k_x} = \hbar\frac{\partial H^{(a)}(\vec{\mathcal{K}},\mathcal{E})}{\partial\vec{\mathcal{K}}_x}. \text{ etc.} \tag{31.81}$$

From Eq. (30.20) this can be written as a lattice Weyl transform $\mathcal{W}$ of the form,

$$\begin{aligned}
&\frac{\partial H_{\alpha\beta}\left(\vec{\mathcal{K}},\mathcal{E}\right)}{\partial\vec{\mathcal{K}}_x}\\
&= \mathcal{W}\left\{\left(\frac{\partial}{\partial\vec{\mathcal{K}}^{\alpha}_x} + \frac{\partial}{\partial\vec{\mathcal{K}}^{\beta}_x}\right)\langle\alpha,\vec{\mathcal{K}},\mathcal{E}|\hat{H}|\beta,\vec{\mathcal{K}},\mathcal{E}\rangle\right\},\\
&= \mathcal{W}\left\{\left\langle\alpha,\frac{\partial}{\partial\vec{\mathcal{K}}_x}\vec{\mathcal{K}},\mathcal{E}\right|\hat{H}|\beta,\vec{\mathcal{K}},\mathcal{E}\rangle + \left\langle\alpha,\vec{\mathcal{K}},\mathcal{E}\right|\hat{H}\left|\beta,\frac{\partial}{\partial\vec{\mathcal{K}}_x}\vec{\mathcal{K}},\mathcal{E}\right\rangle\right\},\\
&= \mathcal{W}\left\{E_{\beta}(\vec{\mathcal{K}},\mathcal{E})\left\langle\alpha,\frac{\partial}{\partial\vec{\mathcal{K}}_x}\vec{\mathcal{K}},\mathcal{E}\right||\beta,\vec{\mathcal{K}},\mathcal{E}\rangle + E_{\alpha}(\vec{\mathcal{K}},\mathcal{E})\langle\alpha,\vec{\mathcal{K}},\mathcal{E}|\left|\beta,\frac{\partial}{\partial\vec{\mathcal{K}}_x}\vec{\mathcal{K}},\mathcal{E}\right\rangle\right\},\\
&= \mathcal{W}\left\{(E_{\beta}(\vec{\mathcal{K}},\mathcal{E}) - E_{\alpha}(\vec{\mathcal{K}},\mathcal{E}))\left\langle\alpha,\frac{\partial}{\partial\vec{\mathcal{K}}_x}\vec{\mathcal{K}},\mathcal{E}\right||\beta,\vec{\mathcal{K}},\mathcal{E}\rangle\right\},
\end{aligned} \tag{31.82}$$

where $\langle\alpha,\frac{\partial}{\partial\vec{\mathcal{K}}_x}\vec{\mathcal{K}},\mathcal{E}|$ symbolically denotes derivative with respect to $\vec{\mathcal{K}}_x$ of the state vector $\langle\alpha,\vec{\mathcal{K}},\mathcal{E}|$ labeled by the three quantum labels. Likewise for $|\beta,\frac{\partial}{\partial\vec{\mathcal{K}}_x}\vec{\mathcal{K}},\mathcal{E}\rangle$.

We also have

$$\frac{\partial(-iG^<_{\beta\alpha}(\vec{\mathcal{K}},\mathcal{E}))}{\partial\mathcal{K}_y} = \mathcal{W}\left\{\left(\frac{\partial}{\partial\vec{\mathcal{K}}^\beta_y}+\frac{\partial}{\partial\vec{\mathcal{K}}^\alpha_y}\right)\langle\beta,\vec{\mathcal{K}},\mathcal{E}|(\hat{\rho})|\alpha,\vec{\mathcal{K}},\mathcal{E}\rangle\right\}, \tag{31.83}$$

where $\hat{\rho}$ is the density matrix operator. From Eq. (30.10), we take the time dependence of $\langle\beta,\vec{\mathcal{K}},\mathcal{E}|(\hat{\rho})|\alpha,\vec{\mathcal{K}},\mathcal{E}\rangle$ to be given by $\langle\beta,\vec{\mathcal{K}},\mathcal{E}|\hat{\rho}(0)|\alpha,\vec{\mathcal{K}},\mathcal{E}\rangle e^{i\omega_{\alpha\beta}t}$.

We have[2]

$$\frac{\partial(-iG^<_{\beta\alpha}(\vec{\mathcal{K}},\mathcal{E}))}{\partial\mathcal{K}_y} = \mathcal{W}\left\{\begin{array}{l}\langle\beta,\frac{\partial}{\partial\vec{\mathcal{K}}_y}\vec{\mathcal{K}},\mathcal{E}|(\hat{\rho}_0)|\alpha,\vec{\mathcal{K}},\mathcal{E}\rangle\\ +\langle\beta,\vec{\mathcal{K}},\mathcal{E}|\hat{\rho}_0|\alpha,\frac{\partial}{\partial\vec{\mathcal{K}}_y}\vec{\mathcal{K}},\mathcal{E}\rangle\end{array}\right\}e^{i\omega_{\alpha\beta}t}. \tag{31.84}$$

The density matrix operator $\hat{\rho}_0$ is of the form,

$$\hat{\rho}_o = \sum_m \rho_m |m\rangle\langle m|, \tag{31.85}$$

$$\hat{\rho}_o|m\rangle = \rho_m|m\rangle = f(E_m)|m\rangle, \tag{31.86}$$

$$\langle m|\hat{\rho}_o|n\rangle = \rho_{mm} = f(E_n)\delta_{mn} \text{ or } f(E_m)\delta_{mn}, \tag{31.87}$$

where the weight function is the Fermi-Dirac function,

$$\rho_0^m = f(E_m). \tag{31.88}$$

Hence

$$\begin{aligned}\hat{\rho}_o|\alpha,\vec{\mathcal{K}},\mathcal{E}\rangle &= \sum_\gamma |\gamma,\vec{\mathcal{K}},\mathcal{E}\rangle\rho_0^\gamma\langle\gamma,\vec{\mathcal{K}},\mathcal{E}||\alpha,\vec{\mathcal{K}},\mathcal{E}\rangle,\\ &= |\alpha,\vec{\mathcal{K}},\mathcal{E}\rangle f(E_\alpha).\end{aligned} \tag{31.89}$$

Similarly,

$$\begin{aligned}\langle\beta,\vec{\mathcal{K}},\mathcal{E}|(\hat{\rho}_0) &= \langle\beta,\vec{\mathcal{K}},\mathcal{E}|\sum_\gamma|\gamma,\vec{\mathcal{K}},\mathcal{E}\rangle\rho_0\langle\gamma,\vec{\mathcal{K}},\mathcal{E}|,\\ &= f(E_\beta)\langle\beta,\vec{\mathcal{K}},\mathcal{E}|.\end{aligned} \tag{31.90}$$

Hence

$$\frac{\partial(-iG^<_{\beta\alpha}(\vec{\mathcal{K}},\mathcal{E}))}{\partial\mathcal{K}_y} = \mathcal{W}\left\{\begin{array}{l}\langle\beta,\frac{\partial}{\partial\vec{\mathcal{K}}_y}\vec{\mathcal{K}},\mathcal{E}||\alpha,\vec{\mathcal{K}},\mathcal{E}\rangle\rho_0^\alpha\\ \langle\beta,\vec{\mathcal{K}},\mathcal{E}||\alpha,\frac{\partial}{\partial\vec{\mathcal{K}}_y}\vec{\mathcal{K}},\mathcal{E}\rangle\rho_0^\beta\end{array}\right\}e^{i\omega_{\alpha\beta}t}. \tag{31.91}$$

---

[2]Here we use the definition of Green's function without the factor $\hbar$, following traditional treatments, i.e., $\rho(1,2) = -iG^<(1,2)$.

Shifting the first derivative to the right, we have

$$\frac{\partial(-iG^{<}_{\beta\alpha}(\vec{\mathcal{K}},\mathcal{E}))}{\partial\mathcal{K}_y}$$
$$= \mathcal{W}\left\{\begin{array}{c} -\langle\beta,\vec{\mathcal{K}},\mathcal{E}||\alpha,\frac{\partial}{\partial\vec{\mathcal{K}}_y}\vec{\mathcal{K}},\mathcal{E}\rangle f\left(E_\alpha\right) \\ \langle\beta,\vec{\mathcal{K}},\mathcal{E}||\alpha,\frac{\partial}{\partial\vec{\mathcal{K}}_y}\vec{\mathcal{K}},\mathcal{E}\rangle f\left(E_\beta\right)\end{array}\right\} e^{i\omega_{\alpha\beta}t},$$
$$= \mathcal{W}\left[\left\{\left(f\left(E_\beta\right)-f\left(E_\alpha\right)\right)\langle\beta,\vec{\mathcal{K}},\mathcal{E}|\left|\alpha,\frac{\partial}{\partial\vec{\mathcal{K}}_y}\vec{\mathcal{K}},\mathcal{E}\right\rangle\right\} e^{i\omega_{\alpha\beta}t}\right]. \tag{31.92}$$

Therefore

$$\frac{\partial H_{\alpha\beta}(\vec{\mathcal{K}},\mathcal{E})}{\partial\vec{\mathcal{K}}_x}\frac{(-iG^{<}_{\beta\alpha}(\vec{\mathcal{K}},\mathcal{E}))}{\partial\vec{\mathcal{K}}_y}$$
$$= \left\{\begin{array}{c} \mathcal{W}\left[\left(-(E_\beta(\vec{\mathcal{K}},\mathcal{E})-E_\alpha(\vec{\mathcal{K}},\mathcal{E}))\right)\left\{\left\langle\alpha,\frac{\partial}{\partial\vec{\mathcal{K}}''_x}\vec{\mathcal{K}},\mathcal{E}\right||\beta,\vec{\mathcal{K}},\mathcal{E}\rangle\right\}\right] \\ \times\mathcal{W}\left[\left\{\left(f\left(E_\beta\right)-f\left(E_\alpha\right)\right)\langle\beta,\vec{\mathcal{K}},\mathcal{E}|\left|\alpha,\frac{\partial}{\partial\vec{\mathcal{K}}_y}\vec{\mathcal{K}},\mathcal{E}\right\rangle\right\} e^{i\omega_{\alpha\beta}t}\right]\end{array}\right\}.$$
$$. \tag{31.93}$$

We have,

$$\frac{\partial H(\vec{\mathcal{K}},\mathcal{E})}{\partial\mathcal{K}_x}\frac{\partial(-iG^{<(b)}(\vec{\mathcal{K}},\mathcal{E}))}{\partial\mathcal{K}_y}$$
$$= \mathcal{W}\left[\left\{\begin{array}{c} \sum_{\alpha,\beta}-(E_\beta(\vec{\mathcal{K}},\mathcal{E})-E_\alpha(\vec{\mathcal{K}},\mathcal{E})) \\ \times\left\{\left\langle\alpha,\frac{\partial}{\partial\mathcal{K}_x}\vec{\mathcal{K}},\mathcal{E}\right||\beta,\vec{\mathcal{K}},\mathcal{E}\rangle\right\}\left\{\langle\beta,\vec{\mathcal{K}},\mathcal{E}|\left|\alpha,\frac{\partial}{\partial\mathcal{K}_y}\vec{\mathcal{K}},\mathcal{E}\right\rangle\right\} \\ \times e^{i\omega_{\alpha\beta}t}\ \left(f\left(E_\beta\right)-f\left(E_\alpha\right)\right)\end{array}\right\}\right]. \tag{31.94}$$

Similarly, we have

$$\frac{\partial H_{\alpha\beta}(\vec{\mathcal{K}},\mathcal{E})}{\partial\vec{\mathcal{K}}_y}\frac{(-iG^{<}_{\beta\alpha}(\vec{\mathcal{K}},\mathcal{E}))}{\partial\vec{\mathcal{K}}_x}$$
$$= \mathcal{W}\left[\left\{\begin{array}{c} \sum_{\alpha,\beta}-(E_\beta(\vec{\mathcal{K}},\mathcal{E})-E_\alpha(\vec{\mathcal{K}},\mathcal{E})) \\ \times\left\{\left\langle\alpha,\frac{\partial}{\partial\mathcal{K}_y}\vec{\mathcal{K}},\mathcal{E}\right||\beta,\vec{\mathcal{K}},\mathcal{E}\rangle\right\}\left\{\langle\beta,\vec{\mathcal{K}},\mathcal{E}|\left|\alpha,\frac{\partial}{\partial\mathcal{K}_x}\vec{\mathcal{K}},\mathcal{E}\right\rangle\right\} \\ \times e^{i\omega_{\alpha\beta}t}\ \left(f\left(E_\beta\right)-f\left(E_\alpha\right)\right)\end{array}\right\}\right]. \tag{31.95}$$

Therefore we obtain

$$\left[\frac{\partial H_{\alpha\beta}(\vec{\mathcal{K}},\mathcal{E})}{\partial\vec{\mathcal{K}}_x}\frac{(-iG^{<}_{\beta\alpha}(\vec{\mathcal{K}},\mathcal{E}))}{\partial\vec{\mathcal{K}}_y}-\frac{\partial H_{\alpha\beta}(\vec{\mathcal{K}},\mathcal{E})}{\partial\vec{\mathcal{K}}_y}\frac{(-iG^{<}_{\beta\alpha}(\vec{\mathcal{K}},\mathcal{E}))}{\partial\vec{\mathcal{K}}_x}\right]$$

$$= \mathcal{W}\left[\left\{\begin{array}{c}\sum\limits_{\alpha,\beta} -\left(E_\beta\left(\vec{\mathcal{K}},\mathcal{E}\right)-E_\alpha\left(\vec{\mathcal{K}},\mathcal{E}\right)\right)\\ \times\left[\begin{array}{c}\left\{\left\langle\alpha,\frac{\partial}{\partial\mathcal{K}_x}\vec{\mathcal{K}},\mathcal{E}\right|\left|\beta,\vec{\mathcal{K}},\mathcal{E}\right\rangle\right\}\\ \times\left\{\left\langle\beta,\vec{\mathcal{K}},\mathcal{E}\right|\left|\alpha,\frac{\partial}{\partial\mathcal{K}_y}\vec{\mathcal{K}},\mathcal{E}\right\rangle\right\}\\ -\left\{\left\langle\alpha,\frac{\partial}{\partial\mathcal{K}_y}\vec{\mathcal{K}},\mathcal{E}\right|\left|\beta,\vec{\mathcal{K}},\mathcal{E}\right\rangle\right\}\\ \times\left\{\left\langle\beta,\vec{\mathcal{K}},\mathcal{E}\right|\left|\alpha,\frac{\partial}{\partial\mathcal{K}_x}\vec{\mathcal{K}},\mathcal{E}\right\rangle\right\}\end{array}\right]\end{array}\right\}\\ \times e^{i\omega_{\alpha\beta}t}\left(f\left(E_\beta\right)-f\left(E_\alpha\right)\right)\right]. \tag{31.96}$$

Therefore from Eq. (31.79), namely,

$$(\sigma_{yx})_{\alpha\alpha} = -\left(\frac{e^2}{h}\right)\frac{1}{(2\pi\hbar)}\int\int\int d\vec{\mathcal{K}}_x d\vec{\mathcal{K}}_y dt$$
$$\times\left[\frac{\partial^{(a)}}{\partial\vec{\mathcal{K}}_x}\frac{\partial^{(b)}}{\partial\vec{\mathcal{K}}_y}-\frac{\partial^{(a)}}{\partial\vec{\mathcal{K}}_y}\frac{\partial^{(b)}}{\partial\vec{\mathcal{K}}_x}\right]\{H_{\alpha\beta}(\vec{\mathcal{K}},\mathcal{E})(-iG^{<}_{\beta\alpha}(\vec{\mathcal{K}},\mathcal{E}))\}, \tag{31.97}$$

we have

$$(\sigma_{yx})_{\alpha\alpha}$$
$$=\left\{\begin{array}{c}-\frac{e^2}{h}\frac{1}{(2\pi\hbar)}\int\int\int d\vec{\mathcal{K}}_x d\vec{\mathcal{K}}_y dt\\ \times\sum\limits_{\alpha,\beta}\left[\begin{array}{c}(E_\beta(\vec{\mathcal{K}},\mathcal{E})-E_\alpha(\vec{\mathcal{K}},\mathcal{E}))\\ \times\left\{\begin{array}{c}\langle\alpha,\frac{\partial}{\partial\mathcal{K}_x}\vec{\mathcal{K}},\mathcal{E}||\beta,\vec{\mathcal{K}},\mathcal{E}\rangle\langle\beta,\vec{\mathcal{K}},\mathcal{E}||\alpha,\frac{\partial}{\partial\mathcal{K}_y}\vec{\mathcal{K}},\mathcal{E}\rangle\\ -\langle\alpha,\frac{\partial}{\partial\mathcal{K}_y}\vec{\mathcal{K}},\mathcal{E}||\beta,\vec{\mathcal{K}},\mathcal{E}\rangle\langle\beta,\vec{\mathcal{K}},\mathcal{E}||\alpha,\frac{\partial}{\partial\mathcal{K}_x}\vec{\mathcal{K}},\mathcal{E}\rangle\end{array}\right\}\\ \times e^{i\omega_{\alpha\beta}t}\left(f\left(E_\beta\right)-f\left(E_\alpha\right)\right)\end{array}\right]\end{array}\right\}. \tag{31.98}$$

The time integral of the RHS amounts to taking zero-order time dependence [zero electric field] of the rest of the integrand, then we have for the remaining time-dependence, explicitly integrated as

$$\int_{-\infty}^{0} dt\exp i\omega_{\alpha\beta}t = \left.\frac{\exp\exp i\omega_{\alpha\beta}t}{i\omega_{\alpha\beta}}\right|_{\tau=-\infty}^{\tau=0},$$
$$= \left.\frac{\exp\left(i\left(\omega_{\alpha\beta}-i\eta\right)\tau\right)}{i\omega_{\alpha\beta}}\right|_{\tau=-\infty}^{\tau=0} = \frac{1}{i\omega_{\alpha\beta}}. \tag{31.99}$$

Thus eliminating the time integral we finally obtain,

$$(\sigma_{yx})_{\alpha\alpha} = i\frac{e^2}{h}\frac{1}{(2\pi\hbar)}\int\int d\vec{\mathcal{K}}_x d\vec{\mathcal{K}}_y \times\sum_{\alpha,\beta}\left[\begin{array}{c}(f(E_\beta)-f(E_\alpha))\left(-\frac{\hbar\omega_{\alpha\beta}}{\omega_{\alpha\beta}}\right)\\ \times\left\{\begin{array}{c}\langle\alpha,\frac{\partial}{\partial\vec{\mathcal{K}}_x}\vec{\mathcal{K}},\mathcal{E}||\beta,\vec{\mathcal{K}},\mathcal{E}\rangle\langle\beta,\vec{\mathcal{K}},\mathcal{E}||\alpha,\frac{\partial}{\partial\vec{\mathcal{K}}_y}\vec{\mathcal{K}},\mathcal{E}\rangle\\ -\langle\alpha,\frac{\partial}{\partial\vec{\mathcal{K}}_y}\vec{\mathcal{K}},\mathcal{E}||\beta,\vec{\mathcal{K}},\mathcal{E}\rangle\langle\beta,\vec{\mathcal{K}},\mathcal{E}||\alpha,\frac{\partial}{\partial\vec{\mathcal{K}}_x}\vec{\mathcal{K}},\mathcal{E}\rangle\end{array}\right\},\end{array}\right] \quad (31.100)$$

$$(\sigma_{yx})_{\alpha\alpha} = i\frac{e^2}{h}\frac{1}{(2\pi)}\int\int dk_x dk_y\sum_{\alpha,\beta}(f(E_\alpha)-f(E_\beta)) \times\left\{\begin{array}{c}\langle\alpha,\frac{\partial}{\partial k_x}\vec{\mathcal{K}},\mathcal{E}||\beta,\vec{\mathcal{K}},\mathcal{E}\rangle\langle\beta,\vec{\mathcal{K}},\mathcal{E}||\alpha,\frac{\partial}{\partial k_y}\vec{\mathcal{K}},\mathcal{E}\rangle\\ -\langle\alpha,\frac{\partial}{\partial k_y}\vec{\mathcal{K}},\mathcal{E}||\beta,\vec{\mathcal{K}},\mathcal{E}\rangle\langle\beta,\vec{\mathcal{K}},\mathcal{E}||\alpha,\frac{\partial}{\partial k_x}\vec{\mathcal{K}},\mathcal{E}\rangle\end{array}\right\} \quad (31.101)$$

Taking the Fourier transform of both sides, we obtain

$$\langle\hat{j}_y(\omega)\rangle = i\frac{e^2}{h}\left|\vec{F}\right|\frac{\delta(\omega)}{(2\pi)}\int\int dk_x dk_y\sum_{\alpha,\beta}(f(E_\alpha)-f(E_\beta)) \times\left\{\begin{array}{c}\langle\alpha,\frac{\partial}{\partial k_x}\vec{\mathcal{K}},\mathcal{E}||\beta,\vec{\mathcal{K}},\mathcal{E}\rangle\langle\beta,\vec{\mathcal{K}},\mathcal{E}||\alpha,\frac{\partial}{\partial k_y}\vec{\mathcal{K}},\mathcal{E}\rangle\\ -\langle\alpha,\frac{\partial}{\partial k_y}\vec{\mathcal{K}},\mathcal{E}||\beta,\vec{\mathcal{K}},\mathcal{E}\rangle\langle\beta,\vec{\mathcal{K}},\mathcal{E}||\alpha,\frac{\partial}{\partial k_x}\vec{\mathcal{K}},\mathcal{E}\rangle\end{array}\right\}, \quad (31.102)$$

$$\langle\hat{j}_y(\omega)\rangle = i\frac{e^2}{h}\left|\vec{F}\right|\frac{\delta(\omega)}{(2\pi)}\int\int dk_x dk_y \times\sum_{\alpha}\left\{\begin{array}{c}-\langle\alpha,\frac{\partial}{\partial k_x}\vec{\mathcal{K}},\mathcal{E}|\sum_\beta|\beta,\vec{\mathcal{K}},\mathcal{E}\rangle f(E_\beta)\langle\beta,\vec{\mathcal{K}},\mathcal{E}||\alpha,\frac{\partial}{\partial k_y}\vec{\mathcal{K}},\mathcal{E}\rangle\\ +\langle\alpha,\frac{\partial}{\partial k_y}\vec{\mathcal{K}},\mathcal{E}|\sum_\beta|\beta,\vec{\mathcal{K}},\mathcal{E}\rangle f(E_\beta)\langle\beta,\vec{\mathcal{K}},\mathcal{E}||\alpha,\frac{\partial}{\partial k_x}\vec{\mathcal{K}},\mathcal{E}\rangle\end{array}\right\} + \left\{\begin{array}{c}\langle\alpha,\frac{\partial}{\partial k_x}\vec{\mathcal{K}},\mathcal{E}|\sum_\beta|\beta,\vec{\mathcal{K}},\mathcal{E}\rangle\langle\beta,\vec{\mathcal{K}},\mathcal{E}||\alpha,\frac{\partial}{\partial k_y}\vec{\mathcal{K}},\mathcal{E}\rangle\\ -\langle\alpha,\frac{\partial}{\partial k_y}\vec{\mathcal{K}},\mathcal{E}|\sum_\beta|\beta,\vec{\mathcal{K}},\mathcal{E}\rangle\langle\beta,\vec{\mathcal{K}},\mathcal{E}||\alpha,\frac{\partial}{\partial k_x}\vec{\mathcal{K}},\mathcal{E}\rangle\end{array}\right\} f(E_\alpha). \quad (31.103)$$

We can show that the first curly bracketed term in Eq. (31.103) contributes to zero.

Proof. Assuming Hermitian matrix for $\hat{\rho}$, consider

$$-\left\{\left\langle \alpha, \frac{\partial}{\partial k_x}\vec{\mathcal{K}}, \mathcal{E}\right| |\beta, \vec{\mathcal{K}}, \mathcal{E}\rangle f\left(E_\beta\right) \langle \beta, \vec{\mathcal{K}}, \mathcal{E}| \left|\alpha, \frac{\partial}{\partial k_y}\vec{\mathcal{K}}, \mathcal{E}\right\rangle\right\}^\dagger$$

$$= -\left\langle \alpha, \frac{\partial}{\partial k_y}\vec{\mathcal{K}}, \mathcal{E}\right| |\beta, \vec{\mathcal{K}}, \mathcal{E}\rangle f\left(E_\beta\right) \langle \beta, \vec{\mathcal{K}}, \mathcal{E}| \left|\alpha, \frac{\partial}{\partial k_x}\vec{\mathcal{K}}, \mathcal{E}\right\rangle. \qquad (31.104)$$

Therefore,

$$\sum_{\alpha,\beta}\left\{\begin{array}{l} -\langle \alpha, \frac{\partial}{\partial k_x}\vec{\mathcal{K}}, \mathcal{E}||\beta, \vec{\mathcal{K}}, \mathcal{E}\rangle f\left(E_\beta\right) \langle \beta, \vec{\mathcal{K}}, \mathcal{E}||\alpha, \frac{\partial}{\partial k_y}\vec{\mathcal{K}}, \mathcal{E}\rangle \\ +\langle \alpha, \frac{\partial}{\partial k_y}\vec{\mathcal{K}}, \mathcal{E}||\beta, \vec{\mathcal{K}}, \mathcal{E}\rangle f\left(E_\beta\right) \langle \beta, \vec{\mathcal{K}}, \mathcal{E}||\alpha, \frac{\partial}{\partial k_x}\vec{\mathcal{K}}, \mathcal{E}\rangle \end{array}\right\}$$

$$= 2i\,\mathrm{Im}\sum_{\alpha,\beta}\left\langle \alpha, \frac{\partial}{\partial k_y}\vec{\mathcal{K}}, \mathcal{E}\right| |\beta, \vec{\mathcal{K}}, \mathcal{E}\rangle f\left(E_\beta\right) \langle \beta, \vec{\mathcal{K}}, \mathcal{E}| \left|\alpha, \frac{\partial}{\partial k_x}\vec{\mathcal{K}}, \mathcal{E}\right\rangle. \qquad (31.105)$$

Assuming $\langle \alpha, \frac{\partial}{\partial k_y}\vec{\mathcal{K}}, \mathcal{E}|$ and $|\alpha, \frac{\partial}{\partial k_x}\vec{\mathcal{K}}, \mathcal{E}\rangle$ are expanded in terms of the orthogonal basis states, then since $\hat{\rho}$ is a projector, we must have,

$$\mathrm{Im}\sum_{\alpha,\beta}\left\langle \alpha, \frac{\partial}{\partial k_y}\vec{\mathcal{K}}, \mathcal{E}\right| |\beta, \vec{\mathcal{K}}, \mathcal{E}\rangle f\left(E_\beta\right) \langle \beta, \vec{\mathcal{K}}, \mathcal{E}| \left|\alpha, \frac{\partial}{\partial k_x}\vec{\mathcal{K}}, \mathcal{E}\right\rangle = 0, \qquad (31.106)$$

being a diagonal Hermitian density matrix. □

Taking the limit $\omega \Longrightarrow 0$ and summing over the states β, we readily obtain the total conductivity, σ_{yx}, sum over the α band index,

$$\sigma_{yx} = \frac{e^2}{h}\sum_\alpha f\left(E_\alpha\right)\frac{i}{(2\pi)}\int\int dk_x dk_y \left[\begin{array}{l} \left\langle \alpha, \frac{\partial}{\partial k_x}\vec{\mathcal{K}}, \mathcal{E}\right| \left|\alpha, \frac{\partial}{\partial k_y}\vec{\mathcal{K}}, \mathcal{E}\right\rangle \\ -\langle \alpha, \frac{\partial}{\partial k_y}\vec{\mathcal{K}}, \mathcal{E}||\alpha, \frac{\partial}{\partial k_x}\vec{\mathcal{K}}, \mathcal{E}\rangle \end{array}\right], \qquad (31.107)$$

where,

$$\vec{\mathcal{K}} = \vec{p} + e\vec{F}t \Longrightarrow \vec{p} = \hbar\vec{k}, \qquad (31.108)$$

in both Eqs. (31.102) and (31.107). Note the transformation from $\vec{\mathcal{K}}$ to $\vec{p}$ integration has a Jacobian unity. Equation (31.107) is the same expression that can obtained to derive the integer quantum Hall effect from Kubo formula [69].

We now prove that for each statevector, $\left|\alpha, \vec{k}\right\rangle$, the expression,

$$\frac{1}{(2\pi)}\int\int dk_x dk_y\; f(E_\alpha(\vec{k}))i\left[\left\langle \frac{\partial}{\partial k_x}\alpha, \vec{k}\right| \frac{\partial}{\partial k_y}|\alpha, \vec{k}\rangle - \left\langle \frac{\partial}{\partial k_y}\alpha, \vec{k}\right| \frac{\partial}{\partial k_x}\left|\alpha, \vec{k}\right\rangle\right], \qquad (31.109)$$

is the winding number around the occupied contour in the Brillouin zone. First we can rewrite the terms within the square bracket multiplied by i as

$$i\left[\left\langle\frac{\partial}{\partial k_x}\alpha,\vec{k}\right|\frac{\partial}{\partial k_y}\left|\alpha,\vec{k}\right\rangle-\left\langle\frac{\partial}{\partial k_y}\alpha,\vec{k}\right|\frac{\partial}{\partial k_x}\left|\alpha,\vec{k}\right\rangle\right]$$
$$=i\left\langle\frac{\partial}{\partial\vec{k}}\alpha,\vec{k}\right|\times\frac{\partial}{\partial\vec{k}}\left|\alpha,\vec{k}\right\rangle=\nabla_{\vec{k}}\times\left\langle\alpha,\vec{k}\right|i\frac{\partial}{\partial\vec{k}}\left|\alpha,\vec{k}\right\rangle. \quad (31.110)$$

The last term indicates the operation of the curl of the Berry connection which is related to the quantization of Hall conductivity. This quantization is due to the uniqueness of the parallel-transported wavefunction, which may also have bearing on the self-consistent Bohr–Sommerfeld quantization [13]. To understand this, first we discuss how the phase of the wavefunction relates to the Berry connection and Berry curvature. This is given in Appendix G.1.

At low temperature, we can just write Eq. (31.107) as,

$$\sigma_{yx}=\frac{e^2}{2\pi\hbar}\frac{1}{(2\pi)}\sum_{\alpha}\int\int_{\text{occupiedBZ}}dk_xdk_y\left[\nabla_{\vec{k}}\times\left\langle\alpha,\vec{k}\right|i\frac{\partial}{\partial\vec{k}}\left|\alpha,\vec{k}\right\rangle\right]_{\text{plane}},$$
$$=\frac{e^2}{2\pi\hbar}\frac{1}{(2\pi)}\sum_{\alpha}\oint dk_c\left[\left\langle\alpha,\vec{k}\right|i\frac{\partial}{\partial k_c}\left|\alpha,\vec{k}\right\rangle\right]_{\text{contour}}. \quad (31.111)$$

Now, from Sec. G.1 of the Appendix [for more details see Chapter 13 of this book [12]],

$$i\Delta\phi_{\text{total}}=-\oint dk_c\left[\left\langle\alpha,\vec{k}\right|\frac{\partial}{\partial k_c}\left|\alpha,\vec{k}\right\rangle\right]_{\text{contour}},$$
$$\Delta\phi_{\text{total}}=\oint dk_c\left[\left\langle\alpha,\vec{k}\right|i\frac{\partial}{\partial k_c}\left|\alpha,\vec{k}\right\rangle\right]_{\text{contour}},$$
$$\frac{\Delta\phi_{\text{total}}}{2\pi}=\frac{1}{2\pi}\oint dk_c\left[\left\langle\alpha,\vec{k}\right|i\frac{\partial}{\partial k_c}\left|\alpha,\vec{k}\right\rangle\right]_{\text{contour}}=n_\alpha, \quad (31.112)$$

where n_α is the winding number or the Chern number. Therefore

$$(\sigma_{yx})_{\alpha\alpha}=\frac{e^2}{2\pi\hbar}\frac{1}{(2\pi)}\sum_{\alpha}\int\int_{\text{occupiedBZ}}dk_xdk_y\left[\nabla_{\vec{k}}\times\left\langle\alpha,\vec{k}\right|i\frac{\partial}{\partial\vec{k}}\left|\alpha,\vec{k}\right\rangle\right]_{\text{plane}},$$
$$=\frac{e^2}{2\pi\hbar}\frac{1}{(2\pi)}\sum_{\alpha}\oint dk_c\left[\left\langle\alpha,\vec{k}\right|i\frac{\partial}{\partial k_c}\left|\alpha,\vec{k}\right\rangle\right]_{\text{contour or Wilson loop over BZ}},$$
$$=\left(\frac{e^2}{h}\right)\sum_{\alpha}\frac{\Delta\phi_{\text{total}}}{2\pi}=\sum_{\alpha}\left(\frac{e^2}{h}\right)n_\alpha, \quad (31.113)$$

over all occupied bands α, where $n_\alpha \in \mathbb{Z}$ is the topological first Chern (or *winding*) number for each band. Thus the the Hall conductivity is quantized in units of $\frac{e^2}{h}$ as derive from Eq. (30.40) of the new method used here.

To touch base with a time-dependent perturbation of the Kubo current-current correlation(KCCC) used by TKNN, we recall that in this particular approach, a time varying electric field is indirectly used. We give a derivation of the KCCC formula from our new quantum transport formalism in Section 5.3 of Part1.

31.7.2 *Recap on IQHE*

By reverting to its equivalent matrix element expression and integrating with respect to time, we obtain

$$\sigma_{yx} = \frac{e^2}{h}\sum_\alpha f\left(E_\alpha\right)\frac{i}{(2\pi)}\int\int dk_x dk_y \begin{bmatrix} \langle\alpha, \frac{\partial}{\partial k_x}\vec{\mathcal{K}}, \mathcal{E}||\alpha, \frac{\partial}{\partial k_y}\vec{\mathcal{K}}, \mathcal{E}\rangle \\ -\langle\alpha, \frac{\partial}{\partial k_y}\vec{\mathcal{K}}, \mathcal{E}||\alpha, \frac{\partial}{\partial k_x}\vec{\mathcal{K}}, \mathcal{E}\rangle \end{bmatrix}. \tag{31.114}$$

At low temperature, $f\left(E_\alpha\right) \simeq 1$, so we can just write Eq. (31.114) as,

$$\begin{aligned}\sigma_{yx} &= \frac{e^2}{2\pi\hbar}\frac{1}{(2\pi)}\sum_\alpha\int\int_{\text{occupiedBZ}} dk_x dk_y \left[\nabla_{\vec{k}} \times \left\langle \alpha, \vec{k}\right| i\frac{\partial}{\partial \vec{k}}\left|\alpha, \vec{k}\right\rangle\right]_{\text{plane}}, \\ &= \frac{e^2}{2\pi\hbar}\frac{1}{(2\pi)}\sum_\alpha\oint dk_c \left[\left\langle \alpha, \vec{k}\right| i\frac{\partial}{\partial k_c}\left|\alpha, \vec{k}\right\rangle\right]_{\text{contour}}. \end{aligned} \tag{31.115}$$

where $\langle\alpha, \vec{k}|i\frac{\partial}{\partial\vec{k}}|\alpha, \vec{k}\rangle$ is the Berry connection in a band and $\oint dk_c\left[\langle\alpha, \vec{k}|i\frac{\partial}{\partial k_c}|\alpha, \vec{k}\rangle\right]$ is the Berry phase. Thus, Eq. (31.114) yields

$$\sigma_{yx} = \sum_\alpha \frac{e^2}{h} n_\alpha. \tag{31.116}$$

Chapter 32

Quantum Superfield Transport Theory of FQHE

32.1 Introduction

In previous chapters, we make use of the gapped energy-band structure of solids under external electric field to derive the integer quantum Hall effect (IQHE) of Chern insulator. We employ the real-time superfield and lattice Weyl transform non-equilibrium Green's function (SFLWT-NEGF) quantum transport formalism of Part 3 [3, 11] to the first-order gradient expansion to derive the topological Chern number of the IQHE for 2D systems, as an integral multiple of quantum conductance, also known as the minimal contact conductance in mesoscopic physics [12].

We find that the quantization of Hall effect occurs strictly not to first order in the electric field *per se* but rather to first-order gradient expansion in the non-equilibrium quantum transport equation. The Berry connection and Berry curvature is the fundamental physics [13] behind the exact quantization of Hall conductance in units of $\frac{e^2}{h}$, which also happens to coincide with the source and drain *contact* conductance per spin in a closed circuit of mesoscopic quantum transport [12].

In Chapter 31, we have shown that the $(p.q; E.t)$ phase-space is renormalized to that of $(\vec{\mathcal{K}}, \mathcal{E})$ phase-space, where in the absence of magnetic fields,

$$\vec{\mathcal{K}} = \vec{p} + e\vec{F}t, \tag{32.1}$$

and

$$\mathcal{E} = E_0 + e\vec{F} \cdot \vec{q}. \tag{32.2}$$

There, we treat the uniform electric field, $\vec{F}$, in the x-direction, and the Hall current is in the y-direction. In the presence of magnetic field, we then have

$$\vec{\mathcal{K}} = \vec{p} + e\vec{F}t + \frac{e}{c}\vec{A}_m, \tag{32.3}$$

where $\vec{A}_m$ is the Maxwell's vector potential.

We have identified the topological invariant in $(\vec{\mathcal{K}}, \mathcal{E})$ phase-space non-equilibrium quantum transport equation leading to IQHE. Moreover, the formula is also applicable to gapped Landau-level structure of a free electron gas in intense magnetic field [73] since the variable $\vec{\mathcal{K}}$ can incorporates the Maxwell's vector potential, $\frac{e}{c}\vec{A}_m$, and its corresponding parallel transport, if present. Note that the change of variables from $\vec{\mathcal{K}}$ to $\vec{p}$ in the integration over the whole Brillouin zone has a Jacobian unity.

It was shown in Chapter 31 that the correct expression for the IQHE conductivity given by

$$\sigma_{yx} = \left(\frac{e^2}{h}\right)\sum_\alpha \frac{\Delta\phi_{\text{total}}}{2\pi} = \sum_\alpha \left(\frac{e^2}{h}\right) n_\alpha. \tag{32.4}$$

We now turn our attention to a macroscopic treatment of fractional quantum Hall effect (FQHE), which may have bearings particularly on the widely separated and occupied lowest sub-band of the Hofstadter butterfly energy spectrum, to be discussed in the last chapter. Remarkably, in FQHE the conductance become a fraction of quantum conductance.

32.1.1 *Fractional quantum Hall effect*

Here, we give a simple explanation for FQHE conductance given by

$$\sigma_{yx} = \left(\frac{e^2}{h}\right) v, \tag{32.5}$$

where v is a fraction with co-primes numerator/denominator in a hierarchy of *scaling* factors. We are essentially referring to the scaling of the lowest *occupied* Landau level in intense magnetic fields, where there is a lowest and highest Landau levels, widely separated from the other sub-bands, in the Hofstadter butterfly magnetic spectrum. At some values of the very intense magnetic fields, these are the only two remaining sub-bands within a zero field energy band. We caution that the number 2 is a prime number and should not be considered an even number. Moreover, prime integers *exclude* many odd integers. All other non-prime integers can be decomposed into products of prime numbers, this we refer to as *higher-order* scaling, i.e., higher-order power of primes. Thus, if one looks for a fundamental hierarchy of scaling patterns, only *fundamental* prime number scaling can represent a primary hierarchy based on *fixed* scaling reference level, namely, the *occupied* lowest Landau level (LLL) at very high magnetic fields.

Although even denominators for v are speculated in the literature, these do not belong to a fundamental hierarchy of scaling in our present analysis for reducibility or factorization reason into power of primes, and therefore represent *higher-order* scaling or successive scaling by co-prime numbers which losses the original fixed scaling reference. Moreover, although factorizable scaling factors can appear in the measurements, these do not belong to a fundamental hierarchy and we believe that these *higher-order* type of scaling should appear weak or less resolved in the measurements. This is discussed in more detail in Section 32.3.1.

32.2 Effects of magnetic fields

For free electrons, the presence of increasing magnetic field has two effects; firstly it creates the sharp Landau energy levels in magnetic fields, $E_n = \hbar\omega_c(n + 1/2)$, $n = 1, 2, 3, \ldots$, where ω_c equals the cyclotron frequency, $\frac{eB}{m}$, and their separations, and secondly, it increases the degeneracy of Landau levels and corresponding density of states. For Bloch energy bands, increasing magnetic field will result in decreasing the number of sub-bands and [242, 243] the flattening of the magnetic sub-bands of the remaining Landau levels.

We can show that the idea of flux attachment is inherent in fully occupied as well as in partially occupied Landau levels, which results in scaling factor closely related to the scaling k-factor in Chern–Simons gauge theory. This is shown to be crucial in giving a simple explanation of FQHE, using non-equilibrium quantum transport in the lattice Weyl transform formalism [11, 12] used in previous chapters [244].

32.2.1 *Magnetic sub-bands*

32.2.1.1 *Low finite fields*

The sharp highly degenerate Landau levels for free electrons in a magnetic field is broadened into magnetic energy sub-bands by the presence of periodic potential at the atomic lattice sites. At values of the magnetic fields where the effect of the periodic atomic lattice sites dominates, the effect of the magnetic fields can be described simply in terms of the dynamics of the Bloch energy bands [24, 239]. For example, for finite magnetic fields, this is manifested in de Haas–van Alphen effects [245, 246] due to Landau orbits at the Fermi surface, and magnetic breakdown between orbits in complicated Fermi surfaces in metals [247, 248]. For smaller magnetic fields, the Bloch-band dynamics is manifested in such phenomena as paramagnetic and diamagnetic susceptibility. Indeed, interband coupling in narrow-gap alloyed semimetals and semiconductors is responsible for the giant diamagnetism of graphite, bismuth and Bi-Sb alloys [8, 9].

32.2.1.2 *Large magnetic fields*

However, when the effect of large magnetic fields dominates over the effect of periodic atomic lattice sites, then the dynamics is now dictated by the magnetic sub-bands. New length scales and time scales now competes the dynamics. These are the following, (ν is an integer or a fraction at very intense field).

Cyclotron frequency	$\omega_c = \frac{eB}{m^*c}$
Magnetic length	$l_B = \sqrt{\frac{\hbar c}{eB}}$
Quantum flux	$\phi_o = \frac{2\pi\hbar c}{e}$
Hall conductivity	$\sigma_{yx} = \frac{e^2}{2\pi\hbar}\nu$
Lattice constant	a

32.2.2 *Commutation relations*

The major quantum mechanical operator in phase space is now played by

$$\vec{K} = \vec{p} + \frac{e}{c}\vec{A}_m, \tag{32.6}$$

which obeys the commutation relation

$$[K_x, K_y] = \frac{e\hbar}{ic}B_z. \tag{32.7}$$

This is more transparent if written as

$$[K_x, K_y] = \frac{e\hbar}{ic}B_z = \frac{e\hbar^2}{ic\hbar}B_z = -i\frac{\hbar^2}{l_B^2}, \tag{32.8}$$

indicating that l_B characterizes the wavelengths in a magnetic sub-bands. For most magnetic field strengths, $l_B > a$ where a is the atomic sites lattice constant. This clearly supports the idea of breakup into magnetic sub-bands. However, for intense magnetic fields, when $l_B = a$ we recover the original whole energy bandwidth.

32.2.2.1 *The von Neumann lattice in phase space*

Equation (32.7) suggest the following relation:

$$\Delta K_x \Delta K_y \succeq \frac{1}{2}\frac{\hbar^2}{l_B^2} = \frac{1}{2}\frac{e\hbar}{c}B_z, \tag{32.9}$$

as a consequence of Heisenberg's uncertainty principle. Thus for the minimal value, of $\Delta K_x \Delta K_y = \frac{1}{2}\frac{\hbar^2}{l_B^2}$, a von Neumann lattice in 2D phase space of (K_x, K_y) has a lattice constant of the order l_B. For our purpose in what follows, it is important to note that $\Delta K_x \Delta K_y \simeq \delta K_x \delta K_y$ is proportional to the perpendicular magnetic field, B_z. Therefore, we may also write a Poisson differential operator of the form inversely proportional to the magnetic field,

$$\frac{\partial^{(a)}}{\partial K_x}\frac{\partial^{(b)}}{\partial K_y} \sim \frac{1}{B_z}. \tag{32.10}$$

This equation means that if we increase the magnetic field by a factor k, then from Eq. (32.10), we can scale the Poisson partial derivatives by $\frac{1}{k}$, resulting in $\frac{1}{k}\left(\frac{\partial^{(a)}}{\partial K_x}\frac{\partial^{(b)}}{\partial K_y}\right) = \left(\frac{\partial^{(a)}}{\partial K_x'}\frac{\partial^{(b)}}{\partial K_y'}\right)$ where the primes embodied the updated magnetic field. This observation is crucial in our macroscopic *scaling* analysis of FQHE.

32.3 Landau level degeneracy

Classically, the Landau-level (LL) degeneracy may be approximated by

$$\frac{A}{\pi r^2} = \frac{R^2}{r^2}, \tag{32.11}$$

where r is the classical radius of tiny Landau orbits in a uniform magnetic fields and the system area is given by $A = \pi R^2$. Equation (32.11) has the dimensional units of total flux divided by the dimensional units of "quantum" flux, $\frac{(\pi R^2)B}{\pi\left(\frac{EL}{e}\right)} \Rightarrow\sim \frac{\Phi}{\phi_o}$, where E stands for units of energy. In this classical analysis, it is thus *implied* that the magnetic flux is equally divided or attached with each tiny cyclotron orbits of the same energy for a fully occupied occupied Landau level.

Quantum mechanically, the more accurate expression for the Landau-level degeneracy, N_{LL} is $N_{\mathrm{LL}} = \frac{\Phi}{\left(\frac{2\pi\hbar c}{e}\right)} = \frac{\Phi}{\phi_o}$. This can be inferred simply by a Bohr-Sommerfeld quantization condition [13] which amounts to counting of Planck states ("pixels" of action) in phase space using Berry's curvature and connection, i.e., magnetic field and vector potential, respectively. In Gaussian units, we have,

$$\begin{aligned}
N_{\rm LL} &= \frac{1}{2\pi\hbar}\iint \vec{\nabla}\times\vec{\mathcal{K}}\cdot d\vec{a},\\
&= \frac{1}{2\pi\hbar}\iint \vec{\nabla}\times\frac{e}{c}\vec{A}\cdot d\vec{a},\\
&= \frac{1}{2\pi\hbar}\frac{|e|}{c}\iint \vec{B}\cdot d\vec{a},\\
&= \frac{\Phi}{\phi_o}\ \epsilon\ \mathbb{Z}, \quad \text{(the same for all Landau levels)}\\
&= \frac{1}{2\pi\hbar}\oint \frac{e}{c}\vec{A}\cdot d\vec{q} = N_{\rm LL}\ \epsilon\ \ \mathbb{Z},
\end{aligned} \tag{32.12}$$

where

$$\vec{K} = \vec{p} + \frac{e}{c}(\vec{A}_m + \vec{F}ct), \tag{32.13}$$

[249], where $\vec{F}$ is the uniform electric field. Here, Φ is the total magnetic flux, and $\phi_o = \frac{hc}{|e|}$ is the quantum flux which one may consider as *quantum mechanically attached* to each quasiparticle (note that instead of classical tiny cyclotron orbits, we now refer these as *quasiparticles*) in a fully occupied LLL.[1] This is actually a precursor of the concept of *flux attachment* afforded by the Chern–Simons U(1) gauge theory in 2-dimensions. Moreover, this is also realization of the BS quantization condition discussed in Chapter 13, as given in Eq. (32.12) [13]. The resulting quantization of orbital motion leads to edge states and IQHE under uniform magnetic fields. The general analysis of edge states marks the works of Laughlin [71, 250], and Halperin [72].

32.3.0.2 *Flux attachments*

From Eq. (32.12), we have for a *fully* occupied LLL at $v = 1$ of the experiments,

$$\frac{1}{2\pi\hbar}\iint \vec{\nabla}\times\vec{K}\cdot d\vec{a} = \frac{\Phi}{\phi_o} = N_o = N_{\rm LLL}, \tag{32.14}$$

[1]In Hofstadter butterfly spectrum, the LLL or sub-band is generally widely separated from a group of bunching sub-bands.

where as before, N_o is the total number of available electrons, e.g., for a given gate voltage bias of a Si MOSFET or GaAs MESFET heterostructures, and $N_{\rm LLL}$ is the degeneracy of the LLL.

For fractionally occupied, say $\frac{1}{3}$ of LL at higher magnetic fields or higher degeneracies, $\frac{N_{\rm LLL}}{N_{\rm LL}} = \frac{N_{\rm O}}{N_{LL}} = \frac{1}{3}$, so we can simply re-express Eq. (32.14), on the spirit of the flux-attachment concept as

$$\frac{k}{2\pi\hbar}\iint \vec{\nabla}\times\vec{K}\cdot d\vec{a} = \frac{\Phi}{\phi_o} = kN_o = N_{\rm LL}, \tag{32.15}$$

where we have indicated the scaling factor by k, e.g., $k = 3$.

32.3.0.3 *Either flux attachments or fractional charge*

Equation (32.15) means that naively, either $3\phi_o = \frac{3hc}{|e|}$ is attach to each quasiparticle or that $|e|$ in $\phi_o = \frac{hc}{|e|}$ is $\frac{|e|}{3}$, a fractional charge. These two alternatives carry different physical meanings, flux attachment means the number of electrons is not enough to fully occupy the new density of states or LL degeneracy, whereas fractional charge means the new quasiparticles *fully* occupy the new Landau level degeneracy, $N_{\rm LL}$, but with fractional charge.

32.3.1 *Hierarchy of primes for the scaling factor k*

Since the magnetic flux of the LLL is our reference point for scaling by a factor k, in order to form a hierarchy of scaling, this cannot be expressed as product of prime numbers, otherwise we will eventually be scaling a different magnetic flux after the first factor of the product. Thus, factorizable numbers and fractions cannot represent as members of a fundamental hierarchy of scaling factor k. Thus, for integer k this must be a prime number, whereas for fractions the numerator and denominator of k must also be a co-prime numbers. This scaling should constitute the principal scaling hierarchy, i.e., first-order in primes.

A second scaling hierarchy would constitute prime numbers for integers and prime-number numerators only for k. The third type of scaling hierarchy will constitute prime number denominators only for k. Since all these scaling k are factorizable, we consider these as *higher-order* scaling hierarchy and to be weakly observed. The fourth scaling hierarchy is for both numerators and denominators to be factorizable. This we expect to be rarely seen in the experiments.

32.3.1.1 *Higher probability for co-prime number for k*

Therefore, from the fundamental scaling hierarchies, we expect the dominance of co-prime number for k in the experiments. We summarize these statements by the following table which give some examples of the values of k and its inverse,

$k \geq 1$	
k	$\nu = \frac{1}{k}$
1	$\nu = 1$
2	$\nu = \frac{1}{2}$
3	$\nu = \frac{1}{3}$
5	$\nu = \frac{1}{5}$
7	$\nu = \frac{1}{7}$
$\frac{3}{2}$	$\nu = \frac{2}{3}$
$\frac{5}{3}$	$\nu = \frac{3}{5}$
$\frac{7}{2}$	$\nu = \frac{2}{7}$
$\frac{7}{3}$	$\nu = \frac{3}{7}$
$\frac{7}{5}$	$\nu = \frac{5}{7}$
..	...

,

$k < 1$	
k	$\nu = \frac{1}{k}$
$\frac{2}{3}$	$\nu = \frac{3}{2}$
$\frac{2}{5}$	$\nu = \frac{5}{2}$
$\frac{2}{7}$	$\nu = \frac{7}{2}$
$\frac{3}{5}$	$\nu = \frac{5}{3}$
$\frac{3}{7}$	$\nu = \frac{7}{3}$
$\frac{5}{7}$	$\nu = \frac{7}{5}$
..	..
..	..
..	..
..	...

. (32.16)

We remark that the number 2 in $\nu = \frac{1}{2}$ is a prime number and should not be considered an even number. The whole number in $\nu = 2$ in Eq. (32.16) simply means two LL are filled by the reference population of the LLL, $N_o = N_{\text{LLL}}$, and corresponds to the re-emergence of IQHE-like for a restructured sub-band. Moreover, prime integers exclude many odd integers. Thus, it appears that the assumption in the literature that the denominator of υ is given by the expression, $(2n + 1)$, in a hierarchy is misleading, this claim is simply borne out of the dominance of prime number numerators for k (2 is a prime number). Furthermore, although even denominators for υ are speculated in the literature, these cannot be members of a fundamental scaling hierarchy for the same reason that these can be decomposed into products of prime numbers. The entries of Eq. (32.16) actually exist as experimental values [251–253], see Figs. 32.1 and 32.2.

32.4 Relation to Chern–Simons gauge theories

We write the Chern–Simons Lagrangian density for U(1) gauge theory for 2D system of manifold, M, as

$$\mathcal{L}_{\text{CS}} = \gamma \varepsilon^{\mu\lambda\nu} A_\mu \partial_\lambda A_\nu - A_\mu J^\mu, \tag{32.17}$$

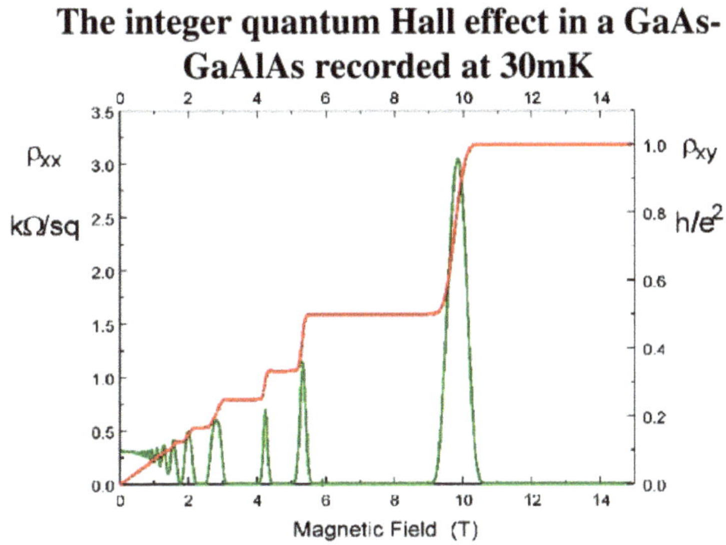

Fig. 32.1. The experimental IQHE in GaAs/GaAlAs.

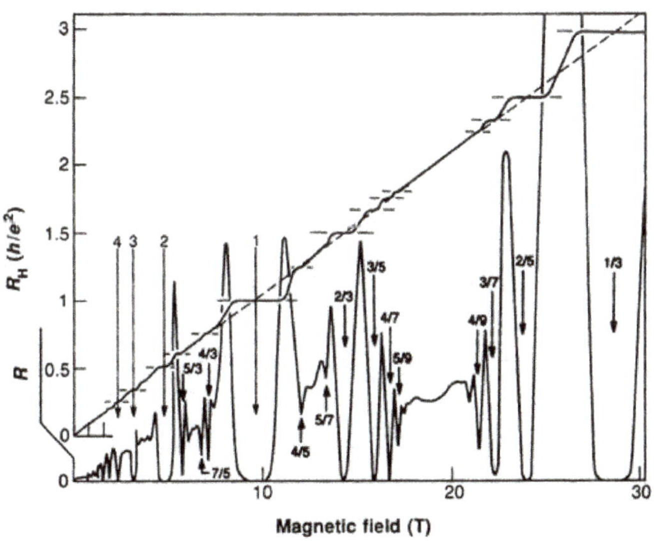

Fig. 32.2. The dashed diagonal line represents the classical Hall resistance and the full drawn diagonal stepped curve the experimental results. The magnetic fields causing the steps are marked with arrows. The step first discovered by Stormer and Tsui at the highest value of the magnetic field and the steps earlier discovered by von Klitzing (integers) with a weaker magnetic field.. Here we consider the $v = 1$ as our reference in calculating the $v = (1/3)$ by employing our novel approach.

where $J^\mu = (\rho, \vec{J})$, ρ is the charge density and $\vec{J}$ is the current density. Later, we will associate the parameter γ with our scaling parameter. Equation (32.17) is often referred to as the Maxwell Chern–Simons theory. The equation of motion is obtained by variation with respect to A_μ

$$\frac{\delta \mathcal{L}_{\text{CS}}}{\delta A_\mu} = \gamma \varepsilon^{\lambda\nu} \partial_\lambda A_\nu - J^\mu = 0. \tag{32.18}$$

This gives

$$\gamma \int_M \nabla \times \vec{A} = \int_M J^0 = \int_M \rho. \tag{32.19}$$

For a fully occupied LLL, we may equate the following,

$$\gamma = \frac{1}{2\pi\hbar}\frac{e}{c}, \tag{32.20}$$

$$\nabla \times \vec{A} \Longrightarrow \nabla \times \frac{e}{c}\vec{A}_m = \nabla \times \vec{\mathcal{K}}, \tag{32.21}$$

$$\int_M \rho = N_{\text{LL}}, \tag{32.22}$$

where,

$$\vec{K} = P + \frac{e}{c}\vec{A} + \vec{F}ct, \tag{32.23}$$

where $\vec{F}$ is the uniform electric field [249]. Thus, Eq. (32.19) reduces to that of Eq. (32.14). The relation of Eq. (32.15) with Chern–Simons U(1) gauge theory thus becomes clear. We re-write Eq. (32.15) with $k = 3$ as

$$\frac{k}{2\pi\hbar}\iint \vec{\nabla} \times \frac{e}{c}\vec{A}_m \cdot d\vec{a} = \frac{\Phi}{\phi_o} = N_{\text{LL}}, \tag{32.24}$$

where $\vec{A}_m$ is the vector potential. Equation (32.24) clearly shows that the magnetic field is scaled by an integer factor k to yield a larger number, N_{LL}, of degenerate states in a Landau level compared to our reference fully occupied LLL, resulting in fractional filling. Equation (32.17), with $k\gamma = k\left(\frac{e}{2\pi\hbar c}\right)$, is basically in the Chern–Simons form of U(1) gauge theory.

32.4.1 *Experiments in IQHE and FQHE*

In 1980, Klaus von Klitzing, working with Si MOSFET samples developed by Michael Pepper and Gerhard Dorda, made the unexpected discovery, see Fig. 32.1, that the Hall resistance was exactly quantized [73].

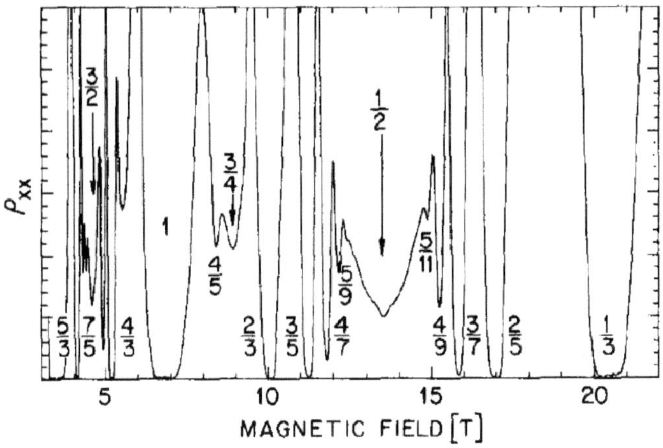

Fig. 32.3. Graph showing the diagonal resistivity. ρ_{xx}, at $T = 80mK$ showing strong transport anomalies at $\eta = \frac{1}{2}$ and $\eta = \frac{3}{2}$. Reproduced from Ref. [252].

Other experiments are done in GaAs MESFETs. In 1982, a group of physicists lead by Tsui and Stormer, working on GaAs MESFET heterostructures, discovered $\nu = \frac{1}{3}$ state [251]. This means they found a state in which the FQHE conductivity was given by

$$\sigma_{yx} = \frac{e^2}{2\pi\hbar}v = \left(\frac{e^2}{h}\right)v, \tag{32.25}$$

as shown in Fig. 32.2. In general, it was discovered over time that there were several states with fractional values that had plateaus corresponding to values of quantum Hall resistivity that correspond to different ν's. This was confusing how these states could occur since $\nu = 1$ should be the lowest possible state given the scheme found by the Landau levels above. At that time, it was considered strange since some fractions would occur and others would not. For example there was $\nu = \frac{1}{3}$ but not $\nu = \frac{1}{2}$. There's $\nu = \frac{3}{5}$, but there is no $\nu = \frac{3}{4}$. In general, it was found at first that no state had an even denominator.

However, more later experiments have presented evidence for even denominator, see Fig. 32.3, in some special cases [252]. More detailed experiments were done [253, 254], see Figs. 32.4 and 32.5.

Our analysis is not based on variational wavefunction microscopic approaches, for example, using Laughlin wavefunctions [250], More-Read wavefunctions [255], or Jain wavefunctions [78, 256, 257], but is based on macroscopic phase-space analysis reminiscent of the macroscopic Landau-Ginzburg analysis of phase transitions in matter via symmetry breaking, employing minimal description using order parameters. Here, our *order*

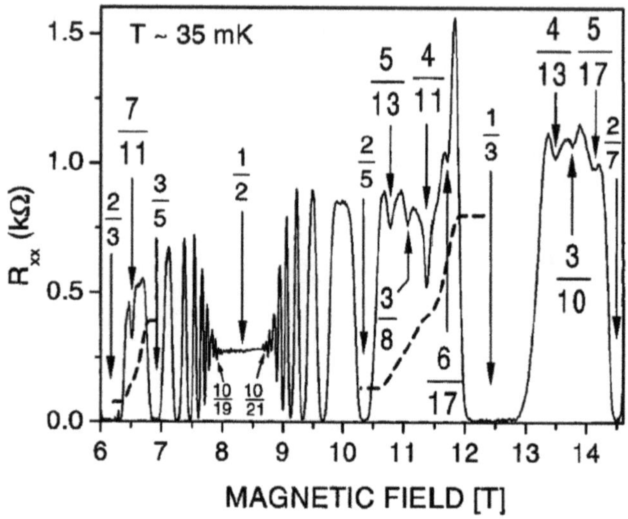

Fig. 32.4. In the present analysis, $\nu = \frac{10}{21}, \frac{3}{8}$, and $\frac{3}{10}$, which are weakly resolved in the figure, do not belong to a hierarchy of scaling factors, since the denominators are factorizable and not prime numbers. Reproduced from Ref. [253].

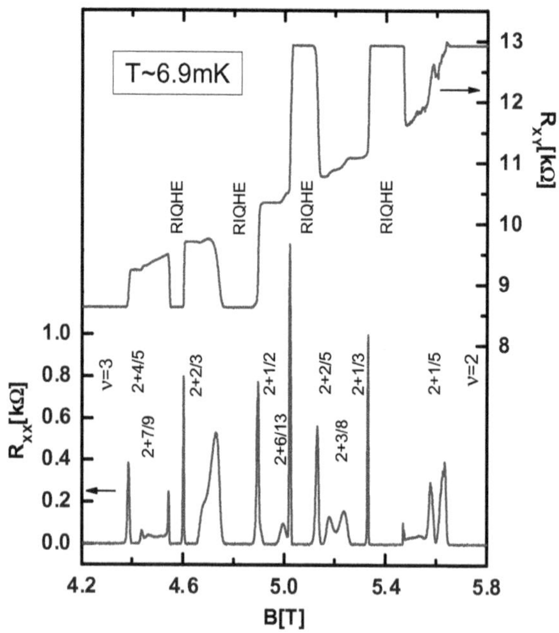

Fig. 32.5. A quantum Hall data showing plateaus in the second Landau level. Note that many of the plateaus are labeled RIQHE (reentrant IQHE). Note that only prime number denominators are resolved. Reproduced from Ref. [254]

parameter is the magnetic flux, $\frac{\Phi}{\phi_o}$, or more precisely the prime number *scaling factor* of the magnetic flux of a fully occupied LLL for a given geometrical 2D feature size, i.e., channel area, of either Si MOSFET or GaAs MESFET heterostructures used in the experiments [251–253]. Further experiments in GaAs/AlGaAs MESFET revealed strong dominance of prime number denominators for the filling factor ν, see Figs. 32.4 and 32.5. Another interesting approach to FQHE, which makes use of Bohr–Sommerfeld quantization, was given by Jacak [79, 98]. This approach may be related to our use here of Maxwell Chern–Simons U(1) gauge theory, which may also related to the Bohr–Sommerfeld quantization [13].

32.5 Topological invariant in phase space and FQHE

The following transport analysis give a macroscopic account of the experimental results [251–253] with the fully occupied LLL as a *reference point*, i.e., the point $v = 1$ moving towards successive v's< 1 with increasing magnetic fields, or moving backward of successive v's> 1 with decreasing magnetic fields. In particular, we will focus on $v = \frac{1}{3}$ or $k = 3$ as a particular case since this looks like a very defined state in the experiments, although the following analysis holds for any values of k.

For one widely separated sub-band at high magnetic fields, the topologically invariant result for σ_{yx} of Eq. (31.79) will thus come out, for $k = 3$, to be,

$$\left(\frac{e^2}{h}\right)\frac{1}{(2\pi\hbar)}\int\int\int dk_x dk_y dt\left(\frac{1}{3}\right)\left[\frac{\partial^{(a)}}{\partial\vec{\mathcal{K}}_x}\frac{\partial^{(b)}}{\partial\vec{\mathcal{K}}_y}-\frac{\partial^{(a)}}{\partial\vec{\mathcal{K}}_y}\frac{\partial^{(b)}}{\partial\vec{\mathcal{K}}_x}\right]$$
$$\times\, H^{(a)}\left(\vec{\mathcal{K}}',\mathcal{E}\right)\left(-iG^{<(b)}\left(\vec{\mathcal{K}}',\mathcal{E}\right)\right), \tag{32.26}$$

where $\vec{\mathcal{K}}'$ comes from the actual lattice Weyl transformation with updated magnetic field. The factor $\left(\frac{1}{3}\right)$ comes from the relation of Eq. (32.10), with $k = 3$. Note that the integrals have to do with the counting of the number of occupied states, moderated by the energy-dependent Wigner distribution, $-iG^{<(b)}(\vec{\mathcal{K}}',\mathcal{E})$, of particle population in phase space. We expect the influence of the product $H^{(a)}(\vec{\mathcal{K}}',\mathcal{E})(-iG^{<(b)}(\vec{\mathcal{K}}',\mathcal{E}))$ in the integrand is weakly affected by the magnetic field since the magnetic field does no work on the LLL.

Thus, in place of the previous expression for fully occupied LLL for $k = 1$, we now have, for any value of k,

$$\sigma_{yx} = \left(\frac{1}{k}\right) \frac{e^2}{(2\pi\hbar)^2} \int\int\int dp_x dp_y dt \left[\frac{\partial^{(a)}}{\partial\vec{\mathcal{K}}_x}\frac{\partial^{(b)}}{\partial\vec{\mathcal{K}}_y} - \frac{\partial^{(a)}}{\partial\vec{\mathcal{K}}_y}\frac{\partial^{(b)}}{\partial\vec{\mathcal{K}}_x}\right]$$
$$\times H^{(a)}(\vec{\mathcal{K}}', \mathcal{E})(-iG^{<(b)}(\vec{\mathcal{K}}', \mathcal{E})). \tag{32.27}$$

32.5.0.1 *Magnetic sub-band representation of magnetic quantum states*

Now the common representation of $\vec{\mathcal{K}}$ and $\vec{\mathcal{K}}'$ of the differential operator in Eq. (32.26) is the $\vec{p}$ of their magnetic sub-bands. Therefore this will yield the same result if we write as

$$\left[\frac{\partial^{(a)}}{\partial p_x}\frac{\partial^{(b)}}{\partial p_y} - \frac{\partial^{(a)}}{\partial p_y}\frac{\partial^{(b)}}{\partial p_x}\right] H^{(a)}(\vec{\mathcal{K}}', \mathcal{E})(-iG^{<(b)}(\vec{\mathcal{K}}', \mathcal{E})), \tag{32.28}$$

since

$$\frac{\partial^{(a)}}{\partial\vec{\mathcal{K}}_x} \equiv \frac{\partial\vec{\mathcal{K}}_x}{\partial p_x}\frac{\partial^{(a)}}{\partial\vec{\mathcal{K}}_x} = \frac{\partial^{(a)}}{\partial p_x}, \tag{32.29}$$

and so on.

Therefore, by going over the analogous procedure used before [244], following Eq. (32.26) we finally obtain,

$$\sigma_{yx} = \left(\frac{e^2}{h}\right) v \quad \left(v = \frac{1}{3}\right). \tag{32.30}$$

Generalizing, we have

$$\sigma_{yx} = \left(\frac{e^2}{h}\right) v, \quad \left(v = \frac{1}{k}\right). \tag{32.31}$$

For convenience, the derivation is similar to the derivation of IQHE, except for the focus on magnetic fields and with only one widely separated subband at high-magnetic fields window. In Eqs. (32.30) and (32.31), there is no summation over occupied levels since we are here mainly concerned with scaling hierarchy of fractionally occupied LLL at very intense magnetic field as a function of magnetic fields beyond the field at the fully occupied LLL. This LLL sub-band is widely separated from the other sub-bands in line with Hofstadter butterfly spectrum.

Chapter 33

Part 7: Concluding Remarks

33.1 On IQHE

We have identified topological invariant in the effective $(\vec{p}, \vec{q}, E, t)$-phase space quantum transport given by

$$\frac{1}{(2\pi\hbar)} \sum_{\alpha.\beta} \int\int\int dk_x dk_y dt \left[\frac{\partial^{(h)}}{\partial k_x} \frac{\partial^{(g)}}{\partial k_y} - \frac{\partial^{(h)}}{\partial k_y} \frac{\partial^{(g)}}{\partial k_x} \right] H^{(a)}_{\alpha\beta}(\vec{\mathcal{K}}, \mathcal{E}) G^{<(b)}_{\beta\alpha}(\vec{\mathcal{K}}, \mathcal{E}), \tag{33.1}$$

an integral which give results in $\mathbb{Z}$ manifold, the so-called first Chern numbers. This is explicitly demonstrated here by reverting to its equivalent matrix elements expression. Moreover, the conventional linearity in the electrical field strength *per se* is not a necessary and sufficient condition to prove the IQHE, but rather it is the first-order gradient expansion in the real-time SFLWT-NEGF quantum transport equation.

In general, nonlinearity in the electric field is still present in the variable $(\mathcal{K}, \mathcal{E})$ in the integrand of Eq. (30.40) based on the assumption of weak electric field and hence weak time dependence. This aspect of our theory cannot arise from purely linear-response theory of equilibrium system. This is one of the virtues of non-equilibrium quantum transport approach. There seems to be experimental evidence on this nonlinearity [258] of quantum Hall effect.

The fundamental physics underlying IQHE is the Berry connection and Berry curvature [13], which is related to the geometrical origin of the old Bohr–Sommerfeld quantization. There have been other attempts to formulate topological invariants and IQHE [259] in phase space, but the results were not explicitly reduced to an expression in terms of the physics of Berry connection, curvature and Chern number, i.e., within the general quantization procedure first enunciated by the Bohr–Sommerfeld quantization [13], in a manner similar to was done in reducing Eq. (33.1) to Eq. (31.113), or to their explicit geometrical equivalence, such as, e.g., eigensheaves of the Hecke operators in geometric Langlands program given by Ikeda [260]. Moreover, the incorporation of varying potential and varying magnetic fields [259] needs self-consistency with Maxwell's equations, which can only be attained through large-scale and iterative numerical simulations incorporating nonlinear transport feedback [11], when IQHE is viewed as a *bona fide* transport problem.

Therefore, we cannot compare our results with the published expressions of other authors [259] for a couple of specific reasons, (a) there is no transparency given towards Berry connection and Berry curvature as indicated above, (b) Weyl transformation is as old as quantum mechanics, and other phase space formulation are based on the old Moyal expansion. The formalism adopted in this chapter is physically based. The discrete phase space is rigorously based on the modular arithmetic of finite field [22] and generalizes Wigner's and other phase space formulation of quantum mechanics by adapting to the problem at hand, specifically in terms of appropriate eigenfunctions for the lattice position and crystal momentum, namely Wannier function (conventional, electric and/or magnetic) and corresponding Bloch functions (conventional, electric and/or magnetic) leading to a renormalized discrete phase space on finite fields [12, 22].

It is worth mentioning that there is other more recent formulation of discrete Weyl–Wigner formulation of quantum mechanics [19–21]. It should be mentioned at the outset that these formulations are not explicitly based on modular arithmetic of finite field modulo prime number of discrete lattice points, which could potentially create some ambiguities in the use of discrete lattice and continuous momentum variables (which we have shown in Part 1 to be non-existent for translational symmetric domains), a bijective deficiency [7]. A naïve discretization of continuous space is incapable of mapping the solid-state lattice sites in a bijective manner and could potentially have arithmetical ambiguities due to lack of closure property. The Buot's formulation of discrete lattice points in a crystalline solid obeying the Born–von Karman boundary condition is fundamentally based on a finite field represented by a finite prime modulus number of lattice points, in all directions on a "torus" obeying modular arithmetic, closed under addition, multiplication, and division, where division by all non-zero numbers is possible if and only if the modulus is a prime number. In the terminology of abstract algebra, the ability to perform division means that modular arithmetic modulo a prime number forms a field or, more specifically, a finite field. This aspect of Buot formalism have been discussed by Gibbons *et al.* [22]. This closure property for a finite number of discrete points only afforded by the modular arithmetic of finite fields [7], that settled the theoretical aspects of discrete quantum mechanics, is missing in all the recent papers on discrete Weyl–Wigner formulation [19–21]. A more complete discussions on various aspects of discrete lattice formulations is covered by Part 1 and in more details in the author's book [12] and references therein.

The key to the straightforward simplicity and power of Buot' formalism lies in the use of appropriate basis states for lattice position and crystal momentum that make up the phase space, akin to a *renormalization of* phase space. This can even deal with a lattice system of only two discrete points yielding the well-known Hadamard transformation for qubits. In contrast, other phase space formalism [19] is unable to treat two-site systems [7].

In the presence of uniform magnetic field, basically the canonical momentum in Eq. (31.102) will strictly incorporate the magnetic field, with Jacobian unity in the integrand, through the vector potential [11]. Thus, the formalism immediately transfers to that of free electron gas under intense magnetic fields, upon which the beautiful pioneering experiments of von Klitzing, Dorda, and Pepper [73] was performed. This interchangeability is one of the advantages of our approach. It also appears that electron–electron interaction [11] which does not break the symmetry of Eqs. (F.40) and (30.13) can be treated in similar manner. The real-time SFLWT-NEGF multi-spinor quantum transport equations are also able to predict various entanglements leading to different topological phases of low-dimensional and nanostructured gapped condensed matter systems in Part 6.

Our new approach to IQHE is a *non-equilibrium quantum transport based*, in contrast with linear-response theory of equilibrium system employed so far. It seems more appropriate for treating non-equilibrium transport of the IQHE. It incorporates residual nonlinearity in $\vec{K}$ and bypasses the use of linear response theory and Kubo formula for the current–current correlation which is based on time-dependent perturbation of equilibrium system.

We believe that this is an entirely new viewpoint, namely, the use of real-time SFLWT-NEGF quantum transport formalism to yield the transport-based topological invariants of condensed matter systems in phase space that is transparent to the real physics of quantization [13]. In Chapter 28, the real-time SFLWT-NEGF multi-spinor quantum transport equations lead to prediction of various entanglements of different topological phases of low-dimensional and nanostructured gapped condensed matter systems.

33.2 Covariant derivatives and Wilson lines

We have shown that the direct route to generalized Peierls phase factor or Wilson lines for crystalline solid under uniform electric fields is embodied in the use of covariant derivatives (representing minimal coupling in U(1) gauge theory) in the finite translation operators. The present calculation thus readily leads to the phase-space topological invariant in non-equilibrium quantum transport equation which yields the IQHE of electrical conductivity upon reverting to equivalent matrix elements expression [244].

33.3 On FQHE

Our analysis of FQHE parallels that of our treatment of IQHE. The only new theoretical ingredient that we need is Eq. (32.12) for the magnetic field, which is equivalent to the Chern–Simons U(1) gauge theory equation of motion, Eq. (32.24). The k-parameter essentially serves as our order parameter analogous to Landau–Ginzburg order parameter. This naturally brings in the flux attachment concept that is concomitant to new statistics of quasi-particles, the so-called anyons. Physically, this means that instead of talking about filled gapped energy levels in IQHE, we are treating a fractionally filled widely-separated gapped energy level at high-field window of the magnetic field, specifically using a fully filled LLL as the point of reference in our FQHE scaling analysis. *This will naturally produced hierarchies of scaling factors, k.* We observe that experiments are done on heterostructured semiconductors, e.g., Si MOSFET or GaAs MESFET, where electrons can be considered 2D Bloch electrons in magnetic sub-bands.

We have shown that the scaling k-factor in Chern–Simons gauge theory corresponds to the scaling of the magnetic fields in FQHE. This leads us to account for the flattening or deformation of the magnetic sub-band of the LLL upon increasing the magnetic field. Following similar procedure using the non-equilibrium quantum transport in a lattice Weyl transform formalism, we determined the exact expression for the Hall conductivity in FQHE found in experiments.

Since the magnetic flux of the LLL is our reference point for scaling by a factor k, this cannot be expressed as product of numbers within the fundamental hierarchy, otherwise we will eventually be scaling a different magnetic flux after the first factor of the product of primes. Thus, the integer k this must be a prime number, whereas for fractions the numerator and denominator of k must also be co-prime numbers. Indeed, the experiments clearly show that the value of the magnetic flux scaling number k has a prime number numerator, or FQHE scaling fraction ν with prime number denominator. These are what is expected to have higher probability of occurring, or is well-structured, and is mostly seen in experiments. Prime integers exclude many odd integers. Thus, it appears that the assumption in the literature that the denominator of v is given by the expression, $(2n+1)$, in a fundamental scaling hierarchy is misleading since prime numbers exclude a lot of odd numbers. Furthermore, although even denominators for v are speculated in the literature, these cannot be members of a fundamental hierarchy for reason of factorizability. Therefore, these prime numbers in our theory must be the controlling parameters of stable FQHE conductance expected to be well-resolved hierarchy in the experiments [254].

Part 8

Chern–Simons Theory of Hofstadter Butterfly Spectrum

Chapter 34

Chern–Simons Theory, IQHE and FQHE Regions in Hofstadter Butterfly

When electrons are subjected to both a magnetic field and a periodic electrostatic potential, 2D systems of electrons exhibit a complex energy spectrum, known as the Hofstadter's butterfly [99]. This complex spectrum results from an incommensurate/commensurate interplay between two characteristic length scales, defined by the magnetic lattice with lattice constant given by the magnetic length, l_B, and the periodic electrostatic potential with lattice constant denoted by a. The magnetic length l_B has the physical meaning of the smallest size of a circular orbit in a magnetic field which is allowed by the uncertainty principle. Following Hofstadter [99] and Wannier [261], we assume the square lattice model with the two quantizing fields which perhaps yield the first fractal-like energy spectrum discovered in physics.

The Schrödinger difference equation studied by Hofstadter [99] is known as the Harper's equation and can be written as

$$u\,(m+1) + u\,(m-1) + 2\cos\left(m\left(\frac{2\pi}{n}\right) - v\right)u\,(m) = \epsilon u\,(m), \quad (34.1)$$

where m and n are integers. We shall see that n denotes the number of magnetic sub-bands in what follows. Obviously, this equation has periodicity in $\Phi = \frac{2\pi}{n}$. The periodicity or the recovery to the original band width has to do with the magnetic length scales approaching the periodic length scale as the magnetic field is increased, thus recovering the original bandwidth. The energy spectrum obtained [99] is shown in Figs. 34.1 and 34.2. Here, the left and right side boundaries represent the original energy bandwidths.

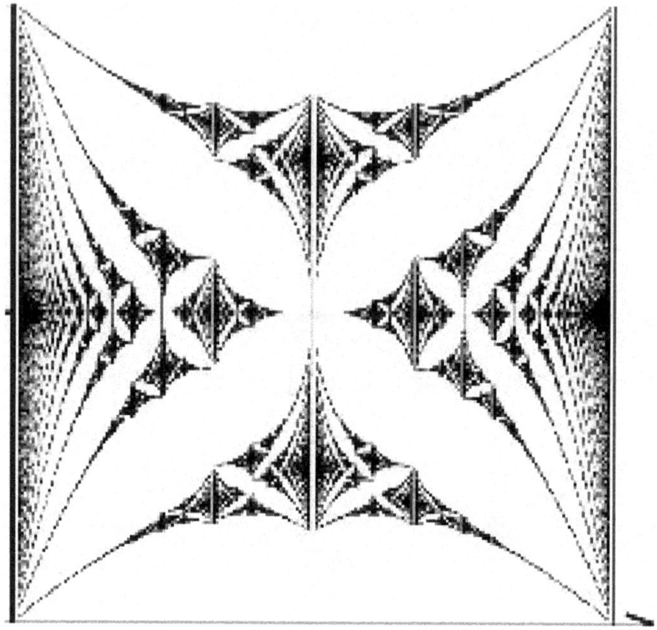

Fig. 34.1. The Hofstadter buttery energy spectrum

One characteristic feature is the wide band gap between topmost sub-band and lowest sub-band. This wide sub-band gap appears to allow the field to restructure the separated lower and upper central regions. Indeed, these two sub-bands are restructured left and right of the middle of the abscissa.

To explain the above complex spectrum, Wannier [261], using counting arguments, derived the so-called Diophantine equation to map out the energy sub-band gaps of the spectrum. Here, we propose a simpler counting argument.

Here, we employ the Maxwell Chern–Simons (CS) gauge theory [262, 263] to start the counting arguments for sub-band gap mapping. First we set the basic variables in analyzing the complex spectrum using CS gauge theory. Acknowledging the symmetry and periodicity of the complex energy spectrum shown in Fig. 34.1, we then draw the lines defining the bottom and top of the sub-bands, i.e., the sub-band gaps. Our results is depicted in Figs. 34.2 and 34.3. These figures will be discussed in more detail in what follows.

More of the lines are drawn in Fig. 34.3 to show how the rays, defining different values, i.e., counting of the sub-bands by lines that emanate from the four corners of the spectrum.

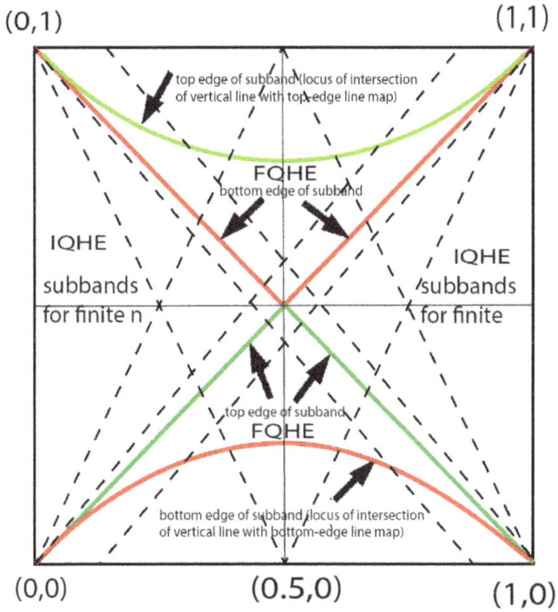

Fig. 34.2. The lines in the figure map the center of the sub-band gaps. There are essentially four regions of the spectrum, separated by the diagonal lines. Remarkably, the diagonal lines represent wide sub-band gaps in Fig. 34.1. The central lower and upper regions of the spectrum is counted as the first lower sub-band and the first upper sub-band, respectively. These two sub-bands clearly show up in the middle of the spectrum. However, to the left and right of the middle, these two sub-bands clearly undergo restructuring reminiscent of FQHE. Therefore, we label these two, lower and upper central regions, of the spectrum as the FQHE regions.

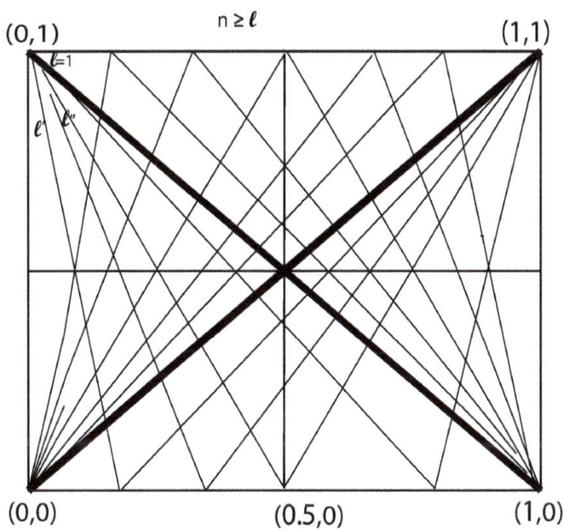

Fig. 34.3. The figure shows how the lines emanates from the four corners of the spectrum.

34.1 Maxwell Chern–Simons gauge theories

We write the CS Lagrangian density for U(1) gauge theory for 2D system of manifold, M, as

$$\mathcal{L}_{CS} = \gamma \varepsilon^{\mu\lambda\nu} A_\mu \partial_\lambda A_\nu - A_\mu J^\mu, \tag{34.2}$$

where $J^\mu = (\rho, \vec{J})$, ρ is the charge density and $\vec{J}$ is the current density. Later, we will associate the parameter γ with our magnetic parameters. Equation (34.2) is often referred to as the Maxwell CS theory [263]. The equation of motion is obtained by variation with respect to A_μ

$$\frac{\delta \mathcal{L}_{CS}}{\delta A_\mu} = \gamma \varepsilon^{\lambda\nu} \partial_\lambda A_\nu - J^\mu = 0. \tag{34.3}$$

This gives

$$\gamma \int_M \nabla \times \vec{A} \cdot da = \int_M J^0 = \int_M \rho, \tag{34.4}$$

where $\int_M \rho$ is equal to the integrated density of states of each sub-band. In general, each sub-band have the same integrated density of states, reflecting the same degeneracy for each Landau levels in free electrons case.

As in Chapter 32, we now show that the total density of states is an integer, equal to the magnetic flux divided by the quantum flux, which is equal to the degeneracy of a Landau level in a magnetic field. We do this by first counting the number of states in magnetic phase-space. This is given by [262]

$$\begin{aligned}\mathcal{N} &= \frac{1}{2\pi\hbar} \iint \vec{\nabla} \times \vec{\mathcal{K}} \cdot d\vec{a}, \\ &= \frac{1}{2\pi\hbar} \iint \vec{\nabla} \times \frac{e}{c} \vec{A} \cdot d\vec{a},\end{aligned} \tag{34.5}$$

where $\vec{\mathcal{K}} = \vec{P} + \frac{e}{c}\vec{A} + \vec{F}ct$, [249] Φ is the total magnetic flux, and $\vec{F}$ is the uniform electric field. We will ignore the electric field from now on. Therefore, the Landau level degeneracy, $\mathcal{N}$, is

$$\begin{aligned}\mathcal{N} &= \frac{1}{2\pi\hbar} \frac{e}{c} \iint \vec{B} \cdot d\vec{a}, \\ &= \frac{\Phi}{2\pi \left(\frac{\hbar c}{e}\right)} = \frac{\Phi}{\phi_o},\end{aligned} \tag{34.6}$$

where $\left(\frac{2\pi\hbar c}{e}\right) = \phi_o$ is the quantum flux. For a square lattice, the number of quantum states in one Bloch band is equal to the number of lattice sites, N. We denote by n the number of magnetic sub-band formed from a Bloch band under a uniform magnetic field. Since the magnetic field does no work or does not impart energy on the systems, we can assume that the number of quantum states in a Bloch band is conserved when uniform magnetic field is applied. Thus for n sub-bands, we have,

$$\begin{aligned} N &= \frac{n}{2\pi\hbar} \iint \vec{\nabla} \times \frac{e}{c} \vec{A} \cdot d\vec{a} = n \frac{Na^2}{2\pi \left(l_B\right)^2}, \\ &= n \frac{\Phi}{\phi_o}. \end{aligned} \tag{34.7}$$

In other words,

$$n = \frac{N}{\left(\frac{\Phi}{\phi_o}\right)} = \frac{N}{\mathcal{N}}, \tag{34.8}$$

where the Landau level (LL) degeneracy $\mathcal{N}$ is,

$$\mathcal{N} = \left(\frac{\Phi}{\phi_o}\right). \tag{34.9}$$

Thus, we can measure the whole Bloch band in units of the LL degeneracy to give us the number of sub-bands. We end up with

$$n \frac{Na^2}{2\pi \left(l_B\right)^2} = N, \tag{34.10}$$

where $\mathcal{N} = \frac{Na^2}{2\pi (l_B)^2}$ is the degeneracy of each sub-band, a is the lattice constant of a 2D square lattice and $l_B = \sqrt{\frac{\hbar c}{eB}}$. Putting $N = 1$ or canceling on both sides of Eq. (34.10), we have effectively normalized our Bloch band width to unity.

We redefine our flux variable as

$$\Phi = \frac{a^2}{2\pi \left(l_B\right)^2} = \frac{1}{n}, \quad 0 \leq \frac{1}{n} \leq 1. \tag{34.11}$$

Then we have

$$n = \frac{1}{\frac{a^2}{2\pi (l_B)^2}} = \frac{1}{\Phi}. \tag{34.12}$$

In Eq. (34.12), the number of sub-bands is determined in units of Φ. We see that our normalized Bloch band is now measured in units of Φ. We see that when $\Phi = 1$, the magnetic lattice coincides with the periodic atomic potential lattice, and hence one expects that a single Bloch band is restored, $E(k)$ is periodic in k and in reduced Brillouin zone or one band, $0 \le k \le \frac{2\pi}{a} = \frac{2\pi}{l_B}$ with l_B the magnetic length, i.e., consisting of only one sub-band equal to the whole Bloch band width.

34.2 Sub-band gap mapping

In tracing the first gap from the top of the Bloch band width, we reckon that the *bottom* of the first gap from the *topmost* first sub-band must lie at values of one value of Φ down in the Bloch energy-band scale. This will define the diagonal line given by the linear equation,

$$W_{\text{bottop}} = 1 - \Phi, \tag{34.13}$$

starting from $(0,1)$ coordinate, where W is the "y-coordinate" representing the points in the normalized zero-field Bloch band width. Other sub-bands *bottom* boundaries will be defined by lines given by the linear equation,

$$W_t = 1 - t\Phi, \quad t \le n, \; t \text{ integer}, \tag{34.14}$$

starting from $(0,1)$ coordinate, where n is the total number of sub-bands.

On the other hand, *top* boundary of the *lowest* sub-band must lie after a one count of Φ. This define another diagonal line of a linear equation,

$$W_{\text{toplow}} = \Phi, \tag{34.15}$$

stating from $(0,0)$ coordinate. These diagonal lines are depicted in Figs. 34.1 and 34.2. Other sub-bands top boundaries will be defined by lines given by the linear equation,

$$W_l = l\Phi, \qquad l \le n, \qquad l \text{ integer}, \tag{34.16}$$

stating from $(0,0)$ coordinate.

34.2.1 *Reflection symmetry at* $\Phi = \frac{1}{2}$

The Hofstadter butterfly spectrum is somewhat reminiscent of a Cantor set, which is an example of *fractal* string. In Fig. 34.4, the horizontal line may represent the zero-field Bloch band width. By tentatively identifying the Hofstadter butterfly spectrum with the cantor set, we see that the limiting

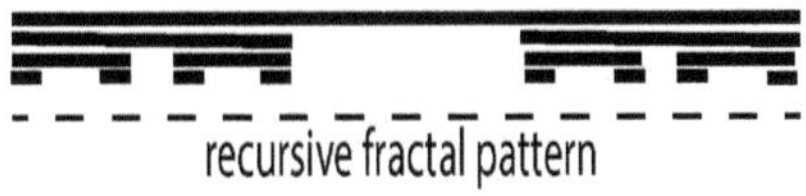

Fig. 34.4. Schematic construction of a Cantor set. Reproduced from Wikipedia.

lowest number of sub-bands is $n \geq 2$. Physically of course, the lower integer value $n = 1$ just coincide with the original zero-field Bloch band width. Just like the Cantor set the gap between the limiting two sub-bands is the largest in the set. This is also physically meaningful since at this magnetic field strength the flux is large, or the degeneracy is large and concentrated in the two sub-bands. Moreover, the cyclotron frequency spacing, $\hbar\omega = \frac{eB}{mc}$, is also large and the two sub-bands are located symmetrically near the middle of the zero-field Bloch band width. Since $n = \infty$ and $n = 1$ are physically identical, giving the whole zero-field Bloch band width at this limiting cases, while also considering the periodicity implied by Eq. (34.1) of period in $\frac{2\pi}{n}$ ($n = \infty$ and $n = 1$), we expect reflection symmetry of the integer number of sub-bands at $\frac{2\pi}{n} = \pi$ or $n = 2$. We can therefore identify $n = 1$ with $n = \infty$, and trace the gaps back toward $n = 2$, with decreasing integer n as a complete mirror image of the left portion of the flux corresponding to $n = 2$. Thus, although reminiscent of Cantor set this cyclic characteristic, among others, of the Hofstadter butterfly spectrum then gives a diametrical contrast to the iterative construction of the Cantor set.

34.2.2 *IQHE and FQHE mapping*

One most notable characteristics of the Hofstadter butterfly is the *reconstruction* of the topmost sub-band as well as that of the lowest sub-band. Although the left and right portion of the spectrum can be characterize as regions of IQHE regions [7, 244, 264], the widely separated top and bottom regions of the spectrum can be identified with the phenomenon of FQHE, since obviously this involves the *reconstruction of a single sub-band* [262] at high magnetic fields . However, in FQHE, only the lowest sub-band is considered relevant as being the *occupied* lowest sub-band. This reconstruction is evidenced by the measurements [251–253] and CS theoretical explanation of the FQHE given in Chapter 32 [262]. These are all indicated in Figs. 34.2 and 34.3.

Chapter 35

Part 8: Concluding Remarks

We need to say something about the profile of the *top* boundary of the *topmost* sub-band, as well as the *bottom* boundary of the *lowest* sub-band. The bottom profile of the lowest sub-band boundary must of the form of a convex curve with endpoints connecting the point $(0,0)$ and $(1,0)$. Similarly, the profile of the top boundary of the topmost sub-band must be a concave curve with endpoints at $(0,1)$ and $(1,1)$. This the main reason for the butterfly profile of the spectrum. This is reminiscent of the Landau levels shifting the lowest energies to higher energies, and the highest energy to lower energies in the presence of magnetic flux, reflecting the increase of $\hbar\omega = \frac{eB}{mc}$ without changing the total energies, since the magnetic field does not work on the system. This aspect of the spectrum is indicated in Figs. 34.2 and 34.3.

Appendix A

Decoupling Degrees of Freedom in CMP and QFT

Another important transformations which has to do with separating degrees of freedom are the so-called decoupling transformations. We will only present here the basic principle which goes back to Foldy and Wouthuysen [66, 75, 265], decoupling positive and negative energy states of the relativistic Dirac electrons. The technique is an essentially iterative procedure to all orders in decoupling by adding to the transformed Hamiltonian at each order of the calculation a *suitable* decoupling perturbation. Similar technique has been employed in the calculation of the magnetic susceptibility of interacting Bloch fermions in solids [18]. More recently, this technique appears in somewhat similar form in the literature which address suppressing decoherence in quantum information systems [266].

A.1 Decoupling transformation for Dirac energy bands

The decoupling of energy bands of 4×4 Hamiltonian of the relativistic Dirac equation was initiated by Foldy–Wouthuysen transformation. This work perhaps marks the birth of dressing transformation in quantum physics, i.e., diagonalization or removal of off-diagonal terms to arbitrary order.

The Dirac equation is given by,

$$i\frac{\partial \Psi}{\partial t} = (\beta m + \vec{\alpha} \cdot \vec{p})\, \Psi, \tag{A.1}$$

where p is the momentum operator, α and β are the well-known Dirac matrices (in the usual representation with β diagonal) and units in which

$\hbar = c = 1$. Restoring dimensional units of c and $\hbar$ we have

$$i\hbar\frac{\partial\Psi}{\partial t} = \left(\beta mc^2 + c\vec{\alpha}\cdot\vec{p}\right)\Psi \tag{A.2}$$

where

$$\beta = \begin{pmatrix} I_2 & 0 \\ 0 & -I_2 \end{pmatrix} = \gamma^o, \text{ “even” since diagonal only} \tag{A.3}$$

$$\alpha_\mu = \begin{pmatrix} 0 & \sigma_\mu \\ \sigma_\mu & 0 \end{pmatrix}, \text{ “odd” since diagonal is zero} \tag{A.4}$$

Notice:

$$\begin{aligned}\frac{1+\beta}{2} &= \frac{1}{2}\begin{pmatrix} I_2 & 0 \\ 0 & I_2 \end{pmatrix} + \frac{1}{2}\begin{pmatrix} I_2 & 0 \\ 0 & -I_2 \end{pmatrix} \\ &= \begin{pmatrix} I_2 & 0 \\ 0 & 0 \end{pmatrix}\begin{pmatrix} \Phi' \\ X' \end{pmatrix} = \begin{pmatrix} \Phi' \\ 0 \end{pmatrix},\end{aligned} \tag{A.5}$$

$$\begin{aligned}\frac{1-\beta}{2} &= \frac{1}{2}\begin{pmatrix} I_2 & 0 \\ 0 & I_2 \end{pmatrix} - \frac{1}{2}\begin{pmatrix} I_2 & 0 \\ 0 & -I_2 \end{pmatrix} \\ &= \begin{pmatrix} 0 & 0 \\ 0 & I_2 \end{pmatrix}\begin{pmatrix} \Phi' \\ X' \end{pmatrix} = \begin{pmatrix} 0 \\ X' \end{pmatrix}.\end{aligned} \tag{A.6}$$

Therefore,

$$\frac{1+\beta}{2}\Psi + \frac{1-\beta}{2}\Psi = \Psi. \tag{A.7}$$

Therefore,

$$i\frac{\partial}{\partial t}\Psi = H'\Psi, \tag{A.8}$$

$$\begin{aligned} i\frac{\partial}{\partial t}\begin{pmatrix} \Phi' \\ X' \end{pmatrix} &= \sqrt{p^2+m^2}\begin{pmatrix} I_2 & 0 \\ 0 & -I_2 \end{pmatrix}\begin{pmatrix} \Phi' \\ X' \end{pmatrix} \\ &= \begin{pmatrix} \sqrt{p^2+m^2}\Phi' \\ -\sqrt{p^2+m^2}X' \end{pmatrix}, \text{ this is a decoupled bands!,} \end{aligned} \tag{A.9}$$

or

$$i\frac{\partial}{\partial t}\Phi' = \sqrt{p^2+m^2}\Phi', \tag{A.10}$$

$$i\frac{\partial}{\partial t}X' = -\sqrt{p^2+m^2}X', \tag{A.11}$$

$$\beta\alpha_\mu = \begin{pmatrix} I_2 & 0 \\ 0 & -I_2 \end{pmatrix}\begin{pmatrix} 0 & \sigma_\mu \\ \sigma_\mu & 0 \end{pmatrix} = \begin{pmatrix} 0 & \sigma_\mu \\ -\sigma_\mu & 0 \end{pmatrix}, \tag{A.12}$$

$$\alpha_\mu\beta = \begin{pmatrix} 0 & \sigma_\mu \\ \sigma_\mu & 0 \end{pmatrix}\begin{pmatrix} I_2 & 0 \\ 0 & -I_2 \end{pmatrix} = \begin{pmatrix} 0 & -\sigma_\mu \\ \sigma_\mu & 0 \end{pmatrix}, \tag{A.13}$$

$$\beta\alpha_\mu\beta = \begin{pmatrix} 0 & \sigma_\mu \\ -\sigma_\mu & 0 \end{pmatrix}\begin{pmatrix} I_2 & 0 \\ 0 & -I_2 \end{pmatrix} = \begin{pmatrix} 0 & -\sigma_\mu \\ -\sigma_\mu & 0 \end{pmatrix} = -\alpha_\mu, \tag{A.14}$$

$$\{\alpha_\mu, \beta\} = 0, \tag{A.15}$$

$$\beta^2 = \begin{pmatrix} I_2 & 0 \\ 0 & -I_2 \end{pmatrix}\begin{pmatrix} I_2 & 0 \\ 0 & -I_2 \end{pmatrix} = I, \tag{A.16}$$

$$\vec{\alpha}\cdot\vec{\alpha} = \sum_\mu \begin{pmatrix} 0 & \sigma_\mu \\ \sigma_\mu & 0 \end{pmatrix}\begin{pmatrix} 0 & \sigma_\mu \\ \sigma_\mu & 0 \end{pmatrix} = \sum_\mu \begin{pmatrix} \sigma_\mu^2 & 0 \\ 0 & \sigma_\mu^2 \end{pmatrix} = \sum_{\mu=1}^{3} \begin{pmatrix} I_2 & 0 \\ 0 & I_2 \end{pmatrix}, \tag{A.17}$$

$$\sigma_x^2 = \begin{pmatrix} 0 & 1 \\ 1 & 0 \end{pmatrix}\begin{pmatrix} 0 & 1 \\ 1 & 0 \end{pmatrix} = \begin{pmatrix} 1 & 0 \\ 0 & 1 \end{pmatrix}, \tag{A.18}$$

$$\sigma_y^2 = \begin{pmatrix} 0 & -i \\ i & 0 \end{pmatrix}\begin{pmatrix} 0 & -i \\ i & 0 \end{pmatrix} = \begin{pmatrix} 1 & 0 \\ 0 & 1 \end{pmatrix}, \tag{A.19}$$

$$\sigma_z^2 = \begin{pmatrix} 1 & 0 \\ 0 & -1 \end{pmatrix}\begin{pmatrix} 1 & 0 \\ 0 & -1 \end{pmatrix} = \begin{pmatrix} 1 & 0 \\ 0 & 1 \end{pmatrix}, \tag{A.20}$$

$$\sigma_x\sigma_y = i\sigma_z, \tag{A.21}$$

$$\sigma_y\sigma_z = i\sigma_x, \tag{A.22}$$

$$\sigma_x\sigma_z = -i\sigma_y. \tag{A.23}$$

Thus

$$H_{Dirac} = \begin{pmatrix} mc^2 & c\vec{p}\cdot\vec{\sigma} \\ c\vec{p}\cdot\vec{\sigma} & -mc^2 \end{pmatrix}. \tag{A.24}$$

The eigenfunction obeys

$$\begin{aligned} i\frac{\partial\psi}{\partial t} &= (\beta m + \vec{\alpha}\cdot\vec{p})\,\psi \\ &= E\psi. \end{aligned} \tag{A.25}$$

The eigenfunctions are of the the form of Bloch function $\sim u(\vec{p})\,e^{-\vec{p}.\vec{x}}$. For each value of $\vec{p}$ there are 4 linearly independent spinors $u(\vec{p})$

$$u(p) = \begin{pmatrix} u_{+\uparrow} \\ u_{+\downarrow} \\ u_{-\uparrow} \\ u_{-\downarrow} \end{pmatrix}, \tag{A.26}$$

corresponding to the two energy eigenvalues $\pm\sqrt{m^2+p^2}$ and two eigenvalues ± 1 for the z-component related to spin. Equation (A.1) contains odd operators, specifically the components of α. It is possible to perform canonical transformation which remove all odd operators. This is the Foldy–Wouthuysen transformation.

If S is an Hermitian operator, then the transformation given by,

$$\Psi' = e^{iS}\Psi, \tag{A.27}$$

$$\begin{aligned} H' &= e^{iS}He^{-iS}, \\ &= e^{iS}\left(i\frac{\partial}{\partial t}\right)e^{-iS}, \end{aligned} \tag{A.28}$$

with S be a non-explicitly time-dependent operator leaves the Hamiltonian in the form,

$$\left(i\frac{\partial}{\partial t}\right)\Psi' = H'\Psi'. \tag{A.29}$$

Let S be a non-explicitly time-dependent operator,

$$S = -i\left(\frac{1}{2m}\right)\beta\vec{\alpha}\cdot\vec{p}\,\omega\left(\frac{\vec{p}}{m}\right), \tag{A.30}$$

where the function[1] $\omega\left(\frac{\vec{p}}{m}\right)$ is to be determined such that H' is free of odd operators.

[1]Interpretation:
$T(\vec{p})\,\Psi(x) = \frac{1}{(2\pi\hbar)^3}\iint d\vec{p}d\vec{x}T(p')\exp\left[\frac{i}{\hbar}p'\cdot(x-x')\,\Psi(x')\right]$.

We have, $He^{-iS} = e^{iS}H$, then

$$H' = e^{iS}He^{-iS} = e^{2iS}H$$
$$= e^{2iS}H = \cos(2S + i\sin 2S)\ H. \quad \text{(A.31)}$$

Since

$$2iS = \left(\frac{1}{m}\right)\beta\vec{\alpha}\cdot\vec{p}\,\omega\left(\frac{\vec{p}}{m}\right) = \Omega\ \beta\vec{\alpha}\cdot\vec{p}, \quad \text{(A.32)}$$

$$\Omega = \left[\left(\frac{1}{m}\right)\omega\left(\frac{\vec{p}}{m}\right)\right], \quad \text{(A.33)}$$

$$H' = \exp\left(\left(\frac{1}{m}\right)\beta\vec{\alpha}\cdot\vec{p}\,\omega\left(\frac{\vec{p}}{m}\right)\right)H$$
$$= \begin{Bmatrix} 1 + \Omega\beta\vec{\alpha}\cdot\vec{p} - \Omega^2\frac{(\vec{\alpha}\cdot\vec{p})^2}{2} - \frac{\Omega^3}{3!}(\vec{\alpha}\cdot\vec{p})^2(\beta\vec{\alpha}\cdot\vec{p}) \\ +\frac{\Omega^4(\vec{\alpha}\cdot\vec{p})^4}{4!} + \frac{\Omega^5(\vec{\alpha}\cdot\vec{p})^4}{5!}(\beta\vec{\alpha}\cdot\vec{p}) - \cdots \end{Bmatrix}H$$
$$= \left[1 - \Omega^2\frac{(\vec{\alpha}\cdot\vec{p})^2}{2} + \frac{\Omega^4(\vec{\alpha}\cdot\vec{p})^4}{4!} - \frac{\Omega^6(\vec{\alpha}\cdot\vec{p})^6}{6!} + \cdots\right]H$$
$$+ \begin{bmatrix} \Omega\beta\vec{\alpha}\cdot\vec{p} - \frac{\Omega^3}{3!}(\vec{\alpha}\cdot\vec{p})^2(\beta\vec{\alpha}\cdot\vec{p}) \\ +\frac{\Omega^5(\vec{\alpha}\cdot\vec{p})^4}{5!}(\beta\vec{\alpha}\cdot\vec{p}) - \frac{\Omega^7(\vec{\alpha}\cdot\vec{p})^6}{6!}(\beta\vec{\alpha}\cdot\vec{p}) + \cdots \end{bmatrix}H$$
$$= \left\{Cos(p\Omega) + \sin(p\Omega)\frac{\beta\vec{\alpha}\cdot\vec{p}}{p}\right\}H. \quad \text{(A.34)}$$

We have after multiplying by H, and using Eqs. (A.12) and (A.14), we obtain the two terms,

$$\left\{Cos(p\Omega)\beta m + \sin(p\Omega)\frac{\beta\vec{\alpha}\cdot\vec{p}}{p}\beta m\right\} = \left\{Cos(p\Omega)\beta m - \sin(p\Omega)\frac{\vec{\alpha}\cdot\vec{p}}{p}m\right\}, \quad \text{(A.35)}$$

$$\left\{Cos(p\Omega)\vec{\alpha}\cdot\vec{p} + \sin(p\Omega)\frac{\beta\vec{\alpha}\cdot\vec{p}}{p}\vec{\alpha}\cdot\vec{p}\right\} = \{pCos(p\Omega)\vec{\alpha}\cdot\vec{p} + \sin(p\Omega)\beta p\}, \quad \text{(A.36)}$$

Grouping the above terms, we obtained by separating the odd terms, $\vec{\alpha}\cdot\vec{p}$,

$$H' = \beta\{m\cos(p\Omega) + p\sin(p\Omega)\} + \frac{\vec{\alpha}\cdot\vec{p}}{p}[pCos(p\Omega) - m\sin(p\Omega)]. \quad \text{(A.37)}$$

Therefore, free of odd terms if

$$[p\cos(p\Omega) - m\sin(p\Omega)] = 0, \tag{A.38}$$

$$\begin{aligned}\frac{p}{m} &= \frac{\sin(p\Omega)}{\cos(p\Omega)} = \tan^{-1}(p\Omega)\\ &= \tan^{-1}\left(\frac{p}{m}\omega\left(\frac{\vec{p}}{m}\right)\right).\end{aligned} \tag{A.39}$$

$$\begin{aligned}H' &= \beta\left\{m\cos(p\Omega) + p\sin(p\Omega)\right\}\\ &= \beta\left\{m\frac{m}{\sqrt{p^2+m^2}} + p\frac{p}{\sqrt{p^2+m^2}}\right\}\\ &= \beta\sqrt{p^2+m^2} = \beta E_p,\end{aligned} \tag{A.40}$$

$$\cos(p\Omega) = \frac{m}{\sqrt{p^2+m^2}}. \tag{A.41}$$

Now the powers of

$$(\beta\vec{\alpha}\cdot\vec{p})^2 = (\beta\vec{\alpha}\cdot\vec{p})(\beta\vec{\alpha}\cdot\vec{p}) = (\beta\vec{\alpha}\beta\cdot\vec{p})(\vec{\alpha}\cdot\vec{p}) = -(\vec{\alpha}\cdot\vec{p})^2,$$

$$(\beta\vec{\alpha}\cdot\vec{p})^3 = (\beta\vec{\alpha}\cdot\vec{p})(\beta\vec{\alpha}\cdot\vec{p})(\beta\vec{\alpha}\cdot\vec{p}) = -(\vec{\alpha}\cdot\vec{p})^2(\beta\vec{\alpha}\cdot\vec{p}), \tag{A.42}$$

$$(\beta\vec{\alpha}\cdot\vec{p})^4 = (\beta\vec{\alpha}\cdot\vec{p})(\beta\vec{\alpha}\cdot\vec{p})(\beta\vec{\alpha}\cdot\vec{p})(\beta\vec{\alpha}\cdot\vec{p}) = (\vec{\alpha}\cdot\vec{p})^4, \tag{A.43}$$

$$(\beta\vec{\alpha}\cdot\vec{p})^5 = (\vec{\alpha}\cdot\vec{p})^4(\beta\vec{\alpha}\cdot\vec{p}), \tag{A.44}$$

$$(\beta\vec{\alpha}\cdot\vec{p})^6 = (\vec{\alpha}\cdot\vec{p})^4(\beta\vec{\alpha}\cdot\vec{p})^2 = -(\vec{\alpha}\cdot\vec{p})^6, \tag{A.45}$$

$$(\beta\vec{\alpha}\cdot\vec{p})^{2n} = (-1)^{\frac{2n}{2}}(\vec{\alpha}\cdot\vec{p})^{2n}, \tag{A.46}$$

$$(\beta\vec{\alpha}\cdot\vec{p})^{2n+1} = (-1)^{\frac{2n}{2}}(\vec{\alpha}\cdot\vec{p})^{2n}(\beta\vec{\alpha}\cdot\vec{p}). \tag{A.47}$$

We have to show that indeed, $\{H, S\} = 0$.

$$\begin{aligned}\{H,S\} &= HS + SH\\ &= (\beta m + \vec{\alpha}\cdot\vec{p})\left[-i\left(\frac{1}{2m}\right)\beta\vec{\alpha}\cdot\vec{p}\,\omega\left(\frac{\vec{p}}{m}\right)\right]\\ &\quad + \left[-i\left(\frac{1}{2m}\right)\beta\vec{\alpha}\cdot\vec{p}\,\omega\left(\frac{\vec{p}}{m}\right)\right](\beta m + \vec{\alpha}\cdot\vec{p})\\ &= \left\{\begin{array}{c}\left[-i\left(\frac{\beta}{2}\right)\beta\vec{\alpha}\cdot\vec{p}\,\omega\left(\frac{\vec{p}}{m}\right)\right]\\ +\left[-i\left(\frac{\vec{\alpha}\cdot\vec{p}\beta}{2m}\right)\vec{\alpha}\cdot\vec{p}\,\omega\left(\frac{\vec{p}}{m}\right)\right]\end{array}\right\}\end{aligned}$$

$$+\left\{\begin{array}{l}\left[-i\left(\frac{\beta}{2}\right)\vec{\alpha}\cdot\vec{p}\,\beta\omega\left(\frac{\vec{p}}{m}\right)\right]\\+\left[-i\left(\frac{\beta\vec{\alpha}\cdot\vec{p}}{2m}\right)\,\vec{\alpha}\cdot\vec{p}\omega\left(\frac{\vec{p}}{m}\right)\right]\end{array}\right\}$$

$$=0,\ \text{since}\ \{\beta\vec{\alpha}\cdot\vec{p}+\vec{\alpha}\cdot\vec{p}\,\beta\}=0. \tag{A.48}$$

A.1.1 *Effective band theory of magnetic susceptibility of relativistic Dirac fermions*

In this section, we formulate the magnetic susceptibility of relativistic Dirac fermions analogous to energy-band dynamics of crystalline solids. The Hamiltonian of free relativistic Dirac fermions is of the form

$$\mathcal{H}=\beta\Delta+c\vec{\alpha}\cdot\vec{P}. \tag{A.49}$$

We designate quantum operators in capital letters and their corresponding eigenvalues in small letters. The equation for the eigenfunctions and eigenvalues is

$$\mathcal{H}b_\lambda(x,p)=E_\lambda(p)b_\lambda(x,p), \tag{A.50}$$

where $E_\lambda(p)=\pm E(p)$, and $E_\lambda(\vec{q}\,'-q)=\frac{1}{(2\pi\hbar)^3}\int d\vec{p}\;e^{(\frac{i}{\hbar})\vec{p}\cdot(\vec{q}\,'-q)}E_\lambda(p)$, λ labels the band index: $\pm$ spin band for positive energy states and $\pm$ spin band for negative energy states.

$$E_\lambda(p)=\pm\sqrt{(cp)^2+(mc^2)^2}. \tag{A.51}$$

The doubly degenerate bands is reminiscent of the Kramer conjugates in bismuth and Bi-Sb alloys. The localized function $a_\lambda(\vec{x}-\vec{q}\,')$ is the "Wannier function" for relativistic Dirac fermions, defined below.

In the absence of magnetic field we may define the Wannier function and Bloch function of a relativistic Dirac fermions as

$$b_\lambda(x,p)=\frac{1}{(2\pi\hbar)^{\frac{3}{2}}}e^{(\frac{i}{\hbar})\vec{p}\cdot\vec{x}}\;u_\lambda(\vec{p}), \tag{A.52}$$

$$a_\lambda(\vec{x}-\vec{q})=\frac{1}{(2\pi\hbar)^{\frac{3}{2}}}\int d\vec{p}\;e^{(\frac{i}{\hbar})\vec{p}\cdot\vec{q}}\;b_\lambda(x,p), \tag{A.53}$$

where $b_\lambda(x,p)$ is the Bloch function, and $a_{\lambda(\vec{x}-\vec{q})}$ the corresponding Wannier function. $u_\lambda(\vec{p})$ is a four-component function. The $u_\lambda(\vec{p})$'s are related to the $u_\lambda(0)$'s by a unitary transformation, S, which also transforms the Dirac Hamiltonian into an *even* form, i.e., no longer have interband terms or the negative and positive energy states are decoupled. This is equivalent to the transformation from Kohn–Luttinger basis to Bloch functions in $\vec{k}\cdot\vec{p}$ theory. We have

$$S = \frac{E+\beta\mathcal{H}}{\sqrt{2E(E+\Delta)}}, \tag{A.54}$$

which can be written in matrix form as

$$S = \begin{pmatrix} \sqrt{\frac{(E+\Delta)}{2E}} & \frac{c\{\vec{\sigma}\cdot\vec{p}\}^*}{\sqrt{2E(E+\Delta)}} \\ -\frac{c\{\vec{\sigma}\cdot\vec{p}\}^*}{\sqrt{2E(E+\Delta)}} & \sqrt{\frac{(E+\Delta)}{2E}} \end{pmatrix}, \tag{A.55}$$

where the entries are 2×2 matrices, $\Delta = mc^2$, and all matrix elements may be viewed as matrix elements of S between the $u_\lambda(0)$'s, which are the spin functions in the Pauli representation. The transformed Hamiltonian is

$$\mathcal{H} = S\mathcal{H}S^\dagger = \beta E(\vec{P}). \tag{A.56}$$

The $a_\lambda(\vec{x}-\vec{q})$ is not a δ-function because of the dependence of $u_\lambda(\vec{p})$ on $\vec{p}$; it is spread out over a region of the order of the Compton wavelength, $\frac{\hbar}{mc}$, of the electron and no smaller, as pointed out first by Newton and Wigner [267], Foldy and Wouthuysen [66] and by Blount [268].

The Weyl correspondence for the momentum and coordinate operator giving the correct dynamics of quasiparticles is given by the prescription that the momentum operator $\vec{P}$ and coordinate operator $\vec{Q}$ be defined with the aid of the Wannier function and the Bloch function as

$$\vec{P}b_\lambda(x,p) = \vec{p}b_\lambda(x,p),$$
$$\vec{Q}a_\lambda(\vec{x}-\vec{q}) = \vec{q}a_\lambda(\vec{x}-\vec{q}),$$

and the uncertainty relation follows in the formalism:

$$[Q_i, P_j] = i\hbar\delta_{ij}. \tag{A.57}$$

From Eq. (A.50), we have

$$\frac{1}{(2\pi\hbar)^{\frac{3}{2}}}\int dq\ e^{(-\frac{i}{\hbar})\vec{p}\cdot\vec{q}}\mathcal{H}a_\lambda(\vec{x}-\vec{q}) = E_\lambda(p)\frac{1}{(2\pi\hbar)^{\frac{3}{2}}}\int dq\ e^{(-\frac{i}{\hbar})\vec{p}\cdot\vec{q}}a_\lambda(\vec{x}-\vec{q}), \tag{A.58}$$

$$\mathcal{H}a_\lambda(\vec{x}-\vec{q}\,') = \int dq\ E_\lambda(\vec{q}\,'-q)a_\lambda(\vec{x}-\vec{q}).$$

These relations allows us to transform the "bare" Hamiltonian operator to an "effective Hamiltonian" expressed in terms of the $\vec{P}$ operator and the $\vec{Q}$

operator. This is conveniently done by the use of the "lattice" Weyl transform ("lattice" Weyl transform and Weyl transform will be used interchangeably for infinite translationally invariant system including crystalline solids). Thus, any operator $A(\vec{P},\vec{Q})$ which is a function of $\vec{P}$ and $\vec{Q}$ can be obtained from the matrix elements of the "bare" operator, A^b_{op}, between the Wannier functions or between the Bloch functions as

$$A(\vec{P},\vec{Q}) = \sum_{\lambda\lambda'} \int d\vec{v}\, d\vec{u}\; a_{\lambda\lambda'}(\vec{u},\vec{v}) \exp\left[\left(-\frac{i}{\hbar}\right)\left(\vec{Q}\cdot\vec{u} + \vec{P}\cdot\vec{v}\right)\right] \Omega_{\lambda\lambda'}, \tag{A.59}$$

$$a_{\lambda\lambda'}(\vec{u},\vec{v}) = h^{-8} \int d\vec{p}\, d\vec{q}\; a_{\lambda\lambda'}(\vec{p},\vec{q}) \exp\left[\left(-\frac{i}{\hbar}\right)(\vec{q}\cdot\vec{u} + \vec{p}\cdot\vec{v})\right], \tag{A.60}$$

$$\begin{aligned} a_{\lambda\lambda'}(\vec{p},\vec{q}) &= \int d\vec{v} e^{\frac{i}{\hbar}\vec{p}\cdot\vec{v}} \left\langle \vec{q} - \frac{1}{2}\vec{v}, \lambda \right| A^b_{op} \left| \vec{q} + \frac{1}{2}\vec{v}, \lambda' \right\rangle \\ &= \int d\vec{u} e^{\frac{i}{\hbar}\vec{q}\cdot\vec{u}} \left\langle \vec{p} + \frac{1}{2}\vec{u}, \lambda \right| A^b_{op} \left| \vec{p} - \frac{1}{2}\vec{u}, \lambda' \right\rangle, \end{aligned} \tag{A.61}$$

where $|\vec{p},\lambda\rangle$ and $|\vec{q},\lambda\rangle$ are the state vectors representing the Bloch functions and Wannier functions, respectively, and

$$\Omega_{\lambda\lambda'} = \int d\vec{p}\; |\vec{p},\lambda\rangle\langle\vec{p},\lambda'|, \tag{A.62}$$

$$= \int d\vec{q} |\vec{q},\lambda\rangle\langle\vec{q},\lambda'|. \tag{A.63}$$

A.1.2 *Dressing of conjugate variables in energy-band quantum dynamics*

A few more words about $\vec{Q}$ and $\vec{P}$. The use of $\vec{Q}$, conjugate to the operator $\vec{P}$ of the Hamiltonian in even form, is preferred in the band-dynamical formalism. [24] The reason we now associate $\vec{Q}$ with the operator $\vec{P}$ of the Hamiltonian in even form is that this momentum operator now belongs to the respective bands (each of infinite width) of the *decoupled* Dirac Hamiltonian. This operator is now analogous to the crystal momentum operator in crystalline solids. For the original Dirac Hamiltonian $\dot{x} = c$ [from Eq. (A.49)] leading to a complex *zitterbewegung* motion in x-space, whereas for the Hamiltonian in even form $\dot{Q} = v$ [from Eq. (A.56)], c is the speed of light and v the velocity of a wave packet in the classical limit, and thus Q is more closely related to the band dynamics of fermions than x. Moreover, on the cognizance that the continuum is the limit when the lattice constant of an array of lattice points goes to zero, there is a more compelling fundamental basis for using the lattice-position operator Q. Since quantum mechanics is the mathematics of measurement processes, [102] the most probable measured values of the positions are the lattice-point coordinates. Indeed, these lattice points, or atomic sites, are where the electrons spend some time in crystalline solids. Therefore the *lattice points* and *crystal momentum* are clearly the *observables* of the theory and q and p constitute the eigenvalues of the lattice-point position operator Q and crystal momentum operator P, respectively. Thus, Q is considered here as the generalized position operator in quantum theory for describing energy-band quantum dynamics, canonical conjugate to "crystal" momentum operator $\vec{P}$ of the Hamiltonian in even form. Although the "bare" operator x can still be used as position operator it only unnecessarily renders very complicated and almost intractable resulting expressions, [74, 269] since this does not directly reflect the appropriate observables in band dynamics as first enunciated by Newton and Wigner[267] and by Wannier several decades ago.[24] Thus, in understanding the dynamics of Dirac relativistic quantum mechanics succinctly, position space should be defined at discrete points q which are eigenvalues of the operator Q.

A.1.3 *The even form of Dirac Hamiltonian in a uniform magnetic field*

The Dirac Hamiltonian for an electron with anomalous magnetic moment in a magnetic field is

$$\mathcal{H}_{op} = \vec{\alpha}\cdot\vec{\Pi}_{op} + \beta mc^2 - \frac{1}{2}(g-2)\mu_B\beta\vec{\sigma}\cdot\vec{B}, \tag{A.64}$$

where

$$\vec{\Pi}_{op} = c\vec{P}_{op} - e\vec{A}\left(\vec{Q}_{op}\right), \tag{A.65}$$

$$\mu_B = \frac{e\hbar}{2mc}. \tag{A.66}$$

The transformed Hamiltonian in even form $\mathcal{H}'_B$ is given by Eriksen and Kolsrud [270]

$$\mathcal{H}'_B = \beta\left[m^2c^4 + \Pi^2 - e\hbar c\left(1+\lambda'\right)\vec{\sigma}\cdot\vec{B} + \beta\left(\frac{\lambda' e\hbar}{2mc}\right)\sigma\cdot\left(B\times\Pi - \Pi\times B\right)\right]^{\frac{1}{2}}, \tag{A.67}$$

where $\lambda' = \frac{1}{2}\left(g-2\right)$, and

$$\tilde{\Pi} = cP - eA(Q) - eA(r)$$

$$= cP - eA(Q+r), \tag{A.68}$$

$$A(Q+r) = \frac{1}{2}B \times (Q+r),$$

$$r = \beta\left(\frac{\lambda'\hbar}{mc}\right)\sigma. \tag{A.69}$$

The above Hamiltonian can be written as

$$\begin{aligned}
\mathcal{H}'_B &= \beta\left[m^2c^4 + \Pi^2 - e\hbar c\,(1+\lambda')\,\vec{\sigma}\cdot\vec{B} - 2\left(\frac{1}{2}B \times r \cdot \Pi\right)\right]^{\frac{1}{2}} \\
&= \beta\left[m^2c^4 + \Pi^2 - e\hbar c\,(1+\lambda')\,\vec{\sigma}\cdot\vec{B} - 2A\,(r)\cdot\Pi\right]^{\frac{1}{2}} \\
&= \beta\left[m^2c^4 + \tilde{\Pi}^2 - e\hbar c\,(1+\lambda')\,\vec{\sigma}\cdot\vec{B} - A^2\,(r)\right]^{\frac{1}{2}} \\
&= \beta\left[m^2c^4 + \tilde{\Pi}^2 - e\hbar c\,(1+\lambda')\,\vec{\sigma}\cdot\vec{B} - \left(\frac{\lambda' e\hbar}{2mc}\right)^2 B^2\right]^{\frac{1}{2}}.
\end{aligned} \tag{A.70}$$

A.1.4 *Translation operator under uniform magnetic fields*

In the presence of a uniform magnetic field, magnetic Wannier Functions, $A_\lambda(x-q)$, and magnetic Bloch functions, $B_\lambda(x,p)$, exist. This is proved by using symmetry arguments. In general, these two basis functions are complete and span all the eigensolutions of the magnetic Hamiltonian belonging to a band index λ. The magnetic Wannier Functions $A_\lambda(x-q)$ and magnetic Bloch functions $B_\lambda(x,p)$ are related by similar unitary transformation in the absence of magnetic field, namely,

$$B_\lambda(x,p) = \frac{1}{(2\pi\hbar)^{\frac{3}{2}}} e^{(\frac{i}{\hbar})\vec{p}\cdot\vec{x}}\, u_\lambda(\vec{p}), \tag{A.71}$$

$$A_{\lambda(\vec{x}-\vec{q})} = \frac{1}{(2\pi\hbar)^{\frac{3}{2}}} \int d\vec{p}\; e^{(\frac{i}{\hbar})\vec{p}\cdot\vec{q}}\, B_\lambda(x,p), \tag{A.72}$$

where $\vec{p}$ and $\vec{q}$ are quantum labels.

Under a uniform magnetic fields, we have for a translation operator, $T_M(q)$, obeying the relation,

$$\begin{aligned}
\nabla_r T_M(q) &= [P, T_M(q)] \\
&= \frac{ie}{\hbar c} A(q) T_M(q).
\end{aligned} \tag{A.73}$$

Therefore,

$$T_M(q) = \exp\left(\frac{-ie}{\hbar c} A(r) \cdot q\right) C(q), \tag{A.74}$$

where $C_0(q)$ is an operator which do not depend explicitly on r. Since $T_M(q)$ is a translation operator by amount q leads us to write

$$C_0(q) = \exp(-q \cdot \nabla r), \qquad \text{a pure displacement operator by amount} - q. \tag{A.75}$$

Equation (A.73) means that $[P, T_M(q)]$ is diagonal if $T_M(q)$ is diagonal, and therefore they have the same eigenfunctions and the same quantum label. Therefore displacement operator in a translationally symmetric system under a uniform magnetic field acquire the so-called *Peierls phase factor.*

Clearly, bringing the wave packet or Wannier function around a closed loop, or around plaquette in the tight-binding limit, would acquire a phase equal to the magnetic flux through the area defined by the loop. This is the so-called Bohm–Aharonov effect or Berry phase. Thus, the concept of Berry phase has actually been floating around in the theory of band dynamics since the time of Peierls. Berry [67] has brilliantly generalized the concept to parameter-dependent Hamiltonians even in the absence of magnetic field through the so-called *Berry connection*, *Berry curvature*, and *Berry flux.*

The magnetic translation operator generates all magnetic Wannier functions belonging to band index λ from a given magnetic Wannier function centered at the origin, $A_\lambda^0(r-0)$, as

$$\begin{aligned} A_\lambda(r-q) &= T_M(q)\, A_\lambda^0(r-0) \\ &= \exp\left(\frac{-ie}{\hbar c} A(r) \cdot q\right) A_\lambda^0(r-q). \end{aligned} \tag{A.76}$$

We also have the following relation,

$$T_M(q)\, T_M(\rho) = \exp\left(\frac{ie}{\hbar c} A(q) \cdot \rho\right) T_M(q+\rho), \tag{A.77}$$

$$\begin{aligned} [T_M(q), T_M(\rho)] &= \exp\left(\frac{ie}{\hbar c} A(q) \cdot \rho\right) T_M(q+\rho) - \exp\left(\frac{ie}{\hbar c} A(\rho) \cdot q\right) T_M(\rho+q) \\ &= 2i \sin\left(\frac{e}{\hbar c} A(q) \cdot \rho\right) T_M(q+\rho). \end{aligned} \tag{A.78}$$

Moreover, we have,

$$\mathcal{H}B_\lambda(x,p) = E_\lambda\left(p - \frac{e}{c}A(q)\right)B_\lambda(x,p),$$

$$\mathcal{H}A_\lambda(\vec{x}-\vec{q}') = \int dq\ e^{i\frac{e}{c}A(q')\cdot q}E_\lambda(\vec{q}'-q)A_\lambda(\vec{x}-\vec{q}), \tag{A.79}$$

and the lattice Weyl transform of any operator, A_{op}, is

$$a_{\lambda\lambda'}(p,q) = \int d\vec{v}e^{\frac{i}{\hbar}\vec{p}\cdot\vec{v}}\left\langle A_\lambda\left(\vec{q}-\frac{1}{2}\vec{v}\right)\right| A_{op}\left|A_{\lambda'}\left(\vec{q}+\frac{1}{2}\vec{v}\right)\right\rangle. \tag{A.80}$$

The Weyl transform of the Hamiltonian operator is easily calculated using Eqs. (A.79) and (A.80). The reader is referred to Refs. [5, 6] for details of the derivation. Applying Eq. (A.80) to the even form of the Dirac Hamiltonian, we have

$$\begin{aligned} h'_B(\vec{p},\vec{q})_{\lambda\lambda'} &= \int d\vec{v}e^{\frac{i}{\hbar}\vec{p}\cdot\vec{v}}\left\langle A_\lambda\left(\vec{q}-\frac{1}{2}\vec{v}\right)\right|\mathcal{H}'_B\left|A_{\lambda'}\left(\vec{q}+\frac{1}{2}\vec{v}\right)\right\rangle \\ &= \int d\vec{v}\exp\left[\frac{i}{\hbar}\left(p-\frac{e}{c}A(q)\right)\cdot v\right]\tilde{E}_\lambda(v;B)\,\delta_{\lambda\lambda'} \\ &= E_\lambda\left(\vec{p}-\frac{e}{c}A(q);B\right)\delta_{\lambda\lambda'}. \end{aligned} \tag{A.81}$$

A.1.5 *The function* $E_\lambda(\vec{p}-\frac{e}{c}A(q);B)\delta_{\lambda\lambda'}$

The function $E_\lambda(\vec{p}-\frac{e}{c}A(q);B)$ is the Weyl transform of $\beta[\mathcal{H}^2]^{\frac{1}{2}}$, where the matrix β served to designate the four bands. In order to calculate χ we only need the knowledge of $E_\lambda(\vec{p}-\frac{e}{c}A(q);B)$ as an expansion up to second order in the coupling constant e and after a change of variable [this is effected by setting $A(q)=0, p=\hbar k$ in the expansion], we obtain the expression of $E_\lambda(\vec{p}-\frac{e}{c}A(q);B)|_{A(q)=0}$, where the dependence in the field B is beyond the vector potential,

$$E_\lambda\left(\vec{k};B\right) = E_\lambda\left(\vec{k};B\right) + BE_\lambda^{(1)}\left(\vec{k}\right) + B^2E_\lambda^{(2)}\left(\vec{k}\right) + \cdots. \tag{A.82}$$

The function $E_\lambda(\vec{p}-\frac{e}{c}A(q);B)|_{A(q)=0}$ which includes the anomalous magnetic moment of the electron is obtained as

$$E_\lambda(k;B) = \beta\left\{\begin{array}{c} E - \frac{ec}{2E}\vec{L}_{c.m.}\cdot\vec{B} - \frac{(1+\lambda')}{2E}e\hbar c\vec{\sigma}\cdot\vec{B} - \frac{(1+\lambda')^2}{8E^3}\left(e\hbar c\vec{\sigma}\cdot\vec{B}\right)^2 \\ +\frac{(e\hbar c)^2\epsilon^2}{8E^5}B^2\left[1+\left(\frac{\lambda' E}{mc^2}\right)^2\right] + O\left(e^3\right) \end{array}\right\}, \tag{A.83}$$

where

$$\vec{L}_{c.m.} = \beta\left(\frac{\lambda'\hbar}{mc}\right)\vec{\sigma}\times\vec{p}, \tag{A.84}$$

$$\epsilon^2 = m^2c^4 + c^2\hbar^2k_z^2, \tag{A.85}$$

$$E\left(\vec{k}\right) = \left[m^2c^4 + c^2\hbar^2k^2\right]^{\frac{1}{2}}. \tag{A.86}$$

The term, $\vec{L}_{c.m.}$, is a magnetodynamic effect, i.e., due to hidden average angular momentum $\vec{L}_{c.m.}$ of a moving electron. Thus, the introduction of the Pauli anomalous term in $\mathcal{H}$ at the outset endows a rigid-body behavior to the electron, and its angular momentum about the origin $\vec{L}_0$ is

$$\vec{L}_0 = \vec{L}_{MO} + \vec{L}_{c.m.}, \tag{A.87}$$

where $\vec{L}_{MO}$ is the angular momentum about the origin of the system of charge concentrated as a point at the center of mass and $\vec{L}_{c.m.}$ is the average angular momentum of the system, as a spread-out distribution of charge about the center of mass. Thus,

$$\vec{L}_o = \vec{q}\times\vec{p} + \left\langle\sum_i \vec{r_i}\times\vec{p_i}\right\rangle, \tag{A.88}$$

$$\left\langle\sum_i \vec{r_i}\times\vec{p_i}\right\rangle = \beta\left(\frac{\lambda'\hbar}{mc}\right)\vec{\sigma}\times\vec{p}, \tag{A.89}$$

$$\begin{aligned} M &= -\left[2E_\lambda^{(2)}\left(\vec{k}\right)B\right]_{sp} \\ &= -\frac{(e\hbar c)^2\epsilon^2}{4\left[E_\lambda\left(\vec{k}\right)\right]^5}\left[1+\left(\frac{\lambda' E}{mc^2}\right)^2\right]B. \end{aligned} \tag{A.90}$$

The induced magnetic moment due to a distribution of electric charge is

$$M = -\frac{Be^2\langle r^2\rangle}{4mc^2}, \tag{A.91}$$

where $\langle r^2\rangle$ is the average of the square of the spatial spread of the distribution normal to the magnetic field. Equating Eqs. (A.90) with (A.91) we obtain

$$\langle r^2\rangle = \frac{mc^2(\hbar c)^2\epsilon^2}{\left[E_\lambda(k)\right]^5}\left[1+\left(\frac{\lambda' E}{mc^2}\right)^2\right]. \tag{A.92}$$

For positive energy states $E_\lambda(k) = (c^2\hbar^2k^2 + m^2c^4)^{\frac{1}{2}}$ and in the non-relativistic limit, Eq. (A.92) reduces to

$$\langle r^2 \rangle = (1+\lambda'^2)\left(\frac{\hbar}{mc}\right)^2, \tag{A.93}$$

and thus the effective spread of the electron at rest, and for $\lambda' = 0$, is precisely equal to the Compton wavelength.

A.2 Magnetic susceptibility of Dirac fermions

The magnetic susceptibility is given by

$$\chi = -\frac{1}{48\pi^3}\left(\frac{e}{\hbar c}\right)^2 \sum_\lambda \int d\vec{k} \left\{ \frac{\partial^2 E_\lambda\left(\vec{k};0\right)}{\partial k_x^2} \frac{\partial^2 E_\lambda\left(\vec{k};0\right)}{\partial k_y^2} - \left(\frac{\partial^2 E_\lambda\left(\vec{k};0\right)}{\partial k_x \partial k_y}\right)^2 \right\} \frac{\partial f\left(E_\lambda\right)}{\partial E_\lambda}$$
$$- \left(\frac{1}{2\pi}\right)^3 \sum_\lambda \int d\vec{k} \left[E_\lambda^{(1)}(k)\right]^2 \frac{\partial f\left(E_\lambda\right)}{\partial E_\lambda} - \left(\frac{1}{2\pi}\right)^3 \sum_\lambda \int d\vec{k}\, 2E_\lambda^{(2)}(k)\, f\left(E_\lambda\right). \tag{A.94}$$

Using the following change of variable of integration,

$$(\hbar c)^3 \int d\vec{k} = \int_{-\infty}^{\infty} d\eta \int_0^{2\pi} d\phi E\left(\vec{k}\right) dE\left(\vec{k}\right), \tag{A.95}$$

where

$$\eta = \hbar c k_z, \tag{A.96}$$

we obtain for the positive energy states the expression for χ which can be divided into more physically meaningful terms as

$$\chi = \chi_{LP} + \chi_P + \chi_{sp} + \chi_g + \chi_{MD}, \tag{A.97}$$

where

$$\chi_{LP} = \frac{1}{24\pi^3}\left(\frac{e}{\hbar c}\right)^2 \int_{-\infty}^{\infty} d\eta \int_\epsilon^\infty \frac{\epsilon^2}{E^3} \frac{\partial f(E)}{\partial E} dE, \tag{A.98}$$

$$\chi_P = -\frac{(1+\lambda')^2}{8\pi^2}\left(\frac{e}{\hbar c}\right)^2 \int_{-\infty}^{\infty} d\eta \int_\epsilon^\infty \frac{1}{E} \frac{\partial f(E)}{\partial E} dE, \tag{A.99}$$

$$\chi_{sp} = -\frac{1}{8\pi^2}\left(\frac{e^2}{\hbar c}\right) \int_{-\infty}^{\infty} d\eta \int_\epsilon^\infty \frac{\epsilon^2}{E^4} \left[1 + \left(\frac{\lambda' E}{mc^2}\right)^2\right] f(E)\, dE, \tag{A.100}$$

$$\chi_g = \frac{(1+\lambda')\,2}{8\pi^2}\left(\frac{e^2}{\hbar c}\right)\int_{-\infty}^{\infty} d\eta \int_{\epsilon}^{\infty} \frac{f(E)}{E^2} dE, \tag{A.101}$$

$$\chi_{MD} = -\frac{\lambda'^2}{8\pi^2}\left(\frac{e^2}{\hbar c}\right)\int_{-\infty}^{\infty} d\eta \int_{\epsilon}^{\infty} \frac{(E^2-\epsilon^2)}{(mc^2)^2 E}\frac{\partial f(E)}{\partial E} dE, \tag{A.102}$$

where

$$\left(\frac{ec}{2E}\vec{L}_{c.m.}\right)_z^2 = \left(\frac{\lambda' e\hbar c}{2mc^2}\right)\frac{(E^2-\epsilon^2)}{E^2}, \tag{A.103}$$

$$\vec{B} = B\frac{\vec{z}}{|\vec{z}|}. \tag{A.104}$$

The total susceptibility for the positive energy states is

$$\chi = \frac{1}{(2\pi)^2}\left(\frac{e^2}{\hbar c}\right)\left[(1+\lambda')^2 - \frac{1}{3}\right]\int_0^{\infty} d\eta\, \frac{f(\epsilon)}{\epsilon}$$
$$+ \frac{1}{(2\pi)^2}\left(\frac{e^2}{\hbar c}\right)\left(\frac{\lambda'}{mc^2}\right)^2\int_0^{\infty} d\eta\, G(\epsilon-\mu), \tag{A.105}$$

where

$$G(\epsilon-\mu) = k_B T \ln\left\{1+\exp\left[-\frac{(\epsilon-\mu)}{k_B T}\right]\right\}$$
$$= \int_{\epsilon}^{\infty} f(E)\, dE. \tag{A.106}$$

The contributions of the holes is obtained by replacement of $f(\epsilon)$ and $G(\epsilon-\mu)$ in Eq. (A.105) by $(1-f(-\epsilon))$ and $G(\epsilon+\mu)$, respectively.

The relative importance of terms that made up χ at $T=0$ of Dirac fermions is summarized in Table A.1, where n is the electron density, $k_F = (3\pi^2 n)^{\frac{1}{3}}$, $\eta_F = \hbar c k_F$ and $E_F = (\Delta^2+\eta_F^2)^{\frac{1}{2}}$, is summarized below.

Table A.1. Relative importance of terms at $T = 0$.

Various contributions to χ_{Dirac} at $T=0$	Non-relativistic, $\frac{\eta_F}{\Delta} \ll 1$	Ultrarelativistic, $\frac{\eta_F}{\Delta} \gg 1$
$\chi_{LP} = -\frac{1}{12\pi^2}\left(\frac{e^2}{\hbar c}\right)\frac{1}{E_F^3}\left(\frac{\eta_F^3}{3}+\Delta^2\eta_F\right)$	$-\frac{1}{12\pi^2}\left(\frac{e}{mc^2}\right)k_F$	$-\frac{1}{12\pi^2}\left(\frac{e^2}{\hbar c}\right)\frac{1}{3}$
$\chi_P = \frac{1}{4\pi^2}(1+\lambda')^2\left(\frac{e^2}{\hbar c}\right)\frac{\eta_F}{E_F}$	$\frac{1}{4\pi^2}(1+\lambda')^2\left(\frac{e}{mc^2}\right)k_F$	$\frac{1}{4\pi^2}(1+\lambda')^2\left(\frac{e^2}{\hbar c}\right)$
$\chi_{MD} = -\frac{\lambda'^2}{4\pi^2}\left(\frac{e^2}{\hbar c}\right)\frac{1}{\Delta^2}\left[\frac{1}{E_F}\left(\frac{\eta_F^3}{3}+\Delta^2\eta_F\right)-\eta_F E_F\right]$	$\Longrightarrow 0$	$\frac{\lambda'^2}{4\pi^2}\left(\frac{e^2}{\hbar c}\right)\frac{2}{3}\left(\frac{\eta_F}{\Delta}\right)^2$
$\chi_{\text{spread}} = -\frac{1}{12\pi^2}\left(\frac{e^2}{\hbar c}\right)\sinh^{-1}\left(\frac{\eta_F}{\Delta}\right)-\chi_{LP}$		$-\frac{1}{12\pi^2}\left(\frac{e^2}{\hbar c}\right)\left[\ln\frac{2\eta_F}{\Delta}-\frac{1}{3}\right]$
$-\frac{\lambda'^2}{4\pi^2}\left(\frac{e^2}{\hbar c}\right)\frac{1}{\Delta^2}\left[\frac{\eta_F(\eta_F^2+\Delta^2)^{\frac{1}{2}}}{2}+\frac{\Delta^2}{2}\sinh^{-1}\left(\frac{\eta_F}{\Delta}\right)\right]$		
$+\frac{\lambda'^2}{4\pi^2}\left(\frac{e^2}{\hbar c}\right)\frac{1}{E_F}\left(\frac{\eta_F^3}{3}+\Delta^2\eta_F\right)\frac{1}{\Delta^2}$	$\Longrightarrow 0$	$-\frac{\lambda'^2}{4\pi^2}\left(\frac{e^2}{\hbar c}\right)\left[\frac{1}{6}\left(\frac{\eta_F}{\Delta}\right)^2+\frac{1}{2}\ln\frac{2\eta_F}{\Delta}\right]$
$\chi_g = \frac{1}{4\pi^2}(1+\lambda')^2\left(\frac{e^2}{\hbar c}\right)\left[\sinh^{-1}\left(\frac{\eta_F}{\Delta}\right)-\frac{\eta_F}{E_F}\right]$	$\Longrightarrow 0$	$\frac{1}{4\pi^2}(1+\lambda')^2\left(\frac{e^2}{\hbar c}\right)\left[\ln\frac{2\eta_F}{\Delta}-1\right]$

Appendix B

Displacement Operator under Uniform High External Electric Fields

To complement Section A.1.4, we give the translation operator for uniform electric field case, $\mathcal{H} = \mathcal{H} - e\vec{F} \cdot \vec{x}$. We have for the displacement operator, $T_E(q)$, obeying the relation,

$$i\hbar \dot{T}_E(q) = [T_E(q), \mathcal{H}], \tag{B.1}$$

$$\dot{T}_E(q) = \frac{ie}{\hbar} F \cdot q \, T_E(q). \tag{B.2}$$

Therefore

$$T_E(q) = C_o(q, \tau) \exp\left(\frac{ie}{\hbar}\right) Ft \cdot q, \tag{B.3}$$

where $C_0(q, \tau)$ is an operator which do not depend explicitly on time, t. $T_E(q)$, being a displacement operator in space and time lead us to write the operator

$$C_0(q, \tau) = \exp\left(q \cdot \frac{\partial}{\partial r} + \tau \frac{\partial}{\partial t}\right). \tag{B.4}$$

$T_E(q)$ plays critical role similar to $T_M(q)$ for establishing the phase space quantum transport dynamics at very high electric fields, where we consider realistic transport problems as time-dependent many-body problems. For zero field case we are dealing with biorthogonal Wannier functions and Bloch functions because the Hamiltonian is no longer Hermitian due to the presence of energy variable, z, in the self-energy. This means that $[T_E(q), \mathcal{H}]$ is diagonal in the bilinear expansion if $T_E(q)$ is diagonal. The eigenfunction of the "lattice" translation operator $T_E(q)$ must then be labeled by a wavenumber $\vec{k}$ which is varying in time as

$$\vec{k} = \vec{k}_0 + \frac{e\vec{F}}{\hbar}t, \tag{B.5}$$

and $\mathcal{H}$ is also diagonal in $\vec{k}$. Similarly, the energy variable, z, in the Hamiltonian must also vary as

$$z = z_0 + e\vec{F} \cdot \vec{q}. \tag{B.6}$$

Similar developments for translationally invariant many-body system subjected to a uniform electric field allows us to define the corresponding electric Bloch functions and electric Wannier functions, in a unifying manner for both magnetic and electric fields. This electric-field version allows us to derive the quantum transport equation of the particle density at very high electric fields. This is discussed in another communication dealing with quantum transport in many-body systems [11], this will not be given here since it is beyond the present scope. A more general displacement operator is discussed in Chapter 31.

Appendix C

Discrete Quantum Mechanics of Two-State Systems: Qubits

Now let us take the smallest field of prime number consisting of only two elements, obeying modular arithmetic, more recently referred to as quantum bit or "qubit". Then the eigenvalues q, may be designated by the number 0 and 1, and the eigenvalues p designated by $\frac{2\pi\hbar n}{N}$, $\{n = 0, 1 \text{ and } N = 2\}$. Observe that the product of the form $pq = \frac{2\pi\hbar nm}{N}$, where n and m are integers, the number N must be a prime number such as the number 2. In this discrete quantum mechanics, we may still designate the *Wannier function* basis states by the Dirac notation $|q\rangle$, in this case by $|0\rangle$ and $|1\rangle$ for $N = 2$.

In the orthonormal *Wannier function* basis states the matrix elements of X, Z, and Y, are calculated as follows:

$$\langle q''| X_{q'} |q\rangle = \delta_{q'',q+q'}, \tag{C.1}$$

$$\langle q''| Z_{p'} |q\rangle = \exp\left\{\frac{i}{\hbar}p' \cdot q\right\} \delta_{q'',q}, \tag{C.2}$$

$$\begin{aligned} \langle q''| Y_{q'p'} |q\rangle &= \exp\left\{\frac{i}{\hbar}\frac{p' \cdot q'}{2}\right\} \langle q''| X_{q'} Z_{p'} |q\rangle, \\ &= \exp\left\{\frac{i}{\hbar}\frac{p' \cdot q'}{2}\right\} \exp\left\{\frac{i}{\hbar}p' \cdot q\right\} \delta_{q'',q+q'}. \end{aligned} \tag{C.3}$$

Substituting the possible values of q and p, namely, $\{0, 1\}$ and $\{\frac{2\pi\hbar 0}{2}, \frac{2\pi\hbar 1}{2}\}$, respectively, and using modular arithmetic, we have

$$\begin{aligned} X_0 &= Z_0 = Y_{0,0} \\ &= \begin{pmatrix} 1 & 0 \\ 0 & 1 \end{pmatrix} = I, \end{aligned} \tag{C.4}$$

$$X_1 = \begin{pmatrix} 0 & 1 \\ 1 & 0 \end{pmatrix} = \sigma_x, \tag{C.5}$$

$$Z_1 = \begin{pmatrix} 1 & 0 \\ 0 & -1 \end{pmatrix} = \sigma_z, \tag{C.6}$$

$$Y_{01'} = \begin{pmatrix} 1 & 0 \\ 0 & -1 \end{pmatrix} = Z_1 = \sigma_z, \tag{C.7}$$

$$Y_{10} = \begin{pmatrix} 0 & 1 \\ 1 & 0 \end{pmatrix} = X_1 = \sigma_x, \tag{C.8}$$

$$Y_{1,1} = \begin{pmatrix} 0 & -i \\ i & 0 \end{pmatrix} = \sigma_y, \tag{C.9}$$

where σ_x, σ_y, and σ_z are the well-known Pauli matrices, the transformation matrices for particles with spin $\frac{1}{2}$.

Likewise, in the orthonormal *Bloch function* basis states the matrix elements of X, Z, and Y, can be calculated. First let us note the shift operator for the *Bloch function* basis states obeys the equation,

$$\exp\left\{\frac{i}{\hbar}p' \cdot Q\right\} |p\rangle = |p + p'\rangle . \tag{C.10}$$

Thus in the *Bloch function* basis, the shift operator, $\tilde{X}_{p'}$, and phase operator, $\tilde{Z}_{q'}$, obey the following equations:

$$\tilde{X}_{p'} |p\rangle = \exp\left\{\frac{i}{\hbar}p' \cdot Q\right\} |p\rangle = |p + p'\rangle , \tag{C.11}$$

$$\tilde{Z}_{q'} |p\rangle = \exp\left\{-\frac{i}{\hbar}q' \cdot P\right\} |p\rangle = \exp\left\{-\frac{i}{\hbar}q' \cdot p\right\} |p\rangle . \tag{C.12}$$

Therefore, the operators, $\exp\left\{\frac{i}{\hbar}p' \cdot Q\right\}$ and $\exp\left\{-\frac{i}{\hbar}q' \cdot P\right\}$, exchange role as the shift operator and phase operator between the *Wannier function* basis space and the *Bloch function* basis space. The product of $\tilde{X}_{p'}$ and $\tilde{Z}_{q'}$ is given by

$$\begin{aligned} \tilde{X}_{p'}\tilde{Z}_{q'} &= \exp\left\{\frac{i}{\hbar}p' \cdot Q\right\} \exp\left\{-\frac{i}{\hbar}q' \cdot P\right\}, \\ &= \exp\left\{\frac{i}{\hbar}\frac{p' \cdot q'}{2}\right\} \exp\left\{-\frac{i}{\hbar}\left(q' \cdot P - p' \cdot Q\right)\right\}. \end{aligned} \tag{C.13}$$

We again identify the corresponding operator $\tilde{Y}_{p',q'}$ the expression

$$\tilde{Y}_{p',q'} = \exp\left\{-\frac{i}{\hbar}\left(q' \cdot P - p' \cdot Q\right)\right\}, \tag{C.14}$$

whose matrix elements must now be calculated using the *Bloch function* basis states. In the orthonormal *Bloch function* basis states the matrix elements of $\tilde{X}$, $\tilde{Z}$, and $\tilde{Y}$, are calculated as follows:

$$\langle p''|\, \tilde{X}_{p'}\, |p\rangle = \delta_{p'',p+p'}, \tag{C.15}$$

$$\langle p''|\, \tilde{Z}_{q'}\, |p\rangle = \exp\left\{-\frac{i}{\hbar} q' \cdot p\right\} \delta_{p'',p}, \tag{C.16}$$

$$\begin{aligned} \langle p''|\, \tilde{Y}_{p'q'}\, |p\rangle &= \exp\left\{-\frac{i}{\hbar}\frac{p' \cdot q'}{2}\right\} \langle p''|\, \tilde{X}_{p'} \tilde{Z}_{q'}\, |p\rangle\,, \\ &= \exp\left\{-\frac{i}{\hbar}\frac{p' \cdot q'}{2}\right\} \exp\left\{-\frac{i}{\hbar} q' \cdot p\right\} \delta_{p'',p+p'}. \end{aligned} \tag{C.17}$$

Again, substituting the possible values of q and p, namely, $\{0, 1\}$ and $\{\frac{2\pi\hbar 0}{2}, \frac{2\pi\hbar 1}{2}\}$, respectively, and using modular arithmetic, we have,

$$\begin{aligned} \tilde{X}_0 = \tilde{Z}_0 &= \tilde{Y}_{0,0} \\ &= \begin{pmatrix} 1 & 0 \\ 0 & 1 \end{pmatrix} = I, \end{aligned} \tag{C.18}$$

$$\tilde{X}_1 = \begin{pmatrix} 0 & 1 \\ 1 & 0 \end{pmatrix} = \sigma_x, \tag{C.19}$$

$$\tilde{Z}_1 = \begin{pmatrix} 1 & 0 \\ 0 & -1 \end{pmatrix} = \sigma_z, \tag{C.20}$$

$$\tilde{Y}_{01'} = \begin{pmatrix} 0 & 1 \\ 1 & 0 \end{pmatrix} = X_1 = \sigma_x, \tag{C.21}$$

$$\tilde{Y}_{10} = \begin{pmatrix} 1 & 0 \\ 0 & -1 \end{pmatrix} = Z_1 = \sigma_z, \tag{C.22}$$

$$\begin{aligned} \tilde{Y}_{1,1} &= \exp\left\{-\frac{i}{\hbar}\frac{p' \cdot q'}{2}\right\} \exp\left\{-\frac{i}{\hbar} q' \cdot p\right\} \delta_{p'',p+p'}, \\ &= \begin{pmatrix} 0 & i \\ -i & 0 \end{pmatrix} = -\sigma_y = \sigma_y^T. \end{aligned} \tag{C.23}$$

C.1 Pauli matrix elements as components of two-state basis operators

It is worth pointing out that the Pauli matrices can also be represented by the components of two-state basis operators. [158] A two-state eigenvector space, denoted by basis $|\uparrow\rangle$ and $|\downarrow\rangle$, has the following complete set of four *basis operators*, $\hat{b}_{\downarrow\uparrow} = |\downarrow\rangle\langle\uparrow|$, $\hat{b}^{\dagger}_{\downarrow\uparrow} = |\uparrow\rangle\langle\downarrow|$, $\hat{d}_{\uparrow\uparrow} = |\uparrow\rangle\langle\uparrow|$, $\hat{d}_{\downarrow\downarrow} = |\downarrow\rangle\langle\downarrow|$. The Pauli spin matrix operators are obtained as simple linear combinations of the above basis operators as, $\hat{S}_o = \frac{\hbar}{2}(\hat{d}_{\uparrow\uparrow} + \hat{d}_{\downarrow\downarrow})$, $\hat{S}_x = \frac{\hbar}{2}(\hat{b}^{\dagger}_{\downarrow\uparrow} + \hat{b}_{\downarrow\uparrow})$, $\hat{S}_y = \frac{\hbar}{2i}(\hat{b}^{\dagger}_{\downarrow\uparrow} - \hat{b}_{\downarrow\uparrow})$, $\hat{S}_z = \frac{\hbar}{2}(\hat{d}_{\uparrow\uparrow} - \hat{d}_{\downarrow\downarrow})$, obeying the quantum-mechanical spin angular-momentum commutation relations, namely, $[\hat{S}_j, \hat{S}_k] = i\hbar\epsilon_{jkl}\hat{S}_l,\ \ i,j,k \neq o$. In *matrix form*, these are represented by the well-known Pauli matrix, $\vec{\sigma}$, namely, $\hat{S}_x = \frac{\hbar}{2}\sigma_x$, $\hat{S}_y = \frac{\hbar}{2}\sigma_y$, $\hat{S}_z = \frac{\hbar}{2}\sigma_z$. Note that operator $\hat{S}_o = \frac{\hbar}{2}I$ is represented by the identity operator (by virtue of the completeness relation for the two eigenvectors; statistically referred to as a pure state) multiplied by $\frac{\hbar}{2}$. Any 2×2 matrix in "spin" space can be expanded in terms of the Pauli and identity matrix operators.

C.1.1 *Discrete Fourier transform: Generalized Hadamard matrix*

The transformation from the *Wannier function*, $|q\rangle$, to the corresponding *Bloch function*, $|p\rangle$, for the two-state system we are considering, yields the Hadamard matrix which is well-known in the quantum computing community. Consider the identity, $|p\rangle = \sum_q \langle q|\, p\rangle\ |q\rangle$ where the $\langle q|\, p\rangle$ is the transformation function. For discrete quantum mechanics, this is given by the discrete Fourier transform function,

$$\langle q|\, p\rangle = \frac{1}{\sqrt{N}} \exp\left(-\frac{i}{\hbar} p \cdot q\right). \tag{C.24}$$

Upon substituting the possible values of q and p, namely, $\{0, 1\}$ and $\{\frac{2\pi\hbar 0}{2}, \frac{2\pi\hbar 1}{2}\}$, respectively, for our two-state system, we obtained the matrix for $\langle q|\, p\rangle$

$$\langle q|\, p\rangle = \frac{1}{\sqrt{2}} \begin{pmatrix} 1 & 1 \\ 1 & -1 \end{pmatrix} \equiv H, \tag{C.25}$$

which is known in the literature as the Hadamard matrix, H, which is really a "two-state discrete Fourier transform".

Noting that the *Bloch function* basis states can be obtained from the *Wannier function* basis states through the Hadamard transformation function matrix, we may thus write

$$\begin{pmatrix} \left|p = \frac{2\pi\hbar}{2}0\right\rangle \\ \left|p = \frac{2\pi\hbar}{2}1\right\rangle \end{pmatrix} = \frac{1}{\sqrt{2}} \begin{pmatrix} 1 & 1 \\ 1 & -1 \end{pmatrix} \begin{pmatrix} |q = 0\rangle \\ |q = 1\rangle \end{pmatrix}, \tag{C.26}$$

which give the *Bloch function* as a linear combination of the *Wannier functions*,

$$\left|p = \frac{2\pi\hbar}{2}0\right\rangle = \frac{1}{\sqrt{2}}\left(|0\rangle + |1\rangle\right), \tag{C.27}$$

$$\left|p = \frac{2\pi\hbar}{2}1\right\rangle = \frac{1}{\sqrt{2}}\left(|0\rangle - |1\rangle\right). \tag{C.28}$$

It follows that the operator in the *Bloch function* representation are related to the operator in the *Wannier function* representation through the Hadamard matrix. Indeed, we have the following relations:

$$\frac{1}{\sqrt{2}}\begin{pmatrix} 1 & 1 \\ 1 & -1 \end{pmatrix}\begin{pmatrix} 0 & -i \\ i & 0 \end{pmatrix}\frac{1}{\sqrt{2}}\begin{pmatrix} 1 & 1 \\ 1 & -1 \end{pmatrix} = \begin{pmatrix} 0 & i \\ -i & 0 \end{pmatrix} = \tilde{Y}_{1,1} = -\sigma_y = \sigma_y^T, \tag{C.29}$$

$$\frac{1}{\sqrt{2}}\begin{pmatrix} 1 & 1 \\ 1 & -1 \end{pmatrix}\begin{pmatrix} 0 & 1 \\ 1 & 0 \end{pmatrix}\frac{1}{\sqrt{2}}\begin{pmatrix} 1 & 1 \\ 1 & -1 \end{pmatrix} = \begin{pmatrix} 1 & 0 \\ 0 & -1 \end{pmatrix} = \tilde{Z}_1 = \sigma_z, \tag{C.30}$$

$$\frac{1}{\sqrt{2}}\begin{pmatrix} 1 & 1 \\ 1 & -1 \end{pmatrix}\begin{pmatrix} 1 & 0 \\ 0 & -1 \end{pmatrix}\frac{1}{\sqrt{2}}\begin{pmatrix} 1 & 1 \\ 1 & -1 \end{pmatrix} = \begin{pmatrix} 0 & 1 \\ 1 & 0 \end{pmatrix} = \tilde{X}_1 = \sigma_x. \tag{C.31}$$

Equations (C.30) and (C.31) reflect the exchange of the roles of the shift and phase operator when viewed from the *Wannier function* basis space to the *Bloch function* basis states.

C.2 Eigenfunctions and eigenvalues of X_1, Z_1, and $Y_{1,1}$

It is clear that the eigenfunctions of σ_z or $\exp\left\{\frac{i}{\hbar}p' \cdot Q\right\}$ are the *Wannier functions* $|0\rangle$ and $|1\rangle$, with eigenvalues 1 and -1, respectively. Similarly the eigenfunctions of σ_x or $\exp\left\{-\frac{i}{\hbar}q' \cdot P\right\}$ are the *Bloch functions* $\left|p = \frac{2\pi\hbar}{2}0\right\rangle = \frac{1}{\sqrt{2}}\left(|0\rangle + |1\rangle\right)$ and $\left|p = \frac{2\pi\hbar}{2}1\right\rangle = \frac{1}{\sqrt{2}}\left(|0\rangle - |1\rangle\right)$, with eigenvalues 1 and -1, respectively. However, the eigenfunctions of $\exp\left\{-\frac{i}{\hbar}\left(q' \cdot P - p' \cdot Q\right)\right\}$, which yield $\sigma_y \Rightarrow \exp\left\{-\frac{i}{\hbar}\left(1 \cdot P - 1 \cdot Q\right)\right\} = iX_1Z_1$ in the *Wannier function* representation, must be a linear combination of the *Wannier function* basis states, namely, $|0\rangle$ and $|1\rangle$. To find this combination it is easier to directly diagonalize the σ_y Pauli matrix by a transformation matrix P, i.e.,

$$P^{-1}\sigma_y P = D, \tag{C.32}$$

where D is a diagonal matrix. The columns of P gives the eigenvectors of σ_y. Indeed, one can readily verify the the matrix P given by

$$P = \frac{1}{\sqrt{2}} \begin{pmatrix} 1 & 1 \\ i & -i \end{pmatrix}, \tag{C.33}$$

diagonalizes σ_y according to Eq. (C.32), with eigenfunctions $|T_1\rangle = \frac{1}{\sqrt{2}} (|0\rangle + i\,|1\rangle)$ and $|T_{-1}\rangle = \frac{1}{\sqrt{2}} (|0\rangle - i\,|1\rangle)$, with eigenvalues 1 and -1, respectively.

C.2.1 *Quantum states of a two-level system: Bloch sphere*

Because of the surging interest in two-level system as qubit in quantum computing, we will discuss this in more detail here. The normalized quantum state of a discrete system is given by $|\Psi\rangle = \sum_q \psi(q)\,|q\rangle$, where $\sum_q |\psi(q)|^2 = 1$ in the *Wannier function* basis states, or given by $|\Psi\rangle = \sum_p \psi(p)\,|p\rangle$, where $\sum_p |\psi(p)|^2 = 1$ in the *Bloch function* basis states. Thus for the two-level system, we have $|\Psi\rangle = \alpha\,|0\rangle + \beta\,|1\rangle$ where $|\alpha|^2 + |\beta|^2 = 1$. Since α and β are complex numbers, we may thus write

$$|\Psi\rangle = r_\alpha e^{i\phi_\alpha}\,|0\rangle + r_\beta e^{i\phi_\beta}\,|1\rangle\,, \tag{C.34}$$

with four real parameters r_α, r_β, ϕ_α, and ϕ_β. We can choose the coefficient of $|0\rangle$ to be real and positive by factoring out $e^{i\phi_\alpha}$. Since the global phase has no observable consequences, we may write this as a three-parameter expression

$$\begin{aligned} |\Psi\rangle &= r_\alpha\,|0\rangle + r_\beta e^{i(\phi_\beta - \phi_\alpha)}\,|1\rangle\,, \\ &= r_\alpha\,|0\rangle + r_\beta e^{i\phi}\,|1\rangle\,. \end{aligned} \tag{C.35}$$

By writing $r_\beta e^{i\phi} = x + iy$, the normalization constraint becomes $r_\alpha^2 + |(x+iy)|^2 = 1 = r_\alpha^2 + x^2 + y^2$ which is the equation of a unit sphere in real 3D space with Cartesian coordinates (x, y, r_α). In terms of spherical polar coordinates for the parameter space these Cartesian coordinates a re given by $x = r\sin\theta\cos\phi$, $y = r\sin\theta\sin\phi$, and $r_\alpha \Rightarrow z = r\cos\theta$, where $r \equiv 1$ by virtue of the normalization constraint on $|\Psi\rangle$. Therefore, we have

$$\begin{aligned} |\Psi\rangle &= z\ \ |0\rangle + \sin\theta\,(\cos\phi + i\sin\phi)\ |1\rangle\,, \\ &= \cos\theta\ \ |0\rangle + e^{i\phi}\sin\theta\ \ |1\rangle\,. \end{aligned} \tag{C.36}$$

Now we have just two parameters, θ and ϕ, describing points on the surface of a unit sphere. Note that for $\theta = \frac{\pi}{4}$, we obtain the Hadamard transformation from the *Wannier function* basis states to the *Bloch function* basis states for the choices of $\phi = 0$ and $\phi = \pi$. Moreover, we also obtain the eigenfunctions of σ_y for the choice of $\phi = \frac{\pi}{2}$ and $\phi = -\frac{\pi}{2}$, for the eigenvalues 1 and -1, respectively. Note that all these eigenfunctions are orthonormal and are defined by opposite points on the horizontal plane cutting the hemisphere in the parameter space at $\theta = \frac{\pi}{4}$. Moreover, quantum states defined by points on the hemisphere for $\theta = 0$ [$|0\rangle$ *Wannier function* basis state] and those defined by $\theta = \frac{\pi}{2}$ [$e^{i\phi}|1\rangle \equiv |1\rangle$ *Wannier function* basis state] are orthonormal. In general two quantum states defined by (θ, ϕ) and $(\frac{\pi}{2} - \theta, \phi + \pi)$, respectively, i.e., points in the Euclidean hemisphere differing in parameter θ by 90° are orthogonal. We have,

$$|\Psi\rangle = \cos\theta\ |0\rangle + e^{i\phi}\sin\theta\ |1\rangle\,, \tag{C.37}$$

$$|\chi\rangle = \cos\left(\frac{\pi}{2} - \theta\right)\ |0\rangle + e^{i(\phi+\pi)}\sin\left(\frac{\pi}{2} - \theta\right)\ |1\rangle\,. \tag{C.38}$$

The dot product of their corresponding vectors is

$$\begin{aligned}\langle\chi\,|\Psi\rangle &= \cos\theta\cos\left(\frac{\pi}{2} - \theta\right) - \sin\theta\sin\left(\frac{\pi}{2} - \theta\right), \\ &= \cos\left(\theta + \frac{\pi}{2} - \theta\right) = \cos\left(\frac{\pi}{2}\right) = 0,\end{aligned} \tag{C.39}$$

demonstrating the orthogonality of the quantum states defined by these points on the Euclidean hemisphere.

Range of the parameters We notice that $|\Psi\rangle = |0\rangle$ for $\theta = 0$ and $|\Psi\rangle = e^{i\phi}\ |1\rangle$ (equivalent to $|1\rangle$) for $\theta = \frac{\pi}{2}$. This suggests that the range $0 \le \theta \le \frac{\pi}{2}$ generates all the quantum states of the two-level system as far as this parameter is concerned. Since the surface of the Euclidean sphere is defined in the range of θ, $0 \le \theta \le \pi$, this means that it is only necessary to consider the upper hemisphere in the parameter space. We expect that corresponding points in the lower hemisphere are equivalent to the points in the upper hemisphere. Indeed, the point corresponding to parameter coordinates (θ, ϕ) lies in the opposite point of the sphere defined by the coordinates $(\pi - \theta, \phi + \pi)$. For this opposite point on the sphere, we have,

$$\begin{aligned}|\Psi'\rangle &= \cos(\pi - \theta)\ |0\rangle + e^{i(\phi+\pi)}\sin(\pi - \theta)\ |1\rangle\,, \\ &= -\Big(\cos\theta\ |0\rangle + e^{i\phi}\sin\theta\ |1\rangle\Big)\,, \\ &= -\,|\Psi\rangle\,,\end{aligned} \tag{C.40}$$

indicating that the opposite points in the sphere only differ by a global phase factor of $e^{i\pi}$, which has no observable consequences, and therefore are equivalent.

C.2.2 *Bloch sphere*

The Bloch sphere is a special construct obtained by scaling the parameter θ by a factor of 2, i.e., $\theta_B = 2\theta$. This maps the points on the Euclidean hemisphere onto points on a Bloch sphere. We have,

$$|\Psi\rangle = \cos\frac{\theta_B}{2}\ |0\rangle + e^{i\phi}\sin\frac{\theta_B}{2}\ |1\rangle\,, \tag{C.41}$$

where $0 \leq \theta_B \leq \pi$, $0 \leq \phi \leq 2\pi$ are the range of the coordinates of the points in the Bloch sphere. The scaling $\theta_B = 2\theta$ is a one-to-one mapping of the points in the Euclidean hemisphere to the Bloch sphere except for the points on the equator of the Euclidean sphere which is mapped to a single point $\theta_B = \pi$, the "south pole" on the Bloch sphere. But as we have remarked above, all the points on the equator of the Euclidean sphere, $e^{i\phi}\ |1\rangle$, are equivalent, and indeed the phase, $e^{i\phi}$, becomes meaningless on the "south pole" of the Bloch sphere.

It is worthwhile mentioning at this point that shift operator X_1 or its matrix representation σ_x interchanges the "south" and "north" pole of the Bloch sphere and may be viewed as a 180° rotation of the Bloch sphere about the x-axis. To see this, note that rotation by angle π of the Bloch sphere about the x-axis would send any vector from the first quadrant to the fourth quadrant, or from the second to third quadrant, with the accompanying change of parameters, $\phi \Rightarrow -\phi$ and $\theta_B \Rightarrow \pi - \theta_B$. Hence Eq. (C.41) becomes

$$\begin{aligned}
|\Psi\rangle' &= \cos\left(\frac{\pi-\theta_B}{2}\right)\ |0\rangle + e^{-i\phi}\sin\left(\frac{\pi-\theta_B}{2}\right)\ |1\rangle\,,\\
&= \sin\left(\frac{\theta_B}{2}\right) |0\rangle + e^{-i\phi}\cos\left(\frac{\theta_B}{2}\right)\ |1\rangle\,,\\
&\Rightarrow e^{i\phi}\sin\left(\frac{\theta_B}{2}\right) |0\rangle + \cos\left(\frac{\theta_B}{2}\right)\ |1\rangle\,,
\end{aligned} \tag{C.42}$$

where the last line is obtained by multiplying by a global factor $e^{i\phi}$. Since X_1 is Hermitian, this correspond to the "contravariant" transformation of the component vector or wavefunction,

$$\begin{pmatrix} 0 & 1 \\ 1 & 0 \end{pmatrix} \begin{pmatrix} \cos\frac{\theta_B}{2} \\ e^{i\phi}\sin\frac{\theta_B}{2} \end{pmatrix} = \begin{pmatrix} e^{i\phi}\sin\frac{\theta_B}{2} \\ \cos\frac{\theta_B}{2} \end{pmatrix}. \tag{C.43}$$

The phase operator Z_1 or its matrix representation σ_z rotates the Bloch sphere about the z-axis by angle π, as can be seen from

$$\begin{pmatrix} 1 & 0 \\ 0 & -1 \end{pmatrix} \begin{pmatrix} \cos\frac{\theta_B}{2} \\ e^{i\phi}\sin\frac{\theta_B}{2} \end{pmatrix} = \begin{pmatrix} \cos\frac{\theta_B}{2} \\ e^{i(\phi+\pi)}\sin\frac{\theta_B}{2} \end{pmatrix}, \tag{C.44}$$

since π-rotation about the z-axis only changes the angle ϕ by π. The $Y_{1,1}$ operator or its matrix representation given by σ_y rotates the Bloch sphere about the y-axis by an angle π, which transform the vector from the first quadrant to the second quadrant, or from the third to the fourth quadrant, with the accompanying change of parameters, $\phi \Rightarrow \pi - \phi$, and $\theta_B \Rightarrow \pi - \theta_B$. Thus the new quantum state is

$$\begin{aligned} |\Psi\rangle' &= \cos\left(\frac{\pi - \theta_B}{2}\right) |0\rangle + e^{i(\pi-\phi)} \sin\left(\frac{\pi - \theta_B}{2}\right) |1\rangle, \\ &\Rightarrow e^{i\phi} \sin\left(\frac{\theta_B}{2}\right) |0\rangle - \cos\left(\frac{\theta_B}{2}\right) |1\rangle. \end{aligned} \tag{C.45}$$

Since $Y_{1,1}$ is Hermitian, this correspond to the "contravariant" transformation of the component vector or wavefunction,

$$\begin{pmatrix} 0 & -i \\ i & 0 \end{pmatrix} \begin{pmatrix} \cos\frac{\theta_B}{2} \\ e^{i\phi}\sin\frac{\theta_B}{2} \end{pmatrix} = i \begin{pmatrix} -e^{i\phi}\sin\frac{\theta_B}{2} \\ \cos\frac{\theta_B}{2} \end{pmatrix} \Rightarrow \begin{pmatrix} e^{i\phi}\sin\frac{(\theta_B+\pi)}{2} \\ \cos\frac{(\theta_B+\pi)}{2} \end{pmatrix}. \tag{C.46}$$

Moreover, successive application of any one of the three operators, X_1, Z_1, $Y_{1,1}$ for even number of times yields identity, whereas when applied for odd number of times yields the same operator.

C.2.3 *Quantum states on opposite points of Bloch sphere*

We note that $\theta = \frac{\pi}{4}$ is mapped to $\theta_B = \frac{\pi}{2}$. Thus the two orthonormal *Bloch function* basis states are now defined on opposite points of the Bloch sphere, as well as the two orthonormal eigenfunctions of the σ_y Pauli matrix. Indeed the relation given by Eq. (C.39) now reads, using θ_B

instead of θ,

$$\langle \chi \,|\Psi\rangle = \cos\frac{\theta_B}{2}\cos\left(\frac{\pi}{2}-\frac{\theta_B}{2}\right) - \sin\frac{\theta_B}{2}\sin\left(\frac{\pi}{2}-\frac{\theta_B}{2}\right),$$
$$= \cos\left(\frac{\pi}{2}\right) = 0, \tag{C.47}$$

which shows that quantum states defined by opposite points on the Bloch sphere are orthogonal.

C.2.4 *Exponential map*

In Section C.2.2, we have shown that the operators X_1, Z_1, and $Y_{1,1'}$ correspond to rotation of the Bloch sphere by a specific angle π about the axes x, z, and y, respectively. In this section, we are interested in rotation of the Bloch sphere about the axes x, z, and y, by an arbitrary angle. Furthermore, we are also interested in the rotation of the Bloch sphere by arbitrary angle about an arbitrary axis of rotation. To do this end we employ the exponential map to parametrize by an arbitrary angle the X_1, Z_1, and $Y_{1,1'}$ rotation operators about their respective axis of rotation.

We can proceed in either of the two ways, namely, obtained the exponential map of the operators themselves, or obtained exponential map through the matrix representation of these operators. We should mention that exponential map is a fundamental construction in the theory of Lie groups.

Case. Exponential map using the operators. We note the following:

$$X_1 = \exp\left\{-\frac{i}{\hbar}P\right\},$$
$$(X_1)^{2n} = I,$$
$$(X_1)^{2n+1} = \exp\left\{-\frac{i}{\hbar}P\right\} = X_1$$

$$Z_1 = \exp\left\{\frac{i}{\hbar}Q\right\},$$
$$(Z_1)^{2n} = I,$$
$$(Z_1)^{2n+1} = \exp\left\{\frac{i}{\hbar}Q\right\} = Z_1$$

$$\begin{aligned}
Y_{1,1'} &= \exp\left\{-\frac{i}{\hbar}(P-Q)\right\}, \\
\left(Y_{1,1'}\right)^{2n} &= I, \\
\left(Y_{1,1'}\right)^{2n+1} &= \exp\left\{-\frac{i}{\hbar}(P-Q)\right\} = Y_{1,1'}
\end{aligned}$$

where we have used modular arithmetic for the prime field of two elements. Now the exponential of an operator A parametrized by θ is given by

$$\exp\{i\theta A\} = I + i\theta A - \frac{(\theta A)^2}{2!} - i\frac{(\theta A)^3}{3!} + \frac{(\theta A)^4}{4!} + i\frac{(\theta A)^5}{5!} + \cdots.$$

For the special case that $A^{2n} = I$, and $A^{2n+1} = A$, we have

$$\begin{aligned}
\exp\{i\theta A\} &= I + i\theta A - \frac{\theta^2 I}{2!} - i\frac{\theta^3 A}{3!} + \frac{\theta^4 I}{4!} + i\frac{\theta^5 A}{5!} + \cdots. \\
&= \cos(\theta)\, I + i\sin(\theta)\, A
\end{aligned}$$

Therefore, we have for arbitrary angle θ_B the following rotation operator on the Bloch sphere

$$R_x(\theta_B) = \exp\left\{-i\frac{\theta_B}{2}X_1\right\} = \cos\left(\frac{\theta_B}{2}\right) I - i\sin\left(\frac{\theta_B}{2}\right) X_1, \quad \text{(C.48)}$$

$$R_y(\theta_B) = \exp\left\{-i\frac{\theta_B}{2}Y_{1,1'}\right\} = \cos\left(\frac{\theta_B}{2}\right) I - i\sin\left(\frac{\theta_B}{2}\right) Y_{1,1'}, \quad \text{(C.49)}$$

$$R_z(\theta_B) = \exp\left\{-i\frac{\theta_B}{2}Z_1\right\} = \cos\left(\frac{\theta_B}{2}\right) I - i\sin\left(\frac{\theta_B}{2}\right) Z_1. \quad \text{(C.50)}$$

From Eqs. (C.48)–(C.50), the matrix representation can readily be obtained by using σ_x, σ_y, and σ_z.

Case. It is also interesting to see the exponential map in terms of the operators Q and P. Then the exponential map looks like an exponential of an exponential. We have

$$\begin{aligned}
R_x(\theta) &= \exp\left\{-i\frac{\theta_B}{2}X_1\right\} \\
&= \exp\left\{-i\frac{\theta_B}{2}\exp\left\{-\frac{i}{\hbar}P\right\}\right\} \\
&= \exp\left\{-i\frac{\theta_B}{2}\cos\left(\frac{1}{\hbar}P\right)\right\}\exp\left\{i\frac{\theta_B}{2}i\sin\left(\frac{1}{\hbar}P\right)\right\}. \quad \text{(C.51)}
\end{aligned}$$

We make use of the Jacobi-Anger identity to evaluate the first factor. We have

$$\exp\left\{-i\frac{\theta_B}{2}\cos\left(\frac{1}{\hbar}P\right)\right\} = \sum_{n=-\infty}^{\infty} i^n J_n\left(\frac{\theta_B}{2}\right) e^{in\frac{1}{\hbar}P}.$$

We can rewrite the right-hand side as

$$\sum_{n=-\infty}^{\infty} i^n J_n\left(\frac{\theta_B}{2}\right) e^{in\frac{1}{\hbar}P} = \sum_{n=-\infty}^{\infty} i^{2n} J_{2n}\left(\frac{\theta_B}{2}\right) e^{i2n\frac{1}{\hbar}P} + \sum_{n=-\infty,n\neq 0}^{\infty} i^{2n+1} J_{2n+1}\left(\frac{\theta_B}{2}\right) e^{i(2n+1)\frac{1}{\hbar}P}.$$

Using modular arithmetic, $e^{i2n\frac{1}{\hbar}P} = I$ for all n, and $e^{i(2n+1)\frac{1}{\hbar}P} = e^{i\frac{1}{\hbar}P}$ for all n. Therefore, we have

$$\sum_{n=-\infty}^{\infty} i^{2n} J_{2n}\left(\frac{\theta_B}{2}\right) e^{i2n\frac{1}{\hbar}P} = \sum_{n=-\infty}^{\infty} (-1)^n J_{2n}\left(\frac{\theta_B}{2}\right) I$$

$$= \left\{J_0\left(\frac{\theta_B}{2}\right) + 2\sum_{n=1}^{\infty}(-1)^n J_{2n}\left(\frac{\theta_B}{2}\right)\right\} I.$$

The expression within the curly bracket is the well-known expression for $\cos\left(\frac{\theta_B}{2}\right)$ in terms of the Bessel function of the first kind, namely,

$$\cos\left(\frac{\theta_B}{2}\right) = J_0\left(\frac{\theta_B}{2}\right) - 2J_2\left(\frac{\theta_B}{2}\right) + J_4\left(\frac{\theta_B}{2}\right) - J_6\left(\frac{\theta_B}{2}\right) + \cdots.$$

$$\sum_{n=-\infty,n\neq 0}^{\infty} i^{2n+1} J_{2n+1}\left(\frac{\theta_B}{2}\right) e^{i(2n+1)\frac{1}{\hbar}P}$$

$$= \sum_{n=-1}^{-\infty} i^{-(2n+1)} J_{-(2n+1)}\left(\frac{\theta_B}{2}\right) e^{i\frac{1}{\hbar}P} + \sum_{n=1}^{\infty} i^{2n+1} J_{2n+1}\left(\frac{\theta_B}{2}\right) e^{i\frac{1}{\hbar}P}$$

$$= \sum_{n=-1}^{-\infty} i^{-(2n+1)} (-1)^{(2n+1)} J_{(2n+1)}\left(\frac{\theta_B}{2}\right) e^{i\frac{1}{\hbar}P} + \sum_{n=1}^{\infty} i^{2n+1} J_{2n+1}\left(\frac{\theta_B}{2}\right) e^{i\frac{1}{\hbar}P}$$

$$= i\left\{2\sum_{n=1}^{\infty}(-1)^n J_{(2n+1)}\left(\frac{\theta_B}{2}\right)\right\} e^{i\frac{1}{\hbar}P},$$

the expression within the curly bracket of the last line is the well-known expression of $\sin\left(\frac{\theta_B}{2}\right)$ in terms of the Bessel functions as

$$\sin\left(\frac{\theta_B}{2}\right) = 2J_1\left(\frac{\theta_B}{2}\right) - 2J_3\left(\frac{\theta_B}{2}\right) + 2J_5\left(\frac{\theta_B}{2}\right) - \cdots .$$

Thus the first factor of Eq. (C.51) yields

$$\begin{aligned}\exp\left\{-i\frac{\theta_B}{2}\cos\left(\frac{1}{\hbar}P\right)\right\} &= \cos\left(\frac{\theta_B}{2}\right)I - i\sin\left(\frac{\theta_B}{2}\right)e^{i\frac{1}{\hbar}P}\\ &= \cos\left(\frac{\theta_B}{2}\right)I - i\sin\left(\frac{\theta_B}{2}\right)X_1.\end{aligned}$$

We are left with second factor of Eq. (C.51)

$$\begin{aligned}&\exp\left\{i\left(\frac{\theta_B}{2}i\right)\sin\left(\frac{1}{\hbar}P\right)\right\}\\ &= \sum_{n=-\infty}^{\infty} J_n\left(i\frac{\theta_B}{2}\right)e^{i\frac{n}{\hbar}P}\\ &= \sum_{n=-\infty}^{\infty} J_{2n}\left(i\frac{\theta_B}{2}\right)e^{i\frac{2n}{\hbar}P} + \sum_{n=-\infty, n\neq 0}^{\infty} J_{2n+1}\left(i\frac{\theta_B}{2}\right)e^{i\frac{2n+1}{\hbar}P}\\ &= \left\{J_0\left(i\frac{\theta_B}{2}\right) + 2\sum_{n=1}^{\infty} J_{2n}\left(i\frac{\theta_B}{2}\right)\right\}I + \sum_{n=-\infty, n\neq 0}^{\infty} J_{2n+1}\left(i\frac{\theta_B}{2}\right)e^{i\frac{2n+1}{\hbar}P}.\end{aligned}$$

The expression in the curly bracket is a well-known sum formula for the Bessel functions [Abramowitz and Stegun, 1972, p. 363] which is equal to identity, i.e.,

$$\left\{J_0\left(i\frac{\theta_B}{2}\right) + 2\sum_{n=1}^{\infty} J_{2n}\left(i\frac{\theta_B}{2}\right)\right\} = 1$$

We are left with the last term which evaluates to zero as

$$\begin{aligned}
&\sum_{n=-\infty,n\neq 0}^{\infty} J_{2n+1}\left(i\frac{\theta_B}{2}\right) e^{i\frac{2n+1}{\hbar}P} \\
&= \sum_{n=1}^{\infty} (-1)^{2n+1} J_{(2n+1)}\left(i\frac{\theta_B}{2}\right) e^{i\frac{-1}{\hbar}P} + \sum_{n=1}^{\infty} J_{(2n+1)}\left(i\frac{\theta_B}{2}\right) e^{i\frac{1}{\hbar}P} \\
&= \sum_{n=1}^{\infty} J_{(2n+1)}\left(i\frac{\theta_B}{2}\right)\left(e^{i\frac{1}{\hbar}P} - e^{i\frac{-1}{\hbar}P}\right) = 0
\end{aligned}$$

We could have readily concluded in Eq. (C.51) that $\exp\left\{i\left(\frac{\theta_B}{2}i\right)\sin\left(\frac{1}{\hbar}P\right)\right\} = I$, since $\sin\left(\frac{1}{\hbar}P\right) = \frac{(e^{i\frac{1}{\hbar}P} - e^{i\frac{-1}{\hbar}P})}{2} \equiv 0$ by virtue of modular arithmetic for the coefficient of the operator P, where $-1 = 1\,(\mathrm{mod}\ 2)$. Therefore we end with the same result as before,

$$R_x(\theta_B) = \cos\left(\frac{\theta_B}{2}\right) I - i\sin\left(\frac{\theta_B}{2}\right) X_1.$$

Similar procedure can be followed for the rotation about the y and z axes, yielding

$$R_y(\theta_B) = \cos\left(\frac{\theta_B}{2}\right) I - i\sin\left(\frac{\theta_B}{2}\right) Y_{1,1}$$

$$R_z(\theta_B) = \cos\left(\frac{\theta_B}{2}\right) I - i\sin\left(\frac{\theta_B}{2}\right) Z_1$$

For $\theta_B = \pi$, $R_x(\pi) = -iX_1$, $R_y(\pi) = -iY_{1,1}$, and $R_z(\pi) = -iZ_1$. Although the presence of $-i$ occurs here as a *non-significant* global factor, the unitary operator that exactly yield X_1, Z_1, and $Y_{1,1'}$ is obtain for the exponential mapping of the whole quantum operators, X, Z, and Y, as we will show below. But first let us examine rotation about an arbitrary axis.

C.2.5 *Rotation about an arbitrary axis in real 3D space*

The parametrization of the rotation the Bloch sphere about an axis in real space by an arbitrary angle $\theta = \frac{\theta_B}{2}$ can likewise be carried out by the exponential map of the projection on a unit vector in real three dimensions, $\hat{n} = (n_x, n_y, n_z)$, multiplied by $\frac{\theta_B}{2}$, of either X_1, Z_1, and $Y_{1,1'}$ or σ_x,

σ_y, and σ_z. The expression for the rotation operator about the $\hat{n}$-axis is given by

$$R_{\hat{n}}\left(\theta_B\right) = \exp\left\{i\frac{\theta_B}{2}\hat{n}\cdot\vec{\sigma}\right\}, \tag{C.52}$$

where $\vec{\sigma}$ denotes the three component vector $\left(X_1, Y_{1,1'}, Z_1\right)$ of Pauli operators. One can readily show that

$$\begin{aligned}(\hat{n}\cdot\vec{\sigma})^2 &= \left(n_x X_1 + n_y Y_{1,1'} + n_z Z_1\right)^2, \\ &= \left(n_x^2 + n_y^2 + n_z^2\right) I, \\ &= I,\end{aligned} \tag{C.53}$$

since the cross products anticommute. The expression for $R_{\hat{n}}\left(\theta_B\right)$ is therefore similar to the expressions of Eqs. (C.48)–(C.50), giving

$$R_{\hat{n}}\left(\theta_B\right) = \cos\left(\frac{\theta_B}{2}\right) I + i\sin\left(\frac{\theta_B}{2}\right)\hat{n}\cdot\vec{\sigma}. \tag{C.54}$$

Note that for $\theta_B = \pi$, we obtain the Hadamard transformation from the *Wannier function* basis states to the *Bloch function* basis states for the choices of $\phi = 0$ and $\phi = \pi$.

C.2.6 *Arbitrary unitary operator for a qubit: General spin Hamiltonian*

So far, we have only given an exponential map of the shift and phase operators and a portion of their products, X_1, Z_1, and $Y_{1,1'}$ out of the whole quantum operators X, Z, and Y. This is the treatment following traditional discussions the literature. However, a complete unitary operator should be an exponential map of $(I + X_1)$, $(I + Z_1)$, and $\left(I + Y_{1,1'}\right)$, and for arbitrary axis of rotation, $(n_0 I + \hat{n}\cdot\vec{\sigma})$, where I is the identity matrix. Thus, for example, the most general density-matrix operator for two-level system and spin $\frac{1}{2}$ Hamiltonian is of the form $\mathcal{O}_{\text{'spin'}} = \alpha + \beta\,\hat{n}\cdot\vec{\sigma}$.

Denoting the arbitrary unitary operator for a two-level system, also known as a qubit, as $U_{\hat{n}}\left(\theta_B\right)$, then Eq. (C.52) must now read as

$$U_{\hat{n}}\left(\theta_B\right) = \exp\left\{i\frac{\theta_B}{2}\left[n_0 I + \hat{n}\cdot\vec{\sigma}\right]\right\}, \tag{C.55}$$

where $\tilde{n} = (n_0, n_x, n_y, n_z)$. Since I commutes with X_1, Z_1, and $Y_{1,1'}$, we can write,

$$\begin{aligned} U_{\tilde{n}}(\theta_B) &= \exp\left\{i\frac{\theta_B}{2}n_0 I\right\}\exp\left\{i\frac{\theta_B}{2}\hat{n}\cdot\vec{\sigma}\right\}, \\ &= \exp\left\{i\frac{\theta_B}{2}n_0\right\}R_{\hat{n}}(\theta_B). \end{aligned} \tag{C.56}$$

Similarly, we have,

$$U_x(\theta_B) = \exp\left\{i\frac{\theta_B}{2}n_0\right\}R_x(\theta_B), \tag{C.57}$$

$$U_y(\theta_B) = \exp\left\{i\frac{\theta_B}{2}n_0\right\}R_y(\theta_B), \tag{C.58}$$

$$U_z(\theta_B) = \exp\left\{i\frac{\theta_B}{2}n_0\right\}R_z(\theta_B). \tag{C.59}$$

By letting $n_0 = 1$ and $\theta_B = \pi$, we indeed have exactly, $U_x(\pi) = X_1$, $U_y(\pi) = Y_{1,1}$, and $U_z(\pi) = Z_1$.

It is also through the unitary operator $U_{\tilde{n}}(\theta_B)$ that one can exactly recover the Hadamard transformation matrix, which is a two-state discrete Fourier transformation between the *Wannier function* basis states and *Bloch function* basis states, or contravariant transformation between components (wavefunction) based on *Wannier function* basis states and based on *Bloch function* basis states. Thus for $n_0 = 1$, $\theta_B = \pi$, and $\hat{n} = \left(\frac{1}{\sqrt{2}}, 0, \frac{1}{\sqrt{2}}\right)$, we have

$$U_{\tilde{n}}(\pi) = \exp\left\{i\frac{\pi}{2}\right\}R_{\hat{n}}(\pi)\Big|_{\hat{n}=\left(\frac{1}{\sqrt{2}},0,\frac{1}{\sqrt{2}}\right)}, \tag{C.60}$$

$$\begin{aligned} U_{\tilde{n}}(\pi) &= \exp\left\{i\frac{\pi}{2}\right\}\left\{\cos\left(\frac{\pi}{2}\right)I - i\sin\left(\frac{\pi}{2}\right)\left[\frac{1}{\sqrt{2}}(X_1 + Z_1)\right]\right\}, \\ &= \exp\left\{i\frac{\pi}{2}\right\}\left\{-i\left[\frac{1}{\sqrt{2}}\begin{pmatrix}1 & 1\\ 1 & -1\end{pmatrix}\right]\right\}, \\ &= \left[\frac{1}{\sqrt{2}}\begin{pmatrix}1 & 1\\ 1 & -1\end{pmatrix}\right]. \end{aligned} \tag{C.61}$$

The general unitary transformation $U_{\tilde{n}}(\theta_B)$ constitutes the Lie group $SU(2)$. The Lie group $SU(2)$ is a continuous map and represents a universal covering of the 3D rotation group $SO(3)$. This can be seen from the following relations: $U_{\tilde{n}}(0) = I$, $U_{\tilde{n}}(2\pi) = (-1)^{n_0+1} I$, $U_{\tilde{n}}(4\pi) = I$. If $n_0 = 0$, $U_{\tilde{n}}(2\pi) = -I$. Since, in general, we need 4π rotation to identically restore the state of a two-level system, $SU(2)$ is a double covering of $SO(3)$ in group theoretical terminology.

Controlling unitaries and tensor product of unitaries are the basis of quantum computing. Recall that finite transformations can be built as a product of many infinitesimal ones

$$U(R) = \mathrm{Lim}_{N \Rightarrow \infty} \left(1 + i\frac{\theta}{N}Q\right)^N \Rightarrow e^{i\theta Q}. \tag{C.62}$$

An *infinitesimal* rotation by $d\theta$ about the axis specified by the unit vector $\hat{n} = (n_1, n_2, n_3)$ can thus be expressed as

$$R(\hat{n}, d\theta) = I + i\frac{d\theta \hat{n} \cdot \vec{J}}{\hbar}, \tag{C.63}$$

where $\vec{J} = (J_1, J_2, J_3)$ is the angular momentum vector. A *finite* rotation is expressed as

$$R(\hat{n}, \theta) = e^{\frac{i}{\hbar}\theta \hat{n} \cdot \vec{J}}. \tag{C.64}$$

Rotations about different axes do not commute since distinct components of the angular momentum do not commute

$$[J_k, J_l] = i\varepsilon_{klm} J_m, \tag{C.65}$$

where ε_{klm} is the totally antisymmetric tensor with $\varepsilon_{123} = 1$, and Einstein summation convention for the indices is used. In representation theory, an irreducible representation is a nontrivial representation with no nontrivial subrepresentations. The defining representation of the rotation group is three dimensional but the simplest nontrivial irreducible representation is two dimensional, namely,

$$J_k = \frac{1}{2}\delta_k, \tag{C.66}$$

where δ_k is one of the Pauli matrices,

$$\delta_1 = \begin{pmatrix} 0 & 1 \\ 1 & 0 \end{pmatrix}, \quad \delta_2 = \begin{pmatrix} 0 & -i \\ i & 0 \end{pmatrix}, \quad \delta_3 = \begin{pmatrix} 1 & 0 \\ 0 & -1 \end{pmatrix}. \tag{C.67}$$

Appendix D

Matrix Equations for Bloch Electrons with Spin

In the presence of spin degree of freedom, Eqs. (21.6), (21.13), (21.14), and (21.15) become matrix equations. For the conduction-electron band, Eq. (21.6), we have the matrix equation with spin indices given by

$$i\hbar\left(\frac{\partial}{\partial t_1}+\frac{\partial}{\partial t_2}\right)\begin{pmatrix} G^{e-h,<}_{cc,\uparrow\uparrow}(12) & G^{e-h,<}_{cc,\uparrow\downarrow}(12) \\ G^{e-h,<}_{cc,\downarrow\uparrow}(12) & G^{e-h,<}_{cc,\downarrow\downarrow}(12) \end{pmatrix}$$

$$= \begin{pmatrix} \begin{bmatrix} v_{c,\uparrow\sigma}(1\xi)\, G^{<}_{cc,\sigma\uparrow}(\xi 2) \\ -G^{<}_{cc,\uparrow\sigma}(1\xi)\; v^{T}_{c,\sigma\uparrow}(\xi 2) \end{bmatrix} & \begin{bmatrix} v_{c,\uparrow\sigma}(1\xi)\, G^{<}_{cc,\sigma\downarrow}(\xi 2) \\ -G^{<}_{cc,\uparrow\sigma}(1\xi)\; v^{T}_{c,\sigma\downarrow}(\xi 2) \end{bmatrix} \\ \begin{bmatrix} v_{c,\downarrow\sigma}(1\xi)\, G^{<}_{cc,\sigma\uparrow}(\xi 2) \\ -G^{<}_{cc,\downarrow\sigma}(1\xi)\; v^{T}_{c,\sigma\uparrow}(\xi 2) \end{bmatrix} & \begin{bmatrix} v_{c,\downarrow\sigma}(1\xi)\, G^{<}_{cc,\sigma\downarrow}(\xi 2) \\ -G^{<}_{cc,\downarrow\sigma}(1\xi)\; v^{T}_{c,\sigma\downarrow}(\xi 2) \end{bmatrix} \end{pmatrix}$$

$$+ \begin{pmatrix} \begin{bmatrix} \Sigma^{r}_{cc,\uparrow\sigma}(1\xi)\, G^{<}_{cc,\sigma\uparrow}(\xi 2) \\ -G^{<}_{cc,\uparrow\sigma}(1\xi)\; \Sigma^{a}_{cc,\sigma\uparrow}(\xi 2) \end{bmatrix} & \begin{bmatrix} \Sigma^{r}_{cc,\uparrow\sigma}(1\xi)\, G^{<}_{cc,\sigma\downarrow}(\xi 2) \\ -G^{<}_{cc,\uparrow\sigma}(1\xi)\; \Sigma^{a}_{cc,\sigma\downarrow}(\xi 2) \end{bmatrix} \\ \begin{bmatrix} \Sigma^{r}_{cc,\downarrow\sigma}(1\xi)\, G^{<}_{cc,\sigma\uparrow}(\xi 2) \\ -G^{<}_{cc,\downarrow\sigma}(1\xi)\; \Sigma^{a}_{cc,\sigma\uparrow}(\xi 2) \end{bmatrix} & \begin{bmatrix} \Sigma^{r}_{cc,\downarrow\sigma}(1\xi)\; G^{<}_{cc,\sigma\downarrow}(\xi 2) \\ -G^{<}_{cc,\downarrow\sigma}(1\xi)\; \Sigma^{a}_{cc,\sigma\downarrow}(\xi 2) \end{bmatrix} \end{pmatrix}$$

$$+ \begin{pmatrix} \begin{bmatrix} \Sigma^{<}_{cc,\uparrow\sigma}(1\xi)\; G^{a}_{cc,\sigma\uparrow}(\xi 2) \\ -G^{r}_{cc,\uparrow\sigma}(1\xi)\; \Sigma^{<}_{cc,\sigma\uparrow}(\xi 2) \end{bmatrix} & \begin{bmatrix} \Sigma^{<}_{cc,\uparrow\sigma}(1\xi)\; G^{a}_{cc,\sigma\downarrow}(\xi 2) \\ -G^{r}_{cc,\uparrow\sigma}(1\xi)\; \Sigma^{<}_{cc,\sigma\downarrow}(\xi 2) \end{bmatrix} \\ \begin{bmatrix} \Sigma^{<}_{cc,\downarrow\sigma}(1\xi)\; G^{a}_{cc}(\xi 2) \\ -G^{r}_{cc,\downarrow\sigma}(1\xi)\; \Sigma^{<}_{cc,\sigma\uparrow}(\xi 2) \end{bmatrix} & \begin{bmatrix} \Sigma^{<}_{cc,\downarrow\sigma}(1\xi)\; G^{a}_{cc,\sigma\downarrow}(\xi 2) \\ -G^{r}_{cc,\downarrow\sigma}(1\xi)\; \Sigma^{<}_{cc,\sigma\downarrow}(\xi 2) \end{bmatrix} \end{pmatrix}$$

$$+\begin{pmatrix} \begin{bmatrix} \Delta^{e-h,r}_{hh,cv,\uparrow\sigma}(1\xi)\, g^{e-h,<}_{ee,vc,\sigma\uparrow}(\xi 2) \\ -g^{e-h,<}_{hh,cv,\uparrow\sigma}(1\xi)\;\Delta^{e-h,a}_{ee,vc,\sigma\uparrow}(\xi 2) \end{bmatrix} & \begin{bmatrix} \Delta^{e-h,r}_{hh,cv,\uparrow\sigma}(1\xi)\, g^{e-h,<}_{ee,vc,\sigma\downarrow}(\xi 2) \\ -g^{e-h,<}_{hh,cv,\uparrow\sigma}(1\xi)\;\Delta^{e-h,a}_{ee,vc,\sigma\downarrow}(\xi 2) \end{bmatrix} \\ \begin{bmatrix} \Delta^{e-h,r}_{hh,cv,\downarrow\sigma}(1\xi)\, g^{e-h,<}_{ee,vc,\sigma\uparrow}(\xi 2) \\ -g^{e-h,<}_{hh,cv,\downarrow\sigma}(1\xi)\;\Delta^{e-h,a}_{ee,vc,\sigma\uparrow}(\xi 2) \end{bmatrix} & \begin{bmatrix} \Delta^{e-h,r}_{hh,cv,\downarrow\sigma}(1\xi)\, g^{e-h,<}_{ee,vc,\sigma\downarrow}(\xi 2) \\ -g^{e-h,<}_{hh,cv,\downarrow\sigma}(1\xi)\;\Delta^{e-h,a}_{ee,vc,\sigma\downarrow}(\xi 2) \end{bmatrix} \end{pmatrix}$$

$$+\begin{pmatrix} \begin{bmatrix} \Delta^{e-h,<}_{hh,cv,\uparrow\sigma}(1\xi)\; g^{e-h,a}_{ee,vc,\sigma\uparrow}(\xi 2) \\ -g^{e-h,r}_{hh,cv,\uparrow\sigma}(1\xi)\;\Delta^{e-h,<}_{ee,vc,\sigma\uparrow}(\xi 2) \end{bmatrix} & \begin{bmatrix} \Delta^{e-h,<}_{hh,cv,\uparrow\sigma}(1\xi)\; g^{e-h,a}_{ee,vc,\sigma\downarrow}(\xi 2) \\ -g^{e-h,r}_{hh,cv,\uparrow\sigma}(1\xi)\;\Delta^{e-h,<}_{ee,vc,\sigma\downarrow}(\xi 2) \end{bmatrix} \\ \begin{bmatrix} \Delta^{e-h,<}_{hh,cv,\downarrow\sigma}(1\xi)\; g^{e-h,a}_{ee,vc,\sigma\uparrow}(\xi 2) \\ -g^{e-h,r}_{hh,cv,\downarrow\sigma}(1\xi)\;\Delta^{e-h,<}_{ee,vc,\sigma\uparrow}(\xi 2) \end{bmatrix} & \begin{bmatrix} \Delta^{e-h,<}_{hh,cv,\downarrow\sigma}(1\xi)\; g^{e-h,a}_{ee,vc,\sigma\downarrow}(\xi 2) \\ -g^{e-h,r}_{hh,cv,\downarrow\sigma}(1\xi)\;\Delta^{e-h,<}_{ee,vc,\sigma\downarrow}(\xi 2) \end{bmatrix} \end{pmatrix}, \tag{D.1}$$

where the repeated spin subscript, σ, indicates the use of Einstein summation convention over the spin degrees of freedom.

For the hole band, we have the corresponding matrix equation determined from Eq. (21.13) as

$$i\hbar\left(\frac{\partial}{\partial t_1}+\frac{\partial}{\partial t_2}\right)\begin{pmatrix} G^{e-h,<}_{vv,\uparrow\uparrow}(12) & G^{e-h,<}_{vv,\uparrow\downarrow}(12) \\ G^{e-h,<}_{vv,\downarrow\uparrow}(12) & G^{e-h,<}_{vv,\downarrow\downarrow}(12) \end{pmatrix}$$

$$=-\begin{pmatrix} \begin{bmatrix} v_{vv,\uparrow\sigma}(1\xi)\, G^{e-h,<}_{vv,\sigma\uparrow}(\xi 2) \\ -G^{e-h,<}_{vv,\uparrow\sigma}(1\xi)\, v^{T}_{vv,\sigma\uparrow}(\xi 2) \end{bmatrix} & \begin{bmatrix} v_{vv,\uparrow\sigma}(1\xi)\, G^{e-h,<}_{vv,\sigma\downarrow}(\xi 2) \\ -G^{e-h,<}_{vv,\uparrow\sigma}(1\xi)\, v^{T}_{vv,\sigma\downarrow}(\xi 2) \end{bmatrix} \\ \begin{bmatrix} v_{vv,\downarrow\sigma}(1\xi)\, G^{e-h,<}_{vv,\sigma\uparrow}(\xi 2) \\ -G^{e-h,<}_{vv,\downarrow\sigma}(1\xi)\, v^{T}_{vv,\sigma\uparrow}(\xi 2) \end{bmatrix} & \begin{bmatrix} v_{vv,\downarrow\sigma}(1\xi)\, G^{e-h,<}_{vv,\sigma\downarrow}(\xi 2) \\ -G^{e-h,<}_{vv,\downarrow\sigma}(1\xi)\, v^{T}_{vv,\sigma\downarrow}(\xi 2) \end{bmatrix} \end{pmatrix}$$

$$-\begin{pmatrix} \begin{bmatrix} \Sigma^{e-h,r}_{vv,\uparrow\sigma}(1\xi)\, G^{e-h,<}_{vv,\sigma\uparrow}(\xi 2) \\ -G^{e-h,<}_{vv,\uparrow\sigma}(1\xi)\, \Sigma^{e-h,a}_{vv,\sigma\uparrow}(\xi 2) \end{bmatrix} & \begin{bmatrix} \Sigma^{e-h,r}_{vv,\uparrow\sigma}(1\xi)\, G^{e-h,<}_{vv,\sigma\downarrow}(\xi 2) \\ -G^{e-h,<}_{vv,\uparrow\sigma}(1\xi)\, \Sigma^{e-h,a}_{vv,\sigma\downarrow}(\xi 2) \end{bmatrix} \\ \begin{bmatrix} \Sigma^{e-h,r}_{vv,\downarrow\sigma}(1\xi)\, G^{e-h,<}_{vv,\sigma\uparrow}(\xi 2) \\ -G^{e-h,<}_{vv,\downarrow\sigma}(1\xi)\, \Sigma^{e-h,a}_{vv,\sigma\uparrow}(\xi 2) \end{bmatrix} & \begin{bmatrix} \Sigma^{e-h,r}_{vv,\downarrow\sigma}(1\xi)\, G^{e-h,<}_{vv,\sigma\downarrow}(\xi 2) \\ -G^{e-h,<}_{vv,\downarrow\sigma}(1\xi)\, \Sigma^{e-h,a}_{vv,\sigma\downarrow}(\xi 2) \end{bmatrix} \end{pmatrix}$$

$$
-\begin{pmatrix}
\begin{bmatrix}
\Sigma^{e-h,<}_{vv,\uparrow\sigma}(1\xi)\,G^{e-h,a}_{vv,\sigma\uparrow}(\xi 2) \\
-G^{e-h,r}_{vv,\uparrow\sigma}(1\xi)\,\Sigma^{e-h,<}_{vv,\sigma\uparrow}(\xi 2)
\end{bmatrix} &
\begin{bmatrix}
\Sigma^{e-h,<}_{vv,\uparrow\sigma}(1\xi)\,G^{e-h,a}_{vv,\sigma\downarrow}(\xi 2) \\
-G^{e-h,r}_{vv,\uparrow\sigma}(1\xi)\,\Sigma^{e-h,<}_{vv,\sigma\downarrow}(\xi 2)
\end{bmatrix} \\
\begin{bmatrix}
\Sigma^{e-h,<}_{vv,\downarrow\sigma}(1\xi)\,G^{e-h,a}_{vv,\sigma\uparrow}(\xi 2) \\
-G^{e-h,r}_{vv,\downarrow\sigma}(1\xi)\,\Sigma^{e-h,<}_{vv,\sigma\uparrow}(\xi 2)
\end{bmatrix} &
\begin{bmatrix}
\Sigma^{e-h,<}_{vv,\downarrow\sigma}(1\xi)\,G^{e-h,a}_{vv,\sigma\downarrow}(\xi 2) \\
-G^{e-h,r}_{vv,\downarrow\sigma}(1\xi)\,\Sigma^{e-h,<}_{vv,\sigma\downarrow}(\xi 2)
\end{bmatrix}
\end{pmatrix}
$$

$$
-\begin{pmatrix}
\begin{bmatrix}
\Delta^{e-h,r}_{ee,vc,\uparrow\sigma}(1\xi)\,g^{e-h,<}_{hh,cv,\sigma\uparrow}(\xi 2) \\
-g^{e-h,<}_{ee,vc,\uparrow\sigma}(1\xi)\,\Delta^{e-h,a}_{hh,cv,\sigma\uparrow}(\xi 2)
\end{bmatrix} &
\begin{bmatrix}
\Delta^{e-h,r}_{ee,vc,\uparrow\sigma}(1\xi)\,g^{e-h,<}_{hh,cv,\sigma\downarrow}(\xi 2) \\
-g^{e-h,<}_{ee,vc,\uparrow\sigma}(1\xi)\,\Delta^{e-h,a}_{hh,cv,\sigma\downarrow}(\xi 2)
\end{bmatrix} \\
\begin{bmatrix}
\Delta^{e-h,r}_{ee,vc,\downarrow\sigma}(1\xi)\,g^{e-h,<}_{hh,cv,\sigma\uparrow}(\xi 2) \\
-g^{e-h,<}_{ee,vc,\downarrow\sigma}(1\xi)\,\Delta^{e-h,a}_{hh,cv,\sigma\uparrow}(\xi 2)
\end{bmatrix} &
\begin{bmatrix}
\Delta^{e-h,r}_{ee,vc,\downarrow\sigma}(1\xi)\,g^{e-h,<}_{hh,cv,\sigma\downarrow}(\xi 2) \\
-g^{e-h,<}_{ee,vc,\downarrow\sigma}(1\xi)\,\Delta^{e-h,a}_{hh,cv,\sigma\downarrow}(\xi 2)
\end{bmatrix}
\end{pmatrix}
$$

$$
-\begin{pmatrix}
\begin{bmatrix}
\Delta^{e-h,<}_{ee,vc,\uparrow\sigma}(1\xi)\,g^{e-h,a}_{hh,cv,\sigma\uparrow}(\xi 2) \\
-g^{e-h,r}_{ee,vc,\uparrow\sigma}(1\xi)\,\Delta^{e-h,<}_{hh,cv,\sigma\uparrow}(\xi 2)
\end{bmatrix} &
\begin{bmatrix}
\Delta^{e-h,<}_{ee,vc,\uparrow\sigma}(1\xi)\,g^{e-h,a}_{hh,cv,\sigma\downarrow}(\xi 2) \\
-g^{e-h,r}_{ee,vc,\uparrow\sigma}(1\xi)\,\Delta^{e-h,<}_{hh,cv,\sigma\downarrow}(\xi 2)
\end{bmatrix} \\
\begin{bmatrix}
\Delta^{e-h,<}_{ee,vc,\downarrow\sigma}(1\xi)\,g^{e-h,a}_{hh,cv,\sigma\uparrow}(\xi 2) \\
-g^{e-h,r}_{ee,vc,\downarrow\sigma}(1\xi)\,\Delta^{e-h,<}_{hh,cv,\sigma\uparrow}(\xi 2)
\end{bmatrix} &
\begin{bmatrix}
\Delta^{e-h,<}_{ee,vc,\downarrow\sigma}(1\xi)\,g^{e-h,a}_{hh,cv,\sigma\downarrow}(\xi 2) \\
-g^{e-h,r}_{ee,vc,\downarrow\sigma}(1\xi)\,\Delta^{e-h,<}_{hh,cv,\sigma\downarrow}(\xi 2)
\end{bmatrix}
\end{pmatrix}.
\tag{D.2}
$$

D.1 Spin-dependent pairing Green's functions

For the "pairing" between electron and holes, we have from Eq. (21.14)

$$
i\hbar\left(\frac{\partial}{\partial t_1}+\frac{\partial}{\partial t_2}\right)
\begin{pmatrix}
g^{e-h,<}_{hh,cv,\uparrow\uparrow}(12) & g^{e-h,<}_{hh,cv,\uparrow\downarrow}(12) \\
g^{e-h,<}_{hh,cv,\downarrow\uparrow}(12) & g^{e-h,<}_{hh,cv,\downarrow\downarrow}(12)
\end{pmatrix}
$$

$$
=\begin{pmatrix}
\begin{bmatrix}
v_{cc,\uparrow\sigma}(\xi 2)\,g^{e-h,<}_{hh,cv,\sigma\uparrow}(1\xi) \\
-g^{e-h,<}_{hh,cv,\uparrow\sigma}(\xi 2)\,v^{T}_{vv,\sigma\uparrow}(1\xi)
\end{bmatrix} &
\begin{bmatrix}
v_{cc,\uparrow\sigma}(\xi 2)\,g^{e-h,<}_{hh,cv,\sigma\downarrow}(1\xi) \\
-g^{e-h,<}_{hh,cv,\uparrow\sigma}(\xi 2)\,v^{T}_{vv,\sigma\downarrow}(1\xi)
\end{bmatrix} \\
\begin{bmatrix}
v_{cc,\downarrow\sigma}(\xi 2)\,g^{e-h,<}_{hh,cv,\sigma\uparrow}(1\xi) \\
-g^{e-h,<}_{hh,cv,\downarrow\sigma}(\xi 2)\,v^{T}_{vv,\sigma\uparrow}(1\xi)
\end{bmatrix} &
\begin{bmatrix}
v_{cc,\downarrow\sigma}(\xi 2)\,g^{e-h,<}_{hh,cv,\sigma\downarrow}(1\xi) \\
-g^{e-h,<}_{hh,cv,\downarrow\sigma}(\xi 2)\,v^{T}_{vv,\sigma\downarrow}(1\xi)
\end{bmatrix}
\end{pmatrix}
$$

$$+\begin{pmatrix} \begin{bmatrix} \Sigma^{e-h,r}_{cc,\uparrow\sigma}(1\xi)\, g^{e-h,<}_{hh,cv,\sigma\uparrow}(\xi 2) \\ -G^{e-h,<}_{cc,\uparrow\sigma}(1\xi)\, \Delta^{e-h,a}_{hh,cv,\sigma\uparrow}(\xi 2) \end{bmatrix} & \begin{bmatrix} \Sigma^{e-h,r}_{cc,\uparrow\sigma}(1\xi)\, g^{e-h,<}_{hh,cv,\sigma\downarrow}(\xi 2) \\ -G^{e-h,<}_{cc,\uparrow\sigma}(1\xi)\, \Delta^{e-h,a}_{hh,cv,\sigma\downarrow}(\xi 2) \end{bmatrix} \\ \begin{bmatrix} \Sigma^{e-h,r}_{cc,\downarrow\sigma}(1\xi)\, g^{e-h,<}_{hh,cv,\sigma\uparrow}(\xi 2) \\ -G^{e-h,<}_{cc,\downarrow\sigma}(1\xi)\, \Delta^{e-h,a}_{hh,cv,\sigma\uparrow}(\xi 2) \end{bmatrix} & \begin{bmatrix} \Sigma^{e-h,r}_{cc,\downarrow\sigma}(1\xi)\, g^{e-h,<}_{hh,cv,\sigma\downarrow}(\xi 2) \\ -G^{e-h,<}_{cc,\downarrow\sigma}(1\xi)\, \Delta^{e-h,a}_{hh,cv,\sigma\downarrow}(\xi 2) \end{bmatrix} \end{pmatrix}$$

$$+\begin{pmatrix} \begin{bmatrix} \Sigma^{e-h,<}_{cc,\uparrow\sigma}(1\xi)\, g^{e-h,a}_{hh,cv,\sigma\uparrow}(\xi 2) \\ -G^{e-h,r}_{cc,\uparrow\sigma}(1\xi)\, \Delta^{e-h,<}_{hh,cv,\sigma\uparrow}(\xi 2) \end{bmatrix} & \begin{bmatrix} \Sigma^{e-h,<}_{cc,\uparrow\sigma}(1\xi)\, g^{e-h,a}_{hh,cv,\sigma\downarrow}(\xi 2) \\ -G^{e-h,r}_{cc,\uparrow\sigma}(1\xi)\, \Delta^{e-h,<}_{hh,cv,\sigma\downarrow}(\xi 2) \end{bmatrix} \\ \begin{bmatrix} \Sigma^{e-h,<}_{cc,\downarrow\sigma}(1\xi)\, g^{e-h,a}_{hh,cv,\sigma\uparrow}(\xi 2) \\ -G^{e-h,r}_{cc,\downarrow\sigma}(1\xi)\, \Delta^{e-h,<}_{hh,cv,\sigma\uparrow}(\xi 2) \end{bmatrix} & \begin{bmatrix} \Sigma^{e-h,<}_{cc,\downarrow\sigma}(1\xi)\, g^{e-h,a}_{hh,cv,\sigma\downarrow}(\xi 2) \\ -G^{e-h,r}_{cc,\downarrow\sigma}(1\xi)\, \Delta^{e-h,<}_{hh,cv,\sigma\downarrow}(\xi 2) \end{bmatrix} \end{pmatrix}$$

$$+\begin{pmatrix} \begin{bmatrix} -\Delta^{e-h,r}_{hh,cv,\uparrow\sigma}(1\xi)\, G^{e-h,>T}_{vv,\sigma\uparrow}(\xi 2) \\ +g^{e-h,<}_{hh,cv,\uparrow\sigma}(1\xi)\, \Sigma^{e-h,rT}_{vv,\sigma\uparrow}(\xi 2) \end{bmatrix} & \begin{bmatrix} -\Delta^{e-h,r}_{hh,cv,\uparrow\sigma}(1\xi)\, G^{e-h,>T}_{vv,\sigma\downarrow}(\xi 2) \\ +g^{e-h,<}_{hh,cv,\uparrow\sigma}(1\xi)\, \Sigma^{e-h,rT}_{vv,\sigma\downarrow}(\xi 2) \end{bmatrix} \\ \begin{bmatrix} -\Delta^{e-h,r}_{hh,cv,\downarrow\sigma}(1\xi)\, G^{e-h,>T}_{vv,\sigma\uparrow}(\xi 2) \\ +g^{e-h,<}_{hh,cv,\downarrow\sigma}(1\xi)\, \Sigma^{e-h,rT}_{vv,\sigma\uparrow}(\xi 2) \end{bmatrix} & \begin{bmatrix} -\Delta^{e-h,r}_{hh,cv,\downarrow\sigma}(1\xi)\, G^{e-h,>T}_{vv,\sigma\downarrow}(\xi 2) \\ +g^{e-h,<}_{hh,cv,\downarrow\sigma}(1\xi)\, \Sigma^{e-h,rT}_{vv,\sigma\downarrow}(\xi 2) \end{bmatrix} \end{pmatrix}$$

$$+\begin{pmatrix} \begin{bmatrix} -\Delta^{e-h,<}_{hh,cv,\uparrow\sigma}(1\xi)\, G^{e-h,rT}_{vv,\sigma\uparrow}(\xi 2) \\ +g^{e-h,r}_{hh,cv,\uparrow\sigma}(1\xi)\, \Sigma^{e-h,>T}_{vv,\sigma\uparrow}(\xi 2) \end{bmatrix} & \begin{bmatrix} -\Delta^{e-h,<}_{hh,cv,\uparrow\sigma}(1\xi)\, G^{e-h,rT}_{vv,\sigma\downarrow}(\xi 2) \\ +g^{e-h,r}_{hh,cv,\uparrow\sigma}(1\xi)\, \Sigma^{e-h,>T}_{vv,\sigma\downarrow}(\xi 2) \end{bmatrix} \\ \begin{bmatrix} -\Delta^{e-h,<}_{hh,cv,\downarrow\sigma}(1\xi)\, G^{e-h,rT}_{vv,\sigma\uparrow}(\xi 2) \\ +g^{e-h,r}_{hh,cv,\downarrow\sigma}(1\xi)\, \Sigma^{e-h,>T}_{vv,\sigma\uparrow}(\xi 2) \end{bmatrix} & \begin{bmatrix} -\Delta^{e-h,<}_{hh,cv,\downarrow\sigma}(1\xi)\, G^{e-h,rT}_{vv,\sigma\downarrow}(\xi 2) \\ +g^{e-h,r}_{hh,cv,\downarrow\sigma}(1\xi)\, \Sigma^{e-h,>T}_{vv,\sigma\downarrow}(\xi 2) \end{bmatrix} \end{pmatrix}, \tag{D.3}$$

and for its non-equilibrium (reverse) "conjugate" process,[1] the matrix equation derived from Eq. (21.15), is

$$i\hbar\left(\frac{\partial}{\partial t_1}+\frac{\partial}{\partial t_2}\right)\begin{pmatrix} g^{e-h,<}_{ee,vc,\uparrow\uparrow}(12) & g^{e-h,<}_{ee,vc,\uparrow\downarrow}(12) \\ g^{e-h,<}_{ee,vc,\downarrow\uparrow}(12) & g^{e-h,<}_{ee,vc,\downarrow\downarrow}(12) \end{pmatrix}$$

$$=\begin{pmatrix} \begin{bmatrix} v_{vv,\uparrow\sigma}(1\xi)\, g^{e-h,<}_{ee,vc,\sigma\uparrow}(\xi 2) \\ -g^{e-h,<}_{ee,vc,\uparrow\sigma}(1\xi)\, v^{T}_{c,\sigma\uparrow}(\xi 2) \end{bmatrix} & \begin{bmatrix} v_{vv,\uparrow\sigma}(1\xi)\, g^{e-h,<}_{ee,vc,\sigma\downarrow}(\xi 2) \\ -g^{e-h,<}_{ee,vc,\uparrow\sigma}(1\xi)\, v^{T}_{c,\sigma\downarrow}(\xi 2) \end{bmatrix} \\ \begin{bmatrix} v_{vv,\downarrow\sigma}(1\xi)\, g^{e-h,<}_{ee,vc,\sigma\uparrow}(\xi 2) \\ -g^{e-h,<}_{ee,vc,\downarrow\sigma}(1\xi)\, v^{T}_{c,\sigma\uparrow}(\xi 2) \end{bmatrix} & \begin{bmatrix} v_{vv,\downarrow\sigma}(1\xi)\, g^{e-h,<}_{ee,vc,\sigma\downarrow}(\xi 2) \\ -g^{e-h,<}_{ee,vc,\downarrow\sigma}(1\xi)\, v^{T}_{c,\sigma\downarrow}(\xi 2) \end{bmatrix} \end{pmatrix}$$

[1]For highly non-equilibrium situation, the reverse process is not simply the Hermitian conjugate of the "forward" process.

$$
+\begin{pmatrix}
\begin{bmatrix} -\Sigma^{e-h,aT}_{vv,\uparrow\sigma}(1\xi)\, g^{e-h,<}_{ee,vc,\sigma\uparrow}(\xi 2) \\ +G^{e-h,>T}_{vv,\uparrow\sigma}(1\xi)\, \Delta^{e-h,a}_{ee,vc,\sigma\uparrow}(\xi 2) \end{bmatrix} & \begin{bmatrix} -\Sigma^{e-h,aT}_{vv,\uparrow\sigma}(1\xi)\, g^{e-h,<}_{ee,vc,\sigma\downarrow}(\xi 2) \\ +G^{e-h,>T}_{vv,\uparrow\sigma}(1\xi)\, \Delta^{e-h,a}_{ee,vc,\sigma\downarrow}(\xi 2) \end{bmatrix} \\
\begin{bmatrix} -\Sigma^{e-h,aT}_{vv,\downarrow\sigma}(1\xi)\, g^{e-h,<}_{ee,vc,\sigma\uparrow}(\xi 2) \\ +G^{e-h,>T}_{vv,\downarrow\sigma}(1\xi)\, \Delta^{e-h,a}_{ee,vc,\sigma\uparrow}(\xi 2) \end{bmatrix} & \begin{bmatrix} -\Sigma^{e-h,aT}_{vv,\downarrow\sigma}(1\xi)\, g^{e-h,<}_{ee,vc,\sigma\downarrow}(\xi 2) \\ +G^{e-h,>T}_{vv,\downarrow\sigma}(1\xi)\, \Delta^{e-h,a}_{ee,vc,\sigma\downarrow}(\xi 2) \end{bmatrix}
\end{pmatrix}
$$

$$
+\begin{pmatrix}
\begin{bmatrix} -\Sigma^{e-h,>T}_{vv,\uparrow\sigma}(1\xi)\, g^{e-h,a}_{ee,vc,\sigma\uparrow}(\xi 2) \\ +G^{e-h,aT}_{vv,\uparrow\sigma}(1\xi)\, \Delta^{e-h,<}_{ee,vc,\sigma\uparrow}(\xi 2) \end{bmatrix} & \begin{bmatrix} -\Sigma^{e-h,>T}_{vv,\uparrow\sigma}(1\xi)\, g^{e-h,a}_{ee,vc,\sigma\downarrow}(\xi 2) \\ +G^{e-h,aT}_{vv,\uparrow\sigma}(1\xi)\, \Delta^{e-h,<}_{ee,vc,\sigma\downarrow}(\xi 2) \end{bmatrix} \\
\begin{bmatrix} -\Sigma^{e-h,>T}_{vv,\downarrow\sigma}(1\xi)\, g^{e-h,a}_{ee,vc,\sigma\uparrow}(\xi 2) \\ +G^{e-h,aT}_{vv,\downarrow\sigma}(1\xi)\, \Delta^{e-h,<}_{ee,vc,\sigma\uparrow}(\xi 2) \end{bmatrix} & \begin{bmatrix} -\Sigma^{e-h,>T}_{vv,\downarrow\sigma}(1\xi)\, g^{e-h,a}_{ee,vc,\sigma\downarrow}(\xi 2) \\ +G^{e-h,aT}_{vv,\downarrow\sigma}(1\xi)\, \Delta^{e-h,<}_{ee,vc,\sigma\downarrow}(\xi 2) \end{bmatrix}
\end{pmatrix}
$$

$$
+\begin{pmatrix}
\begin{bmatrix} \Delta^{e-h,r}_{ee,vc,\uparrow\sigma}(1\xi)\, G^{e-h,<}_{cc,\sigma\uparrow}(\xi 2) \\ -g^{e-h,<}_{ee,vc,\uparrow\sigma}(1\xi)\, \Sigma^{e-h,a}_{cc,\sigma\uparrow}(\xi 2) \end{bmatrix} & \begin{bmatrix} \Delta^{e-h,r}_{ee,vc,\uparrow\sigma}(1\xi)\, G^{e-h,<}_{cc,\sigma\downarrow}(\xi 2) \\ -g^{e-h,<}_{ee,vc,\uparrow\sigma}(1\xi)\, \Sigma^{e-h,a}_{cc,\sigma\downarrow}(\xi 2) \end{bmatrix} \\
\begin{bmatrix} \Delta^{e-h,r}_{ee,vc,\downarrow\sigma}(1\xi)\, G^{e-h,<}_{cc,\sigma\uparrow}(\xi 2) \\ -g^{e-h,<}_{ee,vc,\downarrow\sigma}(1\xi)\, \Sigma^{e-h,a}_{cc,\sigma\uparrow}(\xi 2) \end{bmatrix} & \begin{bmatrix} \Delta^{e-h,r}_{ee,vc,\downarrow\sigma}(1\xi)\, G^{e-h,<}_{cc,\sigma\downarrow}(\xi 2) \\ -g^{e-h,<}_{ee,vc,\downarrow\sigma}(1\xi)\, \Sigma^{e-h,a}_{cc,\sigma\downarrow}(\xi 2) \end{bmatrix}
\end{pmatrix}
$$

$$
+\begin{pmatrix}
\begin{bmatrix} \Delta^{e-h,<}_{ee,vc,\uparrow\sigma}(1\xi)\, G^{e-h,a}_{cc,\sigma\uparrow}(\xi 2) \\ -g^{e-h,r}_{ee,vc,\uparrow\sigma}(1\xi)\, \Sigma^{e-h,<}_{cc,\sigma\uparrow}(\xi 2) \end{bmatrix} & \begin{bmatrix} \Delta^{e-h,<}_{ee,vc,\uparrow\sigma}(1\xi)\, G^{e-h,a}_{cc,\sigma\downarrow}(\xi 2) \\ -g^{e-h,r}_{ee,vc,\uparrow\sigma}(1\xi)\, \Sigma^{e-h,<}_{cc,\sigma\downarrow}(\xi 2) \end{bmatrix} \\
\begin{bmatrix} \Delta^{e-h,<}_{ee,vc,\downarrow\sigma}(1\xi)\, G^{e-h,a}_{cc,\sigma\uparrow}(\xi 2) \\ -g^{e-h,r}_{ee,vc,\downarrow\sigma}(1\xi)\, \Sigma^{e-h,<}_{cc,\sigma\uparrow}(\xi 2) \end{bmatrix} & \begin{bmatrix} \Delta^{e-h,<}_{ee,vc,\downarrow\sigma}(1\xi)\, G^{e-h,a}_{cc,\sigma\downarrow}(\xi 2) \\ -g^{e-h,r}_{ee,vc,\downarrow\sigma}(1\xi)\, \Sigma^{e-h,<}_{cc,\sigma\downarrow}(\xi 2) \end{bmatrix}
\end{pmatrix}. \tag{D.4}
$$

D.2 Semiconductor Bloch equations for spin-magnetization vector

The twelve transport equations for the components of the multi-band spin magnetization transport equations can be written is more compact form as spin-vector equations, a generalization of Eq. (21.1) to multi-band spin-magnetization quantum transport equations. This corresponds to a generalized semiconductor Bloch equations for spin systems. The results can be summarized as four-coupled *vector* equations for the spin magnetization-distribution functions,

$$
\begin{aligned}
& i\hbar\left(\frac{\partial}{\partial t_1}+\frac{\partial}{\partial t_2}\right)\vec{S}^{<}_{cc} \\
&= \frac{1}{2}\left\{\left[\bar{H}_c, \vec{S}^{<}_{cc}\right]+\left[\vec{\mathcal{B}}_{cc}, S^{<}_{cc,o}\right]+i\left[\vec{\mathcal{B}}_{cc}\times\vec{S}^{<}_{cc}-\vec{S}^{<}_{cc}\times\vec{\mathcal{B}}_{cc}\right]\right\}
\end{aligned}
$$

$$
\begin{aligned}
&+\frac{1}{2}\left[\begin{array}{c}+\left[\bar{\Sigma}^{r}_{cc}\vec{S}^{<}_{cc}-\ \vec{S}^{<}_{cc}\bar{\Sigma}^{a}_{cc}\right]+\left[\vec{\Xi}^{r}_{cc}\ S^{<}_{cc,o}-S^{<}_{cc,o}\vec{\Xi}^{a}_{cc}\right]\\ +i\left[\vec{\Xi}^{r}_{cc}\times\vec{S}^{<}_{cc}-\vec{S}^{<}_{cc}\times\vec{\Xi}^{a}_{cc}\right]\end{array}\right]\\
&+\frac{1}{2}\left[\begin{array}{c}\left[\bar{\Sigma}^{<}_{cc}\vec{S}^{a}_{cc}-\vec{S}^{r}_{cc}\bar{\Sigma}^{<}_{cc}\right]+\left[\vec{\Xi}^{<}_{cc}S^{a}_{cc,o}-S^{r}_{cc,o}\vec{\Xi}^{<}_{cc}\right]\\ +i\left[\vec{\Xi}^{<}_{cc}\times\vec{S}^{a}_{cc}-\vec{S}^{r}_{cc}\times\vec{\Xi}^{<}_{cc}\right]\end{array}\right]\\
&+\frac{1}{2}\left[\begin{array}{c}\frac{1}{i\hbar}\left[\bar{\Delta}^{r}_{hh,cv}\vec{S}^{<}_{vc}-\vec{S}^{<}_{cv}\bar{\Delta}^{a}_{ee,vc}\right]+\frac{1}{i\hbar}\left[\vec{\delta}^{r}_{hh,cv}S^{<}_{vc,o}-S^{<}_{cv,o}\vec{\delta}^{a}_{ee,vc}\right]\\ +\frac{1}{\hbar}\left[\vec{\delta}^{r}_{hh,cv}\times\vec{S}^{<}_{vc}-\vec{S}^{<}_{cv}\times\vec{\delta}^{a}_{ee,vc}\right]\end{array}\right]\\
&+\frac{1}{2}\left[\begin{array}{c}+\left[\bar{\Delta}^{<}_{hh,cv}\ \vec{S}^{a}_{vc}-\vec{S}^{r}_{cv}\bar{\Delta}^{<}_{ee,vc}\right]+\left[\vec{\delta}^{<}_{hh,cv}\ S^{a}_{vc,o}-S^{r}_{cv,o}\vec{\delta}^{<}_{ee,vc}\right]\\ +i\left[\vec{\delta}^{<}_{hh,cv}\times\vec{S}^{a}_{vc}-\vec{S}^{r}_{cv}\times\vec{\delta}^{<}_{ee,vc}\right]\end{array}\right],
\end{aligned}
\tag{D.5}
$$

$$
\begin{aligned}
&i\hbar\left(\frac{\partial}{\partial t_1}+\frac{\partial}{\partial t_2}\right)\vec{S}^{<}_{cv}\\
&=\frac{1}{2}\left[\begin{array}{c}\left[\bar{H}_c\vec{S}^{<}_{cv}-\vec{S}^{<}_{cv}\bar{H}_v\right]+\left[\vec{\mathcal{B}}_{cc}S^{<}_{cv,o}-S^{<}_{cv,o}\vec{\mathcal{B}}_{vv}\right]\\ +i\left[\vec{\mathcal{B}}_{cc}\times\vec{S}^{<}_{cv}-\vec{S}^{<}_{cv}\times\vec{\mathcal{B}}_{vv}\right]\end{array}\right]\\
&\times\frac{1}{2}\left[\begin{array}{c}\left[\bar{\Sigma}^{r}_{cc}\vec{S}^{<}_{cv}-\vec{S}^{<}_{cc}\bar{\Delta}^{a}_{hh,cv}\right]+\left[\vec{\Xi}^{r}_{cc}S^{<}_{cv,o}-S^{<}_{cc,o}\vec{\delta}^{a}_{hh,cv}\right]\\ +i\left[\vec{\Xi}^{r}_{cc}\times\vec{S}^{<}_{cv}-\vec{S}^{<}_{cc}\times\vec{\delta}^{a}_{hh,cv}\right]\end{array}\right]\\
&\times\frac{1}{2}\left[\begin{array}{c}\left[\bar{\Sigma}^{<}_{cc}\vec{S}^{a}_{cv}-\vec{S}^{r}_{cc}\bar{\Delta}^{<}_{hh,cv}\right]+\left[\vec{\Xi}^{<}_{cc}S^{a}_{cv,o}-S^{r}_{cc,o}\ \vec{\delta}^{<}_{hh,cv}\right]\\ +i\left[\vec{\Xi}^{<}_{cc}\times\vec{S}^{a}_{cv}-\vec{S}^{r}_{cc}\times\vec{\delta}^{<}_{hh,cv}\right]\end{array}\right]\\
&\times\frac{1}{2}\left[\begin{array}{c}-\left[\bar{\Delta}^{r}_{hh,cv}\vec{S}^{>T}_{vv}-\vec{S}^{<}_{cv}\bar{\Sigma}^{rT}_{\ \ vv}\right]-\left[\vec{\delta}^{r}_{hh,cv}S^{>T}_{vv,o}-S^{<}_{cv,o}\vec{\Xi}^{rT}_{\ \ vv}\right]\\ +i\left[\vec{S}^{<}_{cv}\times\vec{\Xi}^{rT}_{\ \ vv}-\vec{\delta}^{r}_{hh,cv}\times\vec{S}^{,>T}_{vv}\right]\end{array}\right]\\
&\times\frac{1}{2}\left[\begin{array}{c}-\left[\bar{\Delta}^{<}_{hh,cv}\vec{S}^{rT}_{vv}-\vec{S}^{r}_{cv}\bar{\Sigma}^{>T}_{\ \ vv}\right]-\left[\vec{\delta}^{<}_{hh,cv}S^{rT}_{vv,o}-S^{r}_{cv,o}\vec{\Xi}^{>T}_{\ \ vv}\right]\\ +i\left[\vec{S}^{r}_{cv}\times\vec{\Xi}^{>T}_{\ \ vv}-\vec{\delta}^{<}_{hh,cv}\times\vec{S}^{rT}_{vv}\right]\end{array}\right],
\end{aligned}
\tag{D.6}
$$

$$
\begin{aligned}
&i\hbar\left(\frac{\partial}{\partial t_1}+\frac{\partial}{\partial t_2}\right)\vec{S}^{<}_{vc}\\
&=\frac{1}{2}\left[\begin{array}{c}\left[\bar{H}_v\vec{S}^{<}_{vc}-\vec{S}^{<}_{vc}\bar{H}_c\right]+\left[\vec{\mathcal{B}}_{vv}S^{<}_{vc,o}-S^{<}_{vc,o}\vec{\mathcal{B}}_{cc}\right]\\ +i\left[\vec{\mathcal{B}}_{vv}\times\vec{S}^{<}_{vc}-\vec{S}^{<}_{vc}\times\vec{\mathcal{B}}_{cc}\right]\end{array}\right]
\end{aligned}
$$

$$
+\frac{1}{2}\left[\begin{array}{c}-\left[\bar{\Sigma}^{a,T}_{vv}\vec{S}^{<}_{vc}-\vec{S}^{>T}_{vv}\bar{\Delta}^{a}_{ee,vc}\right]-\left[\vec{\Xi}^{a,T}_{vv}S^{<}_{vc,o}-S^{>T}_{vv,o}\vec{\delta}^{a}_{ee,vc}\right]\\ +i\left[\vec{S}^{>T}_{vv}\times\vec{\delta}^{a}_{ee,vc}-\vec{\Xi}^{a,T}_{vv}\times\vec{S}^{<}_{vc}\right]\end{array}\right]
$$

$$
+\frac{1}{2}\left[\begin{array}{c}-\left[\bar{\Sigma}^{>T}_{vv}\;\vec{S}^{a}_{vc}-\vec{S}^{aT}_{vv}\bar{\Delta}^{<}_{ee,vc}\right]-\left[\vec{\Xi}^{>T}_{vv}S^{a}_{vc,o}-S^{aT}_{vv,o}\vec{\delta}^{<}_{ee,vc}\right]\\ +i\left[\vec{S}^{aT}_{vv}\times\vec{\delta}^{<}_{ee,vc}-\vec{\Xi}^{>T}_{vv}\times\vec{S}^{a}_{vc}\right]\end{array}\right]
$$

$$
+\frac{1}{2}\left[\begin{array}{c}\left[\bar{\Delta}^{r}_{ee,vc}\vec{S}^{<}_{cc}-\vec{S}^{<}_{vc}\bar{\Sigma}^{a}_{cc}\right]+\left[\vec{\delta}^{r}_{ee,vc}S^{<}_{cc,o}-S^{<}_{vc,o}\vec{\Xi}^{a}_{cc}\right]\\ +i\left[\vec{\delta}^{r}_{ee,vc}\times\vec{S}^{<}_{cc}-\vec{S}^{<}_{vc}\times\vec{\Xi}^{a}_{cc}\right]\end{array}\right]
$$

$$
+\frac{1}{2}\left[\begin{array}{c}\left[\bar{\Delta}^{<}_{ee,vc}\vec{S}^{a}_{cc}-\vec{S}^{r}_{vc}\bar{\Sigma}^{<}_{cc}\right]+\left[\vec{\delta}^{<}_{ee,vc}\;S^{a}_{cc,o}-S^{r}_{vc,o}\vec{\Xi}^{<}_{cc}\right]\\ +i\left[\vec{\delta}^{<}_{ee,vc}\;\times\vec{S}^{a}_{cc}-\vec{S}^{r}_{vc}\times\vec{\Xi}^{<}_{cc}\right]\end{array}\right], \tag{D.7}
$$

$$
i\hbar\left(\frac{\partial}{\partial t_1}+\frac{\partial}{\partial t_2}\right)\vec{S}^{<}_{vv}
$$

$$
-\,-\frac{1}{2}\left\{\left[\bar{H}_{v},\vec{S}^{<}_{vv}\right]+\left[\vec{\mathcal{B}}_{vv},S^{<}_{vv,o}\right]+i\left[\vec{\mathcal{B}}_{vv}\times\vec{S}^{<}_{vv}-\vec{S}^{<}_{vv}\times\vec{\mathcal{B}}_{vv}\right]\right\}
$$

$$
-\frac{1}{2}\left[\begin{array}{c}\left[\bar{\Sigma}^{r}{}_{vv}\vec{S}^{<}_{vv}-\vec{S}^{<}_{vv}\bar{\Sigma}^{a}{}_{vv}\right]+\left[\vec{\Xi}^{r}{}_{vv}S^{<}_{vv,o}-S^{<}_{vv,o}\vec{\Xi}^{a}{}_{vv}\right]\\ +i\left[\vec{\Xi}^{r}{}_{vv}\times\vec{S}^{<}_{vv}\;-\vec{S}^{<}_{vv}\times\vec{\Xi}^{a}{}_{vv}\right]\end{array}\right]
$$

$$
-\frac{1}{2}\left[\begin{array}{c}\left[\bar{\Sigma}^{<}_{vv}\vec{S}^{a}_{vv}-\vec{S}^{r}_{vv}\bar{\Sigma}^{<}{}_{vv}\right]+\left[\vec{\Xi}^{<}{}_{vv}S^{a}_{vv,o}-S^{r}_{vv,o}\vec{\Xi}^{<}{}_{vv}\right]\\ +i\left[\vec{\Xi}^{<}_{vv}\times\vec{S}^{a}_{vv}-\vec{S}^{r}_{vv}\times\vec{\Xi}^{<}{}_{vv}\right]\end{array}\right]
$$

$$
-\frac{1}{2}\left[\begin{array}{c}\frac{1}{i\hbar}\left[\bar{\Delta}^{r}_{ee,vc}\vec{S}^{<}_{cv}-\vec{S}^{<}_{vc}\bar{\Delta}^{a}_{hh,cv}\right]+\frac{1}{i\hbar}\left[\vec{\delta}^{r}_{ee,vc}S^{<}_{cv,o}-S^{<}_{vc,o}\vec{\delta}^{a}_{hh,cv}\right]\\ +\frac{1}{\hbar}\left[\vec{\delta}^{r}_{ee,vc}\times\vec{S}^{<}_{cv}\;-\vec{S}^{<}_{vc}\times\vec{\delta}^{a}_{hh,cv}\right]\end{array}\right]
$$

$$
-\frac{1}{2}\left[\begin{array}{c}\left[\bar{\Delta}^{<}_{ee,vc}\vec{S}^{a}_{cv}-\vec{S}^{r}_{vc}\bar{\Delta}^{<}_{hh,cv}\right]+\left[\vec{\delta}^{<}_{ee,vc}S^{a}_{cv,o}-S^{r}_{vc,o}\vec{\delta}^{<}_{hh,cv}\right]\\ +i\left[\vec{\delta}^{<}_{ee,vc}\times\vec{S}^{a}_{cv}\;-\vec{S}^{r}_{vc}\times\vec{\delta}^{<}_{hh,cv}\right]\end{array}\right]. \tag{D.8}
$$

D.3 Pseudo-spin correlation functions

In addition to the twelve multi-band magnetization transport equations of the Pauli–Dirac spin components, we still have four more equations for the $S^{<}_{\alpha\beta,o}$ making a total of 16 equations. By the same token as was done for the 2×2 real-spin matrices, we can separate the 2×2 matrix containing $S^{<}_{\alpha\beta,o}$ into a term independent of the band indices, the total charge in the system, plus the pseudo-spin magnetization terms. To do this, we need to revert to the electron picture as discussed before. Then, the additional magnetization transport equations has to do with pseudo-spin magnetization by virtue of our two-band model, i.e., the presence of two-band discrete quantum labels v and c.

From Eqs. (21.21), (21.22), (21.29), and (21.30), we have the following four transport equations for the Pauli–Dirac spin scalar $S^<_{\alpha\beta,o}$, where the subscripts α and β have the range on the set $\{v,c\}$,

$$
\begin{aligned}
i\hbar&\left(\frac{\partial}{\partial t_1}+\frac{\partial}{\partial t_2}\right)S^<_{cc,o}\\
&=\frac{1}{2}\left[\bar{H}_c,S^<_{cc,o}\right]+\frac{1}{2}\left[\vec{\mathcal{B}}_{cc}\cdot\vec{S}^<_{cc}-\vec{S}^<_{cc}\cdot\vec{\mathcal{B}}_{cc}\right]\\
&\quad+\frac{1}{2}\left[\bar{\Sigma}^r_{cc}S^<_{cc,o}-\ S^<_{cc,o}\bar{\Sigma}^a_{cc}\right]+\frac{1}{2}\left[\vec{\Xi}^r_{cc}\cdot\vec{S}^<_{cc}-\vec{S}^<_{cc}\cdot\vec{\Xi}^a_{cc}\right]\\
&\quad+\frac{1}{2}\left[\bar{\Sigma}^<_{cc}S^a_{cc,o}-\ S^r_{cc,o}\bar{\Sigma}^<_{cc}\right]+\frac{1}{2}\left[\vec{\Xi}^<_{cc}\cdot\vec{S}^a_{cc}-\vec{S}^r_{cc}\cdot\vec{\Xi}^<_{cc}\right]\\
&\quad+\frac{1}{2}\left[\bar{\Delta}^r_{hh,cv}S^<_{vc,o}-\ S^<_{cv,o}\bar{\Delta}^a_{ee,vc}\right]+\frac{1}{2}\left[\vec{\delta}^r_{hh,cv}\cdot\vec{S}^<_{vc}-\vec{S}^<_{cv}\cdot\vec{\delta}^a_{ee,vc}\right]\\
&\quad+\frac{1}{2}\left[\bar{\Delta}^<_{hh,cv}S^a_{vc,o}-S^r_{cv,o}\bar{\Delta}^<_{ee,vc}\right]+\frac{1}{2}\left[\vec{\delta}^<_{hh,cv}\cdot\vec{S}^a_{vc}-\vec{S}^r_{cv}\cdot\vec{\delta}^<_{ee,vc}\right],
\end{aligned}
\tag{D.9}
$$

$$
\begin{aligned}
i\hbar&\left(\frac{\partial}{\partial t_1}+\frac{\partial}{\partial t_2}\right)S^<_{cv,o}\\
&=\frac{1}{2}\left[\vec{\mathcal{B}}_{cc}\cdot\vec{S}^<_{cv}-\vec{S}^<_{cv}\cdot\vec{\mathcal{B}}_{vv}\right]+\frac{1}{2}\left[\bar{H}_cS^<_{cv,o}-S^<_{cv,o}\bar{H}_v\right]\\
&\quad+\frac{1}{2}\left[\vec{\Xi}^r_{cc}\cdot\ \vec{S}^<_{cv}-\vec{S}^<_{cc}\cdot\vec{\delta}^a_{hh,cv}\right]+\frac{1}{2}\left[\bar{\Sigma}^r_{cc}S^<_{cv,o}-S^<_{cc,o}\bar{\Delta}^a_{hh,cv}\right]\\
&\quad+\frac{1}{2}\left[\vec{\Xi}^<_{cc}\cdot\vec{S}^a_{cv}-\vec{S}^r_{cc}\cdot\vec{\delta}^<_{hh,cv}\right]+\frac{1}{2}\left[\bar{\Sigma}^<_{cc}S^a_{cv,o}-S^r_{cc,o}\bar{\Delta}^<_{hh,cv}\right]\\
&\quad-\frac{1}{2}\left[\bar{\Delta}^r_{hh,cv}\ S^{>T}_{vv,o}-S^<_{cv,o}\bar{\Sigma}^{rT}_{vv}\right]-\frac{1}{2}\left[\vec{\delta}^r_{hh,cv}\cdot\vec{S}^{>T}_{vv}-\vec{S}^<_{cv}\cdot\vec{\Xi}^{rT}_{vv}\right]\\
&\quad-\frac{1}{2}\left[\bar{\Delta}^<_{hh,cv}S^{rT}_{vv,o}-S^r_{cv,o}\bar{\Sigma}^{>T}_{vv}\right]-\frac{1}{2}\left[\vec{\delta}^<_{hh,cv,z}\cdot\vec{S}^{rT}_{z,vv}-\vec{S}^r_{z,cv}\cdot\vec{\Xi}^{>T}_{vv,z}\right],
\end{aligned}
\tag{D.10}
$$

$$
\begin{aligned}
i\hbar & \left(\frac{\partial}{\partial t_1}+\frac{\partial}{\partial t_2}\right) S^<_{vc,o} \\
&= \frac{1}{2}\left[\left[\bar{H}_v S^<_{vc,o} - S^<_{vc,o}\bar{H}_c\right] + \frac{1}{2}\left[\vec{\mathcal{B}}_{vv}\cdot\vec{S}^<_{vc} - \vec{S}^<_{vc}\cdot\vec{\mathcal{B}}_{cc}\right]\right] \\
&\quad -\frac{1}{2}\left[\bar{\Sigma}^{aT}_{vv}\, S^<_{vc,o} - S^{>T}_{vv,o}\bar{\Delta}^a_{ee,vc}\right] - \frac{1}{2}\left[\vec{\Xi}^{aT}_{vv}\cdot\vec{S}^<_{vc} - \vec{S}^{>T}_{vv}\cdot\vec{\delta}^a_{ee,vc}\right] \\
&\quad -\frac{1}{2}\left[\bar{\Sigma}^{>T}_{vv} S^a_{vc,o} - S^{aT}_{vv,o}\bar{\Delta}^<_{ee,vc}\right] - \frac{1}{2}\left[\vec{\Xi}^{>T}_{vv}\cdot\vec{S}^a_{vc} - \vec{S}^{aT}_{vv}\cdot\vec{\delta}^<_{ee,vc}\right] \\
&\quad +\frac{1}{2}\left[\bar{\Delta}^r_{ee,vc}\, S^<_{cc,o} - S^<_{vc,o}\bar{\Sigma}^a_{cc}\right] + \frac{1}{2}\left[\vec{\delta}^r_{ee,vc}\cdot\vec{S}^<_{cc} - \vec{S}^<_{vc}\cdot\vec{\Xi}^a_{cc}\right] \\
&\quad +\frac{1}{2}\left[\bar{\Delta}^<_{ee,vc} S^a_{cc,o} - S^r_{vc,o}\bar{\Sigma}^<_{cc}\right] + \frac{1}{2}\left[\vec{\delta}^<_{ee,vc,z}\cdot\vec{S}^a_{z,cc} - \vec{S}^r_{z,vc}\cdot\vec{\Xi}^<_{cc,z}\right],
\end{aligned}
\tag{D.11}
$$

$$
\begin{aligned}
i\hbar & \left(\frac{\partial}{\partial t_1}+\frac{\partial}{\partial t_2}\right) S^<_{vv,o} \\
&= \frac{1}{2}\left[\left[\bar{H}_v S^<_{vv,o} - S^<_{vv,o}\bar{H}_v\right] + \frac{1}{2}\left[\vec{\mathcal{B}}_{vv}\cdot\vec{S}^<_{vv} - \vec{S}^<_{vv}\cdot\vec{\mathcal{B}}_{vv}\right]\right] \\
&\quad +\frac{1}{2}\left[\bar{\Sigma}^r{}_{vv} S^<_{vv,o} - S^<_{vv,o}\bar{\Sigma}^a{}_{vv}\right] + \frac{1}{2}\left[\vec{\Xi}^r{}_{vv}\cdot\vec{S}^<_{vv} - \vec{S}^<_{vv}\cdot\vec{\Xi}^a{}_{vv}\right] \\
&\quad +\frac{1}{2}\left[\bar{\Sigma}^<{}_{vv} S^a_{vv,o} - S^r_{vv,o}\bar{\Sigma}^<{}_{vv}\right] + \frac{1}{2}\left[\vec{\Xi}^<{}_{vv}\cdot\vec{S}^a_{vv} - \vec{S}^r_{vv}\cdot\vec{\Xi}^<{}_{vv}\right] \\
&\quad +\frac{1}{2}\left[\bar{\Delta}^r_{ee,vc} S^<_{cv,o} - S^<_{vc,o}\bar{\Delta}^a_{hh,cv}\right] + \frac{1}{2}\left[\vec{\delta}^r_{ee,vc}\cdot\vec{S}^<_{cv} - \vec{S}^<_{vc}\cdot\vec{\delta}^a_{hh,cv}\right] \\
&\quad +\frac{1}{2}\left[\bar{\Delta}^<_{ee,vc} S^a_{cv,o} - S^r_{vc,o}\bar{\Delta}^<_{hh,cv}\right] + \frac{1}{2}\left[\vec{\delta}^<_{ee,vc}\cdot\vec{S}^a_{cv} - \vec{S}^r_{vc}\cdot\vec{\delta}^<_{hh,cv}\right].
\end{aligned}
\tag{D.12}
$$

Note that Eqs. (D.5)–(D.8) and Eqs. (D.9)–(D.12) yield 16 coupled transport equations. By separating $S^<_{\alpha\beta,o}$ into the total charge in the system plus the pseudo-spin, the multi-band spin magnetization quantum transport equations nonlinearly incorporates the effects of pseudo-spin magnetization. We have already perform the sign change of Eq. (21.11) to obtain the electron picture of Eq. (D.12) above. What remains to be done for Eqs. (D.10) and (D.11) in order to revert to electron picture is to replaced all the transposed quantities, $F^{\alpha T}_{vv}$ with their equivalent expressions given in Table 21.2 or Table E.2.

D.3.1 *Spin-independent contributions*

We observe that in Eqs. (D.9)–(D.12) each equation contains spin-independent terms and corresponding terms which involved spin-vector dot products expressing the summation (i.e., the process of "integrating out") of the real-spin degree of freedom. We now calculate the two groups of spin-independent contribution.

Spinless two-level atom and flat-band limits: To gain some understanding into these equations, we examine Eqs. (D.9)–(D.12) by first retaining only the spin-independent terms, i.e., for the moment ignoring the vector dot product portions.

We have the resulting transport equations for the scalar $S^<_{\alpha\beta,o}$ involving only the spin-independent terms of Eqs. (D.9)–(D.12),

$$\begin{aligned}
& i\hbar\left(\frac{\partial}{\partial t_1}+\frac{\partial}{\partial t_2}\right)S^<_{cc,o}\\
&= \frac{1}{2}\left[\bar{H}_c, S^<_{cc,o}\right] + \frac{1}{2}\left[\bar{\Delta}^r_{hh,cv}S^<_{vc,o} - S^<_{cv,o}\bar{\Delta}^a_{ee,vc}\right]\\
&\quad + \frac{1}{2}\left[\bar{\Delta}^<_{hh,cv}S^a_{vc,o} - S^r_{cv,o}\bar{\Delta}^<_{ee,vc}\right] + \frac{1}{2}\left[\bar{\Sigma}^r_{cc}S^<_{cc,o} - S^<_{cc,o}\bar{\Sigma}^a_{cc}\right]\\
&\quad + \frac{1}{2}\left[\bar{\Sigma}^<_{cc}S^a_{cc,o} - S^r_{cc,o}\bar{\Sigma}^<_{cc}\right],
\end{aligned} \tag{D.13}$$

$$\begin{aligned}
& i\hbar\left(\frac{\partial}{\partial t_1}+\frac{\partial}{\partial t_2}\right)S^<_{cv,o}\\
&= \frac{1}{2}\left[\bar{H}_cS^<_{cv,o} - S^<_{cv,o}\bar{H}_v\right] - \frac{1}{2}\left[\bar{\Delta}^r_{hh,cv}\ S^{>T}_{vv,o} + S^<_{cc,o}\bar{\Delta}^a_{hh,cv}\right]\\
&\quad - \frac{1}{2}\left[\bar{\Delta}^<_{hh,cv}S^{rT}_{vv,o} + S^r_{cc,o}\bar{\Delta}^<_{hh,cv}\right] + \frac{1}{2}\left[\bar{\Sigma}^r_{cc}S^<_{cv,o} + S^<_{cv,o}\bar{\Sigma}^{rT}_{vv}\right]\\
&\quad + \frac{1}{2}\left[\bar{\Sigma}^<_{cc}S^a_{cv,o} + S^r_{cv,o}\bar{\Sigma}^{>T}_{vv}\right],
\end{aligned} \tag{D.14}$$

$$\begin{aligned}
& i\hbar\left(\frac{\partial}{\partial t_1}+\frac{\partial}{\partial t_2}\right)S^<_{vc,o}\\
&= \frac{1}{2}\left[\bar{H}_vS^<_{vc,o} - S^<_{vc,o}\bar{H}_c\right] + \frac{1}{2}\left[\bar{\Delta}^r_{ee,vc}\ S^<_{cc,o} + S^{>T}_{vv,o}\bar{\Delta}^a_{ee,vc}\right]\\
&\quad + \frac{1}{2}\left[\bar{\Delta}^<_{ee,vc}S^a_{cc,o} + S^{aT}_{vv,o}\bar{\Delta}^<_{ee,vc}\right] - \frac{1}{2}\left[\bar{\Sigma}^{aT}_{vv}\ S^<_{vc,o} + S^<_{vc,o}\bar{\Sigma}^a_{cc}\right]\\
&\quad - \frac{1}{2}\left[\bar{\Sigma}^{>T}_{vv}S^a_{vc,o} + S^r_{vc,o}\bar{\Sigma}^<_{cc}\right],
\end{aligned} \tag{D.15}$$

$$i\hbar\left(\frac{\partial}{\partial t_1}+\frac{\partial}{\partial t_2}\right)S^{<}_{vv,o}$$

$$=\frac{1}{2}\left[\bar{H}_v S^{<}_{vv,o}-S^{<}_{vv,o}\bar{H}_v\right]+\frac{1}{2}\left[\bar{\Delta}^{r}_{ee,vc}S^{<}_{cv,o}-S^{<}_{vc,o}\bar{\Delta}^{a}_{hh,cv}\right]$$

$$+\frac{1}{2}\left[\bar{\Delta}^{<}_{ee,vc}S^{a}_{cv,o}-S^{r}_{vc,o}\bar{\Delta}^{<}_{hh,cv}\right]+\frac{1}{2}\left[\bar{\Sigma}^{r}{}_{vv}S^{<}_{vv,o}-S^{<}_{vv,o}\bar{\Sigma}^{a}_{vv}\right]$$

$$+\frac{1}{2}\left[\bar{\Sigma}^{<}_{vv}S^{a}_{vv,o}-S^{r}_{vv,o}\bar{\Sigma}^{<}_{vv}\right]. \tag{D.16}$$

To gain insights in these equations, as representing the pseudo-spin part of the spin magnetization transport equations, let us reduce these equations to a two-level flat energy bands or a two-level atomic system. Let us take the zero of energy in the middle of the band or energy gap, and let $\bar{H}_c=-\bar{H}_v=\frac{\hbar\omega_o}{2}$. We will also ignore terms arising from the particle self-energies, Σ, as well as terms involving $\bar{\Delta}^{\lessgtr}$. We identify the interband matrix elements as $\bar{\Delta}^{r}_{ee,vc}=\langle v|\,\mathcal{H}_I\,|c\rangle$, and $\bar{\Delta}^{a}_{hh,cv}=\langle c|\,\mathcal{H}_I\,|v\rangle$, so that $\bar{\Delta}^{a}_{hh,vc}=\bar{\Delta}^{r}_{ee,vc}$, and by virtue of the locality of space–time dependence. The above four equations reduce to the following expressions:

$$i\hbar\frac{\partial}{\partial t}S^{<}_{cc,o}=\frac{1}{2}\left[\bar{\Delta}^{r}_{hh,cv}S^{<}_{vc,o}-S^{<}_{cv,o}\bar{\Delta}^{r}_{ee,vc}\right], \tag{D.17}$$

$$i\hbar\frac{\partial}{\partial t}S^{<}_{cv,o}=\frac{1}{2}\left[\hbar\omega_o S^{<}_{cv,o}+\bar{\Delta}^{r}_{hh,cv}\ \left(S^{<}_{vv,o}-S^{<}_{cc,o}\right)\right], \tag{D.18}$$

$$i\hbar\frac{\partial}{\partial t}S^{<}_{vc,o}=\frac{1}{2}\left[-\hbar\omega_o S^{<}_{vc,o}+\bar{\Delta}^{r}_{ee,vc}\ \left(S^{<}_{cc,o}-S^{<}_{vv,o}\right)\right], \tag{D.19}$$

$$i\hbar\frac{\partial}{\partial t}S^{<}_{vv,o}=\frac{1}{2}\left[\bar{\Delta}^{r}_{ee,vc}S^{<}_{cv,o}-S^{<}_{vc,o}\bar{\Delta}^{r}_{hh,cv}\right], \tag{D.20}$$

where we made use of the relation[2]:

$$S^{>T}_{vv,o}=-S^{<}_{vv,o}, \tag{D.21}$$

in our two-level atomic or flat-band limiting case.[3]

[2]This relation is used to bring Eqs. (D.18) and (D.19) to the electron picture. In fact this substitution is trivial by virtue of the locality in space and time so that the correlation functions involved are proportional to the absolute value of the field variables.

[3]In the electron picture, the right-hand sides of Eqs. (D.17) and (D.20) have opposite sign, whereas in the electron–hole picture they have the same sign or have the same rate of change as discussed before.

The pseudo-spin correlation functions are derived by first expressing the 2×2 matrix in the band indices into a spin-canonical form as

$$\begin{pmatrix} S^<_{cc,o} & S^<_{cv,o} \\ S^<_{vc,o} & S^<_{vv,o} \end{pmatrix} = \frac{1}{2}\left(S_{o,o}I + \vec{S}_o \cdot \vec{\sigma}\right),$$

$$= \frac{1}{2}\begin{pmatrix} S_{o,o} + S_{z,o} & S_{x,o} - iS_{y,o} \\ S_{x,o} + iS_{y,o} & S_{o,o} - S_{z,o} \end{pmatrix}, \tag{D.22}$$

where,

$$S_{x,o} = \left(S^<_{vc,o} + S^<_{cv,o}\right), \tag{D.23}$$

$$iS_{y,o} = \left(S^<_{vc,o} - S^<_{cv,o}\right), \tag{D.24}$$

$$S_{z,o} = \left(S^<_{cc,o} - S^<_{vv,o}\right), \tag{D.25}$$

$$S_{o,o} = \left(S^<_{cc,o} + S^<_{vv,o}\right), \tag{D.26}$$

where we drop the "<" superscript in the pseudo-spin correlation functions, $S_{j,o}$. We note that $S_{o,o}$ represents the trace of the original 4×4 spin matrix for the two bands. Thus, $S_{o,o}$ represents the total charge of the system which may vary in space and time, with the caveat that the background positive charge have to be subtracted from the charge represented by the correlation density $S_{o,o}$ to obtain the net charge.

Similarly, we write the "energy"-matrix in terms of the Pauli matrix as

$$\begin{pmatrix} \bar{H}_c & \bar{\Delta}^r_{hh,cv} \\ \bar{\Delta}^r_{ee,vc} & \bar{H}_v \end{pmatrix} = \frac{1}{2}\left(\breve{H}I + \vec{\mathcal{B}} \cdot \vec{\sigma}\right),$$

$$= \frac{1}{2}\begin{pmatrix} \breve{H} + \mathcal{B}_z & \mathcal{B}_x - i\mathcal{B}_y \\ \mathcal{B}_x + i\mathcal{B}_y & \breve{H} - \mathcal{B}_z \end{pmatrix}, \tag{D.27}$$

where we defined $\vec{\mathcal{B}}$ and $\breve{H}$ as

$$\mathcal{B}_x = \left(\bar{\Delta}^r_{ee,vc} + \bar{\Delta}^r_{hh,cv}\right), \tag{D.28}$$

$$i\mathcal{B}_y = \left(\bar{\Delta}^r_{ee,vc} - \bar{\Delta}^r_{hh,cv}\right), \tag{D.29}$$

$$\mathcal{B}_z = \left(\bar{H}_c - \bar{H}_v\right) = \hbar\omega_o, \tag{D.30}$$

$$\breve{H} = \left(\bar{H}_c + \bar{H}_v\right) = 0. \tag{D.31}$$

Therefore, we have the transport equations for the pseudo-spin correlation functions,

$$\begin{aligned}
i\hbar\frac{\partial}{\partial t}S_{x,o} &= \frac{1}{2}\left[-\hbar\omega_o S^<_{vc,o} + \bar{\Delta}^r_{ee,vc}\ \left(S^<_{cc,o} - S^<_{vv,o}\right)\right] \\
&\quad + \frac{1}{2}\left[\hbar\omega_o S^<_{cv,o} + \bar{\Delta}^r_{hh,cv}\left(S^<_{vv,o} - S^<_{cc,o}\right)\right] \\
&= \frac{i}{2}\left(\mathcal{B}_y S_{z,o} - \mathcal{B}_z S_{y,o}\right),
\end{aligned} \tag{D.32}$$

$$\begin{aligned}
i\hbar\frac{\partial}{\partial t}iS_{y,o} &= \frac{1}{2}\left[-\hbar\omega_o S^<_{vc,o} + \bar{\Delta}^r_{ee,vc}\ \left(S^<_{cc,o} - S^<_{vv,o}\right)\right] \\
&\quad + \frac{1}{2}\left[-\hbar\omega_o S^<_{cv,o} - \bar{\Delta}^r_{hh,cv}\ \left(S^<_{vv,o} - S^<_{cc,o}\right)\right] \\
&= \frac{1}{2}\left[\mathcal{B}_x S_{z,o} - \mathcal{B}_z S_{x,o}\right],
\end{aligned} \tag{D.33}$$

$$\begin{aligned}
i\hbar\frac{\partial}{\partial t}S_{z,o} &= \frac{1}{2}\left[\bar{\Delta}^r_{hh,cv} S^<_{vc,o} - S^<_{cv,o}\bar{\Delta}^r_{ee,vc}\right] \\
&\quad + \frac{1}{2}\left[-\bar{\Delta}^r_{ee,vc} S^<_{cv,o} + S^<_{vc,o}\bar{\Delta}^r_{hh,cv}\right] \\
&= \frac{i}{2}\left(\mathcal{B}_x S_{y,o} - \mathcal{B}_y S_{x,o}\right).
\end{aligned} \tag{D.34}$$

In vector equation form, the pseudo-spin magnetization equation is,

$$\frac{\partial}{\partial t}\vec{S}_o = \frac{1}{\hbar}\vec{\mathcal{B}}\times\frac{1}{2}\vec{S}_o, \tag{D.35}$$

The above equation can also be written as

$$\begin{aligned}
\frac{\partial}{\partial t}\left(\frac{\hbar}{2}\vec{S}_o\right) &= \frac{mc}{2e\hbar}\vec{\mathcal{B}}\times\frac{e}{mc}\left(\frac{\hbar}{2}\vec{S}_o\right), \\
&= \vec{B}' \times \vec{\mu}_B
\end{aligned} \tag{D.36}$$

where

$$\frac{e}{mc}\left(\frac{\hbar}{2}\vec{S}_o\right) = \vec{\mu}_B \tag{D.37}$$

corresponds to the magnetic moment of the pseudo-spin (pseudo-Bohr magneton), and

$$\frac{mc}{2e\hbar}\vec{\mathcal{B}} = \vec{B}', \tag{D.38}$$

has the units of the magnetic field, $\vec{B}'$. Thus we have realized a pseudo-spin angular momentum $\frac{\hbar}{2}\vec{S}_o$, with effective magnetic field, $\vec{B}'$, determined by the interband terms in Δ's and energy gap, Eq. (D.31). Equations (D.35) and (D.36) are also known as the Bloch equations. Note that $\vec{S}_o$ rotates about the z-axis with frequency $\frac{\mathcal{B}_z}{\hbar} = \omega_o$ [138] in the counterclockwise sense.

Indeed, the pseudo-spin angular momentum has eigenvalues $\pm\frac{1}{2}\hbar$ by virtue of the two discrete quantum-energy labels, since from Eq. (D.26) we have,

$$S_{z,o} = 1 \text{ for } \; S_{vv,o} = 0 \text{ for the excited state,} \tag{D.39}$$

$$S_{z,o} = -1 \text{for } S_{cc,o} = 0 \text{ for the unexcited or ground state.} \tag{D.40}$$

D.3.2 *Full spin-independent contributions*

We now consider the full spin-independent terms of Eqs. (D.13)–(D.16), The pseudo-spin correlation functions are given by Eq. (D.26) We expressed in spin-canonical form the following matrix in band indices:

$$\begin{pmatrix} \bar{H}_c + \mathrm{Re}\bar{\Sigma}^r_{cc} & \mathrm{Re}\bar{\Delta}^r_{hh,cv} \\ \mathrm{Re}\bar{\Delta}^r_{ee,vc} & \bar{H}_v + \mathrm{Re}\bar{\Sigma}^r_{vv} \end{pmatrix} = \frac{1}{2}\begin{pmatrix} \bar{H} + \beta_z & \beta_x - i\beta_y \\ \beta_x + i\beta_y & \bar{H} - \beta_z \end{pmatrix}, \tag{D.41}$$

$$\begin{pmatrix} \mathrm{Im}\bar{\Sigma}^r_{cc} & \mathrm{Im}\bar{\Delta}^r_{hh,cv} \\ \mathrm{Im}\bar{\Delta}^r_{ee,vc} & \mathrm{Im}\bar{\Sigma}^r_{vv} \end{pmatrix} = \frac{1}{2}\begin{pmatrix} \bar{\zeta} + \zeta_z & \zeta_x - i\zeta_y \\ \zeta_x + i\zeta_y & \bar{\zeta} - \zeta_z \end{pmatrix}. \tag{D.42}$$

Note that we have separated the real and imaginary components of Eq. (D.27) in Eqs. (D.41) and (D.42).

$$\begin{pmatrix} \bar{\Sigma}^<_{cc} & \bar{\Delta}^<_{hh,cv} \\ \bar{\Delta}^<_{ee,vc} & \bar{\Sigma}^<_{vv} \end{pmatrix} = \frac{1}{2}\begin{pmatrix} \bar{\zeta}^< + \zeta^<_z & \zeta^<_x - i\zeta^<_y \\ \zeta^<_x + i\zeta^<_y & \bar{\zeta}^< - \zeta^<_z \end{pmatrix}. \tag{D.43}$$

Note the use of calligraphic $\mathcal{B}$ in Eq. (D.27) in contrast to the use of the Greek β in Eq. (D.41). The resulting calculation is tedious but the results can be expressed in vector form as

$$i\hbar\left(\frac{\partial}{\partial t_1} + \frac{\partial}{\partial t_2}\right)\vec{S}_o$$
$$= \frac{1}{4}\left[\bar{H}, \vec{S}_o\right] + \frac{1}{4}\left[\vec{\beta}, S_{o,o}\right] + \frac{i}{4}\left[\vec{\beta} \times \vec{S}_o - \vec{S}_o \times \vec{\beta}\right]$$

$$
\begin{aligned}
&+\frac{i}{4}\left[\left\{\bar{\zeta},\vec{S}_o\right\}+\left\{\vec{\zeta},S_{o,o}\right\}\right]-\frac{1}{4}\left\{\vec{\zeta}\times\vec{S}_o+\vec{S}_o\times\vec{\zeta}\right\}\\
&+\frac{1}{4}\left[\vec{\zeta}^{<},\ \mathrm{Re}S^r_{o,o}\right]+\frac{1}{4}\left[\bar{\zeta}^{<},\ \mathrm{Re}\vec{S}^r_o\right]\\
&+\frac{i}{4}\left[\vec{\zeta}^{<}\times\mathrm{Re}\vec{S}^r_o-\mathrm{Re}\vec{S}^r_o\times\vec{\zeta}^{<}\right]\\
&-\frac{i}{4}\left[\left\{\bar{\zeta}^{<},\ \mathrm{Im}\vec{S}^r_o\right\}+\left\{\vec{\zeta}^{<},\ \mathrm{Im}S^r_{o,o}\right\}\right]\\
&+\frac{1}{4}\left\{\vec{\zeta}^{<}\times\mathrm{Im}\vec{S}^r_o+\mathrm{Im}\vec{S}^r_o\times\vec{\zeta}^{<}\right\},
\end{aligned}
\tag{D.44}
$$

where we use the following defined relation:

$$
\begin{aligned}
iS^r_{y,o}&=S^r_{vc,o}-S^r_{cv,o},\\
&=i\left(\mathrm{Re}S^r_{y,o}+i\mathrm{Im}S^r_{y,o}\right).
\end{aligned}
\tag{D.45}
$$

D.3.2.1 *The equation for the scalar $S_{o,o}$*

We have the equation for $S_{o,o}$ representing the total charge is given by

$$
\begin{aligned}
i\hbar\left(\frac{\partial}{\partial t_1}+\frac{\partial}{\partial t_2}\right)S_{o,o}=&\frac{1}{4}\left[\bar{H},S_{o,o}\right]+\frac{1}{4}\left[\bar{\zeta}^{<},\mathrm{Re}S^r_{o,o}\right]\\
&+\frac{i}{4}\left[\{\bar{\zeta},S_{o,o}\}\right]-\frac{i}{4}\left\{\bar{\zeta}^{<},\mathrm{Im}S^r_{o,o}\right\}\\
&+\frac{1}{4}\left[\vec{\beta}^{\cdot psp},\vec{S}_o\right]+\frac{i}{4}\left\{\vec{\zeta}^{\cdot psp},\vec{S}_o\right\}\\
&+\frac{1}{4}\left[\vec{\zeta}^{<\cdot psp},\mathrm{Re}\vec{S}^r_o\right]-\frac{i}{4}\left\{\vec{\zeta}^{<\cdot psp},\mathrm{Im}\vec{S}^r_o\right\},
\end{aligned}
\tag{D.46}
$$

where the superscript $^{\cdot psp}$ means taking the dot product with respect to the pseudo-spin vector components. One observes that the first four terms in the right-hand side of Eq. (D.46) exactly correspond to the terms in the equation of the particle or charge correlation function in system without the spin degree of freedom.

D.3.3 *Spin-dependent contributions*

The spin-dependent terms ignored in Section D.3.1 represent the direct coupling of pseudo-spin to real (Pauli–Dirac) spins. We extract from Eqs. (D.9)–(D.12) the terms involving the dot products of real-spin vectors for their contributions to the pseudo-spin magnetization transport equations. Since the real-spin degree of freedom is essentially integrated out, what is left are the discrete band indices. First, we expressed all 2×2 quantities in the band indices into their pseudo-spin canonical matrix form. We make use of the expressions already given by Eqs. (D.22), (D.26), (D.27), and (D.31) for the pseudo-spin canonical form of the scalar quantities. The vector quantities must also be expressed into their pseudo-spin canonical matrix form in the band indices producing dyadics.

We have for the real-spin vector dot product portion of the pseudo-spin equations, where the dot product can be written for convenience, using the Einstein summation convention for the Pauli–Dirac spin vector dot product,

$$
\begin{aligned}
i\hbar\left(\frac{\partial}{\partial t_1}+\frac{\partial}{\partial t_2}\right) S^<_{cc,o}
&= \frac{1}{2}\left[\mathcal{B}_{cc,i}\, S^<_{cc,i} - S^<_{cc,i}\, \mathcal{B}_{cc,i}\right] \\
&\quad + \frac{1}{2}\left[\Xi^r_{cc,i}\, S^<_{cc,i} - S^<_{cc,i}\, \Xi^a_{cc,i}\right] + \frac{1}{2}\left[\Xi^<_{cc,i}\, S^a_{cc,i} - S^r_{cc,i}\, \Xi^<_{cc,i}\right] \\
&\quad + \frac{1}{2}\left[\delta^r_{hh,cv,i}\, S^<_{vc,i} - S^<_{cv,i}\, \delta^a_{ee,vc,i}\right] + \frac{1}{2}\left[\delta^<_{hh,cv,i}\, S^a_{vc,i} - S^r_{cv,i}\, \delta^<_{ee,vc,i}\right],
\end{aligned} \tag{D.47}
$$

$$
\begin{aligned}
i\hbar\left(\frac{\partial}{\partial t_1}+\frac{\partial}{\partial t_2}\right) S^<_{cv,o}
&= \frac{1}{2}\left[\left(\mathcal{B}_{cc,i} + \Xi^r_{cc,i}\right)\, S^<_{cv,i} - S^<_{cv,i}\, \left(\mathcal{B}_{vv,i} - \Xi^{rT}_{vv,i}\right)\right] \\
&\quad - \frac{1}{2}\left[S^<_{cc,i}\, \delta^a_{hh,cv,i} + \delta^r_{hh,cv,i}\, S^{>T}_{vv,i}\right] + \frac{1}{2}\left[\Xi^<_{cc,i}\, S^a_{cv,i} + S^r_{cv,i}\, \Xi^{>T}_{vv,i}\right] \\
&\quad - \frac{1}{2}\left[\delta^<_{hh,cv,i}\, S^{rT}_{vv,i} + S^r_{cc,i}\, \delta^<_{hh,cv,i}\right],
\end{aligned} \tag{D.48}
$$

$$
\begin{aligned}
i\hbar\left(\frac{\partial}{\partial t_1}+\frac{\partial}{\partial t_2}\right) S^<_{vc,o}
&= \frac{1}{2}\left[\left(\mathcal{B}_{vv,i} - \Xi^{aT}_{vv,i}\right)\, S^<_{vc,i} - S^<_{vc,i}\, \left(\mathcal{B}_{cc,i} + \Xi^a_{cc,i}\right)\right] \\
&\quad + \frac{1}{2}\left[\delta^r_{ee,vc,i}\, S^<_{cc,i} + S^{>T}_{vv,i}\, \delta^a_{ee,vc,i}\right] - \frac{1}{2}\left[\Xi^{>T}_{vv,i}\, S^a_{vc,i} + S^r_{z,vc,i}\, \Xi^<_{cc,i}\right] \\
&\quad + \frac{1}{2}\left[\delta^<_{ee,vc,i}\, S^a_{cc,i} + S^{aT}_{vv,i}\, \delta^<_{ee,vc,i}\right],
\end{aligned} \tag{D.49}
$$

$$
\begin{aligned}
&i\hbar\left(\frac{\partial}{\partial t_1}+\frac{\partial}{\partial t_2}\right)S^{<}_{vv,o}\\
&=\frac{1}{2}\left[\mathcal{B}_{vv,i}\,S^{<}_{vv,i}-S^{<}_{vv,i}\,\mathcal{B}_{vv,i}\right]\\
&+\frac{1}{2}\left[\Xi^{r}_{vv,i}\,S^{<}_{vv,i}-S^{<}_{vv,i}\,\Xi^{a}_{vv,i}\right]+\frac{1}{2}\left[\Xi^{<}{}_{vv,i}\,S^{a}_{vv,i}-S^{r}_{vv,i}\,\Xi^{<}{}_{vv,i}\right]\\
&+\frac{1}{2}\left[\delta^{r}_{ee,vc,i}\,S^{<}_{cv,i}-S^{<}_{vc,i}\,\delta^{a}_{hh,cv,i}\right]+\frac{1}{2}\left[\delta^{<}_{ee,vc,i}\,S^{a}_{cv,i}-S^{r}_{vc,i}\,\delta^{<}_{hh,cv,i}\right].
\end{aligned}
\tag{D.50}
$$

We have the following pseudo-spin canonical matrix form for the real-spin vectors labeled by the discrete band indices,

$$
\begin{pmatrix}\mathcal{B}_{cc,i}+\mathrm{Re}\Sigma^{r}_{cc,i} & \mathrm{Re}\delta^{r}_{cv,i}\\ \mathrm{Re}\delta^{r}_{vc,i} & \mathcal{B}_{vv,i}+\mathrm{Re}\Sigma^{r}_{vv,i}\end{pmatrix}=\frac{1}{2}\begin{pmatrix}\bar{\mathcal{B}}_i+\mathcal{B}_{z,i} & \mathcal{B}_{x,i}-i\mathcal{B}_{y,i}\\ \mathcal{B}_{x,i}+i\mathcal{B}_{y,i} & \bar{\mathcal{B}}_i-\mathcal{B}_{z,i}\end{pmatrix},
\tag{D.51}
$$

$$
\begin{pmatrix}\mathrm{Im}\ \Sigma^{r}_{cc,i} & \mathrm{Im}\ \delta^{r}_{hh,cv,i}\\ \mathrm{Im}\ \delta^{r}_{ee,vc,i} & \mathrm{Im}\ \Sigma^{r}_{vv,i}\end{pmatrix}=\frac{1}{2}\begin{pmatrix}\bar{\zeta}_i+\zeta_{z,i} & \zeta_{x,i}-i\zeta_{y,i}\\ \zeta_{x,i}+i\zeta_{y,i} & \bar{\zeta}_i-\zeta_{z,i}\end{pmatrix},
\tag{D.52}
$$

$$
\begin{pmatrix}\Sigma^{<}_{cc,i} & \delta^{<}_{hh,cv,i}\\ \delta^{<}_{ee,vc,i} & \Sigma^{<}_{vv,i}\end{pmatrix}=\frac{1}{2}\begin{pmatrix}\bar{\zeta}^{<}_i+\zeta^{<}_{z,i} & \zeta^{<}_{x,i}-i\zeta^{<}_{y,i}\\ \zeta^{<}_{x,i}+i\zeta^{<}_{y,i} & \bar{\zeta}^{<}_i-\zeta^{<}_{z,i}\end{pmatrix},
\tag{D.53}
$$

$$
\begin{aligned}
S_{\alpha\beta,i}&=\begin{pmatrix}S^{<}_{cc,i} & S^{<}_{cv,i}\\ S^{<}_{vc,i} & S^{<}_{vv,i}\end{pmatrix},\\
&=\frac{1}{2}\begin{pmatrix}\bar{S}_i+S_{z,i} & S_{x,i}-iS_{y,i}\\ S_{x,i}+iS_{y,i} & \bar{S}_i-S_{z,i}\end{pmatrix}.
\end{aligned}
\tag{D.54}
$$

Equation (D.54) defines pseudo-spin vector for each component of the real spin vector index by "i". We will also make use of the definition

$$
\begin{aligned}
iS^{r}_{y,i}&=i\left(\mathrm{Re}S^{r}_{y,i}+i\mathrm{Im}S^{r}_{y,i}\right),\\
&=S^{r}_{vc,i}-S^{r}_{cv,i}.
\end{aligned}
\tag{D.55}
$$

The results for the pseudo-spin magnetization equation from the contributions of real-spin vector dot product terms in Eqs. (D.47–D.50) can also be written as a pseudo-spin vector equation similar to the scalar-contribution given by Eq. (D.44) as

$$\begin{aligned}
i\hbar&\left(\frac{\partial}{\partial t_1}+\frac{\partial}{\partial t_2}\right)\vec{S}_o \\
&=\frac{1}{4}\left[\bar{\mathcal{B}}_i,\vec{S}_i\right]+\frac{1}{4}\left[\vec{\mathcal{B}}_i,\bar{S}_i\right]+\frac{i}{4}\left[\vec{\mathcal{B}}_i\times\vec{S}_i-\vec{S}_i\times\vec{\mathcal{B}}_i\right] \\
&\quad+\frac{i}{4}\left[\left\{\bar{\zeta}_i,\vec{S}_i\right\}+\left\{\vec{\zeta}_i,\bar{S}_i\right\}\right]-\frac{1}{4}\left\{\vec{\zeta}_i\times\vec{S}_i+\vec{S}_i\times\vec{\zeta}_i\right\} \\
&\quad+\frac{1}{4}\left[\vec{\zeta}_i^{<},\mathrm{Re}\bar{S}_i^{r}\right]+\frac{1}{4}\left[\bar{\zeta}_i^{<},\mathrm{Re}\vec{S}_i^{r}\right] \\
&\quad+\frac{i}{4}\left[\vec{\zeta}_i^{<}\times\mathrm{Re}\vec{S}_i^{r}-\mathrm{Re}\vec{S}_i^{r}\times\vec{\zeta}_i^{<}\right] \\
&\quad-\frac{i}{4}\left[\left\{\bar{\zeta}_i^{<},\mathrm{Im}\vec{S}_i^{r}\right\}+\left\{\vec{\zeta}_i^{<},\mathrm{Im}\bar{S}_i^{r}\right\}\right] \\
&\quad+\frac{1}{4}\left\{\vec{\zeta}_i^{<}\times\mathrm{Im}\vec{S}_i^{r}+\mathrm{Im}\vec{S}_i^{r}\times\vec{\zeta}_i^{<}\right\}.
\end{aligned}\tag{D.56}$$

D.3.3.1 *The equation for the scalar $S_{o,o}$*

Similarly the contribution to the equation for $S_{o,o}$ goes as,

$$\begin{aligned}
i\hbar&\left(\frac{\partial}{\partial t_1}+\frac{\partial}{\partial t_2}\right)S_{o,o} \\
&=\frac{1}{4}\left[\bar{\mathcal{B}}_i,\bar{S}_i\right]+\frac{1}{4}\left[\vec{\mathcal{B}}_i^{\cdot psp},\vec{S}_i\right] \\
&\quad+\frac{1}{4}i\left[\left\{\bar{\zeta}_i,\bar{S}_i\right\}\right]+\frac{1}{4}i\left\{\vec{\zeta}_i^{\cdot psp},\vec{S}_i\right\} \\
&\quad+\frac{1}{4}\left[\bar{\zeta}_i^{<},\mathrm{Re}\bar{S}_i^{r}\right]-\frac{1}{4}i\left\{\bar{\zeta}_i^{<},\mathrm{Im}\bar{S}_i^{r}\right\} \\
&\quad+\frac{1}{4}\left[\vec{\zeta}_i^{<\cdot psp},\mathrm{Re}\vec{S}_i^{r}\right]-\frac{1}{4}i\left\{\vec{\zeta}_i^{<\cdot psp},\mathrm{Im}\vec{S}_i^{r}\right\},
\end{aligned}\tag{D.57}$$

where the superscripts $^{\cdot psp}$ indicates the process of taking the pseudo-spin vector dot products or the process of "integrating out" the pseudo-spin degree of freedoms, aside from the dot product of the Pauli–Dirac spin vector indicated by the Einstein summation convention for the repeated "i" index.

D.4 Non-equilibrium pseudo-spin equations

The non-equilibrium pseudo-spin vector transport equations result by adding Eqs. (D.44) and (D.56)

$$
\begin{aligned}
i\hbar &\left(\frac{\partial}{\partial t_1}+\frac{\partial}{\partial t_2}\right)\vec{S}_o \\
&= \frac{1}{4}\left[\bar{H},\vec{S}_o\right]+\frac{1}{4}\left[\bar{\zeta}^{<},\mathrm{Re}\vec{S}_o^{r}\right]+\frac{i}{4}\left\{\bar{\zeta},\vec{S}_o\right\}-\frac{i}{4}\left\{\bar{\zeta}^{<},\mathrm{Im}\vec{S}_o^{r}\right\} \\
&\quad +\frac{1}{2}\left[\vec{\beta},S_{o,o}\right]+\frac{1}{4}\left[\vec{\zeta}^{<},\mathrm{Re}S_{o,o}^{r}\right]+\frac{i}{4}\left\{\vec{\zeta},S_{o,o}\right\}-\frac{i}{4}\left\{\vec{\zeta}^{<},\mathrm{Im}S_{o,o}^{r}\right\} \\
&\quad +\frac{i}{4}\left[\vec{\beta}\times\vec{S}_o-\vec{S}_o\times\vec{\beta}\right]+\frac{i}{4}\left[\vec{\zeta}^{<}\times\mathrm{Re}\vec{S}_o^{r}-\mathrm{Re}\vec{S}_o^{r}\times\vec{\zeta}^{<}\right] \\
&\quad -\frac{1}{4}\left\{\vec{\zeta}\times\vec{S}_o+\vec{S}_o\times\vec{\zeta}\right\}+\frac{1}{4}\left\{\vec{\zeta}^{<}\times\mathrm{Im}\vec{S}_o^{r}+\mathrm{Im}\vec{S}_o^{r}\times\vec{\zeta}^{<}\right\} \\
&\quad +\frac{1}{4}\left[\bar{\mathcal{B}}_i,\vec{S}_i\right]+\frac{1}{4}\left[\bar{\zeta}_i^{<},\mathrm{Re}\vec{S}_i^{r}\right]+\frac{i}{4}\left\{\bar{\zeta}_i,\vec{S}_i\right\}-\frac{i}{4}\left\{\bar{\zeta}_i^{<},\mathrm{Im}\vec{S}_i^{r}\right\} \\
&\quad +\frac{1}{4}\left[\vec{\mathcal{B}}_i,\bar{S}_i\right]+\frac{1}{4}\left[\vec{\zeta}_i^{<},\mathrm{Re}\bar{S}_i^{r}\right]+\frac{i}{4}\left\{\vec{\zeta}_i,\bar{S}_i\right\}-\frac{i}{4}\left\{\vec{\zeta}_i^{<},\mathrm{Im}\bar{S}_i^{r}\right\} \\
&\quad +\frac{i}{4}\left[\vec{\mathcal{B}}_i\times\vec{S}_i-\vec{S}_i\times\vec{\mathcal{B}}_i\right]+\frac{i}{4}\left[\vec{\zeta}_i^{<}\times\mathrm{Re}\vec{S}_i^{r}-\mathrm{Re}\vec{S}_i^{r}\times\vec{\zeta}_i^{<}\right] \\
&\quad -\frac{1}{4}\left\{\vec{\zeta}_i\times\vec{S}_i+\vec{S}_i\times\vec{\zeta}_i\right\}+\frac{1}{4}\left\{\vec{\zeta}_i^{<}\times\mathrm{Im}\vec{S}_i^{r}+\mathrm{Im}\vec{S}_i^{r}\times\vec{\zeta}_i^{<}\right\}.
\end{aligned}
\tag{D.58}
$$

$$
\begin{bmatrix}
\frac{i}{4}\left[\vec{\mathcal{B}}_i\times\vec{S}_{o,i}-\vec{S}_{o,i}\times\vec{\mathcal{B}}_i\right] \\
+\frac{i}{4}\left[\vec{\zeta}_i^{<}\times\mathrm{Re}\vec{S}_{o,i}^{r}-\mathrm{Re}\vec{S}_{o,i}^{r}\times\vec{\zeta}_i^{<}\right] \\
-\frac{1}{4}\left\{\vec{\zeta}_i\times\vec{S}_{o,i}+\vec{S}_{o,i}\times\vec{\zeta}_i\right\} \\
+\frac{1}{4}\left\{\vec{\zeta}_i^{<}\times\mathrm{Im}\vec{S}_{o,i}^{r}+\mathrm{Im}\vec{S}_{o,i}^{r}\times\vec{\zeta}_i^{<}\right\}
\end{bmatrix}.
$$

The corresponding absolute scalar[4] equation is obtained by adding Eqs. (D.46) and (D.57),

$$
\begin{aligned}
i\hbar\left(\frac{\partial}{\partial t_1}+\frac{\partial}{\partial t_2}\right)S_{o,o}
&=\frac{1}{4}\left[\bar{H},S_{o,o}\right]+\frac{1}{4}\left[\bar{\zeta}^{<},\mathrm{Re}S^{r}_{o,o}\right]+\frac{i}{4}\left[\{\bar{\zeta},S_{o,o}\}\right]-\frac{i}{4}\left\{\bar{\zeta}^{<},\mathrm{Im}S^{r}_{o,o}\right\}\\
&+\frac{1}{4}\left[\vec{\beta}^{\cdot psp},\vec{S}_o\right]+\frac{1}{4}\left[\vec{\zeta}^{<\cdot psp},\mathrm{Re}\vec{S}^{r}_{o}\right]+\frac{i}{4}\left\{\vec{\zeta}^{\cdot psp},\vec{S}_o\right\}-\frac{i}{4}\left\{\vec{\zeta}^{<\cdot psp},\mathrm{Im}\vec{S}^{r}_{o}\right\}\\
&+\frac{1}{4}\left[\bar{\mathcal{B}}_i,\bar{S}_i\right]+\frac{1}{4}\left[\bar{\zeta}^{<}_i,\mathrm{Re}\bar{S}^{r}_{i}\right]+\frac{i}{4}\left\{\bar{\zeta}_i,\bar{S}_i\right\}-\frac{i}{4}\left\{\bar{\zeta}^{<}_i,\mathrm{Im}\bar{S}^{r}_{i}\right\}\\
&+\frac{1}{4}\left[\vec{\mathcal{B}}^{\cdot psp}_i,\vec{S}_i\right]+\frac{1}{4}\left[\vec{\zeta}^{<\cdot psp}_i,\mathrm{Re}\vec{S}^{r}_{i}\right]\\
&+\frac{i}{4}\left\{\vec{\zeta}^{\cdot psp}_i,\vec{S}_i\right\}-\frac{i}{4}\left\{\vec{\zeta}^{<\cdot psp}_i,\ \mathrm{Im}\vec{S}^{r}_{i}\right\}. \qquad \text{(D.59)}
\end{aligned}
$$

Equations (D.5)–(D.8) with (D.58) and (21.60) constitute the non-equilibrium multi-band magnetization quantum transport equations. These 16 equations for the components of the spin vectors are our main results on multi-band quantum transport of electrons with Dirac spin.

[4] $S_{o,o}$ represents the trace of the 4×4 matrix of Eq. (21.16) and hence independent of the Pauli–Dirac and pseudo-spin degrees of freedom. The time variation does not in general vanish by virtue of the non-locality of the space and time arguments of the two-point correlations involved in its equation.

Appendix E

Product of 2 × 2 Matrices

Theorem E.1. *Any product of* 2×2 *matrices given by*

$$AB = \begin{pmatrix} a_{11} & a_{12} \\ a_{21} & a_{22} \end{pmatrix} \begin{pmatrix} b_{11} & b_{12} \\ b_{21} & b_{22} \end{pmatrix}, \tag{E.1}$$

can be expressed in terms of 2×2 *identity and Pauli matrices as*

$$AB = \frac{1}{4}\{S_o\,(ab)\,I + \vec{V}\,(ab) \cdot \vec{\sigma}\}, \tag{E.2}$$

where $S_o\,(ab)$ *is the scalar-transforming operator on product* AB *and* $\vec{V}\,(ab)$ *is the vector-transforming operator on* AB*. Using the Einstein summation convention for the repeated indices, we have,*

$$S_o\,(ab) = \left[\bar{a}\bar{b} + a_k b_k\right], \tag{E.3}$$

$$V_k\,(ab) = \left[\bar{a} b_k + a_k \bar{b} + i\epsilon_{klm}\,(a_l b_m)\right]. \tag{E.4}$$

Here, the scalar and the vector components of $\vec{a}$ *and* $\vec{b}$ *are defined by*

$$\bar{a} = a_{11} + a_{22}, \quad \bar{b} = b_{11} + b_{22}, \tag{E.5}$$

$$a_x = a_{12} + a_{21}, \quad b_x = b_{12} + b_{21}, \tag{E.6}$$

$$ia_y = a_{12} - a_{21}, \quad ib_y = b_{12} - b_{21}, \tag{E.7}$$

$$a_z = a_{11} - a_{22}, \quad b_z = b_{11} - b_{22}. \tag{E.8}$$

Proof. Expressing each matrix in terms of the identity and Pauli matrices, we have

$$
\begin{aligned}
AB &= \frac{1}{2}\begin{pmatrix} \bar{a}+a_z & a_x - ia_y \\ a_x + ia_y & \bar{a}-a_z \end{pmatrix} \frac{1}{2}\begin{pmatrix} \bar{b}+b_z & b_x - ib_y \\ b_x + ib_y & \bar{b}-b_z \end{pmatrix}, \\
&= \frac{1}{4}\begin{pmatrix} (\bar{a}+a_z)(\bar{b}+b_z)+(a_x-ia_y)(b_x+ib_y) \\ (a_x+ia_y)(\bar{b}+b_z)+(\bar{a}-a_z)(b_x+ib_y) \end{pmatrix.} \\
&\qquad \begin{matrix} (\bar{a}+a_z)(b_x-ib_y)+(a_x-ia_y)(\bar{b}-b_z) \\ (a_x+ia_y)(b_x-ib_y)+(\bar{a}-a_z)(\bar{b}-b_z) \end{matrix}\Bigg).
\end{aligned} \tag{E.9}
$$

Upon evaluating the last line, we end up with

$$
AB = \frac{1}{4}\left\{ \begin{matrix} [\bar{a}\bar{b}+\vec{a}\cdot\vec{b}]I + [\bar{a}\vec{b}+\bar{b}\vec{a}+i(\vec{a}\times\vec{b})]_z\sigma_z \\ +[(\bar{a}\vec{b}+\vec{a}\bar{b})+i(\vec{a}\times\vec{b})]_x\sigma_x \\ +[(\bar{a}\vec{b}+\vec{a}\bar{b})+i(\vec{a}\times\vec{b})]_y\sigma_y, \end{matrix} \right\} \tag{E.10}
$$

□

Lemma E.1. *The theorem can be generalized to any binary operation of spinor matrices.*

$$
\begin{aligned}
S_o(a\otimes b) &= [\bar{a}\otimes\bar{b}+\vec{a}\otimes\cdot\vec{b}], \\
&= \frac{1}{4}[\bar{a}\otimes\bar{b}+a_k\otimes b_k],
\end{aligned} \tag{E.11}
$$

$$
\vec{S}(a\otimes b) = \frac{1}{4}[\bar{a}\otimes\vec{b}+\bar{b}\otimes\vec{a}+i(\vec{a}\times\vec{b})_\otimes], \tag{E.12}
$$

$$
S_k(ab) = \frac{1}{4}[\bar{a}\otimes b_k + a_k\otimes\bar{b} + i\epsilon_{klm}a_l \circledast b_m], \tag{E.13}
$$

where $\circledast$ will generally be a different binary operation from that of $\otimes$. This will become clear in the example below.

E.1 Examples

The Pauli–Dirac spinor form of the commutator of A and B is given by

$$
[A,B] = \frac{1}{4}([\bar{a},\vec{b}]+[\vec{a},\bar{b}]+i[\vec{a}\times\vec{b}-\vec{b}\times\vec{a}]), \tag{E.14}
$$

$$
[A,B]_k = \frac{1}{4}([\bar{a},b_k]+[a_k,\bar{b}]+i\epsilon_{kk_1k_2}\{a_{k_1},b_{k_2}\}), \tag{E.15}
$$

where the *bar* indicates total independence with respect to the spin degrees of freedom, $\{a_{k_1}, b_{k_2}\}$ is the anti-commutation of a_{k_1} and b_{k_2}. Note the change from commutator to anti-commutator for the term containing one Levi-Civita symbol. To include dependence on another spin degrees of freedom, it is convenient to arrange the component of the next spin as the succeeding subscripts, such as dyadic tensors $a_{k,l}$ and $b_{k,l}$ where the index l corresponds to the next spin variable, such as valley isospin, for example. Instead of using the bar, quantities independent of two spins will be designated by $a_{o,o}$ and $b_{o,o}$. In all instances, the theorem on binary product of 2×2 matrices can be successively applied. Thus, the fully entangled "torque" [coming from the last term of Eq. (E.15)] now becomes,

$$i\epsilon_{ll_1l_2} i\epsilon_{kk_1k_2} \left[a_{k_1,l_1}, b_{k_2,l_2}\right] = -\epsilon_{ll_1l_2}\epsilon_{kk_1k_2} \left[a_{k_1,l_1}, b_{k_2,l_2}\right]. \tag{E.16}$$

Upon adding dependence on one more pseudospin degrees of freedom, such the pseudospin arising either from the symmetry at low energies of the band indices at Dirac points in graphene or from layer pseudospin in bilayer graphene, the entanglement of the torques becomes

$$-i\epsilon_{mm_1m_2}\epsilon_{ll_1l_2}\epsilon_{kk_1k_2} \left\{a_{k_1,l_1,m_1}, b_{k_2,l_2,m_2}\right\}, \tag{E.17}$$

which is a third rank tensor. Assuming we have considered only the band pseudospin in Eq. (E.17), if one further consider the pseudospin from layer degrees of freedom in bilayer graphene, for example, then the fully entangled "torque" becomes

$$\epsilon_{nn_1n_2}\epsilon_{mm_1m_2}\epsilon_{ll_1l_2}\epsilon_{kk_1k_2} \left[a_{k_1,l_1,m_1,n_1} b_{k_2,l_2,m_2,n_2}\right]. \tag{E.18}$$

Note that starting from the commutation of $[A, B]$, the dependence in spin degrees of freedom results in the alternating commutator and anti-commutator of the spin tensor components in the expression for entangled torques, with commutator for even number of $i\epsilon_{ijk}$ (*Levi-Civita* symbols) and anti-commutator for odd number of $i\epsilon_{ijk}$. Had we started with $\{A, B\}$, then there will be alternating anti-commutator and commutator instead as the number of spin degrees of freedom grows.

These considerations have important significance in formulating the non-equilibrium spin quantum transport equations. However, for economy of space, we will only consider the Pauli–Dirac spin, valley spin or isospin, and pseudospins either due to symmetry of band indices at low energies at the Dirac points [208] or due to bilayer pseudospin. Here we only have eight coupled transport equations to consider. On the other hand, by including any other fourth pseudospin degrees of freedom we will double the number of coupled quantum transport equations to sixteen. Although sixteen coupled spin quantum transport equations are not treated here, the derivation is straightforward following the procedures exemplified here. From the general non-equilibrium coupled spin quantum transport equations various limiting cases appropriate to the problem at hand can be obtained as given in the text.

E.2 Local and non-local terms

As discussed in Section 29.1.2 of Part6, the commutator rigorously incorporates the quantum non-locality in phase space or quantum-diffusive motion of transport kinetics in phase space,[1] and exhibit quantum tunneling across abrupt barriers when all powers of $\hbar$ are taken into account by using the integral kernel [11], whereas the anti-commutator describes local or non-diffusive events in $(\vec{p},\vec{q})$ space as typified by cyclotron-orbit current tied to orbit center or decay and growth of a phase-space distribution function tied to a point $(\vec{p},\vec{q})$ in phase space. Moreover, the exact integral form of the commutator and anti-commutator exhibits non-local kernel and local kernel of integration, respectively. See also the discussion in Section E.7.

E.3 Binary product of "cube matrices"

The result of any binary product of matrices characterized by Pauli–Dirac spin indices, valley or isospin indices, and energy bands or pseudospin indices can be classified into *eight* tensorial groups, obtained by *successive iteration* of the spinor transformation of 2 × 2 matrices. We obtained the following expressions, where as before repeated indices are summed.

E.3.1 *Total scalars or tensors of rank zero*

Here we have $\{k = 0 \text{ or } \sum_k,\ l = 0 \text{ or } \sum_l,\ m = 0 \text{ or } \sum_m\}$, i.e., three 2-degrees of freedom, yielding $2^3 = 8$ terms,

$$S_{o,o,o} = \left(\frac{1}{4}\right)^3 \left\{ \begin{array}{c} \left[\bar{a}_{o,o,o}\bar{b}_{o,o,o} + a_{k,o,o}b_{k,o,o} + \bar{a}_{o,o,m}\bar{b}_{o,o,m}\right] \\ +\bar{a}_{o,l,o}\bar{b}_{o,l,o} + a_{k,l,o}b_{k,l,o} + \bar{a}_{o,l,m}\bar{b}_{o,l,m} \\ +a_{k,o,m}b_{k,o,m} + a_{k,l,m}b_{k,l,m} \end{array} \right\}. \tag{E.19}$$

[1] This is evident even in the semi-classical approximation leading to Boltzmann transport equation, i.e., within the first power of Poisson bracket.

E.3.2 *Dirac-spin vectors and torques*

Here we have [$l = 0$ or $\sum_l$, $m = 0$ or $\sum_m$] on each factor yielding $2 \times 2^2 = 8$ vectors, and two 2-degrees of freedom for the torques, $2^2 = 4$ torques

$$V_{k,o,o} = \left(\frac{1}{4}\right)^3 \begin{bmatrix} \bar{a}_{o,o,o}b_{k,o,o} + a_{k,o,o}\bar{b}_{o,o,o} + i\epsilon_{kk_1k_2}\left(a_{k_1,o,o}b_{k_2,o,o}\right) \\ +\bar{a}_{o,l,o}b_{k,l,o} + a_{k,l,o}\bar{b}_{o,l,o} + i\epsilon_{kk_1k_2}\left(a_{k_1,l,o}b_{k_2,l,o}\right) \\ +\bar{a}_{o,o,m}b_{k,o,m} + a_{k,o,m}\bar{b}_{o,o,m} + i\epsilon_{kk_1k_2}\left(a_{k_1,o,m}b_{k_2,o,m}\right) \\ +\bar{a}_{o,l,m}b_{k,l,m} + a_{k.l,m}\bar{b}_{o,l,m} + i\epsilon_{kk_1k_2}\left(a_{k_1,l,m}b_{k_2,l,m}\right) \end{bmatrix}. \tag{E.20}$$

E.3.3 *Isospin vectors and torques*

Here we have [$k = 0$ or $\sum_k$, $m = 0$ or $\sum_m$] on each factor yielding $2 \times 2^2 = 8$ vectors, and two 2-degrees of freedom for the torques, $2^2 = 4$ torques,

$$V_{o,l,o} = \left(\frac{1}{4}\right)^3 \begin{Bmatrix} \left[\bar{a}_{o,o,o}\bar{b}_{o,l,o} + \bar{a}_{o,l,o}\bar{b}_{o,o,o} + i\epsilon_{ll_1l_2}\left(\bar{a}_{o,l_1,o}\bar{b}_{o,l_2,o}\right)\right] \\ +\bar{a}_{o,o,m}\bar{b}_{o,l,m} + \bar{a}_{o,l,m}\bar{b}_{o,o,m} + i\epsilon_{ll_1l_2}\left(\bar{a}_{o,l_1,m}\bar{b}_{o,l_2,m}\right) \\ +\left[a_{k,o,o}b_{k,l,o} + a_{k,l,o}b_{k,o,o} + i\epsilon_{ll_1l_2}\left(a_{k,l_1,o}b_{k,l_2,o}\right)\right] \\ +a_{k,o,m}b_{k,l,m} + a_{k.l,m}b_{k,o,m} + i\epsilon_{ll_1l_2}\left(a_{k,l_1,m}b_{k,l_2,m}\right) \end{Bmatrix}. \tag{E.21}$$

E.3.4 *Pseudospin vectors and torques*

Similarly, here we have [$k = 0$ or $\sum_k$, $l = 0$ or $\sum_l$] on each factor yielding $2 \times 2^2 = 8$ vectors, and two 2-degrees of freedom for the torques, $2^2 = 4$ torques,

$$V_{o,o,m} = \left(\frac{1}{4}\right)^3 \begin{Bmatrix} \left[\bar{a}_{o,o,o}\bar{b}_{o,o,m} + \bar{a}_{o,o,m}\bar{b}_{o,o,o} + i\epsilon_{mm_1m_2}\left(\bar{a}_{o,o,m_1}\bar{b}_{o,o,m_2}\right)\right] \\ +\left[a_{k,o,o}b_{k,o,m} + a_{k,o,m}b_{k,o,o} + i\epsilon_{mm_1m_2}\left(a_{k,o,m_1}b_{k,o,m_2}\right)\right] \\ +\left[\bar{a}_{o,l,o}\bar{b}_{o,l,m} + \bar{a}_{o,l,m}\bar{b}_{o,l,o} + i\epsilon_{mm_1m_2}\left(\bar{a}_{o,l,m_1}\bar{b}_{o,l,m_2}\right)\right] \\ +\left[a_{k,l,o}b_{k,l,m} + a_{k,l,m}b_{k,l,o} + i\epsilon_{mm_1m_2}\left(a_{k,l,m_1}b_{k,l,m_2}\right)\right] \end{Bmatrix}. \tag{E.22}$$

E.3.5 *Dirac spin-isospin dyadics and entangled torques*

Here we have [$m = 0$ or $\sum_m$, k, l can be distributed in two ways, and l and k can be distributed in two ways] yielding $2^3 = 8$ dyadics, $2^3 = 8$ torques, and $2^1 = 2$ entangled torques from $m = 0$ or $\sum_m$ only.

We have,

$$T_{k,l}=\left(\frac{1}{4}\right)^3\left\{\begin{array}{c}\left[\begin{array}{c}\bar{a}_{o,o,o}b_{k,l,o}+a_{k,l,o}\bar{b}_{o,o,o}+\bar{a}_{o,l,o}b_{k,o,o}+a_{k,o,o}\bar{b}_{o,l,o}\\ +i\epsilon_{kk_1k_2}\left(a_{k_1,o,o}b_{k_2,l,o}\right)+i\epsilon_{kk_1k_2}\left(a_{k_1,l,o}b_{k_2,o,o}\right)\\ +\left[i\epsilon_{ll_1l_2}\left(\bar{a}_{o,l_1,o}b_{k,l_2,o}\right)\right]+i\epsilon_{ll_1l_2}\left(a_{k,l_1,o}\bar{b}_{o,l_2,o}\right)\\ +\frac{1}{2}\left[i\epsilon_{kk_1k_2}i\epsilon_{ll_1l_2}\left(a_{k_1,l_1,o}b_{k_2,l_2,o}\right)\right]\\ +\frac{1}{2}i\epsilon_{kk_1k_2}i\epsilon_{ll_1l_2}\left(a_{k_1,l_1,o}b_{k_2,l_2,o}\right)\end{array}\right]\\ +\left[\begin{array}{c}+\bar{a}_{o,o,m}b_{k,l,m}+a_{k.l,m}\bar{b}_{o,o,m}+\bar{a}_{o,l,m}b_{k,o,m}+a_{k,o,m}\bar{b}_{o,l,m}\\ +i\epsilon_{kk_1k_2}\left(a_{k_1,l,m}b_{k_2,o,m}\right)+i\epsilon_{kk_1k_2}\left(a_{k_1,o,m}b_{k_2,l,m}\right)\\ +i\epsilon_{ll_1l_2}\left(a_{k,l_1,m}\bar{b}_{o,l_2,m}\right)+\left[i\epsilon_{ll_1l_2}\left(\bar{a}_{o,l_1,m}b_{k,l_2,m}\right)\right]\\ +\frac{1}{2}\left[i\epsilon_{kk_1k_2}i\epsilon_{ll_1l_2}\left(a_{k_1,l_1,m}b_{k_2,l_2,m}\right)\right]\\ +\frac{1}{2}\left[i\epsilon_{kk_1k_2}i\epsilon_{ll_1l_2}\left(a_{k_1,l_1,m}b_{k_2,l_2,m}\right)\right]\end{array}\right]\end{array}\right\}. \tag{E.23}$$

E.3.6 *Dirac spin-pseudospin dyadics and entangled torques*

Here we have [$l=0$, or $\sum_l$, k,m can be distributed in two ways, and m and k can be distributed in two ways] yielding $2^3=8$ dyadics, $2^3=8$ torques, and $2^1=2$ entangled torques from $l=0$, or $\sum_l$ only. We have,

$$T_{k,m}=\left(\frac{1}{4}\right)^3\left[\begin{array}{c}\bar{a}_{o,o,o}b_{k,o,m}+a_{k,o,m}\bar{b}_{o,o,o}+\bar{a}_{o,o,m}b_{k,o,o}+a_{k,o,o}\bar{b}_{o,o,m}\\ +i\epsilon_{kk_1k_2}\left(a_{k_1,o,o}b_{k_2,o,m}\right)+i\epsilon_{kk_1k_2}\left(a_{k_1,o,m}b_{k_2,o,o}\right)\\ +i\epsilon_{mm_1m_2}\left(\bar{a}_{o,o,m_1}b_{k,o,m_2}\right)+i\epsilon_{mm_1m_2}\left(a_{k,o,m_1}\bar{b}_{o,o,m_2}\right)\\ +\frac{1}{2}i\epsilon_{kk_1k_2}i\epsilon_{mm_1m_2}\left(a_{k_1,o,m_1}b_{k_2,o,m_2}\right)\\ +\frac{1}{2}i\epsilon_{kk_1k_2}i\epsilon_{mm_1m_2}\left(a_{k_1,o,m_1}b_{k_2,o,m_2}\right)\end{array}\right]$$

$$+\left(\frac{1}{4}\right)^3\left[\begin{array}{c}\bar{a}_{o,l,o}b_{k,l,m}+a_{k.l,m}\bar{b}_{o,l,o}+a_{k,l,o}\bar{b}_{o,l,m}+\bar{a}_{o,l,m}b_{k,l,o}\\ +i\epsilon_{kk_1k_2}\left(a_{k_1,l,o}b_{k_2,l,m}\right)+i\epsilon_{kk_1k_2}\left(a_{k_1,l,m}b_{k_2,l,o}\right)\\ +i\epsilon_{mm_1m_2}\left(\bar{a}_{o,l,m_1}b_{k,l,m2}\right)+i\epsilon_{mm_1m_2}\left(a_{k.l,m_1}\bar{b}_{o,l,m_2}\right)\\ +\frac{1}{2}i\epsilon_{kk_1k_2}i\epsilon_{mm_1m_2}\left(a_{k_1,l,m_1}b_{k_2,l,m_2}\right)\\ +\frac{1}{2}i\epsilon_{kk_1k_2}i\epsilon_{mm_1m_2}\left(a_{k_1,o,m_1}b_{k_2,o,m_2}\right)\end{array}\right]. \tag{E.24}$$

E.3.7 *Isospin-pseudospin dyadics, torques and entangled torques*

Here we have [$k=0$ or $\sum_k$, l,m can be distributed in two ways, and m and l can be distributed in two ways] yielding $2^3=8$ dyadics, $2^3=8$ torques, and $2^1 = 2$ entangled torques from $k = 0$ or $\sum_k$ only. We have,

$$T_{l,m} = \left(\frac{1}{4}\right)^3 \begin{bmatrix} \bar{a}_{o,o,o}\bar{b}_{o,l,m} + \bar{a}_{o,l,m}\bar{b}_{o,o,o} + \bar{a}_{o,l,o}\bar{b}_{o,o,m} + \bar{a}_{o,o,m}\bar{b}_{o,l,o} \\ +i\epsilon_{mm_1m_2}\left(\bar{a}_{o,o,m_1}\bar{b}_{o,l,m_2}\right) + i\epsilon_{mm_1m_2}\left(\bar{a}_{o,l,m_1}\bar{b}_{o,o,m_2}\right) \\ +i\epsilon_{ll_1l_2}\left(\bar{a}_{o,l_1,o}\bar{b}_{o,l_2,m}\right) + i\epsilon_{ll_1l_2}\left(\bar{a}_{o,l_1,m}\bar{b}_{o,l_2,o}\right) \\ +\frac{1}{2}i\epsilon_{ll_1l_2}i\epsilon_{mm_1m_2}\left(\bar{a}_{o,l_1,m_1}\bar{b}_{o,l_2,m_2}\right) \\ +\frac{1}{2}i\epsilon_{ll_1l_2}i\epsilon_{mm_1m_2}\left(\bar{a}_{o,l_1,m_1}\bar{b}_{o,l_2,m_2}\right) \end{bmatrix}$$

$$+\left(\frac{1}{4}\right)^3 \begin{bmatrix} a_{k,l,o}b_{k,o,m} + a_{k,o,m}b_{k,l,o} + a_{k.l,m}b_{k,o,o} + a_{k,o,o}b_{k,l,m} \\ +i\epsilon_{mm_1m_2}\left(a_{k.l,m_1}b_{k,o,m_2}\right) + i\epsilon_{mm_1m_2}\left(a_{k,o,m_1}b_{k,l,m_2}\right) \\ +i\epsilon_{ll_1l_2}\left(a_{k,l_1,m}b_{k,l_2,o}\right) + i\epsilon_{ll_1l_2}\left(a_{k,l_1,o}b_{k,l_2,m}\right) \\ +\frac{1}{2}i\epsilon_{ll_1l_2}i\epsilon_{mm_1m_2}\left(a_{k,l_1,m_1}b_{k,l_2,m_2}\right) \\ +\frac{1}{2}i\epsilon_{ll_1l_2}i\epsilon_{mm_1m_2}\left(a_{k,l_1,m_1}b_{k,l_2,m_2}\right) \end{bmatrix}. \tag{E.25}$$

E.3.8 *Dirac spin-isospin-pseudospin tensors, torques, and entangled 2 and 3 torques*

Here we have [k, l, and m can be considered as three qubits] yielding $2^3 = 8$ tensors of third rank, the number of torques for each spin is $2^2 = 4$ yielding for three spins $3 \times 4 = 12$ torques, the number of entangled torques for each pair of torques is 2 and since there are three pair of torques that can entangle, the results is $3 \times 2 = 6$ entangled torques. Finally we can only have 1 entangled three torques We have the result,

$$T_{k,l,m}$$
$$= \left(\frac{1}{4}\right)^3 \begin{bmatrix} a_{k,l,m}b_{0,0,0} + a_{k,l,0}b_{0,0,m} + a_{k,0,m}b_{0,l,0} + a_{0,l,m}b_{k,0,0} \\ +a_{k,0,0}b_{0,l,m} + a_{0,0,m}b_{k,l,0} + a_{0,l,0}b_{k,0,m} + a_{o,o,o}b_{k.l.m} \end{bmatrix}$$

$$+\left(\frac{1}{4}\right)^3 \begin{bmatrix} \epsilon_{kk_1k_2}a_{k_1,0,0}b_{k_2,l,m} + \epsilon_{kk_1k_2}a_{k_1,l,0}b_{k_2,0,m} \\ +\epsilon_{kk_1k_2}a_{k_1,0,m}b_{k_2,l,0} + \epsilon_{kk_1k_2}a_{k_1,l,m}b_{k_2,0,0} \\ +\epsilon_{ll_1l_2}a_{0,l_1,0}b_{k,l_2,m} + \epsilon_{ll_1l_2}a_{k,l_1,0}b_{0,l_2,m} \\ +\epsilon_{ll_1l_2}a_{0,l_1,m}b_{k,l_2,0} + \epsilon_{ll_1l_2}a_{k,l_1,m}b_{0,l_2,0} \\ +\epsilon_{m;m_1,m_2}a_{0,0.m_1}b_{k,l,m_2} + \epsilon_{m;m_1,m_2}a_{0,l.m_1}b_{k;0,m_2} \\ +\epsilon_{m;m_1,m_2}a_{k,0.m_1}b_{0,l,m_2} + \epsilon_{m;m_1,m_2}a_{k,l.m_1}b_{0,0,m_2} \end{bmatrix}$$

$$+\left(\frac{1}{4}\right)^3\left[\begin{array}{c} i\epsilon_{kk_1k_2}i\epsilon_{ll_1l_2}\left(a_{k_1,l_1,o}b_{k_2,l_2,m}+a_{k_1,l_1,m}b_{k_2,l_2,o}\right)\\ +i\epsilon_{kk_1k_2}i\epsilon_{mm_1m_2}\left(a_{k_1,o,m_1}b_{k_2,l,m_2}+a_{k_1,l,m_1}b_{k_2,o,m_2}\right)\\ +i\epsilon_{ll_1l_2}i\epsilon_{mm_1m_2}\left(a_{o,l_1,m_1}b_{k,l_2,m_2}+a_{k,l_1,m_1}b_{o,l_2,m_2}\right)\\ +i\epsilon_{kk_1k_2}i\epsilon_{ll_1l_2}i\epsilon_{mm_1m_2}\left(a_{k_1,l_1,m_1}b_{k_2,l_2,m_2}\right)\end{array}\right]. \tag{E.26}$$

E.4 Quantum transport equations for $N_s = 3$

The equation for $G^{\gtrless}_{o,o,o}$ has been given in the text. Here we give the rest of the equations. We have for the Pauli–Dirac spin,

$$\begin{aligned} &2^{N_s}i\hbar\left(\frac{\partial}{\partial t_1}+\frac{\partial}{\partial t_2}\right)G^{\gtrless}_{k,o,o}\\ &=\left[\tilde{\mathcal{H}},G^{\lessgtr}\right]_{k,o,o}+\left[\Sigma^{\lessgtr},\mathrm{Re}G^r\right]_{k,o,o}-\frac{i}{2}\left\{\Gamma,G^{\lessgtr}\right\}_{k,o,o}+\frac{i}{2}\left\{\Sigma^{\lessgtr},A\right\}_{k,o,o}, \end{aligned} \tag{E.27}$$

where

$$\begin{aligned} &\left[\tilde{\mathcal{H}},G^{\lessgtr}\right]_{k,o,o}\\ &=\left[H_{o,o,o},G^{\gtrless}_{k,o,o}\right]+\left[H_{o,o,m},G^{\gtrless}_{k,o,m}\right]+i\epsilon_{kk_1k_2}\left\{H_{k_1,o,o},G^{\gtrless}_{k_2,o,o}\right\}\\ &\quad+\left[H_{o,l,o},G^{\gtrless}_{k,l,o}\right]+\left[H_{o,l,m},G^{\gtrless}_{k,l,m}\right]+i\epsilon_{kk_1k_2}\left\{H_{k_1,l,o},G^{\gtrless}_{k_2,l,o}\right\}\\ &\quad+\left[H_{k,o,o},G^{\gtrless}_{o,o,o}\right]+\left[H_{k,o,m},G^{\gtrless}_{o,o,m}\right]+i\epsilon_{kk_1k_2}\left\{H_{k_1,o,m},G^{\gtrless}_{k_2,o,m}\right\}\\ &\quad+\left[H_{k,l,o},G^{\gtrless}_{o,l,o}\right]+\left[H_{k.l,m},G^{\gtrless}_{o,l,m}\right]+i\epsilon_{kk_1k_2}\left\{H_{k_1,l,m},G^{\gtrless}_{k_2,l,m}\right\}, \end{aligned} \tag{E.28}$$

$$\begin{aligned} &\left[\Sigma^{\lessgtr},\mathrm{Re}G^r\right]_{k,o,o}\\ &=\left[\Sigma^{\lessgtr}_{o,o,o},\mathrm{Re}G^r_{k,o,o}\right]+\left[\Sigma^{\lessgtr}_{o,o,m},\mathrm{Re}G^r_{k,o,m}\right]+i\epsilon_{kk_1k_2}\left\{\Sigma^{\lessgtr}_{k_1,o,o},\mathrm{Re}G^r_{k_2,o,o}\right\}\\ &\quad+\left[\Sigma^{\lessgtr}_{o,l,o},\mathrm{Re}G^r_{k,l,o}\right]+\left[\Sigma^{\lessgtr}_{o,l,m},\mathrm{Re}G^r_{k,l,m}\right]+i\epsilon_{kk_1k_2}\left\{\Sigma^{\lessgtr}_{k_1,l,o},\mathrm{Re}G^r_{k_2,l,o}\right\} \end{aligned}$$

$$+\left[\Sigma^{\lessgtr}_{k,o,o}, \mathrm{Re}G^{r}_{o,o,o}\right] + \left[\Sigma^{\lessgtr}_{k,o,m}, \mathrm{Re}G^{r}_{o,o,m}\right]$$
$$+ i\epsilon_{kk_1k_2}\left\{\Sigma^{\lessgtr}_{k_1,o,m}, \mathrm{Re}G^{r}_{k_2,o,m}\right\}$$
$$+\left[\Sigma^{\lessgtr}_{k,l,o}, \mathrm{Re}G^{r}_{o,l,o}\right] + \left[\Sigma^{\lessgtr}_{k.l,m}, \mathrm{Re}G^{r}_{o,l,m}\right] + i\epsilon_{kk_1k_2}\left\{\Sigma^{\lessgtr}_{k_1,l,m}, \mathrm{Re}G^{r}_{k_2,l,m}\right\}, \tag{E.29}$$

$$\left\{\Gamma, G^{\lessgtr}\right\}_{k,o,o}$$
$$= \left\{\Gamma_{o,o,o}, G^{\gtrless}_{k,o,o}\right\} + \left\{\Gamma_{o,o,m}, G^{\gtrless}_{k,o,m}\right\} + i\epsilon_{kk_1k_2}\left[\Gamma_{k_1,o,o}, G^{\gtrless}_{k_2,o,o}\right]$$
$$+ \left\{\Gamma_{o,l,o}, G^{\gtrless}_{k,l,o}\right\} + \left\{\Gamma_{o,l,m}, G^{\gtrless}_{k,l,m}\right\} + i\epsilon_{kk_1k_2}\left[\Gamma_{k_1,l,o}, G^{\gtrless}_{k_2,l,o}\right]$$
$$+ \left\{\Gamma_{k,o,o}, G^{\gtrless}_{o,o,o}\right\} + \left\{\Gamma_{k,o,m}, G^{\gtrless}_{o,o,m}\right\} + i\epsilon_{kk_1k_2}\left[\Gamma_{k_1,o,m}, G^{\gtrless}_{k_2,o,m}\right]$$
$$+ \left\{\Gamma_{k,l,o}, G^{\gtrless}_{o,l,o}\right\} + \left\{\Gamma_{k.l,m}, G^{\gtrless}_{o,l,m}\right\} + i\epsilon_{kk_1k_2}\left[\Gamma_{k_1,l,m}, G^{\gtrless}_{k_2,l,m}\right], \tag{E.30}$$

$$\left\{\Sigma^{\lessgtr}, A\right\}_{k,o,o}$$
$$= \left\{\Sigma^{\lessgtr}_{o,o,o}, A_{k,o,o}\right\} + \left\{\Sigma^{\lessgtr}_{o,o,m}, A_{k,o,m}\right\} + i\epsilon_{kk_1k_2}\left[\Sigma^{\lessgtr}_{k_1,o,o}, A_{k_2,o,o}\right]$$
$$+ \left\{\Sigma^{\lessgtr}_{o,l,o}, A_{k,l,o}\right\} + \left\{\Sigma^{\lessgtr}_{o,l,m}, A_{k,l,m}\right\} + i\epsilon_{kk_1k_2}\left[\Sigma^{\lessgtr}_{k_1,l,o}, A_{k_2,l,o}\right]$$
$$+ \left\{\Sigma^{\lessgtr}_{k,o,o}, A_{o,o,o}\right\} + \left\{\Sigma^{\lessgtr}_{k,o,m}, A_{o,o,m}\right\} + i\epsilon_{kk_1k_2}\left[\Sigma^{\lessgtr}_{k_1,o,m}, A_{k_2,o,m}\right]$$
$$+ \left\{\Sigma^{\lessgtr}_{k,l,o}, A_{o,l,o}\right\} + \left\{\Sigma^{\lessgtr}_{k.l,m}, A_{o,l,m}\right\} + i\epsilon_{kk_1k_2}\left[\Sigma^{\lessgtr}_{k_1,l,m}, A_{k_2,l,m}\right]. \tag{E.31}$$

We have for the valley spin,

$$2^{N_s} i\hbar\left(\frac{\partial}{\partial t_1} + \frac{\partial}{\partial t_2}\right) G^{\gtrless}_{o,l,o}$$
$$= \left[\tilde{\mathcal{H}}, G^{\lessgtr}\right]_{o,l,o} + \left[\Sigma^{\lessgtr}, \mathrm{Re}G^{r}\right]_{o,l,o} - \frac{i}{2}\left\{\Gamma, G^{\lessgtr}\right\}_{o,l,o} + \frac{i}{2}\left\{\Sigma^{\lessgtr}, A\right\}_{o,l,o}, \tag{E.32}$$

where

$$\left[\tilde{\mathcal{H}}, G^{\lessgtr}\right]_{o,l,o}$$
$$= \left[H_{o,o,o}, G^{\gtrless}_{o,l,o}\right] + \left[H_{o,o,m}, G^{\gtrless}_{o,l,m}\right] + i\epsilon_{ll_1l_2}\left\{H_{o,l_1,o}, G^{\gtrless}_{o,l_2,o}\right\}$$

$$+ \left[H_{o,l,o}, G^{\lessgtr}_{o,o,o}\right] + \left[H_{o,l,m}, G^{\gtrless}_{o,o,m}\right] + i\epsilon_{ll_1l_2}\left\{H_{o,l_1,m}, G^{\gtrless}_{o,l_2,m}\right\}$$
$$+ \left[H_{k,o,o}, G^{\gtrless}_{k,l,o}\right] + \left[H_{k,o,m}, G^{\gtrless}_{k,l,m}\right] + i\epsilon_{ll_1l_2}\left\{H_{k,l_1,o}, G^{\gtrless}_{k,l_2,o}\right\}$$
$$+ \left[H_{k,l,o}, G^{\gtrless}_{k,o,o}\right] + \left[H_{k.l,m}, G^{\gtrless}_{k,o,m}\right] + i\epsilon_{ll_1l_2}\left\{H_{k,l_1,m}, G^{\gtrless}_{k,l_2,m}\right\}, \tag{E.33}$$

$$\left[\Sigma^{\lessgtr}, \mathrm{Re}G^{r}\right]_{o,l,o}$$
$$= \left[\Sigma^{\gtrless}_{o,o,o}, \mathrm{Re}G^{r}_{o,l,o}\right] + \left[\Sigma^{\gtrless}_{o,o,m}, \mathrm{Re}G^{r}_{o,l,m}\right] + i\epsilon_{ll_1l_2}\left\{\Sigma^{\gtrless}_{o,l_1,o}, \mathrm{Re}G^{r}_{o,l_2,o}\right\}$$
$$+ \left[\Sigma^{\gtrless}_{o,l,o}, \mathrm{Re}G^{r}_{o,o,o}\right] + \left[\Sigma^{\gtrless}_{o,l,m}, \mathrm{Re}G^{r}_{o,o,m}\right] + i\epsilon_{ll_1l_2}\left\{\Sigma^{\gtrless}_{o,l_1,m}, \mathrm{Re}G^{r}_{o,l_2,m}\right\}$$
$$+ \left[\Sigma^{\gtrless}_{k,o,o}, \mathrm{Re}G^{r}_{k,l,o}\right] + \left[\Sigma^{\gtrless}_{k,o,m}, \mathrm{Re}G^{r}_{k,l,m}\right] + i\epsilon_{ll_1l_2}\left\{\Sigma^{\gtrless}_{k,l_1,o}, \mathrm{Re}G^{r}_{k,l_2,o}\right\}$$
$$+ \left[\Sigma^{\gtrless}_{k,l,o}, \mathrm{Re}G^{r}_{k,o,o}\right] + \left[\Sigma^{\gtrless}_{k.l,m}, \mathrm{Re}G^{r}_{k,o,m}\right] + i\epsilon_{ll_1l_2}\left\{\Sigma^{\gtrless}_{k,l_1,m}, \mathrm{Re}G^{r}_{k,l_2,m}\right\}, \tag{E.34}$$

$$\left\{\Gamma, G^{\lessgtr}\right\}_{o,l,o}$$
$$= \left\{\Gamma_{o,o,o}, G^{\gtrless}_{o,l,o}\right\} + \left\{\Gamma_{o,o,m}, G^{\gtrless}_{o,l,m}\right\} + i\epsilon_{ll_1l_2}\left[\Gamma_{o,l_1,o}, G^{\gtrless}_{o,l_2,o}\right]$$
$$+ \left\{\Gamma_{o,l,o}, G^{\lessgtr}_{o,o,o}\right\} + \left\{\Gamma_{o,l,m}, G^{\gtrless}_{o,o,m}\right\} + i\epsilon_{ll_1l_2}\left[\Gamma_{o,l_1,m}, G^{\gtrless}_{o,l_2,m}\right]$$
$$+ \left\{\Gamma_{k,o,o}, G^{\gtrless}_{k,l,o}\right\} + \left\{\Gamma_{k,o,m}, G^{\gtrless}_{k,l,m}\right\} + i\epsilon_{ll_1l_2}\left[\Gamma_{k,l_1,o}, G^{\gtrless}_{k,l_2,o}\right]$$
$$+ \left\{\Gamma_{k,l,o}, G^{\gtrless}_{k,o,o}\right\} + \left\{\Gamma_{k.l,m}, G^{\gtrless}_{k,o,m}\right\} + i\epsilon_{ll_1l_2}\left[\Gamma_{k,l_1,m}, G^{\gtrless}_{k,l_2,m}\right], \tag{E.35}$$

$$\left\{\Sigma^{\lessgtr}, A\right\}_{o,l,o}$$
$$= \left\{\Sigma^{\gtrless}_{o,o,o}, A_{o,l,o}\right\} + \left\{\Sigma^{\gtrless}_{o,o,m}, A_{o,l,m}\right\} + i\epsilon_{ll_1l_2}\left[\Sigma^{\gtrless}_{o,l_1,o}, A_{o,l_2,o}\right]$$
$$+ \left\{\Sigma^{\gtrless}_{o,l,o}, A_{o,o,o}\right\} + \left\{\Sigma^{\gtrless}_{o,l,m}, A_{o,o,m}\right\} + i\epsilon_{ll_1l_2}\left[\Sigma^{\gtrless}_{o,l_1,m}, A_{o,l_2,m}\right]$$
$$+ \left\{\Sigma^{\gtrless}_{k,o,o}, A_{k,l,o}\right\} + \left\{\Sigma^{\gtrless}_{k,o,m}, A_{k,l,m}\right\} + i\epsilon_{ll_1l_2}\left[\Sigma^{\gtrless}_{k,l_1,o}, A_{k,l_2,o}\right]$$
$$+ \left\{\Sigma^{\gtrless}_{k,l,o}, A_{k,o,o}\right\} + \left\{\Sigma^{\gtrless}_{k.l,m}, A_{k,o,m}\right\} + i\epsilon_{ll_1l_2}\left[\Sigma^{\gtrless}_{k,l_1,m}, A_{k,l_2,m}\right]. \tag{E.36}$$

We have for the pseudospin,

$$2^{N_s} i\hbar \left(\frac{\partial}{\partial t_1} + \frac{\partial}{\partial t_2}\right) G^{\gtrless}_{o,o,m}$$
$$= \left[\tilde{\mathcal{H}}, G^{\lessgtr}\right]_{o,o,m} + \left[\Sigma^{\lessgtr}, \mathrm{Re}G^r\right]_{o,o,m} - \frac{i}{2}\left\{\Gamma, G^{\lessgtr}\right\}_{o,o,m} + \frac{i}{2}\left\{\Sigma^{\lessgtr}, A\right\}_{o,o,m}, \tag{E.37}$$

where

$$\left[\tilde{\mathcal{H}}, G^{\lessgtr}\right]_{o,o,m}$$
$$\begin{aligned}
&= \left[H_{o,o,o}, G^{\gtrless}_{o,o,m}\right] + \left[H_{o,o,m}, G^{\gtrless}_{o,o,o}\right] + i\epsilon_{mm_1m_2}\left\{H_{o,o,m_1}, G^{\gtrless}_{o,o,m_2}\right\}\\
&\quad + \left[H_{o,l,o}, G^{\gtrless}_{o,l,m}\right] + \left[H_{o,l,m}, G^{\gtrless}_{o,l,o}\right] + i\epsilon_{mm_1m_2}\left\{H_{o,l,m_1}, G^{\gtrless}_{o,l,m_2}\right\}\\
&\quad + \left[H_{k,o,o}, G^{\gtrless}_{k,o,m}\right] + \left[H_{k,o,m}, G^{\gtrless}_{k,o,o}\right] + i\epsilon_{mm_1m_2}\left\{H_{k,o,m_1}, G^{\gtrless}_{k,o,m_2}\right\}\\
&\quad + \left[H_{k,l,o}, G^{\gtrless}_{k,l,m}\right] + \left[H_{k,l,m}, G^{\gtrless}_{k,l,o}\right] + i\epsilon_{mm_1m_2}\left\{H_{k,l,m_1}, G^{\gtrless}_{k,l,m_2}\right\},
\end{aligned} \tag{E.38}$$

$$\left[\Sigma^{\lessgtr}, \mathrm{Re}G^r\right]_{o,o,m}$$
$$\begin{aligned}
&= \left[\Sigma^{\lessgtr}_{o,o,o}, \mathrm{Re}G^r_{o,o,m}\right] + \left[\Sigma^{\lessgtr}_{o,o,m}, \mathrm{Re}G^r_{o,o,o}\right] + i\epsilon_{mm_1m_2}\left\{\Sigma^{\lessgtr}_{o,o,m_1}, \mathrm{Re}G^r_{o,o,m_2}\right\}\\
&\quad + \left[\Sigma^{\lessgtr}_{o,l,o}, \mathrm{Re}G^r_{o,l,m}\right] + \left[\Sigma^{\lessgtr}_{o,l,m}, \mathrm{Re}G^r_{o,l,o}\right] + i\epsilon_{mm_1m_2}\left\{\Sigma^{\lessgtr}_{o,l,m_1}, \mathrm{Re}G^r_{o,l,m_2}\right\}\\
&\quad + \left[\Sigma^{\lessgtr}_{k,o,o}, \mathrm{Re}G^r_{k,o,m}\right] + \left[\Sigma^{\lessgtr}_{k,o,m}, \mathrm{Re}G^r_{k,o,o}\right] + i\epsilon_{mm_1m_2}\left\{\Sigma^{\lessgtr}_{k,o,m_1}, \mathrm{Re}G^r_{k,o,m_2}\right\}\\
&\quad + \left[\Sigma^{\lessgtr}_{k,l,o}, \mathrm{Re}G^r_{k,l,m}\right] + \left[\Sigma^{\lessgtr}_{k,l,m}, \mathrm{Re}G^r_{k,l,o}\right] + i\epsilon_{mm_1m_2}\left\{\Sigma^{\lessgtr}_{k,l,m_1}, \mathrm{Re}G^r_{k,l,m_2}\right\},
\end{aligned} \tag{E.39}$$

$$\left\{\Gamma, G^{\lessgtr}\right\}_{o,o,m}$$
$$\begin{aligned}
&= \left\{\Gamma_{o,o,o}, G^{\gtrless}_{o,o,m}\right\} + \left\{\Gamma_{o,o,m}, G^{\gtrless}_{o,o,o}\right\} + i\epsilon_{mm_1m_2}\left[\Gamma_{o,o,m_1}, G^{\gtrless}_{o,o,m_2}\right]\\
&\quad + \left\{\Gamma_{o,l,o}, G^{\gtrless}_{o,l,m}\right\} + \left\{\Gamma_{o,l,m}, G^{\gtrless}_{o,l,o}\right\} + i\epsilon_{mm_1m_2}\left[\Gamma_{o,l,m_1}, G^{\gtrless}_{o,l,m_2}\right]\\
&\quad + \left\{\Gamma_{k,o,o}, G^{\gtrless}_{k,o,m}\right\} + \left\{\Gamma_{k,o,m}, G^{\gtrless}_{k,o,o}\right\} + i\epsilon_{mm_1m_2}\left[\Gamma_{k,o,m_1}, G^{\gtrless}_{k,o,m_2}\right]\\
&\quad + \left\{\Gamma_{k,l,o}, G^{\gtrless}_{k,l,m}\right\} + \left\{\Gamma_{k,l,m}, G^{\gtrless}_{k,l,o}\right\} + i\epsilon_{mm_1m_2}\left[\Gamma_{k,l,m_1}, G^{\gtrless}_{k,l,m_2}\right],
\end{aligned} \tag{E.40}$$

$$
\begin{aligned}
&\left\{\Sigma^{\lessgtr}, A\right\}_{o,o,m} \\
&= \left\{\Sigma^{\lessgtr}_{o,o,o}, A_{o,o,m}\right\} + \left\{\Sigma^{\lessgtr}_{o,o,m}, A_{o,o,o}\right\} + i\epsilon_{mm_1m_2}\left[\Sigma^{\lessgtr}_{o,o,m_1}, A_{o,o,m_2}\right] \\
&\quad + \left\{\Sigma^{\lessgtr}_{o,l,o}, A_{o,l,m}\right\} + \left\{\Sigma^{\lessgtr}_{o,l,m}, A_{o,l,o}\right\} + i\epsilon_{mm_1m_2}\left[\Sigma^{\lessgtr}_{o,l,m_1}, A_{o,l,m_2}\right] \\
&\quad + \left\{\Sigma^{\lessgtr}_{k,o,o}, A_{k,o,m}\right\} + \left\{\Sigma^{\lessgtr}_{k,o,m}, A_{k,o,o}\right\} + i\epsilon_{mm_1m_2}\left[\Sigma^{\lessgtr}_{k,o,m_1}, A_{k,o,m_2}\right] \\
&\quad + \left\{\Sigma^{\lessgtr}_{k,l,o}, A_{k,l,m}\right\} + \left\{\Sigma^{\lessgtr}_{k,l,m}, A_{k,l,o}\right\} + i\epsilon_{mm_1m_2}\left[\Sigma^{\lessgtr}_{k,l,m_1}, A_{k,l,m_2}\right].
\end{aligned}
\tag{E.41}
$$

We have for the entangled Dirac spin and valley spin,

$$
\begin{aligned}
&2^{N_s} i\hbar\left(\frac{\partial}{\partial t_1} + \frac{\partial}{\partial t_2}\right) G^{\gtrless}_{k,l,o} \\
&= \left[\tilde{\mathcal{H}}, G^{\lessgtr}\right]_{k,l,o} + \left[\Sigma^{\lessgtr}, \mathrm{Re}G^r\right]_{k,l,o} - \frac{i}{2}\left\{\Gamma, G^{\lessgtr}\right\}_{k,l,o} + \frac{i}{2}\left\{\Sigma^{\lessgtr}, A\right\}_{k,l,o},
\end{aligned}
\tag{E.42}
$$

where

$$
\begin{aligned}
&\left[\tilde{\mathcal{H}}, G^{\lessgtr}\right]_{k,l,o} \\
&= \left[H_{o,o,o}, G^{\lessgtr}_{k,l,o}\right] + \left[H_{o,o,m}, G^{\lessgtr}_{k,l,m}\right] + i\epsilon_{kk_1k_2}\left\{H_{k_1,o,o}, G^{\lessgtr}_{k_2,l,o}\right\} \\
&\quad + i\epsilon_{ll_1l_2}\left\{H_{o,l_1,o}, G^{\lessgtr}_{k,l_2,o}\right\} + \frac{1}{2} i\epsilon_{kk_1k_2} i\epsilon_{ll_1l_2}\left[H_{k_1,l_1,o}, G^{\lessgtr}_{k_2,l_2,o}\right] \\
&\quad + \left[H_{o,l,o}, G^{\lessgtr}_{k,o,o}\right] + \left[H_{o,l,m}, G^{\lessgtr}_{k,o,m}\right] + i\epsilon_{kk_1k_2}\left\{H_{k_1,l,o}, G^{\lessgtr}_{k_2,o,o}\right\} \\
&\quad + i\epsilon_{ll_1l_2}\left\{H_{k,l_1,o} G^{\lessgtr}_{o,l_2,o}\right\} + \frac{1}{2} i\epsilon_{kk_1k_2} i\epsilon_{ll_1l_2}\left[H_{k_1,l_1,o}, G^{\lessgtr}_{k_2,l_2,o}\right] \\
&\quad + \left[H_{k,l,o}, G^{\lessgtr}_{o,o,o}\right] + \left[H_{k.l,m}, G^{\lessgtr}_{o,o,m}\right] + i\epsilon_{kk_1k_2}\left\{H_{k_1,l,m}, G^{\lessgtr}_{k_2,o,m}\right\} \\
&\quad + i\epsilon_{ll_1l_2}\left\{H_{k,l_1,m}, G^{\lessgtr}_{o,l_2,m}\right\} + \frac{1}{2} i\epsilon_{kk_1k_2} i\epsilon_{ll_1l_2}\left[H_{k_1,l_1,m}, G^{\lessgtr}_{k_2,l_2,m}\right] \\
&\quad + \left[H_{k,o,o}, G^{\lessgtr}_{o,l,o}\right] + \left[H_{k,o,m}, G^{\lessgtr}_{o,l,m}\right] + i\epsilon_{kk_1k_2}\left\{H_{k_1,o,m}, G^{\lessgtr}_{k_2,l,m}\right\} \\
&\quad + i\epsilon_{ll_1l_2}\left\{H_{o,l_1,m}, G^{\lessgtr}_{k,l_2,m}\right\} + \frac{1}{2} i\epsilon_{kk_1k_2} i\epsilon_{ll_1l_2}\left[H_{k_1,l_1,m}, G^{\lessgtr}_{k_2,l_2,m}\right],
\end{aligned}
\tag{E.43}
$$

$$
\begin{aligned}
&\left[\Sigma^{\lessgtr}, \mathrm{Re}G^r\right]_{k,l,o} \\
&= \left[\Sigma^{\lessgtr}_{o,o,o}, \mathrm{Re}G^r_{k,l,o}\right] + \left[\Sigma^{\lessgtr}_{o,o,m}, \mathrm{Re}G^r_{k,l,m}\right] + i\epsilon_{kk_1k_2}\left\{\Sigma^{\lessgtr}_{k_1,o,o}, \mathrm{Re}G^r_{k_2,l,o}\right\}
\end{aligned}
$$

$$
\begin{aligned}
&+ i\epsilon_{ll_1l_2}\left\{\Sigma^{\lessgtr}_{o,l_1,o}, \mathrm{Re}G^r_{k,l_2,o}\right\} + \frac{1}{2}i\epsilon_{kk_1k_2}i\epsilon_{ll_1l_2}\left[\Sigma^{\lessgtr}_{k_1,l_1,o}, \mathrm{Re}G^r_{k_2,l_2,o}\right]\\
&+ \left[\Sigma^{\lessgtr}_{o,l,o}, \mathrm{Re}G^r_{k,o,o}\right] + \left[\Sigma^{\lessgtr}_{o,l,m}, \mathrm{Re}G^r_{k,o,m}\right] + i\epsilon_{kk_1k_2}\left\{\Sigma^{\lessgtr}_{k_1,l,o}, \mathrm{Re}G^r_{k_2,o,o}\right\}\\
&+ i\epsilon_{ll_1l_2}\left\{\Sigma^{\lessgtr}_{k,l_1,o}\mathrm{Re}G^r_{o,l_2,o}\right\} + \frac{1}{2}i\epsilon_{kk_1k_2}i\epsilon_{ll_1l_2}\left[\Sigma^{\lessgtr}_{k_1,l_1,o}, \mathrm{Re}G^r_{k_2,l_2,o}\right]\\
&+ \left[\Sigma^{\lessgtr}_{k,l,o}, \mathrm{Re}G^r_{o,o,o}\right] + \left[\Sigma^{\lessgtr}_{k.l,m}, \mathrm{Re}G^r_{o,o,m}\right] + i\epsilon_{kk_1k_2}\left\{\Sigma^{\lessgtr}_{k_1,l,m}, \mathrm{Re}G^r_{k_2,o,m}\right\}\\
&+ i\epsilon_{ll_1l_2}\left\{\Sigma^{\lessgtr}_{k,l_1,m}, \mathrm{Re}G^r_{o,l_2,m}\right\} + \frac{1}{2}i\epsilon_{kk_1k_2}i\epsilon_{ll_1l_2}\left[\Sigma^{\lessgtr}_{k_1,l_1,m}, \mathrm{Re}G^r_{k_2,l_2,m}\right]\\
&+ \left[\Sigma^{\lessgtr}_{k,o,o}, \mathrm{Re}G^r_{o,l,o}\right] + \left[\Sigma^{\lessgtr}_{k,o,m}, \mathrm{Re}G^r_{o,l,m}\right] + i\epsilon_{kk_1k_2}\left\{\Sigma^{\lessgtr}_{k_1,o,m}, \mathrm{Re}G^r_{k_2,l,m}\right\}\\
&+ i\epsilon_{ll_1l_2}\left\{\Sigma^{\lessgtr}_{o,l_1,m}, \mathrm{Re}G^r_{k,l_2,m}\right\} + \frac{1}{2}i\epsilon_{kk_1k_2}i\epsilon_{ll_1l_2}\left[\Sigma^{\lessgtr}_{k_1,l_1,m}, \mathrm{Re}G^r_{k_2,l_2,m}\right],
\end{aligned}
\tag{E.44}
$$

$$
\begin{aligned}
&\left\{\Gamma, G^{\lessgtr}\right\}_{k,l,o}\\
&= \left\{\Gamma_{o,o,o}, G^{\lessgtr}_{k,l,o}\right\} + \left\{\Gamma_{o,o,m}, G^{\lessgtr}_{k,l,m}\right\} + i\epsilon_{kk_1k_2}\left[\Gamma_{k_1,o,o}, G^{\lessgtr}_{k_2,l,o}\right]\\
&\quad + i\epsilon_{ll_1l_2}\left[\Gamma_{o,l_1,o}, G^{\lessgtr}_{k,l_2,o}\right] + \frac{1}{2}i\epsilon_{kk_1k_2}i\epsilon_{ll_1l_2}\left\{\Gamma_{k_1,l_1,o}, G^{\lessgtr}_{k_2,l_2,o}\right\}\\
&\quad + \left\{\Gamma_{o,l,o}, G^{\lessgtr}_{k,o,o}\right\} + \left\{\Gamma_{o,l,m}, G^{\lessgtr}_{k,o,m}\right\} + i\epsilon_{kk_1k_2}\left[\Gamma_{k_1,l,o}, G^{\lessgtr}_{k_2,o,o}\right]\\
&\quad + i\epsilon_{ll_1l_2}\left[\Gamma_{k,l_1,o}G^{\lessgtr}_{o,l_2,o}\right] + \frac{1}{2}i\epsilon_{kk_1k_2}i\epsilon_{ll_1l_2}\left\{\Gamma_{k_1,l_1,o}, G^{\lessgtr}_{k_2,l_2,o}\right\}\\
&\quad + \left\{\Gamma_{k,l,o}, G^{\lessgtr}_{o,o,o}\right\} + \left\{\Gamma_{k.l,m}, G^{\lessgtr}_{o,o,m}\right\} + i\epsilon_{kk_1k_2}\left[\Gamma_{k_1,l,m}, G^{\lessgtr}_{k_2,o,m}\right]\\
&\quad + i\epsilon_{ll_1l_2}\left[\Gamma_{k,l_1,m}, G^{\lessgtr}_{o,l_2,m}\right] + \frac{1}{2}i\epsilon_{kk_1k_2}i\epsilon_{ll_1l_2}\left\{\Gamma_{k_1,l_1,m}, G^{\lessgtr}_{k_2,l_2,m}\right\}\\
&\quad + \left\{\Gamma_{k,o,o}, G^{\lessgtr}_{o,l,o}\right\} + \left\{\Gamma_{k,o,m}, G^{\lessgtr}_{o,l,m}\right\} + i\epsilon_{kk_1k_2}\left[\Gamma_{k_1,o,m}, G^{\lessgtr}_{k_2,l,m}\right]\\
&\quad + i\epsilon_{ll_1l_2}\left[\Gamma_{o,l_1,m}, G^{\lessgtr}_{k,l_2,m}\right] + \frac{1}{2}i\epsilon_{kk_1k_2}i\epsilon_{ll_1l_2}\left\{\Gamma_{k_1,l_1,m}, G^{\lessgtr}_{k_2,l_2,m}\right\},
\end{aligned}
\tag{E.45}
$$

$$
\begin{aligned}
&\left\{\Sigma^{\lessgtr}, A\right\}_{k,l,o}\\
&= \left\{\Sigma^{\lessgtr}_{o,o,o}, A_{k,l,o}\right\} + \left\{\Sigma^{\lessgtr}_{o,o,m}, A_{k,l,m}\right\} + i\epsilon_{kk_1k_2}\left[\Sigma^{\lessgtr}_{k_1,o,o}, A_{k_2,l,o}\right]\\
&\quad + i\epsilon_{ll_1l_2}\left[\Sigma^{\lessgtr}_{o,l_1,o}, A_{k,l_2,o}\right] + \frac{1}{2}i\epsilon_{kk_1k_2}i\epsilon_{ll_1l_2}\left\{\Sigma^{\lessgtr}_{k_1,l_1,o}, A_{k_2,l_2,o}\right\}\\
&\quad + \left\{\Sigma^{\lessgtr}_{o,l,o}, A_{k,o,o}\right\} + \left\{\Sigma^{\lessgtr}_{o,l,m}, A_{k,o,m}\right\} + i\epsilon_{kk_1k_2}\left[\Sigma^{\lessgtr}_{k_1,l,o}, A_{k_2,o,o}\right]
\end{aligned}
$$

$$
\begin{aligned}
&+ i\epsilon_{ll_1l_2}\left[\Sigma^{\lessgtr}_{k,l_1,o}A_{o,l_2,o}\right] + \frac{1}{2}i\epsilon_{kk_1k_2}i\epsilon_{ll_1l_2}\left\{\Sigma^{\lessgtr}_{k_1,l_1,o}, A_{k_2,l_2,o}\right\} \\
&+ \left\{\Sigma^{\lessgtr}_{k,l,o}, A_{o,o,o}\right\} + \left\{\Sigma^{\lessgtr}_{k.l,m}, A_{o,o,m}\right\} + i\epsilon_{kk_1k_2}\left[\Sigma^{\lessgtr}_{k_1,l,m}, A_{k_2,o,m}\right] \\
&+ i\epsilon_{ll_1l_2}\left[\Sigma^{\lessgtr}_{k,l_1,m}, A_{o,l_2,m}\right] + \frac{1}{2}i\epsilon_{kk_1k_2}i\epsilon_{ll_1l_2}\left\{\Sigma^{\lessgtr}_{k_1,l_1,m}, A_{k_2,l_2,m}\right\} \\
&+ \left\{\Sigma^{\lessgtr}_{k,o,o}, A_{o,l,o}\right\} + \left\{\Sigma^{\lessgtr}_{k,o,m}, A_{o,l,m}\right\} + i\epsilon_{kk_1k_2}\left[\Sigma^{\lessgtr}_{k_1,o,m}, A_{k_2,l,m}\right] \\
&+ i\epsilon_{ll_1l_2}\left[\Sigma^{\lessgtr}_{o,l_1,m}, A_{k,l_2,m}\right] + \frac{1}{2}i\epsilon_{kk_1k_2}i\epsilon_{ll_1l_2}\left\{\Sigma^{\lessgtr}_{k_1,l_1,m}, A_{k_2,l_2,m}\right\}.
\end{aligned} \tag{E.46}
$$

We have for the entangled Dirac spin and pseudospin,

$$
\begin{aligned}
&2^{N_s}i\hbar\left(\frac{\partial}{\partial t_1} + \frac{\partial}{\partial t_2}\right)G^{\gtrless}_{k,o,m} \\
&= \left[\tilde{\mathcal{H}}, G^{\lessgtr}\right]_{k,o,m} + \left[\Sigma^{\lessgtr}, \mathrm{Re}G^r\right]_{k,o,m} - \frac{i}{2}\left\{\Gamma, G^{\lessgtr}\right\}_{k,o,m} + \frac{i}{2}\left\{\Sigma^{\lessgtr}, A\right\}_{k,o,m},
\end{aligned} \tag{E.47}
$$

where

$$
\begin{aligned}
&\left[\tilde{\mathcal{H}}, G^{\lessgtr}\right]_{k,o,m} \\
&= \left[H_{o,o,o}, G^{\lessgtr}_{k,o,m}\right] + \left[H_{o,o,m}, G^{\lessgtr}_{k,o,o}\right] + i\epsilon_{kk_1k_2}\left\{H_{k_1,o,o}, G^{\lessgtr}_{k_2,o,m}\right\} \\
&\quad + i\epsilon_{mm_1m_2}\left\{H_{o,o,m_1}, G^{\lessgtr}_{k,o,m_2}\right\} + \frac{1}{2}i\epsilon_{kk_1k_2}i\epsilon_{mm_1m_2}\left[H_{k_1,o,m_1}G^{\lessgtr}_{k_2,o,m_2}\right] \\
&\quad + \left[H_{o,l,o}, G^{\lessgtr}_{k,l,m}\right] + \left[H_{o,l,m}, G^{\lessgtr}_{k,l,o}\right] + i\epsilon_{kk_1k_2}\left\{H_{k_1,l,o}, G^{\lessgtr}_{k_2,l,m}\right\} \\
&\quad + i\epsilon_{mm_1m_2}\left\{H_{o,l,m_1}, G^{\lessgtr}_{k,l,m2}\right\} + \frac{1}{2}i\epsilon_{kk_1k_2}i\epsilon_{mm_1m_2}\left[H_{k_1,l,m_1}, G^{\lessgtr}_{k_2,l,m_2}\right] \\
&\quad + \left[H_{k,o,o}, G^{\lessgtr}_{o,o,m}\right] + \left[H_{k,o,m}, G^{\lessgtr}_{o,o,o}\right] + i\epsilon_{kk_1k_2}\left\{H_{k_1,o,m}, G^{\lessgtr}_{k_2,o,o}\right\} \\
&\quad + i\epsilon_{mm_1m_2}\left\{H_{k,o,m_1}, G^{\lessgtr}_{o,o,m_2}\right\} + \frac{1}{2}i\epsilon_{kk_1k_2}i\epsilon_{mm_1m_2}\left[H_{k_1,o,m_1}, G^{\lessgtr}_{k_2,o,m_2}\right] \\
&\qquad \times \left[H_{k,l,o}, G^{\lessgtr}_{o,l,m}\right] + \left[H_{k.l,m}, G^{\lessgtr}_{o,l,o}\right] + i\epsilon_{kk_1k_2}\left\{H_{k_1,l,m}, G^{\lessgtr}_{k_2,l,o}\right\} \\
&\quad + i\epsilon_{mm_1m_2}\left\{H_{k.l,m_1}, G^{\lessgtr}_{o,l,m_2}\right\} + \frac{1}{2}i\epsilon_{kk_1k_2}i\epsilon_{mm_1m_2}\left[H_{k_1,o,m_1}, G^{\lessgtr}_{k_2,o,m_2}\right],
\end{aligned} \tag{E.48}
$$

$$
\begin{aligned}
&\left[\Sigma^{\lessgtr}, \mathrm{Re}G^r\right]_{k,o,m} \\
&= \left[\Sigma^{\lessgtr}_{o,o,o}, \mathrm{Re}G^r_{k,o,m}\right] + \left[\Sigma^{\lessgtr}_{o,o,m}, \mathrm{Re}G^r_{k,o,o}\right] + i\epsilon_{kk_1k_2}\left\{\Sigma^{\lessgtr}_{k_1,o,o}, \mathrm{Re}G^r_{k_2,o,m}\right\}
\end{aligned}
$$

$$
\begin{aligned}
&+ i\epsilon_{mm_1m_2}\left\{\Sigma^{\lessgtr}_{o,o,m_1}, \mathrm{Re}G^r_{k,o,m_2}\right\} + \frac{1}{2}i\epsilon_{kk_1k_2}i\epsilon_{mm_1m_2}\left[\Sigma^{\lessgtr}_{k_1,o,m_1}\mathrm{Re}G^r_{k_2,o,m_2}\right]\\
&\quad\times\left[\Sigma^{\lessgtr}_{o,l,o}, \mathrm{Re}G^r_{k,l,m}\right] + \left[\Sigma^{\lessgtr}_{o,l,m}, \mathrm{Re}G^r_{k,l,o}\right] + i\epsilon_{kk_1k_2}\left\{\Sigma^{\lessgtr}_{k_1,l,o}, \mathrm{Re}G^r_{k_2,l,m}\right\}\\
&+ i\epsilon_{mm_1m_2}\left\{\Sigma^{\lessgtr}_{o,l,m_1}, \mathrm{Re}G^r_{k,l,m2}\right\} + \frac{1}{2}i\epsilon_{kk_1k_2}i\epsilon_{mm_1m_2}\left[\Sigma^{\lessgtr}_{k_1,l,m_1},\ \mathrm{Re}G^r_{k_2,l,m_2}\right]\\
&+ \left[\Sigma^{\lessgtr}_{k,o,o}, \mathrm{Re}G^r_{o,o,m}\right] + \left[\Sigma^{\lessgtr}_{k,o,m}, \mathrm{Re}G^r_{o,o,o}\right] + i\epsilon_{kk_1k_2}\left\{\Sigma^{\lessgtr}_{k_1,o,m}, \mathrm{Re}G^r_{k_2,o,o}\right\}\\
&+ i\epsilon_{mm_1m_2}\left\{\Sigma^{\lessgtr}_{k,o,m_1},\ \mathrm{Re}G^r_{o,o,m_2}\right\} + \frac{1}{2}i\epsilon_{kk_1k_2}i\epsilon_{mm_1m_2}\left[\Sigma^{\lessgtr}_{k_1,o,m_1},\ \mathrm{Re}\ G^r_{k_2,o,m_2}\right]\\
&\quad\times\left[\Sigma^{\lessgtr}_{k,l,o}, \mathrm{Re}G^r_{o,l,m}\right] + \left[\Sigma^{\lessgtr}_{k.l,m}, \mathrm{Re}G^r_{o,l,o}\right] + i\epsilon_{kk_1k_2}\left\{\Sigma^{\lessgtr}_{k_1,l,m},\ \mathrm{Re}\ G^r_{k_2,l,o}\right\}\\
&+ i\epsilon_{mm_1m_2}\left\{\Sigma^{\lessgtr}_{k.l,m_1},\ \mathrm{Re}\ G^r_{o,l,m_2}\right\} + \frac{1}{2}i\epsilon_{kk_1k_2}i\epsilon_{mm_1m_2}\\
&\quad\times\left[\Sigma^{\lessgtr}_{k_1,o,m_1},\ \mathrm{Re}\ G^r_{k_2,o,m_2}\right],
\end{aligned}
\tag{E.49}
$$

$$
\begin{aligned}
&\left\{\Gamma, G^{\lessgtr}\right\}_{k,o,m}\\
&= \left\{\Gamma_{o,o,o}, G^{\lessgtr}_{k,o,m}\right\} + \left\{\Gamma_{o,o,m}, G^{\lessgtr}_{k,o,o}\right\} + i\epsilon_{kk_1k_2}\left[\Gamma_{k_1,o,o}, G^{\lessgtr}_{k_2,o,m}\right]\\
&\quad+ i\epsilon_{mm_1m_2}\left[\Gamma_{o,o,m_1}, G^{\lessgtr}_{k,o,m_2}\right] + \frac{1}{2}i\epsilon_{kk_1k_2}i\epsilon_{mm_1m_2}\left\{\Gamma_{k_1,o,m_1}G^{\lessgtr}_{k_2,o,m_2}\right\}\\
&\quad+ \left\{\Gamma_{o,l,o}, G^{\lessgtr}_{k,l,m}\right\} + \left\{\Gamma_{o,l,m}, G^{\lessgtr}_{k,l,o}\right\} + i\epsilon_{kk_1k_2}\left[\Gamma_{k_1,l,o}, G^{\lessgtr}_{k_2,l,m}\right]\\
&\quad+ i\epsilon_{mm_1m_2}\left[\Gamma_{o,l,m_1}, G^{\lessgtr}_{k,l,m_2}\right] + \frac{1}{2}i\epsilon_{kk_1k_2}i\epsilon_{mm_1m_2}\left\{\Gamma_{k_1,l,m_1}, G^{\lessgtr}_{k_2,l,m_2}\right\}\\
&\quad+ \left\{\Gamma_{k,o,o}, G^{\lessgtr}_{o,o,m}\right\} + \left\{\Gamma_{k,o,m}, G^{\lessgtr}_{o,o,o}\right\} + i\epsilon_{kk_1k_2}\left[\Gamma_{k_1,o,m}, G^{\lessgtr}_{k_2,o,o}\right]\\
&\quad+ i\epsilon_{mm_1m_2}\left[\Gamma_{k,o,m_1}, G^{\lessgtr}_{o,o,m_2}\right] + \frac{1}{2}i\epsilon_{kk_1k_2}i\epsilon_{mm_1m_2}\left\{\Gamma_{k_1,o,m_1}, G^{\lessgtr}_{k_2,o,m_2}\right\}\\
&\quad+ \left\{\Gamma_{k,l,o}, G^{\lessgtr}_{o,l,m}\right\} + \left\{\Gamma_{k.l,m}, G^{\lessgtr}_{o,l,o}\right\} + i\epsilon_{kk_1k_2}\left[\Gamma_{k_1,l,m}, G^{\lessgtr}_{k_2,l,o}\right]\\
&\quad+ i\epsilon_{mm_1m_2}\left[\Gamma_{k.l,m_1}, G^{\lessgtr}_{o,l,m_2}\right] + \frac{1}{2}i\epsilon_{kk_1k_2}i\epsilon_{mm_1m_2}\left\{\Gamma_{k_1,o,m_1}, G^{\lessgtr}_{k_2,o,m_2}\right\},
\end{aligned}
\tag{E.50}
$$

$$
\begin{aligned}
&\left\{\Sigma^{\lessgtr}, A\right\}_{k,o,m}\\
&= \left\{\Sigma^{\lessgtr}_{o,o,o}, A_{k,o,m}\right\} + \left\{\Sigma^{\lessgtr}_{o,o,m}, A_{k,o,o}\right\} + i\epsilon_{kk_1k_2}\left[\Sigma^{\lessgtr}_{k_1,o,o}, A_{k_2,o,m}\right]\\
&\quad+ i\epsilon_{mm_1m_2}\left[\Sigma^{\lessgtr}_{o,o,m_1}, A_{k,o,m_2}\right] + \frac{1}{2}i\epsilon_{kk_1k_2}i\epsilon_{mm_1m_2}\left\{\Sigma^{\lessgtr}_{k_1,o,m_1}A_{k_2,o,m_2}\right\}
\end{aligned}
$$

$$
\begin{aligned}
&+\left\{\Sigma^{\lessgtr}_{o,l,o}, A_{k,l,m}\right\}+\left\{\Sigma^{\lessgtr}_{o,l,m}, A_{k,l,o}\right\}+i\epsilon_{kk_1k_2}\left[\Sigma^{\lessgtr}_{k_1,l,o}, A_{k_2,l,m}\right]\\
&+i\epsilon_{mm_1m_2}\left[\Sigma^{\lessgtr}_{o,l,m_1}, A_{k,l,m_2}\right]+\frac{1}{2}i\epsilon_{kk_1k_2}i\epsilon_{mm_1m_2}\left\{\Sigma^{\lessgtr}_{k_1,l,m_1}, A_{k_2,l,m_2}\right\}\\
&+\left\{\Sigma^{\lessgtr}_{k,o,o}, A_{o,o,m}\right\}+\left\{\Sigma^{\lessgtr}_{k,o,m}, A_{o,o,o}\right\}+i\epsilon_{kk_1k_2}\left[\Sigma^{\lessgtr}_{k_1,o,m}, A_{k_2,o,o}\right]\\
&+i\epsilon_{mm_1m_2}\left[\Sigma^{\lessgtr}_{k,o,m_1}, A_{o,o,m_2}\right]+\frac{1}{2}i\epsilon_{kk_1k_2}i\epsilon_{mm_1m_2}\left\{\Sigma^{\lessgtr}_{k_1,o,m_1}, A_{k_2,o,m_2}\right\}\\
&+\left\{\Sigma^{\lessgtr}_{k,l,o}, A_{o,l,m}\right\}+\left\{\Sigma^{\lessgtr}_{k.l,m}, A_{o,l,o}\right\}+i\epsilon_{kk_1k_2}\left[\Sigma^{\lessgtr}_{k_1,l,m}, A_{k_2,l,o}\right]\\
&+i\epsilon_{mm_1m_2}\left[\Sigma^{\lessgtr}_{k.l,m_1}, A_{o,l,m_2}\right]+\frac{1}{2}i\epsilon_{kk_1k_2}i\epsilon_{mm_1m_2}\left\{\Sigma^{\lessgtr}_{k_1,o,m_1}, A_{k_2,o,m_2}\right\}.
\end{aligned}
\tag{E.51}
$$

We have for the entangled valley spin and pseudospin,

$$
\begin{aligned}
&2^{N_s}i\hbar\left(\frac{\partial}{\partial t_1}+\frac{\partial}{\partial t_2}\right)G^{\gtrless}_{o,l,m}\\
&=\left[\tilde{\mathcal{H}}, G^{\lessgtr}\right]_{o,l,m}+\left[\Sigma^{\lessgtr}, \mathrm{Re}G^r\right]_{o,l,m}-\frac{i}{2}\left\{\Gamma, G^{\lessgtr}\right\}_{o,l,m}+\frac{i}{2}\left\{\Sigma^{\lessgtr}, A\right\}_{o,l,m},
\end{aligned}
\tag{E.52}
$$

where

$$
\begin{aligned}
&\left[\tilde{\mathcal{H}}, G^{\lessgtr}\right]_{o,l,m}=\\
&=\left[H_{o,o,o}, G^{\lessgtr}_{o,l,m}\right]+\left[H_{o,o,m}, G^{\lessgtr}_{o,l,o}\right]+i\epsilon_{mm_1m_2}\left\{H_{o,o,m_1}, G^{\lessgtr}_{o,l,m_2}\right\}\\
&\quad+i\epsilon_{ll_1l_2}\left\{H_{o,l_1,o}, G^{\lessgtr}_{o,l_2,m}\right\}+\frac{1}{2}i\epsilon_{ll_1l_2}i\epsilon_{mm_1m_2}\left[H_{o,l_1,m_1}, G^{\lessgtr}_{o,l_2,m_2}\right]\\
&\quad+\left[H_{o,l,o}, G^{\lessgtr}_{o,o,m}\right]+\left[H_{o,l,m}, G^{\lessgtr}_{o,o,o}\right]+i\epsilon_{mm_1m_2}\left\{H_{o,l,m_1}, G^{\lessgtr}_{o,o,m_2}\right\}\\
&\quad+i\epsilon_{ll_1l_2}\left\{H_{o,l_1,m}, G^{\lessgtr}_{o,l_2,o}\right\}+\frac{1}{2}i\epsilon_{ll_1l_2}i\epsilon_{mm_1m_2}\left[H_{o,l_1,m_1}, G^{\lessgtr}_{o,l_2,m_2}\right]\\
&\quad+\left[H_{k,l,o}, G^{\lessgtr}_{k,o,m}\right]+\left[H_{k.l,m}, G^{\lessgtr}_{k,o,o}\right]+i\epsilon_{mm_1m_2}\left\{H_{k.l,m_1}, G^{\lessgtr}_{k,o,m_2}\right\}\\
&\quad+i\epsilon_{ll_1l_2}\left\{H_{k,l_1,m}, G^{\lessgtr}_{k,l_2,o}\right\}+\frac{1}{2}i\epsilon_{ll_1l_2}i\epsilon_{mm_1m_2}\left[H_{k,l_1,m_1}, G^{\lessgtr}_{k,l_2,m_2}\right]\\
&\quad+\left[H_{k,o,o}, G^{\lessgtr}_{k,l,m}\right]+\left[H_{k,o,m}, G^{\lessgtr}_{k,l,o}\right]+i\epsilon_{mm_1m_2}\left\{H_{k,o,m_1}, G^{\lessgtr}_{k,l,m_2}\right\}\\
&\quad+i\epsilon_{ll_1l_2}\left\{H_{k,l_1,o}, G^{\lessgtr}_{k,l_2,m}\right\}+\frac{1}{2}i\epsilon_{ll_1l_2}i\epsilon_{mm_1m_2}\left[H_{k,l_1,m_1}, G^{\lessgtr}_{k,l_2,m_2}\right],
\end{aligned}
\tag{E.53}
$$

$$
\begin{aligned}
&\left[\Sigma^{\lessgtr}, \mathrm{Re}G^{r}\right]_{o,l,m} \\
&= \left[\Sigma^{\lessgtr}_{o,o,o}, \mathrm{Re}G^{r}_{o,l,m}\right] + \left[\Sigma^{\lessgtr}_{o,o,m}, \mathrm{Re}G^{r}_{o,l,o}\right] + i\epsilon_{mm_1m_2}\left\{\Sigma^{\lessgtr}_{o,o,m_1}, \mathrm{Re}G^{r}_{o,l,m_2}\right\} \\
&\quad + i\epsilon_{ll_1l_2}\left\{\Sigma^{\lessgtr}_{o,l_1,o}, \mathrm{Re}G^{r}_{o,l_2,m}\right\} + \frac{1}{2}i\epsilon_{ll_1l_2}i\epsilon_{mm_1m_2}\left[\Sigma^{\lessgtr}_{o,l_1,m_1}, \mathrm{Re}G^{r}_{o,l_2,m_2}\right] \\
&\quad + \left[\Sigma^{\lessgtr}_{o,l,o}, \mathrm{Re}G^{r}_{o,o,m}\right] + \left[\Sigma^{\lessgtr}_{o,l,m}, \mathrm{Re}G^{r}_{o,o,o}\right] + i\epsilon_{mm_1m_2}\left\{\Sigma^{\lessgtr}_{o,l,m_1}, \mathrm{Re}G^{r}_{o,o,m_2}\right\} \\
&\quad + i\epsilon_{ll_1l_2}\left\{\Sigma^{\lessgtr}_{o,l_1,m}, \mathrm{Re}G^{r}_{o,l_2,o}\right\} + \frac{1}{2}i\epsilon_{ll_1l_2}i\epsilon_{mm_1m_2}\left[\Sigma^{\lessgtr}_{o,l_1,m_1}, \mathrm{Re}G^{r}_{o,l_2,m_2}\right] \\
&\quad + \left[\Sigma^{\lessgtr}_{k,l,o}, \mathrm{Re}G^{r}_{k,o,m}\right] + \left[\Sigma^{\lessgtr}_{k.l,m}, \mathrm{Re}G^{r}_{k,o,o}\right] + i\epsilon_{mm_1m_2}\left\{\Sigma^{\lessgtr}_{k.l,m_1}, \mathrm{Re}G^{r}_{k,o,m_2}\right\} \\
&\quad + i\epsilon_{ll_1l_2}\left\{\Sigma^{\lessgtr}_{k,l_1,m}, \mathrm{Re}G^{r}_{k,l_2,o}\right\} + \frac{1}{2}i\epsilon_{ll_1l_2}i\epsilon_{mm_1m_2}\left[\Sigma^{\lessgtr}_{k,l_1,m_1}, \mathrm{Re}G^{r}_{k,l_2,m_2}\right] \\
&\quad + \left[\Sigma^{\lessgtr}_{k,o,o}, \mathrm{Re}G^{r}_{k,l,m}\right] + \left[\Sigma^{\lessgtr}_{k,o,m}, \mathrm{Re}G^{r}_{k,l,o}\right] + i\epsilon_{mm_1m_2}\left\{\Sigma^{\lessgtr}_{k,o,m_1}, \mathrm{Re}G^{r}_{k,l,m_2}\right\} \\
&\quad + i\epsilon_{ll_1l_2}\left\{\Sigma^{\lessgtr}_{k,l_1,o}, \mathrm{Re}G^{r}_{k,l_2,m}\right\} + \frac{1}{2}i\epsilon_{ll_1l_2}i\epsilon_{mm_1m_2}\left[\Sigma^{\lessgtr}_{k,l_1,m_1}, \mathrm{Re}G^{r}_{k,l_2,m_2}\right],
\end{aligned}
\tag{E.54}
$$

$$
\begin{aligned}
&\left\{\Gamma, G^{\lessgtr}\right\}_{o,l,m} \\
&= \left\{\Gamma_{o,o,o}, G^{\lessgtr}_{o,l,m}\right\} + \left\{\Gamma_{o,o,m}, G^{\lessgtr}_{o,l,o}\right\} + i\epsilon_{mm_1m_2}\left[\Gamma_{o,o,m_1}, G^{\lessgtr}_{o,l,m_2}\right] \\
&\quad + i\epsilon_{ll_1l_2}\left[\Gamma_{o,l_1,o}, G^{\lessgtr}_{o,l_2,m}\right] + \frac{1}{2}i\epsilon_{ll_1l_2}i\epsilon_{mm_1m_2}\left\{\Gamma_{o,l_1,m_1}, G^{\lessgtr}_{o,l_2,m_2}\right\} \\
&\quad + \left\{\Gamma_{o,l,o}, G^{\lessgtr}_{o,o,m}\right\} + \left\{\Gamma_{o,l,m}, G^{\lessgtr}_{o,o,o}\right\} + i\epsilon_{mm_1m_2}\left[\Gamma_{o,l,m_1}, G^{\lessgtr}_{o,o,m_2}\right] \\
&\quad + i\epsilon_{ll_1l_2}\left[\Gamma_{o,l_1,m}, G^{\lessgtr}_{o,l_2,o}\right] + \frac{1}{2}i\epsilon_{ll_1l_2}i\epsilon_{mm_1m_2}\left\{\Gamma_{o,l_1,m_1}, G^{\lessgtr}_{o,l_2,m_2}\right\} \\
&\quad + \left\{\Gamma_{k,l,o}, G^{\lessgtr}_{k,o,m}\right\} + \left\{\Gamma_{k.l,m}, G^{\lessgtr}_{k,o,o}\right\} + i\epsilon_{mm_1m_2}\left[\Gamma_{k.l,m_1}, G^{\lessgtr}_{k,o,m_2}\right] \\
&\quad + i\epsilon_{ll_1l_2}\left[\Gamma_{k,l_1,m}, G^{\lessgtr}_{k,l_2,o}\right] + \frac{1}{2}i\epsilon_{ll_1l_2}i\epsilon_{mm_1m_2}\left\{\Gamma_{k,l_1,m_1}, G^{\lessgtr}_{k,l_2,m_2}\right\} \\
&\quad + \left\{\Gamma_{k,o,o}, G^{\lessgtr}_{k,l,m}\right\} + \left\{\Gamma_{k,o,m}, G^{\lessgtr}_{k,l,o}\right\} + i\epsilon_{mm_1m_2}\left[\Gamma_{k,o,m_1}, G^{\lessgtr}_{k,l,m_2}\right] \\
&\quad + i\epsilon_{ll_1l_2}\left[\Gamma_{k,l_1,o}, G^{\lessgtr}_{k,l_2,m}\right] + \frac{1}{2}i\epsilon_{ll_1l_2}i\epsilon_{mm_1m_2}\left\{\Gamma_{k,l_1,m_1}, G^{\lessgtr}_{k,l_2,m_2}\right\},
\end{aligned}
\tag{E.55}
$$

$$
\begin{aligned}
&\left\{\Sigma^{\lessgtr}, A\right\}_{o,l,m} \\
&= \left\{\Sigma^{\lessgtr}_{o,o,o}, A_{o,l,m}\right\} + \left\{\Sigma^{\lessgtr}_{o,o,m}, A_{o,l,o}\right\} + i\epsilon_{mm_1m_2}\left[\Sigma^{\lessgtr}_{o,o,m_1}, A_{o,l,m_2}\right]
\end{aligned}
$$

$$+i\epsilon_{ll_1l_2}\left[\Sigma^{\lessgtr}_{o,l_1,o}, A_{o,l_2,m}\right] + \frac{1}{2}i\epsilon_{ll_1l_2}i\epsilon_{mm_1m_2}\left\{\Sigma^{\lessgtr}_{o,l_1,m_1}, A_{o,l_2,m_2}\right\}$$
$$+\left[\Sigma^{\lessgtr}_{o,l,o}, A_{o,o,m}\right] + \left[\Sigma^{\lessgtr}_{o,l,m}, A_{o,o,o}\right] + i\epsilon_{mm_1m_2}\left\{\Sigma^{\lessgtr}_{o,l,m_1}, A_{o,o,m_2}\right\}$$
$$+i\epsilon_{ll_1l_2}\left\{\Sigma^{\lessgtr}_{o,l_1,m}, A_{o,l_2,o}\right\} + \frac{1}{2}i\epsilon_{ll_1l_2}i\epsilon_{mm_1m_2}\left[\Sigma^{\lessgtr}_{o,l_1,m_1}, A_{o,l_2,m_2}\right]$$
$$+\left\{\Sigma^{\lessgtr}_{k,l,o}, A_{k,o,m}\right\} + \left\{\Sigma^{\lessgtr}_{k.l,m}, A_{k,o,o}\right\} + i\epsilon_{mm_1m_2}\left[\Sigma^{\lessgtr}_{k.l,m_1}, A_{k,o,m_2}\right]$$
$$+i\epsilon_{ll_1l_2}\left[\Sigma^{\lessgtr}_{k,l_1,m}, A_{k,l_2,o}\right] + \frac{1}{2}i\epsilon_{ll_1l_2}i\epsilon_{mm_1m_2}\left\{\Sigma^{\lessgtr}_{k,l_1,m_1}, A_{k,l_2,m_2}\right\}$$
$$+\left[\Sigma^{\lessgtr}_{k,o,o}, A_{k,l,m}\right] + \left[\Sigma^{\lessgtr}_{k,o,m}, A_{k,l,o}\right] + i\epsilon_{mm_1m_2}\left\{\Sigma^{\lessgtr}_{k,o,m_1}, A_{k,l,m_2}\right\}$$
$$+i\epsilon_{ll_1l_2}\left\{\Sigma^{\lessgtr}_{k,l_1,o}, A_{k,l_2,m}\right\} + \frac{1}{2}i\epsilon_{ll_1l_2}i\epsilon_{mm_1m_2}\left[\Sigma^{\lessgtr}_{k,l_1,m_1}, A_{k,l_2,m_2}\right]. \quad \text{(E.56)}$$

We have for the entangled Dirac spin, valley spin, and pseudospin,

$$2^{N_s}i\hbar\left(\frac{\partial}{\partial t_1}+\frac{\partial}{\partial t_2}\right)G^{\gtrless}_{k,l,m}$$
$$=\left[\tilde{\mathcal{H}}, G^{\lessgtr}\right]_{k,l,m} + \left[\Sigma^{\lessgtr}, \mathrm{Re}G^r\right]_{k,l,m} - \frac{i}{2}\left\{\Gamma, G^{\lessgtr}\right\}_{k,l,m} + \frac{i}{2}\left\{\Sigma^{\lessgtr}, A\right\}_{k,l,m}, \quad \text{(E.57)}$$

where

$$\left[\tilde{\mathcal{H}}, G^{\lessgtr}\right]_{k,l,m}$$
$$=\begin{bmatrix} \left[H_{k,l,m}, G^{\lessgtr}_{0,0,0}\right] + \left[H_{k,l,0}, G^{\lessgtr}_{0,0,m}\right] + \left[H_{k,0,m}, G^{\lessgtr}_{0,l,0}\right] \\ + \left[H_{0,l,m}, G^{\lessgtr}_{k,0,0}\right] \\ + \left[H_{k,0,0}, G^{\lessgtr}_{0,l,m}\right] + \left[H_{0,0,m}, G^{\lessgtr}_{k,l,0}\right] + \left[H_{0,l,0}, G^{\lessgtr}_{k,0,m}\right] \\ + \left[H_{o,o,o}, G^{\lessgtr}_{k.l.m}\right] \end{bmatrix}$$
$$+\begin{bmatrix} \epsilon_{kk_1k_2}\left\{H_{k_1,0,0}, G^{\lessgtr}_{k_2,l,m}\right\} + \epsilon_{kk_1k_2}\left\{H_{k_1,l,0}, G^{\lessgtr}_{k_2,0,m}\right\} \\ + \epsilon_{kk_1k_2}\left\{H_{k_1,0,m}, G^{\lessgtr}_{k_2,l,0}\right\} + \epsilon_{kk_1k_2}\left\{H_{k_1,l,m}, G^{\lessgtr}_{k_2,0,0}\right\} \\ + \epsilon_{ll_1l_2}\left\{H_{0,l_1,0}, G^{\lessgtr}_{k,l_2,m}\right\} + \epsilon_{ll_1l_2}\left\{H_{k,l_1,0}, G^{\lessgtr}_{0,l_2,m}\right\} \\ + \epsilon_{ll_1l_2}\left\{H_{0,l_1,m}, G^{\lessgtr}_{k,l_2,0}\right\} + \epsilon_{ll_1l_2}\left\{H_{k,l_1,m}, G^{\lessgtr}_{0,l_2,0}\right\} \\ + \epsilon_{m;m_1,m_2}\left\{H_{0,0.m_1}, G^{\lessgtr}_{k,l,m_2}\right\} + \epsilon_{m;m_1,m_2}\left\{H_{0,l.m_1}, G^{\lessgtr}_{k,0,m_2}\right\} \\ + \epsilon_{m;m_1,m_2}\left\{H_{k,0.m_1}, G^{\lessgtr}_{0,l,m_2}\right\} + \epsilon_{m;m_1,m_2}\left\{H_{k,l.m_1}, G^{\lessgtr}_{0,0,m_2}\right\} \end{bmatrix}$$

$$+\left[\begin{array}{c} i\epsilon_{kk_1k_2} i\epsilon_{ll_1l_2}\left(\left[H_{k_1,l_1,o}, G^{\lessgtr}_{k_2,l_2,m}\right] + \left[H_{k_1,l_1,m}, G^{\lessgtr}_{k_2,l_2,o}\right]\right) \\ + i\epsilon_{kk_1k_2} i\epsilon_{mm_1m_2}\left(\left[H_{k_1,o,m_1}, G^{\lessgtr}_{k_2,l,m_2}\right] + \left[H_{k_1,l,m_1}, G^{\lessgtr}_{k_2,o,m_2}\right]\right) \\ + i\epsilon_{ll_1l_2} i\epsilon_{mm_1m_2}\left(\left[H_{o,l_1,m_1}, G^{\lessgtr}_{k,l_2,m_2}\right] + \left[H_{k,l_1,m_1}, G^{\lessgtr}_{o,l_2,m_2}\right]\right) \\ + i\epsilon_{kk_1k_2} i\epsilon_{ll_1l_2} i\epsilon_{mm_1m_2}\left\{H_{k_1,l_1,m_1}, G^{\lessgtr}_{k_2,l_2,m_2}\right\} \end{array}\right], \tag{E.58}$$

$$\left[\Sigma^{\lessgtr}, \mathrm{Re}G^r\right]_{k,l,m}$$

$$= \left[\begin{array}{c} \left[\Sigma^{\lessgtr}_{k,l,m},\ \mathrm{Re}G^r_{0,0,0}\right] + \left[\Sigma^{\lessgtr}_{k,l,0},\ \mathrm{Re}G^r_{0,0,m}\right] + \left[\Sigma^{\lessgtr}_{k,0,m},\ \mathrm{Re}G^r_{0,l,0}\right] \\ + \left[\Sigma^{\lessgtr}_{0,l,m},\ \mathrm{Re}G^r_{k,0,0}\right] \\ + \left[\Sigma^{\lessgtr}_{k,0,0},\ \mathrm{Re}G^r_{0,l,m}\right] + \left[\Sigma^{\lessgtr}_{0,0,m},\ \mathrm{Re}G^r_{k,l,0}\right] + \left[\Sigma^{\lessgtr}_{0,l,0}\ \mathrm{Re}G^r_{k,0,m}\right] \\ + \left[\Sigma^{\lessgtr}_{o,o,o},\ \mathrm{Re}G^r_{k.l.m}\right] \end{array}\right]$$

$$+\left[\begin{array}{c} \epsilon_{kk_1k_2}\left\{\Sigma^{\lessgtr}_{k_1,0,0},\ \mathrm{Re}G^r_{k_2,l,m}\right\} + \epsilon_{kk_1k_2}\left\{\Sigma^{\lessgtr}_{k_1,l,0},\ \mathrm{Re}G^r_{k_2,0,m}\right\} \\ +\epsilon_{kk_1k_2}\left\{\Sigma^{\lessgtr}_{k_1,0,m},\ \mathrm{Re}G^r_{k_2,l,0}\right\} + \epsilon_{kk_1k_2}\left\{\Sigma^{\lessgtr}_{k_1,l,m},\ \mathrm{Re}G^r_{k_2,0,0}\right\} \\ +\epsilon_{ll_1l_2}\left\{\Sigma^{\lessgtr}_{0,l_1,0},\ \mathrm{Re}G^r_{k,l_2,m}\right\} + \epsilon_{ll_1l_2}\left\{\Sigma^{\lessgtr}_{k,l_1,0},\ \mathrm{Re}G^r_{0,l_2,m}\right\} \\ +\epsilon_{ll_1l_2}\left\{\Sigma^{\lessgtr}_{0,l_1,m},\ \mathrm{Re}G^r_{k,l_2,0}\right\} + \epsilon_{ll_1l_2}\left\{\Sigma^{\lessgtr}_{k,l_1,m},\ \mathrm{Re}G^r_{0,l_2,0}\right\} \\ +\epsilon_{m;m_1,m_2}\left\{\Sigma^{\lessgtr}_{0,0.m_1},\ \mathrm{Re}G^r_{k,l,m_2}\right\} + \epsilon_{m;m_1,m_2}\left\{\Sigma^{\lessgtr}_{0,l.m_1},\ \mathrm{Re}G^r_{k,0,m_2}\right\} \\ +\epsilon_{m;m_1,m_2}\left\{\Sigma^{\lessgtr}_{k,0.m_1},\ \mathrm{Re}G^r_{0,l,m_2}\right\} + \epsilon_{m;m_1,m_2}\left\{\Sigma^{\lessgtr}_{k,l.m_1},\ \mathrm{Re}G^r_{0,0,m_2}\right\} \end{array}\right]$$

$$+\left[\begin{array}{c} i\epsilon_{kk_1k_2} i\epsilon_{ll_1l_2}\left(\left[\Sigma^{\lessgtr}_{k_1,l_1,o},\ \mathrm{Re}G^r_{k_2,l_2,m}\right] + \left[\Sigma^{\lessgtr}_{k_1,l_1,m},\ \mathrm{Re}G^r_{k_2,l_2,o}\right]\right) \\ +i\epsilon_{kk_1k_2} i\epsilon_{mm_1m_2}\left(\left[\Sigma^{\lessgtr}_{k_1,o,m_1},\ \mathrm{Re}G^r_{k_2,l,m_2}\right]\right. \\ \left. + \left[\Sigma^{\lessgtr}_{k_1,l,m_1},\ \mathrm{Re}G^r_{k_2,o,m_2}\right]\right) \\ +i\epsilon_{ll_1l_2} i\epsilon_{mm_1m_2}\left(\left[\Sigma^{\lessgtr}_{o,l_1,m_1},\ \mathrm{Re}G^r_{k,l_2,m_2}\right] + \left[\Sigma^{\lessgtr}_{k,l_1,m_1},\ \mathrm{Re}G^r_{o,l_2,m_2}\right]\right) \\ +i\epsilon_{kk_1k_2} i\epsilon_{ll_1l_2} i\epsilon_{mm_1m_2}\left\{\Sigma^{\lessgtr}_{k_1,l_1,m_1},\ \mathrm{Re}G^r_{k_2,l_2,m_2}\right\} \end{array}\right], \tag{E.59}$$

$$\{\Gamma, G^{\lessgtr}\}_{k,l,m} =$$

$$= \begin{bmatrix} \left\{\Gamma_{k,l,m},\ G^{\lessgtr}_{0,0,0}\right\} + \left\{\Gamma_{k,l,0},\ G^{\lessgtr}_{0,0,m}\right\} + \left\{\Gamma_{k,0,m}\ ,G^{\lessgtr}_{0,l,0}\right\} \\ + \left\{\Gamma_{0,l,m}\ ,G^{\lessgtr}_{k,0,0}\right\} \\ + \left\{\Gamma_{k,0,0},\ G^{\lessgtr}_{0,l,m}\right\} + \left\{\Gamma_{0,0,m},\ G^{\lessgtr}_{k,l,0}\right\} + \left\{\Gamma_{0,l,0},\ G^{\lessgtr}_{k,0,m}\right\} \\ + \left\{\Gamma_{o,o,o},\ G^{\lessgtr}_{k.l.m}\right\} \end{bmatrix}$$

$$+ \begin{bmatrix} \epsilon_{kk_1k_2}\left[\Gamma_{k_1,0,0},\ G^{\lessgtr}_{k_2,l,m}\right] + \epsilon_{kk_1k_2}\left[\Gamma_{k_1,l,0}\ ,G^{\lessgtr}_{k_2,0,m}\right] \\ +\epsilon_{kk_1k_2}\left[\Gamma_{k_1,0,m},\ G^{\lessgtr}_{k_2,l,0}\right] + \epsilon_{kk_1k_2}\left[\Gamma_{k_1,l,m}\ ,G^{\lessgtr}_{k_2,0,0}\right] \\ +\epsilon_{ll_1l_2}\left[\Gamma_{0,l_1,0},\ G^{\lessgtr}_{k,l_2,m}\right] + \epsilon_{ll_1l_2}\left[\Gamma_{k,l_1,0}\ ,G^{\lessgtr}_{0,l_2,m}\right] \\ +\epsilon_{ll_1l_2}\left[\Gamma_{0,l_1,m},\ G^{\lessgtr}_{k,l_2,0}\right] + \epsilon_{ll_1l_2}\left[\Gamma_{k,l_1,m}\ ,G^{\lessgtr}_{0,l_2,0}\right] \\ +\epsilon_{m;m_1,m_2}\left[\Gamma_{0,0.m_1}\ ,G^{\lessgtr}_{k,l,m_2}\right] + \epsilon_{m;m_1,m_2}\left[\Gamma_{0,l.m_1},\ G^{\lessgtr}_{k,0,m_2}\right] \\ +\epsilon_{m;m_1,m_2}\left[\Gamma_{k,0.m_1}\ ,G^{\lessgtr}_{0,l,m_2}\right] + \epsilon_{m;m_1,m_2}\left[\Gamma_{k,l.m_1},\ G^{\lessgtr}_{0,0,m_2}\right] \end{bmatrix}$$

$$+ \begin{bmatrix} i\epsilon_{kk_1k_2}i\epsilon_{ll_1l_2}\left(\left\{\Gamma_{k_1,l_1,o},\ G^{\lessgtr}_{k_2,l_2,m}\right\} + \left\{\Gamma_{k_1,l_1,m},\ G^{\lessgtr}_{k_2,l_2,o}\right\}\right) \\ +i\epsilon_{kk_1k_2}i\epsilon_{mm_1m_2}\left(\left\{\Gamma_{k_1,o,m_1},\ G^{\lessgtr}_{k_2,l,m_2}\right\} + \left\{\Gamma_{k_1,l,m_1},\ G^{\lessgtr}_{k_2,o,m_2}\right\}\right) \\ +i\epsilon_{ll_1l_2}i\epsilon_{mm_1m_2}\left(\left\{\Gamma_{o,l_1,m_1},\ G^{\lessgtr}_{k,l_2,m_2}\right\} + \left\{\Gamma_{k,l_1,m_1},\ G^{\lessgtr}_{o,l_2,m_2}\right\}\right) \\ +i\epsilon_{kk_1k_2}i\epsilon_{ll_1l_2}i\epsilon_{mm_1m_2}\left[\Gamma_{k_1,l_1,m_1},\ G^{\lessgtr}_{k_2,l_2,m_2}\right] \end{bmatrix}, \tag{E.60}$$

$$\{\Sigma^{\lessgtr}, A\}_{k,l,m}$$

$$= \begin{bmatrix} \left\{\Sigma^{\lessgtr}_{k,l,m},\ A_{0,0,0}\right\} + \left\{\Sigma^{\lessgtr}_{k,l,0},\ A_{0,0,m}\right\} + \left\{\Sigma^{\lessgtr}_{k,0,m},\ A_{0,l,0}\right\} \\ + \left\{\Sigma^{\lessgtr}_{0,l,m}\ A_{k,0,0}\right\} \\ + \left\{\Sigma^{\lessgtr}_{k,0,0},\ A_{0,l,m}\right\} + \left\{\Sigma^{\lessgtr}_{0,0,m},\ A_{k,l,0}\right\} + \left\{\Sigma^{\lessgtr}_{0,l,0},\ A_{k,0,m}\right\} \\ + \{\Sigma^{\lessgtr}_{o,o,o},\ A_{k.l.m}\} \end{bmatrix}$$

$$
+\begin{bmatrix}
\epsilon_{kk_1k_2}\left[\Sigma^{\lessgtr}_{k_1,0,0},\ A_{k_2,l,m}\right]+\epsilon_{kk_1k_2}\left[\Sigma^{\lessgtr}_{k_1,l,0},\ A_{k_2,0,m}\right]\\
+\epsilon_{kk_1k_2}\left[\Sigma^{\lessgtr}_{k_1,0,m}\ ,A_{k_2,l,0}\right]+\epsilon_{kk_1k_2}\left[\Sigma^{\lessgtr}_{k_1,l,m},\ A_{k_2,0,0}\right]\\
+\epsilon_{ll_1l_2}\left[\Sigma^{\lessgtr}_{0,l_1,0}\ ,A_{k,l_2,m}\right]+\epsilon_{ll_1l_2}\left[\Sigma^{\lessgtr}_{k,l_1,0},\ A_{0,l_2,m}\right]\\
+\epsilon_{ll_1l_2}\left[\Sigma^{\lessgtr}_{0,l_1,m}\ ,A_{k,l_2,0}\right]+\epsilon_{ll_1l_2}\left[\Sigma^{\lessgtr}_{k,l_1,m},\ A_{0,l_2,0}\right]\\
+\epsilon_{m;m_1,m_2}\left[\Sigma^{\lessgtr}_{0,0.m_1}\ ,A_{k,l,m_2}\right]+\epsilon_{m;m_1,m_2}\left[\Sigma^{\lessgtr}_{0,l.m_1},\ A_{k,0,m_2}\right]\\
+\epsilon_{m;m_1,m_2}\left[\Sigma^{\lessgtr}_{k,0.m_1},\ A_{0,l,m_2}\right]+\epsilon_{m;m_1,m_2}\left[\Sigma^{\lessgtr}_{k,l.m_1},\ A_{0,0,m_2}\right]
\end{bmatrix}
$$

$$
+\begin{bmatrix}
i\epsilon_{kk_1k_2}i\epsilon_{ll_1l_2}\left(\left\{\Sigma^{\lessgtr}_{k_1,l_1,o},\ A_{k_2,l_2,m}\right\}+\left\{\Sigma^{\lessgtr}_{k_1,l_1,m},\ A_{k_2,l_2,o}\right\}\right)\\
+i\epsilon_{kk_1k_2}i\epsilon_{mm_1m_2}\left(\left\{\Sigma^{\lessgtr}_{k_1,l_1,o},\ A_{k_2,l_2,m}\right\}+\left\{\Sigma^{\lessgtr}_{k_1,l_1,m},\ A_{k_2,l_2,o}\right\}\right)\\
+i\epsilon_{ll_1l_2}i\epsilon_{mm_1m_2}\left(\left\{\Sigma^{\lessgtr}_{k_1,l_1,o},\ A_{k_2,l_2,m}\right\}+\left\{\Sigma^{\lessgtr}_{k_1,l_1,m},\ A_{k_2,l_2,o}\right\}\right)\\
+i\epsilon_{kk_1k_2}i\epsilon_{ll_1l_2}i\epsilon_{mm_1m_2}\left[\Sigma^{\lessgtr}_{k_1,l_1,m_1},\ A_{k_2,l_2,m_2}\right]
\end{bmatrix}. \tag{E.61}
$$

E.5 Comparison with multiband quantum transport equations

Here we make a comparison for $N_s = 2$ with multi-band quantum spin transport equations of Buot *et al.* [224]. Note that Ref. [224] essentially calculates Dirac spin semiconductor Bloch equations (DSSBEs) of the spin magnetization quantum transport equations (SMQTEs) in the electron–hole picture. We immediately see that the expression of $G^{<}_{k,o}$, i.e., pseudo-spin independent Pauli Dirac spin transport equations, is identical to the expression of the transport equation of a conduction-band spin transport of $\vec{S}_{cc}$ of Ref. [224], upon using the correspondence of the symbols used in Ref. [224] to the symbols here, i.e., $\Xi_{\gamma\gamma} \Longrightarrow \Sigma_{\gamma\gamma}$, $\delta^{\gtrless}_{\alpha\beta} \Longrightarrow \Sigma^{\gtrless}_{\alpha\beta}$, and $S^{\gtrless} \Longrightarrow G^{\gtrless}$.

By transforming the DSSBEs to the electron picture using Tables E.1 and E.2 (a–b), and using the following relations:

$$iA\,(1,2) = -2i\mathrm{Im}G^r = -\left(G^{>}\,(1,2) - G^{<}\,(1,2)\right), \tag{E.62}$$

$$i\Gamma\,(1,2) = -2i\mathrm{Im}\Sigma^r = -\left(\Sigma^{>}\,(1,2) - \Sigma^{<}\,(1,2)\right), \tag{E.63}$$

in combining the equations to obtain the equations for the components of the magnetization density tensors, the resulting equations exactly reproduce the equations for $N_s = 2$ here. The virtue of doing the calculations from the DSSBEs is that it give us an explicit expression for the sort of terms containing two Levi-Civita tensors. The calculation shows the equivalence, for $m = x$-component of pseudospin with k an arbitrary component of Dirac spin, of the expression,

$$\left(\vec{\Sigma}_y^< \times \vec{A}_z - \vec{\Sigma}_z^< \times \vec{A}_y\right) + \left(\vec{A}_y \times \vec{\Sigma}_z^< - \vec{A}_z \times \vec{\Sigma}_y^<\right)$$

$$= \left\{ \begin{array}{l} \epsilon_{kk_1k_2}\epsilon_{mm_1m_2}\vec{\Sigma}^<_{k_1,m_1}\vec{A}_{k_2m_2} \\ +\epsilon_{kk_1k_2}\epsilon_{mm_1m_2}\vec{A}_{k_1,m_1}\vec{\Sigma}^<_{k_2m_2} \end{array} \right\},$$

$$= \epsilon_{kk_1k_2}\epsilon_{mm_1m_2}\left\{\vec{\Sigma}^<_{k_1,m_1}, \vec{A}_{k_2m_2}\right\}. \tag{E.64}$$

Note that the $\epsilon_{kk_1k_2}\epsilon_{mm_1m_2}\left\{\vec{\Sigma}^<_{k_1,m_1}, \vec{A}_{k_2m_2}\right\}$ and similar other terms are completely symmetric in the simultaneous exchange of $k_1$and k_2 and m_1 and m_2, respectively. The last line of Eq. (E.64) holds for arbitrary components, k and m. Note that we have generalized the spinor Hamiltonian, Eq. (28.12) in the text, to account for the spinor $\mathrm{Re}\Sigma^r$ [167, 226] and/or magnetic field and/or Dresselhaus and/or Rashba spin-orbit coupling.

E.6 Conversion map from electron–hole picture to electron picture

We have the mapping between electron picture and electron–hole picture using Table E.1 [12].

We also have Table E.2, which can also be similarly applied to the self-energies.

Table E.1. Electron to electron–hole picture.

electron picture	$\langle e - \text{field}\rangle$	$\langle e - h\ \text{field}\rangle$	**e − h** picture
$-i\hbar G^<_{vv}(12)$	$\langle\psi_v^\dagger(2)\psi_v(1)\rangle$	$\langle\phi_v(2)\phi_v^\dagger(1)\rangle$	$i\hbar G^{h,>T}_{vv}(12)$
$-i\hbar G^<_{vc}(12)$	$\langle\psi_c^\dagger(2)\psi_v(1)\rangle$	$\langle\psi_c^\dagger(2)\phi_v^\dagger(1)\rangle$	$-i\hbar g^{e-h,<}_{ee,vc}(12)$
$-i\hbar G^<_{cv}(12)$	$\langle\psi_v^\dagger(2)\psi_c(1)\rangle$	$\langle\phi_v(2)\psi_c(1)\rangle$	$-i\hbar g^{e-h,<}_{hh,cv}(12)$
$-i\hbar G^<_{cc}(12)$	$\langle\psi_c^\dagger(2)\psi_c(1)\rangle$	$\langle\psi_c^\dagger(2)\psi_c(1)\rangle$	$-i\hbar G^{e-h,<}_{cc}(12)$
$i\hbar G^{>T}_{vv}(12)$	$\langle\psi_v(2)\psi_v^\dagger(1)\rangle$	$\langle\phi_v^\dagger(2)\phi_v(1)\rangle$	$-i\hbar G^{h,<}_{vv}(12)$
$i\hbar G^>_{vv}(12)$	$\langle\psi_v(1)\psi_v^\dagger(2)\rangle$	$\langle\phi_v^\dagger(1)\phi_v(2)\rangle$	$-i\hbar G^{h,<T}_{vv}(12)$

Table E.2. Similar relations for self-energies.

(a)		(b)	
electron picture	**e − h** picture	**electron** picture	**e − h** picture
$G^r_{vv}(12)$	$-G^{e-h,aT}_{vv}(12)$	$G^a_{vv}(12)$	$-G^{e-h,rT}_{vv}(12)$
$G^r_{vc}(12)$	$g^{e-h,r}_{ee,vc}(12)$	$G^a_{vc}(12)$	$g^{e-h,a}_{ee,vc}(12)$
$G^r_{cv}(12)$	$g^{e-h,r}_{hh,cv}(12)$	$G^a_{cv}(12)$	$g^{e-h,a}_{hh,cv}(12)$
$G^r_{cc}(12)$	$G^{e-h,r}_{cc}(12)$	$G^a_{cc}(12)$	$G^{e-h,a}_{cc}(12)$

E.7 Transport equations in phase space

The transport equations in phase space are obtained by applying the "lattice" Weyl transformation (although continuum approximation is often employed here, this is not essential and we use the word "lattice" when referring to solid-state problems [3, 11, 12, 14] of the propagator equations for the respective excitations by using the following identities:

E.7.1 *"Lattice" Weyl transform of differential operators*

$$i\hbar\left(\frac{\partial}{\partial t_1}+\frac{\partial}{\partial t_2}\right)Y(t_1,t_2)\Leftrightarrow i\hbar\frac{\partial}{\partial t}Y(E,t), \tag{E.65}$$

$$\left(\frac{\partial^2}{\partial t_1^2}-\frac{\partial^2}{\partial t_2^2}\right)Y(t_1,t_2)\Leftrightarrow \frac{2i}{\hbar}E\frac{\partial}{\partial t}Y(E,t), \tag{E.66}$$

$$\left(\nabla_1^2-\nabla_2^2\right)Y(r_1,r_2)\Leftrightarrow \frac{2i}{\hbar}p\cdot\nabla_q Y(p,q). \tag{E.67}$$

E.7.2 *Lattice Weyl transform of product of arbitrary operators*

E.7.2.1 *Poisson bracket expansion*

In terms of differential "Poisson bracket" operator,

$$AB(p,q)=\exp\frac{\hbar}{i}\left(\frac{\partial^{(a)}}{\partial p}\cdot\frac{\partial^{(b)}}{\partial q}-\frac{\partial^{(a)}}{\partial q}\cdot\frac{\partial^{(b)}}{\partial p}\right)a(p,q)\ b(p,q). \tag{E.68}$$

Thus, we are lead to the lattice Weyl transform of a commutator, $[A, B](p, q)$, and an anti-commutator, $\{A, B\}(p, q)$ as

$$[A, B](p, q) = \cos cos\Lambda \left[a(p, q)\ b(p, q) - b(p, q)\, a(p, q)\right]$$
$$-i \sin \Lambda \left[a(p, q)\ b(p, q) + b(p, q)\, a(p, q)\right], \qquad \text{(E.69)}$$
$$\{A, B\}(p, q) = \cos cos\Lambda \left[a(p, q)\ b(p, q) + b(p, q)\, a(p, q)\right]$$
$$-i \sin \Lambda \left[a(p, q)\ b(p, q) - b(p, q)\, a(p, q)\right], \qquad \text{(E.70)}$$

where $\Lambda = \frac{\hbar}{2}\left(\frac{\partial^{(a)}}{\partial p} \cdot \frac{\partial^{(b)}}{\partial q} - \frac{\partial^{(a)}}{\partial q} \cdot \frac{\partial^{(b)}}{\partial p}\right)$. Note that so far no semi-classical approximation is involved in the derivation.

E.7.2.2 *Integral representation*

In terms of integral operators, we have,

$$AB(p, q) = \frac{1}{(\hbar^4)^2} \int dp'\, dq'\ K_{\bar{A}}^{+}(p, q; p'q')\ b(p', q'),$$
$$= \frac{1}{(\hbar^4)^2} \int dp'\, dq'\ a(p', q')\ K_{\bar{B}}^{-}(p, q; p'q'). \qquad \text{(E.71)}$$

where integral kernels are defined by

$$K_{\bar{Y}}^{\pm}(p, q; p'q') = \int dudv \exp\left\{\frac{i}{\hbar}\left[(p - p') \cdot v + (q - q') \cdot u\right]\right\} y$$
$$\times \left(p \pm \frac{u}{2}, q \mp \frac{v}{2}\right). \qquad \text{(E.72)}$$

Therefore, in terms of integral operators,

$$[A, B](p, q) = \frac{1}{(\hbar^4)^2} \int dp'\, dq'$$
$$\times \left[K_{\bar{A}}^{+}(p, q; p'q')\ b(p', q') - b(p', q')\, K_{\bar{A}}^{-}(p, q; p'q')\right], \qquad \text{(E.73)}$$

$$\{A, B\}(p, q) = \frac{1}{(\hbar^4)^2} \int dp'\, dq'\ \left[K_{\bar{A}}^{+}(p, q; p'q')\ b(p', q') + b(p', q')\right.$$
$$\left.\times K_{\bar{A}}^{-}(p, q; p'q')\right]. \qquad \text{(E.74)}$$

E.7.2.3 *Local and non-local integral kernels*

The above expressions simplify considerably when the lattice Weyl transform $a(p,q)$ and $b(p,q)$ are scalar functions. We have for the integral representations,

$$[A,B](p,q) = \frac{1}{(\hbar^4)^2}\int dp'\, dq' du dv \exp\left\{\frac{i}{\hbar}\left[(p-p')\cdot v + (q-q')\cdot u\right]\right\}$$
$$\times \begin{bmatrix} a\left(p+\frac{u}{2}, q-\frac{v}{2}\right) \\ -a\left(p-\frac{u}{2}, q+\frac{v}{2}\right)\end{bmatrix} b\left(p', q'\right), \tag{E.75}$$

$$\{A,B\}(p,q) = \frac{1}{(\hbar^4)^2}\int dp'\, dq' du dv \exp\left\{\frac{i}{\hbar}\left[(p-p')\cdot v + (q-q')\cdot u\right]\right\}$$
$$\times \begin{bmatrix} a\left(p+\frac{u}{2}, q-\frac{v}{2}\right) \\ +a\left(p-\frac{u}{2}, q+\frac{v}{2}\right)\end{bmatrix} b\left(p', q'\right). \tag{E.76}$$

Thus, in its integral forms, which are better suited for carrying numerical simulations, Eq. (E.75) exhibits a *non-local* kernel for the commutator, $[A,B](p,q)$, whereas Eq. (E.76) shows an *averaging* type of kernel tied at point (p,q) or *local* kernel for the anti-commutator $\{A,B\}(p,q)$ to a leading order. Indeed, it is precisely the non-local character of the kernel in Eq. (E.75) that is entirely responsible for quantum-tunneling transport in resonant tunneling diodes [11, 14] which exhibit an entirely quantum nonlocality characteristics.

Appendix F

Spatio-Temporal Translation Operators (STTO): Action Principle Viewpoint

Note that displacement operators are characteristically exponential operators amenable to Fourier series expansion. Remarkably, phase factors are generally acquired due to displacement or motion in parameter space in the presence of electric or magnetic fields. This is ubiquitous in solid-state physics (e.g., Peierls phase factor in magnetic fields) before the Berry connection become mainstream and fashionable. To begin, we write our bare Hamiltonian of electrons as

$$H = H_o - e\vec{F} \cdot \vec{r}, \tag{F.1}$$

where H_o is the periodic Hamiltonian in the absence of the electric field, $\vec{F}$. In what follows, we will prove the following results: (a) displacement operators in space and time carry their respective "Peierls" phase factors, (b) the time derivative of momentum operator is equal to $e\vec{F} \cdot \vec{q}$, and (c) gradient of the energy operator equals $e\vec{F}t$. Specifically we want to show that the displacement operator,

$$\hat{T}(-q) = \exp\left[-\frac{i}{\hbar}\left((\vec{q} \cdot -i\hbar\nabla_{\vec{r}})\right)\right], \tag{F.2}$$

acquires a phase factor in the presence of electric field, and is given by

$$\begin{aligned} \tilde{T}(-\vec{q}) &= \exp\left[-\frac{i}{\hbar}\left(\left(e\vec{F} \cdot \vec{q}\right)t + (\vec{q} \cdot -i\hbar\nabla_{\vec{r}})\right)\right], \\ &= \exp\left[-\frac{i}{\hbar}\left(\left(e\vec{F} \cdot \vec{q}\right)t + \left(\vec{q} \cdot \hat{P}\right)\right)\right], \end{aligned} \tag{F.3}$$

where we indicate by $\tilde{T}$ the translation operator $\hat{T}$ with a phase factor.

Similarly, the time displacement operator,

$$\hat{T}(t) = \exp\left[-\frac{i}{\hbar}\left(\left(t \cdot i\hbar\frac{\partial}{\partial t}\right)\right)\right], \tag{F.4}$$

also acquires a phase factor and is given by

$$\tilde{T}(t) = \exp\left[-\frac{i}{\hbar}\left(\left(e\vec{F}\cdot\vec{q}\right)t + \left(t \cdot i\hbar\frac{\partial}{\partial t}\right)\right)\right], \tag{F.5}$$

where $\tilde{T}(t)$ differs from $\hat{T}(t)$ by a phase factor, and $i\hbar\frac{\partial}{\partial t}$ is the energy operator of the system. In other words, displacement in space by lattice vector $-q$ acquires phase factors equal to $\exp(-\frac{i}{\hbar}e\vec{F}t\cdot\vec{q})$. Similarly, displacement in time by t acquires phase factor equal to $\exp(-\frac{i}{\hbar}(e\vec{F}\cdot\vec{q})t)$. In the absence of the electric field these phase factors give unity, e.g., $\tilde{T}) \Longrightarrow_{\vec{F}\to 0} \hat{T}(t)$, reduces to ordinary translation operator.

The physics behind these phase factors is dictated by a selfconsistent translation of local functions in space and time under a uniform electric field, $\vec{F}$. These phases are explained in more detail in the next section. For efficient bookkeeping and for taking lattice Weyl transform, it is usually more convenient to attach these phase factors to displaced local functions themselves. This yields a generalization of Peierls phase factor, well-known for solid-state problems in magnetic fields. The derivation goes as follows.

F.1 Derivation of the STTO based on action principle

In Hamiltonian–Lagrangian mechanics, $\mathcal{L} = K.E. - V$, where $K.E.$ is the kinetic energy and

$$H = \sum_{q_i} \dot{q}_i \frac{\partial \mathcal{L}}{\partial \dot{q}_i} - \mathcal{L}. \tag{F.6}$$

The action is given by

$$\int dt\mathcal{L} = \int dt\left(\sum_{q_i} p_i\dot{q}_i - H\right), \tag{F.7}$$

which gives the Euler–Lagrange classical equation of motion given by the stationary action principle,

$$\frac{d}{dt}\frac{\partial \mathcal{L}}{\partial \dot{q}_i} - \frac{\partial \mathcal{L}}{\partial q_i} = 0, \tag{F.8}$$

which is the classical Newton force equation.

For our single particle case, a translation in space and time must have inherent particle dynamics. The classical action of the displacement in space and time then becomes operator as

$$p \cdot q - Ht \Longrightarrow \hat{P} \cdot q - \hat{\xi} t. \tag{F.9}$$

Hence, we define translation operator in space and time, which commute, i.e., $[T(q), T(t)] = 0$, as

$$\hat{T}(q)\hat{T}(t) = \exp\frac{i}{\hbar}(\hat{P} \cdot q)\exp\frac{i}{\hbar}(-\hat{\xi}t) = \exp\frac{i}{\hbar}(\hat{P} \cdot q - \hat{\xi}t), \tag{F.10}$$

where $\hat{P}$ is the momentum operator,

$$-i\hbar\nabla_{\mathbf{q}}\psi = \hat{P}\psi, \tag{F.11}$$

and ξ is the energy operator, explicitly given by

$$\hat{\xi} = i\hbar\frac{\partial}{\partial t} \tag{F.12}$$

since in the Schrödinger equation,

$$i\hbar\frac{\partial}{\partial t}\psi = \hat{H}\psi. \tag{F.13}$$

In Eq. (F.10), the Baker–Campbell–Hausdorff formula has been applied, since $\hat{T}(\vec{q})$ and $\hat{T}(t)$ commute and therefore equal to the right side of the equation. The phase of the spatio-temporal translation operator mirrors that of the Lagrangian, $\mathcal{L}$, in Hamilton classical mechanics, Eq. (F.7), where the stationary action gives the classical equation of motion, Eq. (F.8). In quantum mechanics, this stationary action translates into stationary quantum phase. With respect to our translation operator, we have indeed a classical (i.e., using eigenvalues) equation of motion which gives the classical distance (position) q given by the group velocity multiplied by the time traveled, $\vec{q} = \vec{v}_g t$ where

$$\vec{v}_g = \frac{\partial \mathcal{E}}{\partial \vec{p}} \tag{F.14}$$

is the group velocity. Moreover, if we denote

$$L = \int_0^t (p \cdot \dot{q} - \mathcal{E})\, dt, \tag{F.15}$$

then we have a stationary action principle, $\dot{\mathbf{p}} = e\mathbf{F}$ which describes the the impulse imparted on the particle. Thus, our proper choice of translation operators in space and time, Eq. (F.10) has inherent particle dynamics built into it. Moreover, the phase of the Schrödinger wavefunction is also proportional to L, for describing a traveling wave packet, again justifying our spatio-temporal translation operator, Eq. (F.10).

F.1.1 *Peierls phase factors*

F.1.1.1 *Non-locality in coordinates*

The nature of the derivatives of exponential displacement operator is determined, e.g., by the following operation:

$$i\hbar\frac{\partial}{\partial t}\tilde{T}\left(q\right)W_\lambda(r-0)=i\hbar\frac{\partial\phi}{\partial t}\tilde{T}\left(q\right)W_\lambda\left(r-0\right), \tag{F.16}$$

$$\frac{\partial}{\partial t}\tilde{T}\left(q\right)=\frac{\partial\phi}{\partial t}\tilde{T}\left(q\right), \tag{F.17}$$

$$\frac{d\tilde{T}\left(q\right)}{\tilde{T}\left(q\right)}=d\phi, \tag{F.18}$$

where specifically $\hat{T}\left(-q\right)W_\lambda\left(r-0\right)$ is a translation of the center coordinate of a localized Wannier function centered in the origin to another lattice point q, yielding $W_\lambda\left(r-q\right)$. Therefore $\hat{T}\left(-q\right)$ resembles the physical process of transferring a localized function centered at the origin to a localized function centered at another lattice point q. Here we define ϕ as undetermined for the moment as

$$\phi=-\frac{i}{\hbar}(f(q,t)+(\vec{q}\cdot\hat{P})), \tag{F.19}$$

where $f\left(q,t\right)$ is to be determined. Note that the presence of $f(q,t)$ is needed for selfconsistency in the presence of electric field. Now consider the second term in the exponent, namely,

$$\begin{aligned}\phi_2&=-\frac{i}{\hbar}\vec{q}\cdot\hat{P}=-\frac{i}{\hbar}\vec{q}\cdot\left(-i\hbar\frac{\partial}{\partial\vec{r}}\right),\\&=(-q)\cdot\frac{\partial}{\partial\vec{r}}\end{aligned} \tag{F.20}$$

leading to Fourier series expansion of the translation, $\hat{T}\left(-q\right)$. Then, we obtain

$$-i\hbar\frac{\partial\phi_2}{\partial t}=[H,\phi_2]=\frac{i}{\hbar}[H,((-\vec{q})\cdot\hat{P})]=e\vec{F}\cdot(-\vec{q}), \tag{F.21}$$

since $-\frac{i}{\hbar}f\left(q,t\right)$ commutes with the Hamiltonian. Therefore, we have

$$\frac{d\tilde{T}\left(-q\right)}{\tilde{T}\left(-q\right)}=d\phi=\frac{i}{\hbar}e\vec{F}\cdot(-\vec{q})\ dt. \tag{F.22}$$

Therefore,

$$\begin{aligned}\tilde{T}(-q) &= \exp\frac{i}{\hbar}[(e\vec{F}\cdot(-\vec{q})\ \Delta t) + \hat{P}\cdot(-\vec{q})],\\ &= \exp\frac{i}{\hbar}[\hat{P} + e\vec{F}t]\cdot(-\vec{q}) \qquad \text{(F.23)}\end{aligned}$$

This means that a displacement $(-\vec{q})$ of localized function acquires a phase factor given by

$$Peierls\ phase\ factor = \exp\left[\frac{i}{\hbar}\left(e\vec{F}t\cdot(-\vec{q})\right)\right]. \qquad \text{(F.24)}$$

We may thus write

$$-i\hbar\frac{\partial}{\partial t}\tilde{T}(-\vec{q},t) = [H,\tilde{T}(-\vec{q},t)] = e\vec{F}\cdot(-\vec{q})\ \tilde{T}(-\vec{q},t). \qquad \text{(F.25)}$$

We can also deduce from Eq. (F.21) the relation for the momentum operator,

$$\begin{aligned}\frac{i}{\hbar}[H,[(-\vec{q})\cdot -i\hbar\nabla_{\vec{r}}]] &= e\vec{F}\cdot(-\vec{q}) = \frac{\partial\hat{P}}{\partial t}\cdot(-q), \Longrightarrow \frac{\partial\hat{P}}{\partial t} = e\vec{F},\\ &\Longrightarrow \hat{P} = \hat{P}_o + e\vec{F}t. \qquad \text{(F.26)}\end{aligned}$$

F.1.1.2 *Simultaneous eigenvalues for $\hat{H}$ and $\tilde{T}(-\vec{q},t)$*

One very important conclusion is implied in Eq. (F.25), which we rewrite here for convenience

$$[\hat{H},\tilde{T}(-\vec{q},t)] = -e\vec{F}\cdot\vec{q}\,\tilde{T}(-\vec{q},t). \qquad \text{(F.27)}$$

What the above relation means is that if $\tilde{T}(-\vec{q},t)$ is diagonal then $[H,\tilde{T}(-\vec{q},t)]$ is also diagonal. But if $\tilde{T}(-\vec{q},t)$ is diagonal, then $\hat{H}$ is also diagonal with the same eigenvalues. The eigenfunction of $\tilde{T}(-\vec{q},t)$ is labeled by the quantum label $\vec{\mathcal{K}} = \vec{p}_o + e\vec{F}t$. This implies that $\hat{H}$ is also diagonal in $\vec{\mathcal{K}}$. The electric Bloch function labeled by $B\left(k_o + \frac{e}{\hbar}Ft, ..\right)$ is the eigenfunction of $\tilde{T}(-\vec{q},t)$ as well as that of the renormalized

$$H_{\text{renormalized}}(\hat{K},\hat{Q}) \Longleftrightarrow E_n(\vec{\mathcal{K}}) - e\vec{F}\cdot\vec{q}, \qquad \text{(F.28)}$$

where the double pointed arrow denotes lattice Weyl correspondence, Eq. (F.28), using the localized electric Wannier function.

F.1.1.3 *Non-locality in time*

Because non-local arguments in time acquires phase factor also, the energy variable of the theory now varies with $e\vec{F} \cdot \vec{q}$, as discussed next. From Eq. (F.10) for a displacement in time

$$\hat{T}(t) = \exp \frac{i}{\hbar}(-\hat{\xi}t) \tag{F.29}$$

or

$$\tilde{T}(t) \equiv \mathcal{F} \exp\left(\frac{-i}{\hbar}\left(i\hbar t \frac{\partial}{\partial t}\right)\right) = \mathcal{F} \exp\left(t \frac{\partial}{\partial t}\right) \tag{F.30}$$

where,

$$\mathcal{F} = \exp\left[-\frac{i}{\hbar}\left(f_t(q,t)\right)\right], \tag{F.31}$$

is a phase factor to be determined

$$\phi_t(q) = -\frac{i}{\hbar}\left(f_t(q,t) + t\, i\hbar \frac{\partial}{\partial t'}\right). \tag{F.32}$$

Therefore,

$$-i\hbar \frac{\partial \tilde{T}(t)}{\partial \vec{q}} = -i\hbar \frac{\partial \phi_t(q)}{\partial \vec{q}} \tilde{T}(t), \tag{F.33}$$

$$\begin{aligned} -i\hbar \frac{\partial \phi_t(q)}{\partial \vec{q}} &= -\frac{i}{\hbar}\left[\vec{\mathcal{K}}(t'), (t)\, i\hbar \frac{\partial}{\partial t'}\right] = \left[\vec{p} + \frac{e}{c}\vec{A}_o, t\nabla_{t'}\right], \\ &= [\vec{p} + e\vec{F}t', t\nabla_{t'}] = -e\vec{F}t. \end{aligned} \tag{F.34}$$

Therefore

$$\begin{aligned} -i\hbar \frac{\partial \phi_t(q)}{\partial \vec{q}} &= -e\vec{F}t, \\ \frac{\partial \phi_t(q)}{\partial \vec{q}} &= -\frac{i}{\hbar} e\vec{F}t. \end{aligned} \tag{F.35}$$

Thus, we obtained,

$$\frac{\partial \tilde{T}(t)}{\partial \vec{q}} = \left(-\frac{i}{\hbar} e\vec{F}t\right) \tilde{T}(t), \tag{F.36}$$

and hence,

$$d \ln \tilde{T}(t) = \left(-\frac{i}{\hbar} e\vec{F}t\right) \cdot d\vec{q}, \tag{F.37}$$

$$\ln \tilde{T}(t) = \left(-\frac{i}{\hbar} e\vec{F}t\right) \cdot \Delta\vec{q}. \tag{F.38}$$

Hence a displacement in time carries a phase factor given by $\exp(-\frac{i}{\hbar}(e\vec{F}t)\cdot\Delta\vec{q})$ and

$$\tilde{T}(t) = \exp\left\{-\frac{i}{\hbar}\left[(e\vec{F}t)\cdot\Delta\vec{q}+t\left(i\hbar\frac{\partial}{\partial t'}\right)\right]\right\} \tag{F.39}$$

Therefore,

$$-i\hbar\frac{\partial}{\partial\vec{q}}\tilde{T}\left(\vec{q},t\right) = [\vec{\mathcal{K}}(t),\tilde{T}(\vec{q},t)] = (-e\vec{F}t)\ \tilde{T}(\vec{q},t) \tag{F.40}$$

Now from second line of Eq. (F.34), we have

$$-\frac{i}{\hbar}\left[\vec{\mathcal{K}}\left(t'\right),(t)\,i\hbar\frac{\partial}{\partial t'}\right] = \left[\vec{p}+e\vec{F}t',(t)\,\frac{\partial}{\partial t'}\right] = -e\vec{F}t, \tag{F.41}$$

$$[\vec{\mathcal{K}}\left(t'\right),\hat{\xi}] = -i\hbar e\vec{F} = -i\hbar\frac{\partial\hat{\xi}}{\partial q}, \tag{F.42}$$

which leads to to the expression for in Eq. (F.31) for f_t in the phase factor as

$$\frac{\partial\hat{\xi}}{\partial\vec{q}} = e\vec{F} \Longrightarrow f_t = e\vec{F}\cdot\vec{q}. \tag{F.43}$$

All these results, Eq. (F.23) and Eq. (F.39), lead to the time-dependent wave vector,

$$\hbar\vec{k} = \hbar\vec{k}_o + \hbar e\vec{F}t \tag{F.44}$$

and to the position-dependent energy,

$$\mathcal{E} = E_o + e\vec{F}\cdot\vec{q}, \tag{F.45}$$

respectively. Again, for for taking lattice Weyl transform, it is more convenient to attach these phase factor to the displaced local functions themselves. This allows us to generalize the Peierls phase factor to space and time displacements, originally well-known for solid-state problems in magnetic fields.

Appendix G

Hall Current Density and Current Through a Discrete Periodic Lattice

The definition of the current density given by the first row or first line of Eq. (30.39) of the text, reads

$$\frac{a^2}{(2\pi\hbar)^2}\iint dK_x\, dK_y \left(\frac{e}{a^2}\frac{\partial \mathcal{E}}{\partial K_y}\right)(-iG^<(\vec{\mathcal{K}},\mathcal{E})). \tag{G.1}$$

One observes that the above expression embodies the intuitive concept of a current density $j(E)$ as a product of group velocity along the Hall current direction, represented by $v_g = \frac{\partial \mathcal{E}}{\partial K_y}$ multiplied by areal density of particles, represented by

$$\rho = \frac{1}{a^2}(-iG^<(\vec{\mathcal{K}},\mathcal{E})), \tag{G.2}$$

multiplied by the electric charge, e, and summed over all the 2D variables, K_x and K_y with proper counting of states. It can easily be shown that Eq. (G.1) has a corresponding discrete lattice expression. Indeed, we can readily use the expression for the *directional* electric current through the lattice derived in detail in Section 21.3 of Ref. [12], namely, Eq. (21.16) in Section 21.3 of the author's book, which give the current through the lattice as

$$\begin{aligned} j(E,t) &= \frac{-e}{\hbar}\sum_{\vec{q},\vec{q}''}(\vec{q}-\vec{q}'')\left\langle \vec{q}\right| H_o \left|\vec{q}''\right\rangle G^<(\vec{q}'',\vec{q},E,t), \\ &= \frac{e}{\hbar}\sum_{q,q''}(\vec{q}''-\vec{q})\left\langle \vec{q}\right| H_o \left|\vec{q}''\right\rangle G^<(\vec{q}'',\vec{q},E,t). \end{aligned} \tag{G.3}$$

The above equation involves summation over all the lattice points, q and q'', i.e., the expression is really a trace. Now a trace involving products of matrix elements is also a trace involving products of their respective lattice Weyl transforms [refer to Chapter 5]. In what follows we will treat steady-state condition only. The lattice Weyl transform of

$$\left(\vec{q}'' - \vec{q}\right)\left\langle \vec{q}\right| H_o \left|\vec{q}''\right\rangle \tag{G.4}$$

can be calculated as follows. We have,

$$\begin{aligned}
\mathcal{W}\left[\left(\vec{q}'' - \vec{q}\right)\left\langle \vec{q}\right| H_o \left|\vec{q}''\right\rangle\right] &= \sum_{\vec{v}} e^{\left(\frac{2i}{\hbar}\right)\vec{p}\cdot\vec{v}} \left((\tilde{q}+\vec{v}) - (\tilde{q}-\vec{v})\right)\left\langle \tilde{q}-\vec{v}\right| H_o \left|\tilde{q}+\vec{v}\right\rangle, \\
&= \sum_{\vec{v}} e^{\left(\frac{2i}{\hbar}\right)\vec{p}\cdot\vec{v}} \left(2\vec{v}\right)\left\langle \tilde{q}-\vec{v}\right| H_o \left|\tilde{q}+\vec{v}\right\rangle, \\
&= \frac{\partial}{\partial\vec{p}} \sum_{\vec{v}} e^{\left(\frac{2i}{\hbar}\right)\vec{p}\cdot\vec{v}} \left\langle \tilde{q}-\vec{v}\right| H_o \left|\tilde{q}+\vec{v}\right\rangle.
\end{aligned} \tag{G.5}$$

However, for crystalline solid the matrix element of the energy band depends in the difference of positions $\vec{q}$ which is equivalent to saying that the matrix element $\left\langle \tilde{q}-\vec{v}\right| H_o \left|\tilde{q}+\vec{v}\right\rangle \Longrightarrow H_o\left(2\vec{v}\right)$. Therefore, we have

$$\mathcal{W}\left(\vec{q}'' - \vec{q}\right)\left\langle \vec{q}\right| H_o \left|\vec{q}''\right\rangle = \frac{\partial}{\partial\vec{p}} E\left(\vec{p}\right). \tag{G.6}$$

In the presence of uniform electric or magnetic fields, we then have,

$$\frac{\partial}{\partial\vec{p}} E\left(\vec{p}\right) \Longrightarrow \frac{\partial}{\partial\vec{\mathcal{K}}} \mathcal{E}(\vec{\mathcal{K}}). \tag{G.7}$$

Similarly, the lattice Weyl transform of $G^<\left(\vec{q}'', \vec{q}, E, t\right)$ is

$$W\left[-iG^<\left(\vec{q}'', \vec{q}, t, t'\right)\right] = -iG^<(\vec{p}, \vec{q}, E, t), \tag{G.8}$$

which goes into

$$-iG^<\left(\vec{p}, \vec{q}, E, t\right) \Longrightarrow -iG^<(\vec{\mathcal{K}}, \mathcal{E}), \tag{G.9}$$

in the presence of electric or magnetic field. Using the proper counting of states in phase space for 2D systems in the taking of the trace, we finally arrive at

$$j_y = \frac{a^2}{(2\pi\hbar)^2} \iint dK_x dK_y \left(\frac{e}{a^2}\frac{\partial\mathcal{E}}{\partial K_y}\right)\left(-iG^<(\vec{\mathcal{K}}, \mathcal{E})\right), \tag{G.10}$$

which is what we want to derive from the current density through the lattice, Eq. (G.3). Thus, this equation is valid for both lattice and continuum by simply performing lattice Weyl transformation of the appropriate quantities from discrete to continuum and vice versa.

G.1 Phase change of wavefunction under parallel transport

To relate to the phase of the wavefunction, we recall that for parallel transport [for more details see Chapter 13 of author's book [12]],

$$\begin{aligned} \frac{\partial \psi_\alpha(\vec{k})}{\partial t} &= \left(-\Gamma^j_{\vec{k},\beta} \frac{d\vec{k}}{dt} \right) \psi_\beta(\vec{k}), \\ &= \left(-\langle \alpha, \vec{k} | \frac{\partial}{\partial \vec{k}} | \beta, \vec{k} \rangle \frac{d\vec{k}}{dt} \right) \psi_\beta \left(\vec{k} \right). \end{aligned} \tag{G.11}$$

In the adiabatic case, this becomes (assuming energy band α is far remove from the other bands),

$$\frac{\partial \psi_\alpha(\vec{k})}{\partial t} = \left(-\langle \alpha, \vec{k} | \frac{\partial}{\partial \vec{k}} | \alpha, \vec{k} \rangle \cdot \frac{d\vec{k}}{dt} \right) \psi_\alpha(\vec{k}). \tag{G.12}$$

Thus,

$$\frac{d \ln \psi_\alpha(\vec{k})}{dt} = \left(-\langle \alpha, \vec{k} | \frac{\partial}{\partial \vec{k}} | \alpha, \vec{k} \rangle \cdot \frac{d\vec{k}}{dt} \right), \tag{G.13}$$

$$d \ln \psi_\alpha(\vec{k}) = -\langle \alpha, \vec{k} | \frac{\partial}{\partial \vec{k}} | \alpha, \vec{k} \rangle \cdot d\vec{k}, \tag{G.14}$$

$$i d\phi = i \langle \alpha, \vec{k} | i \frac{\partial}{\partial \vec{k}} | \alpha, \vec{k} \rangle \cdot d\vec{k}, \tag{G.15}$$

where $id\phi$ is the change of phase of the wavefunction along a curve in $\vec{k}$-space (Brillouin zone). Around a closed curve the total change of phase must be a multiple of 2π, i.e., $\Delta\phi = 2\pi n$ ($n \in \mathbb{Z}$) for the wavefunction to return to its original state. We can write

$$\begin{aligned} &\frac{i}{(2\pi)} \int\!\!\int dk_x dk_y \; f(E_\alpha(\vec{k})) \begin{bmatrix} \left\langle \frac{\partial}{\partial k_x} \alpha, \vec{k} \right| \frac{\partial}{\partial k_y} \left| \alpha, \vec{k} \right\rangle \\ - \left\langle \frac{\partial}{\partial k_y} \alpha, \vec{k} \right| \frac{\partial}{\partial k_x} \left| \alpha, \vec{k} \right\rangle \end{bmatrix}, \\ &= \frac{1}{(2\pi)} \int\!\!\int dk_x dk_y \; f(E_\alpha(\vec{k})) \left[\nabla_{\vec{k}} \times \langle \alpha, \vec{k} | i \frac{\partial}{\partial \vec{k}} | \alpha, \vec{k} \rangle \right]_{\text{plane}}. \end{aligned} \tag{G.16}$$

This result can also be traced to the self-consistent Bohr–Sommerfeld quantization condition [13].

Bibliography

[1] H. A. Kramers and G. H. Wannier, Statistics of the two-dimensional ferromagnet, *Phys. Rev.* **60**, 252–262 (1941).

[2] F. A. Buot, R. Montecillo, M. D. Nacar, and R. E. S. Otadoy, Nonlinear physics of THz sources: Duality theory in nonequilibrium nanodevice physics, *AIP Conf. Proc.* **1871**, 030001 (2017).

[3] F. A. Buot, Method for calculating $Tr\mathcal{H}^n$ in solid-state theory, *Phys. Rev. B* **10**, 3700 (1974).

[4] F. A. Buot, Formalism of distribution-function method in impurity screening, *Phys. Rev. B* **13**, 977–989 (1976).

[5] F. A. Buot, Discrete phase-space model for quantum mechanics, in *Bell's Theorem, Quantum Theory, and Conceptions of the Universe*, edited by M. Kafatos, Fundamental Physics Series, pages 159–162, Kluwer, 1989.

[6] F. A. Buot, Operator space and discrete phase space methods in quantum transport and quantum computing, *J. Comp. Theor. Nanoscience* **6**, 1864–1926 (2009).

[7] F. A. Buot, Comments on the Weyl-Wigner calculus for lattice models, March 2021, arXiv:2103.10351.

[8] F. A. Buot and J. W. McClure, Theory of diamagnetism of bismuth, *Phys. Rev. B* **6**, 4525 (1972).

[9] F. A. Buot, R. E. S. Otadoy, and K. B. Rivero, Magnetic susceptibility of Dirac fermions, BiSb alloys, interacting Bloch fermions, dilute nonmagnetic alloys, and Kondo alloys, *Physica B* **503**, 69–97 (2017).

[10] F. A. Buot, Foundation of computational nanoelectronics, in *Handbook of Theoretical and Computational Nanotechnology*, edited by M. Rieth and W. Schommers, volume 23, pages 221–310, American Scientific Publishers, 2006.

[11] F. A. Buot and K. L. Jensen, Lattice Weyl-Wigner formulation of exact many-body quantum transport theory and applications to novel quantum-based devices, *Phys. Rev. B* **42**, 9429–9456 (1990).

[12] F. A. Buot, *Nonequilibrium Quantum Transport Physics in Nanosystems*, World Scientific, 2009, and references therein.

[13] F. A. Buot, A. R. Elnar, G. Maglasang, and R. E. S. Otadoy, On quantum Hall effect, Kosterlitz-Thouless phase transition, Dirac magnetic monopole, and Bohr–Sommerfeld quantization, *J. Phys. Communi.* **5**, 025007 (2021), See also arXiv:2001.06993v3.

[14] K. L. Jensen and F. A. Buot, Numerical simulation of intrinsic bistability and high-frequency current oscillations in resonant tunneling structures, *Phys. Rev. Lett.* **66**, 1078 (1991).
[15] J. Schwinger, *Quantum Kinematics and Dynamics*, Benjamin, 1970, chapter 4.
[16] F. A. Buot, Direct construction of path integrals in the lattice-space multiband dynamics of electrons in a solid, *Phys. Rev. A* **33**, 2544–2562 (1986).
[17] J. Klauder, *Coherent States: Applications in Physics and Mathematical Physics*, World Scientific Publishers, 1985.
[18] F. A. Buot, Magnetic susceptibility of interacting free and Bloch electrons, *Phys. Rev. B* **14**, 3310–3328 (1976).
[19] I. V. Fialkovsky and M. A. Zubkov, Precise Wigner-Weyl calculus for lattice models, *Nucl. Phys. B* **954**, 114999 (2020), See also arXiv:2103.10351.
[20] M. Ligabo, Torus as phase space: Weyl quantization, dequantization, and Wigner formalism, *J. Math. Phys.* **57**, 082110 (2016).
[21] P. Kasperkovitz and M. Peev, Weyl-Wigner formalisms for toroidal geometries, *Annals of Phys.* **230**, 21–51 (1994).
[22] K. S. Gibbons, M. J. Hoffmann, and W. K. Wooters, Discrete phase space based on finite fields, *Phys. Rev. A* **70**, 062101 (2004).
[23] F. A. Buot, A. R. Elnar, G. Maglasang, and C. M. Galon, A mechanical implementation and diagrammatic calculation of entangled basis states, 2021, arXiv:2112.10291.
[24] G. H. Wannier, Dynamics of Band Electrons in Electric and Magnetic Fields, *Rev. Mod. Phys.* **34**, 645 (1962).
[25] R. L. Stratonovich, *Topics in Theory of Random Noise*, volume 1, Gordon and Breach, Philadelphia, 1963.
[26] L. Tonks, The complete equation of state of one, two and three-dimensional gases of hard elastic spheres, *Phys. Rev.* **50**, 955–963 (1936).
[27] M. Girardeau, Relationship between systems of impenetrable bosons and fermions in one dimension, *J. Math. Phys.* **1**, 516 (1960).
[28] E. H. Lieb and W. Liniger, Exact analysis of an interacting Bose gas. I. The general solution and the ground state, *Phys. Rev.* **130**(6), 1605–1616 (1963).
[29] M. K. Gaillard, P. D. Grannis, and F. J. Sciulli, The standard model of particle physics, *Rev. Mod. Phys.* **71**(6), S96 (1999).
[30] N. Borstnik, Fermionization, Number of families, 2018, arXiv:1805.06774.
[31] S. Bera, B. Chakrabarti, A. Gammal, M. Tsatsos, M. Lekala, B. Chatterjee, C. Leveque, and A. Lode, Sorting fermionization from crystallization in many-Boson Wavefunctions, *Sci. Rep.* **9**, 17873 (2019).
[32] E. Miranda, Introduction to Bosonization, *Brazi. J. Phys.* **33**(1), 3 (2003).
[33] D. Senechal, An introduction to Bosonization, Aug 1999, arXiv:cond-mat/9908262.
[34] J. Wilson, N. Malvania, Y. Le, Y. Zhang, M. Rigol, and D. Weiss, Observation of dynamical fermionization, 2019, arXiv:1908.05364.
[35] V. Zatloukal, Jordan-Schwinger map in the theory of angular momentum, https://www.zatlovac.eu/lecturenotes/Jordan-Schwinger.pdf.
[36] Knowen, Jordan-Wigner transform, https://knowen.org/nodes/947.
[37] S. Kivelson and D. Rokhsart, Bogoliubov quasiparticles, spinons, and spin-charge decoupling in superconductors, *Phys. Rev. B* **41**(1), 11693 (1990).
[38] T. Senthil, A. Vishwanath, L. Balents, S. Sachdev, and M. P. A. Fisher, Quantum criticality beyond the Landau–Ginzburg–Wilson paradigm, *Science* **303**, 1490 (2004).

[39] T. Senthil, A. Vishwanath, L. Balents, S. Sachdev, and M. P. A. Fisher, Quantum criticality beyond the Landau–Ginzburg–Wilson paradigm, *Phys. Rev. B* **70**(1), 144407 (October 2004).
[40] U. Cambridge, Discovery about behavior of building block of nature could lead to computer revolution, https://www.eurekalert.org/news-releases/560825, July 2009.
[41] Y. Jompol, C. Ford, J. Griffiths, I. Farrier, G. Jones, D. Anderson, D. Ritchie, T. Silk, and A. Schofieldl, Probing spin-charge separation in a Tomonaga-Luttinger liquid, *Science* **325**(5940), 597–601 (July 2009), See also arXiv:1002.2782 (2010).
[42] S. Demokritov, V. Demidov, O. Dzyapko, G. Melkov, A. Serga, B. Hillebrands, and A. Slavin, Bose-Einstein condensation of quasi-equilibrium magnons at room temperature under pumping, *Nature* **443**(7110), 430–433 (September 2006), see also arXiv:1002.2782 (2010).
[43] T. Brauner, Spontaneous symmetry breaking and Nambu–Goldstone bosons in quantum many-Body systems, 2010, arXiv:1001.5212.
[44] H. Leutwyler, Phonons as Goldstone bosons, 1996, arXiv:hep-ph/9609466.
[45] Schubell, Being in two places at once: Spin-charge separation, https://guava.physics.uiuc.edu/~nigel/courses/569/Essays_Fall2010/Files/Schubel.pdf, 2010.
[46] S. Tomonaga, Remarks on Bloch's method of sound waves applied to many-fermion problems, *Prog. Theor. Phys.* **5**, 544 (1950).
[47] J. M. Luttinger, An exactly soluble model of a many-fermion system, *J. Math. Phys.* **4**, 1154 (1963).
[48] F. D. M. Haldane, General relation of correlation exponents and spectral properties of one-dimensional fermi systems: Application to the anisotropic S=1/2 Heisenberg chain, *Phys. Rev. Lett.* **45**, 1358 (1980), See also, **47**, 1840 (1981); *J. Phys. C* **14**, 2585 (1981).
[49] J. van den Brink, W. Stekelenburg, D. Khomskii, G. Sawatzky, and K. Kugel, Spin and orbital excitations in magnetic insulators with Jahn-Teller ions, *Phys. Rev. B* **58**, 10276–10282 (1998), See also, **47**, 1840 (1981); *J. Phys. C* **14**, 2585 (1981).
[50] J. Schlappa, K. Wohlfeld, K. Zhou, M. Mourigal, M. Haverkort, V. N. Strocov, *et al.*, Spin-orbital separation in the quasi-one-dimensional Mott insulator Sr2CuO3, *Nature* **485**, 82–85 (2012).
[51] M. Halperin, A. Jevicki, and P. Senjanovic, Field theories in terms particle-string variables: Spin, internal symmetries and arbitrary dimension, *Phys. Rev. D* **16**, 2476 (October 1977), See Appendix.
[52] F. A. Berezin, Non-Wiener functional integrals, *Theor. Math. Phys.* **6**, 141 (1971).
[53] J. Zinn-Justin, *Quantum Field Theory and Critical Phenomena*, Oxford University Press, 2002.
[54] D. E. Soper, Construction of the functional-integral representation for fermion Green functions, *Phys. Rev. D* **18**, 4590 (December 1978).
[55] W. Guttinger, Quantum field theory in the light of distribution analysis, *Phys. Rev.* **89**(5), 1004 (1953).
[56] P. D. Drummond and M. D. Reid, Retrocausal model of reality for quantum fields, *Phys. Rev. Res.* **2**, 033266 (2020).
[57] P. D. Drummond, Time evolution with symmetric stochastic action, *Phys. Rev. Res.* **3**, 013240 (2021).
[58] S. Friederich, Introducing the Q-based interpretation of quantum theory, 2021, arXiv:2106.13502.
[59] V. I. Man'ko, L. Rosa, and P. Vitale, Probability representation in quantum field theory, 1998, arXiv:hep-th/9806164.

[60] I. Raptis, Third' quantization of vacuum Einstein gravity and free Yang-Mills theories, 2006, arXiv:gr-qc/0606021v4.
[61] S. B. Giddings and A. Strominger, Baby universe, third quantization and the cosmological constant, *Nuclear Phys. B* **321**, 481–508 (1989).
[62] T. Prosen, Third quantization: a general method to solve master equations for quadratic open Fermi systems, 2008, arXiv:0801.1257v1.
[63] T. H. Seligman and T. Prosen, Third quantization, 2010, arXiv:1011.0625v1.
[64] M. Bouhmadi-Lopez, M. Kramer, J. Morais, and S. Robles-Perez, The third quantization: To tunnel or not to tunnel?, *Galaxies* **6**, 21 (2018), Conference Report.
[65] W. E. Brittin and W. R. Chappell, The Wigner distribution function and second quantization in phase space, *Rev. Mod. Phys.* **34**, 620 (1962).
[66] L. L. Foldy and S. A. Wouthuysen, On the Dirac theory of spin-(1/2) particles and its non-relativistic limit, *Phys. Rev.* **78**, 29–36 (1950).
[67] M. V. Berry, Quantal phase factors accompanying adiabatic changes, *Proc. R. Soc. London* **392**(1802), 45–57 (1984).
[68] P. A. M. Dirac, Quantized singularities in the electromagnetic field, *Proc. R. Soc. London* **133**, 50–72 (1931).
[69] D. J. Thouless, M. Kohmoto, M. P. Nightingale, and M. den Nijs, Quantized Hall conductance in a two-dimensional periodic potential, *Phys. Rev. Lett.* **49**(6), 405 (1982).
[70] P. Streda, Quantised Hall effect in a two-dimensional periodic potential, *J. Phys. C: Solid State Phys.* **15**, L1299 (1982).
[71] R. B. Laughlin, Quantized Hall conductivity in two dimension, *Phys. Rev. B* **23**, 5632 (1981).
[72] B. I. Halperin, Quantized Hall conductance, current-carrying edge states, and the existence of extended states in a two-dimensional disordered potential, *Phys. Rev. B* **25**, 2185 (1982).
[73] K. von Klitzing, G. Dorda, and M. Pepper, A new method for high-accuracy determination of the fine–structure constant based on quantized Hall resistance, *Phys. Rev. Lett.* **45**, 494 (1980).
[74] L. G. Suttorp and S. R. D. Groot, Covariant equation of motion for a charged particle with magnetic dipole moment, *Il Nouvo Cimento* **65**, 245–274 (1970).
[75] F. A. Buot, Weyl transform and the magnetic susceptibility of a relativistic Dirac electron gas, *Phys. Rev. A* **8**, 1570 (1973).
[76] N. D. Mermin and H. Wagner, Absence of ferromagnetism or antiferromagnetism in one-dimensional or two dimensional isotropic Heisenberg models, *Phys. Rev. Lett.* **17**, 1133–1136 (1966).
[77] J. E. Jacak, Explanation of an unexpected occurrence of fractional quantum Hall effect states in monolayer graphene, *J. Phys. Condens. Matter* **31**, 475601 (2019).
[78] J. K. Jain, Composite-fermion approach for the fractional quantum Hall effect, *Phys. Rev. Lett.* **63**, 199 (1989).
[79] J. E. Jacak, Magnetic flux quantum in 2D correlated states of multiparticle charged system, *New J. Phys.* **22**, 093027 (2020).
[80] A. S. Davydov, *Quantum Mechanics*, pages 253, 435, Pergamon Press, 1965.
[81] J. Sniatycki, Bohr-Sommerfeld conditions in geometric quantization, *Rep. Math. Phys.* **7**(6837), 303–311 (1975).
[82] K. Wu and Y. Zhang, Topological quantum field via Chern–Simons theory, part1, 2011, arXiv:1112.0373v1.
[83] A. Moll, Exact Bohr-Sommerfeld conditions for the quantum periodic Benjamin-Ono equation, symmetry, integrability and geometry: Methods and applications, *SIGMA* **15**(6837), 098 (2019).

[84] F. D. M. Haldane, Nonlinear field theory of large-spin Heisenberg antiferromagnets: Semiclassically quantized solitons of one-dimensional easy-axis Neel state, *Phys. Rev. Lett.* **50**(6837), 1153 (1983).
[85] R. Jackiw, Dirac's magnetic monopole (again), 2002, arXiv:hep-th/0212058.
[86] J. Kosterlitz and D. Thouless, Ordering, metastability and phase transition in two dimensional systems, *J. Phys. C: Solid State Phys.* **6**(7), 1181 (1973).
[87] J. M. Kosterlitz and D. J. Thouless, Long-range order and metastability in two dimensional solids and superfluids (Application of dislocation theory), *J. Phys. C: Solid State Phys.* **5**(11), L124 (1972).
[88] A. Polyakov, Interaction of goldstone particles in two dimensions. Applications to ferromagnets and massive Yang-Mills fields, *Phys. Lett. B* **59**, 1–108 (1975).
[89] E. Witten, Chiral symmetry, the (1/N) expansion and the SU(N) Thirring model, *Nucl. Phys. B* **145**, 110–118 (1978).
[90] D. Nelson and B. Halperin, Dislocation-mediated melting in two dimensions, *Phys. Rev. B* **19**, 2457 (1979).
[91] X.-G. Wen, Zoo of quantum-topological phases of matter, 2017, arXiv:1610.03911v3.
[92] L. M. C. Armas, Revisiting the quantum Hall effect in the Haldane model, https://app.uff.br/riuff/bitstream/handle/1/6284/tese.pdf?sequence=1, Thesis, Universidade Federal Fluminense (2017).
[93] F. D. M. Haldane, Model for a quantum Hall effect without Landau levels: Condensed-matter realization of the parity anomaly, *Phys. Rev. Lett.* **61**, 2015–2018 (October 1988).
[94] W. Goff, F. Gaitan, and M. Stone, Berry phase and the Wess-Zumino effective action in ^{3}He-A, *Phys. Lett. A* **136**(7,8), 433 (1989).
[95] P. S. Hsin, A. Kapustin, and R. Thorngren, Berry phase in quantum field theory: Diabolical points and boundary phenomena, 2020, arxiv.org/abs/2004.10758.
[96] T. Tochishita, M. Mizui, and H. Kuratsuji, Semiclassical quantization for the motion of guiding center using the coherent state path integral, 1996, arXiv:cond-mat/9512174v2.
[97] N. Sansonetto, Monodromy and the Bohr-Sommerfeld geometric quantization, *J. Geom. Symm. Phys.* **20**(7,8), 97–106 (2010).
[98] J. E. Jacak, Application of path-integral quantization to indistinguishable particle systems topologically confined by a magnetic field, *Phys. Rev. A* **97**, 012108 (2018).
[99] D. R. Hofstadter, Energy levels and wave functions of Bloch electrons in rational and irrational magnetic fields, *Phys. Rev. B* **14**(7,8), 2239–2249 (1976).
[100] G. Wannier, A result not dependent on rationality for Bloch electrons in a magnetic field, *Phys. Status Solidi* **88**(7,8), 757 (1978).
[101] Y. Iwasaki, Y. Hashimoto, T. Nakamura, and S. Katsumoto, Observation of Conductance Fluctuation due to zitterbewegung in InAs 2-dimensional electron gas, *J. Phys. Conf. Series* **864**, 012054 (2017).
[102] J. Schwinger, Brownian motion of a quantum oscillator, *J. Math. Phys.* **2**, 1018 (1961).
[103] F. A. Buot, Mesoscopic physics and nanoelectronics: Nanoscience and nanotechnology, *Phys. Report* **234**, 73 (1993).
[104] C. Piermarocchi, P. Chen, Y. S. Dale, and L. J. Sham, Theory of fast quantum control of exciton dynamics in semiconductor quantum dots, *Phys. Rev. B* **65**, 075307 (2002).
[105] T. D. Ladd, J. R. Goldman, F. Yamaguchi, and Y. Yamamoto, All-Silicon quantum computer, *Phys. Rev. Lett.* **89**, 017901 (2002).

[106] L. V. Keldysh, Diagram technique for nonequilibrium processes, *Soviet Phys. JETP* **20**, 1018 (1965), See also V. Korenman, *Annals Phys.* **39**, 72 (1966); P. Danielewicz, *Annals Phys.* **152**, 239 (1984).
[107] L. P. Kadanoff and G. Baym, *Quantum Statistical Mechanics*, Benjamin, NY, 1962.
[108] V. Korenman, Nonequilibrium quantum statistics; Application to the laser, *Annals Phys.* **39**, 72–126 (1966), See also P. Danielewicz, *Annals Phys.* **152**, 239 (1984).
[109] G. D. Mahan, Quantum transport equation for electric and magnetic fields, *Phys. Report* **145**, 251 (1987).
[110] D. S. Chemla and J. Shah, Many-body and correlation effects in semiconductors, *Nature* **411**(6837), 549 (2001).
[111] F. A. Buot and A. K. Rajagopal, High-frequency behavior of quantum-based devices: Equivalent-circuit, nonperturbative-response, and phase-space analyses, *Phys. Rev. B* **48**, 17217 (1993), See also, *Mater. Sc. Engng. B* **35**, 303 (1995).
[112] F. A. Buot, P. Zhao, H. L. Cui, D. Woolard, K. L. Jensen, and C. M. Krowne, Emitter quantization and double hysteresis in resonant-tunneling structures: A nonlinear model of charge oscillation and current bistability, *Phys. Rev. B* **61**, 5644 (2000).
[113] W. R. Frensley, Wigner function model of a resonant-tunneling semiconductor device, *Phys. Rev. B* **36**, 1570 (1987), See also, *Mater. Sci. Engng. B* **35**, 303 (1995).
[114] J. R. Barker, D. W. Lowe, and S. Murray, A Wigner function approach to transport and switching in sub-micron heterostructures, in *The Physics of Submicron Structures*, edited by H. L. Grubin, K. Hess, G. J. Iafrate, and D. K. Ferry, page 277, Plenum Press, NY, 1984.
[115] N. C. Kluksdahl, A. M. Kriman, D. K. Ferry, and C. Ringhofer, Self-consistent study of the resonant-tunneling diode, *Phys. Rev. B* **39**(11), 7720–7735 (1989).
[116] F. A. Buot and K. L. Jensen, Intrinsic high-frequency oscillations and equivalent circuit model in the negative differential resistance region of resonant tunneling devices, *COMPEL* **10**, 241 (1991).
[117] F. A. Buot, General theory of quantum distribution function transport equations: Superfluid systems and ultrafast dynamics of optically excited semiconductors, *La Rivista del Nuovo Cimento* **20**(9), 1–75 (1997).
[118] M. Schmutz, Real-time Green's functions in many body problems, *Z. Physik B* **30**, 97–106 (1978).
[119] T. Arimatsu and H. Umezawa, General structure of non-equilibrium thermo field dynamics, *Prog. Theor. Phys.* **77**, 53 (1987).
[120] J. Frenkel, *Wave Mechanics, Advanced General Theory*, pages 253, 435, Clarendon, Oxford, 1934.
[121] E. Gerjouy, A. R. Rao, and L. Spruch, A unified formulation of the construction of variational principles, *Rev. Mod. Phys.* **55**, 725 (1983).
[122] R. Balian and M. Veneroni, Static and dynamic variational principles for expectation values of observables, *Annals Phys.* **187**, 29 (1988).
[123] R. Jackiw and A. Kerman, Time-dependent variational principle and the effective action, *Phys. Lett. A* **71**(2-3), 158–162 (1979).
[124] A. K. Rajagopal and F. A. Buot, A nonequilibrium time-dependent functional theory based on Liouvillian quantum field dynamics, *Int. J. Quantum Chem.* **56**, 389 (1995).
[125] A. K. Rajagopal and F. A. Buot, Nonequilibrium time-dependent functional theory for coupled interacting fields, *Phys. Rev. B* **51**, 1883 (1995).

[126] A. K. Rajagopal and F. A. Buot, Generalized Functional theory of interacting coupled Liouvillean fields in condensed matter, in *Density Functional Theory II: Relativistic and Time-Dependent Extensions*, edited by R. F. Nalewajski, volume 181, pages 159–162, Springer-Verlag, 1996.

[127] T. Kato, T. Kobayashi, and M. Namiki, Formal theory of Green functions, *Prog. Theor. Phys.* **15**(3), (1960).

[128] C. De Dominicis and P. C. Martin, Stationary entropy principle and renormalization in normal and superfluid systems. I. Algebraic formulation, *J. Math. Phys.* **5**(1), 14–30 (1964).

[129] Y. Nambu, Quasi-particles and Gauge invariance in the theory of superconductivity, *Phys. Rev.* **117**(2–3), 648 (1960).

[130] T. Morita and K. Hiroike, A new approach to the theory of classical fluids. III, *Prog. Theor. Phys.* **25**, 537–578 (1961).

[131] W. Kohn and L. J. Sham, Self-consistent equations including exchange and correlation effects, *Phys. Rev. A* **40**, 1133 (1965).

[132] F. A. Buot, Quantum superfield renormalization procedure, unpublished, 1996.

[133] G. Lindblad, On the generators of quantum dynamical semigroups, *Commun. Math. Phys.* **48**(2), 119–130 (1976).

[134] T. R. Kirkpatrick and J. R. Dorfman, Transport in a dilute but condensed nonideal Bose gas: Kinetic equations, *J. Low Temp. Phys.* **58**(2), 301–331 (1985).

[135] A. G. Aronov, Y. M. Gal'perin, V. L. Gurevich, and V. I. Kozub, The Boltzmann-equation description of transport in superconductors, in *Nonequilibrium Superconductivity*, edited by D. N. Langenberg and A. I. Larkin, volume 12, page 325, North Holland, NY, 1986.

[136] A. L. Fetter and J. D. Walecka, *Quantum Theory of Many-Particle Systems*, page 183, McGraw Hill, 1971.

[137] F. A. Buot, Generalized semiconductor Bloch equations, *J. Comp. Theor. Nanosci.* **1**, 144–168 (2004).

[138] L. Mandel and E. Wolf, *Optical Coherence and Quantum Optics*, Cambridge University Press, NY, 1995.

[139] A. Stahl and I. Balslev, *Electrodynamics of Semiconductor Band Edge*, Springer Tracts in Modern Physics, Springer-Verlag, NY, 1987.

[140] F. A. Buot, An interband tunnel high-frequency source, *J. Phys. D: Appl. Phys.* **30**, 3016 (1997).

[141] J. M. Ziman, *Electrons and Phonons*, Clarendon Press, Oxford, 1979.

[142] C. Horie and J. A. Krumhansl, Boltzmann equation in a phonon system, *Phys. Rev. A* **136**, 1397 (1964).

[143] F. A. Buot, On the relaxation rate spectrum of phonons, *J. Phys. C: Solid State Phys.* **5**, 5 (1972).

[144] D. Benin, Phonon viscosity and wide-angle phonon scattering in superfluid helium, *Phys. Rev. B* **11**, 145 (1975).

[145] M. Bonitz, T. Bornath, D. Kremp, M. Schlanges, and W. D. Kraeft, Quantum kinetic theory for laser plasmas. Dynamical screening in strong fields, in *Lectures in Theoretical Physics*, edited by W. E. Brittin, volume IX C, page 469, Gordon and Breach, NY, 1967.

[146] P. C. K. Kwok, Green's function method in lattice dynamics, in *Solid State Physics*, edited by F. Seitz, D. Turnbull, and H. Ehrenreich, volume 20, pages 213–303, Elsevier, NY, 1968.

[147] F. Jahnke and S. W. Koch, Many-body theory for semiconductor microcavity lasers, *Phys. Rev. A* **52**, 1712 (1995).

[148] R. J. Stoner and H. J. Maris, Kapitza conductance and heat flow between solids at temperatures from 50 to 300 K, *Phys. Rev. B* **48**, 16373 (1993).

[149] N. Perrin, Determination of the thermal boundary resistance in the transport approach, *J. Low Temp. Phys.* **31**, 257 (1978).

[150] M. L. Huberman and A. W. Overhauser, Electronic Kapitza conductance at a diamond-Pb interface, *Phys. Rev. B* **50**, 2865 (1994).

[151] A. Sergeev, Inelastic electron-boundary scattering in thin films, *Physica B* **263–264**, 217 (1999).

[152] W. A. Little, Kapitza resistance between helium and metals in the normal and superconducting states, *Phys. Rev.* **123**, 435 (1961).

[153] F. W. Sheard, R. M. Bowley, and G. A. Toombs, Microscopic theory of the Kapitza resistance at a solid-liquid 4He interface, *Phys. Rev. A* **8**, 3135 (1973).

[154] G. D. Mahan, Kapitza thermal resistance between a metal and a nonmetal, *Phys. Rev. B* **79**, 075408 (2009).

[155] W. L. Barnes, A. Dereux, and T. W. Ebbesen, Surface plasmon subwavelength optics, *Nature* **424**, 824–830 (2003).

[156] A. V. Zayats and I. I. Smolyaninov, Near-field photonics: surface plasmon polaritons and localized surface plasmons, *J. Opt. A: Pure Appl. Opt.* **5**, S16 (2003).

[157] S. A. Wolf, J. Lu, M. R. Stan, E. Chen, and D. M. Treger, The promise of nanomagnetics and spintronics for future logic and universal memory, *Proc. IEEE* **98**(12), 2155–2168 (2010).

[158] F. A. Buot, R. A. Loberternos, D. L. Villarin, and R. E. S. Otadoy, Spin magnetization quantum transport equations: Coupled nonequilibrium spin and charge correlations, *Quantum Matter* **1**, 1–7 (2013).

[159] H. C. Torrey, Bloch equations with diffusion terms, *Phys. Rev.* **104**, 563 (1956).

[160] M. I. D'yakonov and V. I. Perel', Current-induced spin orientation of electrons in semiconductors, *Phys. Lett. A* **35**, 459–460 (1971).

[161] T. L. Gilbert, A phenomenological theory of damping in ferromagnetic materials, *IEEE Trans. Magneti.* **40**(6), 3443 (2004).

[162] S. S. L. Zhang and S. Zhang, Interplay between magnetization dynamics and spin transport, *IEEE Trans. Magnet.* **46**(6), 2297 (2010).

[163] R. F. L. Evans, D. Hinzke, U. Atxitic, U. Nowak, R. W. Cantrell, and O. C. Fesenko, Wannier-based calculation of the orbital magnetization in crystals, *Phys. Rev. B* **85**(6), 01443 (2012).

[164] L. Berger, Side-jump mechanism for the Hall effect of ferromagnets, *Phys. Rev. B* **2**, 4559 (1970).

[165] S. Zhang, Extraordinary Hall effect in magnetic multilayers, *Phys. Rev. B* **51**, 3632 (1995).

[166] C. L. Canedy, X. W. Li, and G. Xiao, Large magnetic moment enhancement and extraordinary Hall effect in Co/Pt superlattices, *Phys. Rev. B* **62**, 508 (2000).

[167] A. K. Rajagopal and M. Mochena, Spin-orbit interactions in the many-body theory of magnetic electron systems, *Phys. Rev. B* **57**, 11582 (1998).

[168] V. P. Zhukov and E. V. Chulkov, The femtosecond dynamics of electrons in metals, *Phys.-Usp.* **52**, 105–136 (2009).

[169] M. D. Mower, G. Vignale, and I. Tokatly, D'yakonov–Perel' spin relaxation for degenerate electrons in the electron–hole liquid, October 2010, arXiv:1010.5544v1.

[170] I. Zutic, J. Fabian, and S. D. Sarma, Spintronics: Fundamentals and applications, *Rev. Mod. Phys.* **76**, 323 (2004).

[171] F. A. Buot, Discrete phase space and quantum superfield theory in nanosystem quantum transport, *J. Comp. Theor. Nanosci.* **4**, 1037 (2007).

[172] F. A. Buot, Quantum superfield theory and lattice weyl transform in nonequilibrium quantum transport physics, *Quantum Matter* **2**(4), 247–288 (2013).

[173] A. Sakurai and Y. Tanimura, Self-excited current oscillations in a resonant tunneling diode described by a model based on the Caldeira — Leggett Hamiltonian, *New J. Phys.* **16**, 015002 (2014).

[174] F. A. Buot, Generalized semiconductor Bloch equations, *J. Comp. Theor. Nanosci.* **1**, 144–168 (2004).

[175] Y. D. Jung and W. P. Hong, Electron spin-interaction effects on the ponderomotive magnetization and radiation in degenerate quantum plasmas, *Phys. Plasmas* **20**, 084502 (2013).

[176] G. Brodin, M. Marklund, J. Zamanian, and M. Stefan, Spin and magnetization effects in plasmas, 2010, arXiv:1010.0572v1.

[177] P. A. Andreev, Hydrodynamic and kinetic models for spin-1/2 electron-positron quantum plasmas: Annihilation interaction, helicity conservation, and wave dispersion in magnetized plasmas, 2014, arXiv:1404.4899v1.

[178] M. N. Baibich, J. M. Broto, A. Fert, F. N. V. Dau, F. Petroff, P. Etinne, G. Creuzet, A. Friederich, and J. Chazelas, Giant magnetoresistance of (001)Fe/(001)Cr magnetic superlattices, *Phys. Rev. Lett.* **61**, 2472 (1988).

[179] S. Datta and B. Das, Electronic analog of the electro-optic modulator, *App. Phys. Lett.* **56**(7), 665–667 (1990).

[180] D. C. Ralph and M. D. Stiles, Spin transfer torques, *J. Magn. Magn. Mater.* **320**(7), 1190 (2008).

[181] J. C. Slonczewski, Current-driven excitation of magnetic multilayers, *J. Magn. Magn. Mater.* **159**(1-2), L1–L7 (1996).

[182] M. Johnson and J. Byers, Charge and spin diffusion in mesoscopic metal wires and at ferromagnet/nonmagnet interfaces, *Phys. Rev. B* **67**, 125112 (2003).

[183] C. Gros and G. Y. Chitov, The spin-SAF transition in NaV2O5 induced by spin-pseudospin coupling, *Europhys. Lett.* **69**(3), 447 (2005).

[184] A. Shipiro, P. M. Levy, and S. Zhang, Self-consistent treatment of nonequilibrium spin torques in magnetic multilayers, *Phys. Rev. B* **67**, 104430 (2003).

[185] A. Moreo, M. Daghofer, A. Nicholson, and E. Dagotto, Interband pairing in multiorbital systems, *Phys. Rev. B* **80**, 104507 (2009).

[186] E. Gubankova, W. V. Liu, and F. Wilczek, Breached pairing superfluidity: Possible realization in QCD, *Phys. Rev. Lett.* **91**, 032001 (2003).

[187] W. V. Liu and F. Wilczek, Interior gap superfluidity, *Phys. Rev. Lett.* **90**, 047002 (2003).

[188] F. A. Buot, R. A. Loberternos, R. E. S. Otadoy, and D. L. Villarin, Nonequilibrium spin magnetization quantum transport equations, 2002, arXiv:1112.4220v1.

[189] F. A. Buot, General theory of quantum distribution function transport equations: Superfluid systems and ultrafast dynamics of optically excited semiconductors, *La Rivista del Nuovo Cimento* **20**, 1–75 (1997).

[190] D. K. Singh, B. Kamble, and A. Singh, Spin-charge and spin-orbital coupling effects on spin dynamics in ferromagnetic manganites, *J. Phys. Condens. Matter* **22**, 396001 (2010).

[191] F. Buot, C. Scott, I. Mack, and K. Sieger, Fundamental physical limits of computation: Rigid coupling model, broken symmetry and self-induced transmission in a computer medium, in *Molecular Electronic Devices: Proceedings of the 3rd International Symposium on Molecular Electronic Devices*, edited by F. Carter, R. Sietkowski, and H. Wohltjen, North Holland, Amsterdam, 1989.

[192] J. Lin, M. Marcolli, H. Ooguri, and B. Stoica, Locality of gravitational systems from entanglement of conformal field theories, *Phys. Rev. Lett.* **114**, 221601 (2015).

[193] M. Sarovar, A. Ishizaki, G. R. Fleming, and K. B. Whaley, Quantum entanglement in photosynthetic light-harvesting complexes, *Nature Phys.* **6**, 462 (2010).

[194] A. Einstein, B. Podolsky, and N. Rosen, Can quantum-mechanical description of physical reality be considered complete?, *Phys. Rev.* **47**, 777–80 (1935).

[195] S. Hill and W. K. Wooters, Entanglement of a pair of quantum bits, *Phys. Rev. Lett.* **78**, 5022 (1977).

[196] W. K. Wooters, Entanglement of formation of an arbitrary state of two qubits, *Phys. Rev. Lett.* **80**, 2245 (1988).

[197] J. Kim, Entanglement of formation and monogamy of multi-party quantum entanglement, *Sci. Rep.* **11**, 2364 (2021).

[198] F. A. Buot, K. B. Rivero, and R. E. S. Otadoy, Generalized nonequilibrium quantum transport of spin and pseudospins: Entanglements and topological phases, *Physica B* **559**, 42–61 (2019), See also, F. A. Buot, Multi-Spinor nonequilibrium quantum transport: Entanglement-induced delocalization and topological insulators, in *Topics in Nanoscience* I, edited by W. Schommers, World Scientific, 2022.

[199] E. Madarassay, Decay of soliton-like perturbations into vortex-anti vortex pairs, August 2009, arXiv:0807.3893v3[cond-mat.other].

[200] T. H. Solomon, M. S. Paoletti, and M. E. Schwartzr, Experimental studies of advection-reaction-diffusion systems, in *Chaos, Complexity and Transport*, edited by C. Chandre, X. Leoncini, and G. Zaslavsky, World Scientific, Singapore, 2008.

[201] U. Tokyo, How spacetime is built by quantum entanglement, https://phys.org/news/2015-05-spacetime-built-quantum-entanglement.html, May 2015.

[202] J. Maldacena, Entanglement and the geometry of spacetime, https://www.ias.edu/ideas/2013/maldacena-entanglement, May 2013.

[203] J. S. Bell, On the Einstein-Podolsky-Rosen paradox, *Physics* **1**, 195–200 (1964), reprinted in *Speakable and Unspeakable in Quantum Mechanics*, 2nd edition, pages 14–21, Cambridge, Cambridge University Press, (2004).

[204] J. M. Kosterlitz, Nobel lecture: Topological defects and phase transitions, *Rev. Mod. Phys.* **89**, 040501 (2017).

[205] F. D. M. Haldane, Nobel lecture: Topological quantum matter, *Rev. Mod. Phys.* **89**, 040502 (2017).

[206] X. Xu, W. Yao, D. Xiao, and T. Heinz, Spin and pseudospins in layered transition metal dichalcogenides, *Nature Phys.* **10**, 343 (2014).

[207] M. S. Fuhrer and A. R. Hamilton, Chasing the exciton condensate, physics viewpoint, *Physics* **9**, 80 (2016).

[208] M. Kindermann, Pseudospin entanglement and Bell test in graphene, July 2008, arXiv:0807.0217v2.

[209] X. Liu, K. Watanabe, T. Taniguchi, B. L. Halperin, and P. Kim, Quantum Hall drag of exciton condensate in graphene, *Nat. Phys.* **13**, 746 (2017).

[210] I. Sodemann, H. Chen, and A. H. MacDonald, Theory of transport phenomena in coherent quantum hall bilayers, October 2014, arXiv: 1411.0008v1.

[211] B. Altshuler, Anderson localization at 1D edge of 2D topological insulator, *Phys. Rev. Lett.* **111**, 08641 (2013).

[212] J. Maciejko, C. Liu, Y. Oreg, X.-L. Qi, C. Wu, and S. C. Zhang, Kondo effect in helical edge liquid of the quantum spin Hall state, *Phys. Rev. Lett.* **102**, 256803 (2009).

[213] A. van Oudenaarden, S. J. K. Vardy, and J. E. Mooij, One-dimensional localization of quantum vortices in disordered Josephson junction arrays, *Phys. Rev. Lett.* **77**, 4257 (1996).

[214] E. Babaev, J. Jaykka, and M. Speight, Magnetic field delocalization and flux inversion in fractional vortices in two-component superconductors, 2009, arXiv:0903.3339v2.

[215] C. J. O. Reichhardt and M. B. Hastings, Do Vortices Entangle?, *Phys. Rev. Lett.* **92**, 157002 (2004).

[216] J. L. Ge, T. R. Wu, M. Gao, Z. B. Bai, L. Cao, X. F. Wang, Y. Y. Qin, and F. Q. Song, Weak localization of bismuth cluster-decorated graphene and its spin-orbit interaction, *Front. Phys.* **12**, 127210 (2017).

[217] W. Knapp, G. E. Pikus, F. G. Pikus, S. V. Iordanskii, V. Mosser, K. Zekentes, and Y. B. Lyanda-Geller, Weak antilocalization and spin precession in quantum wells, 1996, arXiv:cond-mat/9602068v1.

[218] M. Kammermeier, P. Wenk, J. Schiliemann, S. Heedt, and T. Schapers, Weak (anti)localization in tubular semiconductor nanowires with spin-orbit coupling, *Phys. Rev. B* **93**, 205306 (2016).

[219] S. I. Dorozhkin, A. A. Kapustin, and S. S. Murzin, Observation of crossover from weak localization to antilocalization in the temperature dependence of the resistance of a two-dimensional system with spin-orbit interaction, *JETP Lett.* **97**, 149–154 (2013).

[220] A. Kitaev, Unpaired Majorana fermions in quantum wires, 2020, arXiv:cond-mat/0010440.

[221] M. Buttiker, Edge state without magnetic fields, *Science* **325**, 278 (2009).

[222] F. Wilczek, Basic notions of entropy and entanglement, Lecture notes, March 2014.

[223] G. Baym and T. Ozawa, Condensation of bosons with Rashba-Dresselhaus spin-orbit coupling, *J. Phys.: Conf. Series* **529**, 012006 (2014).

[224] F. A. Buot, R. A. Narandan, and R. E. S. Otadoy, Nonequilibrium multi-band spin quantum transport: Spin, pseudospin, and charge coupled equations, *Quantum Matter* **5**, 155–181 (2016).

[225] F. A. Buot, Y. Jiang, and A. Fedoseyev, Novel quantum hydrodynamic equations and quantum-hierarchy decoupling scheme, *Phys. Rev. E* **66**, 066119 (2002).

[226] A. Rajagopal and M. Mochena, Broken-symmetry-adapted Green's function theory of condensed matter: Towards a vector spin density functional theory, July 1999, arXiv:cond-mat/9907089.

[227] S. Murakami, Two-dimensional topological insulators and their edge states, *J. Phys.: Conf. Series* **302**, 012019 (2011).

[228] E. D. K. Finck, J. Eisenstein, L. Pfeifer, and K. West, Exciton transport and Andreev reflection in a bilayer quantum Hall system, *Phys. Rev. Lett.* **106**, 236807 (2011).

[229] O. D. Jefimenko, Torque exerted by a moving electric charge on a stationary electric charge distribution, *J. Phys. A: Math. Gen.* **35**, 5305–5314 (2002).

[230] M. I. D'yakonov and V. I. Perel', Possibility of orienting electron spins with current, *JETP Lett.* **13**(11), 657–660 (1971).

[231] R. C. Iotti, F. Dolcini, and F. Rossi, Wigner-function formalism applied to semiconductor quantum devices: Need for nonlocal scattering models, *Phys. Rev. B* **96**, 115420 (2017).

[232] R. C. Iotti and F. Rossi, Microscopic theory of energy dissipation and decoherence in solid-state quantum devices: Need for nonlocal scattering models, *Entropy* **20**, 726 (2018).

[233] P. J. Newton, R. Mansell, S. N. Holmes, M. Myronov, and C. H. W. Barnes, Weak localization and weak antilocalization in doped germanium epilayers, *App. Phys. Lett.* **110**, 062101 (2017).
[234] M. Dzero, J. Xia, V. Galitski, and P. Coleman, Topological Kondo insulators, *Ann. Rev. Cond. Matter Phys.* **7**(1), 249–280 (2016).
[235] P.-Y. Chang, O. Erten, and P. Coleman, Mobius Kondo insulators, *Nature Phys.* **13**, 794–798 (2017).
[236] S. Diehl, E. Rico, M. A. Baranov, and P. Zoller, Topology by dissipation in atomic quantum wires, *Nat. Phys.* **7**, 971 (2011).
[237] X. Chen, L. Zhang, and H. Guo, Valley caloritronics and its realization by graphene nanoribbons, *Phys. Rev. B* **92**, 155427 (2015).
[238] H. Adachi, K. Uchida, E. Saitoh, and S. Maekawa, Theory of the spin Seebeck effect, *Rep. Prog. Phys.* **76**, 036501 (2013).
[239] F. A. Buot, Zener Effect, in *Encyclopedia of Electrical and Electronics Engineering*, edited by J. Webster, volume 23, pages 669–688, John Wiley and Sons, 1999.
[240] S. Mandelstam, Gauge independent formulation of electrodynamics, *Annals Phys.* **19**, 1 (1962).
[241] T. T. Wu and C. N. Yang, Concept of non-integrable phase factors and global formulation of gauge fields, *Phys. Rev. D* **12**, 3845 (1975).
[242] K. F. Berggren and D. J. Newson, Magnetic depopulation of electronic subbands in low-dimensional semiconductor systems and their influence on the electrical resistivity and Hall effect, *Semicond. Sci. Technol.* **1**, 327 (1986).
[243] M. Wilkinson, Von Neumann lattices of Wannier functions for Bloch electrons in magnetic fields, *Proc. R. Soc. London* **403**, 145–166 (1956).
[244] F. A. Buot, Nonequilibrium superfield and lattice Weyl transform approach to quantum Hall effect, 2020, arXiv:2001.06993.
[245] W. J. de Haas and P. M. van Alphen, The dependence of the susceptibility of diamagnetic metals upon the field, *Proc. Acad. Sci. Amst.* **33**, 1106–1118 (1930).
[246] V. G. Peschanskii and Y. A. Kolesnichenko, On the 60th anniversary of the Lifshitz-Kosevich theory, *Low Temp. Phys.* **40**(4), 267 (2014).
[247] F. A. Buot, P. L. Li, and J. O. Strom-Olsen, The influence of scatterings on magnetic breakdown, *J. Low Temp. Phys.* **22**, 535–556 (1976).
[248] A. Alexandradinata and L. Glazman, Modern theory of magnetic breakdown, 2017, arXiv:1708.09387v1.
[249] F. A. Buot, On quantum Hall effect: Covariant derivatives, Wilson lines, gauge potentials, lattice Weyl transforms, and Chern numbers, February 2021, arXiv:2106.16238v1.
[250] R. B. Laughlin, Anomalous quantum Hall effect: An incompressible quantum fluid with fractionally charged excitations, *Phys. Rev. Lett.* **6**(18), 1395–1398 (1983).
[251] H. Stormer, Two-dimensional electron correlation in high magnetic fields, *Physica B* **177**(14), 401–408 (1992).
[252] R. Willett, J. P. Eisenstein, H. L. Stormer, D. C. Tsui, A. C. Gossard, and J. H. English, Observation of an even-denominator quantum number in the fractional quantum Hall effect, *Phys. Rev. Lett.* **59**, 1776 (1987).
[253] W. Pan, H. L. Stormer, D. C. Tsui, L. N. Pfeiffer, K. W. Baldwin, and K. W.West, Fractional quantum Hall effect of composite fermions, *Phys. Rev. Lett.* **90**, 016801 (2003).
[254] A. Kumar, G. A. Csathy, M. J. Manfra, L. N. Pfeiffer, and K. W. West, Nonconventional odd-denominator fractional quantum Hall states in the second Landau level, *Phys. Rev. Lett.* **105**, 246808 (2010).

[255] G. Moore and N. Read, Nonabelians in the fractional quantum Hall effect, *Nucl. Phys. B* **360**, 362 (1991).
[256] J. K. Jain, Theory of the fractional quantum Hall effect, *Phys. Rev. B* **41**, 7653 (1990).
[257] J. K. Jain, Microscopic theory of the fractional quantum Hall effect, *Adv. Phys.* **41**, 105–146 (1992).
[258] N. B. Schade, D. I. Schuster, and S. R. Nagel, A nonlinear, geometric Hall effect without magnetic field, *PNAS* **116**(49), 24475–24479 (2019).
[259] M. A. Zubkov and X. Wu, Topological invariant in terms of the Green functions for quantum Hall effect in the presence of varying magnetic field, *Annals Phys.* **430**, 168170 (2020).
[260] K. Ikeda, Quantum Hall effect and Langlands program, 2017, arXiv:1708.00419v2.
[261] G. H. Wannier, A result not dependent on rationality for Bloch electrons in a magnetic field, *Phys. Stat. Sol. B* **88**, 757 (1978).
[262] F. A. Buot, G. Maglasang, A. R. F. Elnar, and C. Galon, On Fractional Quantum Hall Effect (FQHE): A Chern–Simons and nonequilibrium quantum transport Weyl transform approach, 2021, arXiv: 2107.13970.
[263] T. V. Mechelen and Z. Jacob, Viscous Maxwell–Chern–Simons theory for topological electromagnetic phases of matter, *Phys. Rev. B* **102**, 155425 (2020).
[264] F. A. Buot, On the quantization of Hall effect in electrical conductivity: A nonequilibrium quantum superfield and lattice Weyl transform transport approach, *AIP Conf. Proc.* **2286**, 030007 (2020).
[265] A. J. Silenko, General method of the relativistic Foldy-Wouthuysen tansformation and proof of validity of the F-W Hamiltonian, February 2015, arXiv:quant-ph/9809071v2.
[266] L. Viola, E. Knill, and S. Lloyd, Dynamical decoupling of open quantum systems, 1999, ArXiv:1501.02052v2.
[267] T. D. Newton and E. P. Wigner, Localised state for elementary systems, *Rev. Mod. Phys.* **21**, 400–406 (1949).
[268] E. I. Blount, Formalisms of band theory, in *Solid State Physics*, edited by F. Seitz and D. Turnball, volume 13 of *Solid State Physics*, pages 335–343, Academic, NY, 1961.
[269] E. I. Blount, Motion of an electron in a perturbed periodic potential, *Phys. Rev.* **126**, 1636 (1962).
[270] E. Eriksen and M. Kolsrud, Canonical transformation of Dirac's equation to even form. Expansion in terms of external fields, *Il Nouvo Cimento* **18**, 1 (1960).

Index

O

P

Q

R

S

T

www.ingramcontent.com/pod-product-compliance
Ingram Content Group UK Ltd.
Pitfield, Milton Keynes, MK11 3LW, UK
UKHW052113170926
8624IPUK00026B/372

9 789811 264719